P. Colinet, J.C. Legros, M.G. Velarde

Nonlinear Dynamics
of Surface-Tension-Driven
Instabilities

P. Colinet, J.C. Legros, M.G. Velarde

Nonlinear Dynamics of Surface-Tension-Driven Instabilities

With a Foreword by I. Prigogine

Berlin · Weinheim · New York · Chichester · Brisbane · Singapore · Toronto

Authors:
Dr. Pierre Colinet, Chargé de Recherches, Fonds National de la Recherche Scientifique;
Professor Dr. Jean Claude Legros, Université Libre de Bruxelles,
Service de Chimie Physique E. P., C. P. 165/62, Bruxelles, Belgium
pcolinet@ulb.ac.be, jclegros@ulb.ac.be
Professor Dr. Manuel G. Velarde, Universitad Complutense de Madrid, Instituto Pluridisciplinar,
Madrid, Spain
mvelarde@eucmax.sim.ucm.es

With 189 figures and 14 tables

1st edition

Library of Congress Card No: applied for

Die Deutsche Bibliothek – CIP Cataloguing-in-Publication-Data
A catalogue record for this publication is available from Die Deutsche Bibliothek

ISBN 3-527-40291-8

© Wiley-VCH Verlag Berlin GmbH, Berlin (Federal Republic of Germany), 2001

Printed on non-acid paper.

Printing: Strauss Offsetdruck, D-69509 Mörlenbach
Bookbinding: J. Schäffer GmbH & Co. KG., D-67269 Grünstadt

Printed in the Federal Republic of Germany.

Wiley-VCH Verlag Berlin GmbH
Bühringstrasse 10, D-13086 Berlin
Federal Republic of Germany

Contents

Preface

The development of Nonlinear Physics has not yet ceased to surprise us. Progresses achieved during the last decades have allowed physicists to understand many key aspects of dissipative structures in systems far from equilibrium. The emergence of macroscopic order in systems disordered at the microscopic scale, the spontaneous breaking of spatio-temporal symmetries, and highly non-relaxational and stochastic dynamics are features common to virtually all natural phenomena. Today, dissipation as well as global invariance properties such as conservation laws appear as central concepts in the theory of pattern formation, without which our task of identifying the sources of such a universality would have been hopeless.

The Brussels School has been playing a pioneering role in the study of dissipative structures far from equilibrium, since the decisive works of Th. De Donder on chemically reacting systems. His illuminating theory of affinity, including the derivation of their nonlinear relationships with chemical reaction rates, were some of the essential milestones on the way to a theory of non-equilibrium complex processes. Since then, the study of open systems exchanging matter and energy with their environment has allowed Thermodynamics to be reconsidered in an enlarged sense, encompassing the study of systems ordering spontaneously under the action of an external constraint. In this sense, the active and cross-fertilizing development of Non-Equilibrium Thermodynamics and Nonlinear Science has allowed more and more diverse and real-life macroscopic phenomena to be considered from a unified point of view.

Still, there remains sufficient potential for surprise and fascination. The study of particular nonlinear systems continuously leads to the discovery of new dynamical regimes and transitions, which stimulate further developments of the theory of complex systems. In Hydrodynamics, the paradigm of pattern-forming system has for long been the buoyancy-induced Bénard instability of a fluid confined between plates at different temperatures. Even though the important role of surface tension effects in Bénard experiments was quoted as early as at the end of the 1950s, it is only far later that surface-tension-driven flows have been studied in their own right.

Accordingly, the influence of various physico-chemical processes at interfaces is now described in an abundant literature, which a newcomer to the field might find quite complex to sort out. In their book, Pierre Colinet, Jean Claude Legros and Manuel G. Velarde very nicely describe the progress accomplished in that field over recent years, and provide the reader with a complete and pedagogic view of the diversity of phenomena specific to surface-tension-driven, rather than buoyancy-driven, convection. I believe that the presentation of the text, written in the modern language of Nonlinear Physics, should allow scientists of various fields to discover the fascinating and unexpected features of hydrodynamic instabilities in systems with interfaces.

I. Prigogine, Bruxelles, December 2000.

Preamble

One century after the discovery of cellular convective structures by Bénard, thermal convection in fluid layers still remains a subject of intensive research and of new developments. This phenomenon is considered today as a key example of hydrodynamic nonlinear system displaying pattern formation, complex dynamics and transition to turbulence. Most of the effort however has been focused on buoyancy-induced patterns, even though it has been known since the end of the 1950s that Bénard hexagonal cells were induced by surface tension gradients, not by buoyancy. Accordingly, there exists no comprehensive monograph about surface-tension-driven instabilities, despite the even wider variety of pattern and wave-forming phenomena observed in this situation. Motivated by a renewed interest in recent years, the present book aims to fill this gap by providing a coherent view of past and current research about complex dynamics of convection influenced by physico-chemical processes at fluid interfaces.

The phenomenon of Bénard convection is now understood in terms of an instability of the motionless state of the system, triggered by the joint action of buoyancy and surface tension forces. In a wider context, nearly-periodic structures such as hexagons, rolls or waves have been predicted and observed in various fields, such as in chemically reactive systems, in solidification, in biology, in lasers, in geology, in meteorology, ... At first, it became apparent that many macroscopic features of these dissipative structures are common to the bifurcating states observed in all these fields, and essentially independent of the specific details of the system under consideration.

Later, Nonlinear Science has provided tools to explain how the symmetry properties and the nature of bifurcating modes indeed determine the possible spatio-temporal structures near instability onset. While pattern formation is now well understood at this general level, the detailed analysis of particular physical set-ups remains essential as far as the mechanisms leading to selection between dissipative or weakly-dissipative structures is concerned. In writing this book, our first goal has been to provide an overview of the wide variety of steady patterns, waves and other more complex phenomena observed in multiphase and multicomponent fluid systems. Various realistic situations are thus considered, including effects such as heat and mass transfer through interfaces, surface deformability, phase change, thermal and mechanical couplings at interfaces, non-isothermal diffusion effects in multicomponent mixtures, non-Boussinesq effects, ... Several theories which compare successfully with experiments are described, though attention is also drawn to still unexplained experimental data, which are likely to remain a long-lived challenge for future developments in the field of surface-tension-driven convection.

We have thus devoted significant space to detailed "First Principle" analyzes of realistic experimental set-ups. It is indeed our opinion that, nowadays, the progress accomplished in Nonlinear Science should allow a variety of complex processes to be considered, starting from a system of thermo-hydrodynamic equations and boundary conditions including all the effects relevant to a particular experiment. The existing tools could possibly be applied to industrially-relevant situations, although in many cases of interest, even the basic equations (or more exactly the boundary conditions) applying in the strongly nonlinear regime are not yet firmly established, and compared with experimental situations. This is particularly the case for problems involving mass transfer through interfaces, such as evaporation, solute transfer through an interface, ... where non-equilibrium kinetic relations leading to the rate of transfer through the interface are still subject to intensive discussions.

Hence, the mathematical tools presented here are first explained at a general level, without reference to any particular set-up. For instance, using some operational formulation, we describe and compare several techniques which allow the center-manifold reduction of the dynamics of an infinite-dimensional system to a finite-dimensional system of evolution equations in the vicinity of bifurcation points. Extensive use is also made of symmetry arguments, which allow to obtain quite a detailed (though not necessarily complete) picture of the possible spatio-temporal behaviors in the vicinity of a bifurcation point. Using these general results, the methods are systematically applied to convection problems involving complicated boundary conditions at interfaces, and multiphase systems. In view of the above discussion of effects specific to interfacial mass transfer, most of the examples studied refer to the more usual (and simpler) case of heat transfer, still leaving doors open for future extensions to mass-transfer systems. Some account is also given of modern numerical techniques which allow three-dimensional time-dependent situations to be analyzed in detail, including the strongly nonlinear convection domain.

The selection of physical systems is based either on the illustration of specific mathematical difficulties, or on the identification of the minimal set of ingredients necessary to account for a particular behavior. For instance, some of the systems studied are characterized by so many parameters, that an exhaustive quantitative analysis of all possible behaviors would probably justify forming subjects of books in themselves. Rather than doing this, we have tried instead to isolate some basic instability modes, and to explain their mechanisms.

Coming back to a more general point of view, we have also tried to emphasize the link with nonlinear behaviors observed in other fields, outside the realm of Fluid Dynamics. As expected, such analogies exist, and are more quantitatively established by showing that behaviors observed in quite different fields are described by the same nonlinear equations. Hence, the focus is also placed on the derivation of several classical equations of Mathematical Physics from the original coupled Navier–Stokes and energy or mass balance equations, complemented by realistic boundary conditions. Whenever possible, the emphasis is placed on rigorous asymptotic derivations of these equations. This includes systems of coupled real and complex Ginzburg–Landau equations, Newell–Whitehead–Segel equations, nonlinear Schrödinger equations, and long-wavelength approximations of convection problems or nonlinear phase equations, leading for instance to Kuramoto–Sivashinsky or dissipation-modified Korteweg–de Vries equations. In turn, some of these generic equations are characterized by complex dynamical regimes, such as strongly nonlinear oscillations, spatially quasiperiodic and chaotic patterns, weakly dissipative solitons and cnoidal waves, or turbulent behavior.

In addition to providing a summarized account of the classical literature on the subject of Bénard convection, the text also covers recent experimental, theoretical and numerical results, and stresses the essential differences that surface-tension-driven instabilities present with respect to their purely buoyancy-driven counterpart. Among these are the possibility of various types of oscillatory behaviors at the onset of convection, the mode mixing and resonance between transverse and longitudinal waves, the particular non-variational dynamics of hexagonal patterns and their defects, resonant wave patterns, multiple bifurcations, spatially quasiperiodic and chaotic patterns, soliton and shock-like waves, and the peculiarities of the transitions to turbulent dynamics occurring at a large driving force. While the emphasis is on spatially extended systems, results are also provided about the influence of lateral confinement.

We believe that the text should be found useful by students who want to gain familiarity and begin research in this vast and fascinating field, by physicists interested about how physico-chemical processes at interfaces affect pattern and wave formation in hydrodynamical systems, as well as by chemical engineers desiring to apply some of the mathematical techniques presented here in view of their application to industrial and microgravity-relevant processes. Finally, applied mathematicians might also wish to further improve rigor in the methodologies or to consider in greater mathematical detail some of the nonlinear equations presented in various parts of the book.

Although a preliminary knowledge of Fluid Mechanics and perturbation techniques is certainly an advantage, sufficient material has been gathered to provide the reader with a progressive and complete insight into this vast field. The first introductory chapter describes nonlinear dissipative structures at a general level, with several illustrations of recent findings in the field of surface-tension-driven convection. Chapter 2 contains a succinct derivation of thermo-hydrodynamic equations and boundary conditions prevailing at interfaces. Chapter 3 is devoted to linear stability analyzes and identification of basic instability modes. The weakly nonlinear theories are presented in Chapter 4 (monotonic instabilities) and Chapter 5 (oscillatory instabilities). In Chapter 6, experimental and theoretical results are presented about solitonic and shock-like surface waves, while Chapter 7 is devoted to selected examples of multiple bifurcations. Finally, recent results on strongly nonlinear surface-tension-driven convection and transitions to interfacial turbulence are presented in Chapter 8.

We hope that the contents of our book correctly reflect the numerous collaborations and fruitful discussions we had with so many colleagues and friends. We especially wish to express our deep gratitude to those who helped us improving our text with some of their results : M. Bestehorn, J. Bragard, P. Dauby, K. Eckert, M. Hennenberg, D. Johnson, E.L. Koschmieder, G. Lebon, H. Linde, S.W. Morris, A.A. Nepomnyashchy, R. Narayanan, A.Ye. Rednikov, M.F. Schatz, D. Schwabe, H.L. Swinney, A. Thess, U. Thiele, W. Tokaruk, S.J. VanHook, and A. Wierschem. All these collaborations and this book would not have been possible without the continuous support received from the European Union, the European Space Agency, the Federal Office for Scientific, Technical and Cultural Affairs (Belgium), the Fonds National de la Recherche Scientifique (Belgium), and the Ministry of Science and Technology (Spain). This text is part of the training effort done by the authors in the frame of the ICOPAC (Interfacial Convection and Phase Change) network sponsored by the European Union, and of the InterUniversity Poles of Attraction Programme sponsored by the Belgian Federal Office for Scientific, Technical and Cultural Affairs.

P. Colinet, J.C. Legros and M.G. Velarde.

1 Introduction

1.1 Equilibrium versus non-equilibrium states and instability phenomena

When a system initially at thermodynamic equilibrium is progressively driven away from it by some externally imposed constraint, the homogeneous equilibrium state is first continuously replaced by an inhomogeneous non-equilibrium state (called the "Thermodynamic Branch", although hereafter we will most often refer to it as a reference state). Such non-equilibrium states are usually described by an extension of Equilibrium Thermodynamics known as Thermodynamics of Irreversible Processes [1], typically leading to expressions of the fluxes and of the entropy production rate as a function of the gradients present throughout the system. These regimes may be seen as the linear response of the system to the imposed driving gradients (the control parameters), such linearity being guaranteed at least in a close vicinity of equilibrium. Note that even though the system as a whole may be in a non-equilibrium state, the deviation from equilibrium remains sufficiently small for the system to be described as an ensemble of infinitesimal volume elements (though they remain macroscopic in a thermodynamic sense) locally at equilibrium, but corresponding to infinitesimally different values of the state variables (local equilibrium hypothesis [2, 3]). These volume elements thus exchange energy and matter through their fictive boundaries, generating fluxes of these quantities throughout the system.

Like the equilibrium state, reference states observed in situations of weak non-equilibrium possess the property of asymptotic stability. Namely, due to dissipative mechanisms (e.g. heat conduction and viscosity in the case of fluid mechanics), fluctuations (thermodynamic, or of external origin) around the reference state are damped after a certain time. However, when the non-equilibrium constraint is increased further, not only nonlinearities of various origins may cause deviations with respect to linear response, but the ability to dissipate the fluctuations may possibly decrease. Above a certain critical constraint, some fluctuations may grow in time rather than decay, thus indicating the loss of stability of the reference non-equilibrium state. This instability phenomenon, or bifurcation, is generally accompanied by the emergence of new stable states (such as other steady states, oscillatory or even more complex behavior).

It will be useful in the following to employ some concepts of dynamical systems theory [4], and in particular the notion of a phase space. By suitably defining coordinate axes in such abstract space (usually infinite-dimensional), the instantaneous state of the system may be represented as a point, eventually describing some trajectory if the system evolves in time. A bifurcation phenomenon corresponds to a qualitative change in the structure of the phase space of the system. A steady reference state is a fixed point in the phase space, which becomes unstable above the critical value of the constraint. This means that in its neighborhood, not all trajectories are attracted to it as time goes on. In fact, contrary to the situation prevailing before the transition, there now exist some directions in phase space along which the trajectory gets away from the fixed point to tend to another fixed point (steady state), to a limit cycle (oscillatory behavior), or to a more complex attractor (e.g. quasi-periodic or chaotic behavior). Note that for the dissipative systems we consider here, the dimensionality of the various kinds of attractors in phase space is much lower than the dimension of this space, because of the property of contraction of volumes in phase space. Owing to this property, the attracting states reached by the system after a sufficiently long time are in most cases independent of the initial conditions, at least if some more complicated situations such as multistability (i.e. several solutions simultaneously stable) are excluded for the moment.

It is rather remarkable that a number of general macroscopic properties are common to the bifurcating states observed in fields apparently as different as fluid mechanics, chemistry, nonlinear optics, solid-state physics, biology, or even population dynamics or economics [5, 6, 7, 8, 9]. The new states created at the bifurcation points indeed display macroscopic features that are essentially independent of the specific details of the system under consideration (although these are important in the mechanism giving rise to the instability). For instance, patterned states (defined for the moment as near-periodic variations of typical quantities throughout the system) as well as periodic waves are observed in lasers, convection, solidification, chemical reactions, meteorology, aggregation of amoebas, ... Oscillatory behaviors are observed in population dynamics, in traffic dynamics, in semiconductor devices, in electronic circuits, in neurodynamics, ...

In most of these situations, the newly formed states emerge from the cooperative behavior of the 'microscopic" entities composing the system (interacting atoms or molecules, electrons, photons, coupled oscillators, biological entities such as amoebas, insects, animals or humans, ...), which are able to self-organize themselves on time and length scales that are macroscopic by comparison with the microscopic scales of their individual interactions [5, 6, 7]. The basic reason for this appearance of strongly correlated structures is linked to the fact that the growth (or decay) rate of the fluctuations depends on their spatial structure (i.e. typically on some length scales). A simplified view is that near the threshold, only a very limited number of fluctuations (hereafter also denoted as modes) are able to extract energy from the reference state fast enough to overcome dissipation. It is intuitive that such "fittest" fluctuations (those with a positive growth rate) will strongly determine the appearance of structures above the threshold of instability. The spatial structure of these bifurcating modes is in turn determined by the symmetries underlying the physical system considered [10, 11]. For instance, for systems that are translationally invariant along some directions (like Bénard convection), it appears that the basic modes (normal modes) are characterized by a periodic dependence upon the coordinates along these directions.

To illustrate the above general concepts, consider the example of buoyancy-induced convection. When a horizontal layer of fluid is confined between rigid plates maintained at different temperatures (here we consider the case where the lower plate is hotter), fluid at the bottom has a lower density than fluid at the top (due to thermal expansion), which is a potentially unstable situation in the field of gravity. As long as the temperature difference does not exceed a certain critical value (depending on the liquid depth as well as on properties of the fluid), heat is propagated by conduction from the bottom to the top plate and no motions occur (this is the reference state introduced above). When the temperature difference exceeds this critical value, convective motions set in and organize themselves in a system of convective rolls, with hot fluid rising at some horizontal locations and cooler fluid descending at some other places (Fig. 1.1).

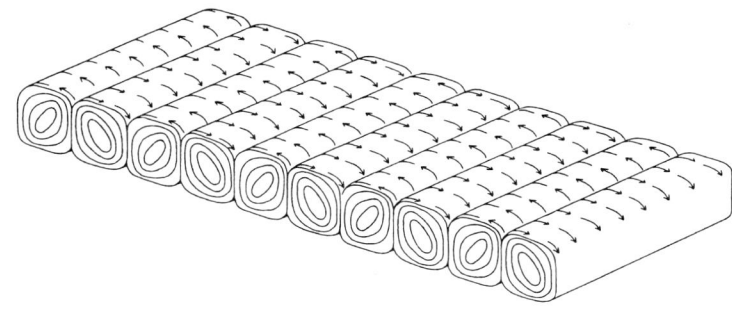

Figure 1.1: Roll-shaped convective cells observed in Rayleigh–Bénard convection. The fundamental unit of the pattern consists in two rolls rotating in opposite directions. The horizontal period is approximately twice the depth of the fluid layer (Velarde and Normand, 1980).

In some other situations, the motions are organized in the form of an hexagonal pattern [Figs 1.2(left) and 1.3], as observed in the experiments of Bénard [12] around 1900. In Fig. 1.2(left), the flow is visualized using aluminum flakes aligning with the local velocity vector. Bright zones correspond to motions along the top free surface, while darker zones correspond to upwards motion (center of the hexagons) and downwards motions (periphery of the hexagons). Note that it was realized only in 1958 that the main source of instability in the experiments of Bénard was associated with surface tension variation with temperature (Marangoni effect), rather than with buoyancy. As the surface tension generally decreases with temperature, tangential stresses induce motions from hot to cold at the free surface. This is the prototype system we shall study in great detail throughout this book, in a variety of configurations.

In relation to the discussion made above, it is worth mentioning that such kinds of striped (rolls) or hexagonal patterns are observed in a variety of other systems, such as reaction-diffusion systems [13, 14] [see Fig. 1.2(right)], solidification patterns [15, 16, 17], optical systems [18, 19], ... Some examples of geological pattern formation (permafrost cellular patterns) are also described in [20].

Buoyancy-driven convection is also at the origin of the formation of cloud streets (see Fig. 1.4) at the boundaries of convection rolls in the atmosphere (reaching sizes of several hundred meters), even though their specific form can vary widely according to the relative importance of two instability mechanisms, namely convection and shear instability [21].

A beautiful example of thermo-hydrodynamic instability is observed when a salty lake has

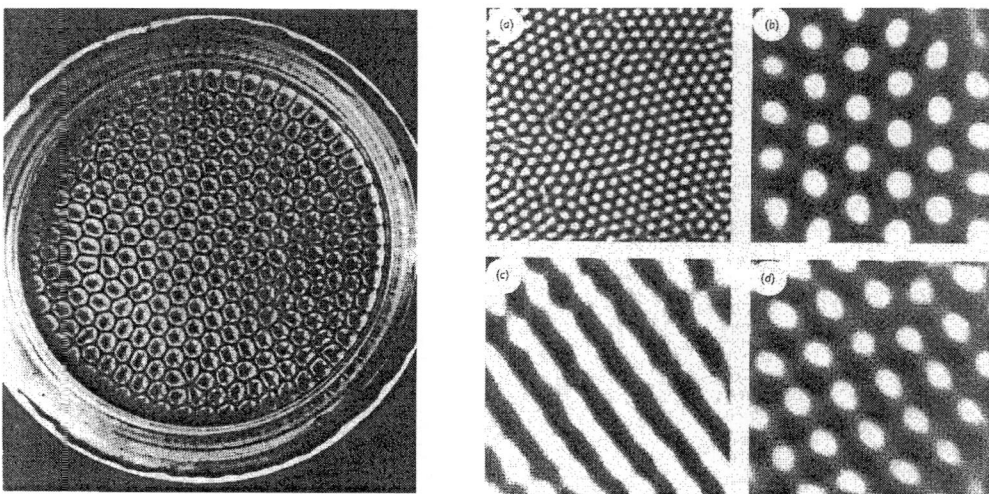

Figure 1.2: Left: Top view of an hexagonal convection pattern in Marangoni–Bénard convection (Koschmieder and Pallas, 1974). Courtesy of E.L. Koschmieder. Right: Stationary chemical patterns in a continuously fed unstirred reactor (Ouyang and Swinney, 1991): (a), (b) hexagons; (c) stripes; (d) mixed state. Courtesy of H.L. Swinney.

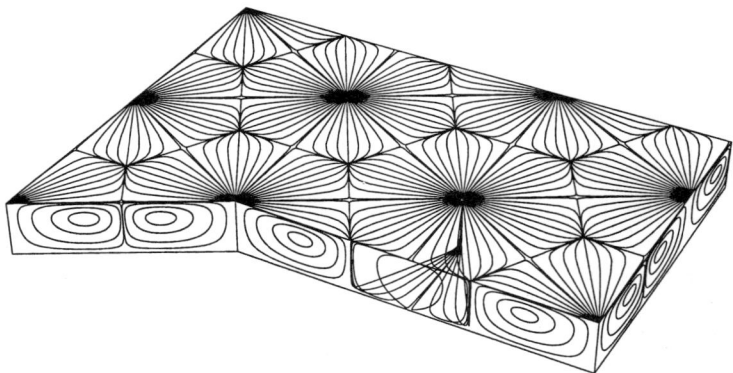

Figure 1.3: Reconstruction of particle paths for up-hexagons (observed in Marangoni–Bénard convection in moderate and large Prandtl number fluids). Fluid motion is upwards at the center of hexagons, where the free surface temperature is maximal.

dried out, such as in Fig. 1.5. There, it is likely that at the final stages of evaporation, the increasing concentration of minerals has led to preferential precipitation at the colder periphery of hexagonal (and distorted polygonal) convection cells.

Owing to its apparent simplicity and fundamental interest, buoyancy-driven Rayleigh–Bénard convection has been extensively studied in the past (for reviews, see [22, 23, 24, 25,

Figure 1.4: Cloud streets, due to the combined action of buoyancy (Rayleigh–Bénard instability) and wind shear (Kelvin–Helmholtz instability).

Figure 1.5: The 12106 km^2 Salar de Uyuni, Bolivia's largest salt pan, in the Daniel Campos province. Polygonal patterns formed by mineral deposited during lake evaporation. Similar patterns are also found near San Pedro de Atacama in Chile. Courtesy of Nathalie Dubois.

26, 27]), as a model hydrodynamic system displaying pattern formation and nonlinear dynamics. We now turn to some recently obtained results illustrating features more specific to the presence of interfaces and surface tension effects.

1.2   Phenomenology of surface-tension-driven instabilities

When a *thin* layer of liquid presenting a free surface is heated from below, the variation of surface tension with temperature can result in an instability of the motionless conductive state of the system, generally leading to hexagonal patterns such as those represented in Fig.1.2(left). This occurs for layer depths typically of the order of 1 cm or less. Otherwise, the buoyancy-driven mechanism is generally dominant, leading to rolls such as shown in Fig.1.1. The basic properties of such hexagonal patterns have been studied experimentally by Koschmieder and collaborators [28]. Chapters 3 and 4 are specifically devoted to explanations of mechanisms and determinations of threshold values of temperature gradients (§3.3), nonlinear stability properties (§4.4), and cell size selection and defects in the hexagonal structures (§4.5).

Here, we merely wish to illustrate some more recent experimental findings, the theoretical description of which is also undertaken in the next chapters.

While these hexagonal patterns are typically observed in some range above the instability threshold, Eckert (née Nitschke) et al. [29, 30] have recently provided evidence of a transition from hexagonal to *square* patterns when the temperature gradient is increased (see Fig. 4.33 of Chapter 4 and Fig. 1.6).

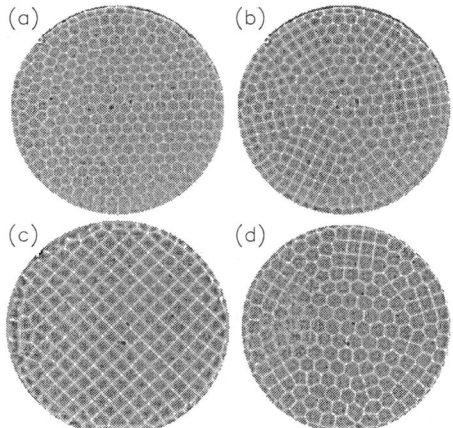

Figure 1.6: Shadowgraph images illustrating the secondary instability leading to square patterns in surface-tension-driven convection, in a layer of silicone oil (viscosity 7 cSt, liquid depth 0.711 mm). The convective pattern changes with increasing supercriticality, measured by a parameter ϵ defined in the main text, from hexagons at $\epsilon = 1.61$ (a); through a mixed state at $\epsilon = 3.9$ (b); and to a square pattern at $\epsilon = 7.22$ (c). As ϵ is then decreased, hexagons reappear in the pattern by $\epsilon = 3.5$ (d). Note the larger size of hexagons in (d) than in (a). For fixed ϵ, all the patterns are time-independent. From [31]. Courtesy of M. Schatz and H.L. Swinney.

In Fig. 1.6 and some of the following ones, the supercriticality parameter ϵ is defined as $\epsilon = \Delta T / \Delta T_c - 1$, where ΔT is the temperature drop (e.g. between the bottom plate and the interface), while ΔT_c is its value at the threshold of the first instability of the motionless state.

The hexagon-square transition, which has motivated active theoretical research (see §4.4.3), has also been studied in details by Thiele and Eckert [32] (see also §4.5.4), Eckert and Thess [33], and Schatz et al. [31]. Moreover, Tokaruk et al. [34] have also very recently provided evidence for a square pattern in *two-layer* surface-tension-driven instabilities, i.e. in a situation where the upper phase is also a liquid, rather than ambient air.

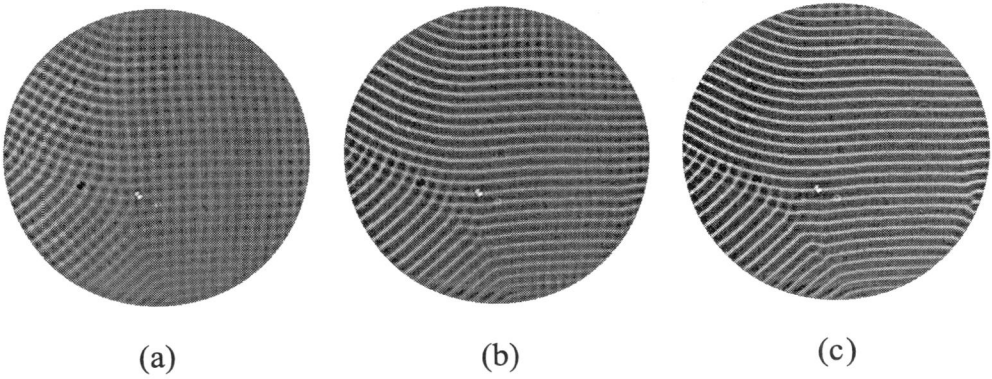

Figure 1.7: Squares-rolls transition observed in two-layer Bénard convection in a Fluorinert FC-104 (bottom layer, 1.08 mm thick)–water (top layer, 2.98 mm thick) system, at increasing thermal constraint (measured by the supercriticality parameter ϵ). The container is circular, with a radius 38 times larger than the Fluorinert layer depth. (a) the initial square planform, for $\epsilon = 0.56$; (b) squares-rolls transition, for $\epsilon = 0.63$; (c) the secondary roll planform, for $\epsilon = 0.70$. Courtesy of W. Tokaruk and S.W. Morris (unpublished, see also [34]).

In this case, it appears that at increasing ϵ, a transition occurs to a *roll* pattern (see Fig. 1.7). Even though the latter is still unexplained (while it can be shown that squares are indeed predicted very close to the first instability threshold, see §4.4.3), it appears that buoyancy should not be the determinant factor, due to the small liquid thicknesses used in the experiment.

The study of two-layer systems considerably enriches the classical Bénard problem of a single layer in contact with a gas. In particular, due to the couplings between the dynamics of temperature and velocity fields in both layers, a two-layer system heated from below or above may be subject to steady patterns, *oscillatory* structures, or more complex behaviors, which have no counterpart in single-layer systems (see e.g. §3.5, §5.4, §7.2). A related question is the following: for a liquid/gas system, is it possible to achieve a theoretical description of the system by considering the dynamics in the liquid phase only, i.e. *neglecting* dynamical effects in the upper gas phase? This is discussed in some details in §3.2.3 and §3.5, where we will derive the conditions necessary to allow such a reduction.

Recently, the dynamics of surface-tension-driven instabilities in very thin *films* of relatively viscous liquids such as oils has been studied by VanHook et al. [35, 36], both theoretically and experimentally. Some of their experimental findings are represented in Fig. 1.8.

VanHook et al. have provided evidence that sufficiently thin liquid layers heated from below may lead to the formation of dry spots (free surface depressions reaching the bottom

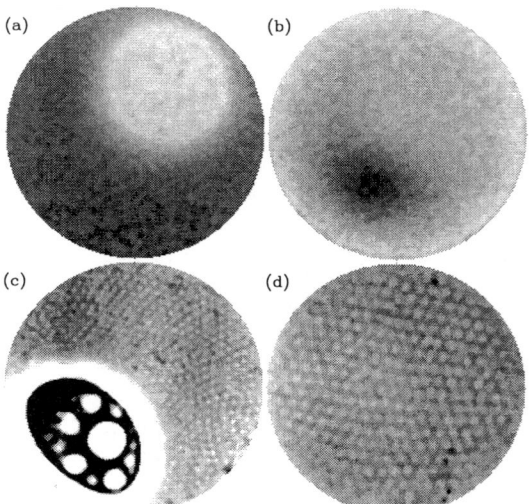

Figure 1.8: Infrared images of the free surface of a thin film of silicone oil 10 cSt in contact with a gas layer approximately twice as thick as the mean oil depth d. The two-layer system is heated from below (metallic plate), and the gas is in contact with a cooled sapphire window. The temperature increases with increasing brightness, so warm depression regions are white (except in c) and cool elevated regions are dark. Each image has its own brightness scale. (a) a localized depression (dry spot) with a helium gas layer, where $d = 0.25$ mm; (b) a localized elevation (high spot) with an air gas layer, where $d = 0.37$ mm; (c) a dry spot with hexagons in the surrounding region, with an air gas layer and $d = 0.25$ mm; (d) hexagons with an air gas layer, where $d = 0.45$ mm. From VanHook et al. [35]. Courtesy of S.J. VanHook and H.L. Swinney.

heated plate), or even high spots (surface elevations, contacting in some cases the upper cooled plate). While it had been theoretically predicted earlier that heated thin films of highly viscous liquids are typically unstable to a large-scale surface deformational mode (see §3.6), and that such instability could lead to the formation of dry spots in the nonlinear regime (see §4.6.3), the work of VanHook et al. not only confirmed these predictions, but also elucidated the role of the gas phase in this instability. In particular, in addition to dry spots, the system may also spontaneously form high spots [Fig. 1.8(b)] and in some range of depths, mixed states of large-scale deformations and short-scale hexagonal patterns [Fig. 1.8(c)]. Such effects have also been theoretically predicted (see §4.6.3 and §4.6.4), using long-wave lubrication-type equations whose general features are discussed in §4.6.

Now, for liquid depths of the order of 1 mm or more (but still without significant influence of buoyancy), Schatz has recently been able to increase the driving constraint sufficiently high above threshold to observe further instabilities of surface-tension-driven patterns. At very high supercriticality ϵ, the convection cells not only acquire irregular polygonal shapes and larger size, but also become time-dependent (Fig. 1.9).

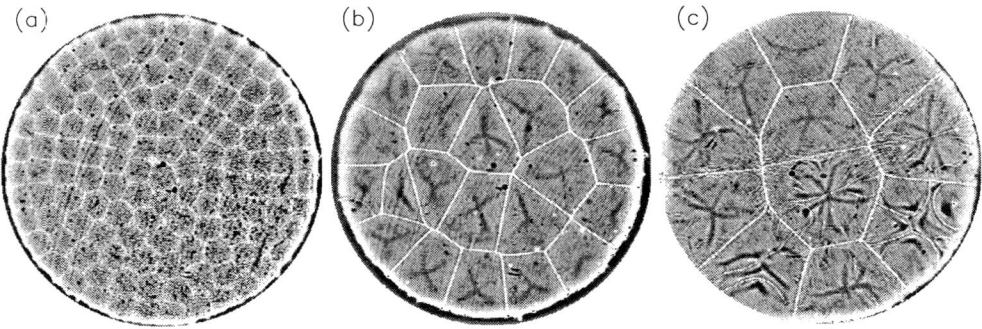

Figure 1.9: Observation of pattern coarsening in Marangoni–Bénard convection as ϵ is increased beyond the onset of squares for $d = 1.22$ mm. The convection cells become both larger and time-dependent with increasing ϵ. (a) $\epsilon = 10.5$; (b) $\epsilon = 18.3$; (c) $\epsilon = 47.8$. Courtesy of M. Schatz (unpublished).

At these high values of ϵ, the boundaries of convection cells, i.e. the coldest regions, become extremely sharp and straight. This will be explained in Chapter 8, where some models of surface-tension-driven convection at very high thermal gradient will be constructed. As the governing thermo-hydrodynamic equations (see Chapter 2) cannot be solved exactly in this case, these models rely on some assumptions, and should in this sense be considered as phenomenological. Nevertheless, they seem to account for time-dependence of the patterns, in qualitative agreement with dynamical processes observed by Schatz [37]. Note that similar time-dependent patterns have also been observed in systems of interest to chemical engineering, i.e. in mass-transfer, rather than heat-transfer, experiments [38, 39, 40]. Still, several aspects of highly supercritical convection patterns, such as the coarsening phenomenon observed in Fig. 1.9, remain unexplained, and should be the subject of future experimental and theoretical work.

Experiments such as those shown in the previous figures concern *spatially extended* systems, i.e. the size of the vessel containing the liquid layer(s) is much larger in horizontal extent than the typical size of the convection cells. When the lateral size of the container is comparable with the size of convection cells, the walls may drastically influence the pattern shape, as illustrated in Fig. 1.10.

Strong confinement by lateral walls not only affects the shape of the convection patterns (see also [42]), but may also induce complex dynamical regimes even near the first instability threshold from the motionless conductive state. This may typically happen when the lateral size of the container is near some particular values, for which two different convective structures (or modes) can be amplified. For reasons explained in Chapter 7, one then speaks of a *codimension-2 bifurcation*. For instance, Johnson and Narayanan [43] have experimentally observed dynamical switchings between two convective modes in circular containers with aspect ratio (radius over depth) equal to 2.5 (see Fig. 7.8 of 7.1). The linear and weakly nonlinear regimes of surface-tension-driven convection in small aspect ratio containers are considered

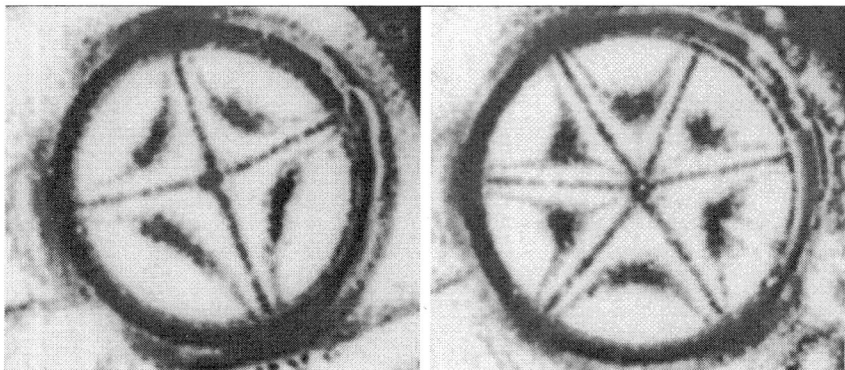

Figure 1.10: Surface-tension-driven patterns observed in conditions of reduced gravity (hence, with negligible buoyancy) during the MAXUS 2 sounding-rocket experiment (1995). Convection is observed, using aluminum flakes, within circular holes in a Polycarbonate solid lid (1.5 mm thick) covering the liquid/gas interface. The liquid is silicone oil 10 cSt, with depth $d = 3$ mm, and left and right figures correspond to circular openings of radii 12 mm and 15 mm, respectively. The flow is upwards at the center of each triangle-like cell, downwards at its periphery and near the contact point with the solid lid. Courtesy of D. Schwabe. From [41].

in §3.3.4 and §7.1.

Finally, surface-tension-driven instabilities can also lead to solitonic nonlinear waves in heat or mass-transfer driven systems, when a single layer of liquid is either heated from *above*, or allowed to *absorb* a surface-active substance. Typical examples of periodic wave trains observed in an annular geometry are shown in Fig. 1.11.

Linde, Velarde and collaborators have considered numerous combinations of liquids and surfactants (see e.g. [44, 45, 46]), and identified some of the observed waves as dissipative *solitons* and *cnoidal* waves (see e.g. [47, 48]). In systems heated from above, or subject to surfactant absorption, a variety of oscillatory instabilities may be triggered, and recent work by Rednikov et al. [49, 50] has allowed some of their generic features to be partly understood (see §3.4, §3.6.2, §6.3). Some relevant experimental findings and related nonlinear theories form the subject of Chapter 6.

1.3   Mathematical description of bifurcation phenomena

Let us consider for illustration the roll mode of Bénard convection represented in Fig. 1.1. For our purpose in the following discussion, a convective mode may be defined as a particular coupling between velocity and temperature fluctuations (with respect to the reference conductive state), made possible by the thermal expansion of the fluid and the associated buoyancy effect, and/or by the variation of surface tension with temperature.

Given the spatial structure of such a mode, the mathematical description of the problem

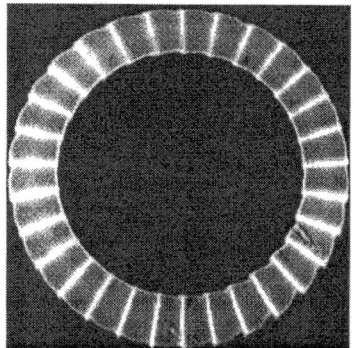 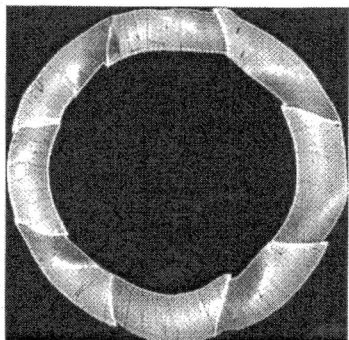

Figure 1.11: Typical two-dimensional (left) and three-dimensional (right) waves observed during absorption of pentane vapor in liquid toluene, visualized using a shadowgraph method in an annular container. The waves propagate in azimuthal directions, clockwise or counter-clockwise, and conserve the same wavelength and phase speed during many turns.

often relies on the calculation of its amplitude, i.e. a multiplicative factor determining the intensity of convection (as well as of the temperature fluctuation). When increasing the temperature difference above the threshold, the amplitude of the unstable modes grows with time, indicating the instability of the reference state, and may be expected to reach a stationary value after some time. One of the purposes of the present text is to describe the evolution of the amplitudes of these near-critical modes via a system of evolution equations (generally having a universal form) called amplitude, or Landau equations. We will now briefly discuss some basic characteristics of such systems of equations.

Above a certain critical value of the control parameter(s), a certain number of modes become unstable, and their evolution and nonlinear interactions determine the spatio-temporal properties of the emerging dissipative structures. The number of bifurcating modes is strongly linked to the size of the system under investigation. In small systems (compared with some correlation length of the fluctuations), the spatial structure at onset is strongly determined by the shape of the boundaries enclosing the system, and typically a small number of modes (and possibly a single one) have a positive growth rate at some slightly supercritical value of the constraint. As the spatial structure of these modes is "frozen" by the small size of the system, the nonlinear problem amounts to determine the temporal evolution of the amplitudes of the most unstable modes. This evolution is generally governed by a low-dimensional system of ordinary differential equations (ODEs) for these amplitudes, which can in principle be obtained via a method known as Center Manifold Reduction [51]. It turns out that the reduction of the infinite-dimensional problem to a low-dimensional dynamical system is possible, because in the neighborhood of the critical point, most of the modes are still strongly damped (large negative growth rate) and are adiabatically slaved to the "master" modes, i.e. those whose growth rate is close to zero (eventually positive). In the phase space of the system, the trajectory remains confined (at least after a short transient) in a subspace of much smaller dimensionality than the full phase space.

For spatially extended systems on the contrary, the number of unstable modes rapidly increases with the control parameter, and the observed behaviors typically involve a large number of degrees of freedom. If the "slaving" approach described above is adopted, the large number of ODEs rapidly makes the method ineffective. A better approach in this case is to consider the lateral boundaries as perturbations of an infinitely large system. This typically leads to systems of a few partial differential equations (PDEs), describing the evolution of a small number of amplitudes (envelope functions, or order parameters), which depend not only on time, but also on some (slow) spatial coordinates. It is worth noting that the amplitude equations obtained are very similar to equations used for describing equilibrium phase transition phenomena (e.g. Ginzburg–Landau equations [52]). Such equations eventually provide a satisfactory description of a number of phenomena observed experimentally, such as selection of the wavelength of the pattern by finite-size effects, evolution of imperfections and defects (dislocations, domain walls, ...), secondary instabilities, ... In fact, the analysis of non-equilibrium transitions in spatially extended systems has strongly benefited from the previous knowledge of equilibrium phase transitions, to which many concepts and vocabulary were borrowed. It is also remarkable that such important similarities in macroscopic behaviors are observed in systems that are described by different "microscopic" equations (typically an extremum principle for equilibrium systems, versus a system of deterministic evolution equations such as Navier–Stokes equations for fluid dynamics problems).

Intuitively, it may be expected that the degree of complexity (spatial or temporal) of the behaviors observed above the instability threshold is directly related to the number of bifurcating modes. Such a view is not entirely correct however, at least for two reasons. First, in rotationally invariant systems such as the Bénard set-up described above, rolls could have any orientation in the horizontal plane, and the description of the weakly nonlinear problem requires to take into account the nonlinear competitions between a large number of roll modes (possibly infinite) with different orientations. It can be shown however that such competition is generally governed by variational principles [53], which basically prevent the system from reaching any time-dependent asymptotic behavior. Second, chaotic behaviors have been predicted in numerous low-dimensional dynamical systems, the first example of which is the Lorenz model [54], describing the nonlinear coupled evolution of only three modes, and providing a qualitative picture of buoyancy-driven convection and transition to turbulence in the Rayleigh–Bénard problem. On the other hand, turbulent phenomena in fluids are known to involve a considerable range of length and time scales [55, 56, 57], and the question of the relevance of low-dimensional chaotic behaviors to such phenomena is still subject to debate. In particular, the transition to turbulence in Bénard convection is known to be strongly linked to secondary instabilities of thermal boundary layers near bounding plates [58, 59, 60], as also discussed in Chapter 8 for the case of surface-tension-driven convection.

1.4   Application-oriented aspects

In addition to the fundamental interest of Bénard instabilities, practical aspects are numerous, in both Rayleigh–Bénard (buoyancy-driven) and Marangoni–Bénard (surface-tension-driven) situations. When buoyancy and surface tension are acting together, we shall speak of

Rayleigh–Marangoni–Bénard convection. Rayleigh–Bénard instability phenomena are at the origin of large-scale convective motions in the Earth's mantle, thereby affecting the drift of continents (plate tectonics). Moreover, buoyancy induces motions in the Earth's atmosphere (and in oceans), which are greatly determinant as far as short- and medium-term weather variations are concerned. In the Sun, buoyancy-induced convection affects the transport of heat from the center to the surface, thus strongly participating in the global solar activity.

In smaller scale laboratory and industrial set-ups, both Rayleigh and Marangoni–Bénard instabilities also yield important modifications of the heat/mass transfer characteristics. In fact, as we shall see later, the onset of Bénard convection always leads to an increase of the total heat/mass transfer (conductive and convective) through the layers. Although this principle cannot be considered as general, it will appear that many nonlinear transitions (observed when increasing the constraint) indeed lead to increasing efficiency of the convective heat/mass transfer. In surface-tension-driven situations, this transfer occurs through interfaces, and important consequences may be expected in techniques such as liquid/liquid extraction, liquid/gas absorption or desorption, distillation, ... In many situations of industrial interest, order of magnitude changes in the rates of transfer from one phase to another have been measured and correlated with empirical relationships (see e.g. [61, 62, 63, 64, 65, 66, 67, 68, 69, 70, 71, 72, 73]). Marangoni–Bénard convection is also important in phase separation processes, or more generally in processes involving phase change (e.g. evaporation [40, 74]) such as in heat exchangers based on the transport of (latent) heat by vapor (heat pipes), in spray drying, in the drying of paint films, in the coating industry, ... While these devices and techniques have mostly been designed and used empirically, it is believed that an understanding of the basic mechanisms could lead to optimization of these processes (which depend on properties such as direction of the transfer and bulk diffusivities [75], interfacial rheology [76], surface chemical reactions [77, 78, 79, 80], ...).

Actually, the increase of the interfacial mass exchange coefficient (i.e. per unit surface) by spontaneous "interfacial turbulence" is not the only factor influencing the overall transfer rates. In many industrial processes, Marangoni flows induce significant interfacial deformation, leading to important variations of the total surface of exchange itself. This is discussed in more details in the thesis of Molenkamp [81], who also provides a rather complete bibliography on possible industrial applications. We also want to point out here the example of thin-film evaporators, used e.g. in the food and pharmaceutical industries, where liquid flowing on a heated plate (at reduced pressure) may organize in rivulet-like structures, with increased transfer rate in regions where the film is thinner, though however with the possible appearance of dry spots [82].

Finally, in crystal growth onboard space laboratories (where the masking effect of buoyancy is drastically reduced), interfacial phenomena are of primary importance, and generally play a significant role as far as the quality of the produced materials is concerned. The interest in studying Marangoni convection in the latter case was motivated by the research started a few decades ago, with the hope that the weightlessness environment could allow growing crystals of better quality than those grown on earth. Instabilities of the so-called floating zone configuration, and of the half-zone (liquid bridge) model, have been extensively studied, in connection with hydrothermal and surface instabilities predicted and observed in systems with interfaces submitted to lateral temperature gradients [83, 84, 85, 86, 87, 88, 89, 90, 91, 92, 93, 94].

Before embarking on detailed descriptions of surface-tension-driven Bénard flows, we point out a number of useful references which might be of interest to the reader, in addition to those already cited. There are classical books on theoretical aspects of hydrodynamic instabilities [95, 96, 97, 98], early theoretical papers on Rayleigh–Marangoni–Bénard instabilities [99, 100, 101, 102, 103, 104], and more experiments on surface-tension-driven convection [105, 106, 107, 108, 109, 110, 111, 112, 113, 114, 115, 116, 117].

2 Balance equations and boundary conditions

Hydrodynamic systems are subject to a variety of instabilities. Their description is most commonly achieved in terms of macroscopic equations expressing the conservation of fundamental quantities such as mass, momentum and energy. A succinct derivation of these equations is presented in this chapter, which should merely be considered as a quick reference guide, and by no means pretends to be exhaustive. After a derivation of equations applying in liquid bulks, we discuss interfacial boundary conditions, i.e. equations expressing the balance of conserved quantities across interfaces separating different phases. These thermo-hydrodynamic balance equations have to be complemented by constitutive and state equations specific to the media or interface under consideration. Some useful concepts of the Thermodynamics of Irreversible Processes will thus also be included for this purpose.

The derivation of conservation equations and constitutive relations presented in this chapter can be found in Glansdorff and Prigogine [2], Aris [182], de Groot and Mazur [188], Haase [3], Platten and Legros [24], and Vidal, Dewel and Borckmans [184]. Although these equations are now well-known and commonly used, completeness here requires their self-consistent derivation from first principles. This will enable us to stress certain assumptions limiting the scope of the present book, as well as to set notation used in later chapters. Moreover, the derivation of interfacial boundary conditions at deformable interfaces is less common in textbooks and often subject to subtle and not so explicitly discussed assumptions. Finally, it is hoped that sufficient material is gathered in this chapter, to allow the reader to apply some of the techniques presented in later chapters to other systems than those specifically studied.

2.1 Thermo-Hydrodynamic equations

Some basic material concerning the description of fluid motion is given in Appendix A. Hereafter, $\vec{x}$ denotes the position vector within the fluid volume, $\vec{v}$ the fluid velocity defined by Eq. (A.1), while d/dt stands for the material derivative, defined by Eq. (A.2).

2.1.1 Mass conservation and continuity equation

If $\rho(\vec{x}, t)$ is the specific mass of the fluid, the total mass in a volume $V(t)$ is

$$m = \iiint_{V(t)} \rho(\vec{x}, t) dV \tag{2.1}$$

The principle of conservation of mass states that in the absence of sources or sinks internal to the volume V, m should not change during fluid motion [$V(t)$ is composed of the same fluid particles]. Applying Reynolds' Transport Theorem (A.11) it follows that

$$\frac{dm}{dt} = \iiint_{V(t)} \left(\frac{\partial \rho}{\partial t} + \vec{\nabla}.(\rho \vec{v}) \right) dV \tag{2.2}$$

where $\vec{\nabla}$ denotes the gradient operator. As Eq. (2.2) is valid for any volume $V(t)$, the integrand must vanish, leading to the local form of the continuity equation

$$\frac{\partial \rho}{\partial t} + \vec{\nabla}.(\rho \vec{v}) = \frac{d\rho}{dt} + \rho \vec{\nabla}.\vec{v} = 0 \tag{2.3}$$

Thus, for an incompressible fluid [see Eq. (A.8)],

$$\vec{\nabla}.\vec{v} = 0 \quad or \quad \frac{d\rho}{dt} = 0 \tag{2.4}$$

are equivalent expressions.

2.1.2   Momentum balance

Newton's law of conservation of the momentum states that the rate of change of linear momentum $\rho \vec{v}$ in a volume $V(t)$ is equal to the sum of forces acting on this volume. Forces may be external or body forces, denoted by $\rho \vec{f}$ per unit volume, and surface or contact forces, which act on the volume $V(t)$ through its bounding surface $S(t)$. The Cauchy Stress Principle states that the force $\vec{t}_{\vec{n}}$ per unit area acting on an element of surface dS with outwards normal $\vec{n}$ is a function of position $\vec{x}$, of time t and of $\vec{n}$. Then, the conservation of momentum reads

$$\frac{d}{dt} \iiint_{V(t)} \rho \vec{v} dV = \iiint_{V(t)} \rho \vec{f} dV + \iint_{S(t)} \vec{t}_{\vec{n}} dS \tag{2.5}$$

Note that the equilibrium of local stresses implies [182] that $\vec{t}_{\vec{n}}$ is related to the direction $\vec{n}$ by a second order tensor $\mathbf{T}$

$$\vec{t}_{\vec{n}} = \mathbf{T}.\vec{n} \quad or \quad (t_{\vec{n}})_i = T_{i,j} n_j \tag{2.6}$$

where summation holds on repeated indices. Using Eqs (A.12), (2.6) and Green's Theorem, the local form of Eq. (2.5) is obtained as

$$\rho \vec{a} = \rho \frac{d\vec{v}}{dt} = \rho \vec{f} + \vec{\nabla}.\mathbf{T} \tag{2.7}$$

where $\vec{a}$ is the acceleration, and the divergence of the stress tensor $\vec{\nabla}.\mathbf{T}$ is a vector with components

$$(\vec{\nabla}.\mathbf{T})_i = \frac{\partial}{\partial x_j} T_{i,j} \tag{2.8}$$

Equation (2.7) is Cauchy's equation of motion, which holds for any continuum of matter, whatever the particular form of the stress tensor $\mathbf{T}$. For the fluids considered here, the stress tensor $\mathbf{T}$ will be taken symmetric, i.e. $T_{i,j} = T_{j,i}$. This results from the law of conservation of angular momentum, valid in situations where there are no couple stresses (nonpolar fluid) nor body torques [182].

When the fluid is at rest, each element of area experiences a stress normal to itself, and $\mathbf{T}$ is diagonal

$$T_{i,j} = -p\delta_{i,j} \tag{2.9}$$

where p can be identified with the thermodynamic pressure.

The stress tensor $\mathbf{T}$ may always be rewritten

$$T_{i,j} = -p\delta_{i,j} + P_{i,j} \tag{2.10}$$

where $P_{i,j}$ is called the viscous stress tensor, which vanishes when the fluid is at rest. It represents the stresses induced by fluid motions, and should thus be related to the velocity gradients. In Eq. (2.10), p may still be considered as the thermodynamic pressure, as long as each fluid element may be assumed in a state of *local equilibrium*, i.e. thermo-hydrodynamic gradients will be considered weak enough[1].

Later, we will come back to the functional form of p and $\mathbf{P}$ which are needed to make the problem self-consistent. The equation for p is called an equation of state, relating the pressure to other intensive variables, while the equation for $\mathbf{P}$, derived in §2.1.6, is called a constitutive equation, thus specifying the type of fluid and flow. In particular, when $\mathbf{P}$ vanishes identically, we have a perfect, ideal, inviscid or Euler fluid flow.

2.1.3 Energy conservation

The First Principle of Thermodynamics (conservation of energy) states that the variation of total energy (kinetic + internal) inside a material volume $V(t)$ is equal to the heat transferred inside $V(t)$ through its bounding surface $S(t)$, plus the work done on $V(t)$. In our notations, this reads

$$\frac{d}{dt} \iiint_{V(t)} \rho \left(\frac{1}{2}v^2 + \epsilon_{\text{int}} \right) dV = \iint_{S(t)} -\vec{q}.\vec{n}dS + \iint_{S(t)} \vec{t}_{\vec{n}}.\vec{v}dS + \iiint_{V(t)} \rho\vec{f}.\vec{v}dV \tag{2.11}$$

where ϵ_{int} is the specific internal energy, $\vec{q}$ is the heat flux and other quantities have been defined earlier. Using Eqs (A.12), (2.6) and Green's Theorem, we obtain the local form

$$\rho\frac{d}{dt}(\frac{1}{2}v^2 + \epsilon_{\text{int}}) = -\vec{\nabla}.\vec{q} + \frac{\partial}{\partial x_j}(T_{i,j}v_i) + \rho\vec{f}.\vec{v} \tag{2.12}$$

Multiplying Eq. (2.7) by $\vec{v}$, we get the rate of change of the kinetic energy, which can then be subtracted from Eq. (2.12). This leads to

$$\rho\frac{d}{dt}\epsilon_{\text{int}} = -\vec{\nabla}.\vec{q} + T_{i,j}\frac{\partial v_i}{\partial x_j} \tag{2.13}$$

---

[1]This means that even though the fluid experiences gradients of concentration, temperature, ... there still exists volume elements in which equilibrium is achieved, i.e. in which thermodynamic functions may be considered as homogeneous. Two nearby fluid elements are thus in equilibrium states corresponding to infinitesimally different values of the state variables. For more detailed explanations, see [2, 3, 24, 182, 183, 184, 188].

Using thermodynamic relationships, the latter equation can be written as an equation for the temperature T. If the assumption of local equilibrium is adopted (this implies that the internal energy and other intensive variables locally depend on the same state variables as in equilibrium), then $\epsilon_{\text{int}} = \epsilon_{\text{int}}(v, T)$ where $v = 1/\rho$ is the specific volume. Accordingly [24, 184], we have

$$\frac{d\epsilon_{\text{int}}}{dt} = c_v \frac{dT}{dt} + \left[T \left(\frac{\partial p}{\partial T} \right)_v - p \right] \frac{dv}{dt} \tag{2.14}$$

where $c_v = (\partial \epsilon_{\text{int}}/\partial T)_v = T(\partial s/\partial T)_v$ is the specific heat at constant volume (s is the specific entropy). Thus, Eq. (2.13) reads, using Eqs (2.14), (2.3) and (2.10),

$$\begin{aligned} \rho c_v \frac{dT}{dt} &= \left[p - T \left(\frac{\partial p}{\partial T} \right)_v \right] (\vec{\nabla}.\vec{v}) - \vec{\nabla}.\vec{q} + T_{i,j} \frac{\partial v_i}{\partial x_j} \\ &= -T \left(\frac{\partial p}{\partial T} \right)_v (\vec{\nabla}.\vec{v}) - \vec{\nabla}.\vec{q} + P_{i,j} \frac{\partial v_i}{\partial x_j} \end{aligned} \tag{2.15}$$

Thus, the first term on the r.h.s. of the second identity is associated with compressibility effects, the second one with heat conduction and the third one is the heat production due to viscous stresses (viscous heating). Note that for an incompressible single-component fluid for which viscous heating is negligible (as is the case for most liquids under usual conditions), the energy equation takes the simpler form

$$\rho c_p \frac{dT}{dt} = \rho c_p \left[\frac{\partial T}{\partial t} + (\vec{v}.\vec{\nabla})T \right] = -\vec{\nabla}.\vec{q} \tag{2.16}$$

where we have also used $c_v \simeq c_p$, valid for liquids. Note that neglecting viscous heating greatly simplifies the energy equation, an approximation which will be further discussed in §3.1.4, when considering the Boussinesq–Oberbeck approximation of thermo-hydrodynamic equations.

2.1.4   Multicomponent liquids

Up to now, only single-component liquids have been considered. As some problems treated in this book involve binary mixtures, we now derive the equations expressing the individual species conservation, as well as the corresponding modifications of the energy equation. In the general case of a mixture of c components $\gamma = 1, ..., c$, the specific mass or partial density of component γ is defined as

$$\rho_\gamma = \lim_{\Delta V \to 0} \frac{\Delta m_\gamma}{\Delta V} \tag{2.17}$$

where Δm_γ is the mass of component γ inside ΔV. If $\Delta m = \Sigma_{\gamma=1}^c \Delta m_\gamma$ is the total mass inside ΔV, the total density or specific mass of the mixture is given by

$$\rho = \lim_{\Delta V \to 0} \frac{\Delta m}{\Delta V} = \sum_{\gamma=1}^c \rho_\gamma \tag{2.18}$$

and the local composition of the mixture is characterized by the mass fractions

$$N_\gamma = \lim_{\Delta V \to 0} \frac{\Delta m_\gamma}{\Delta m} = \frac{\rho_\gamma}{\rho} \tag{2.19}$$

which satisfy $\Sigma_\gamma N_\gamma = 1$.

As, in the absence of chemical reactions, the conservation of mass applies to each component individually, we get a conservation equation for ρ_γ identical to Eq. (2.3). However, if chemical reactions occur, an additional source term accounting for the production of γ is to be added. Considering a single reaction to simplify notation, we then have

$$\frac{\partial \rho_\gamma}{\partial t} + \vec{\nabla}.(\rho_\gamma \vec{v}_\gamma) = w \nu_\gamma M_\gamma \tag{2.20}$$

where $\vec{v}_\gamma$ is the velocity of component γ, w is the volumic rate of the chemical reaction, and ν_γ and M_γ are respectively the stoichiometric coefficient and the molecular weight of species γ.

Now, as $\Sigma_\gamma \nu_\gamma M_\gamma = 0$, Eq. (2.20) summed up on γ reduces to Eq. (2.3), i.e. the continuity equation, provided the fluid velocity is understood as the barycentric velocity

$$\vec{v} = \frac{1}{\rho} \sum_\gamma \rho_\gamma \vec{v}_\gamma \tag{2.21}$$

The deviation of the velocity of component γ with respect to the barycentric velocity $\vec{v}$, i.e.

$$\vec{\Delta}_\gamma = \vec{v}_\gamma - \vec{v} \tag{2.22}$$

is due to non-convective (diffusive) processes to be discussed later on (§2.1.6). Note that the diffusion fluxes will here always be defined relative to the velocity of the center of mass (barycentric system). The reader is referred to [3, 184, 188] for definitions of other reference systems.

Finally, the individual species conservation equation reads

$$\frac{\partial \rho_\gamma}{\partial t} + \vec{\nabla}.(\rho_\gamma \vec{v}) = -\vec{\nabla}.(\rho_\gamma \vec{\Delta}_\gamma) + w \nu_\gamma M_\gamma \tag{2.23}$$

or, using the relation $\rho_\gamma = \rho N_\gamma$ and the continuity equation (2.3),

$$\rho \frac{dN_\gamma}{dt} = -\vec{\nabla}.(\rho_\gamma \vec{\Delta}_\gamma) + w \nu_\gamma M_\gamma \tag{2.24}$$

From the definition (2.21) of $\vec{v}$, it is seen that $\rho \vec{v}$ represents the *total* linear momentum, which guarantees that the momentum conservation law [i.e. Cauchy's equation (2.7)] remains valid for a multicomponent system. However, the external force per unit volume is now given by $\rho \vec{f} = \Sigma_\gamma \rho_\gamma \vec{f}_\gamma$, where $\vec{f}_\gamma$ is the external force per unit mass acting on component γ.

Finally, we consider the modifications of the energy conservation equation (2.15) [or (2.16)]. Keeping the form $\epsilon_{kin} = \rho v^2/2$ of the kinetic energy (see [3, 24, 184] and a discussion of the validity of this approximation in Ref. [188]), $\rho \vec{f}.\vec{v}$ has to be replaced by $\Sigma_\gamma \rho_\gamma \vec{f}_\gamma.\vec{v}_\gamma$

(work done by external forces acting on individual species) in Eq. (2.12). Moreover, a term must be added in case chemical reactions occur, if these produce a variation of the potential energy. Assuming that external forces derive from a potential, i.e. $\vec{f}_\gamma = -\vec{\nabla}\phi_\gamma$, the energy equation (2.12) now reads

$$\rho\frac{d}{dt}\left(\frac{1}{2}v^2 + \epsilon_{\text{int}}\right) = -\vec{\nabla}.\vec{q} + \frac{\partial}{\partial x_j}(T_{i,j}v_i) + \sum_\gamma \rho_\gamma \vec{f}_\gamma.\vec{v}_\gamma - w\sum_\gamma \nu_\gamma M_\gamma \phi_\gamma \quad (2.25)$$

From this equation, we again subtract the rate of change of the kinetic energy. With the definition (2.22), we then obtain

$$\rho\frac{d\epsilon_{\text{int}}}{dt} = -\vec{\nabla}.\vec{q} + T_{i,j}\frac{\partial v_i}{\partial x_j} + \sum_\gamma \rho_\gamma \vec{f}_\gamma.\vec{\Delta}_\gamma - w\sum_\gamma \nu_\gamma M_\gamma \phi_\gamma \quad (2.26)$$

For gravitational forces (no other body forces will be considered in this book), $\phi_\gamma = \phi$ and $\vec{f}_\gamma = \vec{f}$ is independent of γ. Accordingly, the last two terms in Eq. (2.26) vanish, i.e. the energy equation reduces to Eq. (2.13). We will always omit those terms in the following. However, the thermodynamic relationships used for obtaining Eq. (2.15) from Eq. (2.13) are different when multicomponent systems are considered. In such a case, it proves more convenient to evaluate partial derivatives and thermodynamic coefficients at constant pressure (and temperature), rather than at constant volume as was the case for single-component fluids. Introducing the specific enthalpy $h = \epsilon_{\text{int}} + pv$, and using the continuity equation (2.3) and the definition (2.10) of the viscous stress tensor $\mathbf{P}$, Eq. (2.26) yields

$$\rho\frac{dh}{dt} = -\vec{\nabla}.\vec{q} + P_{i,j}\frac{\partial v_i}{\partial x_j} + \frac{dp}{dt} \quad (2.27)$$

Assuming that $h = h(T, p, N_\gamma)$, we have

$$dh = c_p dT + \left(\frac{\partial h}{\partial p}\right)_{T,N_\gamma} dp + h_\gamma dN_\gamma \quad (2.28)$$

where c_p is the specific heat at constant pressure and composition, and h_γ is the specific partial enthalpy of component γ. Using the species conservation equation (2.24), the continuity equation (2.3), and the thermodynamic relationship

$$1 - \rho\left(\frac{\partial h}{\partial p}\right)_{T,N_\gamma} = \rho T\left(\frac{\partial v}{\partial T}\right)_{p,N_\gamma} = \left(\frac{\partial \ln v}{\partial \ln T}\right)_{p,N_\gamma} \quad (2.29)$$

then combining Eqs (2.27) and (2.28) leads to

$$\rho c_p \frac{dT}{dt} = -\vec{\nabla}.\vec{q}^+ + P_{i,j}\frac{\partial v_i}{\partial x_j} + \left(\frac{\partial \ln v}{\partial \ln T}\right)_{p,N_\gamma}\frac{dp}{dt} - w\sum_\gamma \nu_\gamma h_\gamma M_\gamma \quad (2.30)$$

where

$$\vec{q}^+ = \vec{q} - \sum_\gamma \rho_\gamma h_\gamma \vec{\Delta}_\gamma \quad (2.31)$$

is a modified heat flux, accounting for the enthalpy transfer due to diffusive processes. Note that Eq. (2.30) also includes the production of heat due the chemical reaction (last term). As for single-component fluids, a simpler form of the energy equation is obtained for incompressible liquids ($v = 1/\rho =$ constant, $c_p \simeq c_v$), viscous heating can be neglected, and no chemical reactions occur. We then obtain

$$\rho c_p \frac{dT}{dt} = -\vec{\nabla}.\vec{q}^{+}$$
(2.32)

which is the form generally used throughout this book.

2.1.5 Entropy balance

In this section, we outline the derivation of the balance equation for the entropy. This balance reads

$$\rho \frac{ds}{dt} = -\vec{\nabla}.\vec{J}_s + \sigma_s$$
(2.33)

where s is the specific entropy, $\vec{J}_s$ is the entropy flux and σ_s the entropy production per unit mass. The second principle of thermodynamics states that σ_s is semi-definite positive.

If we maintain the local equilibrium hypothesis, the Gibbs relationship is still valid

$$Tds = d\epsilon_{\text{int}} + pdv - \sum_\gamma \mu_\gamma dN_\gamma$$
(2.34)

where μ_γ is the chemical potential of component γ.

The Gibbs identity (2.34) is written in the coordinate system of the fluid particle. Using Eqs (2.3), (2.10), (2.24) and (2.26), and introducing the affinity [1]

$$A = -\sum_\gamma \nu_\gamma \mu_\gamma M_\gamma$$
(2.35)

of the chemical reaction considered, it is found that the entropy evolution is indeed governed by (2.33), with the entropy flux

$$\vec{J}_s = \frac{\vec{q}}{T} - \sum_\gamma (\mu_\gamma T^{-1}) \rho_\gamma \vec{\Delta}_\gamma$$
(2.36)

and the local entropy source

$$\begin{aligned}\sigma_s = \vec{q}.\vec{\nabla} T^{-1} - \sum_\gamma (\rho_\gamma \vec{\Delta}_\gamma)\left[\vec{\nabla}(\mu_\gamma T^{-1}) + T^{-1}\vec{\nabla}\phi_\gamma\right] \\ + T^{-1} P_{i,j}\partial v_i/\partial x_j + T^{-1}w(A - \sum_\gamma \nu_\gamma M_\gamma \phi_\gamma)\end{aligned}$$
(2.37)

where $\vec{f}_\gamma$ has been replaced by $-\vec{\nabla}\phi_\gamma$, i.e. we have again assumed that body forces derive from a potential. This allows a redefinition of the chemical affinity $\tilde{A} = -\Sigma_\gamma \nu_\gamma M_\gamma(\mu_\gamma + \phi_\gamma)$. We also make use of the Gibbs–Helmholtz relation

$$\frac{\partial}{\partial T}\left(\frac{\mu_\gamma}{T}\right)_{p,N_\gamma} = -\frac{h_\gamma}{T^2}$$
(2.38)

from which we have $\vec{\nabla}(\mu_\gamma T^{-1}) = T^{-1}(\vec{\nabla}\mu_\gamma)_T + h_\gamma\vec{\nabla}T^{-1}$, where $(\vec{\nabla}\mu_\gamma)_T$ is the isothermal gradient of μ_γ, i.e. the gradient due to the spatial variation of p and N_γ at constant T. Then, Eq. (2.37) leads [188, 184, 24] to

$$\sigma_s = \vec{q}^{+}.\vec{\nabla}T^{-1} - T^{-1}\sum_\gamma(\rho_\gamma\vec{\Delta}_\gamma)\left[(\vec{\nabla}\mu_\gamma)_T - \vec{f}_\gamma\right] + T^{-1}P_{i,j}\frac{\partial v_i}{\partial x_j} + T^{-1}\tilde{A}w$$

(2.39)

where $\vec{q}^{+} = \vec{q} - \Sigma_\gamma\rho_\gamma h_\gamma\vec{\Delta}_\gamma$ is the heat flux incorporating the contribution of heat transport due to species diffusion, already encountered when discussing the energy equation (2.30) or (2.32).

2.1.6   Generalized forces and fluxes – Phenomenological relations

In the preceding sections, we introduced unknowns of two kinds: field quantities ρ, $\vec{v}$, T and N_γ (or ρ_γ), and fluxes or rates $T_{i,j} = -p\delta_{i,j} + P_{i,j}$, $\vec{q}$ (or $\vec{q}^{+}$), $\rho_\gamma\vec{\Delta}_\gamma$ and w. The latter quantities appear in the balance equations describing the evolution of the former. Consequently, we have more unknowns than equations, even by incorporating the state equation $\rho = \rho(p, T, N_\gamma)$. On the other hand, experiments show that some relationships exist between the fluxes and the gradients of the state variables (i.e. the forces). The derivation of such constitutive relations is one of the goals of the Thermodynamics of Irreversible Processes, whose formalism defines a general form for the entropy production

$$\sigma_s = \sum_i J_i X_i$$

(2.40)

where J_i and X_i denote generalized fluxes and forces, respectively. In fact, different forms of σ_s are possible [e.g. Eqs (2.37) or (2.39)], which can always be written under the general form (2.40). The other possible forms can be obtained by transforming the system of fluxes and forces according to linear transformations

$$J_i' = \sum_j a_{ij}J_j \qquad X_i = \sum_j a_{ji}X_j'$$

(2.41)

such that the entropy production, a measurable quantity, is unchanged

$$\sigma_s = \sum_i J_i'X_i' = \sum_i J_i X_i$$

(2.42)

The arbitrariness existing in the definition of generalized forces and fluxes should thus not lead to differences in physical results, such that the discussion which follows can be based on the form (2.39) of σ_s (see Table 2.1).

The form (2.40) of the entropy production allows some important considerations. At thermodynamic equilibrium, J_i and X_i must vanish simultaneously. If the fluxes J_i are assumed to be analytical functions of the fluxes X_i in the neighborhood of equilibrium $X_i = 0$, a Taylor series expansion limited to the first order in X_i yields

$$J_i = \sum_j L_{ij}X_j$$

(2.43)

Table 2.1: A possible choice for the definition of forces X_i and fluxes J_i [24].

J_i	X_i
$\vec{q}^{+}$	$\vec{\nabla}T^{-1}$
$\rho_\gamma \vec{\Delta}_\gamma$	$-T^{-1}\left[(\vec{\nabla}\mu_\gamma)_T - \vec{f}_\gamma\right]$
$P_{i,j}$	$T^{-1}\partial v_i/\partial x_j$
w	$T^{-1}\tilde{A}$

where the proportionality factors L_{ij} are called phenomenological coefficients, which in general are functions of the local state variables (temperature, pressure, composition), but not of the gradients X_i. The relations (2.43) define the linear domain of Non-Equilibrium Thermodynamics [1].

The determination of phenomenological coefficients L_{ij} and of the domain of validity of the linear relationships (2.43) is in principle possible using Non-Equilibrium Statistical Thermodynamics [188], though these results can also be directly obtained from experiment. It turns out that linear laws described in the next sections, such as the Fourier's law of heat conduction, the Fick's law of species diffusion and the Newton's stress-strain relation are excellent approximations in a large range of experimental conditions. For the chemical reaction rates, the linear domain seems to be restricted to a more narrow domain around equilibrium [1, 2, 3, 184].

Finally, it must be emphasized that the form of the entropy production rate (2.40) is fairly general and can also be obtained from consideration of irreversible processes occurring in heterogeneous (discontinuous) systems, such as systems composed of several homogeneous phases. In this case, linear relations such as (2.43) may also be postulated [3, 184], and allow in principle the description of processes such as phase change, osmosis, ... These phenomena will not be examined here, with the exception of the influence of evaporation on thermocapillary convection, discussed in §3.8.

Properties of the phenomenological coefficients

The phenomenological coefficients L_{ij} appearing in relations (2.43) are not all independent, as they should satisfy three types of conditions:

The Second Principle of Thermodynamics

Combining Eqs (2.40) and (2.43), the Second Principle is expressed as

$$\sigma_s = \sum_{i,j} L_{ij} X_i X_j \geq 0 \tag{2.44}$$

As this inequality should be valid whatever the value (and sign) of the generalized forces X_i, it implies that

$$L_{ii} \geq 0 \qquad\qquad \text{for all } i \qquad\qquad\qquad (2.45)$$

$$4L_{ii}L_{jj} \geq (L_{ij} + L_{ji})^2 \qquad \text{for all } i, j \neq i \qquad\qquad (2.46)$$

Note also that only the symmetric part of the matrix of coefficients L_{ij} contributes to the entropy production [1, 2]. Indeed, if $L_{ij} = L_{ij}^s + L_{ij}^a$ (with $L_{ij}^s = L_{ji}^s$ and $L_{ij}^a = -L_{ji}^a$), we get $\sigma_s = \sum_{i,j} L_{ij}^s X_i X_j$.

Curie's Symmetry Principle for isotropic media

As the sums in Eq. (2.43) run over all generalized forces present in the system considered, couplings are apparently possible between processes with different tensorial character (e.g. scalars and vectors). However, not all these couplings are possible in isotropic media, due to symmetry (Curie's Principle). For example, a non-zero chemical affinity (scalar force) cannot be at the origin of a heat flux with a privileged direction (vectorial flux), as long as the fluid considered is indeed isotropic. More precisely, couplings between processes with an odd difference between their tensorial orders (e.g. a scalar and a vector, or a vector and a second order tensor) are forbidden in isotropic media. Thus, a number of cross-effects are a priori excluded (see Table 2.2), and should be characterized by vanishing coefficients L_{ij}.

Table 2.2: Couplings allowed by the Curie's Principle in isotropic media.

	Velocity gradients	Temperature gradients	Concentration gradients	Affinity of chemical reaction
Momentum flux	Newton's law			weak coupling
Heat flux		Fourier's law	Dufour effect	
Mass flux		Soret effect	Fick's law	
Chemical reaction rate	weak coupling			dependent on reaction scheme

Note that a priori nothing contradicts the possibility of a coupling between chemical reactions (order 0) and momentum transfer (order 2), as here the difference between tensorial orders is even. However, the amplitude of such phenomena appears very weak, and does not seem to have been studied systematically [184].

Onsager's reciprocity relations

On the basis of the microscopic reversibility, i.e. the invariance with respect to time-reversal of the evolution equations at the microscopic level (e.g. the Schrödinger equation, or the Hamilton equations), Onsager showed that the matrix of phenomenological coefficients should be symmetric [188], i.e.

$$L_{ij} = L_{ji} \quad \text{for all } i, j \qquad\qquad\qquad (2.47)$$

This result is fundamental when studying cross-effects such as thermodiffusion (Soret effect) whose magnitude is thus related to that of the Dufour effect (see Table 2.2). As the property (2.47) was derived from a molecular statistical theory of non-equilibrium and cannot be obtained from purely macroscopic (e.g. symmetry) arguments, it is generally asserted that they should merely be considered as a postulate at macroscopic level [184, 185]. An important condition underlying the validity of reciprocity relations (2.47) is that the fluxes and the forces should be linearly independent. Indeed, in cases where linear relationships exist between the forces and also between the fluxes, some arbitrariness exists when selecting the set of J_i's and X_i's, and the Onsager's reciprocity relations are not satisfied for all possible choices. Instructive examples can be found in Haase [3], or Vidal et al. [184]. The reader is also referred to [3, 184] for explanations of the Casimir's generalization of Onsager's reciprocity laws, applying in place of Eq. (2.47) to couplings between processes with different parities with respect to time-reversal. For our purpose, Eq. (2.47) proves to be sufficient.

Phenomenological relation for heat diffusion – Fourier's law

Consider a medium in which the only irreversible process is heat transfer by conduction. Then, the entropy production (2.39) reduces to

$$\sigma_s = \vec{q}.\vec{\nabla}T^{-1} = -\frac{\vec{q}}{T^2}.\vec{\nabla}T \tag{2.48}$$

where $\vec{q}^+ = \vec{q}$ as no species diffusion occurs. Now, according to Eq. (2.43) and Table 2.1, the phenomenological relation between the heat flux $\vec{q}$ and the thermal force $\vec{\nabla}T^{-1}$ is

$$\vec{q} = L_{qq}\vec{\nabla}T^{-1} \tag{2.49}$$

in isotropic media (L_{qq} should be replaced by a tensor in anisotropic media).

The determination of the phenomenological coefficient L_{qq} is usually experimental. Measurements show that in isotropic bodies, heat "flows" from hot to cold regions, according to the Fourier's law

$$\vec{q} = -\lambda\vec{\nabla}T \tag{2.50}$$

where λ is the thermal conductivity of the material (which has units W/m K). Identification of Eqs (2.49) and (2.50) leads to $L_{qq} = T^2\lambda$. Note that $\lambda > 0$ in order for σ_s to be positive [see also Eq. (2.45)].

Heat conduction is an irreversible process for which the domain of validity of the phenomenological law (2.49) is quite extended (up to gradients of order 10^9 K/cm), even though the thermal conductivity cannot be considered as independent of temperature in cases where large temperature differences are involved.

Under normal conditions, liquids have thermal conductivities of the order of $0.1-0.6$ W/m K, while the thermal conductivity of metallic solids is much higher (of order 10^2 W/m K). The thermal conductivity of gases is lower than that of liquids (except for helium which has 0.15 W/m K, comparable to the conductivity of silicone oils). Most often in fluid dynamics, one uses the thermometric conductivity or heat diffusivity $\kappa = \lambda/\rho c_p$, which is of the order of 10^{-3} cm²/s for most liquids, and typically two orders of magnitude larger for gases.

Phenomenological relation for species diffusion – Fick's law

In isothermal multicomponent systems at rest and without chemical reactions, the only dissipative mechanism is mass diffusion. Then, the entropy source term (2.39) reads

$$\sigma_s = -T^{-1} \sum_{\gamma=1}^{c} (\rho_\gamma \vec{\Delta}_\gamma) \left[(\vec{\nabla}\mu_\gamma)_T - \vec{f}_\gamma \right] \tag{2.51}$$

where the sum runs over the c components of the mixture. For isotropic binary mixtures ($c = 2$), it was proposed by Fick in 1855 that a relation formally analogous to Fourier's law (2.50) should hold for each mass flux, $\vec{J}_\gamma = \rho_\gamma \vec{\Delta}_\gamma$, in the presence of a gradient of concentration of γ

$$\vec{J}_\gamma = -D_\gamma \vec{\nabla} \rho_\gamma \tag{2.52}$$

For mixtures of more than two components, Onsager proposed to generalize this relation under the form

$$\vec{J}_\gamma = -\sum_{\xi=1}^{c-1} D_{\gamma\xi} \vec{\nabla} \rho_\xi \tag{2.53}$$

where non-diagonal diffusion coefficients $D_{\gamma\xi}$ account for couplings between the mass fluxes of some components and the mass fraction gradients of others. In this respect, it should be stressed that the matrix of diffusion coefficients $D_{\gamma\xi}$ is in general not symmetric. Indeed, even though the Onsager's reciprocity laws (2.47) apply to the phenomenological coefficients L_{ij}, this does not imply the same property for the diffusion coefficients $D_{\gamma\xi}$, which are in general linear combinations of coefficients L_{ij} [184].

According to Eq. (2.51), the generalized force associated with the mass flux $\vec{J}_\gamma = \rho_\gamma \vec{\Delta}_\gamma$ is $\vec{X}_\gamma = -T^{-1}[(\vec{\nabla}\mu_\gamma)_T - \vec{f}_\gamma]$. In this expression, the external force term $\vec{f}_\gamma$ accounts for forced diffusion, i.e. diffusion created by external forces acting *differently* on each species (e.g. ionic systems submitted to electric fields). In the case of gravity, this contribution vanishes. Now, the isothermal gradient of the chemical potential can be rewritten as $(\vec{\nabla}\mu_\gamma)_T = (\vec{\nabla}\mu_\gamma)_{T,p} + v_\gamma \vec{\nabla} p$, where v_γ is the specific volume of component γ. The contribution proportional to the pressure gradient accounts for pressure diffusion, which may lead (e.g. during centrifugation, i.e. at very high pressure gradients) to a net displacement of component γ in the mixture. In fact, Eqs (2.7) and (2.10) imply that when the fluid is at rest (mechanical equilibrium), the pressure gradient is related to the external forces according to $\vec{\nabla} p = \Sigma_\gamma \rho_\gamma \vec{f}_\gamma$. Hereafter, we will neglect both forced and pressure diffusion (e.g. in the absence of external forces), and concentrate on "ordinary diffusion", associated with the force $-T^{-1}(\vec{\nabla}\mu_\gamma)_{T,p}$ due to gradients of composition.

Restricting the following discussion to binary mixtures, we note that there exists a linear relationship between the fluxes, because

$$\vec{J}_1 + \vec{J}_2 = \rho_1 \vec{\Delta}_1 + \rho_2 \vec{\Delta}_2 = 0 \tag{2.54}$$

and also between the forces $-T^{-1}(\vec{\nabla}\mu_\gamma)_{T,p}$, owing to the Gibbs–Duhem relation

$$N_1(\vec{\nabla}\mu_1)_{T,p} + N_2(\vec{\nabla}\mu_2)_{T,p} = 0 \qquad (2.55)$$

In this case, it is preferable to eliminate one of the forces and one of the fluxes using the latter relationships (this is particularly important for mixtures of more than two components [3, 184]). It is then found that the force conjugate to $\vec{J}_1$ is $-T^{-1}N_2^{-1}(\vec{\nabla}\mu_1)_{T,p} = -T^{-1}N_2^{-1}(\partial\mu_1/\partial N_1)_{T,p}\vec{\nabla}N_1$. The phenomenological expression relating them thus reads (for isotropic fluids)

$$\vec{J}_1 = -\frac{L_{jj}}{TN_2}\left(\frac{\partial\mu_1}{\partial N_1}\right)_{T,p}\vec{\nabla}N_1 \qquad (2.56)$$

which may be rewritten under the form

$$\vec{J}_1 = \rho_1\vec{\Delta}_1 = -\rho D\vec{\nabla}N_1 \qquad (2.57)$$

where D is the isothermal diffusion coefficient. It is related to the phenomenological coefficient L_{jj} by $D = L_{jj}(\partial\mu_1/\partial N_1)_{T,p}/\rho TN_2$. Even though the Second Principle implies that $L_{jj} > 0$, this does not imply a condition on D, as $(\partial\mu_1/\partial N_1)_{T,p}$ can be either positive or negative, the latter case corresponding to an unstable thermodynamic situation [2]. Typical orders of magnitudes of the isothermal diffusion coefficient D are 0.1 to 1 cm²/s for gases, 10^{-5} cm²/s for liquid mixtures and 10^{-8} cm²/s for solid alloys.

Phenomenological relations for fluxes in non-isothermal binary mixtures

In this more general case, the entropy production (in the absence of chemical reactions and at mechanical equilibrium) reads

$$\sigma_s = \vec{q}^+.\vec{\nabla}T^{-1} - T^{-1}\sum_{\gamma=1}^{2}(\rho_\gamma\vec{\Delta}_\gamma)\left[(\vec{\nabla}\mu_\gamma)_T - \vec{f}_\gamma\right] \qquad (2.58)$$

Using Eqs (2.54) and (2.55), and assuming that external forces and pressure gradients have a negligible influence, it is found that the force conjugate to $\vec{J}_1$ has the same form as for an isothermal mixture. Then, for isotropic mixtures, phenomenological relations (2.43) may be written in the form [24, 184, 188]

$$\vec{q}^+ = -\lambda\vec{\nabla}T - \rho N_1\left(\frac{\partial\mu_1}{\partial N_1}\right)_{T,p}TD_F\vec{\nabla}N_1 \qquad (2.59)$$

$$\rho_1\vec{\Delta}_1 = -\rho N_1 N_2 D_T\vec{\nabla}T - \rho D\vec{\nabla}N_1 \qquad (2.60)$$

with the definitions

$$\lambda = \frac{L_{11}}{T^2} \qquad\qquad \text{Thermal conductivity}$$

$$D_F = \frac{L_{12}}{\rho N_1 N_2 T^2} \qquad\qquad \text{Dufour coefficient}$$

$$D_T = \frac{L_{21}}{\rho N_1 N_2 T^2} \qquad\qquad \text{Thermal diffusion coefficient}$$

$$D = \frac{L_{22}}{\rho T N_2} \left(\frac{\partial \mu_1}{\partial N_1}\right)_{T,p} \qquad \text{Isothermal diffusion coefficient}$$

with $D_F = D_T$ in virtue of the Onsager's reciprocity relation $L_{12} = L_{21}$. Furthermore, the second principle implies $L_{11}, L_{22} > 0$ [Eq. (2.45)], and $L_{12}^2 < L_{11}L_{22}$ [Eq. (2.46)]. In addition to the conditions $\lambda > 0$ and $D > 0$ [the latter being valid if the thermodynamic stability condition $(\partial \mu_1/\partial N_1)_{T,p} > 0$ is verified], we thus obtain

$$D_T^2 < \frac{\lambda D}{\rho_1 T N_1 N_2 \, (\partial \mu_1/\partial N_1)_{T,p}} \tag{2.61}$$

The relations (2.59–2.60) provide an example of thermodynamic coupling between molecular processes of different nature (heat and mass diffusion). A steady gradient of temperature in a binary mixture induces a mass flux, which leads to a separation of components of the mixture and ultimately to a stationary concentration gradient (Soret effect). This stationary non-equilibrium state (which is characterized by a steady rate of entropy production) corresponds to the vanishing of the total mass flux $\rho_1 \vec{\Delta}_1$. According to (2.60), this leads to

$$\vec{\nabla} N_1^s = -\frac{D_T}{D} N_1 N_2 \vec{\nabla} T^s \tag{2.62}$$

This stationary Soret separation is usually small in magnitude for typical mixtures (the Soret coefficient D_T/D of gas and liquid mixtures is of order 10^{-5}–$10^{-2}\mathrm{K}^{-1}$ with either sign) but may lead to important modifications of mechanical, electrical and chemical properties of materials which are submitted to high temperature gradients and/or subject to interfacial phenomena. Moreover, in liquids, the Soret effect is at the origin of important qualitative and quantitative modifications of hydrodynamic stability properties [272, 273, 24]. The main reason for this is that even small concentration variations can lead to significant variation of physical properties such as density or surface tension.

We finally note that despite the relation $D_F = D_T$ which indicates that the magnitude of Soret and Dufour effects are related, the latter effect is usually negligible in liquids, as the temperature differences induced by composition gradients still remain too small (of order 10^{-3} K in liquids) to affect physical properties significantly. The Dufour effect becomes more relevant for compressible fluids.

Phenomenological relations for momentum transfer – Newton's law

We now turn to the determination of constitutive relations for the stress tensor $T_{i,j}$, in the simplest case where, as for heat and mass diffusion, linear phenomenological relations are assumed valid. Thus, from the beginning we exclude the possibility of viscoelastic properties (memory effects) and non-Newtonian behavior (nonlinearity of the relation between stress and strain). It was shown that $T_{i,j} = -p\delta_{i,j} + P_{i,j}$ [Eq. (2.10)], which defines the viscous stress tensor $P_{i,j}$ and introduces the hydrodynamic pressure p (which can be identified with the thermodynamic pressure under conditions of local equilibrium and for compressible fluids, while it should be taken as an independent variable in the case of incompressible liquids [182]).

For the isothermal viscous flow of a single-component liquid, using Eq. (A.15), the entropy production (2.39) reads

$$\sigma_s = \frac{P_{i,j}}{T}\frac{\partial v_i}{\partial x_j} = \frac{P_{i,j}}{T}(e_{i,j} + \Omega_{i,j}) \tag{2.63}$$

where $e_{i,j} = e_{j,i}$ is the deformation or rate of strain tensor, while $\Omega_{i,j} = -\Omega_{j,i}$ corresponds to rotation of the fluid as a rigid body. In view of Eq. (2.63), phenomenological relations (2.43) should here take the form $P_{i,j} = L_{i,j,p,q}(e_{p,q} + \Omega_{p,q})$, where $L_{i,j,p,q}$ is a fourth-order tensor. For homogeneous fluids, the tensor $\mathbf{L}$ is independent of position. Moreover, $\mathbf{L}$ must be isotropic if the fluid has no preferred direction. Imposing the symmetry of the viscous stress tensor $P_{i,j}$ (see §2.1.2) and using $e_{p,p} = \vec{\nabla}.\vec{v}$, we finally get [182]

$$P_{i,j} = \alpha\delta_{i,j}(\vec{\nabla}.\vec{v}) + 2\mu e_{i,j} \tag{2.64}$$

which contains only two independent phenomenological coefficients α and μ. This relation defines the Newtonian fluid, which is a particular (linear) case of a Stokesian (nonlinear but isotropic) fluid, for which $P_{i,j} = \alpha\delta_{i,j} + \beta e_{i,j} + \gamma e_{i,k}e_{k,j}$ (where α, β and γ may depend on the invariants of the rate of strain tensor, i.e. they are also functions of the velocity gradients). Thus for a Newtonian fluid (and certainly when the velocity gradients are weak enough), μ is the shear or dynamic viscosity, and relates the shear stress to the velocity gradient (strain).

Note that the mean of the principal stresses is

$$\frac{1}{3}T_{i,i} = -p + \frac{1}{3}P_{i,i} = -p + (\alpha + \frac{2}{3}\mu)(\vec{\nabla}.\vec{v}) \tag{2.65}$$

The Stokes hypothesis consists in identifying this quantity with $-p$, where p is the thermodynamic pressure. This implies $\alpha + \frac{2}{3}\mu = 0$, i.e. the vanishing of the coefficient of bulk viscosity. This hypothesis obviously has no implications for incompressible fluids, for which the viscous stress tensor finally reads

$$P_{i,j} = 2\mu e_{i,j} = \mu\left(\frac{\partial v_i}{\partial x_j} + \frac{\partial v_j}{\partial x_i}\right) \tag{2.66}$$

Dynamic viscosities μ of most liquids are of the order 10^{-2} g/cm s, with the exception of oils whose viscosity can be several orders of magnitude larger. Dynamic viscosities of gases are typically of the order of 10^{-4} g/cm s. In hydrodynamics, one also uses the kinematic viscosity (or momentum diffusivity) $\nu = \mu/\rho$, which is of order 10^{-2} cm^2/s (i.e. 1 cSt) for most liquids except oils (e.g. silicone oils have kinematic viscosities up to several thousands cSt). As the density of gases is much lower than that of liquids, their kinematic viscosities can be of the same order, or even larger.

Phenomenological relations for chemical reaction rates

For systems where chemical reactions are the only irreversible processes, the entropy production (2.39) reads

$$\sigma_s = \sum_{\alpha=1}^{r} \frac{A_\alpha w_\alpha}{T} \tag{2.67}$$

where r is the number of simultaneous chemical reactions, and $\tilde{A}_\alpha = A_\alpha$ when the potential of the external forces is independent of the nature of the chemical species ($\phi_\gamma = \phi$), as is the case for gravitational forces (see also §2.1.5). According to Eq. (2.67), phenomenological relations (2.43) should take the form

$$w_\alpha = \sum_{\beta=1}^{r} l_{\alpha\beta} \frac{A_\beta}{T} \tag{2.68}$$

where $l_{\alpha\beta}$ is a matrix of phenomenological coefficients. However, the domain of validity of such constitutive relations is in general limited to a rather small domain in the vicinity of equilibrium, as revealed by comparison with formulas obtained using kinetic expressions for the reaction rates (such as De Donder relations [1, 184]), which are typically nonlinear in the affinities. Thus in most cases, kinetic expressions are preferred over relations (2.68). However, this is only mentioned for completeness, as chemical reactions will be disregarded in the following.

2.2 Boundary conditions

The bulk equations obtained in §2.1 have to be complemented by an appropriate set of boundary conditions. These can hold either at non-deformable boundaries (§2.2.1), or at deformable fluid interfaces (§2.2.2). The derivation of these interfacial conditions will include the possibility of mass transfer through the interface, i.e. phase change. As will be seen in subsequent chapters, not all instability modes require surface deformation to be taken into account. The interfacial conditions will be greatly simplified in this case.

As the systems considered in this book generally consist in one or several superposed layers of fluids, the boundary conditions are directly expressed in a cartesian system of coordinates (x, y, z), where x and y coordinates are parallel, and z is perpendicular to the plane of the layers.

2.2.1 Non-deformable plane boundaries

Mechanical boundary conditions

On a rigid wall and for a viscous fluid, it is usually assumed that no slip takes place, i.e. all components of velocity vanish

$$\vec{v} = (u, v, w) = 0 \tag{2.69}$$

If the fluid is incompressible and the rigid boundary is horizontal, using the continuity equation $\vec{\nabla}.\vec{v} = 0$ to eliminate horizontal velocity components yields

$$w = \frac{\partial w}{\partial z} = 0 \tag{2.70}$$

When the horizontal boundary is a free surface, the horizontal velocity components do not vanish in general. For an incompressible Newtonian liquid, and in the absence of tangential stress, Eq. (2.66) yields

$$w = \frac{\partial u}{\partial z} + \frac{\partial w}{\partial x} = \frac{\partial v}{\partial z} + \frac{\partial w}{\partial y} = 0 \tag{2.71}$$

or, using the incompressibility condition to eliminate u and v, we obtain

$$w = \frac{\partial^2 w}{\partial z^2} = 0 \tag{2.72}$$

on a stress-free boundary.

Mass transfer boundary conditions

At an impervious surface, the vanishing of the normal velocity ensures that no mass flows through the boundary. This is sufficient for single-component fluids, but for mixtures, the vanishing of normal velocity must apply to every chemical species individually. Alternatively, we may impose the vanishing of the normal barycentric velocity, together with that of diffusive mass fluxes $\rho_\gamma \vec{\Delta}_\gamma.\vec{n}$ [see Eq. (2.22)], where $\vec{n}$ is the normal to the surface. The particular form of these boundary conditions thus depends on the phenomenological law adopted for diffusive fluxes (see §2.1.6). In particular, for an isothermal binary fluid [Eq. (2.57)], the condition for an impervious horizontal wall is

$$\vec{n}.\vec{\nabla}N_1 = \frac{\partial N_1}{\partial z} = 0 \tag{2.73}$$

When the binary fluid is not isothermal, the Soret effect has to be taken into account [see Eq. (2.60)], and this relation should be replaced by

$$N_1 N_2 D_T \frac{\partial T}{\partial z} + D \frac{\partial N_1}{\partial z} = 0 \tag{2.74}$$

Thermal boundary conditions

At an isothermal wall with prescribed temperature T_w, we have

$$T = T_w \tag{2.75}$$

while for a heat-insulated wall, the normal heat flux vanishes. According to Eqs (2.50) or (2.59), for pure fluids (and for liquid mixtures in which Dufour effects can be neglected [188]), we obtain

$$\vec{n}.\vec{\nabla}T = \frac{\partial T}{\partial z} = 0 \tag{2.76}$$

In expressing the condition (2.75), we have assumed the continuity of temperature across the boundary. While this condition is realistic in most situations (especially for solid boundaries), it is not valid when the boundary possesses a non-negligible thermal resistance. Then a difference of temperature may exist across the boundary, and the continuity of the heat flux is often written

$$-\lambda \frac{\partial T}{\partial z} = \alpha(T - T_0) \qquad (2.77)$$

where T is the temperature of the fluid at the boundary, T_0 is the temperature on the other side of the boundary, and α is a heat transfer coefficient. Equation (2.77) is referred to as "Newton's law of cooling" in the absence of radiation. This relation is phenomenological in the sense that it defines the heat transfer coefficient α. Its validity thus depends on the particular situation considered. For example, in the case of a very thin slab of a highly insulating solid material, and when the heat transfer is steady, it may be shown by resolution of the Fourier equation inside the slab, that the relation (2.77) holds with $\alpha = \lambda_s/h$, where λ_s is the thermal conductivity of the slab and h is its thickness. More precisely, the condition for the validity of Eq. (2.77) is that the thickness of the slab is much smaller than the scale of horizontal variations of T (and T_0).

Another situation where Eq. (2.77) is often employed is that of a free surface, e.g. when the liquid is in contact with air. In this case, T_0 is taken as the ambient temperature, i.e. the temperature of the gas at a large distance from the surface. Though Newton's cooling law is mathematically convenient in this situation, it may not be adequate in describing cooling due to convection in the gas. A fortiori, even if such convective phenomena are negligible, the validity of the Newton's law with a constant heat transfer coefficient is not guaranteed, as will be shown in the next chapter by particular examples.

2.2.2 Deformable fluid interfaces

In this section, we consider boundary conditions applying at the interface between two fluids with different thermal and mechanical properties. For simplicity, their derivation will be limited to the case where the unknown position of the interface can be described by its variable height $z = h(x, y, t)$, taken as a single-valued function of the horizontal coordinates x and y. Simplifying further, the boundary conditions will be established in a two-dimensional geometry, though subsequently generalized to three dimensions. The conservation of basic quantities will be expressed in a fixed infinitesimal volume element of size $dx.dz$ whose center $[x, h(x, t)]$ is placed on the interface at time t (see Fig. 2.1). This approach is complementary to that used in deriving bulk equations (see §2.1), and obviously leads to conditions identical to those obtained using a convected volume element [182, 187].

We also neglect any temperature jump across the interface (i.e. we assume that it has a negligible thermal resistance), and assume the no-slip condition to be valid (the tangential velocity is continuous across the interface).

If $\vec{v}_\Sigma(x, t)$ denotes the interface velocity, a kinematic relation between $\vec{v}_\Sigma$ and h is readily obtained by differentiating $z = h(x, t)$ with respect to t, with $w_\Sigma = dz/dt$ and $u_\Sigma = dx/dt$.

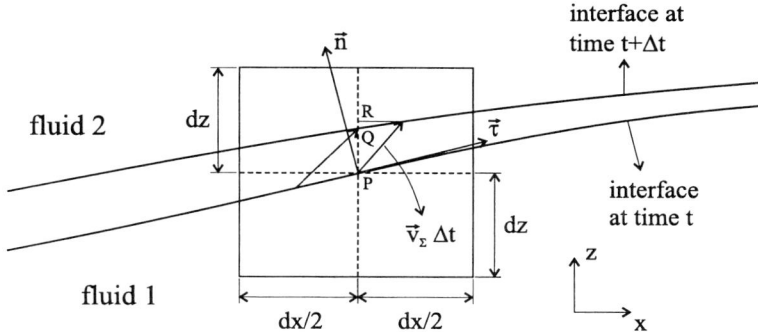

Figure 2.1: Definition of the volume element and position of the interface.

Thus,

$$w_\Sigma = \frac{\partial h}{\partial t} + u_\Sigma \frac{\partial h}{\partial x} \tag{2.78}$$

which actually expresses the equality $|PR|=|PQ|+|QR|$ (see Fig. 2.1). This kinematic relation is straightforwardly generalized to three dimensions[2].

We now consider the conservation of any specific quantity f_i (where $i = 1, 2$ denotes the phase considered). At time t and leading order in dx and dz, the total amount of f inside the volume element is $F(t) = (f_1 + f_2)dxdz$. Note that in this expression, the contribution of the interfacial phase itself (e.g. matter accumulation at the interface) is neglected, referring the reader to [76, 182, 187] for a more general derivation. At time $t + \Delta t$, the interface has moved, and the amount of f inside the fixed volume reads $F(t + \Delta t) = f_1(t + \Delta t)dV_1(t + \Delta t) + f_2(t + \Delta t)dV_2(t + \Delta t)$, with $dV_1(t + \Delta t) = dxdz + dx\Delta t\partial h/\partial t$ and $dV_2(t + \Delta t) = dxdz - dx\Delta t\partial h/\partial t$. Thus, the time variation of the total amount of f inside the volume element is

$$\frac{\partial F}{\partial t} = \lim_{\Delta t \to 0} \frac{F(t + \Delta t) - F(t)}{\Delta t} = (f_1 - f_2)\frac{\partial h}{\partial t}dx \tag{2.79}$$

at the leading order for $dx, dz \to 0$. The time variation of F is due to the flow of f through the faces of the element, plus the amount of f eventually produced inside the element. The latter source terms can be written $Q_1 dxdz + Q_2 dxdz$, where Q_1 and Q_2 are the production of f per unit volume pertaining to each phase. This contribution is of higher order compared with the right-hand side (r.h.s.) of Eq. (2.79), and therefore does not contribute to the boundary condition. Now, denoting by $\vec{v}_i = (u_i, w_i)$ the velocity field in each phase $i = 1, 2$, the flux of f across horizontal faces reads

$$f_1 w_1 dx - f_2 w_2 dx \tag{2.80}$$

---

[2]In three dimensions with $\vec{v}_\Sigma = (u_\Sigma, v_\Sigma, w_\Sigma)$, the kinematic condition reads $w_\Sigma = \partial h/\partial t + u_\Sigma \partial h/\partial x + v_\Sigma \partial h/\partial y$.

which is of first order, and the flux across vertical faces is calculated as

$$[f_1 u_1]_{x-dx/2}(dz - h'\tfrac{dx}{2}) - [f_1 u_1]_{x+dx/2}(dz + h'\tfrac{dx}{2}) + [f_2 u_2]_{x-dx/2}(dz + h'\tfrac{dx}{2})$$
$$-[f_2 u_2]_{x+dx/2}(dz - h'\tfrac{dx}{2}) = -dx h'(f_1 u_1 - f_2 u_2) + h.o.t.$$

$$(2.81)$$

where $h' = \partial h/\partial x$, $[]_{x\pm dx/2}$ denotes a quantity evaluated on the lateral faces of the volume element (non-bracketed quantities are evaluated at x), and h.o.t. stands for higher-order terms.

Combining Eq. (2.79) with Eqs (2.80) and (2.81), and neglecting second-order contributions, we get the balance (or jump) condition

$$f_1\left(w_1 - u_1\frac{\partial h}{\partial x} - \frac{\partial h}{\partial t}\right) = f_2\left(w_2 - u_2\frac{\partial h}{\partial x} - \frac{\partial h}{\partial t}\right)$$

$$(2.82)$$

Using Eq. (2.78) and the definition $\vec{n} = (-h', 1)/N$ of the unit normal to the interface (where $N = (1 + h'^2)^{1/2}$ is the normalization constant), Eq. (2.82) can be rewritten

$$f_1(\vec{v}_1 - \vec{v}_\Sigma).\vec{n} = f_2(\vec{v}_2 - \vec{v}_\Sigma).\vec{n}$$

$$(2.83)$$

which indeed expresses that the flux of f is continuous across the interface. In particular, the mass flux through the interface is given by

$$J = \rho_1(\vec{v}_1 - \vec{v}_\Sigma).\vec{n} = \rho_2(\vec{v}_2 - \vec{v}_\Sigma).\vec{n}$$

$$(2.84)$$

which is also valid in three spatial dimensions[3].

Individual species conservation

For multicomponent mixtures, a conservation equation must be written for every chemical species γ. Denoting by $\rho_{i\gamma} = \rho_i N_{i\gamma}$ ($i = 1, 2$ and $\gamma = 1, ..., c$) the partial densities in each phase, the derivation of the jump condition closely parallels that achieved above for pure fluids. Hence, the mass flux of component γ through the interface is

$$J_\gamma = \rho_{1\gamma}(\vec{v}_{1\gamma} - \vec{v}_\Sigma).\vec{n} = \rho_{2\gamma}(\vec{v}_{2\gamma} - \vec{v}_\Sigma).\vec{n}$$

$$(2.85)$$

where $\vec{v}_{i\gamma}$ denotes the velocity of component γ in phase i. Using the definition $\rho_i\vec{v}_i = \Sigma_\gamma \rho_{i\gamma}\vec{v}_{i\gamma}$ of the barycentric velocity in each phase (see §2.1.4), it is readily checked that the summation of Eq. (2.85) on γ reduces to the total mass flux condition (2.84), with $\sum_\gamma J_\gamma = J$.

Note that in addition to the mass conservation relations (2.84) and (2.85), it is necessary to consider supplementary boundary conditions to make the problem self-consistent. Indeed, just as for bulk conservation equations, constitutive expressions relating the fluxes to the other state variables are needed. When the interface is taken locally at equilibrium, conditions of phase coexistence can be used and typically relate the concentrations in both phases via a certain partition coefficient. For the evaporation of a pure liquid, the equilibrium condition is the

---

[3]In three dimensions, the unit normal reads $\vec{n} = (-h_x, -h_y, 1)/(1 + h_x{}^2 + h_y{}^2)^{1/2}$, where suffixes denote derivatives with respect to the corresponding variable.

Clausius–Clapeyron equation, defining a relation between the saturation pressure and temperature [248]. However, when the interface cannot be considered at equilibrium, rate equations should be used in place of the coexistence relations. Phenomenological rate equations can be derived from the interfacial entropy source [187], though their correct form is still a matter of current research. As an example, evaporative convection will be discussed in §3.8.

Momentum conservation

The jump condition (2.83) may also be applied to express the conservation of the linear momentum $f_i = \rho_i \vec{v}_i$, but new terms should be added accounting for the sum of all external forces acting on the volume element considered. These external forces are the hydrodynamic stresses acting on each face (pressure and viscous stresses), and surface tension forces acting along the interface[4]. Note that body force terms are neglected, being proportional to $dx\,dz$.

If $\mathbf{T_i}$ denotes the stress tensor (2.10) in phase i, $\vec{1}_x$ and $\vec{1}_z$ the unit vectors respectively along x and z, σ the surface tension, and $\vec{\tau}$ the unit tangent to the interface (see Fig. 2.1), the sum of external forces is calculated as

$$
\begin{aligned}
& (dz + h'\tfrac{dx}{2})[\mathbf{T_1}.\vec{1}_x]_{x+dx/2} - (dz - h'\tfrac{dx}{2})[\mathbf{T_1}.\vec{1}_x]_{x-dx/2} - \\
& (dz + h'\tfrac{dx}{2})[\mathbf{T_2}.\vec{1}_x]_{x-dx/2} + (dz - h'\tfrac{dx}{2})[\mathbf{T_2}.\vec{1}_x]_{x+dx/2} + \\
& (\mathbf{T_2}.\vec{1}_z - \mathbf{T_1}.\vec{1}_z)dx + [\sigma\vec{\tau}]_{x+dx/2} - [\sigma\vec{\tau}]_{x-dx/2} \\
& = -dx\,h'(\mathbf{T_2} - \mathbf{T_1}).\vec{1}_x + dx(\mathbf{T_2} - \mathbf{T_1}).\vec{1}_z + dx\tfrac{\partial}{\partial x}(\sigma\vec{\tau}) + h.o.t.
\end{aligned}
\tag{2.86}
$$

Using Eq. (2.78), and dividing by N (= ds/dx where ds is the element of arc length along the surface), the momentum jump condition may be rearranged as

$$
[\rho((\vec{v} - \vec{v}_\Sigma).\vec{n})\vec{v} - \mathbf{T}.\vec{n}]_1^2 = \frac{\partial}{\partial s}(\sigma\vec{\tau})
\tag{2.87}
$$

where $[f]_1^2 = f_2 - f_1$ denotes the discontinuity across the interface.

With the definition (2.84) of the total mass flux J, Eq. (2.87) may also be rewritten

$$
J[\vec{v}]_1^2 - [\mathbf{T}.\vec{n}]_1^2 = \frac{\partial}{\partial s}(\sigma\vec{\tau})
\tag{2.88}
$$

In particular, using Eq. (2.10), the projection of this equation on the normal $\vec{n}$ (normal stress boundary condition) reads

$$
J[\vec{v}.\vec{n}]_1^2 + [p - \mathbf{P}.\vec{n}.\vec{n}]_1^2 = 2\sigma H
\tag{2.89}
$$

where the mean curvature H is defined by

$$
H = \frac{1}{2}\vec{n}.\frac{\partial}{\partial s}\vec{\tau} = \frac{1}{2}\frac{h''}{N^3}
\tag{2.90}
$$

If there is no slip,

$$
[\vec{v}]_1^2.\vec{\tau} = 0
\tag{2.91}
$$

---

[4]Here, we consider the surface fluid to be inviscid, hence the surface stress tensor is diagonal [182, 187, 281]. The equivalent of the pressure p (for bulk fluids) is *minus* the surface tension σ (for an interface).

Therefore, the projection of Eq. (2.88) on the unit tangent $\vec{\tau}$ (tangential stress boundary condition) reads

$$-[\mathbf{T}.\vec{n}.\vec{\tau}]_1^2 = -[\mathbf{P}.\vec{n}.\vec{\tau}]_1^2 = \frac{\partial \sigma}{\partial s} = \frac{1}{N}\frac{\partial \sigma}{\partial T}(T_x + h'T_z) \tag{2.92}$$

In the last equality of Eq. (2.92), the surface tension σ has been assumed to depend on temperature T only, and a suffix is used to denote a derivative with respect to the corresponding variable.

In three dimensions, the r.h.s. of Eqs (2.87) and (2.88) has to be replaced by $\vec{\nabla}_s\sigma - \sigma(\vec{\nabla}_s.\vec{n})\vec{n}$, where $\vec{\nabla}_s$ is the surface nabla operator [182, 187]. Actually, the normal stress condition (2.89) remains valid, with the mean curvature H now given by

$$H = -\frac{1}{2}\vec{\nabla}_s.\vec{n} = \frac{1}{2}\frac{h_{xx}(1+h_y{}^2) + h_{yy}(1+h_x{}^2) - 2h_x h_y h_{xy}}{(1+h_x{}^2+h_y{}^2)^{3/2}} \tag{2.93}$$

Defining two unit tangent vectors by

$$\vec{\tau}_x = \frac{(1,0,h_x)}{(1+h_x{}^2)}, \quad \vec{\tau}_y = \frac{(0,1,h_y)}{(1+h_y{}^2)}, \tag{2.94}$$

the two tangential momentum balance conditions read [182]

$$-[\mathbf{P}.\vec{n}]_1^2.\vec{\tau}_i = \vec{\tau}_i.\vec{\nabla}_s\sigma = \frac{\sigma_i}{(1+h_i{}^2)^{1/2}} = \frac{\partial \sigma}{\partial T}\frac{T_i + h_i T_z}{(1+h_i{}^2)^{1/2}} \quad i = x,y \tag{2.95}$$

Energy conservation

We may again use Eq. (2.83) with $f_i = \rho_i(\epsilon_{i,\text{int}} + v_i^2/2)$, but we must here also take into account the heat flux $\vec{q}$ entering the element, and the work of the stresses acting on the elementary surfaces (as in §2.1.3). The work of external body forces is found to be of second order ($\sim dxdz$), as well as the release of energy due to chemical reactions. We shall also neglect the work of surface tension forces, as well as the variation of energy of the interfacial phase, as order of magnitude analysis shows that these contributions are generally negligible relative to the thermal contributions. As the case of non-isothermal multispecies mass transfer through interfaces is not treated in this text, the following derivation is limited to the case of single-component fluids.

After calculations similar to those presented above, we get

$$[\rho(\epsilon_{\text{int}} + v^2/2)(\vec{v} - \vec{v}_\Sigma).\vec{n}]_1^2 = -[\vec{q}.\vec{n}]_1^2 + [\mathbf{T}.\vec{v}.\vec{n}]_1^2 \tag{2.96}$$

Using Eq. (2.84), this is rewritten as

$$J[\epsilon_{\text{int}} + v^2/2]_1^2 + [\vec{q}.\vec{n}]_1^2 = [\mathbf{T}.\vec{v}.\vec{n}]_1^2 \tag{2.97}$$

This equation may be simplified by first calculating the scalar product of the momentum jump boundary condition (2.88) with the interface velocity $\vec{v}_\Sigma$, and subtracting the result from Eq. (2.97). However, to be coherent with the assumption that surface tension effects are negligible in the energy conservation condition, we also have to neglect the r.h.s. of Eq. (2.88). Then, the result is rearranged as

$$J[\epsilon_{\text{int}} + (\vec{v} - \vec{v}_\Sigma)^2/2]_1^2 + [\vec{q}.\vec{n}]_1^2 = [\mathbf{T}.(\vec{v} - \vec{v}_\Sigma).\vec{n}]_1^2 \tag{2.98}$$

Finally, introducing the specific enthalpy $h = \epsilon_{\text{int}} + pv$ and the latent heat of evaporation (per unit mass) $L = h_2 - h_1$, we obtain

$$JL + J[(\vec{v} - \vec{v}_\Sigma)^2/2]_1^2 + [\vec{q}.\vec{n}]_1^2 = [\mathbf{P}.(\vec{v} - \vec{v}_\Sigma).\vec{n}]_1^2 \tag{2.99}$$

where the decomposition (2.10) of the stress tensor $\mathbf{T}$ has been used. Equation (2.99) is identical to the form considered in [186] and is also valid in three dimensions. The reader is also referred to the more general derivation of interfacial boundary conditions presented in [187], including surface tension effects and material properties of the interface. The reader interested in peculiar interfacial rheologies and other effects specific to mass transfer through interfaces is referred to [76, 119].

Note finally that the interfacial conditions derived above can be greatly simplified when the interface is only weakly deformed. This is discussed in §3.8, where linearized interfacial boundary conditions will be analyzed in some detail. Large (nonlinear) deformations of thin liquid films are discussed in §4.5.3.

3 Instability modes in Bénard layers

In this chapter, we deal with the identification and classification of several instability modes in one or several layers of fluids submitted to transverse temperature or concentration gradients. First, we recall how normal mode analysis is carried out with the buoyancy-driven instability problem, to subsequently apply it to several problems involving surface tension, with increasing degree of complexity. It is hoped that our presentation, which emphasizes the relevant time scales of the various problems, dimensionless parameters, and the underlying physics, will allow the reader to gain a coherent view of the field, and to extend some of the analyzes to the problems of his/her particular interest.

Although reference is in general made to heat transfer systems, most of the mathematical developments also apply to isothermal binary mixtures undergoing mass transfer, provided complications specific to mass transfer through interfaces (adsorption/desorption processes, surface diffusion, viscoelastic behavior, ...) are neglected in a first approximation. Incorporation of such effects would require a generalization of the boundary conditions established in Chapter 2. As their exact form is not firmly established yet, in particular for nonlinear situations involving departures for local equilibrium at the interface, we will not discuss such situations here, referring the reader to the specialized literature on the subject (see e.g. [76, 119, 191] and references therein).

Though the focus will be on planar layers, some of the instability mechanisms described in the following can be transposed to other geometries (e.g. cylindrical volumes, or drops undergoing radial heat or mass transfer). This certainly holds when the typical length scale of interfacial convection is small compared with the size of the fluid volume. When the size of convection cells increases, however, non-trivial modifications of instability modes occur. The influence of confinement will also be investigated in some detail for layers with small lateral extension.

3.1 Reference states and small perturbations

The equations presented in the previous chapter, complemented by appropriate boundary conditions, may admit several solutions for fixed values of the parameters entering the problem. This multiplicity appears due to nonlinearities in both the equations and the boundary conditions. In the balance equations, nonlinearity is brought about by the advective terms [i.e. $(\vec{V}.\vec{\nabla})\vec{V}$ or $(\vec{V}.\vec{\nabla})T$], while for the interfacial boundary conditions (at a deformable interface), it arises because the position of the interface itself is variable and hence is an unknown of the problem. Other nonlinearities may also occur due to variations of fluid properties with temperature or concentration, both in the equations and in the boundary conditions. Apart from surface tension and density dependence on temperature and concentration, which will be assumed linear, we will generally disregard temperature and concentration variations of other fluid properties, a hypothesis generally referred to as the Boussinesq (or Boussinesq–Oberbeck) approximation, and discussed in §3.1.4. Other effects responsible for nonlinearity such as chemical reaction source terms will always be neglected.

The number of possible solutions is a function of the parameters characterizing the problem, such as fluid properties, externally imposed gradients, ... Dimensional analysis, which allows the equations to be written under dimensionless form, reduces the number of independent parameters to a certain "minimal set" (Buckingham's Π–Theorem) of dimensionless numbers, which will be considered as the control parameters of the problem (although this denomination is sometimes reserved for those parameters which directly characterize deviations from equilibrium, such as externally imposed fluxes, ...).

Thus, given a set of control parameters, only the analysis of the relative stability of the possible solutions will allow the determination of which of them will effectively be realized in an experiment. In this chapter, we will be interested in the stability of regimes emerging from thermodynamic equilibrium, when a certain driving constraint is progressively increased. Such "reference" regimes (the Thermodynamic Branch), will here be steady, restricting consideration to external constraints which are time-independent. For instance, in the case of a fluid layer infinite in horizontal extent and heated homogeneously from below (the "Bénard" configuration), our reference solution will be a motionless state, where heat is transported by pure conduction across the layer. When the thermal conductivity of the fluid is constant, this reference steady solution is characterized by a linear temperature profile throughout the layer. Such a high-symmetry state[1] thus depends on the intensity of the constraint, and may or may not be stable against hydrodynamic disturbances depending on the value of the driving gradient. It is the purpose of linear stability theory, to be presented in this chapter, to find the threshold values of the constraints above which the reference solution is unstable to infinitesimal disturbances.

For small deviations from the reference state, the equations governing the evolution of the perturbations can be linearized, which often allows analytical stability criteria (sufficient conditions for instability) to be obtained. More generally, stability should also be tested against finite-amplitude disturbances, which clearly requires nonlinearity to be retained in the equations. Some examples of solutions stable to infinitesimal perturbations, but unstable to finite-amplitude disturbances will be studied in subsequent chapters. It is also possible to determine sufficient conditions for stability to finite-amplitude disturbances, on the basis of nonlinear energy stability theory, briefly considered in the next chapter for the Marangoni–Bénard case.

3.1.1 Rayleigh–Bénard instability – time scales

As announced, let us first consider the Rayleigh–Bénard instability (Fig. 3.1, left). A layer of pure liquid (which will be considered as incompressible and Newtonian) is confined between two horizontal plates maintained at controllable temperatures.

---

[1]In fact, regimes observed near equilibrium or emerging continuously from it are expected to possess all symmetries of the equations and boundary conditions. This may not be the case for states appearing through subsequent

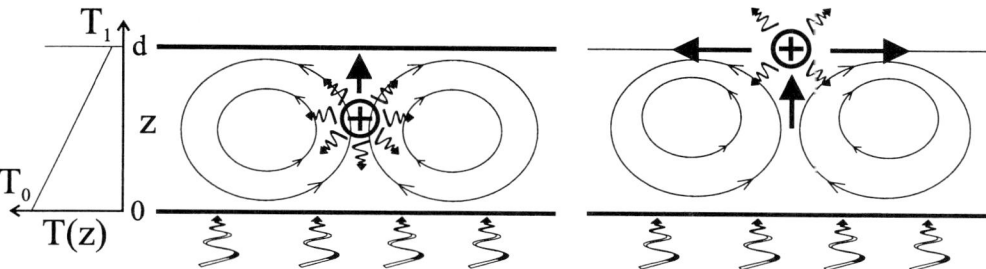

Figure 3.1: Left: sketch of a typical Rayleigh–Bénard experiment: a liquid is confined between rigid plates at different temperatures T_0 and $T_1 < T_0$ (the layer is heated from below). Right: sketch of a typical Marangoni–Bénard experiment (see §3.3.2): in this case, the upper surface is open to air, and the surface tension depends on temperature.

Imposing a certain temperature gradient throughout the system allows the state of the fluid to be varied from thermodynamic equilibrium ($T_0 = T_1 = T_{\text{fluid}}$) to situations of moderate or strong non-equilibrium ($|T_0 - T_1| \geq 0$). What are the qualitative changes occurring in the system when the difference $|T_0 - T_1|$ is progressively increased from zero? At equilibrium, apart from fluctuations of thermodynamic origin, the fluid temperature is homogeneous, and no macroscopic motions occur. When a small temperature difference is imposed across the layer, the steady state reached by the system after a sufficiently long time (more precisely a time on the scale of the thermal relaxation time d^2/κ, where d is the fluid depth, and κ the thermal diffusivity) is a motionless (or rest) state characterized by a linear temperature profile from T_0 to T_1 (see Fig. 3.1). Note that in the following, the thermal conductivity λ is assumed to be independent of temperature, such that the reference state temperature profile remains linear even for finite values of $\Delta T = T_0 - T_1$.

Owing to the thermal expansion of the liquid[2], a density gradient exists along the vertical. When heated from below ($\Delta T > 0$), lighter fluid lies near the bottom of the layer, while heavier fluid is on top. Even though this state corresponds to an equilibrium situation, it is unstably stratified in the gravity field, as the total potential energy could be reduced by a redistribution of the fluid inside the layer. To illustrate this, and to establish the time scale associated with this process, we now consider the fate of temperature or velocity fluctuations in this system. A thermal fluctuation of length scale l will be damped by diffusion after a time of order l^2/κ. Thus we may expect that the fluctuations with the longest lifetime (the "most dangerous" ones) have the size of the layer, i.e. decay on a time scale d^2/κ. We thus define the thermal time scale

$$\tau_{\text{th}} = \frac{d^2}{\kappa} \tag{3.1}$$

Similarly, velocity fluctuations (or more precisely, vorticity fluctuations) decay with a viscous

bifurcations, as seen in following chapters.

[2]Note that thermal expansion is not incompatible with the hypothesis of incompressibility (Boussinesq approximation, see §3.1.4).

time scale

$$\tau_{\text{visc}} = \frac{d^2}{\nu} \tag{3.2}$$

where ν is the kinematic viscosity.

Assume for the moment that the above stabilizing effects are associated with very long time scales (i.e. ν and κ are small). Then, when a localized fluctuation of temperature (say, a hot spot) arises in a certain volume element, an upward force on this fluid particle is induced by buoyancy, as its density is decreased compared with its environment (differential buoyancy force). As molecular diffusion is assumed to be negligible, the fluid particle will set into motion in the upward direction, without friction, and will remain at the same temperature during this motion. The buoyancy force it will experience will then increase, because the particle moves towards colder regions, such that the velocity is increased, ... This is the origin of the amplification (instability) phenomenon. We may also associate a characteristic time scale with the amplification process, by considering an equation of motion of the type $F = m\,a$ for the particle [198] : if z is the height of the element, the net upward buoyancy force per unit volume is given by $g\delta\rho = g\alpha\rho_0\delta T$, where g is gravity, ρ_0 is the mean density, α is the coefficient of linear expansion of the fluid, and δT is the temperature difference between the element and its surroundings. As the fluid temperature at height z is $T = T_0 - \Delta T z/d$, the linear equation of motion is $\rho_0 \ddot{z} = \alpha\rho_0 g\Delta T z/d$, which for $\Delta T > 0$ (heating from below) is solved by an exponential growth $z \sim \exp[t/\tau_{\text{buoy}}]$, where the time constant is the buoyancy time scale

$$\tau_{\text{buoy}} = \sqrt{\frac{d}{\alpha g \Delta T}} \tag{3.3}$$

Note that when heating from the top ($\Delta T < 0$), the simplified equation of motion admits an oscillatory solution. This behavior is not unrealistic, as in this case (stably stratified liquid) it is known [21] that internal waves occur with a (Brunt–Väisälä) frequency of order $1/\tau_{\text{buoy}}$ (see next sections).

Comparing the different time scales given by Eqs (3.1–3.3), it may be expected that instability will develop if the time for a particle to travel over a distance of order d is shorter than the times necessary for the particle either to be slowed down by viscosity, or thermally equilibrated with its environment. Thus, instability would occur typically if $\tau_{\text{buoy}}^2 \ll \tau_{\text{visc}}\tau_{\text{th}}$, which may be rearranged as

$$Ra = \frac{g\alpha\Delta T d^3}{\nu\kappa} \gg 1 \tag{3.4}$$

This equation defines the Rayleigh number Ra, which is the usual measure for the relative importance of destabilizing and stabilizing effects. Typically, the critical value of the Rayleigh number above which instability sets in is of order 10^3. Its actual value depends on the nature of upper and lower plates (e.g. rigid and heat-conducting, or free and poorly-conducting, ...). The above dimensional arguments will now be confirmed by a more rigorous linear stability analysis in the simplified case of a layer of infinite depth.

3.1.2 Dynamics of fluctuations in an infinite stratified medium

First, we do not consider the effect of horizontal walls, thus restricting the analysis to the case of disturbances with an extent much smaller than the actual fluid depth (short-wave analysis). This allows obtaining a simple and quite general view of the behavior of potentially unstable modes.

We consider an incompressible Newtonian fluid, submitted to a constant temperature gradient parallel to the gravity axis (vertical z-coordinate). All fluid properties are assumed to be independent of temperature, except density which decreases according to the linearized state equation

$$\rho = \rho_0 \left[1 - \alpha(T - T_0)\right] \tag{3.5}$$

where ρ_0 is the density at reference temperature T_0. Furthermore, the variation of the density is only taken into account in the body force term of the Navier–Stokes equation (in other places ρ will be assumed equal to ρ_0). These hypotheses are known as the Boussinesq approximation, and will be justified in more details in §3.1.4. It will be seen that even when density varies due to temperature fluctuations, the incompressibility condition may remain valid. i.e. the velocity field $\vec{V}$ is solenoidal:

$$\vec{\nabla} . \vec{V} = 0 \tag{3.6}$$

The Navier–Stokes equation [see Eqs (2.7), (2.10) and (2.66)] then reads

$$\rho_0 \frac{d\vec{V}}{dt} = \mu \Delta \vec{V} - \vec{\nabla}p - \rho g \vec{1}_z \tag{3.7}$$

where $\vec{1}_z$ is the unit vector directed vertically upwards, p is the pressure, μ is the dynamic (or shear) viscosity, and Δ is the Laplacian operator. The equation describing the evolution of the temperature field T is the energy equation (2.16) [for an incompressible fluid with negligible viscous heating], complemented by Fourier's law (2.50). This gives

$$\frac{dT}{dt} = \kappa \Delta T \tag{3.8}$$

where $\kappa = \lambda/\rho_0 c_p$ is the thermal diffusivity.

Note that the non-isothermal single-component fluid problem considered here is formally equivalent to an isothermal two-component mixture problem, where stratification is provided by a gradient of concentration along the vertical, and by the dependency of ρ on the mass fraction N of one of the components. If we ignore chemical reactions between species, the mass diffusion equation (2.24) may be combined with Fick's law for the mass flux (2.57) to give a convection-diffusion equation analogous to Eq. (3.8) :

$$\frac{dN}{dt} = D \Delta N \tag{3.9}$$

once again relying on the Boussinesq approximation (constant density ρ and isothermal diffusion coefficient D). Thus, an equivalence exists between the "thermal" and the "concentrational" problems, under the substitution of T by N or vice versa.

In Eqs (3.7) and (3.8) [or (3.9)], $d/dt = \partial/\partial t + \vec{V}.\vec{\nabla}$ is the material derivative, the second (advective) term of which introduces nonlinearity in the equations. However, in this chapter, we will only be concerned with the evolution of infinitesimally small perturbations added to a steady reference solution, and thus with the linearized form of these terms.

It is seen that Eqs (3.5–3.8) admit the solution

$$\vec{V}_{\text{ref}} = 0, \quad T_{\text{ref}} = T_0 - \beta z, \quad p_{\text{ref}} = p_0 - \rho_0 g \left(z + \alpha\beta\frac{z^2}{2} \right) \tag{3.10}$$

where β is the constant gradient along the vertical (by convention, $\beta > 0$ corresponds to heating from below), and p_0 is a reference pressure.

The stability of this reference motionless state is investigated by adding perturbations of temperature T', velocity $\vec{V}'$ and pressure p', and linearizing the equations with respect to such perturbations. Omitting the primes, Eq. (3.6) is unchanged, while from (3.7) and (3.8), we get

$$\frac{\partial \vec{V}}{\partial t} = \nu\Delta\vec{V} - \frac{1}{\rho_0}\vec{\nabla}p + g\alpha T \vec{1}_z \tag{3.11}$$

where $\nu = \mu/\rho_0$ is the kinematic viscosity, and

$$\frac{\partial T}{\partial t} = \beta W + \kappa\Delta T \tag{3.12}$$

where W is the vertical component of velocity.

We may eliminate the pressure from the Navier–Stokes equation (3.11) and at the same time derive an equation for W and T only by first taking the divergence of Eq. (3.11). Using Eq. (3.6), an equation for p is obtained in the form

$$\Delta p = \rho_0\alpha g\frac{\partial T}{\partial z} \tag{3.13}$$

Then, applying the Laplacian operator to Eq. (3.11), using Eq. (3.13), and projecting on $\vec{1}_z$, we get

$$\frac{\partial}{\partial t}\Delta W = \nu\Delta^2 W + g\alpha\Delta_h T \tag{3.14}$$

where Δ_h is the horizontal Laplacian operator.

Solving the resulting system of Eqs (3.12) and (3.14) is possible by separation of variables. It is then seen that perturbations of temperature and velocity can be written as the superposition of plane waves (normal modes) of the form

$$\begin{pmatrix} W \\ T \end{pmatrix} = e^{\sigma t}e^{i\vec{k}.\vec{r}} \begin{pmatrix} w \\ \theta \end{pmatrix} \tag{3.15}$$

where σ is the growth rate, $\vec{k} = k_x\vec{1}_x + k_y\vec{1}_y + k_z\vec{1}_z$ is a three-dimensional wavevector and $\vec{r} = x\vec{1}_x + y\vec{1}_y + z\vec{1}_z$ is the coordinate vector.

Inserting Eq. (3.15) into (3.12), we get a relation between the constants θ and w:

$$\theta = \frac{\beta}{\sigma + \kappa k^2} w \tag{3.16}$$

Now, it can be seen from the last three equations that non-trivial solutions ($w, \theta \neq 0$) exist if and only if a certain compatibility relation is satisfied. This equation is the dispersion relation of the problem, and relates σ to other quantities, according to

$$(\sigma + \nu k^2)(\sigma + \kappa k^2) = \alpha\beta g \frac{(k^2 - k_z{}^2)}{k^2} \tag{3.17}$$

where $k = |\vec{k}|$. This is a second degree equation for σ, implying that two different modes exist. To examine the stability of the reference state, we should normally consider $\vec{k}$ as a free parameter. However, we will first discuss Eq. (3.17) keeping k and k_z constant.

For $\beta = 0$, it is seen that the two modes are decoupled and damped at rates $\sigma = -\nu k^2$ (viscous mode) and $\sigma = -\kappa k^2$ (thermal mode). For $\beta \neq 0$, both modes are coupled by the buoyancy, due to thermal expansion. For $\beta \to +\infty$, Eq. (3.17) admits the asymptotic behavior

$$\sigma \to \pm \left(\alpha g \beta \frac{k_h^2}{k^2}\right)^{\frac{1}{2}} \quad for \ \beta \to +\infty \tag{3.18}$$

where k_h is the modulus of the horizontal component of $\vec{k}$. In this limit, the effects of thermal and viscous diffusion are negligible, and one of the modes has a positive growth rate, indicating the (Rayleigh–Bénard) instability of the reference state. Apart from a geometric factor k_h/k ($= \sin\phi$, where ϕ is the angle between $\vec{k}$ and the vertical), it is seen that the time scale given by Eq. (3.18) corresponds to the result (3.3) obtained from dimensional analysis. The angular dependency shows that the most unstable modes are those with horizontal wavevectors.

As both roots are negative for $\beta = 0$, and one of them is positive for $\beta \to \infty$, a positive value of β should exist at which the growth rate goes through zero [roots of Eq. (3.17) are purely real for $\beta > 0$]. This is the neutral stability locus, obtained by setting σ to zero in Eq. (3.17). We then obtain

$$\beta_{\vec{k}} = \frac{\nu\kappa k^6}{\alpha g k_h^2} \tag{3.19}$$

where the index $\vec{k}$ means that this threshold value depends on the modulus and orientation of the disturbance wavevector with respect to the vertical.

Thus, the instability condition (for a disturbance with wavevector $\vec{k}$) may be written as $Ra_{\vec{k}} > 1$, where the disturbance Rayleigh number is defined by $Ra_{\vec{k}} = \alpha g \beta k_h^2/\nu\kappa k^6$. Here also, we see that perturbations with wavevectors along the vertical ($k_h = 0$) are not dangerous because their associated Rayleigh number is always zero.

To complete the discussion, it is instructive to analyze the behavior of eigenvalues for $\beta < 0$. In the limit $\beta \to -\infty$, we see that the eigenvalues are imaginary, corresponding to oscillatory perturbations. In reality, for finite negative values of β, the effect of thermal and viscous dissipation will be responsible for damping of the oscillations of temperature and velocity fields. These modes are known in the literature as (Brunt–Väisälä) internal waves, observed in situations of stable density stratification (e.g. in the atmosphere or the ocean [21, 505]). The possibility of sustaining these normally damped modes by another effect (thermocapillarity at a free interface) will be examined later on in this chapter (§3.4).

The complete picture of the behavior of the solutions of Eq. (3.17) is presented in Fig. 3.2. We have

$$\sigma = \frac{-(\nu + \kappa)k^2 \pm \sqrt{k^4(\nu - \kappa)^2 + 4\alpha g \beta k_\mathrm{h}^2 / k^2}}{2} \tag{3.20}$$

from which it is seen that the two real eigenvalues obtained for $\beta > 0$ merge into two complex conjugate eigenvalues at

$$\beta = -\frac{(\nu - \kappa)^2 k^6}{4\alpha g k_\mathrm{h}^2} \tag{3.21}$$

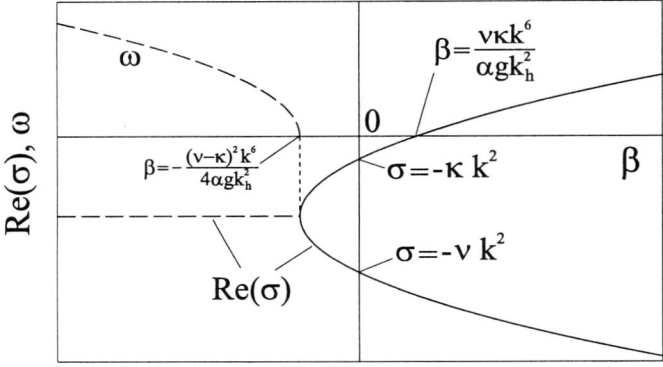

Figure 3.2: General behavior of eigenvalues as a function of the imposed temperature gradient β. Dashed branches correspond to oscillatory modes, full branches to monotonic modes. It has been assumed that $\nu > \kappa$ ($Pr > 1$).

3.1.3 Rayleigh–Bénard instability of a fluid confined between stress-free boundaries

We now return to the case of the Rayleigh–Bénard instability in a liquid layer of depth d comparable to the scale of the disturbances. We thus have to complement the linearized equations (3.6), (3.12) and (3.14) by boundary conditions at top and bottom plates (see §2.2.1), which for simplicity will be both considered as isothermal (imposed temperatures T_0 and T_1), plane and stress-free. Clearly, this case is unrealistic, but it allows direct exploitation of the previous results with straightforward modifications. In fact, such idealized walls have been considered in numerous studies, both in the linear and nonlinear regimes (see e.g. [22, 23, 25]). Still, we will consider a layer unbounded in the lateral direction, thus restricting the analysis to layers with a large aspect ratio L/d (length over depth).

Impermeability of the horizontal boundaries requires the normal (z-component) of velocity to vanish

$$W = 0 \quad at \ z = 0, d \tag{3.22}$$

The absence of tangential stress means, for a Newtonian fluid [Eq. (2.72)], that

$$D^2 W = 0 \quad at \ z = 0, d \tag{3.23}$$

where D will hereafter denote the z-derivative. The boundary condition (3.23) is an equation for W only, which is necessary to solve Eq. (3.14) without having to use the Navier–Stokes equations for horizontal velocities.

Finally, we consider that boundaries are maintained at constant temperature, implying vanishing temperature perturbation

$$T = 0 \quad at \ z = 0, d \tag{3.24}$$

As the layer is infinite in the horizontal directions, normal modes remain plane wave solutions $\exp[i\vec{k}_h.\vec{r}]$ where (in contradistinction with the previous section) $\vec{k}_h$ here denotes a horizontal wavevector, and $\vec{r} = x\vec{1}_x + y\vec{1}_y$ denotes the horizontal coordinate vector. In fact, the decomposition into a complete set of normal modes (Fourier decomposition) is a general characteristic of systems considered in this book, for which the horizontal planform function $\phi(\vec{r})$ (i.e. the horizontal dependency) is a solution of the Helmholtz equation $\Delta_h\phi + k_h^2\phi = 0$. The time dependency is also exponential[3], such that normal modes now read

$$\begin{pmatrix} W \\ T \end{pmatrix} = \exp[\sigma t + i\vec{k}.\vec{r}] \begin{pmatrix} w(z) \\ \theta(z) \end{pmatrix} \tag{3.25}$$

where the index h for the horizontal wavevector is omitted in what follows.

Substituting the normal modes into the balance equations (3.12) and (3.14), we get the system of ordinary differential equations satisfied by $w(z)$ and $\theta(z)$ as

$$\sigma(D^2 - k^2)w = \nu(D^2 - k^2)^2 w - \alpha g k^2 \theta \tag{3.26}$$

$$\sigma\theta = \kappa(D^2 - k^2)\theta + \beta w \tag{3.27}$$

which have to be solved with the boundary conditions

$$w = D^2 w = \theta = 0 \quad at \ z = 0, d \tag{3.28}$$

---

[3]Note that we do not consider the initial-value problem here. This could be done by using a Laplace transform of the differential equations.

The simplicity of the problem arises because only even-order z-derivatives enter equations and boundary conditions (had we used thermally insulating boundary conditions $D\theta = 0$ or realistic rigid plates $w = Dw = 0$, the resolution would have been more complicated, as seen in §3.3.1). Then, it is seen that the vertical dependencies $\sin(n\pi z/d)$ where $n = 1, 2, ...$ satisfy boundary conditions for both temperature and velocity. It is readily checked that the dispersion relation of the infinite problem considered in §3.1.2 remains applicable if the vertical wavenumber k_z is restricted to values $n\pi/d$ with integer n. We thus rewrite Eq. (3.17) as

$$\left[\sigma + \nu\left(k^2 + \frac{n^2\pi^2}{d^2}\right)\right]\left[\sigma + \kappa\left(k^2 + \frac{n^2\pi^2}{d^2}\right)\right] = \alpha\beta g\frac{k^2}{k^2 + n^2\pi^2/d^2} \tag{3.29}$$

In particular, neutral stability ($\sigma = 0$) occurs for

$$\frac{\alpha\beta g}{\nu\kappa} = \frac{(k^2 + n^2\pi^2/d^2)^3}{k^2} \tag{3.30}$$

Defining the dimensionless wavenumber $a = kd$, this is put under dimensionless form as

$$Ra = \frac{\alpha g\beta d^4}{\nu\kappa} = \frac{\alpha g(T_0 - T_1)d^3}{\nu\kappa} = \frac{(a^2 + n^2\pi^2)^3}{a^2} \tag{3.31}$$

which again shows that the dimensionless Rayleigh number Ra is the parameter controlling the stability of the motionless state.

The relation (3.31) is represented in Fig. 3.3, for several values of n. The value of the Rayleigh number rapidly increases with the "vertical wavenumber" n, such that the most dangerous modes are those with $n = 1$.

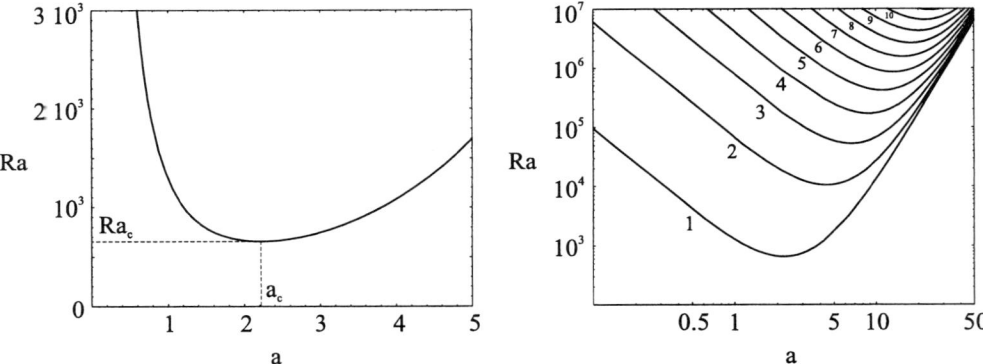

Figure 3.3: Locus of the neutrally stable perturbations for the Rayleigh–Bénard problem with stress-free heat-conducting top and bottom plates. Left: threshold of the most dangerous mode $n = 1$. Right: logarithmic plot of successive thresholds of modes with n cells in the vertical direction.

For every point (a, Ra) below the curve labeled $n = 1$ in this figure, the reference state is stable [i.e. all solutions of the dispersion relation (3.29) with $n = 1, 2, ...$ are real and negative]. If, for a given horizontal wavenumber a, the Rayleigh number is increased above the value $(a^2 + \pi^2)^3/a^2$, one of the eigenvalues becomes positive, indicating the loss of stability of the reference state to this perturbation. As temperature and velocity fields may be written as a superposition of all Fourier components, it is seen that the reference state first becomes unstable when the Rayleigh number exceeds the value corresponding to the minimum of the curve $n = 1$. Exactly at this point, $Ra = Ra_c$, only the perturbation with wavenumber a_c (the critical wavenumber) is neutrally stable, and all the other perturbations are exponentially damped. Hence, a convective structure in the form of rolls with lateral size π/a_c (half a wavelength, see Fig. 3.4) can be expected when the critical Rayleigh number is first exceeded.

The critical conditions are readily computed from Eq. (3.31), which gives

$$Ra_c = \frac{27\pi^4}{4} \simeq 657.5 \quad \text{at} \quad a_c = \frac{\pi}{\sqrt{2}} \simeq 2.221 \tag{3.32}$$

indicating that the lateral size of a convective roll should be close to 1.4 times the depth of the liquid, as seen in Fig. 3.4.

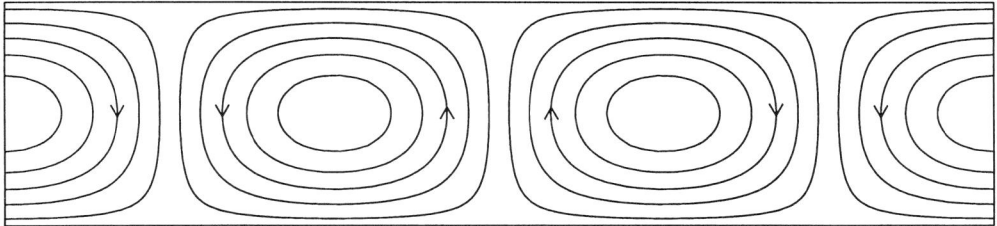

Figure 3.4: Streamlines as obtained from the velocity field $W \sim \sin \pi z \cos a_c x$. The structure is neutrally stable at $Ra = Ra_c$. The figure may also be interpreted as the contour plot of the associated perturbation of temperature $T \sim \sin \pi z \cos a_c x$, after a horizontal shift of $\pi/2a_c$ (half a roll), matching positive temperature perturbations with uprising currents.

3.1.4 The Boussinesq approximation

In the previous sections, we considered effects due to the thermal expansion of a liquid, even though the velocity field was assumed solenoidal ($\vec{\nabla}.\vec{V} = 0$). This approximation, associated with the names of Oberbeck and Boussinesq, may clearly not be valid in all situations. Hence, it is important to determine its limits of validity, and to check in particular whether it is sufficient for the set-ups considered in this book. A related question is the choice of the equation of state for the fluid. In general, in addition to the variation of density with temperature (or concentration), we might consider the variation with pressure, which is clearly non-uniform in convective regimes. Moreover, other fluid properties (e.g. viscosity, thermal conductivity, ...) may also vary with temperature (or concentration) and pressure. In fact, we shall see that these effects may also be neglected, provided suitable *additional* conditions are satisfied.

Since the Boussinesq approximation is used quite generally for a variety of situations including large-scale atmospheric and oceanic flows, the reader is also referred to more detailed discussions of its validity (e.g. [255, 264, 265, 266, 267] and references therein). In particular, the careful analysis of de Boer [266, 267] clearly distinguishes several cases (weakly or strongly heated fluids, shallow or deep fluids, viscous or inertial regimes) and provides a number of higher-order corrections to the Boussinesq equations.

The following developments will be made considering either a single-component fluid submitted to temperature variations, or a two-component isothermal mixture with non-uniform concentration (but no chemical reactions). In the latter case, deviation from the isothermal situation occasioned by the Dufour effect will not be considered, as this effect is negligible in liquids. The formal similarity between thermal and solutal problems may again be exploited here, although in the former case, additional terms need to be taken into account in the energy equation. This is discussed at the end of this section. For the moment, we denote the diffusive field by N and we will generally refer to the case of mass diffusion, where N stands for the mass fraction. According to Eqs (2.3), (2.7), (2.10), (2.24), (2.30), (2.50) and (2.57), the system of equations we consider is

$$\frac{\partial \rho}{\partial t} + \vec{\nabla}.(\rho \vec{V}) = 0 \tag{3.33}$$

$$\rho \left(\frac{\partial \vec{V}}{\partial t} + (\vec{V}.\vec{\nabla})\vec{V} \right) = -\vec{\nabla}p + \vec{\nabla}.\mathbf{P} + \rho \vec{f} \tag{3.34}$$

$$\rho \left(\frac{\partial N}{\partial t} + (\vec{V}.\vec{\nabla})N \right) = \vec{\nabla}.(\rho D \vec{\nabla} N) \tag{3.35}$$

where the viscous stress tensor $\mathbf{P}$ reads [Eq. (2.64)]:

$$P_{i,j} = \alpha \delta_{i,j}(\vec{\nabla}.\vec{V}) + \mu \left(\frac{\partial V_i}{\partial x_j} + \frac{\partial V_j}{\partial x_i} \right) \tag{3.36}$$

We first make the assumption that the equation of state $\rho = \rho(p, N)$ can be linearized in the range of pressures and concentrations (or temperatures) considered:

$$\rho = \rho_0 \left[1 + \gamma(N - N_0) + \chi(p - p_0) \right] \tag{3.37}$$

where $\gamma = \rho_0^{-1}(\partial \rho / \partial N)_{N_0, p_0}$ is the coefficient of solutal (or thermal) expansion, and $\chi = \rho_0^{-1}(\partial \rho / \partial p)_{N_0, p_0}$ is the compressibility coefficient. N_0 and p_0 are respectively reference composition and pressure. For other fluid properties, expressions similar to Eq. (3.37) will be considered. For instance, $\mu = \mu_0 \left[1 + \gamma_\mu(N - N_0) + \chi_\mu(p - p_0) \right]$ for the dynamical viscosity, ..

Our goal in the following is to investigate the limit of negligible compressibility and small, but non-negligible, solutal (or thermal) expansion. More precisely, the scales of variation of N and of p being respectively denoted by θ_N and θ_p, we will consider the limit of small dimensionless "Boussinesq numbers" $\gamma\theta_N$ and $\chi\theta_p$. An essential assumption making the Boussinesq approximation possible is that $\gamma\theta_N \gg \chi\theta_p$, i.e. the variations of density are due to variations of N only, and pressure variations induce negligible density variations. Note that other small numbers must be introduced for the variations of other fluid properties (which will also need to be small for the Boussinesq approximation to be valid).

For $\gamma\theta_N, \chi\theta_p \to 0$ (i.e. $\rho = \rho_0$ constant), it is tempting to conclude that the system of equations (3.33–3.35) admits the motionless solution $\vec{V} = 0$. Eq. (3.34) would then reduce to the hydrostatic condition $\vec{\nabla}p = \rho_0\vec{g}$ (the body force term is assumed to consist of gravity only, i.e. $\vec{f} = \vec{g}$), and Eq. (3.35) would give $\Delta N = 0$ at the steady state (or more generally an unsteady diffusion equation), to be solved with adequate boundary conditions which eventually fix the scale θ_N. However, this view is not necessarily correct, because even though $\gamma\theta_N, \chi\theta_p \ll 1$, these small numbers might be multiplied by large quantities in some terms. This in fact happens for the body force term $\rho\vec{g}$ of the Navier–Stokes equations. For the moment, consider Eq. (3.33) in the limit $\gamma\theta_N, \chi\theta_p \to 0$. If we assume that the time scale of the density variations remains finite (e.g. it is determined by diffusive time scales), substituting Eq. (3.37) and taking the limit, we get the incompressibility relation

$$\vec{\nabla}.\vec{V} = 0 \tag{3.38}$$

With this result, the divergence of the viscous stress tensor Eq. (3.36) reads

$$\vec{\nabla}.\mathbf{P} = \mu\Delta\vec{V} + \frac{\partial\mu}{\partial x_j}\left(\frac{\partial\vec{V}}{\partial x_j} + \vec{\nabla}V_j\right) \tag{3.39}$$

If we now consider Eq. (3.34) with Eqs (3.37–3.39) in the limit $\gamma\theta_N, \chi\theta_p \to 0$ (and similar limits for the quantities $\gamma_\mu\theta_N, \chi_\mu\theta_p$), we get

$$\rho_0\left[\frac{\partial\vec{V}}{\partial t} + (\vec{V}.\vec{\nabla})\vec{V}\right] = -\vec{\nabla}p + \rho_0\vec{g} + \mu_0\Delta\vec{V} + \rho_0\gamma(N - N_0)\vec{g} + \rho_0\chi(p - p_0)\vec{g} \tag{3.40}$$

The two last terms of this equation have not been neglected compared with $\rho_0\vec{g}$, because $\rho_0\vec{g}$ can be exactly balanced by an hydrostatic pressure gradient $\vec{\nabla}p_h = \rho_0\vec{g}$ (we exclude here the case of open flows, such as an inclined layer not bounded in the lateral directions). Generally the two last terms (which may be irrotational, depending on N and p) cannot be compensated by pressure gradients, and can only be balanced by a non-zero velocity field. Now, the last term of the equation (compressibility effect) may be neglected compared with the previous one (solutal or thermal expansion) provided

$$\gamma\theta_N \gg \chi\theta_p \tag{3.41}$$

It is important to realize that the pressure scale θ_p to be considered in this estimate is the scale of the fluctuation of pressure (associated with fluid motions), and not the scale of the hydrostatic pressure contribution. Indeed, the hydrostatic part may always be eliminated by redefining p_h as the solution of

$$\vec{\nabla}p_h = \rho_0\vec{g}[1 + \chi(p_h - p_0)] \tag{3.42}$$

Redefining p as $p + p_h$ and using the estimate (3.41) for the fluctuation p, Eq. (3.40) becomes

$$\rho_0 \left[\frac{\partial \vec{V}}{\partial t} + (\vec{V}.\vec{\nabla})\vec{V} \right] = -\vec{\nabla}p + \mu_0 \Delta \vec{V} + \rho_0 \gamma (N - N_0)\vec{g} \tag{3.43}$$

which is the Navier–Stokes equation in the Boussinesq approximation. Finally, it is straightforward to show that the species conservation equation (3.35) under the same approximations reads

$$\frac{\partial N}{\partial t} + (\vec{V}.\vec{\nabla})N = D_0 \Delta N \tag{3.44}$$

As a preliminary conclusion, we may state that the Boussinesq approximation [Eqs (3.38), (3.43) and (3.44)] is valid in situations where $\gamma \theta_N, \chi \theta_h \ll 1$ (and similar expressions for all other fluid properties) and $\chi \theta_p \ll \gamma \theta_N$. In the first condition, θ_h is the scale of the hydrostatic pressure, while in the second θ_p is the scale of the fluctuations of pressure due to the flow, which is in general much smaller than θ_h. Note that up to now, we have excluded the possibility of open flows (forced convection due to external pressure gradients or gravity forces) and of liquid layers presenting a deformable interface (see end of this section).

For a horizontal layer of depth d, infinite in horizontal extent, and confined between non-deformable boundaries, we may test the above conditions. As p_h is obtained from Eq. (3.42), we get $\theta_h \sim \rho_0 g d$. Considering θ_N as the scale of concentration (or temperature) variations, it appears that for a quasi-steady viscous flow [setting to zero the left-hand side of Eq. (3.43)], scales of pressure and of velocity fluctuations are, respectively, $\theta_p \sim \rho_0 g \gamma \theta_N d$ and $U \sim \gamma \theta_N g d^2 / \nu_0$ (where $\nu_0 = \mu_0 / \rho_0$ is the kinematic viscosity). Note that the scale of velocity U may be rewritten as $U \sim Ra D_0 / d$, where Ra is the Rayleigh number

$$Ra = \gamma \theta_N \frac{g d^3}{\nu_0 D_0} = \gamma \theta_N Ga \tag{3.45}$$

with the Galileo number $Ga = g d^3 / \nu_0 D_0$. Note that the Rayleigh number appears as the product of a Boussinesq number $\gamma \theta_N$ and Ga. Thus, if a situation is considered for which the Rayleigh number is of order unity or higher, the use of the Boussinesq approximation (with $\gamma \theta_N \ll 1$) implies that, simultaneously, $Ga \gg 1$ [200]. The Galileo number is indeed very large for standard experiments (e.g. for a 5 mm thick water layer, $Ga \sim 10^7$ for a thermal problem, while $Ga \sim 10^9$ for a solutal problem). Moreover, the two conditions on the pressure scales reduce to $\chi \rho_0 g d \ll 1$, which is well verified under usual conditions.

Because of the relation Eq. (3.45) between Ra, $\gamma \theta_N$ and Ga, it is particularly important (when using the Boussinesq approximation) to ensure that the ratio $Ra/Ga = \gamma \theta_N$ indeed remains very small. In some specific problems, the Galileo number appears in some terms not necessarily multiplied by the Boussinesq parameter $\gamma \theta_N$. In these cases, it should be remembered that if $Ra = O(1)$ or higher, the Boussinesq approximation requires $Ga \gg 1$. For instance, if a horizontal liquid layer (two-dimensional case) is in contact with a gas, and has a deformable surface, the normal stress boundary condition (2.89) must be considered along the deformed boundary $z = h(x, t)$. Assuming that the system only slightly deviates from the rest state for which the interface is flat (at $z = 0$), we may obtain the boundary condition satisfied by perturbations of pressure p, interface position h and vertical velocity W (see §3.8) as

$$\sigma \frac{\partial^2 h}{\partial x^2} - \rho_0 gh = -p + 2\mu \frac{\partial W}{\partial z} \quad at\ z = 0 \tag{3.46}$$

The second term of the l.h.s. of this expression is proportional to the hydrostatic pressure gradient, which is in general a large quantity. In fact, using the above "buoyancy" scales, $\theta_p \sim \rho_0 g \gamma \theta_N d$ for pressure and $U \sim \gamma \theta_N gd^2/\nu_0$ for velocity, the r.h.s. of Eq. (3.46) is found to be of order $\rho_0 g \gamma \theta_N d$. Thus, in the case where both terms of the l.h.s. are of the same order (typically for a Bond number $Bo = \rho_0 gd^2/\sigma$ of order unity), we get $h \sim \gamma \theta_N d$ which shows that the surface deformation should be neglected if the Boussinesq approximation is used ($\gamma \theta_N \to 0$). Surface deformation thus appears as a non-Boussinesq effect [200] when the driving mechanism for fluid motions is buoyancy. Therefore, including its effect in the presence of buoyancy would require taking into account several terms of the same order which were neglected when taking the limit of vanishing Boussinesq number $\gamma \theta_N$.

Finally, let us now specifically consider the thermal problem, for which additional terms occur in the energy equation [Eq. (2.30)]. First, the production of heat due viscous dissipation reads

$$P_{i,j} \frac{\partial V_i}{\partial x_j} = \mu \left(\frac{\partial V_i}{\partial x_j} + \frac{\partial V_j}{\partial x_i} \right) \frac{\partial V_i}{\partial x_j} \tag{3.47}$$

Using the above estimates for buoyancy-induced flow (induced by a temperature difference θ_T), this is of order $\rho_0 (\gamma \theta_T)^2 g^2 d^2/\nu_0$, which can be neglected against the diffusive term $\lambda_0 \Delta T$ (where λ_0 is the thermal conductivity) provided $Ra \ll c_p/\gamma gd$. This condition is well verified (except at very large Rayleigh numbers, or for large-scale atmospheric flows [265]) owing to the large value of the specific heat c_p. The quantity $Di = \gamma gd/c_p$ is sometimes called the dissipation number [265].

A more critical term in Eq. (2.30) is that associated with expansion effects

$$\left(\frac{\partial \ln v}{\partial \ln T} \right)_p \frac{dp}{dt} = -T\gamma \left(\frac{\partial p}{\partial t} + \vec{V}.\vec{\nabla}p \right) \tag{3.48}$$

in which the pressure gradient should include the hydrostatic contribution. Thus, still in the case of steady viscous buoyancy-induced flow, the order of magnitude of Eq. (3.48) is $\gamma^2 T_0 \theta_T \rho_0 g^2 d^2/\nu_0$, which is negligible compared with $\lambda_0 \theta_T/d^2$ when $d^4 \ll \lambda_0 \nu_0/\gamma^2 T_0 \rho_0 g^2$. For both water and a 100 cSt silicone oil, this leads to $d \ll 2 - 3$ cm. In some situations, it might thus be incorrect to neglect this contribution. Note that de Boer [266] also discusses situations where it is incorrect to neglect the term $\sim \partial p/\partial t$ in Eq. (3.48), for instance for gases in a constant volume vessel (where spatially homogeneous pressure variations cannot be excluded). The latter effect turns out to be negligible for liquids under normal conditions.

Another surprising deviation from the Boussinesq approximation may occur for buoyancy-induced flows in a low-gravity environment [201, 202]. There, due to residual accelerations of magnitude a, a weak flow may exist with an order of magnitude $U \sim \gamma \theta_N ad^2/\nu_0$. However, due to volume expansion caused by mass (or heat) diffusion [an additional term $-\gamma dN/dt$ in the r.h.s. of Eq. (3.38)], another velocity scale $U_d \sim \gamma D_0 \theta_N/d$ exists, which becomes the dominating velocity when $U_{\mathrm{d}} \gg U$. This gives $Ga \ll 1$ (where the Galileo number is here calculated using the reduced acceleration a), and it may thus be expected that the Boussinesq approximation is not valid in this case.

In conclusion, the Boussinesq approximation may break down for several reasons (e.g. surface deformation in the presence of buoyancy, very low gravity, gravity induced channel flow, expansion effects in the energy equation, large-scale flows, ...). Rather than to attempt a general formulation of the necessary conditions for its validity (which depend on the flow regime, on the geometry considered, ...), it appears preferable to rely on a careful case by case analysis. As for the variation of fluid properties with temperature or concentration, it is necessary to check in each case that the corresponding smallness parameters remain small, independently of the smallness of the Boussinesq parameter $\gamma\theta_N$. Usually, the temperature dependence of viscosity appears as the most critical among these effects [268, 269]. In this case, the divergence of the viscous stress tensor should be written as (3.39). The influence of this effect on large-wavelength Bénard convection is considered in §4.5.2.

Note finally that Batchelor [255] considered a more general discussion of the validity of the incompressibility condition $\vec{\nabla}.\vec{V} = 0$, where the scales of pressure and velocity are not a priori related to buoyancy-induced motions. Furthermore, the variation of density in his case is expressed not as a function of variations of temperature and pressure, but rather of pressure and entropy. His estimates are thus expressed as a function of the sound speed c, defined by $c^2 = (\partial\rho/\partial p)_s$, rather than as a function of the isothermal compressibility $\chi = \rho_0^{-1}(\partial\rho/\partial p)_{T_0}$. His conclusions, though of wider applicability, agree with those derived above for buoyancy-induced flows.

3.2   Dimensionless parameters and equations

3.2.1   Dimensionless parameters

Some dimensionless groups have already been defined in the previous sections. For instance, the Prandtl number is

$$Pr = \frac{\nu}{\kappa} \tag{3.49}$$

which, according to Eqs (3.1) and (3.2), is the ratio of characteristic thermal and viscous relaxation times. This is a characteristic of the fluid used, which is particularly important for the dynamical and nonlinear regimes [in contrast, the monotonic neutral stability result Eq. (3.31) is independent of Pr]. Typical values of Pr are presented in Table 3.1.

Most of the liquids used for experiments on Bénard instabilities have a Prandtl number of order 5 or more (especially for highly viscous oils), and only liquid metals are characterized by values of Pr as low as 10^{-2}–10^{-3}. Gases typically have a Pr of order unity.

For binary mixtures, the dimensionless number characterizing the ratio of the mass diffusion relaxation time $\tau_{\mathrm{diff}} = d^2/D$ (D is the isothermal diffusion coefficient) to the viscous time $\tau_{\mathrm{visc}} = d^2/\nu$ is the Schmidt number

$$Sc = \frac{\tau_{\mathrm{diff}}}{\tau_{\mathrm{visc}}} = \frac{\nu}{D} \tag{3.50}$$

Table 3.1: Prandtl number of some fluids.

Fluid	Pr
Water	6
Silicone oil	$5 - 10^4$
Mercury	0.026
Air	0.7

Thus, this is the equivalent of the Prandtl number for systems where the diffusive field is concentration rather than temperature. As D is typically two orders of magnitude lower than the thermal diffusivity κ, the Schmidt number is often very large for typical binary liquid mixtures (see Table 3.2).

Table 3.2: Typical values of the Schmidt number.

Fluid mixture	Sc
Aqueous solutions	$\sim 10^3$
Metallic melts	~ 50

In non-isothermal binary mixtures, both Prandtl and Schmidt numbers need to be considered. The Lewis number may be defined as $Le = Pr/Sc = D/\kappa$, and is equal to the ratio of thermal and solutal relaxation times.

In §3.1.2 and 3.1.3, the Rayleigh number [99] was defined as

$$Ra = \frac{\tau_{\text{visc}}\tau_{\text{th}}}{\tau_{\text{buoy}}^2} = \frac{\alpha g \Delta T d^3}{\nu\kappa} = \frac{\alpha g \beta d^4}{\nu\kappa} \tag{3.51}$$

where τ_{buoy} is the characteristic "buoyancy" time scale defined by Eq. (3.3), and interpreted as the typical time constant of the amplification phenomenon associated with the Rayleigh–Bénard instability, in the absence of stabilizing diffusive effects. The Rayleigh number therefore measures the relative importance of destabilizing over stabilizing factors.

Another important destabilizing mechanism studied here is associated with the surface tension dependence on temperature (or concentration) at a free surface, or at the interface between two liquids. When considering such situations, one is led to introduce the Marangoni number

$$Ma = \frac{(-\partial\sigma/\partial T)\Delta T d}{\mu\kappa} = \frac{(-\partial\sigma/\partial T)\beta d^2}{\mu\kappa} \tag{3.52}$$

where $\partial\sigma/\partial T$ is the surface tension variation with temperature (negative for usual liquids). The Marangoni number measures the importance of the tangential thermocapillary stresses at a free interface. It will appear in the tangential stress boundary condition (see §3.2.3). The Marangoni number may also be defined as the ratio of thermal and viscous time scales to a "thermocapillary" time scale constructed from dimensional reasoning. If a gradient of temperature (say, of order $\beta = \Delta T/d$) exists along the interface, a stress (surface tension gradient) acts tangentially on it, and induces motion of the fluid (from hot to cold regions of the surface if the surface tension decreases with temperature). Thus, the fluid of density ρ will experience an acceleration of order $\sigma_T\beta/\rho d$ [hereafter, σ_T will denote $(-\partial\sigma/\partial T)$], from which the thermocapillary time scale $\tau_{\mathrm{Ma}} = (\rho d^2/\sigma_T\beta)^{1/2}$ can be defined. It is then checked that the Marangoni number defined by Eq. (3.52) is equal to $Ma = \tau_{\mathrm{th}}\tau_{\mathrm{visc}}/\tau_{\mathrm{Ma}}^2$.

Both Rayleigh and Marangoni numbers are proportional to the driving gradient $\beta = \Delta T/d$. In the presence of both effects (e.g. for a layer of fluid heated from below and presenting a free surface), the relative importance of buoyancy and surface tension forces is often characterized by the dynamic Bond number

$$Bo_{\mathrm{d}} = \frac{Ra}{Ma} = \frac{\tau_{\mathrm{Ma}}^2}{\tau_{\mathrm{buoy}}^2} = \frac{\alpha\rho g d^2}{\sigma_T} \tag{3.53}$$

which is independent of β. This quantity is seen to increase with the square of the liquid depth d, thus indicating the dominating effect of buoyancy over surface tension for thick layers. This is physically justified by the increasing role of volume forces compared with surface forces at increasing system size. In terms of time scales, the dominating effect will be the one which most rapidly sets the fluid into motion. Some typical values of Bo_{d} are reproduced in Table 3.3, as a function of the fluid thickness.

Table 3.3: Dynamic Bond number as a function of depth d [cm].

Fluid	Bo_{d}/d^2 [cm^{-2}]
Silicone oil	19
Water	1.9

Note that this Bond number is qualified as "dynamic" by opposition with the static Bond number (or just Bond number), which reads

$$Bo = \frac{\rho g d^2}{\sigma} \tag{3.54}$$

where σ is the surface tension. The Bond number also quantifies the importance of gravity over surface tension forces, but is now connected with the role of equilibrium capillarity rather than thermocapillarity. For small Bo, the fluid behavior is mainly determined by surface tension (e.g. at small d), while gravity dominates for large Bo. The transition between these two regimes occurs at a length scale obtained by setting $Bo = 1$. We then get the capillary length scale $d_{\mathrm{cap}} = (\sigma/\rho g)^{1/2}$, which is of order 2.5 mm for water.

The Galileo number has already been defined as

$$Ga = \frac{g d^3}{\nu\kappa} \tag{3.55}$$

which appears as the ratio $\tau_{\mathrm{th}}\tau_{\mathrm{visc}}/\tau_{\mathrm{grav}}^2$ of thermal and viscous time scales to the square of the gravity time scale $\tau_{\mathrm{grav}} = (d/g)^{1/2}$, i.e. the time needed for a body to travel over a distance

d under the acceleration g. It is instructive to note the relation between the definitions of Ga and of the Rayleigh number (3.51). It was seen when studying the Boussinesq approximation that $Ra/Ga = \alpha\Delta T$, i.e. the Boussinesq parameter. Thus it follows that $\tau_{grav}/\tau_{buoy} = (\alpha\Delta T)^{1/2}$.

We may finally introduce the capillary number (also called the crispation number)

$$Ca = \frac{\mu\kappa}{\sigma d} = \frac{Bo}{Ga} \tag{3.56}$$

from which it can be seen that $Ma = \gamma_\sigma\Delta T Ca^{-1}$ (with $\gamma_\sigma = \sigma^{-1}\sigma_T$), i.e. a relation formally analogous to $Ra = \alpha\Delta T Ga$, but for the surface tension rather than the buoyancy effect. Accordingly, we also have $\tau_{cap}/\tau_{Ma} = (\gamma_\sigma\Delta T)^{1/2}$ where $\tau_{Ma} = (\rho d^3/\sigma_T\Delta T)^{1/2}$ is the time scale associated with the thermocapillary effect, while $\tau_{cap} = (\rho d^3/\sigma)^{1/2}$ is associated with capillary forces acting at a deformable interface (Laplace force in the normal stress boundary condition).

For completeness, we also define the Peclet number

$$Pe = \frac{Ud}{\kappa} \tag{3.57}$$

and the Reynolds number

$$Re = \frac{Ud}{\nu} \tag{3.58}$$

where U is a typical fluid velocity. Defining a convective time scale by $\tau_{conv} = d/U$, it is seen that the Peclet and Reynolds numbers may be rewritten as $Pe = \tau_{th}/\tau_{ccnv}$ and $Re = \tau_{visc}/\tau_{conv}$. We also have $Re = Pe/Pr$.

3.2.2 Dimensionless equations

In this section, we establish the dimensionless equations which will generally be referred to and eventually recalled later on in this book. Denoting the unit of length by d (generally this is taken as the fluid depth, or one of the fluid depths in the case of multilayer systems), the unit time as τ, the unit mass as ρd^3, and the unit temperature as θ, and using the Boussinesq approximation (§3.1.4), we rewrite the Navier–Stokes equations [see Eqs (2.7), (2.10) and (2.66)] as

$$\frac{d\vec{V}}{dt} = \frac{\tau}{\tau_{visc}}\Delta\vec{V} - \frac{\tau^2}{\tau_{th}\tau_{visc}}\vec{\nabla}p + \frac{\tau^2}{d}\frac{\rho}{\rho_0}\vec{g} \tag{3.59}$$

where both terms in the material derivative $d/dt = \partial/\partial t + \vec{V}.\vec{\nabla}$ have the same scale because the unit of velocity is chosen as d/τ. In Eq. (3.59), the unit of pressure is chosen as $\mu\kappa/d^2$, the thermal and viscous time scales are defined by Eqs (3.1) and (3.2), and the notations $\vec{V}$ and p are conserved for dimensionless variables. The energy equation (2.16), written for the dimensionless temperature field T, reads, using the Fourier's law (2.50)

$$\frac{dT}{dt} = \frac{\tau}{\tau_{th}}\Delta T \tag{3.60}$$

Note that some terms in this equation were neglected (viscous heating and expansion effects), an approximation justified at the end of §3.1.4. In order to take into account buoyancy effects, a linearized equation of state will be used in the form $\rho = \rho_0[1 - \alpha(T - T_0)]$, where α is the thermal expansion coefficient, and where isothermal compressibility effects are neglected (see also §3.1.4).

Now, in principle, the unit of time τ can be chosen for convenience, e.g. as $\tau = \tau_{\text{th}}$ or $\tau = \tau_{\text{visc}}$. This will be unimportant (it is always possible to recalculate physical quantities), as long as no limiting values of dimensionless parameters are taken. For example, we might be interested in the limit of small Prandtl number $Pr = \tau_{\text{th}}/\tau_{\text{visc}} \to 0$. In this case, the choice of a thermal time scale $\tau = \tau_{\text{th}}$ leads to the drop of essential terms ($\Delta\vec{V}$ and $\vec{\nabla}p$), resulting in erroneous results (as an indication, it is not possible to satisfy all boundary conditions because the highest-order viscous terms have been dropped). The relevant choice in this limit thus appears to be $\tau = \tau_{\text{visc}}$ (eventually with a more adequate choice of the pressure unit as $\mu\nu/d^2$ and of the temperature unit as $Pr\theta$ [375]), leading for $Pr \to 0$ to disregarding the material derivative term from the energy equation (3.60). Note however that it was also shown in [375] that this viscous scaling breaks down when increasing the constraint (inertial convection), a difficulty which can only be solved by reintroducing the thermal nonlinearity $\vec{V}.\vec{\nabla}T$ in Eq. (3.60). This is further discussed at the end of §4.5.6.

In this book, we will mostly be interested in the case of moderate and high Prandtl number liquids, which represent the vast majority of liquids used for experimentation (apart from liquid metals such as mercury, see Table 3.1). In these situations, the relevant time scale appears to be the thermal scale $\tau = \tau_{\text{th}}$. We then obtain the dimensionless equations

$$\Delta\vec{V} - \vec{\nabla}p - Ga\vec{1}_z[1 + \alpha\theta(T - T_0)] = Pr^{-1}\left[\frac{\partial\vec{V}}{\partial t} + (\vec{V}.\vec{\nabla})\vec{V}\right] \tag{3.61}$$

$$\Delta T = \frac{\partial T}{\partial t} + (\vec{V}.\vec{\nabla})T \tag{3.62}$$

where the z-coordinate has been taken along the vertical ($\vec{1}_z$ is a unit vector directed vertically upwards), and the Galileo number is defined as $Ga = gd^3/\nu\kappa$ (linked to the Rayleigh number by $Ra = \alpha\theta Ga$, see §3.2.1).

The incompressibility condition is unchanged

$$\vec{\nabla}.\vec{V} = 0 \tag{3.63}$$

To conclude this section, we recall that if the liquid is an isothermal binary mixture, and thus concentration of the solute appears as the diffusive scalar field (rather than temperature), we may in principle use the system of dimensionless equations (3.61) to (3.63) with the change of notations $T \to N$ (the mass fraction of the solute), $\kappa \to D$ (the isothermal diffusion coefficient), $\alpha\theta \to \gamma\Gamma$ (where γ is the coefficient of density variation with concentration and Γ a typical concentration scale), and $Pr \to Sc$ (the Schmidt number). The time scale is then associated with species diffusion.

3.2.3 Dimensionless boundary conditions

Here we briefly establish the dimensionless form of the most frequently used boundary conditions, based on time and velocity scales introduced in §3.2.2. We will however omit for the moment the boundary conditions at a deformable interface, which will be established when specifically investigating those cases (see §3.6, §3.8 and §4.5.3). All conditions will here be written along a flat horizontal boundary located at $z = a$, or at an interface with height $h = h_0$ and no phase change.

Mechanical boundary conditions

Rigid boundary

On a rigid boundary, for a viscous fluid, all components of the velocity vanish. For a wall located at height $z = a$, this reads

$$\vec{V}(z = a) = 0 \tag{3.64}$$

Using the incompressibility condition (3.63), an equivalent set of boundary conditions may be obtained for the normal (vertical) component W:

$$W = DW = 0 \quad at \ z = a \tag{3.65}$$

where D denotes the (dimensionless) z-derivative.

Stress-free undeformable boundary

Here, the normal component of velocity vanishes, as well as the tangential stress. This applies at a free surface without Marangoni effect. Again, for a horizontal boundary located at $z = a$ and denoting by (U, V, W) the components of $\vec{V}$, this reads

$$W = DU = DV = 0 \quad at \ z = a \tag{3.66}$$

Using Eq. (3.63), it is seen that equivalent boundary conditions on W are

$$W = D^2W = 0 \quad at \ z = a \tag{3.67}$$

Thermal boundary conditions

Isothermal (perfectly-conducting) boundary

Along a perfectly-conducting boundary, the temperature is uniform, such that

$$T = T_{\mathrm{w}} \quad at \ z = a \tag{3.68}$$

Poorly-conducting boundary

Here, the normal component of the heat flux vanishes. If Fourier's law is assumed, this reads

$$DT = 0 \quad at \ z = a \tag{3.69}$$

Thermally-resistive boundary

The last two cases are particular cases of a more general mixed condition

$$DT + Bi(T - T_0) = 0 \tag{3.70}$$

where Bi is the dimensionless Biot number, which measures the efficiency of heat transfer at the boundary. Comparing Eq. (3.70) with Eqs (3.68) and (3.69), we see that $Bi = 0$ corresponds to the poorly-conducting case, while $Bi \to \infty$ corresponds to a perfectly-conducting boundary. The condition (3.70) will now be discussed in some detail.

This kind of phenomenological boundary condition in principle applies to several situations, though with a variable degree of success (see also §2.2.1). For example in the case of a free surface, heat removal by the gas phase (by conduction or convection) is often described by the phenomenological Newton's cooling law (2.77). Then, rescaling lengths by d as usual, we may identify T_0 with the ambient temperature (i.e. the temperature prevailing sufficiently far from the surface), and the Biot number as

$$Bi = \frac{\alpha d}{\lambda} \tag{3.71}$$

where α here stands for the heat transfer coefficient, and λ for the fluid thermal conductivity. Of course, Eq. (3.70) is a definition of Bi (or α), rather than a rigorous condition expressing energy conservation through an interface (see later). It is often used instead of the exact boundary condition, with a value of Bi determined empirically.

Another situation where Eq. (3.70) may apply is the case of a liquid phase in contact with a purely conductive medium with thermal conductivity λ_w and thickness d_w. This material will be denoted as the passive medium in what follows. Then, if the heat transfer is unidirectional (along z) and quasi-steady in the passive medium (such that the temperature profile is linear in this phase[4]), continuity of the heat flux at the interface requires

$$-\lambda \frac{\partial T}{\partial z} = \lambda_w \frac{(T - T_0)}{d_w} \tag{3.72}$$

where T denotes the temperature at the interface, and $T - T_0$ is the temperature drop across the passive phase. Identifying Eq. (3.72) with (3.70), we get $Bi = \lambda_w d / \lambda d_w$.

However, the use of Eq. (3.72) is certainly not valid for a non-isothermal free surface (e.g. when convection occurs). Indeed, near the threshold of cellular Bénard convection, the temperature at the free interface (say, located at $z = 0$) is well approximated by $T =$

---

[4]Note that this holds provided the thermal relaxation time d_w^2/κ_w of the passive medium is much smaller than that of the liquid.

$T_s + A \cos kx$, where A is a small amplitude measuring the intensity of fluid motions and k is a wavenumber characterizing the periodicity of convection rolls in the horizontal x-direction. It is helpful for some interpretations of further results to examine the adequacy of a law of the form (3.70) for describing the heat transfer in such a convective regime. We consider a quasi-steady situation in the passive gas phase (this is justified, except for large depths d_w, owing to the generally high thermal diffusivity of gaseous media). Then, we have to solve the steady energy equation

$$\Delta T_w = 0 \tag{3.73}$$

with boundary conditions $T_w = T_s + A \cos kx$ at $z = 0$ and $T_w = T_0$ at $z = d_w$. The solution of this problem reads

$$T_w = T_s + \frac{z}{d_w}(T_0 - T_s) + A\frac{\sinh[k(d_w - z)]}{\sinh[kd_w]} \cos[kx] \tag{3.74}$$

and from this relation, we may express the continuity of the heat flux at the free surface as

$$
\begin{aligned}
-\lambda \frac{\partial T}{\partial z}\Big|_0 &= -\lambda_w \frac{\partial T_w}{\partial z}\Big|_0 \\
&= \frac{\lambda_w}{d_w}(T_s - T_0) + A \cos[kx]k\lambda_w \coth[kd_w]
\end{aligned}
\tag{3.75}
$$

Strictly speaking, this relation cannot be cast under the form (3.70) with $T = T_s + A \cos[kx]$ and constant Bi. However, this becomes possible if the Biot number Bi is allowed to depend on the wavenumber k through the relation

$$Bi_k = \frac{\lambda_w}{\lambda} kd \coth[kd_w] \tag{3.76}$$

Then it is seen that for the first term of the right-hand side of Eq. (3.75), the appropriate value of the Biot number is $Bi_{k\to 0} = \lambda_w d / \lambda d_w$, which is the result found above for the horizontally uniform case (zero wavenumber mode). The heat transfer for Fourier components with wavenumber k will be described by a relation of the form (3.70), but with a value of the Biot number given by Eq. (3.76). Note that $Bi_k = Bi_0 kd_w \coth[kd_w]$, i.e. the zero wavenumber Biot number is multiplied by a monotonically increasing function of kd_w.

The boundary condition (3.70) with (3.76) will indeed prove to be useful (and even exact) for describing the linear stage of the Bénard instability (because Fourier components with different wavenumbers are decoupled). However, it is not straightforward, although possible, to extend this boundary condition to nonlinear regimes, for which the dynamics of modes with different wavenumbers (e.g. the harmonics generated by nonlinear interactions) cannot be decoupled. Furthermore, in the case where the upper phase cannot be considered as inert (i.e. the motions cannot be neglected or the dynamics of heat diffusion is not quasi-static), we will rather turn to the exact two-layer approach for which the equations are solved in both phases, subject to the realistic jump conditions established in Chapter 2 and written under dimensionless form hereafter.

Interfacial boundary conditions at a flat interface

At an undeformable flat horizontal interface separating two fluid media with different properties (hereafter indexes 1 and 2 will refer to lower and upper phase respectively), the continuity/jump conditions are:

The non-deformability (or impermeability) condition:

$$W_1 = W_2 = 0 \tag{3.77}$$

which is coherent with Eqs (2.78) and (2.84) when the interface is flat ($h = h_0$) and there is no phase change ($J = 0$).

The no-slip condition:

$$U_1 = U_2, \quad V_1 = V_2 \quad \text{or} \quad DW_1 = DW_2 \tag{3.78}$$

the second condition being obtained from the incompressibility relation (3.63).

The continuity of temperature:

$$T_1 = T_2 \tag{3.79}$$

which follows from the fact that the interface is assumed to have negligible thermal resistance. Note that similarly, Eq. (3.78) holds where there is no resistance to momentum transfer.

The continuity of the heat flux:

$$DT_1 = \frac{\lambda_2}{\lambda_1} DT_2 \tag{3.80}$$

This condition follows from the general energy jump equation (2.99) and Fourier's law (2.50) when $h = h_0$ and $J = 0$.

The continuity of normal and tangential stress: the equation for the linear momentum conservation is given by Eq. (2.88). In general, one should consider its projections on the normal and the tangent to the interface. For an undeformable interface however, only the tangential stress component needs to be used. Indeed, should we want to assume the interface perfectly flat as done in this section, the normal component [Eq. (2.89)] would indicate that in the presence of fluid motions, there exists a jump of pressure across the interface, due to a discontinuity of the normal viscous stresses. In reality, this small pressure jump is compensated by a capillary (Laplace) pressure induced by a small surface deformation. However, this generally turns out to be unimportant, because the pressure is a hidden variable for an incompressible fluid and can be eliminated from the Navier–Stokes equations, as seen later. Thus, in most cases, the normal stress boundary condition will merely serve to check that surface deformation can indeed be neglected. A more detailed discussion of the normal stress condition including surface deformability and phase change is presented in §3.8.

Now, projecting Eq. (2.88) on the tangent directions ($\vec{1}_x$ and $\vec{1}_y$), using Eqs (2.10), (2.66) and the conditions (3.77), leads to the tangential stress boundary conditions

$$-\mu_2 \frac{\partial U_2}{\partial z} + \mu_1 \frac{\partial U_1}{\partial z} = \frac{\partial \sigma}{\partial x} \tag{3.81}$$

$$-\mu_2 \frac{\partial V_2}{\partial z} + \mu_1 \frac{\partial V_1}{\partial z} = \frac{\partial \sigma}{\partial y} \tag{3.82}$$

Differentiating the first of these boundary conditions with respect to x, the second with respect to y, adding the results and using the incompressibility condition (3.63) yields

$$\mu_2 \frac{\partial^2 W_2}{\partial z^2} - \mu_1 \frac{\partial^2 W_1}{\partial z^2} = \Delta_h \sigma \tag{3.83}$$

where Δ_h is the horizontal Laplacian operator. Then, assuming a linear variation of surface tension with temperature $\sigma = \sigma_0 - \sigma_T(T - T_0)$, and using the usual thermal scales, the dimensionless form of this condition reads

$$\frac{\mu_2}{\mu_1} D^2 W_2 - D^2 W_1 = -Ma_1 \Delta_h T \tag{3.84}$$

where $T = T_1 = T_2$ and the Marangoni number is

$$Ma_1 = -\frac{\sigma_T \theta_1 d_1}{\mu_1 \kappa_1} \tag{3.85}$$

Here, θ_1 is the temperature drop across the bottom layer, and as for most liquid interfaces the surface tension decreases with temperature ($\sigma_T = \partial \sigma / \partial T < 0$), the Marangoni number will be positive when heating from below.

Note that when the upper phase is a gas, the dynamical viscosity μ_2 is in general very small compared with μ_1 ($\mu_2/\mu_1 \to 0$). As the velocities induced in the gas layer in the immediate vicinity of the interface are typically of the same order as those in the liquid [because of the no-slip condition (3.78)], we get the condition

$$D^2 W_1 = Ma \, \Delta_h T \tag{3.86}$$

This is the condition applicable to a single-layer problem. Discussions of the validity of the so-called "one-sided" model [186] will be provided later on (§3.5 and §3.8).

3.3 Monotonic modes in one-layer systems

3.3.1 Rayleigh–Bénard Instability

In this section we generalize the analysis presented in §3.1.3 to various types of realistic boundary conditions. These results are classical [24, 25, 95], but are included here for completeness, also allowing to establish linearized equations to be used in later sections, as well as to address some basic features of monotonic instability modes. Although oscillatory instability is not expected here, it is worth emphasizing that damped oscillatory modes are not excluded for negative Rayleigh numbers (see §3.1.2 and 3.1.3), a possibility discussed in Appendix B on the basis of self-adjointness properties of linear operators. Linear energy stability and variational principles are also summarized in Appendix C.

We will study the linear stability of a motionless reference state, in the presence of a constant gradient β of temperature (or concentration) orthogonal to the layer and parallel to gravity ("heating from below", as it is known from §3.1 that monotonic instability modes may only be expected in this situation). The general set-up has already been sketched in Fig. 3.1 of §3.1.1.

The dimensionless Boussinesq equations describing Rayleigh–Bénard convection have been established in §3.2.2, using the thermal time scale $\tau_{\mathrm{th}} = d^2/\kappa$, the length scale d, the pressure scale $\mu\kappa/d^2$, and the temperature scale $\theta = \beta d = T_0 - T_1$ (where T_0 and T_1 are respectively the bottom and top surface temperatures). We reproduce the result [Eqs (3.61–3.63)] as

$$\vec{\nabla}.\vec{V} = 0 \tag{3.87}$$

$$\Delta\vec{V} - \vec{\nabla}p - Ga\vec{1}_z[1 + \alpha\theta(T - \tilde{T}_{\mathrm{r}})] = Pr^{-1}\left(\frac{\partial\vec{V}}{\partial t} + (\vec{V}.\vec{\nabla})\vec{V}\right) \tag{3.88}$$

$$\Delta T = \frac{\partial T}{\partial t} + (\vec{V}.\vec{\nabla})T \tag{3.89}$$

where $\vec{V} = (U, V, W)$, T, and p are respectively the fields of velocity, temperature and pressure, $Pr = \nu/\kappa$ is the Prandtl number, $Ga = gd^3/\nu\kappa \gg 1$ is the Galileo number, $|\alpha\theta| \ll 1$ is the Boussinesq parameter, and the Rayleigh number is the product $Ra = -\alpha\theta Ga = O(1)$. Note that the coefficient of thermal expansion α is negative here. We also introduced the reduced reference (mean) temperature $\tilde{T}_{\mathrm{r}} = (T_0 + T_1)/2\theta$, at which the physical properties are evaluated (e.g. in the calculation of Ga, Ra and Pr).

The motionless state is a steady solution of Eqs (3.87–3.89) with $\vec{V} = 0$, such that

$$\vec{\nabla}p_{\mathrm{ref}} = -Ga\vec{1}_z[1 + \alpha\theta(T_{\mathrm{ref}} - \tilde{T}_{\mathrm{r}})] \tag{3.90}$$

$$\Delta T_{\mathrm{ref}} = 0 \tag{3.91}$$

For the horizontally homogeneous boundary conditions we consider at $z = 0$ and $z = 1$ (where $T = T_0/\theta$ and $T = T_1/\theta$ respectively), the only solution of this system is

$$T_{\mathrm{ref}} = -z + \tilde{T}_{\mathrm{r}} + \frac{1}{2} \tag{3.92}$$

$$p_{\mathrm{ref}} = -Gaz - Ra\frac{z(z-1)}{2} \tag{3.93}$$

$$\vec{V}_{\text{ref}} = 0 \tag{3.94}$$

i.e. T_{ref} is a linear temperature profile (constant dimensionless temperature gradient -1) and p_{ref} is the sum of an hydrostatic pressure $-Gaz$ (i.e. $-\rho g z$ in physical units) and a parabolic correction due to thermal expansion of the fluid. Note that the temperatures T_0 and T_1 at bottom and top plates, respectively, should in principle be calculated from the thermal boundary conditions. However, this is not performed here as it will have no direct influence on later developments.

The linear stability of this motionless state is studied by adding infinitesimal perturbation fields T', $\vec{V}'$ and p' to the solution (3.92–3.94), and linearizing the basic equations with respect to the primed quantities. Then, omitting the primes, we get the system describing the evolution of perturbations as

$$\vec{\nabla}.\vec{V} = 0 \tag{3.95}$$

$$\Delta\vec{V} - \vec{\nabla}p + Ra\vec{1}_z T = Pr^{-1}\frac{\partial\vec{V}}{\partial t} \tag{3.96}$$

$$\Delta T + W = \frac{\partial T}{\partial t} \tag{3.97}$$

This system of equations, linear and homogeneous in the perturbation fields $\vec{V} = (U, V, W)$, T, p, is quite general as it does not depend on boundary conditions. Frequent reference will be made to these equations in the following.

It is possible to reduce the above system to a simpler one, by first taking the divergence of Eq. (3.96) and using Eq. (3.95). We then get an equation for the pressure perturbation

$$\Delta p = Ra\, DT \tag{3.98}$$

where D stands for the dimensionless z-derivative throughout in the following. This allows eliminating p (as well as U and V) from the above system, by applying the Laplacian operator to Eq. (3.96), and using Eq. (3.98). We then obtain

$$\Delta^2 W + Ra\Delta_{\text{h}}T = Pr^{-1}\frac{\partial}{\partial t}\Delta W \tag{3.99}$$

$$\Delta T + W = \frac{\partial T}{\partial t} \tag{3.100}$$

The system (3.99–3.100) only couples W and T, and is equivalent to the system (3.95–3.97). This is especially interesting insofar as the boundary conditions established in §3.2.3 could also be expressed in terms of W and T. Note that it is possible to obtain the result (3.99) in one step, by applying twice the curl operator to Eq. (3.96) and using Eq. (3.95), and the relations $\vec{\nabla} \times \vec{\nabla} \times \vec{a} = -\Delta\vec{a} + \vec{\nabla}(\vec{\nabla}.\vec{a})$ and $\vec{\nabla} \times \vec{\nabla}f = 0$ valid for any $\vec{a}$ and f.

We may now proceed to the linear stability analysis, using the different combinations of boundary conditions established in §3.2.3. As the calculation has been described by several authors (see e.g. [24, 99, 95]), we will simply underline the essential steps hereafter.

By separation of variables, it is possible to show [206] that

$$\left(\begin{array}{c} W \\ T \end{array} \right) = \exp[\sigma t]\, \phi(x,y) \left(\begin{array}{c} w(z) \\ \theta(z) \end{array} \right) \tag{3.101}$$

where the planform function ϕ is a solution of the Helmholtz equation

$$\Delta_h \phi + k^2 \phi = 0 \tag{3.102}$$

in which k^2 is an arbitrary constant. The general solution of this equation may be written as a superposition of plane wave solutions

$$\phi = \exp[i\vec{k}.\vec{r}] = \exp[i(k_x x + k_y y)] \tag{3.103}$$

with wavenumber $k = (k_x^2 + k_y^2)^{1/2}$. Note that we excluded solutions of Eq. (3.101) where $k^2 < 0$, because they lead to solutions not bounded at infinity, and are thus not valid in laterally unbounded geometries.

Thus, solutions of Eqs (3.99–3.100) or Eqs (3.95–3.97) can be written as superpositions of normal modes $\{U, V, W, T, p\} \sim \exp[\sigma t + i\vec{k}.\vec{r}]\{u(z), v(z), w(z), \theta(z), \pi(z)\}$. Using Eqs (3.99–3.100), we readily obtain the system

$$(D^2 - k^2)^2 w - k^2 Ra\theta = \sigma Pr^{-1}(D^2 - k^2)w \tag{3.104}$$

$$(D^2 - k^2)\theta + w = \sigma\theta \tag{3.105}$$

which may be solved exactly by a superposition of solutions of the form $\exp[rz]$, with r given by the roots of a characteristic equation found by substituting D by r, and solving algebraically for w and θ. Alternatively, we may write an equation for w alone by applying $(D^2 - k^2 - \sigma)$ to the first equation and using the second to eliminate θ. Substituting D by r, the characteristic polynomial then reads

$$(r^2 - k^2 - \sigma)(r^2 - k^2 - \sigma Pr^{-1})(r^2 - k^2) + k^2 Ra = 0 \tag{3.106}$$

which leads to 6 solutions r_i, $i = 1, ..., 6$ in general different from each other if $\sigma \neq 0$ and $Pr \neq 1$. Thus, the z-dependencies of temperature and velocity perturbations read

$$w = \sum_{i=1}^{6} c_i \exp[r_i z] \tag{3.107}$$

$$\theta = \sum_{i=1}^{6} c_i' \exp[r_i z] \tag{3.108}$$

and a relation is found between c_i' and c_i by inserting Eqs (3.107–3.108) into one of the equations (3.104–3.105). We then obtain $c_i = c_i'(k^2 + \sigma - r_i^2)$. Thus there remains 6 independent integration constants, which have to be determined from the boundary conditions. For any combination of the linear boundary conditions of §3.2.3, this leads to a 6×6 linear homogeneous system of the form $A.C = 0$, where C is the vector of unknowns c_i and A is a 6×6 matrix. Then, the compatibility relation of the system (which guarantees the existence of non-trivial solutions) is $\det(A) = 0$, which leads to the dispersion relation

$$f(\sigma, k^2, Ra) = 0 \tag{3.109}$$

This relation may in principle be solved for $\sigma = \sigma(Ra, k^2)$. However, this is technically impossible (the reason for this being the complicated dependency of the roots r_i upon σ), and the determinant (3.109) in general has to be calculated numerically, and combined with an iterative method for root finding (e.g. Newton's method).

Even for the neutral stability locus $\sigma = 0$ [where the solutions (3.107–3.108) should be rewritten taking into account the multiplicity of the roots of the characteristic equation $(r^2 - k^2)^3 = k^2 Ra$], the neutral stability condition [which is also the limit of Eq. (3.109) for $\sigma \to 0$] reads

$$f_0(Ra, k^2) = 0 \tag{3.110}$$

and should be numerically solved for Ra (at given k). The only situations where analytical results may be obtained are:

- the heat-conducting stress-free boundary conditions $w = D^2 w = \theta = 0$ at $z = 0, 1$. This case has already been examined in §3.1.3, which lead to the neutral stability curve $Ra = (k^2 + n^2\pi^2)^3/k^2$ with $n = 1, 2, \ldots$ denoting the "vertical wavenumber". Minimization of Ra with respect to k leads to the critical Rayleigh number $Ra_c = 27\pi^4/4$ with $k_c = \pi/2^{1/2}$ and $n = 1$.

- cases where the heat flux is held fixed at both bottom and top boundaries (whatever the mechanical conditions). Although the full analysis must still be performed numerically, it appears that the critical conditions are reached for vanishing wavenumbers ($k_c = 0$). An example of neutral stability curve in this case is represented in Fig. 3.5. Using perturbation methods with the smallness parameter k (long-wave asymptotics), the threshold $Ra_c = O(1)$ may be found analytically [96]. This has also been widely exploited in the nonlinear regime (see e.g. [212, 215]), as also discussed in §4.6.

The numerical results are summarized in Table 3.4. Various combinations of boundary conditions can be constructed among $W = 0$ (undeformable), $DW = 0$ (rigid no-slip condition), $D^2W = 0$ (stress-free "slip" condition, or free surface without significant Marangoni effect), $T = 0$ (heat conducting) and $DT = 0$ (heat insulating).

Note that similarly to the situation encountered in the free-free conducting case, a countable infinite number of solutions of Eq. (3.110) exist at given k (these solutions correspond to different values of the vertical wavenumber n, which represents the number of superposed cells in the vertical direction) for all the combinations considered in Table 3.4. The most dangerous mode always appears to be $n = 1$, leading to the critical values of Ra indicated in the table. However, higher-order modes with $n > 1$ may play an important role in the

Table 3.4: Critical Rayleigh numbers and wavenumbers [24] for various combinations of boundary conditions. The critical mode always corresponds to a vertical wavenumber $n = 1$ (one convective cell in the vertical direction).

$z = 0$	$z = 1$	$z = 0$	$z = 1$	Ra_c	k_c
$W = 0, D^2W = 0$	$W = 0, D^2W = 0$	$T = 0$	$T = 0$	657.511	2.22
$W = 0, D^2W = 0$	$W = 0, D^2W = 0$	$DT = 0$	$T = 0$	384.693	1.76
$W = 0, D^2W = 0$	$W = 0, D^2W = 0$	$DT = 0$	$DT = 0$	120	0
$W = 0, DW = 0$	$W = 0, D^2W = 0$	$T = 0$	$T = 0$	1100.66	2.68
$W = 0, DW = 0$	$W = 0, D^2W = 0$	$DT = 0$	$T = 0$	816.748	2.21
$W = 0, DW = 0$	$W = 0, D^2W = 0$	$T = 0$	$DT = 0$	669.001	2.09
$W = 0, DW = 0$	$W = 0, D^2W = 0$	$DT = 0$	$DT = 0$	320	0
$W = 0, DW = 0$	$W = 0, DW = 0$	$T = 0$	$T = 0$	1707.76	3.12
$W = 0, DW = 0$	$W = 0, DW = 0$	$DT = 0$	$T = 0$	1295.78	2.55
$W = 0, DW = 0$	$W = 0, DW = 0$	$DT = 0$	$DT = 0$	720	0

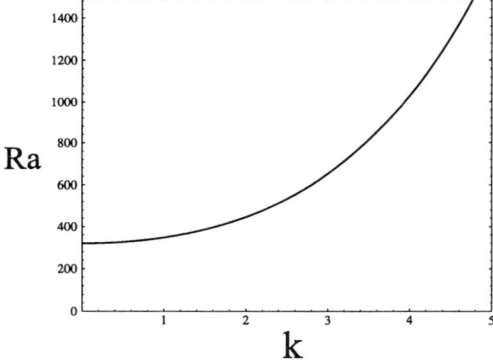

Figure 3.5: The neutral stability curve of the Rayleigh–Bénard problem, in the case where the bottom plate is rigid, the upper boundary is a free surface, and the heat flux crossing them is held fixed. The critical wavenumber is zero, corresponding here to a critical Rayleigh number $Ra_c = 320$.

strongly nonlinear domain, when several different vertical eigenmodes have a positive linear growth rate. It is worth mentioning that this situation has no counterpart in the purely surface-tension-driven cases considered in the following sections.

Another specific aspect of the Rayleigh–Bénard problem is the possibility of establishing certain self-adjointness properties. This is developed in some detail in Appendix B. There it will be shown that for all the systems considered above, the linear problem is self-adjoint with respect to a certain scalar product. Then, it is shown how this property determines the nature of eigenvalues (i.e. the growth rates). In particular, it will be seen that all eigenvalues should be real (no oscillatory modes) for *positive* values of the Rayleigh number, i.e. when heating is from below. On the contrary, nothing equivalent can be concluded when $Ra < 0$ (heating from above), and oscillatory modes are not excluded in this situation. Indeed, oscillatory modes are predicted (internal waves) but they are always damped (as shown for the stress-free heat conducting case in §3.1.3).

Finally, we also briefly explain in Appendix C how a variational formulation of the neutral stability problems considered above can be derived (see also [20, 24, 95]), on the basis of linear energy stability theory. The result is that a sufficient condition for stability of the reference state is $Ra < Ra^*$, where Ra^* is given by

$$\frac{1}{Ra^*} = \max_{k,\theta(z),w(z)} \frac{\int_0^1 dz \left[-k^2(k^2\theta^2 + (D\theta)^2) + 2k^2 w\theta\right]}{\int_0^1 dz \left[k^4 w^2 + 2k^2(Dw)^2 + (D^2w)^2\right]} \tag{3.111}$$

where the maximization is performed on k, and on real functions $\theta(z)$ and $w(z)$ satisfying the selected set of boundary conditions. In fact, it may be shown [20] for all the problems considered in this section that the result of this maximization precisely leads to the critical Rayleigh number Ra_c. This variational principle thus provides a useful tool for using numerical methods based on the Galerkin or finite difference minimization of Eq. (3.111).

3.3.2 Marangoni–Bénard Instability

Now, let us concentrate on the main subject of this book, namely the instability associated with the surface tension dependence upon temperature (or concentration). As surface tension gradients at a free surface act as tangential stresses on the adjacent fluids, a dependence of surface tension on temperature (i.e. thermocapillarity) will provide a coupling of temperature and velocity fluctuations at the level of the free surface. The systems considered will be either one-phase systems (one layer of liquid in contact with an inert gas) treated in this section, or two-phase systems (two layers of "active" fluids in contact at an interface) considered in §3.5.

Similarly to the variation of density with temperature which is at the origin of the Rayleigh effect, the thermocapillary (or Marangoni) effect may be responsible for cellular convective instabilities. The basic mechanism of instability can be explained by considering the evolution of fluctuations of temperature and velocity (see also Fig. 3.1 and the discussion of §3.1.1. for the Rayleigh–Bénard case). If a fluctuation of temperature (say, a hot spot) occurs at the free surface (or interface), a local surface tension gradient is created, directed radially away from the fluctuation. The associated tangential stresses then induce radially divergent surface fluid motion (the interface may be seen as an elastic membrane which relaxes at the point where the disturbance is created). Now, continuity of the fluid requires a vertical ascending flow to take place below the disturbance. If the layer of fluid is heated from below, this uprising fluid, being hotter, will make the free surface temperature increase at the initial location of the disturbance. Consequently, the surface tension at that point decreases, thus increasing the tangential stresses. Typically, amplification of the fluctuations will take place provided the characteristic thermocapillary time $\tau_{Ma} = (\rho d^2/\sigma_T\beta)^{1/2}$ introduced in §3.2.1 is lower than thermal and viscous relaxation times[5]. This leads to the condition that the Marangoni number

---

[5]Note that when $\beta < 0$ (heating from the top), τ_{Ma} is complex. Similarly to the discussion of the buoyant time scale in §3.1, this indicates that oscillatory modes may exist in this situation (dilational surface waves, see §3.4).

[Eq. (3.52)] should be larger than a numerical constant of order unity, to be determined via the linear stability analysis presented below, which parallels the seminal work of Pearson [101].

One-phase systems – Finite-depth case

As the developments made at the beginning of the last section were independent of boundary conditions, we may here directly rewrite the system of equations (3.104–3.105) for vertical z-dependencies of normal modes with $Ra = 0$ (no buoyancy is considered in this section) as

$$(D^2 - k^2)^2 w = \sigma Pr^{-1}(D^2 - k^2)w \tag{3.112}$$

$$(D^2 - k^2)\theta + w = \sigma\theta \tag{3.113}$$

and the boundary conditions must also be written for perturbations w and θ. For a rigid heat conducting plate at the bottom (see §3.2.3), we have

$$w = Dw = \theta = 0 \quad \text{at } z = 0 \tag{3.114}$$

while at the upper flat free surface, we will consider a heat transfer described by Newton's law of cooling, together with the continuity of tangential stress. According to Eqs (3.70) and (3.86), this reads (for normal modes)

$$w = D\theta + Bi\,\theta = D^2 w + k^2 Ma\,\theta = 0 \quad \text{at } z = 1 \tag{3.115}$$

The above system of equations and boundary conditions can be solved analytically for the case of neutral stability ($\sigma = 0$). This is also possible for the general dispersion relation ($\sigma \neq 0$), but the result will not be reproduced here. The simplification (compared with Rayleigh–Bénard instabilities) is due to the decoupling of the Navier–Stokes equation (3.112) from the heat equation (3.113) occurring in the absence of buoyancy. In Marangoni–Bénard instabilities, the coupling between thermal and mechanical modes (without which instability would not occur) is provided by the tangential stress boundary condition [the last of Eq. (3.115)] expressing the equality between thermocapillary stresses due to surface tension variation with temperature, and Newtonian viscous stresses. Moreover, as this relation is a linear function of Ma, it is possible to solve the compatibility relation for Ma, thus directly providing the locus of marginally stable perturbations in the plane (k, Ma).

Restricting the analysis to the neutral stability case, we set $\sigma = 0$, and solve Eq. (3.112) for $w(z)$. Then, expressing the boundary conditions on w at $z = 0, 1$ except for the Marangoni stress boundary condition, we get

$$w(z) = \alpha \left\{ (1 - z)\sinh[kz] + kz(\coth[k]\sinh[kz] - \cosh[kz]) \right\} \tag{3.116}$$

and from Eq. (3.113), the temperature perturbation is found as

$$\theta(z) = \alpha \left\{ \begin{array}{l} f\sinh[kz] + z\cosh[kz]\dfrac{z(1 - k\coth[k]) - 3}{4k} + \\[2mm] z\left(\dfrac{z}{4} - \dfrac{1 - k\coth[k]}{4k^2}\right)\sinh[kz] \end{array} \right\} \tag{3.117}$$

which satisfies $\theta = 0$ at $z = 0$, and where f is a constant found by expressing the boundary condition $D\theta + Bi\,\theta = 0$ at $z = 1$.

Finally, solving the remaining boundary condition $D^2w + k^2Ma\,\theta = 0$ at $z = 1$ with respect to the Marangoni number leads to

$$Ma = \frac{16k(k\cosh[k] + Bi\sinh[k])(2k - \sinh[2k])}{4k^3\cosh[k] + 3\sinh[k] - \sinh[3k]} \tag{3.118}$$

which is the result obtained by Pearson [101], and represented in Fig. 3.6, for several values of the free surface Biot number.

Note that the multiplicative constant α appearing in expressions (3.116–3.117) remains undetermined. This is a general feature of linear stability problems, due to the homogeneous character of the linear equations and boundary conditions. Only nonlinear stability analyzes will allow determination of the value of this amplitude α, as explained in Chapter 4.

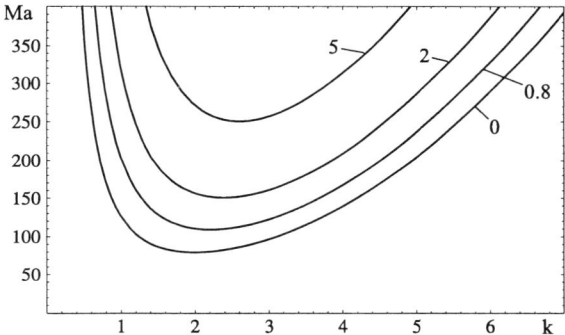

Figure 3.6: Neutral stability locus: Marangoni number Ma as a function of the wavenumber k of the perturbation, for several free surface Biot numbers Bi (indicated on each curve). $Ma > 0$ corresponds to heating the layer from below.

The curves drawn in Fig. 3.6 thus represent the locus of neutrally stable perturbations. At a given Bi, the reference diffusive state is stable against disturbances with a given wavenumber k, provided the Marangoni number is below the value given in Fig. 3.6. Once this threshold is exceeded, the perturbation with wavenumber k has a positive growth rate σ, solution of the full problem Eqs (3.112–3.115). Although the analysis proceeds exactly along the same lines as for neutral stability, only numerical solutions of the dispersion relation $f(\sigma, Ma, k, Bi, Pr) = 0$ will be presented here. In general, this relation admits several solutions for σ at given values of other parameters. However, contrary to what happens for the Rayleigh–Bénard instability (see last section), at a given k and varying Ma at constant values of Bi and Pr, only one eigenvalue solution of this dispersion relation may become positive. This would be totally confirmed by the fact that only one critical value of Ma was found in the neutral stability case, if the possibility of oscillatory onset ($\sigma = i\omega \neq 0$) was excluded. Although no proof exists that the spectral problem is self-adjoint (see Appendices B and C), no oscillatory onset was detected by Vidal and Acrivos [209], when solving the problem (3.112–3.115) numerically.

The highest value of the growth rate σ is represented in Fig. 3.7 for $Bi = 0$ (heat insulating upper surface, which is a good approximation of a liquid/gas interface for sufficiently large gas depth), $Pr \to \infty$ (which is a good approximation for typical liquids, and especially viscous liquids like oils), and for several values of Ma. It is seen that the growth rate presents a maximum at a given k_{max} depending on Ma, and that a range of wavenumbers leads to instability above the critical value $Ma_c \simeq 79.6$. This critical value corresponds to the minimum of the neutral stability curve ($Bi = 0$) of Fig. 3.6, which also defines the critical wavenumber $k_c \simeq 1.99$.

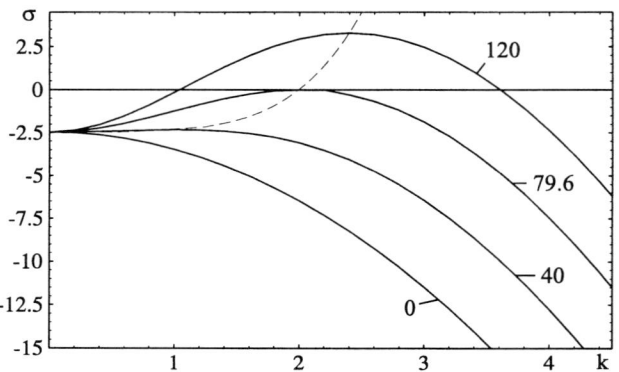

Figure 3.7: The most dangerous eigenvalue as a function of k, for several values of Ma, and fixed values of $Bi = 0$ (insulating free surface) and $Pr \to \infty$.

Another result is the locus of the fastest growing modes (maximal growth rate), represented in Fig. 3.8, for $Bi = 0$ and several values of Pr. Note that the neutral stability result (3.118) is independent of Pr, a general characteristic of neutrally stable monotonic modes.

The critical temperature and velocity perturbations, given by Eqs (3.116–3.117), are depicted in Fig. 3.9. Note the displacement towards the free surface of the maximum of the vertical velocity perturbation, characteristic of surface-tension-driven instabilities. This is also visible on the streamlines given in Fig. 3.10.

Infinite-depth case – Short-wavelength modes

We conclude this section by considering the limiting case $k \to \infty$, corresponding to convective cells with wavelength $\lambda = 2\pi/k \to 0$, i.e. short-wavelength modes. Some of these results will be needed in the following chapters. When $k \to \infty$, we may substitute hyperbolic functions in (3.118) by their asymptotic values. We then simply get the asymptotics of the neutral stability curve as

$$Ma \to 8k(k + Bi) \qquad \text{for} \quad k \to \infty \tag{3.119}$$

Note that Bi was not a priori neglected with respect to k, thus allowing for large values of the surface Biot number Bi. In fact, the result (3.119) can also be obtained considering a

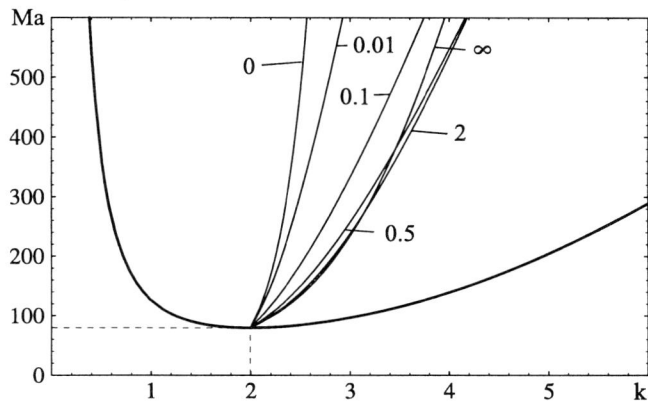

Figure 3.8: The locus of the fastest growing perturbations (wavenumber $k_{\max}$) for $Bi = 0$, and various values of Pr. All curves start at the critical value $Ma_c = 79.6$ and $k_c = 1.99$.

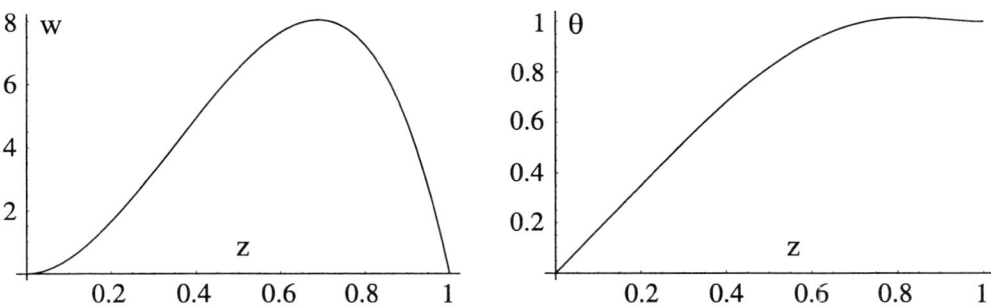

Figure 3.9: Critical perturbations w and θ for the Marangoni–Bénard problem with $Bi = 0$.

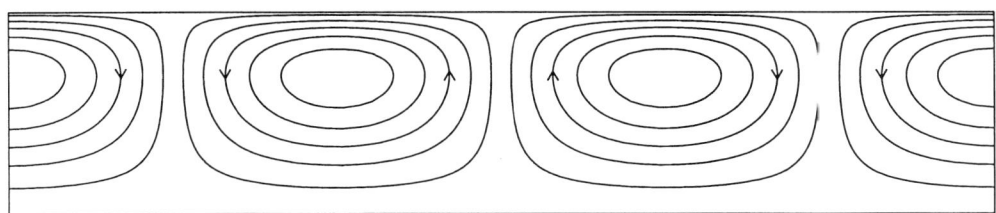

Figure 3.10: Streamlines corresponding to the onset of Marangoni–Bénard convection.

semi-infinite problem and $k = O(1)$ (where the unit length d can here be chosen arbitrarily), because in this case the $O(1)$ size of convective cells is indeed much smaller than the depth. We thus replace the lower boundary conditions by the requirement that fields should remain bounded for $z \to -\infty$. To simplify notations, we will now consider the interface at $z = 0$. It is then straightforward to solve the neutral stability problem (the full dispersion relation will be obtained in Chapter 8) as

$$w = -4k^2(k + Bi)z \exp[kz] \tag{3.120}$$

$$\theta = \left[1 - (k + Bi)z + k(k + Bi)z^2\right] \exp[kz] \tag{3.121}$$

with the Marangoni number indeed given by Eq. (3.119). Thus, the critical Marangoni number is lower than in the finite depth case, because of the absence of friction and thermal dissipation at the bottom wall. In particular, it is seen that whatever the value of Bi, the critical wavenumber is now $k_c = 0$, leading to the conclusion that critical modes are always scaled with the depth of the liquid layer, as large as it can be.

However, this does not prevent the use of this model for obtaining significant conclusions about linear and nonlinear behaviors in the Marangoni–Bénard problem (see e.g. [103]). Furthermore, most of the results will be seen to be qualitatively consistent with those derived, at much greater computational expense, for the finite-depth case.

Note that Eqs (3.120–3.121) may also be obtained, up to a multiplicative constant, from the $k \to \infty$ asymptotics of the finite-depth perturbations Eqs (3.116–3.117). It is also necessary to change the position of the interface from $z = 1$ to $z = 0$. The exponentially decreasing factor $\exp[kz]$ in Eqs (3.120–3.121) indeed indicates that the penetration depth of convective cells is proportional to their wavelength, thus much smaller than the depth.

3.3.3   Combined Rayleigh–Marangoni–Bénard instability

When a layer of liquid lying on a heated rigid surface is in contact with a gas at an undeformable free surface, both Rayleigh–Bénard and Marangoni–Bénard mechanisms may act together in destabilizing the reference diffusive state. This is the possibility considered by Nield [102]. No other results than the final one will be given here, as the linear stability analysis can only be achieved via numerical methods, due to the same difficulties as those encountered in §3.3.1 (in addition to the presence of surface tension effects). The equations to be considered are Eqs (3.104–3.105), with the usual rigid-conducting boundary conditions (3.114) at the bottom, and boundary conditions (3.115) at the top free surface.

The linear results are most conveniently presented in the plane of Rayleigh and Marangoni numbers Ra and Ma. We note that although self-adjointness of the linear operator cannot be proved here (exchange of stability is not valid, see Appendix B), no oscillatory modes are observed when both Rayleigh and Marangoni numbers are positive. This is the quadrant considered here, for which the neutral stability locus is shown in Fig. 3.11.

A nearly straight line is seen to connect the "pure Rayleigh" case $Ra_c^0 = 669$, $k_c = 2.09$ to the "pure Marangoni" case $Ma_c^0 = 79.6$, $k_c = 1.99$ (for $Bi = 0$). Minimization with respect to the wavenumber (critical conditions) is performed at each point of the neutral stability curve, resulting in a slight variation of k_c with $Bo_d = Ra/Ma$. However, the mechanisms being quite similar in their effects, with critical wavenumbers not too different, buoyancy and surface tension couple in a cooperative way to destabilize the diffusive state.

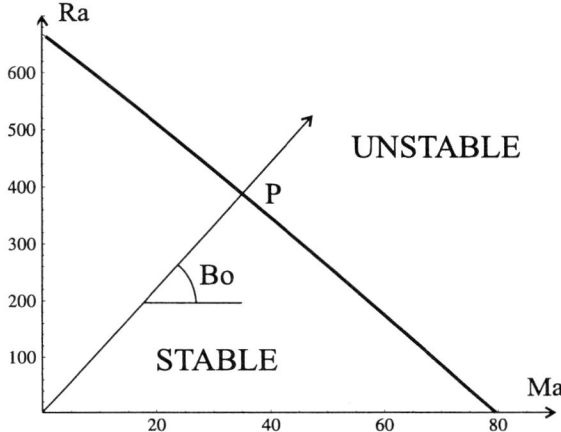

Figure 3.11: The neutral stability boundary for the combined Rayleigh–Marangoni–Bénard instability, in the quadrant $Ra > 0$, $Ma > 0$, after minimization with respect to the wavenumber, and for a zero free surface Biot number. The arrow from the origin represents the theoretical trajectory of the representative point of the system when the driving flux is slowly increased (quasi-stationary conditions). Instability then occurs at P.

Approximating the neutral stability curve by a straight line [102], we get

$$\frac{Ra_c}{Ra_c^0} + \frac{Ma_c}{Ma_c^0} \simeq 1 \tag{3.122}$$

where the "pure" critical Rayleigh and Marangoni numbers Ra_c^0 and Ma_c^0 depend on the Biot number. Even for $Bi > 0$ however, the linear correlation (3.122) remains approximately valid. In Fig. 3.11, the arrow drawn from the origin represents the trajectory followed by a system progressively heated from equilibrium (origin), and sufficiently slowly to allow thermal equilibration to take place at every moment (such that the temperature profile remains linear throughout the layer). The slope of this trajectory is given by the dynamic Bond number Eq. (3.53), i.e.

$$Bo_d = \frac{Ra}{Ma} = \frac{\alpha \rho g d^2}{\sigma_T} \tag{3.123}$$

which is proportional to the square of the depth d, reflecting the increasing importance of the buoyancy mechanism for increasing fluid depths. On the contrary, for shallow layers the thermocapillary effect is dominant over buoyancy. A fortiori, in a microgravity environment ($g \simeq 0$), the Rayleigh effect is negligible, and the instability is mostly surface-tension-driven.

Note that in the paper by Nield [102] (and in this section), only the positive quadrant $Ra > 0$, $Ma > 0$ was considered. This corresponds to a liquid layer heated from below (apart for water below 4 °C, which would have $Ra > 0$ when heated from above, leading to important geological and oceanographic implications [20] and non-Boussinesq effects [216]), and for which the surface tension decreases with temperature. This is not the only quadrant of interest. Even when omitting consideration of quadrants $Ra > 0$, $Ma < 0$ and $Ra < 0$, $Ma > 0$ which correspond to rather exceptional cases (e.g. water below 4 °C heated from above, or anomalous surface tension dependence upon temperature), the remaining quadrant $Ra < 0$, $Ma < 0$ appears unexpectedly rich in behaviors. As this quadrant corresponds to a situation where both surface tension and buoyancy mechanisms are stabilizing, no monotonic instability is obtained. However, this does not rule out the possibility of oscillatory onset. This is considered in §3.4, where it will be seen that such oscillatory modes indeed exist.

3.3.4   Influence of lateral confinement

Up to now, the liquid layer was considered infinite in horizontal extent. The results obtained so far should thus compare satisfactorily with experiments in containers with a large aspect ratio $A = L/d$, where L is the lateral dimension of the container, while d is the fluid depth. For $A = O(1)$ however, the influence of the lateral walls certainly has to be taken into account. In this section, we mostly consider the mathematically convenient cases of stress-free rectangular containers and vorticity-free circular containers. Typical modifications occurring for realistic no-slip walls are considered at the end of the section.

Stress-free rectangular containers

In unbounded layers ($A \to \infty$), the eigenfunctions of the linear problems had periodic dependencies $\exp(i\vec{k}.\vec{r})$ upon the lateral coordinate $\vec{r} = x\vec{1}_x + y\vec{1}_y$. Separation of variables leads in particular to Eqs (3.101–3.103), i.e. both vertical velocity W and temperature perturbation T were found to be proportional to a planform function $\phi(\vec{r})$, solution of the Helmholtz equation $\Delta_{\mathrm{h}}\phi = -k^2\phi$, where Δ_{h} is the horizontal Laplacian, while $k = |\vec{k}|$ is an arbitrary wavenumber.

The results obtained in unbounded layers may actually be straightforwardly applied to containers with $A = O(1)$, provided boundary conditions on the lateral walls are also satisfied by eigenfunctions of Δ_{h}. This is possible only for particular types of containers (stress-free rectangular containers or vorticity-free circular containers) considered hereafter.

Two-dimensional stress-free containers

Referring to Fig. 3.10 representing the streamlines of a particular eigenmode with wavenumber k in the Marangoni–Bénard case, it is seen that no flow occurs through fictive vertical walls separated by a distance $A = n\pi/k$, $n = 1, 2, ...$ (half a wavelength $2\pi/k$ is the size of a convective cell). Thus, given a two-dimensional box with such fictive lateral walls at $x = 0$ and $x = A$, the eigenmodes are given by the normal modes found in the corresponding infinite problem, with wavenumbers restricted to

$$k = n\frac{\pi}{A}, \quad n = 1, 2, ... \tag{3.124}$$

where n represents the number of convective cells in the box.

What is the nature of these fictive walls at $x = 0$ and $x = A$? Recalculating the horizontal velocity U using the incompressibility condition $\partial U/\partial x + DW = 0$, a particular eigenmode with n convective cells may be written

$$\begin{pmatrix} U \\ W \\ T \end{pmatrix} = \exp[\sigma t] \begin{pmatrix} -\frac{A}{n\pi}\sin(n\pi x/A)Dw(z) \\ \cos(n\pi x/A)w(z) \\ \cos(n\pi x/A)\theta(z) \end{pmatrix} \tag{3.125}$$

from which it is seen that at $x = 0$ and $x = A$, not only $U = 0$ (impermeable walls), but $\partial W/\partial x = 0$ (stress-free walls) and $\partial T/\partial x = 0$ (thermally insulating walls). This is the only combination of lateral boundary conditions for which separation of variables may be achieved. Note that Eqs (3.124–3.125) also hold for periodic lateral boundary conditions on the interval $[-A, +A]$, though in this case an arbitrary phase (horizontal translation) could be included in Eq. 3.125.

Even though these walls are unrealistic, they allow the effect of lateral confinement to be considered in a simple way, and have been used by several authors [149. 156, 43, 159, 151, 206, 225], both in linear and nonlinear regimes. They allow in particular to apply any dispersion relation found for a laterally infinite domain to a container with aspect ratio A, just considering discrete values (3.124) of the wavenumber. This is illustrated in Fig. 3.12 for the pure Marangoni–Bénard problem, for which the monotonic instability threshold is given by Pearson's [101] result (3.118).

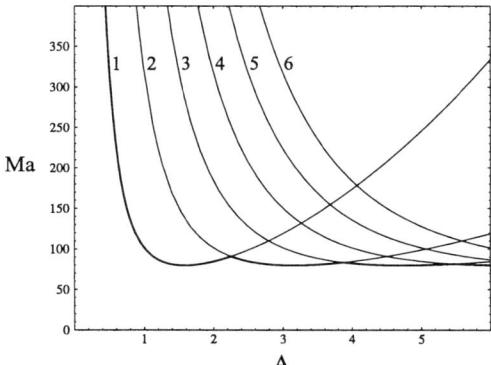

Figure 3.12: Threshold Marangoni numbers Ma for modes with different number n of convective cells in lateral direction (indicated on each curve), as a function of the aspect ratio A of the stress-free/insulating two-dimensional container. Pure Marangoni–Bénard convection is considered, with zero Biot number. The lower (thicker) curve gives the critical Marangoni number as a function of A.

From Fig. 3.12, it is seen that the critical value $Ma_c = 79.6$ is recovered in the limit $A \rightarrow \infty$, as expected. This value is also reached for some particular values of A, when the convection cells just have a value of $k = n\pi/A$ equal to the critical wavenumber $k_c = 1.99$ valid in the infinite case. When varying A around one of these values, convection cells get compressed ($k > k_c$) or elongated ($k < k_c$), and may only be triggered for values of Ma higher than 79.6. Finally, note that there exists particular values of A for which two curves with different n cross. Near these points (named codimension-2 points for reasons explained in Chapter 7), the convective structure expected when crossing the threshold may consist in a mixed state of the two modes, or even in dynamical regimes. The analysis of the competition between unstable modes may only be achieved from the full nonlinear system of equations.

Three-dimensional stress-free rectangular containers

The considerations presented above may be directly generalized to three-dimensional rectangular containers with dimensionless lengths A_x and A_y (scaled by the liquid depth) along x and y respectively. Following [149, 151], we consider here impermeable and stress-free (or slippery) thermally insulated walls, i.e. lateral boundary conditions read

$$U = \partial W/\partial x = \partial V/\partial x = \partial T/\partial x = 0 \quad \text{at} \quad x = 0, A_x,$$
$$V = \partial W/\partial y = \partial U/\partial y = \partial T/\partial y = 0 \quad \text{at} \quad y = 0, A_y \tag{3.126}$$

where the U, V and W are x-, y- and z-components of the velocity field $\vec{V}$, respectively.

Consider for instance the general system (3.99–3.100). Separation of variables allows W and T to be written as a superposition of eigenfunctions of the horizontal Laplacian Δ_h with boundary conditions (3.126) :

$$\begin{pmatrix} W \\ T \end{pmatrix} = \exp(\sigma t)\cos(\frac{n_x\pi x}{A_x})\cos(\frac{n_y\pi y}{A_y})\begin{pmatrix} w(z) \\ \theta(z) \end{pmatrix} \tag{3.127}$$

Thus, the system (3.104–3.105) is here again valid, with the equivalent wavenumber

$$k = \left[\left(n_x\frac{\pi}{A_x}\right)^2 + \left(n_y\frac{\pi}{A_y}\right)^2\right]^{\frac{1}{2}}, \quad n_x, n_y = 1, 2, ... \tag{3.128}$$

The horizontal velocity field $\vec{V}_h = U\vec{1}_x + V\vec{1}_y$ may now be written $\vec{V}_h = k^{-2}\vec{\nabla}_h DW$ (where $\vec{\nabla}_h$ is the horizontal gradient), such as the incompressibility condition $\vec{\nabla}.\vec{V} = 0$ is satisfied[6]. It is then seen that boundary conditions (3.126) on U and V are also satisfied.

Therefore, the dispersion relations found for laterally infinite domains may here again be exploited using the equivalent wavenumber (3.128), for rectangular domains with boundary conditions (3.126). This is illustrated in Fig. 3.13, which shows a map of the most critical mode (lower instability threshold) as a function of A_x and A_y, for the case of pure Marangoni–Bénard convection.

---

[6]At the linear stage, only the potential part $k^{-2}\vec{\nabla}_h DW$ of the horizontal velocity field is coupled to the temperature field, and may therefore contribute to instability. A solenoidal part $\vec{\nabla}_h \times (\Psi\vec{1}_z)$ could be added to $\vec{V}_h$, but leads to eigenmodes which are never linearly amplified. These modes may however play an important role in the nonlinear dynamics, as seen in subsequent chapters.

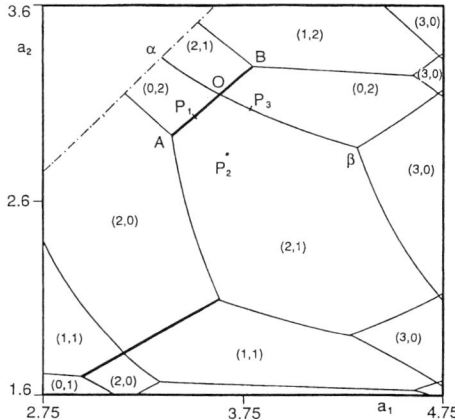

Figure 3.13: Map of the most dangerous mode, denoted by (n_x, n_y), as a function of the aspect ratios A_x and A_y (noted a_1, a_2), for pure Marangoni–Bénard convection [Pearson's dispersion relation (3.118)] with $Bi = 0$. The picture is symmetrical about the line $a_1 = a_2$, up to the exchange of n_x by n_y. From [151].

Vorticity-free circular containers

We now consider Bénard convection in a circular container of aspect ratio $A = R/d$, where R is the radius of the cylinder. Following [149], we consider that there is neither heat flux nor mass flow through the lateral boundary, but we allow the fluid to slip on it, assuming that this occurs with *zero tangential vorticity*. Thus, introducing a cylindrical reference frame (r, ϕ, z) where r and ϕ denote radial and azimuthal coordinates, and defining the fluid velocity as $\vec{V} = (U, V, W)$ in this reference frame, we impose the boundary conditions (at $r = A$)

$$\frac{\partial T}{\partial r} = U = \frac{\partial W}{\partial r} = \frac{\partial (rV)}{\partial r} = 0 \tag{3.129}$$

Apart from this, the system is governed by the usual equations (3.95–3.97) or (3.99–3.100), and upper and lower boundary conditions selected from §3.2.3. It should be pointed out that the method is still applicable for more general multi-layered and multi-component systems considered later.

As in the last section, the horizontal velocity $\vec{V}_h = U\vec{1}_r + V\vec{1}_\phi$ may be assumed purely potential ($\vec{V}_h = \vec{\nabla}_h \Phi$), without loss of generality. Then, the continuity equation (3.95) reads

$$\frac{\partial^2 \Phi}{\partial r^2} + \frac{1}{r}\frac{\partial \Phi}{\partial r} + \frac{1}{r^2}\frac{\partial^2 \Phi}{\partial \phi^2} + DW = 0 \tag{3.130}$$

where D still denotes $\partial/\partial z$. Taking into account the 2π-periodicity in the azimuthal direction, this equation may be solved by separating variables as

$$W = \exp(\sigma t) \exp(im\phi) f(kr) w(z)$$
$$\Phi = k^{-2} \exp(\sigma t) \exp(im\phi) f(r) Dw(z) \tag{3.131}$$

where $m = 0, 1, 2, ...$ is the azimuthal wavenumber, k is arbitrary for the moment, an exponential factor $\exp(\sigma t)$ has been assumed, and the radial function f satisfies

$$f''(y) + \frac{1}{y} f'(y) + (1 - \frac{m^2}{y^2}) f(y) = 0 \tag{3.132}$$

i.e. a Bessel equation whose only solutions not diverging at $y = 0$ are denoted by $J_m(y)$. Then, eigenmodes of the linear problem can be written[7]

$$W = \exp(\sigma t) \exp(im\phi) J_m(kr) w(z)$$
$$U = k^{-1} \exp(\sigma t) \exp(im\phi) J'_m(kr) Dw(z)$$
$$V = imk^{-2} r^{-1} \exp(\sigma t) \exp(im\phi) J_m(kr) Dw(z)$$
$$T = \exp(\sigma t) \exp(im\phi) J_m(kr) \theta(z) \tag{3.133}$$

Substituting these equations into the original problem once again shows that vertical dependencies w and θ satisfy Eqs (3.104–3.105). Lower and upper boundary conditions may be selected as usual from those given in §3.2.3. For a liquid layer on a rigid/conducting plate, with an undeformable upper surface submitted to thermocapillarity, Eqs (3.114) and (3.115) still hold. Therefore, k may again be considered as an equivalent wavenumber. As for rectangular containers, the latter may only assume discrete values due to the lateral boundary conditions. Indeed, from Eqs (3.133), it is seen that all boundary conditions Eqs (3.129) are satisfied provided

$$k = \frac{y_{m,n}}{A} \tag{3.134}$$

where $y_{m,n}$ denotes the nth positive zero of J'_m. Thus, the integer $n = 1, 2, ...$ can be regarded as a radial wavenumber, as seen in Fig. 3.14.

For the problem considered here, if we omit buoyancy ($Ra = 0$), the neutral stability curve ($\sigma = 0$) is given by the Pearson's result (3.118). Using Eq. (3.134), we may readily calculate the critical Marangoni number for different modes (m, n), where m and n respectively denote azimuthal and radial wavenumbers. This is done as a function of the aspect ratio in Fig. 3.15. Note that in this figure, the same characteristics are observed as for rectangular slippery containers, namely $Ma = 79.6$ is recovered for several container sizes, and several codimension-2 points occur.

---

[7] Note that, as for other linear problems considered in this chapter, Eqs (3.133) may be multiplied by an arbitrary *complex* amplitude. In the nonlinear regime, equations governing the evolution of this amplitude factor will be determined, and in particular, its argument, say χ, will play an important role, as it represents the angular position of the pattern inside the cylinder. For example, $\cos(m\phi + \chi) J_m(r)$ would give the same patterns as in Fig. 3.14, though rotated by $-\chi/m$.

m=0 m=1

m=2 m=3

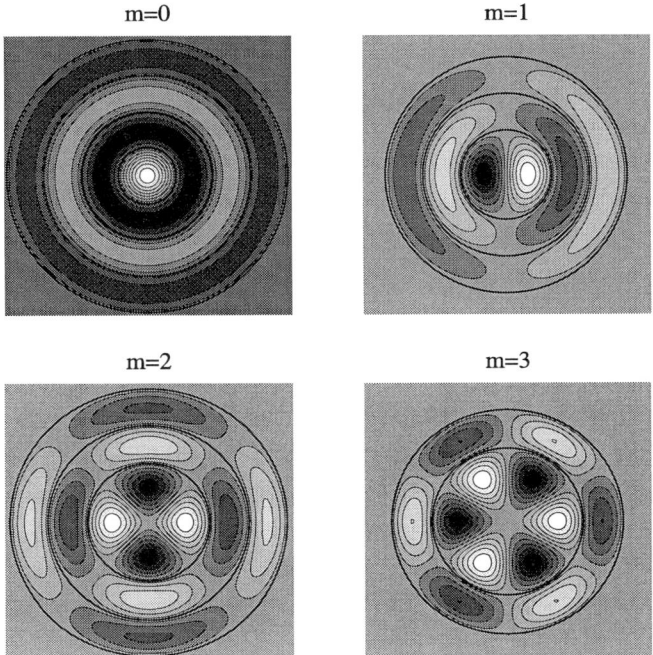

Figure 3.14: Contourplot of the planform function $\cos(m\phi)J_m(r)$, which can be interpreted as the free surface temperature field (black zones are cold, white zones hot) of modes with azimuthal wavenumber m. For each m, structures with different radial wavenumber $n = 1, 2, ...$ are represented (the mode $n = 1$ is the structure inside the circle with smaller radius, the mode $n = 2$ inside the second circle, ...).

No-slip containers

When slippery (or vorticity-free) conditions on the lateral walls are replaced by realistic no-slip conditions, the method of separation of variables is not applicable anymore [257, 208, 206], whatever the thermal boundary conditions. Therefore, only purely numerical methods have been used, such as Galerkin or spectral Tau methods [152, 154, 257], or finite-element methods [225, 258, 259]. It is beyond the scope of the present text to describe these methods in detail, and only essential differences with the results of previous sections will be traced hereafter. Being closer to experiment, only three-dimensional containers (with thermally insulated sidewalls) will be considered.

Davis [257] showed, for buoyancy-driven convection, that convective rolls with axes parallel to the shorter side of the container are always observed at threshold. Though Davis assumed that the axial velocity component of the rolls vanishes identically (which cannot be an exact solution [24]), it turns out that rolls parallel to the shorter side are indeed very often observed in experiments. Moreover, this result still holds even for surface-tension-driven convection, except for boxes with particular shapes (see below).

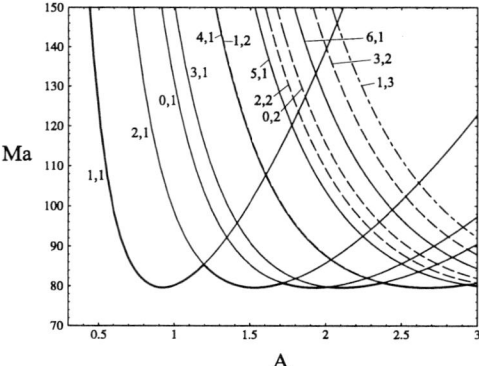

Figure 3.15: Threshold Marangoni numbers Ma for different modes labeled m, n in vorticity-free circular containers, where m and n respectively denote azimuthal and radial wavenumbers. Ma is represented as a function of A, the aspect ratio of the circular container. Pure Marangoni–Bénard convection is considered, with zero Biot number. The lower (thicker) curve gives the critical Marangoni number as a function of A.

An equivalent wavenumber cannot be rigorously defined for rigid boxes. Following Dauby and Lebon [152], instability modes in three-dimensional rectangular containers (with walls, say, at $x = -A_x/2, A_x/2$ and $y = -A_y/2, A_y/2$) may however be subdivided into four classes, characterized by the parity of the eigenfunctions with respect to the coordinates x and y[8]. These will be denoted by EE, EO, OE and OO, the first and second letters referring respectively to the parity (even or odd) of the x- and y-dependencies of the temperature field.

Dauby and Lebon [152] have considered the onset of Bénard convection in rectangular containers with aspect ratios A_x and A_y up to 9. In particular, for square boxes and pure Marangoni convection, their result is presented in Fig. 3.16. Note that no oscillatory onset of instability has been reported in any of the existing studies, even though the principle of exchange of stability cannot be established rigorously.

Although Fig. 3.16 presents some similarities with corresponding figures for slippery containers, clear differences also exist: first, the value $Ma_c = 79.6$ valid for infinite boxes is only recovered when A_x and A_y both tend to infinity, and not for any finite value. Furthermore, Dauby and Lebon [152] have also shown that, contrary to slippery containers, the convective pattern expected at threshold does vary along each of the curves of Fig. 3.16. When increasing aspect ratios, the number of convective cells increases, and the curves present successive minima each time the pattern reaches some optimal shape (i.e. when rolls have a wavelength close to $2\pi/k_c$, where k_c is the critical wavenumber in an infinite layer).

When only one of the two aspect ratios becomes infinite, the other one being fixed, the

---

[8] Actually, as can be seen e.g. from the system (3.95–3.97), the temperature perturbation T and the vertical velocity W should possess identical parity with respect to x and y, while the horizontal component U (V) should possess an opposite parity with respect to x (y) and the same parity with respect to y (x). When speaking about the parity of an eigenmode, we will thus by convention mean the parity of T or W.

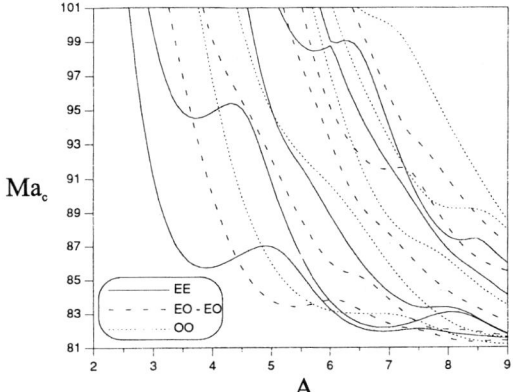

Figure 3.16: Critical Marangoni number as a function of the aspect ratio for square boxes ($A = A_x = A_y$) and $Ra = Bi = 0$. The curves corresponding to several successive eigenvalues and the parity of the eigenmodes is indicated. Courtesy of P. Dauby and G. Lebon.

critical temperature gradient tends to a value which is greater than the corresponding one for an infinite box [257, 152]. In Fig. 3.17, a map of the preferred mode at threshold is presented, still for purely surface-tension-driven convection [152]. Zones of different darkness correspond to different classes of solutions.

Each strip in Fig. 3.17 corresponds to a flow pattern taking the form of convective rolls which are usually parallel to the shorter side of the box[9]. Some exceptions exist however: the horizontal light grey zone (EO) centered on $A_y = 1.6$ corresponds to a unique convection roll parallel to x. Rolls parallel to the longer side of the box also occur in the small EO area near $(A_x, A_y) = (5.6, 5.2)$ (three rolls parallel to x). Other peculiarities of square or nearly square patterns are analyzed in detail in [152].

Dauby et al. have also treated the case of rigid circular containers [154]. Here, the fields are periodic with respect to the azimuthal angle ϕ, and the stability can still be tested separately with respect to disturbances proportional to $\exp(im\phi)$, where m is the azimuthal wavenumber. An example of a linear stability diagram will be given in §7.1, where some particular regimes occurring in the vicinity of a codimension-2 point will be studied in detail. This is motivated by recent experiments of Johnson and Narayanan [43] who have provided evidence of interesting dynamic switchings between patterns with different azimuthal wavenumbers.

---

[9] This result was also obtained by Davis [257], and his Figure 13 of the map of convective modes at threshold is actually very similar to Fig. 3.17 [up to a change of horizontal scale determined by the variation of critical wavenumber from $k_c = 3.12$ (see Table 3.4) to $k_c = 1.99$]. Dauby et al. [152, 154] have also shown that the map of preferred modes only weakly depends on Ra and Bi, up to changes of preferred wavenumbers. Actually, the map can be almost completely explained in terms of two arguments: first, rolls parallel to shorter sides dissipate less kinetic energy than rolls parallel to longer sides [28, 257]. Second, rolls with wavelength close to the critical value in an infinite layer are also preferred.

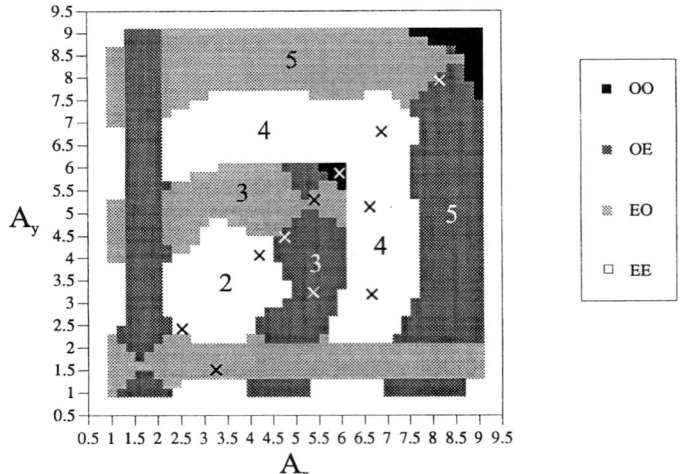

Figure 3.17: Map of convective modes (and their parity) at threshold for $Ra = Bi = 0$. Numbered zones correspond to onset of convection via a given number of rolls parallel to the shorter side of the box. Crosses indicate box sizes for which nonlinear regimes are studied in [152]. Courtesy of P. Dauby and G. Lebon.

Dauby et al. [154] have analyzed both insulating and thermally conducting rigid sidewalls, for which stability results strongly differ.

3.4 Wave modes in one-layer systems

In this section, we re-examine the problem studied in §3.3.3, i.e. a layer of liquid with unde-formable interface in contact with a passive gas, and subject to the joint action of buoyancy and surface tension forces. For a long time since the classical paper of Nield [102], it was admitted that only monotonic modes (real growth rate) could lead to instability in this situa-tion, and a fortiori only in the case of bottom heating (see Fig. 3.11). When the layer is heated from above (corresponding to negative Rayleigh and Marangoni numbers), an unconditionally stable situation was predicted (both effects are "stabilizing"). Recently, Rednikov et al. [50] showed that such a view was incorrect, because the motionless diffusive state may actually become unstable to oscillatory disturbances for sufficiently *negative* values of the Rayleigh and Marangoni numbers[10]. A summary of these findings is presented in this section.

---

[10] In fact, when heating is from above, at least two kinds of waves may exist. The first waves are internal and associated with the name of Brunt–Väisälä, as already mentioned in §3.1. The second kind are surface (dilational) waves, which were first studied by Lucassen [242], though in a slightly different set-up. The latter waves may be explained considering, say, a hot spot at the free surface. As usual the Marangoni effect drives a surface flow away from the spot, which now brings *colder* liquid from the bulk. Because of fluid inertia, the initially hot spot eventually becomes colder than its neighborhood (overshoot), thus reversing the flow and leading to oscillations.

The equations have already been established in previous sections and will not be reproduced here. We consider Eqs (3.104–3.105) for the perturbations of temperature θ and vertical velocity w, with rigid-conducting boundary conditions (3.114) at the bottom plate, and free surface boundary conditions (3.115). A zero Biot number Bi will be assumed for results presented below, corresponding to a heat insulating free surface.

This section is mostly concerned with a numerical analysis of the linear stability problem. The numerical method presented in §3.4.1 can be applied to several problems involving onset of waves in Bénard convection, as is done in the following sections. However, deeper understanding of physical mechanisms involved here can be obtained via boundary-layer analysis [50], which also leads to analytical results valid in some asymptotic limit (§3.4.2). From this refined analysis, some general features may also be extracted (§3.4.3), which will be helpful in considering other wave modes, e.g. when surface deformation is considered (§3.6). In §3.4.4, some hints are provided for designing experiments aimed to observe the waves predicted by Rednikov et al. [50].

3.4.1 General dispersion relation

The full dispersion relation of the problem is found by solving Eqs (3.104) and (3.105) for w and θ, and expressing all boundary conditions (3.114) and (3.115) but the Marangoni stress condition. Then, this last boundary condition provides the compatibility relation under the form $Ma = f(\sigma, k, Ra, Pr)$, where Ma is the Marangoni number, σ is the complex growth rate, k is the wavenumber, while Ra and Pr are the Rayleigh and Prandtl numbers. As Ma is real, we have in fact two relations $Ma = \text{Re}[f(\sigma_R + i\omega, k, Ra, Pr)]$ and $\text{Im}[f(\sigma_R + i\omega, k, Ra, Pr)] = 0$, allowing determination of the amplification rate σ_R and of the frequency ω. As we seek marginal states, we set $\sigma_R = 0$ and solve $\text{Im}[f] = 0$ for the critical frequency ω. Then, $Ma = \text{Re}[f]$ provides the critical Marangoni number. For convenience, we will also use notations $M = -Ma$ and $R = -Ra$.

In the following plots, the Prandtl number will be fixed to $Pr = 6$ (e.g. water). Figure 3.18 presents the marginal curves in the plane (M, k) for some *fixed* values of the (sign-changed) Rayleigh number R. The corresponding values of the marginal frequency are represented in Fig. 3.19. It is seen that the marginal curves appear in the form of "bubbles", the interior of which corresponds to the instability region. The bubble labeled 1 in Fig. 3.18 corresponds to $R = 3 \cdot 10^6$. When decreasing R, the bubble shrinks and moves downwards (see the bubble labeled 2 for $R = 1.5 \cdot 10^6$), and finally collapses at point 3 for $R = 1.21 \cdot 10^6$. Note that these apparently high values of R and M are easily attainable with liquids of small viscosity (e.g. water) and not too small depths (see below). In Figs 3.18 and 3.19, the dashed curves are asymptotic results given hereafter.

Note that these bubbles only appear if the Rayleigh number Ra is *artificially* maintained constant. A more physical representation of results is obtained by keeping the ratio of Rayleigh and Marangoni numbers constant (i.e. constant dynamic Bond number $Bo_d = Ra/Ma$ defined in §3.2.1). Indeed, both Ra and Ma are proportional to the driving gradient, so that their ratio only depends on fluid physical properties (and depth). In this representation, the marginal

---

Both internal and surface waves are usually damped, though it is shown in this section that their interaction may lead to amplification. Besides, if the surface is deformable, the capillary-gravity wave is yet another oscillatory mode. Various aspects of all three types of waves will be discussed later on.

$M \times 10^{-6}$

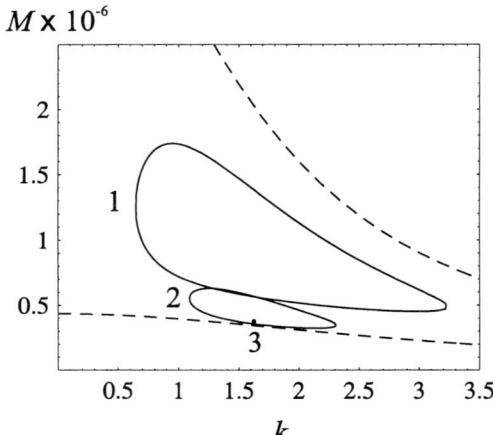

Figure 3.18: Marginal stability "bubbles" (overstability) for the Rayleigh–Marangoni–Bénard problem heated from above and with undeformable interface ($Pr = 6$). The Marangoni number $M = -Ma$ is plotted against the wavenumber k for different values of the Rayleigh number $R = -Ra$. (1) $R = 3 \ 10^6$, (2) $R = 1.5 \ 10^6$, point (3) $R = 1.21 \ 10^6$. The dashed curves are asymptotic results given by Eqs (3.138) [lower curve] and (3.139) [upper curve] and calculated for $R = 3 \ 10^6$. The agreement with numerical results is improved at higher R.

$\omega \times 10^{-3}$

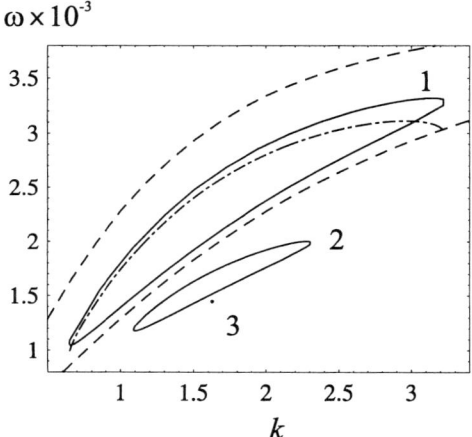

Figure 3.19: The marginal frequency ω along the bubbles of Fig. 3.18, and the asymptotic results (3.136) [lower dashed curve] and (3.137) [upper dashed curve] for $R = 3 \ 10^6$. The dot-dashed curve is a plot of Eq. (3.137) with M taken as the exact numerical value represented in Fig. 3.18.

curves have the usual shape (presenting a minimum at a certain k), even if in general they are limited by left and right asymptotes, as seen in Fig. 3.20. The frequencies corresponding to Fig. 3.20 are represented in Fig. 3.21.

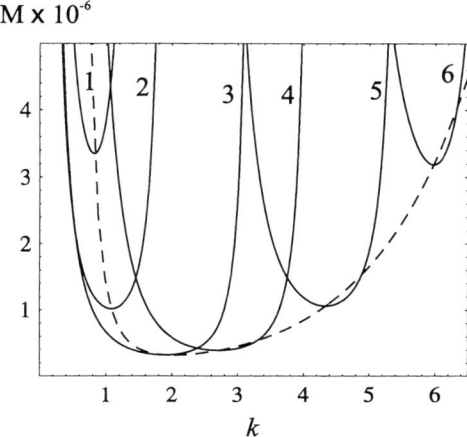

M × 10^{-6}

Figure 3.20: Marginal curves at several fixed values of the dynamic Bond number $Bo_d = R/M$. Curves: (1) $Bo_d = 1.5$, (2) $Bo_d = 2$, (3) $Bo_d = 4$, (4) $Bo_d = 6$, (5) $Bo_d = 10$, (6) $Bo_d = 15$ ($Pr = 6$). The dashed line is the locus of minima (critical conditions).

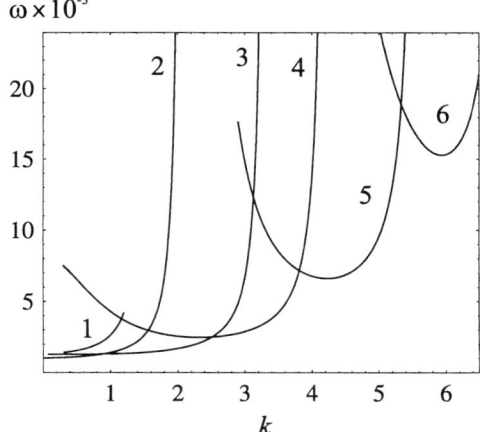

$\omega \times 10^{-3}$

Figure 3.21: The marginal frequency curves corresponding to Fig. 3.20.

For increasing Bond number (e.g. when increasing the liquid depth), it is seen in Figs 3.20 and 3.21 that the critical wavenumber shifts from long to short waves. Noteworthy is that there is a minimal value of $Bo_d = 0.715$ (for $Pr = 6$) below which no instability is possible. At this value, the right asymptote of the marginal curve appears at $k = 0$ and subsequently moves to the right when increasing Bo_d. The left asymptote of the curve appears at $Bo_d = 6.89$ (for $Pr = 6$).

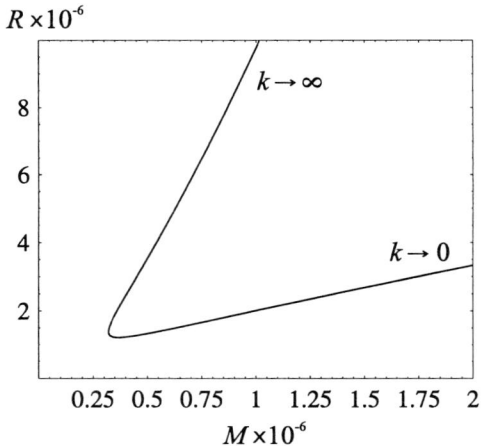

Figure 3.22: The stability diagram in the "Nield's plane" (R, M), when heating is from above, for the most dangerous mode $n = 1$ and at $Pr = 6$. Minimization with respect to the wavenumber is performed along the critical curve.

The locus of the critical points is depicted in Fig. 3.22 using Nield's (R, M) plot. As already noted, the trajectory described by the system progressively heated from above is a line of slope Bo_d passing through the origin. The system thus appears to be greatly stabilized for large values of Bo_d (buoyancy dominates) and absolutely stable for $Bo_d < 0.715$ (surface tension dominates). Thus, contrary to the situation occurring when heating is from below (§3.3.3), instability here definitely requires the simultaneous action of both mechanisms. In this sense, for a given fluid, there is an optimal depth at which the system is most unstable. For water with a mean temperature about 50 °C, we may compute $Pr = 3.57$, $Bo_d = 2.61d^2$ and $M = 20058\Delta Td$, where d is the depth expressed in centimeters and ΔT the imposed temperature difference. For a zero Biot number (which is justified for not too small air gaps above water), critical parameters are represented in Fig. 3.23, as a function of the depth. The minimum ΔT is found to be $(\Delta T_c)_{min} = 6.16$ °C for $d = 1.51$ cm. The period of oscillation is then $T = 7.41$ s, and the wavelength is $\lambda = 4.26$ cm. Further hints for experiments are given in §3.4.4.

The results presented in Figs 3.18–3.23 concern the most dangerous modes. In fact, as in the case of the pure Rayleigh–Bénard instability (e.g. between stress-free boundaries, see §3 1.3), less dangerous modes exist (corresponding to much higher values of the critical temperature gradient) characterized by vertical wavenumbers $n = 2, 3, ...$ (which may here also be seen as the number of convective cells in the vertical direction). The above results all correspond to $n = 1$. It has been numerically observed that such higher-order bubbles appear from

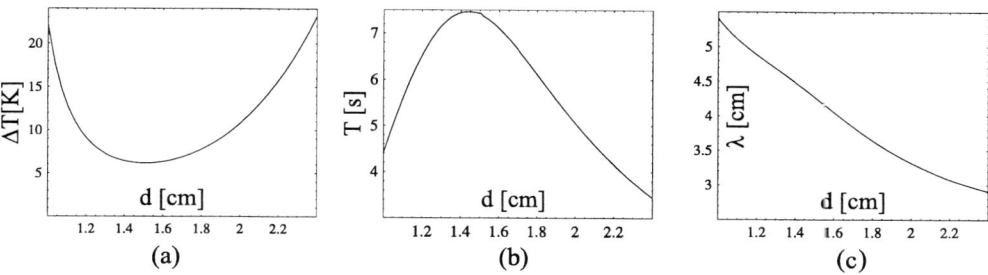

Figure 3.23: Critical temperature difference (a), period of oscillation (b), and wavelength (c) as a function of the depth for water at a mean temperature of $50\,^\circ\text{C}$ ($Pr = 3.57$) and $Bi = 0$.

some points (in the representation used in Fig. 3.18) at increasing values of R (for example, the bubble $n = 2$ appears at $R = 5.29\ 10^7$, the third bubble ($n = 3$) appears at $R = 4.68\ 10^8$, and so on. These higher-order modes also have a minimal value of Bo_d below which stability is guaranteed (in the case $Pr = 6$, for $n = 1$ we had $Bo_d = 0.715$, for $n = 2$ we obtain $Bo_d = 6.44$, ...). For every n, it is possible to repeat the same numerical calculations as for the fundamental mode $n = 1$. The final results for critical conditions of modes $n = 1, 2$, and 3 are represented in Fig. 3.24, where it is seen that higher-order modes are indeed always less dangerous.

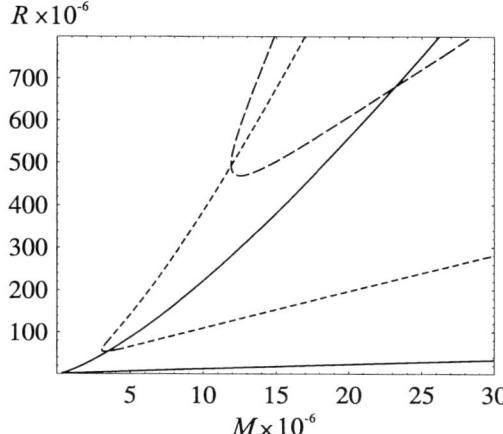

Figure 3.24: The stability diagram in the "Nield's plane" (R, M), when heating is from above, for modes $n = 1$ (full curve), 2 (short-dashed) and 3 (long-dashed), and at $Pr = 6$. Minimization with respect to the wavenumber is performed along each critical curve.

3.4.2   Asymptotic analysis – Internal and surface waves

It was seen in §3.2.1 that the dimensionless Rayleigh and Marangoni numbers could be written as the ratios of characteristic time scales. For instance, $Ra = \tau_{visc}\tau_{th}/\tau_{buoy}^2$ where $\tau_{visc} = d^2/\nu$ and $\tau_{th} = d^2/\kappa$ are respectively the characteristic viscous and thermal time scales, and $\tau_{buoy} = (d/\alpha g \Delta T)^{1/2}$ is the time scale of internal waves (when heating is from above). Similarly, $Ma = \tau_{visc}\tau_{th}/\tau_{Ma}^2$ where $\tau_{Ma} = (\rho d^3/\sigma_T \Delta T)^{1/2}$ is the thermocapillary time scale, characteristic of surface-tension-driven motions. At very large Rayleigh and Marangoni numbers, τ_{buoy} and τ_{Ma} are thus much shorter than viscous and thermal time scales (which will here be assumed of the same order, i.e. $Pr = \nu/\kappa$ is of order unity). In this case, an asymptotic analysis has been carried out [50], choosing as time scales the buoyant time scale τ_{buoy} (and associated velocity scale d/τ_{buoy}). A smallness parameter $\epsilon = (-Pr/Ra)^{1/4}$ then explicitly appears in the linearized equations for the disturbances. We only reproduce the main results of the analysis here, referring the reader to [50] for further details. Note that a similar analysis is presented in §6.3.1, when discussing dilational surface waves triggered by nonlinearity of the reference temperature profile.

In the limit $\epsilon \to 0$, the dissipative (Laplacian) terms (proportional to ϵ^2) disappear from the Navier–Stokes and energy equations. Thus at the leading order, the flow in the bulk of the liquid is ideal and free from thermal dissipation. However, the order of the equations has been reduced in this limit, and not all the boundary conditions can be satisfied. This indicates that boundary layers (of size ϵ) are formed near the upper and lower boundaries (where dissipative effects cannot be neglected). The correct resolution thus requires the definition of stretched vertical coordinates [230]. Expanding the unknown quantities in powers of ϵ, successive approximations are computed by identifying the various powers of ϵ in the equations and boundary conditions.

At the lowest order, the resolution of equations in each zone (the liquid bulk and the boundary layers at bottom and top surfaces), together with asymptotic matching and boundary conditions leads to the lowest-order dispersion relation [50]

$$\sinh\left(\sqrt{\frac{\sigma^2 - PrRa}{\sigma^2}}\,k\right)\left(\sigma^2 - \frac{PrMa\,k^2}{1 + Pr^{1/2}}\right) = 0 \qquad (3.135)$$

in which σ is the growth rate (in units of the usual thermal time scale τ_{th}). At this order, the two factors of Eq. (3.135) lead to purely imaginary growth rates $\sigma = i\omega$ (for negative Ra and Ma). Solving for the frequency ω, the first factor gives

$$\omega^2 = \frac{k^2 Pr\,(-Ra)}{k^2 + n^2\pi^2} \qquad n = 1, 2, \dots \qquad (3.136)$$

which is the frequency of ideal (Brunt–Väisälä) internal waves with vertical wavenumber n in a stably stratified liquid layer. Note that the dimensional form of Eq. (3.136) is identical to Eq. (3.18) and indeed corresponds to the buoyant time scale τ_{buoy}. The second factor of Eq. (3.135) leads to

$$\omega^2 = \frac{k^2 Pr\,(-Ma)}{1 + Pr^{1/2}} \qquad (3.137)$$

which is the frequency of dilational or longitudinal surface waves [231], depending only on the Marangoni number Ma. Note that for k and Pr of order unity, the time scale of these waves is $\tau_{Ma} = (\rho d^3/\sigma_T \Delta T)^{1/2}$, i.e. indeed the thermocapillary time scale. In fact, these surface waves are associated with an intense horizontal velocity field in the surface boundary layer (ϵ^{-1} times larger than other quantities in this asymptotic analysis [50]). In contrast, for internal waves the asymptotic order is the same for all quantities, such that a clear distinction between both types of waves is possible.

Now, both internal and dilational surface waves are undamped at the lowest order (purely imaginary growth constant σ). Hence, it is necessary to compute the next order correction to the dispersion relation, in order to determine the real part of σ. Consequently, although this real part will remain small (of order ϵ) compared with the frequency, this will allow the determination of conditions under which damping or amplification prevails (and thereby of the marginal conditions for both modes).

We state directly the result for the marginal conditions. For internal waves, the real part of the $O(\epsilon)$ correction to the growth rate is found to vanish when

$$Bo_{int} = \frac{Ra_{int}}{Ma_{int}} = \frac{2Pr^{1/2} + 1}{Pr^{1/2}(Pr^{1/2} + 1)}[k^2 + n^2\pi^2] \quad n = 1, 2, ... \tag{3.138}$$

while for the surface wave, we get

$$Bo_{sf} = \frac{Ra_{sf}}{Ma_{sf}} = \frac{1}{Pr^{1/2} + 1}[k^2 + (n - 1/2)^2\pi^2] \quad n = 1, 2, ... \tag{3.139}$$

An important point is that in between these two values of the dynamic Bond number (in fact for given k and n, $Bo_{sf} < Bo_{int}$), there exists a point where the frequencies of both modes are equal (resonance). Identifying Eqs (3.136) and (3.137), we get the resonance condition

$$Bo_{res} = \frac{k^2 + n^2\pi^2}{Pr^{1/2} + 1} \quad n = 1, 2, ... \tag{3.140}$$

and it is seen that at given k and n, $Bo_{sf} < Bo_{res} < Bo_{int}$.

The above results and the following discussions must be understood in the asymptotic limit $Ra, Ma \to \infty$, $Bo = O(1)$. Considering Bo as a free parameter progressively increased from 0 (pure Marangoni effect), instability first occurs to surface waves at the minimal value of Bo_{sf}. According to Eq. (3.139), this is $Bo_{min} = \pi^2/4(Pr^{1/2} + 1)$ at $k = 0, n = 1$. In agreement with numerical results of §3.4.1, this gives $Bo_{min} \simeq 0.715$ for $Pr = 6$. Then, when Bo is increased (at constant k and n), the growth constant of surface waves eventually increases, while that of internal waves remains negative. In the vicinity of the resonance point $Bo = Bo_{res}$ (for given k and n) it is shown in [50] that a continuous transition occurs between the surface wave mode and the internal wave mode, a phenomenon called mode-mixing. There, a higher degree of amplification is predicted for the unstable mixed mode, while the stable mixed mode is more damped (however the real parts of both modes still remain small compared with the frequency). This mode mixing results in the fact that above Bo_{res}, the mode with positive growth rate turns out to be the internal wave mode. Increasing Bo again (at fixed k and n), the amplification rate of the internal wave mode decreases, and passes through zero at Bo_{int}. This does not mean that the system is stable above this value however, because other modes (with different values of k and n) are in fact still unstable. Indeed, it is seen that Eq. (3.138) has no upper bound when maximized with respect to k and n. Thus when Bo is increased, the system remains unstable (in the limit $Ra, Ma \to \infty$), although with respect to modes with increasing k and n. This may also be seen from Figs 3.22 and 3.24 above.

Another representation is obtained at given Bond number Bo and vertical wavenumber n. The resolution of $Bo = Bo_{int}$ and $Bo = Bo_{sf}$ with respect to the horizontal wavenumber k leads to, say, k_{int} and k_{sf} respectively. For $n = 1$, a solution $k_{sf} > 0$ exists if $Bo > \pi^2/4(Pr^{1/2}+1)$, while $k_{int} > 0$ exists if $Bo > \pi^2(2Pr^{1/2}+1)/Pr^{1/2}(Pr^{1/2}+1)$. In this case, it is readily shown that $k_{sf} > k_{int}$. These values of the wavenumber in fact correspond to the left and right asymptotes of marginal stability curves of Fig. 3.20 (limits of the stability boundary for $M \to \infty$, defining the interval of unstable wavenumbers).

3.4.3   General features of mode-mixing phenomena

The problem studied here shares several generic features with another problem treated later on in §3.6, namely the interaction between capillary-gravity waves and dilational waves at a *deformable* interface. Among these, the presence of two different high-frequency modes in the limit of large thermal gradient, whose interaction may lead to instability. In the case considered here, these modes are internal (transverse) waves and surface (longitudinal or dilatioral) waves. In §3.6, it will be seen that dilational surface waves, which do not require essential surface deformation, can also resonantly interact with capillary-gravity waves, which involve motions orthogonal to the interface (and are in this sense transverse waves). Just as here, these waves will be shown to mix near resonance, where a higher degree of amplification/damping is also predicted. Such phenomena were also studied by Earnshaw and McLaughlin [243, 226, 287], though in a situation were the waves are not sustained by a thermal gradient.

The different physical nature of waves considered here and those studied in §3.6 also leads to specific phenomena, such as the possible destabilization of a countable infinite number of modes $n = 1, 2, ...$, as seen in Fig. 3.24. This is apparently a characteristic of buoyancy-induced flows, linked to the vertical structure of perturbation fields. Another difference is the particular form of the stability boundary (which may not always be in the form of bubbles such as those encountered here), and different scalings of critical Marangoni numbers.

Finally, note that dilational surface waves can also be amplified when considering the non-linearity of the basic temperature (or concentration) gradient, or the convective heat transport in the gas phase, as we shall see respectively in §6.3 and §3.5.

3.4.4   Some hints for experiments

As seen below for water, amplification of mixed internal and surface waves may thus be expected for reasonable values of temperature gradients, provided the liquid depth is chosen near the optimal one. However, water is certainly not the most suitable liquid for observing these waves, as its free surface is easily contaminated by impurities, strongly biasing the Marangoni effect [261].

In selecting another liquid, the following facts have to be taken into account. First, it has been shown in [50] that the increase of the Prandtl number stabilizes the diffusive state (higher critical Rayleigh and Marangoni numbers). Hence, low-viscosity liquids are certainly preferable. Moreover, critical temperature gradients will also be lower if the optimal depth d^* is found to be large (Rayleigh and Marangoni are proportional to d^3 and d respectively). As the optimal dynamic Bond number $Bo^* = \alpha \rho g (d^*)^2 / \sigma_T$ is about 4 or larger for Prandtl numbers below 6, liquids with low expansion coefficient α and high surface tension variation with temperature σ_T would also be preferable. Acetone ($Pr = 3.6$) appears to be a good candidate. Indeed, the optimal depth is found to be $d^* \simeq 1$ cm, and the corresponding critical temperature difference is about 6.5 °C. The period of oscillations at onset is then about 3.7 s. Preliminary results on acetone have indeed been recently obtained by A. Ezersky (personal communication, 1999), which are in reasonable agreement with calculated values of critical temperature gradient and frequency.

Furthermore, it is thought that such modes could be partly responsible for some of the waves experimentally observed in mass-transfer systems [46, 228] (see also Chapter 6), but this should be investigated more carefully. Indeed, the waves described above are not the only ones predicted when heating a Bénard layer from the top (or absorbing a surface-tension-lowering solute). In §3.5, it will be seen that waves can also be triggered when the convective heat transport in the *gas phase* is important, i.e. when the gas layer is sufficiently thick. For thin gas layers, which is also a condition for the use of a Biot number (see §3.2.3 and §3.5.2), it has been checked [50] that the wave modes described above are indeed the most dangerous. Other possible wave modes are due to surface deformation, but as will be seen in §3.6, such capillary-gravity waves are not expected when the layer has a thickness of the order of 1 cm, such as for the modes considered here.

3.5 Two-layer systems

In the previous sections, Rayleigh and Marangoni–Bénard instabilities were considered for single phase systems. The gas phase adjacent to the liquid layer was considered as passive, thus neglecting the dynamics of thermal and mechanical perturbations in the gas, and modeling the heat transfer across the free surface by a Newton's cooling law of the form (2.77). However, even if this analysis may be satisfactory in some cases, it remains to determine the conditions under which such a one-sided model applies to realistic set-ups. In the general case, we have to consider the gas phase in the same way as the liquid phase, thus solving continuity, momentum and Navier–Stokes equations in both media.

This two-layer approach will also be necessary when considering two immiscible liquid phases in contact. The analysis is thus developed without making special assumptions for the upper phase (except that it is assumed incompressible, which is reasonable even for gases provided flow velocities are low enough), and an attempt is made to obtain a sufficiently general view of the behavior of two-layer systems, as a function of the relative thermal and mechanical properties of the fluids in contact. The number of dimensionless parameters characterizing the system is increased with respect to the one-phase problems, because in addition to the Marangoni and Prandtl numbers, we will have to consider all the ratios of fluid properties. Note that the Biot number will not appear in the two-layer formulation, because the heat transfer boundary condition through the interface will here be expressed by the general equation for the conservation of energy (which here reduces to the equality of the normal heat fluxes on both sides of the interface). A sketch of the system studied in this section is given in Fig. 3 25.

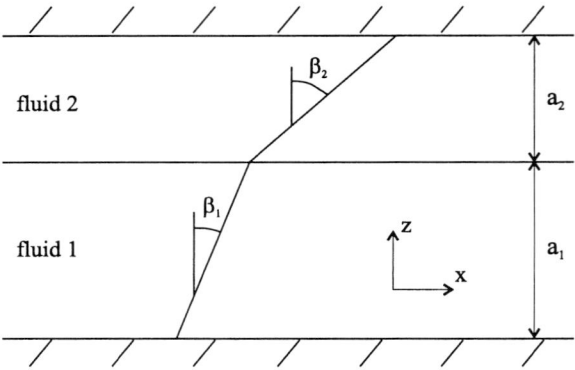

Figure 3.25: Geometry of the two-layer system and reference diffusive state.

Note that the general approach in this book is to isolate specific phenomena and to focus on limiting cases, rather than to attempt a full resolution, which would be at the detriment of physical understanding. Accordingly, we neglect here buoyancy effects in both layers[11], and interface deformations (specific modes linked to the latter effect will be studied in §3.6). This also allows the use of the Boussinesq approximation, as discussed in §3.1.4.

The equations will be written under dimensionless form using thermal scales (as defined in §3.2.2) referring to the phase labeled 1 (the lower one by convention). The scale of length will be taken as the lower fluid depth a_1, the temperature scale as the temperature drop ΔT_1 over the first layer in the reference diffusive state, the time scale as a_1^2/κ_1 (the unit of velocity as κ_1/a_1) and the unit of pressure as $\mu_1\kappa_1/a_1^2$. Hereafter, the symbols ρ_i, μ_i, $\nu_i = \mu_i/\rho_i$, λ_i, and κ_i will respectively denote the density, the dynamic viscosity, the kinematic viscosity, the thermal conductivity and the thermal diffusivity (the indices $i = 1, 2$ referring to lower and upper phases respectively). The dimensionless ratios of these quantities will be denoted by $\rho = \rho_2/\rho_1$, $\mu = \mu_2/\mu_1$, $\nu = \nu_2/\nu_1$, $\lambda = \lambda_2/\lambda_1$, and $\kappa = \kappa_2/\kappa_1$. Similarly, the depth ratio is denoted as $a = a_2/a_1$, and the ratio of the thermal gradients in both layers for the reference state is $\beta = \beta_2/\beta_1 = a^{-1}\Delta T_2/\Delta T_1 = \lambda^{-1}$. Then, the system of dimensionless equations for vertical velocity $w_i(z)$ and temperature perturbation $\theta_i(z)$ reads in the usual way [see Eqs

---

[11] In the simplest case, it may be guessed that if the local Rayleigh numbers defined separately for both layers are very different, one of the layers will drive convection in the other one, and instability modes should not be qualitatively modified compared with those studied in previous sections. However, when local Rayleigh numbers are close to each other, specific kinds of coupling are observed. This is studied in §5.4. We also refer the reader to interesting works [97, 221, 222], on "anticonvection", a specific mechanism due to buoyancy and occurring when the two-layer system is heated *from above*. Other situations involving coupling between buoyancy and surface tension effects in two-layer systems are studied in [43, 159, 223].

(3.104–3.105) already written for normal modes] for layer 1:

$$(D^2 - k^2)^2 w_1 - k^2 Ra_1\theta_1 = \sigma Pr_1^{-1}(D^2 - k^2)w_1 \tag{3.141}$$

$$(D^2 - k^2)\theta_1 + w_1 = \sigma\theta_1 \tag{3.142}$$

where $Pr_1 = \nu_1/\kappa_1$, and $Ra_1 = g\alpha_1\Delta T_1 a_1^3/\nu_1\kappa_1$. Note that the Rayleigh effect is included here for later reference, though we neglect it in this section.

Now, owing to our choice of units, a number of dimensionless ratios appear in the equations in the upper layer. These equations read

$$(D^2 - k^2)^2 w_2 - k^2\alpha\nu^{-1}Ra_1\theta_2 = \sigma\nu^{-1}Pr_1^{-1}(D^2 - k^2)w_2 \tag{3.143}$$

$$(D^2 - k^2)\theta_2 + \beta\kappa^{-1}w_2 = \sigma\kappa^{-1}\theta_2 \tag{3.144}$$

where $\alpha = \alpha_2/\alpha_1$ is the ratio of thermal expansion coefficients, and other ratios have been defined above.

The boundary conditions to be used here have been established in §3.2.3. At the rigid isothermal top and bottom walls, we have

$$w_1 = Dw_1 = \theta_1 = 0 \quad at\ z = -1 \tag{3.145}$$

$$w_2 = Dw_2 = \theta_2 = 0 \quad at\ z = a \tag{3.146}$$

and at the undeformable interface $z = 0$, we have Eqs (3.77–3.80) and Eq. (3.84). In the present notations, these are

$$w_1 = w_2 = 0,\ Dw_1 = Dw_2,\ \theta_1 = \theta_2,\ \lambda D\theta_2 = D\theta_1 \quad at\ z = 0 \tag{3.147}$$

and the tangential stress boundary condition

$$\mu D^2 w_2 - D^2 w_1 = k^2 Ma_1\theta_1 \quad at\ z = 0 \tag{3.148}$$

where $Ma_1 = -\sigma_T\Delta T_1 a_1/\mu_1\kappa_1$ is the Marangoni number defined using properties of the lower phase (and will often be written Ma in the following). This choice of scales will facilitate comparison of results with the one-layer case, and thus allow the determination of the conditions under which the upper phase may be considered as an inert gas.

It is seen from Eqs (3.141–3.148) with $Ra_1 = 0$ and $\beta = \lambda^{-1}$, that in the case of neutral stability ($\sigma = 0$), a compatibility relation should be found between Ma, k^2, a, κ, λ and μ. For overstability ($\sigma = i\omega$), the discussion also depends on additional parameters Pr_1 and ν.

3.5.1 Infinitely deep layers – Short-wavelength analysis

The analysis of the dispersion relation (and in particular the neutral stability condition) considerably simplifies in situations where the influence of the rigid bottom and top boundaries may be neglected. Furthermore it was seen in §3.3.2, related to the one-phase problem, that this situation corresponds to the asymptotic form of the neutral stability curve and of the critical perturbations when the wavenumber is increased to infinity (i.e. perturbations of wavelength much shorter than the depth).

The analysis presented here is similar to the work of Sternling and Scriven [75], who first established the neutral and overstability conditions, in an attempt to provide an explanation of the phenomenon of "interfacial turbulence". Since the 1950s, this name has been used to denote the various forms of spontaneous agitation and convective motions experimentally observed in the vicinity of an interface, formed by initially putting in contact two immiscible phases in thermal or concentrational non-equilibrium with each other. This non-equilibrium leads to transfer of heat (or matter) through the interface, accompanied by the formation of diffusive profiles of temperature (or concentration) in the bulk of the adjacent fluids. After a certain time, these profiles might extend sufficiently deep into the bulks and become sufficiently slow (and linear) for a linear stability analysis of the kind presented here to be valid. It is now currently admitted that such hydrodynamic instability phenomena (due to the Marangoni effect) are indeed responsible for the experimentally observed intensive convective motions and important increases of the overall heat/mass transfer rates through the interface.

Monotonic modes

When $\sigma = 0$, the Stokes equations (3.141) and (3.143) may be readily integrated (in the case $Ra_1 = 0$) to give

$$w_1 = bz \exp[kz], \quad w_2 = bz \exp[-kz] \tag{3.149}$$

where b is an unknown constant, and the boundary conditions $w_1 = w_2 = 0$ and $Dw_1 = Dw_2$ at $z = 0$ have been used, together with boundedness conditions at $z \to -\infty$ and $z \to +\infty$. The energy equations (3.142) and (3.144) are then solved for θ_1 and θ_2. Using the boundary condition $\theta_1 = \theta_2$ at $z = 0$, we obtain

$$\theta_1 = \exp[kz] \left[c + b\frac{z(1 - kz)}{4k^2} \right], \quad \theta_2 = \exp[-kz] \left[c + \frac{b}{\lambda\kappa}\frac{z(1 + kz)}{4k^2} \right] \tag{3.150}$$

where c is another constant, which can be determined from the heat flux continuity $\lambda D\theta_2 = D\theta_1$ at $z = 0$. This gives

$$c = b\frac{(\kappa^{-1} - 1)}{4k^3(1 + \lambda)} \tag{3.151}$$

which is the expression for the amplitude of the temperature perturbation at the interface (up to an arbitrary multiplicative constant b, the amplitude of the velocity perturbation). The surface tension gradient created by this temperature inhomogeneity has to be balanced by the viscous stresses. This is expressed by the tangential stress boundary condition (3.148), whose resolution with respect to Ma gives the neutral stability condition

$$Ma = 8k^2 \frac{(1 + \lambda)(1 + \mu)}{1 - \kappa^{-1}} \tag{3.152}$$

Thus, the neutral stability curve in the (Ma, k) plane is a parabola, with critical wavenumber $k_c = 0$ always. This characteristic is general for infinitely deep Marangoni–Bénard problems (see §3.3.2), and reflects the fact that the size of the critical perturbations should always be comparable to the depth of the liquid layers. Note also that the critical Marangoni number $Ma_c = 0$, thus indicating that an infinitesimally small temperature gradient will be sufficient to excite convective modes with vanishing wavenumber. In this sense, the system studied is always unstable. However, in this case it will be instructive to analyze the full dispersion relation, which will lead to the result that the actual growth rate of these modes is also vanishingly small (e.g. the maximal growth rate for the one-layer zero Biot number case is proportional to the Marangoni number, which indicates that the fastest growing perturbations are amplified with a rate $\sim \sigma_T \beta/\mu$, where β is the imposed thermal gradient).

The result (3.152) also shows the influence of properties ratios μ, λ and κ. The dynamic viscosity ratio μ and thermal conductivity ratio λ are seen to influence only the magnitude of the Marangoni number (the curvature of the parabola), but not its sign. On the contrary, the thermal diffusivity ratio κ is crucial as far as the sign of Ma is concerned. For $\kappa > 1$, monotonic instability occurs only if heating is from below ($Ma > 0$), while heating must be from above when $\kappa < 1$. As "above" and "below" have actually no meaning here (e.g. in a microgravity environment), it is preferable to remember that instability occurs if the liquid with lower thermal diffusivity is hotter. The simplest example is the liquid/gas interface problem, where it is known that instability occurs when the liquid side is heated, due to the much larger thermal diffusivity in the gas. Note that similar conclusions apply for isothermal systems, where a soluble surface active solute (which lowers the surface tension) diffuses through the interface. There, instability occurs when the surfactant is transferred out of the phase with lower diffusion coefficient.

The physical reason of this preference for one of the directions of heating (we now refer to the thermal problem) is connected with the efficiency of heat transport by convection in both phases. This is measured by the Peclet number $Pe = Ud/\kappa$, where U and d are typical velocity and length scales, and κ is the thermal diffusivity. Considering an initially positive temperature fluctuation at the interface, there is upflow in the lower layer and downflow in the upper one at the vertical of the disturbance (because of the coupling of horizontal velocities at the interface). If the upper phase has higher diffusivity, isotherms in this phase are significantly less deformed by convection than in the lower phase [velocities have comparable orders of magnitude in both phases as indicated by Eqs (3.149)]. The lower layer thus contributes more than the upper one to the variation of the interface temperature, and as it is hotter if heating is from below, there is a net increase in temperature at the initial location of the disturbance [see also Eq. (3.151)], leading to the instability of the motionless diffusive state.

We can also investigate the conditions under which this two-layer problem reduces to the single phase formulation considered in §3.3.2. It is seen that the neutral stability condition Eq. (3.119) with $Bi = 0$ is recovered from Eq. (3.152) when $\mu \to 0$, $\lambda \to 0$ and $\kappa \to \infty$. In fact, even though $\kappa = \lambda/\rho c_p$, the last two conditions are not incompatible, at least for liquid/gas systems. The value of $\rho = \rho_{gas}/\rho_{liquid}$ may indeed be sufficiently small for these conditions to be approximately valid (e.g. a water/air system has $\mu \simeq 0.02$, $\lambda \simeq 0.04$, $\kappa \simeq 152$). However, it is seen that the one-layer modeling in terms of the Biot number fails in cases where the above three conditions are not met simultaneously (Bi cannot be identified with a certain function of fluid properties and independent of the wavenumber).

General dispersion relation and wave modes

In the general case $\sigma \neq 0$ (and in general complex), we can solve Eqs (3.141–3.144) proceeding along the same lines as for neutral stability. We will not reproduce intermediate results here, because of their greater complexity, but rather state the final result directly. Denoting the roots (with positive real parts) of the characteristic polynomials by r_1, r_2 (in the heat equations) and s_1, s_2 (in the Navier–Stokes equations), we have

$$r_1^2 = k^2 + \sigma, \quad r_2^2 = k^2 + \sigma \kappa^{-1} \tag{3.153}$$

$$s_1^2 = k^2 + \sigma Pr^{-1}, \quad s_2^2 = k^2 + \sigma \nu^{-1} Pr^{-1} \tag{3.154}$$

and assuming them to be different (no degeneracy), we finally obtain the full dispersion relation

$$Ma = \frac{(k + r_1)(k + r_2)(r_1 + s_1)(r_2 + s_2)(r_1 + \lambda r_2)\left(k(1 + \mu) + s_1 + \mu s_2\right)}{k^2\left[(kr_2 + r_2^2 + ks_2 + r_2 s_2) - \kappa^{-1}(kr_1 + r_1^2 + ks_1 + r_1 s_1)\right]} \tag{3.155}$$

Note that when $\sigma = \sigma_R + i\sigma_I$, the Marangoni number calculated from this expression is in general complex, which is not physical. We will thus only be interested in the points where the imaginary part of this expression is identically zero. This in principle provides a relation for determining the frequency $\omega = \sigma_I$, and the corresponding real part of Eq. (3.155) then leads to the determination of the real part σ_R (growth rate) at given Ma. Unfortunately, it is practically impossible to solve for σ analytically, and the general discussion may only be achieved from a numerical resolution of Eq. (3.155).

We may check that the expression (3.155) leads to the neutral stability condition (3.152) when $\sigma \to 0$ [from Eqs (3.153–3.154) it is seen that $r_1, r_2, s_1, s_2 \to k$ in this limit]. Moreover, the wavenumber k may easily be eliminated from the equation by the changes $\sigma \to \tilde{\sigma} k^2$ and $Ma \to \tilde{M} a k^2$.

Then, defining

$$\tilde{r}_1^2 = 1 + \tilde{\sigma}, \quad \tilde{r}_2^2 = 1 + \tilde{\sigma} \kappa^{-1} \tag{3.156}$$

$$\tilde{s}_1^2 = 1 + \tilde{\sigma} Pr^{-1}, \quad \tilde{s}_2^2 = 1 + \tilde{\sigma} \nu^{-1} Pr^{-1} \tag{3.157}$$

we get

$$\tilde{M}a = \frac{(1 + \tilde{r}_1)(1 + \tilde{r}_2)(\tilde{r}_1 + \tilde{s}_1)(\tilde{r}_2 + \tilde{s}_2)(\tilde{r}_1 + \lambda \tilde{r}_2)(1 + \mu + \tilde{s}_1 + \mu \tilde{s}_2)}{(\tilde{r}_2 + \tilde{s}_2)(1 + \tilde{r}_2) - \kappa^{-1}(\tilde{r}_1 + \tilde{s}_1)(1 + \tilde{r}_1)} \tag{3.158}$$

In the following, we will mainly be interested in the conditions sufficient to have oscillatory onset of instability. Thus, substituting $\tilde{\sigma} = i\tilde{\omega}$ in this expression (marginal conditions), we will try to determine conditions for the curve of $\text{Im}[\tilde{M}a]$ as a function of $\tilde{\omega}$ to present at least one zero. If this happens at $\tilde{\omega} = \tilde{\omega}_0$, we will determine the corresponding critical Marangoni number $\tilde{M}a_0$ from the real part of Eq. (3.158). In fact, this will give a branch of critical oscillatory modes given in the original non-rescaled variables by $Ma = \tilde{M}a_0 k^2$ (thus a parabola as for neutral stability) and $\omega = \tilde{\omega}_0 k^2$. The physical reason for such a simple dependence with respect to the wavenumber k is linked to the absence of any geometric length scale in the infinite-depth problem, and the rescaling made above in fact corresponds to a redefinition of the unit length (and associated time and velocity scales) based on the wavelength of a convective perturbation. Note that we will write all quantities without tilde in the following.

For $\sigma = i\omega$ and $\omega \to 0$, Ma given by Eq. (3.158) is purely real [and equal to Eq. (3.152)]. The sign of the imaginary part for small $\omega > 0$ is thus determined by the slope of the curve of $\text{Im}[Ma]$ versus ω at $\omega = 0$, which can be computed from Eq. (3.158). As $\text{Im}[Ma]$ is a continuous function of ω, the denominator of which cannot vanish (except for very particular values of parameters), it will be possible to decide whether it possesses at least one zero by comparing its sign near the origin and at $\omega \to \infty$. For conciseness, we only give here essential results and some details needed for the discussion (see also [97, 75]).

The properties ratios κ and ν appear to be the crucial parameters. In fact, a sufficient condition for overstability is that κ and ν are simultaneously superior or inferior to unity, whatever the value of Pr, λ and μ. In other cases, the discussion should also depend on numerical values of these parameters. The complete discussion is summarized in Fig. 3.26.

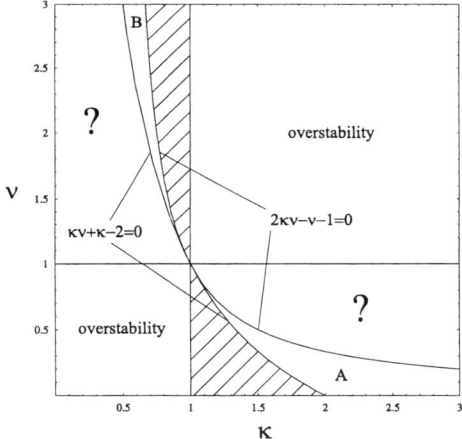

Figure 3.26: Sufficient conditions for overstability in a system of two semi-infinite layers with undeformable interface. Overstability occurs whatever the values of Pr, λ and μ, provided κ and ν are simultaneously superior or inferior to unity. In the dashed regions, overstability also occurs if Pr is sufficiently small. In region A, both μ and Pr should be sufficiently small, while in region B, μ should be sufficiently large and Pr sufficiently small for overstability to occur. In the regions labeled with a question mark, no overstability is predicted, though this cannot be proven rigorously.

When $\nu < 1$ and $1 < \kappa < 2/(1 + \nu)$ [or $\nu > 1$ and $(1 + \nu)/2\nu < \kappa < 1$], i.e. in the dashed regions of Fig. 3.26, a branch of overstability also exists if

$$Pr < Pr^* = \frac{\kappa(1+\lambda)[-1-\nu+2\kappa\nu+\mu(-2+\kappa+\kappa\nu)]}{2\nu(1-\kappa)(1+\mu)[1+2\lambda+\kappa(2+\lambda)]} \tag{3.159}$$

When $\nu < 1$ and $2/(1+\nu) < \kappa < (1+\nu)/2\nu$ (region A), Pr should also be smaller than Pr^* given by Eq. (3.159), but this is only possible provided

$$\mu < \mu^* = (1+\nu-2\kappa\nu)/(\kappa+\kappa\nu-2) \tag{3.160}$$

Finally, when $\nu > 1$ and $2/(1+\nu) < \kappa < (1+\nu)/2\nu$ (region B), overstability certainly occurs if $Pr < Pr^*$ and $\mu > \mu^*$. In other regions (labeled with a question mark), the sign of $\text{Im}[Ma]$ is identical at small and large ω, and overstability branches might only exist by pairs (although a numerical investigation of the relation (3.158) has not indicated any oscillatory onset).

We will now derive some useful results in the limit of infinite Prandtl number. This is motivated because most of the fluids used in experiment (excluding gases and liquid metals) are characterized by moderate and large values of Pr, for which applicability of this analysis may be expected. This is a fortiori true for soluble surfactant mass transfer systems, for which the Prandtl number should be replaced by the Schmidt number, which generally takes very high values (see §3.2.1). In this case however, it should be emphasized that we did not take into account processes specific to mass transfer, such as accumulation of surfactant at the interface, energy barriers, surface diffusion, ... [76, 119, 191, 217], which may not be negligible in all cases of interest.

Asymptotic behavior at large Prandtl (or Schmidt) number

A numerical tracking of the overstability branch at increasing values of Pr shows that the frequency ω increases indefinitely with Pr, tending to the asymptotic behavior $\omega \sim Pr^{2/3}$ at $Pr \rightarrow \infty$. In order to obtain analytical results, we thus assume $\sigma = SPr^{2/3}$ in Eq. (3.158), and let Pr tend to infinity. The asymptotic form of Ma (limited to the first two leading order terms) is then

$$Ma \rightarrow \frac{(1+\lambda\kappa^{-1/2})(1+\mu)S^2 Pr^{4/3}}{(\kappa^{1/2}-1)} + \frac{PrS^{3/2}(1+\lambda\kappa^{-1/2})\left(S^{3/2}(\nu-1)(\mu+\kappa^{1/2})+4\nu(\mu+1)(\kappa-1)\right)}{4\nu(\kappa-1)^2} \tag{3.161}$$

To determine the marginal cases, we set $S = i\Omega$. Thus, the first term in Eq. (3.161) is real, such that Ω has to be found by setting the imaginary part of the second term to zero. Then, reintroducing this value in the first term leads to the critical value of the Marangoni number. After some algebra, we finally get

$$\omega = \Omega Pr^{2/3} = \left[\frac{2\sqrt{2}\nu(\mu+1)(\kappa-1)Pr}{(\mu+\kappa^{1/2})(\nu-1)}\right]^{\frac{2}{3}} \tag{3.162}$$

and

$$Ma = -\frac{4(1 + \lambda\kappa^{-1/2})(1 + \mu)^{7/3}Pr^{4/3}\nu^{4/3}(1 + \kappa^{1/2})(\kappa - 1)^{1/3}}{(\mu + \kappa^{1/2})^{4/3}(\nu - 1)^{4/3}} \qquad (3.163)$$

which have also been checked numerically. Remember that both these expression have to be multiplied by k^2 to return to unscaled variables. We finally note that this result agrees with the numerical finding of Reichenbach and Linde [219] that the critical Marangoni numbers are generally higher for mass transfer systems (for which Pr has to be understood as the Schmidt number Sc) than for heat transfer systems.

Summary – Physical mechanisms and heuristic arguments

Consider the case $\kappa > 1$ and $\nu > 1$, for which at least one branch of oscillatory modes exists, whatever the values of other parameters (Pr, λ and μ). For moderate and large Prandtl numbers, these oscillatory modes occur when heating is from above [see the sign of Eq. (3.163)]. It was also seen that for $\kappa > 1$, monotonic modes are expected when heating is from below. Thus, instability occurs for both directions of heating. This is also the case for $\kappa < 1$ and $\nu < 1$, but the directions of heating are reversed with respect to the previous situation. In both cases, the expressions (3.162–3.163) are asymptotically valid for $Pr \to \infty$ (or $Sc \to \infty$ for mass-transfer systems). In the following, we disregard overstable modes occurring when $\kappa - 1$ and $\nu - 1$ have opposite signs (see above), thus restricting the analysis to fluids with moderate or large Prandtl (or Schmidt) number.

We may also reformulate our conclusions without making use of the notions "above" and "below" (although in the normal field of gravity care should be taken to choose as the bottom phase the one with the higher density, i.e. $\rho = \mu/\nu$ should be lower than unity). When the phase with lower diffusivity is also the one with lower kinematic viscosity, monotonic instability occurs if transfer of heat (or of surfactant) is from this phase to the other one, and overstability occurs when the transfer occurs in the opposite direction (at least for sufficiently large Pr or Sc).

A physical interpretation of the overstability mechanism is the following. Consider that the phase with lower diffusivity is also the one with the lower kinematic viscosity. To simplify, the description is here made in terms of heat transfer. When heating is from the phase with higher diffusivity, the analysis of previous sections shows that the diffusive state is stable to monotonic perturbations. Indeed, considering again an initially positive fluctuation of temperature at a point of the interface, fluid is brought towards that point from both phases, but the colder one (lower diffusivity) is more effective in transporting heat by convection, such that the disturbance is damped. However, we now have to consider the effect of fluid inertia, as stimulated by the fact that overstability will not occur at infinite Prandtl numbers [for which the critical Marangoni number is infinite, as seen from Eq. (3.163)]. Following the initial temperature fluctuation, the flow does not establish instantaneously, but rather needs a certain viscous time of the order of d^2/ν to penetrate (over a depth d) into the adjacent liquid bulks. As the hotter phase possesses the higher kinematic viscosity ν, it will respond faster than the colder one to the temperature perturbation at the interface. During this first stage, the incoming hot fluid is able to produce an increase of the temperature at the point of the interface where the fluctuation occurred, thereby amplifying the surface tension gradient. After a certain time, the flow has established into both phases, and the effect of increased convective heat transport in the cold phase overcomes the inertia effect. Cold fluid brought to the interface makes the temperature decrease, eventually leading to an overshoot of the temperature at the location of the disturbance (i.e. it becomes negative). Then, the situation is reversed (i.e. the process has completed half a period of oscillation), as the negative temperature disturbance produces flow (after a certain time, due to inertia) away from the disturbance, first in the hotter phase. The disturbance is thus further cooled in the first (inertial) stage, then is heated again when flows have established, and so on. An oscillation period is completed when the disturbance becomes positive again, and the initial situation is restored, possibly with a slight global increase of the temperature perturbation. In this case (when the Marangoni number is higher than critical), the oscillations are increasing in amplitude and the diffusive state is progressively replaced by an oscillatory convective structure (the description of which requires nonlinear effects to be taken into account).

This discussion is illustrated on the model example depicted in Fig. 3.27, for $\nu = \mu = 0.1$, $\kappa = 0.4$, $\lambda = 0.5$ and $Pr = 0.01$. Note that the Prandtl number is taken small here, in order to obtain a sufficiently small value of the critical Marangoni number. This allows visualizing the behavior of the temperature fluctuation at the interface, which would not have been possible for too large Marangoni number. The formulas (3.162–3.163) derived for $Pr \to \infty$ are of course not satisfactory for $Pr = 0.01$, although the predicted sign of Ma is correct (overstability occurs when heating from below). In Fig. 3.27, temperature and vertical velocity fluctuations are represented as a function of the height z, for parameters given above and for critical conditions ($Ma = 245.2$ and $\omega = 0.0754$). The vertical line corresponds to the position of the interface, and the bottom hotter layer lies at the left of this line (negative z). The left column (starting from above) represents the first half of the period, starting with a situation where there is a small positive temperature fluctuation at the interface.

The right column (from top to bottom) corresponds to the second half of the period, and is symmetric ($w \to -w$, $\theta \to -\theta$) with respect to the left column. The discussion made above is applicable to Fig. 3.27. In particular, note that the bottom hotter layer responds faster than the colder one to the positive temperature perturbation at the interface (times $T/10$ to $3T/10$). When the flow is established in the top layer, convective transport induces a decrease of the surface temperature (from $3T/10$ to $5T/10$, where overshoot occurred).

The mechanism described above would seem to indicate the onset of some kind of standing wave oscillating structure above threshold. However, it should be emphasized that at the linear stage, nothing can rigorously be said about the ultimate nature of the nonlinear regime. In fact, the solution of the linear problem solved here can be any superposition of waves traveling with phase velocity ω/k in different horizontal directions (assuming that only one wavenumber is selected). In a two-dimensional situation, the general solution (say, considering the surface temperature) can be written

$$T_s = A_L \exp[i(kx + \omega t)] + A_R \exp[i(kx - \omega t)] + c.c. \tag{3.164}$$

with *arbitrary* A_L and A_R (corresponding respectively to the amplitudes of left and right traveling waves), and $c.c.$ denotes the complex conjugate.

In fact, it will be shown in next chapters that among the infinite number of superpositions possible at the linear stage, only a reduced set will lead to lasting small-amplitude oscillating

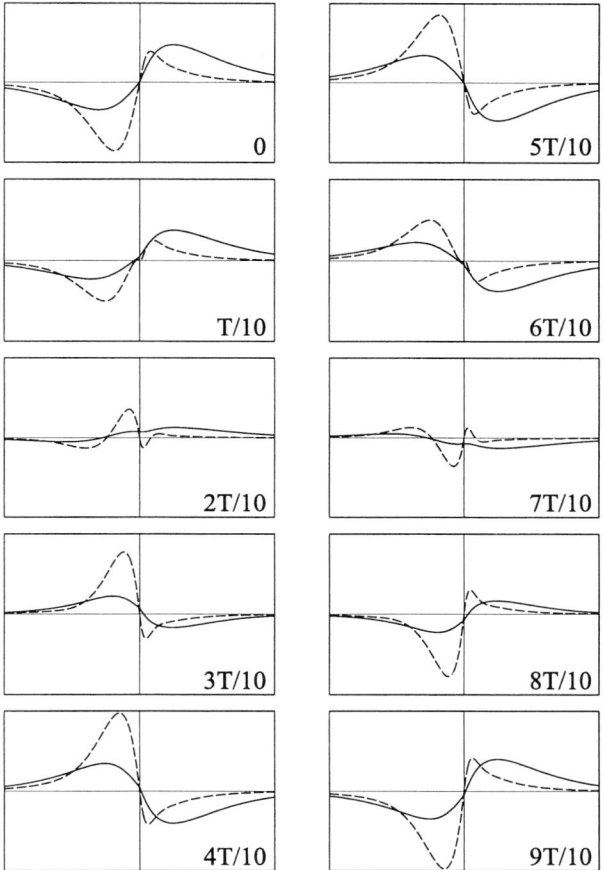

Figure 3.27: Critical (overstable) mode for $\nu = \mu = 0.1$, $\kappa = 0.4$, $\lambda = 0.5$, $Pr = 0.01$. Critical parameters are $Ma = 245.2$ and $\omega = 0.0754$. The full curve is the temperature fluctuation $\mathrm{Re}[\theta(z)\exp(i\omega t)]$, while the dashed one is the vertical velocity $\mathrm{Re}[w(z)\exp(i\omega t)]$, represented as a function of z, and for 10 different times spanning the whole period T.

nonlinear regimes (after all transients have been damped). In a two-dimensional situation, the only constant amplitude solutions are either a pure traveling wave solution (i.e. $A_L = 0$ and $A_R \neq 0$ in Eq. (3.164), or $A_R = 0$ and $A_L \neq 0$, leading respectively to $T_s \sim \cos[kx - \omega t]$ and $T_s \sim \cos[kx + \omega t]$), or a standing wave (i.e. $|A_L| = |A_R| \neq 0$, leading to $T_s \sim \cos[kx]\cos[\omega t]$). Still, nothing can be said about the nonlinear stability of these two solutions without taking nonlinear terms into account. It may also happen that both these solutions are unstable, eventually leading to oscillations of large amplitude, or even more complicated aperiodic or chaotic behaviors.

As a conclusion, the heuristic arguments given above (valid at the linear stage) merely indicate that there is an oscillatory coupling of temperature and velocity fluctuations around the diffusive reference state. The corresponding convective regimes observed above the threshold of instability may only be investigated in the nonlinear domain.

3.5.2   Finite depth effects – General case

We now turn to an analysis of the full system of equations and boundary conditions (3.141–3.148), though always in the case $Ra_1 = 0$. Only the essential steps of the mathematical analysis are presented here, referring to Smith [218] for the neutral stability case, Reichenbach and Linde [219] (who also considered weak surface deformation) and the book of Simanovsky and Nepomnyashchy [97] for further details. In view of the large number of independent parameters characterizing the system, it appears impossible here to provide an exhaustive discussion of all possible cases. Rather, our approach in this section will be to determine classes of systems (distinguished by their values of ν, κ, but now also the depth ratio a) presenting qualitatively similar stability diagrams in the plane (Ma, k).

Neutral stability

Proceeding along the same lines as for the infinite-depth case, we first solve Eqs (3.141) and (3.143) with $\sigma = 0$ for the vertical velocities in both layers, taking into account all boundary conditions but the Marangoni stress condition at $z = 0$. Then, velocities in both layers are determined up to a constant factor which can be set to unity (normalization condition). The result reads

$$w_1(z) = b_1 \sinh[kz] + z\left(\cosh[kz] + d_1 \sinh[kz]\right) \tag{3.165}$$

$$w_2(z) = b_2 \sinh[kz] + z\left(c_2 \cosh[kz] + d_2 \sinh[kz]\right) \tag{3.166}$$

with $b_1 = -k\operatorname{cosech}^2[k]$, $b_2 = -a^2 k \operatorname{cosech}^2[ak]c_2$, $d_1 = F_1(k)/2$, $d_2 = -c_2 F_1(ak)/2$ and $c_2 = F_2(k)/F_2(ak)$. For simplicity, we have defined $F_1(\xi) = \operatorname{cosech}^2[\xi](\sinh[2\xi] - 2\xi)$ and $F_2(\xi) = \xi^2 \operatorname{cosech}^2[\xi] - 1$. It may then be shown that the tangential viscous stress $\left[\mu D^2 w_2 - D^2 w_1\right]_{z=0} < 0$ for every $k > 0$, such that the numerator of Ma given by the thermocapillary stress boundary condition Eq. (3.148) never vanishes.

We now compute the denominator of Ma (surface tension gradient). Solving Eqs (3.142) and (3.144) with (3.165–3.166), we obtain

$$\begin{aligned}\theta_1(z) =\;&p_1 \cosh[kz] + p_2 \sinh[kz]\\ &+ \tfrac{z}{4k^2}\left[\cosh[kz](1 - 2b_1 k - d_1 kz) + \sinh[kz](d_1 - kz)\right]\end{aligned} \tag{3.167}$$

$$\begin{aligned}\theta_2(z) =\;&q_1 \cosh[kz] + q_2 \sinh[kz]\\ &+ \tfrac{\beta}{\kappa}\tfrac{z}{4k^2}\left[\cosh[kz](c_2 - 2b_2 k - d_2 kz) + \sinh[kz](d_2 - c_2 kz)\right]\end{aligned} \tag{3.168}$$

Then, expressing the boundary conditions at $z = -1$, $z = a$ and at the interface, we obtain the integration constants p_1, q_1, p_2 and q_2. Expressions of p_2 and q_2 are too bulky to be reproduced here, but the expression of $p_1 = q_1$ (i.e. the temperature at $z = 0$) simplifies to

$$\theta_1(0) = \frac{F_3(k) - c_2\kappa^{-1}F_3(ak)}{4k^3(\lambda\coth[ak] + \coth[k])} \tag{3.169}$$

with $F_3(\xi) = \xi^3\coth[\xi]\mathrm{cosech}^2[\xi] - 1$. Thus, using the Marangoni condition (3.148) leads to the critical Marangoni number. After some manipulations, the latter can be put under the simple form

$$Ma = P(k)\left(1 + \lambda\frac{\tanh[k]}{\tanh[ak]}\right)\left(1 + \mu\frac{G_2(k)}{G_2(ak)}\right)\left(1 - \kappa^{-1}\frac{G_1(k)}{G_1(ak)}\right)^{--} \tag{3.170}$$

where

$$G_1(\xi) = \frac{F_2(\xi)}{F_3(\xi)} = \frac{\xi^2\mathrm{cosech}^2[\xi] - 1}{\xi^3\coth[\xi]\mathrm{cosech}^2[\xi] - 1} \tag{3.171}$$

$$G_2(\xi) = \frac{F_2(\xi)}{F_1(\xi)} = \frac{\sinh^2[\xi] - \xi^2}{2\xi - \sinh[2\xi]} \tag{3.172}$$

and

$$P(k) = -4k^2\coth[k]\frac{F_1(k)}{F_3(k)} = \frac{4k^2\coth[k]\mathrm{cosech}^2[k](\sinh[2k] - 2k)}{1 - k^3\coth[k]\mathrm{cosech}^2[k]} \tag{3.173}$$

Note that $P(k)$ is equal to the Pearson's result Eq. (3.118) with $Bi = 0$ (zero free surface Biot number). Thus it appears that the critical Marangoni number for the full two-layer problem is given by the Pearson's neutral stability curve for the one-layer problem with zero Biot number, multiplied by three factors, each of them depending on one of the ratios κ, λ and μ (in addition to the depth ratio a and the wavenumber k). From the expression (3.170), we obtain the following interesting results:

A. For $\mu \to 0$, $\lambda \to 0$ and $\kappa \to \infty$, we obtain the zero Biot number Pearson's result $Ma = P(k)$. This was also the case for the infinite-depth problem (see §3.5.1). These limits should thus be understood as conditions under which the upper phase may be considered as an inert insulating gas (one-sided model).

B. For $\mu \to 0$ and $\kappa \to \infty$, we obtain $Ma = P(k)(1 + \lambda\tanh[k]/\tanh[ak])$. Interestingly, this formula corresponds to the Pearson's result (3.118) when the Biot number is set to $Bi = \lambda k\coth[ak]$. This is precisely the result (3.76), obtained by solving the stationary heat equation in the gas phase with a free surface temperature spatially periodic with wavenumber k. Thus it is proved that the one-layer approach in terms of a Biot number depending on k is exact when dynamic and thermal relaxation effects may be neglected in the gas phase. For very thin gas layers or for long-wave modes (in fact in the case $a \ll k^{-1}$), the one-layer approach also proves to be correct with the constant value $Bi = \lambda/a$.

C. When $a = 1$ (equal depths for both layers), Eq. (3.170) simplifies to $Ma = P(k)(1 + \lambda)(1 + \mu)/(1 - \kappa^{-1})$, i.e. the zero Biot number Pearson's result is multiplied by a factor independent of the wavenumber k. As a consequence, the critical conditions are always given by $k_c = 1.99$ and $Ma_c = 79.6(1+\lambda)(1+\mu)/(1-\kappa^{-1})$ (see §3.3.2). In particular, we recover the conclusion of the infinite-depth case, that monotonic instability should occur when heating is from the side with the lower thermal diffusivity.

D. For $\xi \to 0$, $G_1(\xi) \to 5/\xi^2 + O(1)$ and $G_2(\xi) \to -\xi/4 + O(\xi^2)$. As $P(k) \to 80/k^2 + O(1)$ for $k \to 0$, we obtain the critical Marangoni number in the long-wave limit as

$$Ma \to \frac{80}{k^2} \left(1 + \frac{\lambda}{a}\right) \left(1 + \frac{\mu}{a}\right) \left(1 - \frac{a^2}{\kappa}\right)^{-1} \tag{3.174}$$

The sign of this expression is thus determined by the value of a^2/κ, which represents the ratio of thermal relaxation times in both layers. Thus for long-wavelength monotonic modes, destabilization occurs when heating from the side with the longer thermal relaxation time. In contrast, for short-wavelength modes ($k \to \infty$), which are insensitive to the actual fluid depths, Eq. (3.152) predicts monotonic instability when heating from the side with the lower thermal diffusivity. As a consequence, there exist situations where the two-layer problem is monotonically unstable for both directions of heating. For example, if $\kappa > 1$ but $a^2/\kappa > 1$, short-wave modes are destabilized when heating is from below, while long-wave modes are destabilized when heating is from above. As the critical Marangoni number (3.170) cannot vanish, the only possibility is that an asymptote exists at a certain wavenumber k^* (at which the temperature perturbation at the interface vanishes). From Eq. (3.169) or Eq. (3.170), we thus get

$$\kappa G_1(ak^*) = G_1(k^*) \tag{3.175}$$

Some examples of neutral stability curves displaying asymptotes will be presented in the next sections, combining them with results obtained for overstable modes.

Wave modes

The full dispersion relation ($\sigma \neq 0$) may be obtained for the problem (3.141–3.148), but is too cumbersome to be presented here. We will thus rely on a numerical analysis, combining results obtained for the short-wavelength modes, with heuristic arguments based on the experience gained when studying neutral stability.

For neutral stability, it was seen (point D of the last section) that the conditions on the direction of heating obtained for $k \to \infty$ can be transposed to the case $k \to 0$, by simply replacing κ with κ/a^2. The main hypothesis underlying the following discussion is that this is valid for overstability too, i.e. the conditions for overstability obtained for $k \to \infty$ and formulated in terms of κ and ν have to be replaced by the same conditions in terms of κ/a^2 and ν/a^2. This is clearly only a conjecture based on the intuitive idea that the vertical scale of long-wave modes should be associated with the depth, and not with the wavelength. It should thus be cross-checked on several examples. Even if the analysis of given fluid systems will always require a case-by-case analysis, the primary classification presented hereafter will certainly help to get a first reasonable picture of the stability diagram, and thereby also to make a selection of liquids as a function of the desired behavior.

We thus directly refer to the classification provided in Fig. 3.28, in relation to the following discussion. Without loss of generality, we assume $\kappa > 1$ (the case $\kappa < 1$ follows exactly the same lines). This means that for short waves ($k \to \infty$), neutral stability will always occur when heating is from below (positive Marangoni number). Now, we must distinguish two possible cases, according to whether $\nu < 1$ or $\nu > 1$. In the first situation, there are probably no overstable modes in the short-wave limit (see §3.5.1), except in some special cases (e.g. low Prandtl number fluids). On the contrary, when $\nu > 1$, these modes certainly exist, whatever the Prandtl number, and we may draw a parabolic branch of short-wave overstable modes for negative Marangoni number in all the diagrams below the branch $\nu > 1$ of Fig. 3.28.

We now discuss the long-wave modes, for which the depth ratio a is a crucial parameter. For $a^2 < \nu < 1 < \kappa$, long-wave monotonic modes should appear for $Ma > 0$, and the neutral stability curve thus presents no asymptote ($Ma > 0$ for all k). Oscillatory modes should appear in the long-wave limit $k \to 0$, because a^2/ν and a^2/κ are both inferior to one (here we make use of the conjecture mentioned above). If the mechanism of these oscillations is similar to that discussed in §3.5.1 for short-wave modes, it is likely that they will appear when heating is from above ($Ma < 0$). However, the branch of overstability cannot extend up to $k \to \infty$ because oscillatory modes are not expected in this limit. The most reasonable possibility is a vertical asymptote at some finite k, as represented in Fig. 3.28(a). We now discuss the case $\nu < a^2 < \kappa$. Here, the neutral stability curve remains positive for all k and no oscillatory modes are expected, even for $k \to 0$ because $a^2/\nu > 1$ and $a^2/\kappa < 1$. This results in the second stability diagram of the case $\nu < 1$, i.e. Fig. 3.28(b). The last situation for this case is $\nu < 1 < \kappa < a^2$ [Fig. 3.28(c)], for which an important difference is the asymptote occurring in the neutral stability curve (indeed, monotonic long-wave modes now occur for $Ma < 0$ as $a^2/\kappa > 1$). As we now have $a^2/\nu > 1$ and $a^2/\kappa > 1$, long-wave overstable modes should appear for $Ma > 0$. Again, there is no prolongation to $k \to \infty$ so that a possible behavior would be an asymptote in the oscillatory branch at a certain k. Another possibility [see Fig. 3.28(c)] is a branching of the overstability branch on the neutral stability curve. At the branching point, the frequency of the oscillation vanishes. This completes the discussion of the case $\nu < 1$.

The discussion of the second situation $\nu > 1$ (still with $\kappa > 1$) proceeds along the same lines and will not be detailed here. The resulting stability diagrams are presented in Fig. 3.28(d-h). Remark the cases (f, h), corresponding to $a^2 > \nu, \kappa$, where there is an asymptote in the neutral stability curve, short-wave oscillatory modes for $Ma < 0$ and long-wave oscillatory modes for $Ma > 0$. These oscillatory modes may either branch on the neutral stability curve, or present an asymptote at finite k, depending on whether $\nu > \kappa$ or $\nu < \kappa$. The choice adopted (either for branching or for asymptote) in Fig. 3.28 in fact relies on results obtained in another study [49] of the behavior of eigenvalues at large Marangoni number, and which indicates that an asymptote may exist in the overstability branch at a value of k determined by the equation $\nu^{1/2} \tanh[k_o^*] = \tanh[ak_o^*]$. A solution for k_o^* exists either if $a^2 < \nu$ and $\nu < 1$ or if $a^2 > \nu$ and $\nu > 1$. The corresponding sign of the Marangoni number is positive if $\kappa < \nu$,

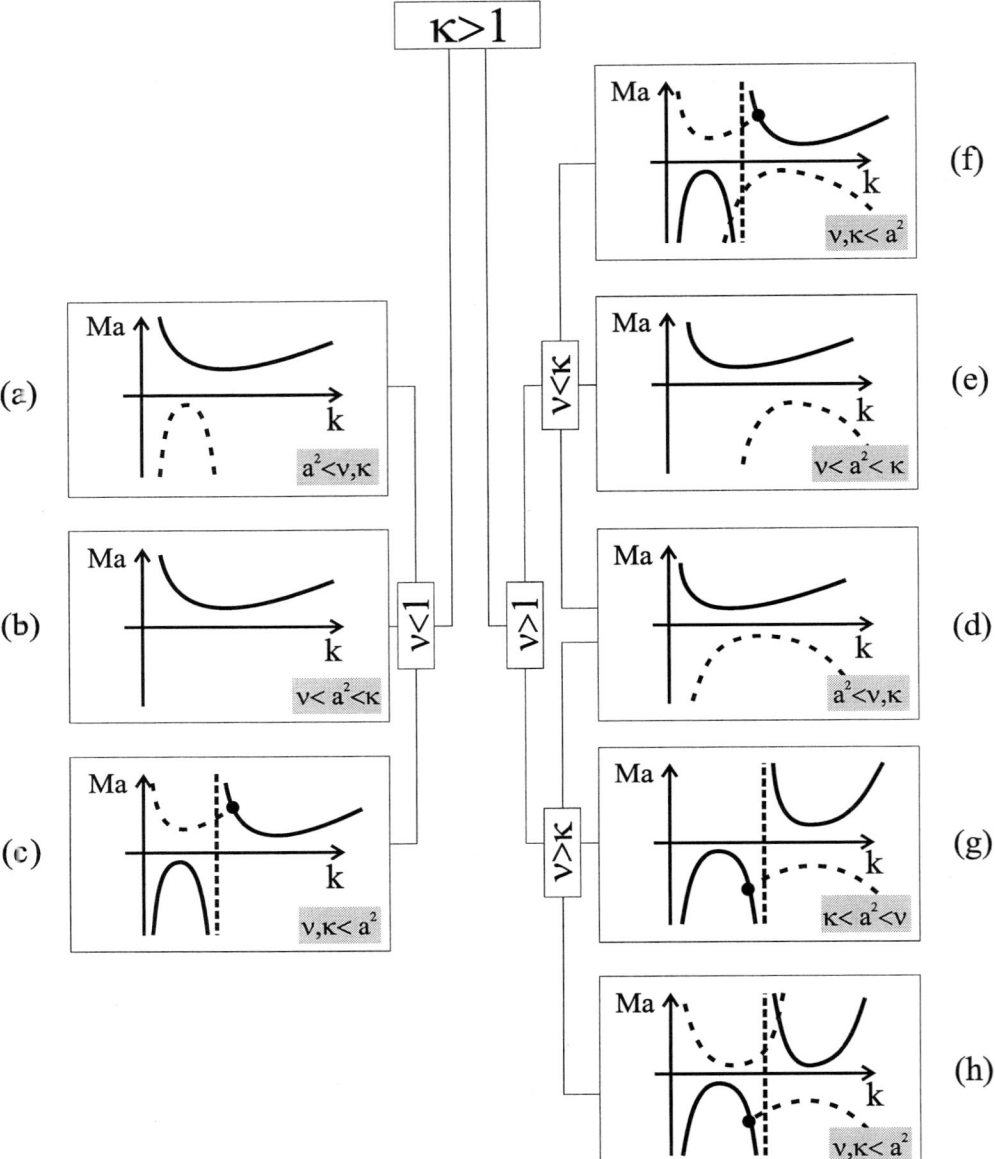

Figure 3.28: Classification of stability diagrams as a function of the heat (or mass) diffusivity ratio κ, the kinematic viscosity ratio ν and the depth ratio a. Neutral stability curves are represented as full curves, while overstability appears as dashed. The behavior of the overstable modes at moderate and low wavenumber k is conjectural. Note that for $\kappa < 1$, this diagram may be applied by changing $<$ to $>$ and vice versa in every inequality, and changing the sign of the Marangoni number.

negative if $\kappa > \nu$.

We proceed now to the full numerical determination of the stability diagrams for several liquid/liquid and liquid/gas systems, illustrating different cases discussed in Fig. 3.28.

A. Water (lower phase) – Silicone oil 10 cSt (upper phase), $a = 4$: this corresponds to parameters $\rho = 0.939$, $\nu = 10.2$, $\mu = 9.54$, $\kappa = 0.655$, $\lambda = 0.222$, $Pr_1 = 5.85$. Here $\kappa < 1$, $\nu > 1$, and $a^2 > \nu, \kappa$. Note that redefining layer 1 as layer 2 and vice versa, this is equivalent to the case $\kappa > 1$, $\nu < 1$, and $a^2 < \nu, \kappa$, i.e. Fig. 3.28(a) where the sign of the Marangoni number should also be changed. Indeed, Fig. 3.28(a) is then in agreement with the exact stability diagram represented in Fig. 3.29.

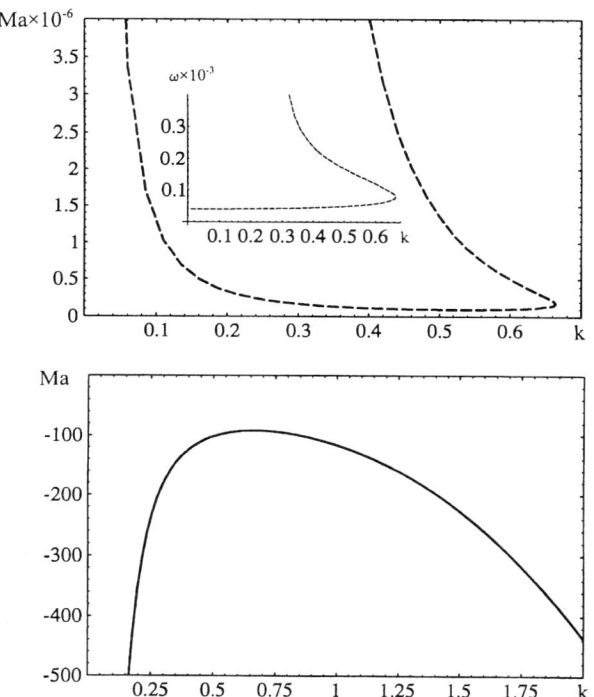

Figure 3.29: Stability diagrams for bottom heating (above) and top heating (below) for the water–silicone oil 10 cSt system ($\rho = 0.939$, $\nu = 10.2$, $\mu = 9.54$, $\kappa = 0.655$, $\lambda = 0.222$, $Pr_1 = 5.85$) with $a = 4$. Full curve: neutral stability, dashed curve: overstability. The inset of the above plot represents the frequency for marginal oscillatory states.

B. Silicone oil 50 cSt (lower phase) – Air (upper phase), $a = 1$: the parameters are $\rho = 0.00122$, $\nu = 0.314$, $\mu = 3.87 \, 10^{-4}$, $\kappa = 180$, $\lambda = 0.146$, $Pr_1 = 405$. This is a liquid/gas case, for which $\kappa > 1$, $\nu < 1$ and $\kappa > a^2 > \nu$. We are thus in the situation of Fig. 3.28(b), as confirmed by the stability diagram presented in Fig. 3.30. Note that for a lower viscosity silicone oil (e.g. 10 cSt), we would have $\nu > 1$ and there should be oscillatory modes when heating from above in the whole range of wavenumbers (however, these modes become unstable only at unrealistically large values of Ma). In Fig. 3.30 we also represented the neutral stability curve for the one-layer system [Pearson's result Eq. (3.118)] with $Bi = 0$ and $Bi = \lambda/a = 0.146$. As commented above (point B), the latter value provides good agreement in the region of long waves. Using $Bi = \lambda k \coth[ak]$ leads to excellent agreement for all k.

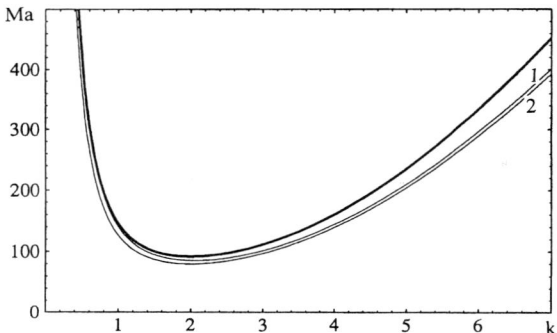

Figure 3.30: Stability diagram for the silicone oil 50 cSt – air system ($\rho = 1.22\,10^{-3}$, $\nu = 0.314$, $\mu = 3.87\,10^{-4}$, $\kappa = 180$, $\lambda = 0.146$, $Pr_1 = 405$). The exact neutral stability curve is represented as a thick line. The thinner lines represent Pearson's result with $Bi = 0$ (curve 2) and with $Bi = \lambda/a = 0.146$ (curve 1, with good agreement in the limit $k \to 0$). Excellent agreement for all k is obtained when using $Bi = \lambda k \coth[ak]$.

C. Fluorinert FC 70 – Silicone oil 20 cSt, variable a : Here, $\rho = 0.485$, $\nu = 1.43$, $\mu = 0.689$, $\kappa = 3.16$, $\lambda = 2.11$, $Pr_1 = 395$. For $a = 1$, the stability diagram is given in Fig. 3.31(a), and corresponds to Fig. 3.28(d), i.e. for $a^2 < \nu, \kappa$. For increasing thickness of the oil layer, at $a = 1.5$ we have $\nu < a^2 < \kappa$, and no long-wave oscillatory modes are expected [according to our conjecture, see also Fig. 3.28(e)]. This is indeed apparent from the stability diagram for this case in Fig. 3.31(b), where it is seen that a vertical asymptote appeared from $k = 0$ in the overstability branch at $Ma < 0$. Note in particular the different behavior of the frequency for Fig. 3.31(b) and Fig. 3.31(a). This asymptote displaces to the right with increasing a, and for $a = 4$ [Fig. 3.31(c)], an asymptote also exists in the neutral stability branch, so that long-wave monotonic modes now also occur when $Ma < 0$. Furthermore, long-wave oscillatory modes also exist when $Ma > 0$, and the global stability diagram indeed corresponds to Fig. 3.28(f). Note however the generally much larger critical values of Ma needed for overstability.

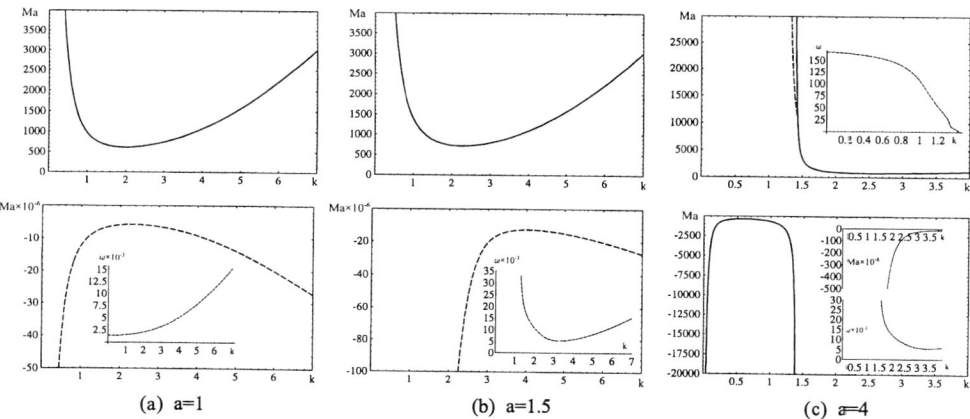

Figure 3.31: The Fluorinert FC 70 - Silicone oil 20 cSt stability diagrams: $\rho = 0.485$, $\nu = 1.43$, $\mu = 0.689$, $\kappa = 3.16$, $\lambda = 2.11$, $Pr_1 = 395$. (a): $a = 1$, (b): $a = 1.5$, (c): $a = 4$. Dashed curves: overstable modes, full curves: neutrally stable modes.

D. Water – n-octane system, variable a: values of parameters are $\rho = 0.705$, $\nu = 0.813$, $\mu = 0.573$, $\kappa = 0.703$, $\lambda = 0.248$, $Pr_1 = 5.85$. This system is qualitatively similar to the previous one, although with a reversed direction of heating. Furthermore, the value of the Prandtl number is smaller, thus allowing more reasonable values for thresholds of oscillatory modes. The resulting diagrams are presented in Fig. 3.32, for cases $a = 1$ and $a = 0.8$, corresponding respectively to Figs 3.28(d) and 3.28(f).

3.6 Influence of surface deformation

In all the previous sections, the deformation of the free surface (or the interface) with respect to a plane situation was neglected. In reality, it is clear that deformations will occur, as a result of hydrodynamic stresses caused by fluid motions. The effects opposing deformation are surface tension and gravity, and there indeed exist conditions under which surface deformation is a second-order effect and can be safely neglected. In this section however, our purpose will be to determine the conditions under which surface deformation plays an essential role, and qualitatively affects the behavior of the Bénard set-up.

3.6.1 One-layer systems

We will first briefly review the case where the liquid layer is heated from below [86, 233, 234]. A neutral stability condition generalizing that obtained by Pearson is easily obtained. This will indicate the negligible role of surface deformation except for very thin layers of viscous liquids, where surface deformation is important for long-wave modes. Then, the opposite case where the layer is heated from above will be considered in more detail. Indeed, several authors

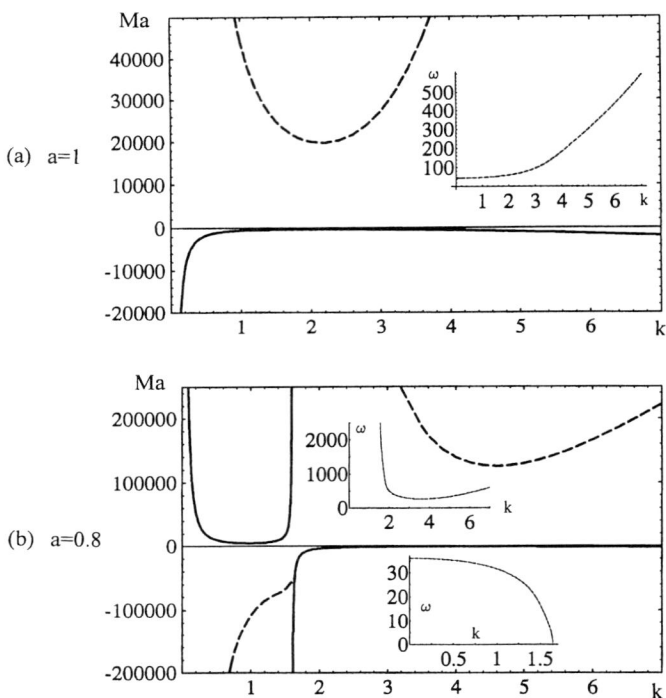

Figure 3.32: The water – n-octane system: values of parameters are $\rho = 0.705$, $\nu = 0.813$, $\mu = 0.573$, $\kappa = 0.703$, $\lambda = 0.248$, $Pr_1 = 5.85$. (a) $a = 1$, (b) $a = 0.8$.

[48, 231, 235, 236, 240] have shown that wave modes may be triggered in this situation, by the combined action of surface deformation and the Marangoni effect. An asymptotic analysis similar to that presented in §3.4 for internal and surface waves will be developed, which will help to clarify the physical nature of these wave modes.

Neutral stability

The system of equations to be considered here has already been established in §3.3.1. As the normal momentum balance boundary condition will now be needed, and since it involves the pressure in the liquid, we will use the general formulation (3.95–3.97), although we here neglect buoyancy effects[12]. Thus, we rewrite the system of equations for the z-dependencies of normal modes (keeping the usual notations) as

$$iku + Dw = 0 \tag{3.176}$$

---

[12]As discussed in §3.1.4 in relation with the Boussinesq approximation, care has to be taken when considering the buoyancy effect in the presence of surface deformation.

$$(D^2 - k^2)w - Dp = Pr^{-1}\sigma w \tag{3.177}$$

$$(D^2 - k^2)u - ikp = Pr^{-1}\sigma u \tag{3.178}$$

$$(D^2 - k^2)\theta + w = \sigma\theta \tag{3.179}$$

where u, w, p and θ are respectively (amplitudes of) perturbations of horizontal velocity, vertical velocity, pressure and temperature.

At the bottom plate $z = 0$, we impose the usual no-slip isothermal boundary conditions

$$w = Dw = \theta = 0 \tag{3.180}$$

The derivation of linearized boundary conditions at the interface $z = 1$ is classical [48, 86, 218, 233, 234, 235, 236], and will not be detailed here. Referring to §3.8 for a more general derivation including evaporation, we here directly state the result, written for normal modes. The amplitude of the surface deformation (in units of the liquid depth) will be denoted by ξ. The linearized kinematic condition [see Eq. (2.78)] reads

$$w = \sigma\xi \tag{3.181}$$

The normal stress balance (already discussed in §3.1.4 in connection with the Boussinesq approximation) is written as

$$-p + 2Dw = -k^2 Ca^{-1}\xi - Ga\xi \tag{3.182}$$

where the capillary number Ca and the Galileo number Ga have been defined earlier [Eqs (3.55) and (3.56)]. The tangential stress boundary condition is obtained as

$$Du + ikw + ikMa(\theta - \xi) = 0 \tag{3.183}$$

Finally, the energy balance at the interface reads

$$D\theta + Bi(\theta - \xi) = 0 \tag{3.184}$$

We first derive the neutral stability condition ($\sigma = 0$) for the closed system (3.176–3.184), proceeding in the usual way. From Eqs (3.176–3.178), it is seen that $(D^2 - k^2)p = 0$, whose general solution is

$$p = a\cosh[kz] + b\sinh[kz] \tag{3.185}$$

We then solve Eq. (3.177) for w, Eq. (3.176) for u, and finally Eq. (3.179) for θ. The results read

$$w = (c + \frac{a}{2}z)\cosh[kz] + (d + \frac{b}{2}z)\sinh[kz] \tag{3.186}$$

$$u = \frac{i}{k}\left[\left(\frac{a}{2} + k(d + \frac{b}{2}z)\right)\cosh[kz] + \left(\frac{b}{2} + k(c + \frac{a}{2}z)\right)\sinh[kz]\right] \qquad (3.187)$$

$$\theta = \left(r + \frac{a - 4dk}{8k^2}z - \frac{b}{8k}z^2\right)\cosh[kz] + \left(q + \frac{b - 4ck}{8k^2}z - \frac{a}{8k}z^2\right)\sinh[kz] \qquad (3.188)$$

and we are left with 6 unknowns a, b, c, d, r and q to be calculated from the boundary conditions (3.180–3.184). Actually, one of them will serve to calculate the surface deformation ξ, which is also unknown. We adopt $d = 1$ as the normalization condition. From Eq. (3.180) valid at $z = 0$, we get $c = r = 0$ and $a = -2k$. Then, from Eq. (3.181) at $z = 1$, we obtain $b = 2(k\coth[k] - 1)$, and from Eq. (3.182), we get the amplitude of the surface deformation

$$\xi = -\frac{2k^2\text{cosech}[k]}{Ga + k^2Ca^{-1}} \qquad (3.189)$$

The expression for q obtained from (3.184) is not reproduced here. Instead we directly state the compatibility condition obtained by solving Eq. (3.183) with respect to Ma. This reads

$$Ma = \frac{16k(k\cosh[k] + Bi\sinh[k])(2k - \sinh[2k])}{4k^3\cosh[k] + 3\sinh[k] - \sinh[3k] - 32k^5\cosh[k](Ga + k^2Ca^{-1})^{-1}} \qquad (3.190)$$

which reduces to the Pearson's relation (3.118) in the limit $Ga \to \infty$, $Ca \to 0$. This limit corresponds to the case of an undeformable interface, as also seen from Eq. (3.189). The combination $Ga + k^2Ca^{-1}$ indicates that the effect of gravity (measured by the Galileo number $Ga = gd^3/\nu\kappa$) is dominant at large scales ($k \to 0$), while that of surface tension (measured by the capillary number $Ca = \mu\kappa/\sigma d$) is stronger at smaller scales. For increasing wavenumber k, the effect of surface deformation quickly becomes negligible, as indicated by the fact that the asymptotic behavior of the threshold Marangoni number for $k \to \infty$ and $Ga\,Ca^{-1} = O(1)$ remains $Ma \to 8k(Bi+k)$, as for the Pearson's analysis in the short-wave limit (see §3.3.2).

On the contrary, the long-wave asymptotic behavior of the neutral stability curve is strongly determined by surface deformation effects. Indeed, it may be computed that the limit at $k \to 0$ is $Ma_0 = \frac{2}{3}(1 + Bi)Ga$, thus strongly differing from the behavior $80(1 + Bi)/k^2$ obtained in the undeformable case. Taking the next order term ($\sim k^2$) into account, we in fact have the Taylor expansion

$$Ma \simeq \frac{2}{3}(1 + Bi)Ga + k^2\left[\frac{2}{3}(1 + Bi)Ca^{-1} + \frac{Ga}{180}(24 - 16Bi - Ga - BiGa)\right] \qquad (3.191)$$

which is the result found by Regnier and Lebon [254]. For normal experimental situations, the Galileo number is very large (e.g. for a 1 mm deep water layer, $Ga \sim 10^5$), while the capillary number is small ($Ca \sim 10^{-6}$ in the same situation). Thus the effects of interface deformation are typically negligible. Some neutral stability curves are represented in Fig. 3.33 for $Bi = 0$.

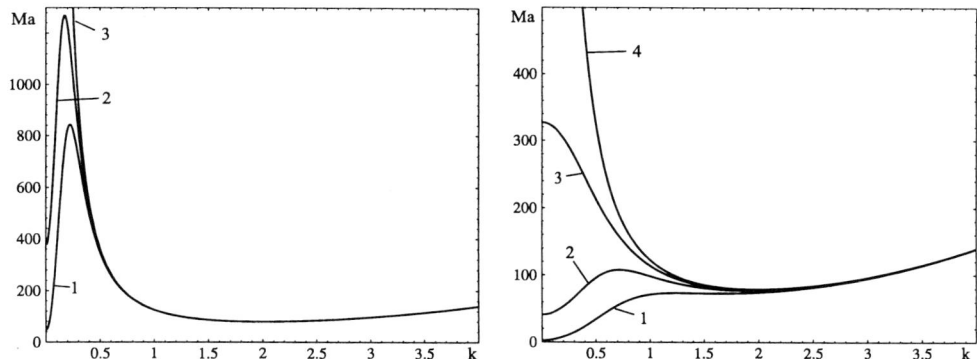

Figure 3.33: Neutral stability curves in the deformable free surface case, for several liquid depths. Left (water): (1) $d = 0.1$ mm, (2) $d = 0.2$ mm, (3) $d = 1$ mm. Right (silicone oil 200 cSt): (1) $d = 0.2$ mm, (2) $d = 0.5$ mm, (3) $d = 1$ mm, (4) $d = 2$ mm.

For very thin layers of highly viscous liquids however, long-wave deformational modes may become dangerous, and a fortiori the most dangerous ones when the long-wave threshold $Ma_0 = \frac{2}{3}Ga$ (for $Bi = 0$) becomes lower than the usual threshold $Ma \simeq 80$ at $k \simeq 2$ for cellular convection. While for water this corresponds to unusually thin liquid films ($d \sim 100$ μm or less), this becomes experimentally accessible with highly viscous liquids such as silicone oils [e.g. for a 200 cSt oil, $Ga < 120$ (i.e. $Ma_0 < 80$ if $Bi = 0$) for $d < 0.6$ mm]. Recent experimental results of VanHook et al. [35, 36] have indicated that the cross-over between the two modes may indeed be observed in ground-based experiments. VanHook et al. have further observed and explained that the long-wave deformational mode generally leads to film rupture (see also [237]). Moreover, an interesting situation occurs when the thresholds of long-wave deformational modes and cellular modes $k \simeq 2$ nearly coincide. In this case, it is possible to derive a system of nonlinear evolution equations taking into account their coupling, resulting in highly complex behavior [238, 239]. These situations are examined in §4.5.3 and §4.5.4.

Overstability – Transverse and longitudinal waves

It was seen in the previous section that the surface deformation has negligible influence when gravity and surface tension effects are strong enough to maintain the interface practically flat, whatever flows and thermal inhomogeneities may exist. This situation corresponds to a high Galileo number and a small capillary number. On the other hand, it is known that in such conditions, a high-frequency transverse mode exists (capillary-gravity waves) which is only slightly damped (i.e. the frequency is much higher than the damping rate). These are the propagating waves commonly observed at the open surface of a liquid, after the latter has been perturbed externally. As shown in this section, such waves can be sustained (and amplified) by the Marangoni effect, when the liquid layer is heated from above. This has been the subject of numerous studies [48, 231, 235, 236, 240, 241]. Note that a second high-frequency wave mode exists when the liquid layer is strongly heated from above, as seen in §3.4. As opposed to the *transverse* capillary-gravity waves which show essential fluid motions *perpendicular* to the interface, this second mode is characterized by intense velocities *along* the interface. Hence, these waves may be called *longitudinal* or *dilational* surface waves.

The first description of dilational waves was proposed by Lucassen [242], who considered a different system, namely a free surface covered by a surfactant[13]. He also discussed the resonance between transverse and longitudinal waves (mode-mixing). Later, Earnshaw [243] (and earlier papers cited therein) also considered this aspect, both theoretically and experimentally, though in a situation where the waves were not amplified by a thermal instability, but rather excited by external means and eventually damped by viscosity.

In fact, the problem studied in this section shares many important features with that considered in §3.4. Here also, instability will be shown to result from the interaction and mode-mixing between two types of high-frequency wave modes (here, capillary-gravity waves and dilational waves, while in §3.4, the modes involved were internal and dilational waves). Many asymptotic features turn out to be similar, although with some important differences, as will become apparent later on.

Numerical analysis of the full dispersion relation

The system of equations and boundary conditions is the same as before [Eq. (3.176–3.184)]. First, the general dispersion relation is solved numerically and used to determine the threshold of marginal oscillatory modes. The numerical method will not be detailed here, as it is similar to that used in §3.4.1. Then, the main results of asymptotic developments at large Galileo and inverse capillary numbers are presented, referring to Rednikov et al. [244] for the complete analysis.

The stability diagrams will be presented in terms of the modified Marangoni number

$$m = -\frac{Ma}{Ga} = \frac{\sigma_T \Delta T}{\rho g d^2} \tag{3.192}$$

which is positive when heating the layer from above (σ_T is negative for normal liquids and ΔT stands for the temperature difference between the bottom rigid surface and the free upper surface).

In this section, we always consider $Bi = 0$ (poorly-conducting upper surface). The Prandtl number is chosen as that of water ($Pr = 6$). In examining the limit of large Ga and Ca^{-1}, it will also be convenient to fix the value of the dynamic Bond number $Bo = GaCa = \rho g d^2/\sigma$, while allowing Ga to tend to infinity. Figure 3.34 presents some marginal curves for oscillatory instability in the plane (m, k). In each case, the inset represents the frequency ω along marginal states.

The instability region is inside the "bags" represented in Fig. 3.34. Each bag is limited in the direction of increasing k by some k^* (called the turning point in what follows). Qualitatively, it is seen that the wavenumber k^* at the turning point increases with Ga, and decreases with Bo. Asymptotic results for k^* are given later.

---

[13]Even though Lucassen longitudinal waves are strongly damped (their relaxation time is of the order of the period of oscillations), many of their features turn out to be similar to those studied in §3.4 and in the following (see also §6.3). In particular, dilational waves are genuinely dissipative. Indeed, being generated by the Marangoni effect, viscosity is essential in their mechanism and they cannot be observed in ideal liquids.

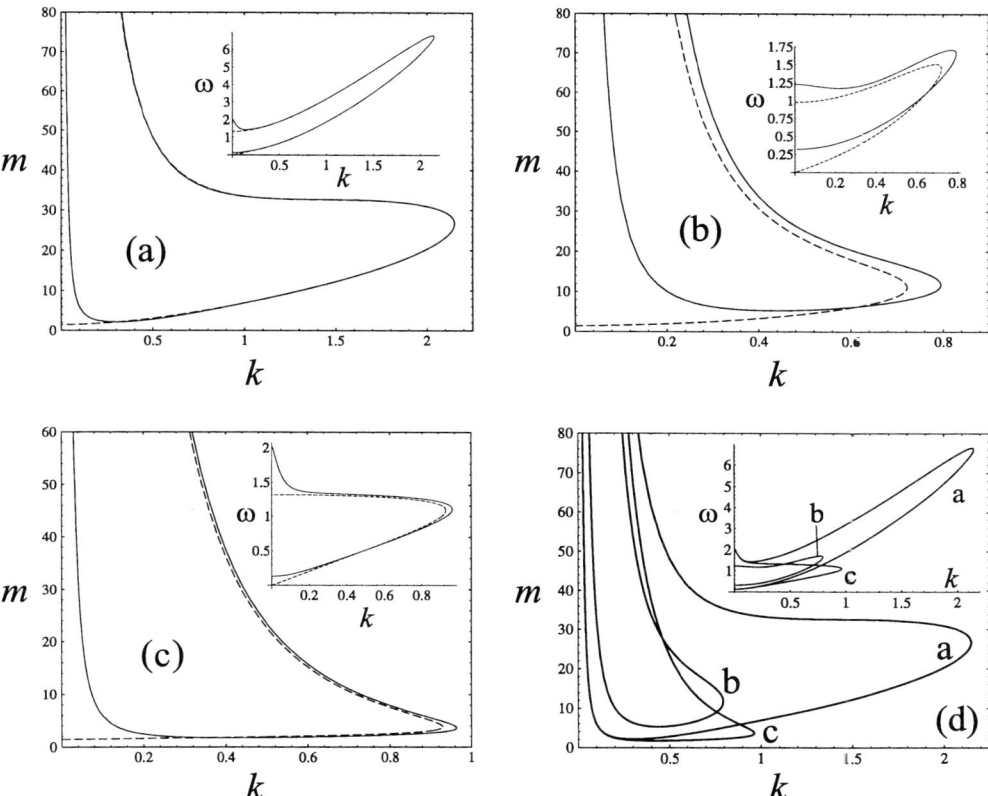

Figure 3.34: Marginal curves for oscillatory instability: modified Marangoni number m as a function of the wavenumber k. The inset gives the corresponding frequency. (a): $Bo = 0.333$, $Ga = 9155$, (b): $Bo = 0.333, Ga = 1536$, (c): $Bo = 25, Ga = 9155$. The full lines are exact numerical results. The dashed lines are asymptotic results (boundary-layer approximation). The graph (d) compares cases (a–c). In each case, instability occurs inside the "bag" formed by the marginal curve.

It is seen that the behavior in the region of long waves, i.e. for $k \rightarrow 0$, is not much influenced by the value of the Bond number [curves a and c in Fig. 3.34(d) coincide in this limit], but is much more sensitive to the value of Ga. As in the previous section, this is attributed to the fact that long waves are much more affected by gravity than by surface tension.

Figure 3.35 clarifies the behavior of the two most dangerous eigenvalues at fixed wavenumber k, while varying the modified Marangoni number m. Both real [Fig. 3.35(a)] and imaginary parts [Fig. 3.35(b)] of $\lambda = \sigma/(PrGa)^{1/2}$ [i.e. λ is the complex growth constant reduced by the gravity time scale $(d/g)^{1/2}$] are represented. One of the eigenvalues has a positive real part (indicating instability) within a certain interval of m, defined by the intersections of the bag of Fig. 3.34(a) with a vertical line at $k = 1$.

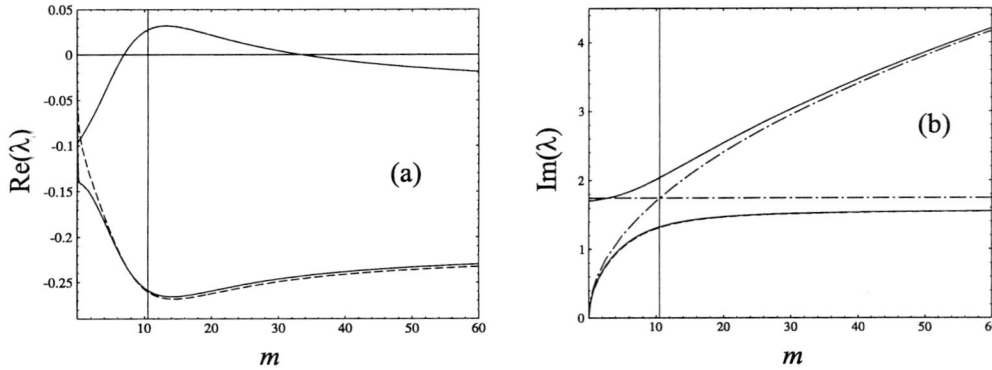

Figure 3.35: Real (a) and imaginary (b) parts of the eigenvalue λ as a function of the modified Marangoni number m, at a fixed wavenumber $k = 1$. $Pr = 6$, $Bo = 0.333$, $Ga = 9155$, corresponding to Fig. 3.34(a). Full lines: exact numerical results. Dashed lines: boundary-layer approximation. Dot-dashed lines: frequencies of the transverse and longitudinal waves in the leading-order approximation [Eqs (3.194) and (3.195)]. Resonance occurs at $m = m_{res} \simeq 10.5$ (vertical line), given by Eq. (3.196).

The numerical calculations also show that when Ga is increased, the upper branch of the marginal "bag" of Fig. 3.34 shifts to larger and larger values of m. The lower part does not vary significantly (in the scale of the modified Marangoni number m) for k of order unity. The turning point k^* rapidly shifts to the region of short waves, and the corresponding Marangoni number m^* at this point grows faster than the upper branch of the bag at $k = O(1)$, leading to the appearance of a minimum in the upper branch [almost visible in Fig. 3.34(a)]. To summarize, we may say that the bag appreciably grows upwards and to the right when increasing Ga.

Analytical results – Boundary-layer approximation

We can already point out some interesting similarities (Rednikov et al. [50, 244]) between the present problem and that studied in §3.4, namely the interaction between internal and longitudinal waves at an undeformable interface. Although differences exist in the shape of the instability domain (a "bubble" in §3.4, and a "bag" here), the important feature is that at given k of order unity, there exist two marginal stability thresholds (the upper and lower branches of the bubbles or bags). In both cases, two oscillatory modes exist (i.e. two pairs of complex conjugate eigenvalues), and referring to Fig. 3.35(b) we see that in the neighborhood of the (resonance) point where their frequencies almost coincide, a higher degree of amplification/damping is predicted (this was also the case in §3.4, although no graph was presented). Moreover, the value of $m = m_{res}$ at which resonance occurs is located in between the lower branch (hereafter denoted by m_{tr}) and the upper branch (denoted by m_{long}) of the instability region.

In the limit of large Galileo number Ga (and small capillary number $Ca = Bo/Ga$), there exists a transverse weakly damped mode (capillary-gravity waves). In fact, as seen in §3.2.1, the Galileo number appears as the ratio $Ga = \tau_{th}\tau_{visc}/\tau_{grav}^2$ of thermal and viscous time scales to the square of the gravity time scale $\tau_{grav} = (d/g)^{1/2}$ (note that we also have $Ca^{-1} = \tau_{th}\tau_{visc}/\tau_{cap}^2$, where $\tau_{cap} = (\rho d^3/\sigma)^{1/2}$ is a capillary time scale). For $Ga, Ca^{-1} \to \infty$, τ_{grav} and τ_{cap} both become much smaller than thermal and viscous time scales, and dissipative effects become negligible (many oscillations occur before significant damping is observed). Note that this discussion is also applicable to §3.4, under the conditions to replace τ_{grav} and τ_{cap} by the scales τ_{buoy} and τ_{Ma} of buoyancy and thermocapillary effects, respectively.

As in §3.4, we construct an asymptotic analysis by first rewriting the equations and boundary conditions in units of the (fast) gravity time scale τ_{grav} and associated velocity scale d/τ_{grav}. The resulting system is presented in [244], and hence only salient results are reproduced hereafter. Once again, the dissipative (Laplacian) terms in Navier–Stokes and energy equations are multiplied by ϵ^2, where $\epsilon = (Pr/Ga)^{1/4} \ll 1$ is the smallness parameter. Thus, dissipative effects are a priori negligible everywhere except near the upper and lower boundaries, where boundary layers of thickness of order ϵ exist, inside which higher-order vertical derivative terms cannot be neglected (the vertical coordinate z must be rescaled with ϵ). Note that thermal and viscous boundary layers have comparable thicknesses, because the Prandtl number Pr is assumed to be of order unity.

i) Leading-order asymptotics for $\epsilon \to 0$

We proceed in the usual way (see Nayfeh [230]), and solve the equations in each zone (the bulk and the top and bottom boundary layers) separately, using matching and boundary conditions to determine the unknown integration constants (the reader is referred to §6.3 for a detailed resolution of a very similar problem). Note that according to the continuity equation $Dw + iku = 0$, the asymptotic expansion for the horizontal velocity u in the free surface boundary layer (where $D \sim \epsilon^{-1}$) should in general be ϵ^{-1} times larger than the vertical velocity w.

The first result is that the leading-order problem admits a non-trivial solution if

$$(\lambda^2 + \omega_0^2)\left(1 + \frac{mk^2\lambda^{-2}}{Pr^{1/2} + 1}\right) = 0 \tag{3.193}$$

which is the lowest-order dispersion relation, allowing the determination of the complex growth rate λ (in units of τ_{grav}^{-1}) as a function of other parameters. In Eq. (3.193), ω_0 is the frequency of ideal capillary-gravity waves, i.e.

$$\omega_0^2 = k\left(1 + \frac{k^2}{Bo}\right)\tanh[k] \tag{3.194}$$

Hence, one of the modes is associated with capillary-gravity waves, while the other one (also purely oscillatory in the leading order) has a frequency found from

$$\lambda^2 = -\frac{mk^2}{Pr^{1/2} + 1} \tag{3.195}$$

which corresponds to Eq. (3.137) up to a change from gravitational to thermal time scales. The second mode is thus the longitudinal (dilational, and sound-like) wave mode already encountered in §3.4, and associated with a strong horizontal velocity along the free surface. In this respect, it is noteworthy that for transverse capillary-gravity waves, the horizontal and vertical velocities in the surface boundary layer are of the same order.

Note that the frequencies of both waves are identical (resonance) at

$$m_{\text{res}} = (Pr^{1/2} + 1)\left(1 + \frac{k^2}{Bo}\right)\frac{\tanh[k]}{k} \tag{3.196}$$

As in §3.4.2, it is necessary to proceed to higher order to determine the dissipative characteristics of the waves, i.e. we will seek a correction to the eigenvalue λ which has a non-vanishing real part.

ii) Next order corrections to the dispersion relation

It is worth noting that a difference with §3.4.2 will appear in what follows. In fact, while the marginal conditions for the transverse mode can be found at order ϵ, this is not the case for longitudinal waves. For the latter, we must proceed to order ϵ^2. Note that this situation is specific to the case considered here, as also shown by the analysis of the transverse and longitudinal waves in the two-layer case [49].

The dispersion relation up to order ϵ^2 reads

$$A_0 + \epsilon A_1 + \epsilon^2 A_2 = 0 \tag{3.197}$$

where A_0 is just the left-hand side of Eq. (3.193), while A_1 and A_2 are functions of k, λ, Pr, m and Bo not reproduced here for conciseness. Rather, results are directly stated for the marginal conditions, obtained by expanding $\lambda = \lambda_0 + \epsilon\lambda_1 + \ldots$

At order ϵ, an expression is found for λ_1, for both transverse and longitudinal waves. For the transverse mode, the condition of marginal stability $\text{Re}[\lambda_1] = 0$ leads to

$$m_{\text{tr}} = \frac{Pr^{1/2}(Pr^{1/2} + 1)}{Pr^{1/2} + (Pr^{1/2} + 1)\cosh^2[k]}\frac{\omega_0^2}{k^2} \tag{3.198}$$

while for longitudinal waves, $\text{Re}[\lambda_1]$ does not vanish.

On the other hand, the expression (3.198) for m_{tr} is not satisfactory, because it indicates an exponentially small threshold for $k \to \infty$. The physical reason is that for such short waves, the influence of the rigid bottom is exponentially small, and viscous dissipation is only due to the weak (next order) dissipation in the bulk. This also justifies the investigation of the next order (ϵ^2) correction to the growth rate [due to the viscous stresses, i.e. in the combination $Du - ikw$ of the tangential stress balance (3.183)].

The threshold of transverse waves is then found as

$$m_{\text{tr}} = \frac{Pr^{1/2}(Pr^{1/2} + 1)}{Pr^{1/2} + (Pr^{1/2} + 1)\cosh^2[k]}\frac{\omega_0^2}{k^2} \times$$
$$\times \left[1 + \epsilon\sqrt{2}\frac{k}{\omega_0^{1/2}}(4\sinh[k]\cosh[k] + 3\tanh[k])\right] \tag{3.199}$$

Figure 3.36(a) shows this marginal curve for $Pr = 6$ and $Bo = 0.333$ (dot-dashed lines) for several values of Ga (or equivalently ϵ). The full lines represent the corresponding numerical results for the lower branch of the bag. A satisfactory agreement is obtained between asymptotic and numerical results. In case 3, full and dot-dashed curves are nearly indistinguishable. Note that the turning point (which is not predicted by the asymptotic results) lies far to the right and above the figure. In Figs 3.34 and 3.35, ϵ is in fact not small enough to obtain satisfactory agreement with exact results. The dashed lines in these plots correspond to another approximation, derived from the full dispersion relation by neglecting only exponentially small terms when $\epsilon \to 0$ (but not polynomial ones). This approximation [244], which in fact amounts to retaining an infinite number of terms in the boundary layer dispersion relation (3.197), obviously leads to results valid at larger ϵ (lower Ga).

The curve 3 in Fig. 3.36(a) presents a minimum at finite k, which may become the absolute minimum defining the critical conditions. For increasing Bo, this minimum becomes deeper and shifts to the region of short waves. For $Bo \gg 1$ and sufficiently small ϵ, the absolute minimum (critical conditions) can be calculated as

$$k_c = \sqrt{Bo/5}, \quad m_c = 7.93\,\epsilon\,Pr^{1/2}Bo^{-1/8} \quad (Ma_c = -7.93Pr^{3/4}Ga^{3/4}Bo^{-1/8})$$
(3.200)

Rewriting these results in dimensional variables, the depth d disappears from the expressions, and the critical wavelength scales with the capillary length. Thus we recover earlier results in this limiting case (see [48, 231, 235]).

We now turn to the critical conditions for the longitudinal modes. Using the next order correction to $\mathrm{Re}[\lambda_1]$ yields

$$m_{\mathrm{long}} = \frac{\epsilon^{-4/3}2^{2/3}}{k^2} \frac{(Pr^{1/2} + 1)^{7/3}}{(7Pr^{1/2} + 3)^{4/3}} \left(1 + \frac{k^2}{Bo}\right)^{4/3}$$
(3.201)

which shows that the critical Marangoni number for the longitudinal mode is asymptotically higher than for the transverse mode. Indeed, m_{long} scales as $\epsilon^{-4/3}$ ($Ma_{\mathrm{long}} \sim Ga^{4/3}$), while for the transverse mode $m_{\mathrm{tr}} \sim \epsilon^0$ ($Ma_{\mathrm{tr}} \sim Ga$). These results agree with the numerical observation that the lower branch of the stability bag is not significantly modified by ϵ (in terms of m), while the upper branch indeed shifts to higher m when ϵ is decreased. This is the reason why in Fig. 3.36(b), the curve 3 (corresponding to $\epsilon = 0.01$) is not represented.

At the resonance point [defined by Eq. (3.196)], $\mathrm{Re}[\lambda_1]$ is found to diverge [because its coefficient in the dispersion relation at order ϵ is proportional to $(1 - m/m_{\mathrm{res}})$]. This situation was also encountered in [50], and indicates that an improved asymptotic analysis is needed in the vicinity of m_{res}. However, this result also shows that the stability status of the modes is exchanged at $m = m_{\mathrm{res}}$. Taking into account that $m_{\mathrm{tr}} < m_{\mathrm{res}} < m_{\mathrm{long}}$, it is seen that the transverse mode is unstable for $m_{\mathrm{tr}} < m < m_{\mathrm{res}}$, and stable for $m < m_{\mathrm{tr}}$ or $m > m_{\mathrm{res}}$. Similarly, the longitudinal mode is unstable for $m_{\mathrm{res}} < m < m_{\mathrm{long}}$, while it is stable outside this interval. As the behavior of the eigenvalues should remain continuous across $m = m_{\mathrm{res}}$ (as also shown in [244] and in Fig. 3.35), it is clear that at resonance a continuous transition occurs between transverse and longitudinal modes (mode mixing). This is confirmed by Fig. 3.35, where it is seen that the numerically determined frequencies tend to be closer to one of the leading-order results (dot-dashed lines) for $m < m_{\mathrm{res}}$ and to the other

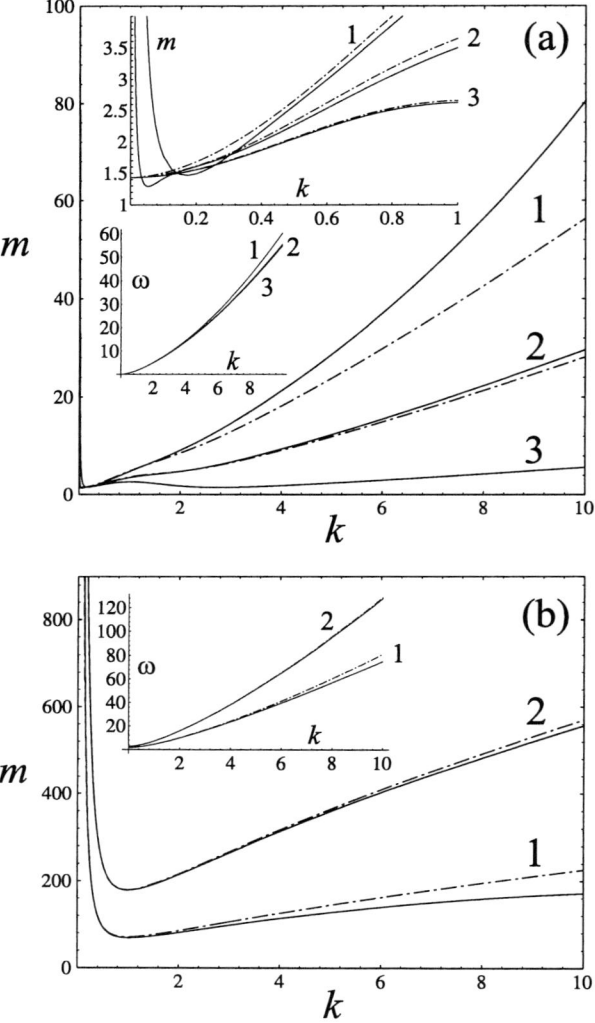

Figure 3.36: Marginal states for (a) transverse and (b) longitudinal waves. Numerical results (full lines) and leading-order asymptotic results [dot-dashed lines, Eqs. (3.199) and (3.201)]. $Pr = 6$, $Bo = 0.333$ and $\epsilon = 0.1$ ($Ga = 60000$) for curves 1, $\epsilon = 0.05$ ($Ga = 960000$) for curves 2, $\epsilon = 0.01$ ($Ga = 6\,10^8$) for curves 3. Insets show frequencies ω along marginal states [Eqs (3.194) and (3.195) as dot-dashed], and in case (a) the behavior at small k.

leading-order approximation for $m > m_{res}$. Moreover, Fig. 3.35(a) shows that a higher degree of amplification/damping is indeed predicted for the two mixed modes in the neighborhood of m_{res}. When decreasing ϵ, the approximation of $Re[\lambda]$ by $Re[\lambda_1]$ is improved and this feature becomes even more apparent.

iii) Asymptotic behavior of the turning point

For decreasing ϵ (increasing Ga), the turning point k^* was seen to shift to the region of short waves ($k^* \to \infty$). In fact, under the assumption that $k^* \gg Bo^{1/2}$, it is possible to show that the asymptotic behavior of the turning point is

$$k^* = 0.73\,10^{-4}GaBo^{-1}, \quad \omega^* = 0.74\,10^{-6}Ga^{3/2}Bo^{-2},$$
$$m^* = 0.28\,10^{-3}GaBo^{-2} \quad (Ma^* = -0.28\,10^{-3}Ga^2Bo^{-2}) \tag{3.202}$$

Thus $k^* \gg Bo^{1/2}$ is satisfied if $Ga \gg Bo^{3/2}$ (in practice $Ga \gg 15000Bo^{3/2}$). Consequently, if we define a Galileo number using the capillary length (instead of the depth d), its value must be sufficiently large for the result (3.202) to be valid. Note that the powers in relations (3.202) are exact, and only the coefficients necessitated (one) numerical calculation.

General behavior of the eigenvalues

The eigenvalue corresponding to the lowest full curve of Fig. 3.35(a) can be numerically tracked to the region of negative m (heating from below). The result is represented in Fig. 3.37, where the inset shows a magnification near the origin.

It is seen that the pair of complex conjugate eigenvalues corresponding to the longitudinal mode when $m < m_{res}$ is converted into a pair of purely real eigenvalues at a certain positive value of m. The most dangerous of these two real eigenvalues may then be destabilized when m passes through zero and becomes sufficiently negative. At this point, the diffusive state becomes unstable to the monotonic Pearson's mode (see §3.3.2), as indicated by the calculation of the corresponding Marangoni number $Ma = mGa$ [this gives $Ma \simeq 125$, which is the Pearson's threshold value (3.118) at $k = 1$]. Thus, in the parameter space, there is a continuous transition from the transverse to the longitudinal wave and then to the mode responsible for Pearson's monotonic instability. The dot-dashed curve represents Eq. (3.195) evaluated at negative m, and shows a satisfactory agreement at large enough $|m|$. Note that the negative root of Eq. (3.195) at $m < 0$ is unphysical [244].

3.6.2 Two-layer systems

When allowing for surface deformation in a two-layer system, neutral stability and overstability results obtained in §3.5 are modified in a way similar to what has been predicted in the last section for one-layer systems. Namely, a monotonic deformational mode also occurs at small wavenumbers, as shown hereafter. Only a brief account of results of Smith [218] is presented here. The main conclusion is that, similarly to what happens for one-layer systems, the deformational mode may become the most dangerous one at small Galileo number, i.e. when sufficiently thin layers of highly viscous liquids are considered.

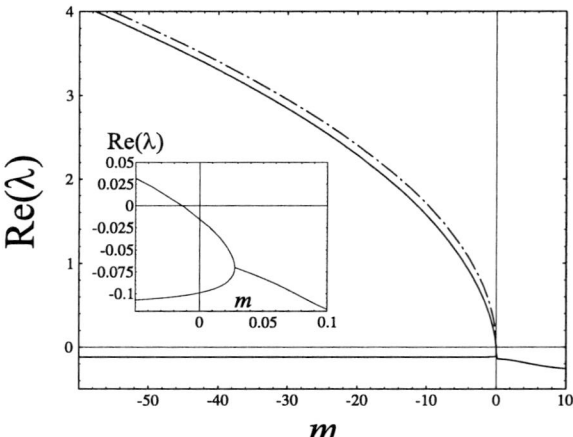

Figure 3.37: Continuation of the lower branch of Fig. 3.35(a) [with the same parameters] to the region of negative m (heating from below). The inset shows details of the behavior near the origin. At the branching point, the two complex conjugate eigenvalues merge giving rise to two purely real eigenvalues. One of them further leads to Pearson's monotonic instability at negative m. The dot-dashed curve is Eq. (3.195) evaluated at negative m.

Mode-mixing and resonance of capillary-gravity waves and surface dilational modes may also occur for two-layer systems, although this case will not be treated here. For an application of techniques similar to those used in §3.4.2 and §3.6.1 to two-layer systems, the reader is referred to the detailed analysis of Rednikov et al. [49]. Other interfacial oscillatory modes due to buoyancy and occurring when the density jump between both layers is small are not considered either, and the reader is referred to works of Renardy et al. [176, 220] for detailed analyzes of these interfacial modes.

The basic equations considered here are Eqs (3.141–3.144) with boundary conditions (3.145–3.146) at the upper ($z = a$) and lower ($z = -1$) plates. Interfacial boundary conditions including small deformability may be written at $z = 0$ after linearization, and read

$$w_1 = w_2 = \sigma\xi \tag{3.203}$$

$$Dw_1 = Dw_2 \tag{3.204}$$

$$\theta_1 = \theta_2 + (\lambda^{-1} - 1)\xi \tag{3.205}$$

$$\lambda D\theta_2 = D\theta_1 \tag{3.206}$$

$$(D^2 - 3k^2 - \sigma Pr_1^{-1})Dw_1 = \mu(D^2 - 3k^2 - \sigma\nu^{-1}Pr_1^{-1})Dw_2 + k^2((1 - \rho)Ga + k^2Ca^{-1})\xi \tag{3.207}$$

$$(D^2 + k^2)w_1 = \mu(D^2 + k^2)w_2 + k^2 Ma_1(\xi + \theta_1) \tag{3.208}$$

where ξ is the amplitude of surface deformation, and the other usual notations apply. The normal momentum balance (3.207) has been written in a form involving only vertical components of the velocity field, using the horizontal components of the Navier–Stokes equations as well as the incompressibility condition. Note that the capillary and Galileo numbers appearing in this relation are still defined by Eqs (3.55) and (3.56), though using properties of the lower layer. Note also the factor $1 - \rho = 1 - \rho_2/\rho_1 > 0$ multiplying Ga in Eq. (3.207).

Referring the reader to the work of Smith [218] for details of the calculation, we just give the result obtained for the neutral stability relation ($\sigma = 0$) in the absence of buoyancy ($Ra_1 = 0$) as

$$Ma\left\{ \frac{1 - \kappa^{-1}\frac{G_1(k)}{G_1(ak)} -}{\frac{8G_3(k)}{(1-\rho)Ga + k^2 Ca^{-1}}\left[1 + \frac{\tanh(k)}{\tanh(ak)}\right]\left[1 - \frac{\mu}{a^2}\frac{G_4(k)}{G_4(ak)} + k^2(1-\mu)G_4(k)\right]} \right\}$$
$$= P(k)\left[1 + \lambda\frac{\tanh(k)}{\tanh(ak)}\right]\left[1 + \mu\frac{G_2(k)}{G_2(ak)}\right] \tag{3.209}$$

where the definitions (3.171–3.173) still apply, while G_3 and G_4 are defined by

$$G_3(\xi) = \frac{\xi^3 \cosh[\xi]\sinh^2[\xi]}{\xi^3 \cosh[\xi] - \sinh^3[\xi]} \tag{3.210}$$

$$G_4(\xi) = \frac{\xi^2 - \sinh^2[\xi]}{\xi^2 \sinh^2[\xi]} \tag{3.211}$$

At large Galileo and small capillary numbers, Eq. (3.209) reduces to Eq. (3.173), as expected when surface deformation is negligible. This certainly holds for large wavenumbers, as the supplementary term in Eq. (3.209) vanishes exponentially when $k \to \infty$. In this limit, the result (3.152) for two semi-infinite layers is recovered.

As for one-layer systems, surface deformation plays a more important role at small wavenumbers. When $k \to 0$, the neutral stability condition indeed reduces to

$$Ma = \frac{2}{3}(1 - \rho)Ga\frac{(1 + \lambda a^{-1})(1 + \mu a^{-1})}{(1 + a^{-1})(1 - \mu a^{-2})} \tag{3.212}$$

whose sign is determined by $1 - \mu a^{-2}$. Consequently, the long-wave deformational mode may be excited by heating from below if $\mu < a^2$, and from above if $\mu > a^2$. Thus, comparing with Eq. (3.174), the same direction of heating is predicted when $\mu < a^2 < \kappa$ (from below) or $\kappa < a^2 < \mu$ (from above). In other cases, an asymptote should exist at a value k^* of the wavenumber which can be approximated as

$$k^* \simeq \left[\frac{120(1 - \mu a^{-2})(1 + a^{-1})}{Ga(1 - \rho)(a^2\kappa^{-1} - 1)}\right]^{\frac{1}{2}} \tag{3.213}$$

in the long wave limit $k^* \to 0$.

3.7   Two-component one-layer systems

The Bénard instability of a binary fluid layer has been extensively studied in the past, both theoretically [162, 163, 165, 166, 167, 232] and experimentally [24, 168, 169, 274, 455, 457], as an example of an hydrodynamic system displaying a Hopf bifurcation (overstability) at onset from the basic conductive state. These oscillations may occur due to the competition between thermal and solutal contributions to the density field (Rayleigh mechanism), or to the surface tension at a free surface (Marangoni mechanism).

Consider first the buoyancy-induced mechanism. A temperature gradient is established by heating the horizontal layer, say from below. On a larger time scale, the Soret effect generates a concentration gradient, also perpendicular to the layer. Depending on the sign of the Soret coefficient, defined by Eq. (2.62), the heavier component may either migrate towards the cold upper surface, or to the hotter bottom plate. In the former case, thermal and solutal fields both contribute to create an unstable density stratification, while in the latter, the solutal contribution rather stabilizes the basic diffusive state, and eventually enters into competition with the destabilizing thermal contribution. In this case, oscillations are possible because the ratio of thermal to solutal relaxation times (the Lewis number) is typically small (of order 10^{-2}).

In this section, we will mainly focus on the effect of surface tension, i.e. the situation where the upper surface of the binary liquid layer is free and in contact with an inert gaseous phase. Linear stability analyzes [232, 279, 463, 464] show that overstability may also occur in this situation, even for purely surface-tension-driven convection (i.e. for very thin layers of liquid, or in microgravity). The stability diagrams (see §3.7.2) are found to be rather similar for buoyancy and surface-tension-driven convection.

Note that the analysis of Marangoni–Bénard convection with Soret effect will be pursued in §5.3, when studying two- and three-dimensional waves in the weakly nonlinear regime.

Using the thickness of the layer d as the unit length, d^2/κ as unit time (κ is the thermal diffusivity), the temperature drop Θ_T over the layer as unit temperature and the Soret-induced concentration drop $\Theta_N = S\Theta_T$ as the unit concentration (S is a Soret coefficient), the Boussinesq equations governing the evolution of perturbations from the diffusive state can be found from the balance equations presented in Chapter 2 (see also [232, 279, 463, 465]) as

$$\vec{\nabla}.\vec{v} = 0 \tag{3.214}$$

$$Pr^{-1}\frac{d\vec{v}}{dt} = -\vec{\nabla}p + \Delta\vec{v} + \vec{1}_z Ra(T + Le\Psi_{\mathrm{Ra}}N) \tag{3.215}$$

$$\frac{dT}{dt} = \Delta T + W \tag{3.216}$$

$$\frac{dN}{dt} = Le\Delta(N - T) + W \tag{3.217}$$

where $\vec{v} = (U, V, W)$ is the velocity field, $\vec{1}_z$ is the unit vector along the vertical, T, N and p are respectively the perturbations of temperature, mass fraction and pressure, $d/dt =$

$\partial/\partial t + \vec{v}.\vec{\nabla}$ is the material derivative, Δ is the Laplacian operator, Pr is the Prandtl number, Le is the Lewis number, Ψ_{Ra} is the Rayleigh separation ratio, and Ra is the Rayleigh number. These dimensionless parameters are defined as

$$Pr = \nu/\kappa, \; Le = D/\kappa, \; Ra = \frac{g\alpha\Theta_T d^3}{\nu\kappa}, \; \Psi_{Ra} = \frac{Ra_S}{Ra} = Le^{-1}\frac{\beta}{\alpha}S \tag{3.218}$$

where ν is the kinematic viscosity, D is the isothermal diffusion coefficient, g is gravity, α and β are respectively the thermal and solutal expansion coefficients, while Ra_S is a solutal Rayleigh number defined by $Ra_S = g\beta\Theta_N d^3/\nu D$, by analogy with Ra.

The boundary conditions to be satisfied at the bottom plate $z = 0$ are:

$$T = \vec{v} = \partial_z(N - T) = 0 \tag{3.219}$$

and at the undeformable interface $z = 1$, we have:

$$W = \partial_z T + BiT = \partial_z(N - T) = 0, \\ \partial_z\vec{v}_h + Ma\vec{\nabla}_h(T + Le\Psi_{Ma}N) = 0 \tag{3.220}$$

where Bi is the Biot number, the index h denotes horizontal components, Ma and Ψ_{Ma} are respectively the Marangoni number and the Marangoni separation ratio, defined here as

$$Ma = \frac{-\sigma_T\Theta_T d}{\mu\kappa}, \; \Psi_{Ma} = Le^{-1}\frac{\sigma_N}{\sigma_T}S \tag{3.221}$$

where μ is the dynamic viscosity, and σ_T and σ_N are respectively the surface tension variations with temperature and mass fraction.

3.7.1 Neutral stability

Linear stability analysis of the reference state is once more carried out via normal modes analysis, i.e. Fourier amplitudes (periodic dependency upon the horizontal coordinate $\vec{r}$) are introduced in the form

$$(\vec{v}, T, N) \sim \exp[i\vec{k}.\vec{r} + \sigma t] \tag{3.222}$$

Assuming small perturbations, linearizing the system (3.214–3.220) leads to an eigenvalue problem for the complex growth rate σ, given the values of all dimensionless numbers. As usual for horizontally isotropic systems, the stability problem only depends on the wavenumber $k = |\vec{k}|$, and can finally be written (using the same notations as in the previous sections, as well as η for the amplitude of the concentration field N) as:

$$(D^2 - k^2 - Pr^{-1}\sigma)(D^2 - k^2)w = k^2 Ra(\theta + Le\Psi_{Ra}\eta) \tag{3.223}$$

$$(D^2 - k^2 - \sigma)\theta + w = 0 \tag{3.224}$$

$$Le(D^2 - k^2)(\eta - \theta) - \sigma\eta + w = 0 \tag{3.225}$$

together with boundary conditions

$$\theta = w = Dw = D\eta - D\theta = 0 \quad \text{at } z = 0 \tag{3.226}$$

$$w = D\theta + Bi\theta = D\eta - D\theta = D^2 w + k^2 Ma(\theta + Le\Psi_{\text{Ma}}\eta) = 0 \quad \text{at } z = 1 \tag{3.227}$$

As stability results are well-known for the Rayleigh–Bénard case (e.g. with rigid-rigid boundary conditions, see [166]), we here restrict the presentation of results to the case $Ra = 0$. Solving the corresponding neutral stability problem ($\sigma = 0$) as in previous sections[14] is possible, and is more easily achieved by defining a combined field

$$\chi(z) = Le\left[\eta(z) - \theta(z)\right] \tag{3.228}$$

which should thus satisfy $(D^2 - k^2)\chi + w = 0$ in the bulk, and boundary conditions $D\chi = 0$ at $z = 0, 1$. The neutral stability condition may finally be written in the form

$$Ma = \frac{P(k, Bi)}{1 + Le\Psi_{\text{Ma}} + \Psi_{\text{Ma}}R(k, Bi)} \tag{3.229}$$

where $P(k, Bi)$ is equal to the Pearson's result (3.118), while $R(k, Bi)$ is defined by

$$R = \frac{\chi(1)}{\theta(1)} = \frac{(Bi + k\coth[k])\left(\cosh[3k] - \cosh[k](1 - 4k^2)\right) - 4k\sinh[k](2 + k^2))}{(k\sinh[3k] - 4k^4\cosh[k] - 3k\sinh[k])} \tag{3.230}$$

which thus quantifies the relative contribution of fields χ and θ to the surface tension gradient. Note that the expression (3.230) turns out to be positive for all k and Bi, and is monotonically decreasing with k. It is also readily seen that $\lim_{k\to\infty} R = 1$, while in the long-wave limit $k \to 0$, $R \simeq 5k^{-2}(1 + Bi)/3$. As the Pearson's function P tends to $P \simeq 80k^{-2}(1 + Bi)$ in this limit, we have $\lim_{k\to 0} Ma = 48/\Psi_{\text{Ma}}$. Expanding (3.229) to the next (curvature) term in fact yields

$$Ma = \frac{48}{\Psi_{\text{Ma}}} + 16k^2 \frac{-9 + (1 + Bi - 9Le)\Psi_{\text{Ma}}}{5(1 + Bi)\Psi_{\text{Ma}}^2} + O(k^4) \quad \text{for } k \ll 1 \tag{3.231}$$

Hence, assuming that $Le < (1 + Bi)/9$, critical conditions may be given by $k_c = 0$, $Ma_c = 48/\Psi_{\text{Ma}}$ provided either $\Psi_{\text{Ma}} > 9/(1 + Bi - 9Le)$, or $\Psi_{\text{Ma}} < 0$. In both cases, this long-wave instability is due to the solutal field alone. As boundaries are impervious to matter, the instability is indeed due to long-wave modes (see also results of §3.3.1 for insulating-insulating cases), and we recover the value $Ma_S = \Psi_{\text{Ma}}/Ma = 48$ for the critical solutal Marangoni number Ma_S.

---

[14]Note that Eqs (3.116) and (3.117) of §3.3.2 are still valid here, while the concentration field η still has to be found from Eq. (3.225) with boundary conditions $D\eta - D\theta = 0$ at $z = 0, 1$. Therefore, it can be expected in advance that the neutral stability relation found here will be similar to the Pearson's relation (3.118), except for an additional term proportional to Ψ_{Ma} in the denominator.

Taking into account that $P(k, Bi) > 0$ and $R(k, Bi) > 1$ it can be shown that the neutral curve (3.229) has no asymptote either for $\Psi_{Ma} > 0$ $(Ma > 0)$, or for $\Psi_{Ma} < -(1 + Le)^{-1}$ $(Ma < 0)$. For $-(1 + Le)^{-1} < \Psi_{Ma} < 0$, an asymptote exists at a wavenumber $k = k^*$, and $Ma < 0$ $(Ma > 0)$ for $k < k^*$ $(k > k^*)$. Hence, instability occurs for both directions of heating. When heating is from above, the solutally-dominated long-wave instability just described occurs when $Ma < 48/\Psi_{Ma} < 0$. When heating is from below, instability occurs first to modes with finite critical wavenumber k_c, and the relation (3.229) must be minimized numerically. Note that this also holds for $0 < \Psi_{Ma} < 9/(1 + Bi - 9Le)$.

The discussion is summarized in Fig. 3.38, which is drawn for $Le = 0.05$ and $Bi = 0.5$. Note that in the neighborhood of $\Psi_{Ma} = 9/(1 + Bi - 9Le)$ (transition from a finite to a zero critical wavenumber), it can be calculated that

$$k_c^2 \simeq \frac{35}{2} \frac{(1 + Bi - 9Le)^2}{169 + 64Bi} \left(-\Psi_{Ma} + \frac{9}{1 + Bi - 9Le} \right) \qquad (3.232)$$

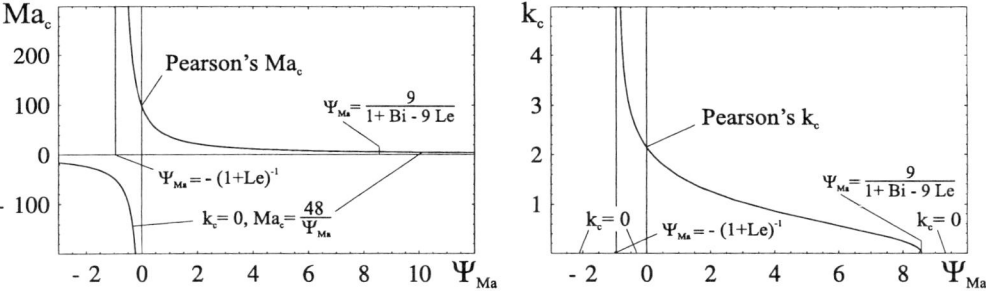

Figure 3.38: Critical Marangoni number (left) and wavenumber (right) for monotonic modes, together with representative points. Parameters used here are $Ra = 0$, $Le = 0.05$ and $Bi = 0.5$.

Hence, for $\Psi_{Ma} > 0$ the critical Marangoni number Ma_c is seen to decrease with increasing solutal contribution. For $\Psi_{Ma} < 0$, thermal and solutal fields compete and an increase of Ma_c is observed at increasing $|\Psi_{Ma}|$. However, oscillations are also possible in this case, as discussed now.

3.7.2 Oscillatory modes and complete stability diagrams

The system of ODEs (3.223–3.225) and boundary conditions (3.226–3.227) can also be solved analytically when $\sigma \neq 0$ (even for $Ra \neq 0$), although interpretation of the results for the branch of marginal oscillatory perturbations $\sigma = i\omega$ is purely numerical. The method used to trace the oscillatory branch has already been described in §3.4.1 and will not be detailed here. Combining these results with neutral stability curves calculated in the last section yields stability diagrams such as presented in Fig. 3.39, for $Le = 0.05$, $Pr = 6$, $Bi = 0.5$.

For negative Marangoni separation ratio Ψ_{Ma} and positive thermal Marangoni number Ma (which for liquids with negative σ_T means that the layer is heated from below while

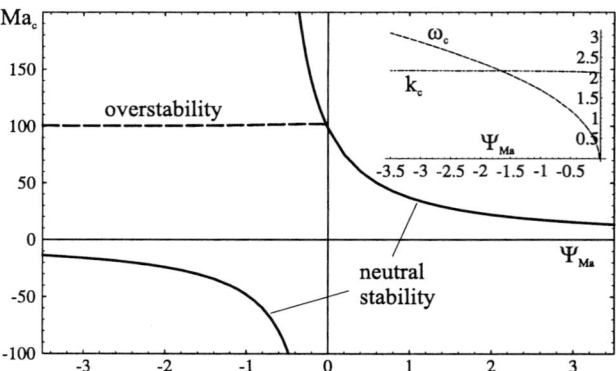

Figure 3.39: Linear stability diagram for $Ra = 0$, $Le = 0.05$, $Pr = 6$, $Bi = 0.5$. Full curves represent neutral stability ($\sigma = 0$), while the dashed curve represents oscillatory onset or overstability ($\sigma = i\omega$). The critical frequency ω_c and critical wavenumber k_c at the oscillatory threshold are shown as inset.

the surfactant migrates towards the cold free surface by Soret effect), the onset is oscillatory provided $\Psi_{\mathrm{Ma}} < \Psi_{\mathrm{CTP}} < 0$, where Ψ_{CTP} corresponds to the (codimension-2) point where the oscillatory branch crosses the neutral stability line. The frequency increases with the absolute value of Ψ_{Ma}. Note that as Ma is minimized with respect to the wavenumber, the point where the frequency vanishes along the Hopf branch does not generally corresponds to a codimension-2 point (see §5.3) in a strict sense [166]. However, this small effect is not visible on Fig. 3.39.

3.7.3 Physical mechanisms and heuristic arguments

For two-dimensional situations, the perturbation fields will be written

$$U(x, z, t) = A \exp[i(k_c x + \omega_c t)]U_c(z) + B \exp[i(k_c x - \omega_c t)]U_c^*(z) + c.c. \quad (3.233)$$

where we have used the notation U for vertical velocity, temperature or concentration perturbations, and U_c for the corresponding component of the critical eigenmode. In Eq. (3.233), A represents the amplitude of the left-traveling wave, while B is the amplitude of the right-traveling wave. Perturbation analysis will be used in Chapter 5 to derive evolution equations for amplitudes $A(t)$ and $B(t)$. It will be seen that apart from the reference solution $A = B = 0$, which is unstable above the critical Marangoni number, the only solutions with steady values of $|A|$ and $|B|$ are traveling waves (TW) for which $A = 0, |B| \neq 0$ (or $|A| \neq 0, B = 0$), and standing waves (SW) for which $|A| = |B| \neq 0$. Introducing these results in Eq. (3.233) leads to

$$U(x, z) \sim |U_c(z)| \cos[k_c x + \omega_c t + \varphi_U(z)] \quad \text{(TW)} \qquad\qquad (3.234)$$

$$U(x, z) \sim |U_c(z)| \cos(k_c x) \cos[\omega_c t + \varphi_U(z)] \quad \text{(SW)} \qquad\qquad (3.235)$$

where $\varphi_U(z) = \arg[U_c(z)]$ is the phase (generally z-dependent) of the critical eigenmode U_c. Note that a suitable redefinition of space and time origins has been used to eliminate arbitrary phases of A and B.

According to the values of Pr, Le, Ψ_{Ma} and Bi, TW or SW could be preferred in a supercritical situation (see §5.3.1). For the moment, in order to illustrate the mechanisms of instability, we have represented both TW (Fig. 3.40) and SW (Fig. 3.41), though using values of A and B calculated for typical values of dimensionless parameters in §5.3.1.

Figure 3.40: Traveling waves reconstructed from the first-order fields for $Ra = 0$, $Pr = 6$, $Le = 0.05$, $Bi = 0.5$, $\Psi_{\text{Ma}} = -0.05$, $(Ma - Ma_c)/Ma_c = 0.05$. Thin full (closed) curves are the streamlines, thin dashed lines are isotherms (hotter at the bottom), while thick dot-dashed lines are isoconcentration lines of the surfactant (more concentrated at the free surface). The whole pattern is drifting (without deformation) to the left.

It is seen from Figs 3.40 and 3.41 that the concentration field is generally more distorted by convection than the thermal field, owing to the small value of $Le = 0.05$. Moreover, an important ingredient of the oscillation mechanism is the phase shift between concentration and temperature fields (also relevant in the buoyancy-driven situation [165]). Consider the standing waves on Fig. 3.41: the velocity field is not represented but may be guessed by the deformation it produces on isotherms (there is almost no phase shift between velocity and temperature fields). In the top left picture, there is upflow in the center. While the thermal field favors this situation, the concentration field opposes it, since upflow brings surfactant-poor liquid to the free surface. Eventually, the surfactant-induced surface tension gradient is strong enough to slow down the motion (2nd picture), and to reverse it (pictures 3 and 4), inducing downflow at the middle. This overshoot is possible because of the slow diffusive evolution of the concentration field when the velocity is weak. This motion is also amplified by the thermal contribution. Amplification occurs from the 3rd to the 5th picture, up to the moment where the intensity of convection has become sufficiently strong to advect the concentration perturbation sufficiently. There, the concentration-induced surface tension gradient overcomes the thermally-induced surface tension gradient, and therefore tends once again to reduce the flow (pictures 6 and 7), and to reverse it (pictures 8 to 10), restoring the initial pattern.

Coming back to Fig. 3.39, obtained for $Le = 0.05$, it can be seen that the threshold Marangoni number for oscillations only weakly depends on the separation ratio Ψ_{Ma}. At larger Le, the Hopf threshold increases with increasing $-\Psi_{\text{Ma}}$, while at lower Le, it generally decreases in some range of separation ratios. This surprising effect appears to be linked to the increasing phase shift between temperature and concentration fields when the Hopf frequency increases.

Note finally the z-dependence of the phase of the concentration field, visible in Fig. 3.41.

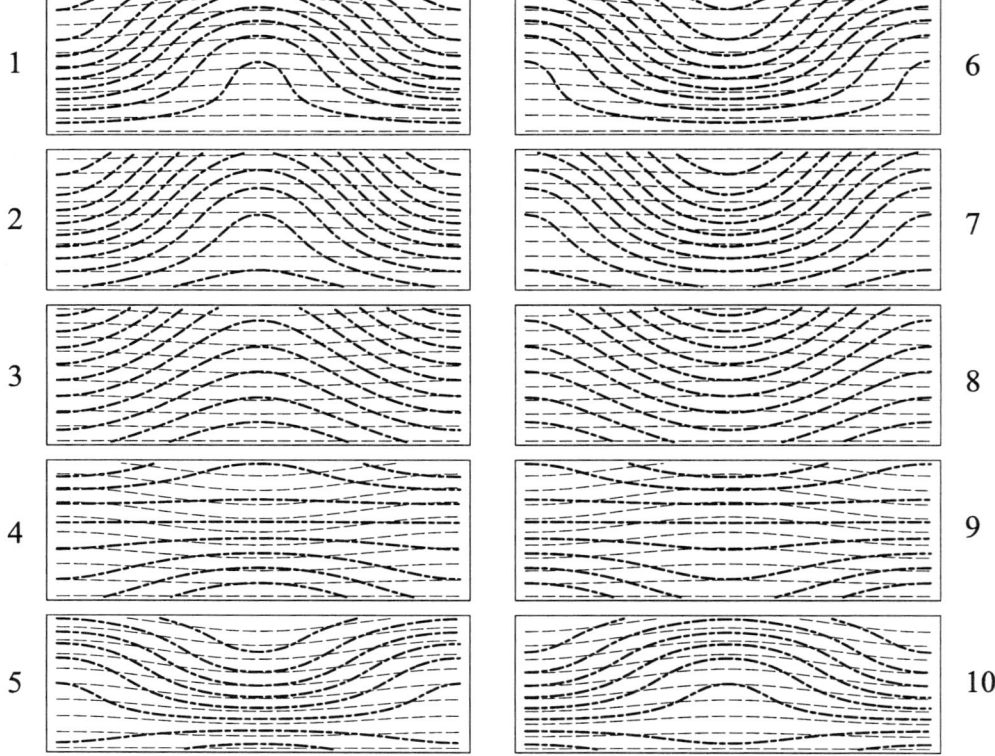

Figure 3.41: Standing waves reconstructed from the first-order fields for $Ra = 0$, $Pr = 6$, $Le = 0.05$, $Bi = 0.5$, $\Psi_{\mathrm{Ma}} = -0.5$, $(Ma - Ma_{\mathrm{c}})/Ma_{\mathrm{c}} = 0.1$. Only isotherms (thin dashed lines) and isoconcentration lines of the surfactant (thick dot-dashed lines) are represented. Ten snapshots sampled over the whole period are depicted, running from top to bottom and left to right.

For instance, when the flow reverses (e.g. pictures 3 to 5), the concentration field reacts faster near the interface than at the bottom. In contrast, the phase of temperature and velocity fields only weakly depend on z, at least for moderate frequencies.

3.8   Influence of evaporation

In this section, we consider a layer of evaporating pure liquid in contact with its own vapor (the gas phase is also pure), and examine the possibility of thermocapillary instabilities. Evaporative convection is important in a number of applications [40, 81], such as drying of paint films, distillation, spray drying, thin-film evaporators, and heat-exchangers. As an example, heat pipes allow to transfer heat efficiently, by evaporating a liquid continuously at a certain end (evaporator), and recondensing it another location (condenser). These devices take advantage of the large amount of energy transported by the vapor phase (latent heat), which allows to reach high efficiencies (and at the same time avoid the use of mechanical parts such as pumps, thus making it suitable for microgravity applications). It is thus important to understand the role of thermocapillarity, as a first step in the idealized configuration considered here, as Marangoni convection leading to increased heat transport could increase performances of the heat exchanger.

From a physical point of view, evaporative convection also raises a number of interesting questions. At thermodynamic equilibrium, the chemical potential is continuous across the (presumably flat) liquid/vapor interface (in addition to the equality of temperatures and of pressures of both phases). These conditions lead to the Clausius–Clapeyron relation [248], i.e. a relation between the saturation pressure and temperature at equilibrium. Now, when imposing a certain heat flux through the layer, the interface may or may not stay under *local* equilibrium conditions. If it does, this means that everywhere along the interface, the local pressure will be equal to the saturation pressure (at the local temperature). As the pressure should remain homogeneous (apart from small hydrodynamic pressure gradients), the only possibility is that the temperature remains constant everywhere along the interface (corresponding to the pressure in the vapor via the Clausius–Clapeyron relation). Thus, no surface tension gradient will occur in this situation, and Marangoni–Bénard instability is not possible.

This fact has been noted in experimental studies [249], which have evidenced other types of instabilities, due to the effect of vapor recoil occurring when the pressure in the vapor phase is very low (see also [246, 250, 251]). This specific instability mode will not be considered here, referring the reader to the detailed theoretical works of Palmer [245] and Burelbach et al. [186]. Rather, we shall focus on the effect of departure from local equilibrium at the interface (i.e. a deviation with respect to the Clausius–Clapeyron relation), which is shown to destabilize the system, possibly leading to Marangoni–Bénard convection. Starting from general equations and boundary conditions including surface deformability, our approach here will differ from the previous sections, in that we will attempt to simplify the starting equations as much as possible before solving them. Actually, the simplest formulation turns out to be formally identical to the Pearson's problem treated in §3.3.2, though including a Biot number for which a value is *calculated* hereafter, as a function of evaporation parameters. Our goal will thus be to identify and discuss the physical assumptions sufficient to obtain this result (moderate evaporation fluxes, weak vapor recoil effect, one-sided model, weak interfacial deformation, weak gas pressure fluctuations, negligible interfacial temperature jump, ...).

It is also worth stressing that the following approach is not applicable to evaporation of a liquid under ambient air. In this case, the gas phase is a mixture of two components (air and vapor). Hence, even though the total pressure in the gas remains quasi-homogeneous, fluctuations of the partial pressure of vapor are allowed (concentration fluctuations in the gas), which should in turn allow temperature fluctuations along the interface, and hence Marangoni effects [253].

3.8.1 Non-equilibrium effects

Consider a layer of a single-component liquid, lying on a heated rigid plate ($z = 0$) and in contact with its own vapor at a free interface whose shape is given by $z = h(x, t)$, in which x

is the horizontal coordinate and t is time. At this interface, evaporation may occur, such that a mass flux $J = \vec{J}.\vec{n}$ is allowed [$\vec{n} = (-h', 1)/N$ is the free surface unit normal pointing to the vapor, $N = [1 + (h')^2]^{1/2}$ is the normalization factor, and a prime denotes a derivative with respect to x]. In several studies of evaporating liquid layers (see e.g. [245, 246, 251]), the mass flux J is assumed to obey the Hertz–Knudsen equation, derived from the kinetic theory of gases [252]. This non-equilibrium law is derived for a flat interface along which the equality of liquid and vapor temperatures is assumed (the vapor is further assumed to be a perfect gas), and reads

$$J = \beta \sqrt{\frac{M}{2\pi RT}} [p_{\mathrm{s}}(T) - p_0(T)] \qquad (3.236)$$

where β is the accommodation coefficient, M is the molecular weight of vapor, $p_{\mathrm{s}}(T)$ is the saturation pressure at surface temperature T, $p_0(T)$ is the vapor pressure just beyond the interface, and R is the universal gas constant.

In this section, in order to start from a more general form of the non-equilibrium mass flux equation which should in particular be valid for a moving interface of arbitrary shape (and for any form of the vapor state equation), we use the thermodynamics of irreversible processes [247, 263]. First, note that the Clausius–Clapeyron equation follows from the equality $\mu_{\mathrm{v}}(p_{\mathrm{v}}, T_{\mathrm{v}}) = \mu_{\mathrm{l}}(p_{\mathrm{l}}, T_{\mathrm{l}})$ between chemical potentials of the liquid and of the vapor. Developing this formula and using classical thermodynamics relations, the Clausius–Clapeyron equation gives the slope of the coexistence curve [248]:

$$\frac{\partial p_{\mathrm{s}}}{\partial T} = \frac{\rho_{\mathrm{v}} \rho_{\mathrm{l}} L}{(\rho_{\mathrm{l}} - \rho_{\mathrm{v}}) T} \qquad (3.237)$$

obtained assuming thermal equilibrium $T_{\mathrm{v}} = T_{\mathrm{l}} (= T)$, and mechanical equilibrium $p_{\mathrm{v}} = p_{\mathrm{l}}$ $(= p_{\mathrm{s}})$. In this relation, ρ_{v} and ρ_{l} are respectively the vapor and liquid densities, and L is the latent heat of evaporation. Out of equilibrium, the equality of chemical potentials no longer holds. Calculating the interfacial entropy source [187], it turns out that the difference of chemical potentials is a generalized thermodynamic force (see §2.1.6) giving rise to a non-equilibrium mass flux across the interface. In the linear domain of irreversible thermodynamics, one can then make use of the phenomenological law:

$$J = K[\mu_{\mathrm{l}}(p_{\mathrm{l}}, T) - \mu_{\mathrm{v}}(p_{\mathrm{v}}, T)] \qquad (3.238)$$

where K is a positive coefficient. The derivation of this phenomenological relation is more rigorously justified in [247]. Note that if the temperature jump is not neglected, then other phenomenological coefficients appear, since there also exists a "thermal force" $\Delta T = T_{\mathrm{l}} - T_{\mathrm{v}}$. In this respect, the reader is referred to the recent works of Ward and Fang [192, 193], who provide careful experimental measurements of temperature jumps as high as 7.8 °C for evaporation of water and other liquids. Ward and Fang also propose a new expression for the evaporation flux, derived on the basis of statistical rate theory, and compare the latter with their experimental findings. The influence of this temperature jump on Marangoni convection has not been studied yet, and as a first approximation, it will be neglected in the following developments. Note that under this conditions, the expression for the evaporation flux derived by Ward and Fang indeed reduces to Eq. (3.238) at small enough $\mu_l - \mu_v$. It also appears that many of the experimental results of Ward and Fang indeed fall in this linear range.

The hydrodynamic definition of the mass flux J (see §2.2.2) is

$$J = \rho_l(\vec{v}_l - \vec{v}_\Sigma).\vec{n} = \rho_v(\vec{v}_v - \vec{v}_\Sigma).\vec{n} \tag{3.239}$$

where $\vec{v}_v$, $\vec{v}_l$ and $\vec{v}_\Sigma$ are respectively the vapor, liquid and interface velocities.

Let $p_s(T)$ be a function defining the equilibrium (saturation) pressure as a function of temperature T [$p_s(T)$ is for instance the Clausius–Clapeyron coexistence curve, or any fitting of experimental points]. Assume that the liquid and vapor state equations are respectively

$$\rho_l = \rho_l^0(p, T), \ \rho_v = \rho_v^0(p, T)$$

and that the inequalities

$$\frac{|p_l - p_s(T)|}{p_s(T)} \ll 1, \ \frac{|p_v - p_s(T)|}{p_s(T)} \ll 1$$

hold everywhere along the liquid/vapor interface. Then, using a Taylor expansion around the point $(p_s(T), T)$, at which chemical potentials are equal [by definition of $p_s(T)$],we get a linearized form of Eq. (3.238) as

$$J = K(T) \left[\frac{p_s(T) - p_v}{\rho_v(T)} - \frac{p_s(T) - p_l}{\rho_l(T)} \right] \tag{3.240}$$

where $\rho_v(T)$, $\rho_l(T)$ and $K(T)$ stand for $\rho_v^0(p_s(T), T)$, $\rho_l^0(p_s(T), T)$ and $K(p_s(T), T)$ respectively.

Note that if $p_v \simeq p_l$ in Eq. (3.240), neglecting ρ_l^{-1} against ρ_v^{-1} we get a form $J = K[p_s(T) - p_v]/\rho_v(T)$, presenting a formal analogy with the Hertz–Knudsen equation (although obtained by different means), and leading to a rough estimate of the coefficient $K = \beta \rho_v(T)(M/2\pi RT)^{1/2}$. Of course, only experimental measurements of K or non-equilibrium rate theories [193] can lead to satisfactory values.

However, the approximation $p_v \simeq p_l$ is only valid for weakly curved interfaces, and when dynamical effects may be neglected in the interfacial momentum balance. In the following, we will prefer to use Eq. (3.240), and consider K, ρ_v and ρ_l as being independent of T (which is valid provided temperature differences involved are sufficiently small). Note that as only a few measurements of K are available, we will either rely on the estimate provided above as a function of the accommodation coefficient β, or discuss the results as a function of K. In general, K might range from $K = 0$ (no evaporation) to $K \to \infty$ (which corresponds to $\mu_v = \mu_l$ for finite J according to Eq. (3.238), i.e. the interface is at local equilibrium).

3.8.2   Rate of evaporation in the reference state

We here compute the rate of evaporation of a liquid layer of depth h lying on a heated rigid plate maintained at a constant temperature T_b (for a given pressure p_g of the vapor). The basic state is assumed as unidimensional (all variables only depend on the vertical coordinate z, and the only non-zero component of the velocity in both liquid and gas phases is the vertical one, denoted by w). The vapor phase is infinitely deep. The solution of the hydrodynamic equations established here is the reference solution whose stability will be studied in §3.8.3.

When evaporation takes place, the relation (3.239) indicates that if the liquid is at rest ($w_l = 0$), the interface moves with a vertical velocity $w_\Sigma = -J/\rho_l$. The second relation (3.239) then indicates that the vertical velocity of the vapor is $w_v = J(1/\rho_v - 1/\rho_l) \simeq J/\rho_v$. This means that the liquid depth will decrease, and that it may finally completely evaporate. In order to obtain a steady reference solution (and a steady mass flux J), we will assume in the following that fresh liquid is injected at the bottom plate $z = 0$, at a rate which exactly matches the rate of liquid evaporation at the free surface $z = h$. This is attempted in order to mimic the steady regime obtained in heat exchanging devices such as heat pipes, where liquid condensed at the condenser is continuously brought back to the evaporator. The rigid bottom boundary may eventually be considered as a porous material, through which liquid is injected at a velocity $w_b = J/\rho_l$. Thus, the whole liquid layer has the vertical velocity $w_l = w_b = J/\rho_l$ (because the fluid is incompressible, such that $dw/dz = 0$), and the relation (3.239) indicates that $w_\Sigma = 0$. The interface is thus fixed in space at the location $z = h$. Note that for sufficiently small evaporation flux (more exactly, at small Peclet number, as seen below), results should not depend on whether the bottom plate is taken as being impervious or permeable, such as is done here, as the depth only varies slowly with time, allowing a quasi-steady approximation to be used [253].

The effective rate of evaporation J is given by the relation (3.240), which is seen to depend on the vapor and liquid pressures p_v and p_l on each side of the interface, and on the interfacial temperature T. Their value may only be obtained by solving the hydrodynamic equations in both liquid and gas phases, with suitable boundary conditions.

In both phases (indices v and l are omitted for simplicity), the full Navier–Stokes equations for incompressible Newtonian fluids read

$$\rho \left[\frac{\partial \vec{v}}{\partial t} + (\vec{v}.\vec{\nabla})\vec{v} \right] = \mu \Delta \vec{v} - \vec{\nabla}p + \rho\vec{g} \tag{3.241}$$

where μ is the dynamic viscosity (assumed to be independent of temperature), and $\vec{g} = g\vec{1}_z$ is the gravity vector ($\vec{1}_z$ is the unit vector along the vertical z-direction). As the system is unidimensional and the velocities are independent of z and t (steady problem) in both phases (the gas is also assumed to be incompressible, which is realistic for the range of velocities involved), the equations (3.241) reduce to an equation for pressure

$$\frac{dp}{dz} = -\rho g \tag{3.242}$$

We also have to consider the energy equation governing the temperature distribution in both phases:

$$\rho c_p \left[\frac{\partial T}{\partial t} + (\vec{v}.\vec{\nabla})T \right] = \lambda \Delta T \tag{3.243}$$

where c_p is the specific heat at constant pressure, and λ is the thermal conductivity, both assumed to be constant in the range of temperatures involved. For unidimensional steady transport, the equation (3.243) reduces to

$$w\frac{dT}{dz} = \kappa\frac{d^2T}{dz^2}$$ (3.244)

where $\kappa = \lambda/\rho c_p$ is the fluid thermal diffusivity.

Equations (3.242) and (3.244) may be directly integrated, to give the distribution of pressure and of temperature in both phases. We obtain $p_v = -\rho_v gz + c_1$, $p_l = -\rho_l gz + c_2$, $T_v = c_3 \exp[w_v z/\kappa_v] + c_4$ and $T_l = c_5 \exp[w_l z/\kappa_l] + c_6$, as a function of 6 integration constants c_i to be determined from boundary conditions.

First of all, for $z \to \infty$ the temperature in the vapor phase must not diverge, which implies $c_3 = 0$. Thus the temperature is constant in the gas ($T_v = c_4 = T_i$ i.e. the temperature at the interface), and all the thermal energy is transported by the gas flow with velocity w_v [247]. Furthermore, we will neglect the barometric pressure variation in the gas phase (because ρ_v is small). Thus $p_v = c_1 = p_g$, i.e. the pressure is constant in the gas phase.

The boundary condition at the rigid conducting plate ($z = 0$) is $T = T_b$, implying $c_5 + c_6 = T_b$. At the interface $z = h$, we have assumed no temperature jump, such that $T_l(z = h) = c_5 \exp[w_l h/\kappa_l] + c_6 = T_i$. From these two relations, we obtain

$$c_5 = \frac{T_i - T_b}{\exp[w_l h/\kappa_l] - 1}, \quad c_6 = T_b - c_5$$ (3.245)

The constant c_2 is obtained by considering the interfacial balance of normal momentum [see Eq. (2.89)], which here reads (hereafter, we generally neglect ρ_l^{-1} against ρ_v^{-1})

$$p_v - p_l + J^2/\rho_v = 0$$ (3.246)

where the last term accounts for the vapor recoil effect, i.e. the fact that there is a difference in velocity (and in normal momentum) between molecules of liquid arriving at the interface, and molecules of vapor leaving it. Note that in general, this effect is safely negligible, except for very high mass fluxes (e.g. when the liquid evaporates under very low pressure [245, 186]). We may retain this effect anyway, such that Eq. (3.246) gives $c_2 = p_g + J^2/\rho_v + \rho_l gh$.

Assuming that the gas pressure is given, we are thus left with two unknowns T_i (the interfacial temperature) and J (the mass flux). The fluid velocity w_l appearing in Eq. (3.245) is linked to the mass flux J by the relation $w_l = J/\rho_l$.

The conservation of energy at the interface [see Eq. (2.99)] expresses that the jump of normal heat fluxes is equal to the heat used for evaporation (latent heat L), plus a small amount of heat transformed into kinetic energy of the leaving vapor molecules (note that the even smaller kinetic energy of arriving liquid molecules is negligible [186]):

$$\lambda_v\frac{dT_v}{dz} - \lambda_l\frac{dT_l}{dz} = J\left(L + \frac{1}{2}w_v^2\right) = J\left[L + \frac{1}{2}\left(\frac{J}{\rho_v}\right)^2\right]$$ (3.247)

This relation allows the interfacial temperature T_i to be computed as a function of the mass flux, using the expressions obtained above for the temperature profiles. We obtain

$$T_i = T_b - \frac{L + \frac{1}{2}(\frac{J}{\rho_v})^2}{c_{pl}}(1 - \exp[-w_l h/\kappa_l])$$ (3.248)

The last boundary condition, providing a second relation between the mass flux and the interfacial temperature, is the phenomenological relation (3.240). With the help of Eq. (3.246), this may be rewritten

$$J = \frac{K}{\rho_v} \left[p_s(T_i) - p_g + \frac{J^2}{\rho_l} \right]$$

(3.249)

where $1/\rho_l$ has again been neglected with respect to $1/\rho_v$.

Relations (3.248) and (3.249) allow in principle T_i and J to be computed, provided that we know the function $p_s(T)$, i.e. the relation between the equilibrium pressure and the interfacial temperature. As an illustration, we may for instance use the Clausius–Clapeyron relation for a perfect gas

$$p_s(T) = p_0 \exp \left[-\frac{L}{R} \left(\frac{1}{T} - \frac{1}{T_0} \right) \right]$$

(3.250)

where (p_0, T_0) is a point lying on the saturation curve (e.g. $T_0 = 100\,°C$ and $p_0 = 1$ atm for pure water). For simplicity, we assume that the difference $T - T_0$ is small, and we use the linearized form of Eq. (3.250)

$$p_s(T) \simeq p_0 + p_T(T - T_0)$$

(3.251)

where $p_T = Lp_0/RT_0^2$ ($\simeq 0.036$ atm/K for pure water).

We may now proceed to the resolution of the system of equations (3.248), (3.249) and (3.251). We first replace (3.248) in (3.251)

$$p_s(T_i) = p_b + p_T \frac{\left[L + \frac{1}{2} \left(\frac{J}{\rho_v} \right)^2 \right]}{c_{pl}} \left(\exp \left[-\frac{Jh}{\rho_l \kappa_l} \right] - 1 \right)$$

(3.252)

where p_b stands for $p_0 + p_T(T_b - T_0)$, i.e. the saturation pressure evaluated at the temperature of the bottom plate.

In the general case, the analytical resolution of Eqs (3.249) and (3.252) with respect to J and T_i appears to be impossible. In order to attempt a graphical resolution, we first solve (3.249) with respect to $p_s(T_i)$:

$$p_s(T_i) = p_g + \frac{\rho_v J}{K} - \frac{J^2}{\rho_l}$$

(3.253)

and both values of $p_s(T_i)$, given by Eqs (3.252) and (3.253), are represented as a function of J in the same diagram (Fig. 3.42). The solution for J is at the intersection of both curves.

Now, it turns out that for the conditions of Fig. 3.42, J is not too large, and an analytical expression can be obtained. First of all, order of magnitude estimates show that we can neglect the term $(J/\rho_v)^2$ with respect to L in Eq. (3.252). This amounts to neglecting the kinetic energy imparted to the molecules of gas leaving the interface, compared with the energy used for evaporation. Another simplification is to neglect the effect of vapor recoil, i.e. J^2/ρ_l compared with other terms in (3.253). This is valid for the range of values of J considered in this analysis (both these effects are only important at very high mass fluxes J, and the latter can lead to specific instabilities [186, 245, 249]).

A more restrictive approximation concerns the exponential term in Eq. (3.252). This term may be linearized provided that

$$Pe = \frac{Jh}{\rho_1 \kappa_1} = \frac{w_1 h}{\kappa_1} \ll 1 \tag{3.254}$$

defining the thermal Peclet number Pe, representing the ratio of the velocity of the fluid (associated with the rate of phase change) to the characteristic thermal velocity. It may also be defined as the ratio of the thermal relaxation time h^2/κ_1 to the convective time h/w_1. Thus, its value will be small provided the rate of phase change is not too high. More precisely, for a water layer of depth $h = 1$ mm, this approximation is valid if J is at most of the order 10^{-3} g/cm^2 s, representing a heat transport of $JL = 2.3$ W/cm^2.

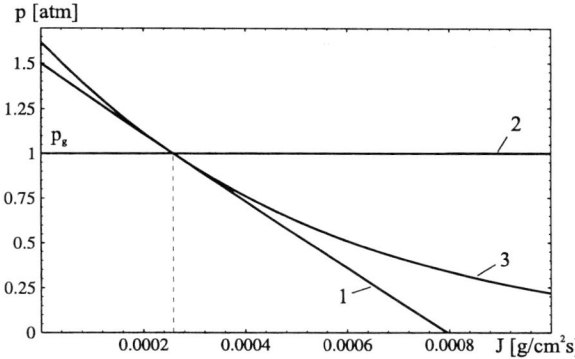

Figure 3.42: Graphical determination of J. $p_g = 1$ atm is the vapor pressure, $p_b = 1.5$ atm is the saturation pressure at the rigid plate temperature $T_b = 387$ K [Eq. (3.251)]. Curves 1 and 2 are given by Eqs (3.252) and (3.253), curve 3 is the Clausius–Clapeyron curve (3.250) in which the temperature of the interface is evaluated by Eq. (3.248). All the parameters are those of water ($T_0 = 373$ K, $p_0 = 1$ atm), with a depth $h = 1$ mm, and an accommodation coefficient $\beta = 0.1$. The supersaturation is $T_b - T_0 = 14$ °C. It is found that $J \simeq 2.6 \, 10^{-4}$ g/cm^2 s ($JL \simeq 0.6$ W/cm^2).

With these approximations, Eq. (3.252) may be written

$$p_s(T_i) \simeq p_b + \frac{p_T L}{c_{pl}}\left(-\frac{Jh}{\rho_1 \kappa_1}\right) = p_b - \frac{p_T L J h}{\lambda_1} \tag{3.255}$$

and Eq. (3.253) reads

$$p_s(T_i) = p_g + \frac{\rho_v J}{K} \tag{3.256}$$

Equalization of these two relations then leads to the mass flux

$$J = \frac{p_b - p_g}{\dfrac{\rho_v}{K} + \dfrac{p_T L h}{\lambda_l}} = \frac{K}{\rho_v} \frac{p_b - p_g}{1 + Bi} \tag{3.257}$$

where $Bi = K p_T L h / \rho_v \lambda_l$ is a dimensionless number characterizing the regime of evaporation. For $Bi \ll 1$, Eq. (3.257) leads to $J = K \rho_v^{-1}(p_b - p_g)$, and hence $p_s(T_i) = p_b$ according to Eq. (3.256), i.e. the interface is at the temperature of the bottom plate, and the rate of evaporation is limited by the magnitude of K (*reaction-limited regime*). On the contrary, for $Bi \gg 1$, we get the relation

$$J = \frac{\lambda_l(p_b - p_g)}{p_T L h} \tag{3.258}$$

and the mechanism limiting the evaporation rate is now heat diffusion (*diffusion-limited regime*). Indeed, it is checked from Eq. (3.255) that $p_s(T_i) = p_g$ in this limit, i.e. the interface is at local equilibrium.

As mentioned earlier, K (and hence, Bi) can be roughly estimated by the formula $K = \beta \rho_v(T)(M/2\pi RT)^{1/2}$. For water, this gives $K \simeq 6\,10^{-10}$ g s/cm^4 (with a value 0.1 of the accommodation coefficient β), and $Bi \sim 800$ for the example considered above, i.e. the evaporation rate is limited by the transport of heat towards the interface, and Eq. (3.258) holds.

Still, in the following, we will adopt the more general expression (3.257) of the reference mass flux J, in part because the actual values of K could be quite different from those estimated with the help of the Hertz–Knudsen relation. Moreover, the value of the accommodation coefficient β is generally found to be significantly reduced by the presence of impurities on the interface, such that it could not be legitimate to neglect non-equilibrium effects (finite K). This holds in particular when the *stability* of the reference state is considered.

3.8.3 Stability of the reference solution

Here, a particular model of the problem will be used. This is known in the literature as the one-sided model [186], for which the dynamics of the gas phase can be decoupled from the dynamics of the liquid phase. We may thus solve the problem (with suitable boundary conditions) in the liquid phase only. This model requires the assumption that the ratios of densities ρ_v/ρ_l, of thermal conductivities λ_v/λ_l and of dynamic viscosities μ_v/μ_l are small compared with unity, which is verified for most liquids in contact with their vapor, far enough from the critical point. We also refer to §3.5.2 for the discussion of the conditions under which a two-phase system without evaporation can be reduced to a one-layer formulation.

As mentioned above, our goal here will be to stress the assumptions necessary to reduce the full problem to the Pearson's problem considered in §3.3.2, with a value of the Biot number actually identical to the parameter $Bi = K p_T L h / \rho_v \lambda_l$ characterizing the cross-over between reaction-limited ($Bi \ll 1$) and heat-diffusion-limited ($Bi \gg 1$) regimes.

Perturbation equations

Linearizing Eqs (3.241) and (3.243) around the evaporative reference state $\vec{v}_{\text{ref}}$, p_{ref} and T_{ref}, and including the incompressibility condition, we have

$$\vec{\nabla} \cdot \vec{v} = 0 \tag{3.259}$$

$$\rho \frac{\partial w}{\partial t} + \rho w_{\text{ref}} Dw = \mu \Delta w - Dp \tag{3.260}$$

$$\rho \frac{\partial u}{\partial t} + \rho w_{\text{ref}} Du = \mu \Delta u - \frac{\partial p}{\partial x} \tag{3.261}$$

$$\frac{\partial T}{\partial t} + w_{\text{ref}} DT + w DT_{\text{ref}} = \kappa \Delta T \tag{3.262}$$

where both components of the Navier–Stokes equation (3.241) have been separated ($\vec{v} = u \vec{1}_x + w \vec{1}_z$), D denotes the z-derivative, and $\kappa = \lambda/\rho c_p$ is the thermal diffusivity. Moreover, buoyancy is assumed to be negligible (thin layers or microgravity). Note that hereafter, subscripts v and l will be omitted when referring to liquid quantities.

On the rigid conducting bottom plate $z = 0$, we have

$$u = w = T = 0 \tag{3.263}$$

i.e. the total temperature is maintained constant at T_{b} and the total velocity is kept at its reference value $\vec{V} = w_{\text{ref}} \vec{1}_z = \vec{1}_z J_{\text{ref}}/\rho_1$ [J_{ref} is the reference value of the mass flux, given by Eq. (3.257), or determined via graphical resolution for higher rates].

At the interface $z = h + \xi(x, t)$, linearization of the mass conservation relation (3.239) yields

$$J = J_{\text{ref}} + J' = \rho_1 w_{\text{ref}} + \rho_1 [w_1(z = h) - w_\Sigma] \tag{3.264}$$

As $J_{\text{ref}} = \rho_1 w_{\text{ref}}$, the perturbed mass flux J' is given by

$$J' = \rho_1 [w_1(z = h) - w_\Sigma] \tag{3.265}$$

and will be denoted by J in the following. Note that an equivalent treatment of the second equality in (3.239) gives

$$J' = \rho_v [w_v(z = h) - w_\Sigma] \tag{3.266}$$

The general form of the momentum balance at the interface reads [Eq. (2.87)]

$$[\![\rho\{(\vec{v} - \vec{v}_\Sigma) \cdot \vec{n}\} \vec{v} + p\vec{n} - \mathbf{P} \cdot \vec{n}]\!] = \frac{\sigma \partial \vec{t}/\partial x + \vec{t} \partial \sigma/\partial x}{N} \tag{3.267}$$

where $[\![x]\!]$ stands for the discontinuity $x_v - x_1$ (v and l denote respectively the vapor and the liquid), $\vec{n} = (-\xi' \vec{1}_x + \vec{1}_z)/N$ and $\vec{t} = (\vec{1}_x + \xi' \vec{1}_z)/N$ are respectively the free surface normal and tangential unit vectors [$N = (1 + \xi'^2)^{1/2}$ is the normalization factor], σ is the surface tension, $\mathbf{P}$ is the viscous stress tensor with components $P_{ij} = \mu(\partial v_i/\partial x_j + \partial v_j/\partial x_i)$. Note that according to Eq. (3.239), $[\![\rho\{(\vec{v} - \vec{v}_\Sigma).\vec{n}\}\vec{v}]\!] = J[\![\vec{v}]\!]$ and $J[\![\vec{v}]\!].\vec{n} \simeq J^2/\rho_v$.

Applying the one-sided model approximation ($\rho_v/\rho_l \to 0$, $\mu_v/\mu_l \to 0$), substituting $p \to p_{ref} + p$, $w \to w_{ref} + w$, and linearizing, the projection of (3.267) on the surface normal reads

$$(p_v)_{ref} - (p_l)_{ref} + p_v - p_l = -(\mathbf{P}_l.\vec{n}).\vec{n} - \frac{J_{ref}^2}{\rho_v} - 2\frac{J_{ref} J}{\rho_v} + \sigma \frac{\partial^2 \xi}{\partial x^2} \tag{3.268}$$

where the relations (3.265–3.266) have been used. This relation must be expressed at $z = h + \xi(x, t)$. Since $\xi(x, t) \ll h$, the Taylor series around $z = h$ can be limited to linear terms in ξ. Using (3.246) at $z = h$, we obtain

$$p_v - p_l = -\rho_l g\xi - 2\mu_l Dw_l - 2\frac{J_{ref} J}{\rho_v} + \sigma \frac{\partial^2 \xi}{\partial x^2} \tag{3.269}$$

We must also project the relation (3.267) on the surface unit tangent $\vec{t}$. This gives, after linearization,

$$\mu_l \left(\frac{\partial u_l}{\partial z} + \frac{\partial w_l}{\partial x} \right) = \sigma_T \frac{\partial}{\partial x}(T + \xi DT_{ref}) \tag{3.270}$$

where σ_T is the surface tension variation with temperature. We may use the continuity relation (3.259) to rewrite (3.270) in the form (expressed at $z = h$)

$$\mu_l \left(\frac{\partial^2 w}{\partial x^2} - \frac{\partial^2 w}{\partial z^2} \right) = \sigma_T \frac{\partial^2 T}{\partial x^2} \tag{3.271}$$

Note that we used the no-slip condition

$$\vec{v}_l.\vec{t} = \vec{v}_v.\vec{t} \tag{3.272}$$

whose perturbed and linearized form is

$$u_l = u_v + \frac{\partial \xi}{\partial x} \frac{J_{ref}}{\rho_v} \tag{3.273}$$

Now, perturbing Eq. (3.240) [neglecting the variations of K, ρ_v and ρ_l with T], and applying the one-sided model approximation, we get the perturbed mass flux as

$$J = K \left[\frac{p_T}{\rho_v}(T + \xi DT_{ref}) + \frac{p_l}{\rho_l} - \frac{p_v}{\rho_v} \right] \tag{3.274}$$

We may eliminate p_v between Eqs (3.269) and (3.274), and get the following relation for the flux J:

$$J = \frac{K}{1 - 2K\rho_v^{-2} J_{ref}} \left[\frac{p_T}{\rho_v}(T + \xi DT_{ref}) - \frac{p_l}{\rho_v} + \frac{\rho_l g}{\rho_v}\xi + 2\frac{\mu_l}{\rho_v}Dw_l - \frac{\sigma}{\rho_v}\frac{\partial^2 \xi}{\partial x^2} \right] \tag{3.275}$$

Finally, the general conservation equation for the energy at the interface [see Eq. (2.99)] reads (in the one-sided model [186], and neglecting the work of viscous stresses) as

$$
J\left[L + \frac{1}{2}\left(\frac{J}{\rho_v}\right)^2\right] = -\lambda_1\,\vec{n}.\vec{\nabla}T_1
\tag{3.276}
$$

whose perturbed and linearized form is

$$
-\lambda_1 DT_1 - \lambda_1 \xi D^2 T_{\text{ref}} = J\left[L + \frac{3}{2}\left(\frac{J_{\text{ref}}}{\rho_v}\right)^2\right]
\tag{3.277}
$$

Dimensionless equations

The equations and boundary conditions can be put under dimensionless form by using the usual thermal scales: h (the unperturbed liquid depth) as unit length, h^2/κ as unit time, κ/h as unit velocity, and $\mu\kappa/h^2$ as unit pressure (fluid quantities without indices are those of the liquid phase). The temperature unit θ is based on the amplitude of the thermal gradient at the interface in the basic state, i.e.

$$
\theta = h|DT_{\text{ref}}(z = h)| = \frac{h|J_{\text{ref}}|\left[L + \frac{1}{2}(J_{\text{ref}}/\rho_v)^2\right]}{\lambda_1}
\tag{3.278}
$$

The dimensionless equations valid in the liquid phase thus read (the same notations are kept for dimensionless velocity, temperature and pressure):

$$
\vec{\nabla}.\vec{v} = 0
\tag{3.279}
$$

$$
Pr(\Delta w - Dp) = \frac{\partial w}{\partial t} + PeDw
\tag{3.280}
$$

$$
Pr(\Delta u - \frac{\partial p}{\partial x}) = \frac{\partial u}{\partial t} + PeDu
\tag{3.281}
$$

$$
\Delta T + w\exp[Pe(z - 1)] = \frac{\partial T}{\partial t} + PeDT
\tag{3.282}
$$

where $Pr = \mu/\rho\kappa$ is the Prandtl number of the liquid, and the Peclet number $Pe = w_{\text{ref}}h/\kappa$ has been defined earlier [Eq. (3.254)].

The boundary conditions are:

At the bottom plate ($z = 0$): the usual conditions

$$
u = w = T = 0
\tag{3.283}
$$

At the interface ($z = 1$): the mass conservation relation – definition of mass flux [Eqs. (3.265) and (3.275)]:

$$J = Pr^{-1}(w - w_\Sigma) = \frac{PeBi}{Pr}\left[(T - \xi) - \Pi_1\left(p - 2Dw - Ga\xi + Ca^{-1}\frac{\partial^2\xi}{\partial x^2}\right)\right]$$

(3.284)

where units of J are taken as μ_1/h. In this relation, the "Biot" number Bi has been defined as

$$Bi = \frac{p_T h}{\rho_v \lambda_1}\frac{K[L + 2^{-1}(J_{\text{ref}}/\rho_v)^2]}{1 - 2KJ_{\text{ref}}/\rho_v^2}$$

(3.285)

while other dimensionless groups appearing in Eq. (3.284) are

$$\Pi_1 = \frac{\mu_1\kappa_1}{p_T\theta h^2} \simeq \frac{\mu_1\lambda_1\kappa_1}{J_{\text{ref}}h^3 p_T L},$$

(3.286)

and the usual capillary number $Ca = \mu_1\kappa_1/\sigma h$ and Galileo number $Ga = gh^3/\nu_1\kappa_1$ (see §3.2.1).

Another boundary condition at $z = 1$ is the normal momentum balance [Eq. (3.269)]

$$p_v - p_1 + 2Dw_1 = Ca^{-1}\frac{\partial^2\xi}{\partial x^2} - Ga\xi - \frac{2Pe}{\rho^* Pr}(w - w_\Sigma)$$

(3.287)

where we introduced the density ratio $\rho^* = \rho_v/\rho_1$.

We also have the tangential stress condition [Eq. (3.271)]

$$D^2 w - \frac{\partial^2 w}{\partial x^2} = Ma\frac{\partial^2}{\partial x^2}(T - \xi)$$

(3.288)

where $Ma = -\sigma_T\theta h/\mu_1\kappa_1 = -\sigma_T\beta h^2/\mu_1\kappa_1$ is the Marangoni number [$\beta = |DT_{\text{ref}}(z = h)|$ is the reference temperature gradient at the interface].

The no-slip condition [Eq. (3.272)] reads

$$u_1 = u_v + \frac{Pe}{\rho^*}\frac{\partial\xi}{\partial x}$$

(3.289)

The thermal boundary condition is obtained from Eq. (3.277), using $D^2 T_{\text{ref}}|_{z=h} = -Pe\theta/h^2$. With (3.275) and the definitions of Bi and Π_1, Ca and Ga, we get

$$DT - Pe\xi = \delta Bi\left\{-(T - \xi) + \Pi_1\left[(p - 2Dw) - Ga\xi + Ca^{-1}\frac{\partial^2\xi}{\partial x^2}\right]\right\}$$

(3.290)

where $\delta = \left[L + 3(J_{\text{ref}}/\rho_v)^2/2\right] / \left[L + (J_{\text{ref}}/\rho_v)^2/2\right] \simeq 1$. The last boundary condition is the kinematic condition, whose dimensionless linearized form reads

$$w_\Sigma = \frac{\partial\xi}{\partial t}$$

(3.291)

The system of equations and boundary conditions (3.279–3.291) is still not closed, essentially because the gas pressure perturbation still appears in the boundary condition (3.287). However, an important limiting case occurs when the capillary number is very small (for a water layer of depth $h = 1$ mm, $Ca = 1.4\,10^{-6}$) and the Galileo number very large (under the same conditions $Ga \simeq 10^5$), such that $\xi \to 0$ according to the boundary condition (3.287), and the interface deformation is suppressed by the combined action of surface tension and gravity (see also §3.6.1). Furthermore, order of magnitude estimates show that $\Pi_1 \sim 10^{-9}$ for the range of mass fluxes considered, such that terms multiplied by Π_1 may be neglected in Eqs (3.284) and (3.290), provided $|p - 2Dw|/|T|$ is not too large (this can be verified a posteriori).

With these approximations, the system of equations (3.279–3.282) is unchanged, while the free surface boundary conditions (3.284–3.291) reduce to:

$$w = PeBiT \tag{3.292}$$

$$D^2 w - \frac{\partial^2 w}{\partial x^2} = Ma\frac{\partial^2 T}{\partial x^2} \tag{3.293}$$

$$DT + BiT = 0 \tag{3.294}$$

and the system is closed in this limit, since no quantities linked to the gas phase appear in these boundary conditions.

It was also seen in the last section that the liquid Peclet number Pe is often small compared with unity. In the limit $Pe \to 0$, the problem reduces to the problem treated linearly by Pearson [101], and presented in §3.3.2. However, in Pearson's analysis, the value of the Biot number Bi is not related to evaporation parameters (because the heat transfer at the free surface is described by a classical Newton's cooling law (2.77), with a phenomenological heat-transfer coefficient α whose value depends on several experimental peculiarities, such as the gas depth and thermal conductivity, gas convection, ...). Here, the relation (3.285) can be used to evaluate the Biot number. Still, this relation also contains a phenomenological coefficient K [defined by Eq. (3.240)], whose order of magnitude may in principle be obtained from the kinetic theory of gases. For water with $h = 1$ mm, and an accommodation coefficient $\beta = 0.1$, we find $Bi \simeq 10^3$, but this value could be significantly reduced by impurities, leading to much lower values of the accommodation coefficient ($\beta = 0.01$ or even $\beta = 0.001$).

Excluding the case of very low vapor pressures, the vapor recoil effects can be safely neglected in Eq. (3.285), leading to

$$Bi = \frac{Kp_ThL}{\rho_v\lambda_l} \tag{3.295}$$

i.e. indeed the dimensionless parameter characterizing the evaporation regime (heat-diffusion-limited or reaction-limited).

As the effective Biot number may vary over a wide range, plots of the Pearson's result (3.118) are extended to higher values of Bi in Fig. 3.43. The critical Marangoni number $Ma_c(Bi)$ is seen to increase drastically with the Biot number. For $Bi \gg 1$, Ma_c is proportional to Bi, while the critical wavenumber tends to a constant close to 3.

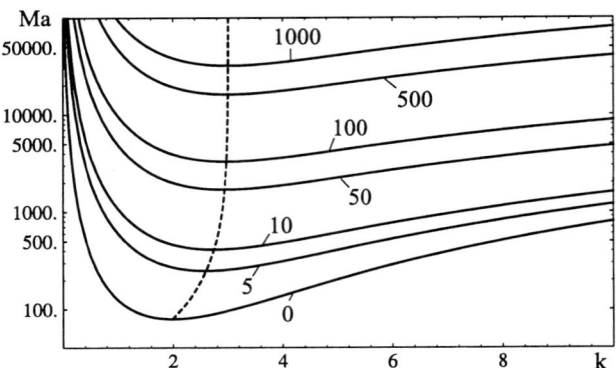

Figure 3.43: Neutral stability curves for various Biot numbers, as given by Eq. (3.118). The critical Marangoni number is given by the minimum of these curves as a function of the wavenumber k (dashed line).

Note that the actual value of the Marangoni number $Ma = -\sigma_T \theta h / \mu_l \kappa_l$ here depends on the evaporation-driven temperature drop θ, i.e. it must be *calculated* using Eq. (3.278). For a given temperature T_b at the bottom plate, and considering the vapor pressure p_v as the control parameter, it is found that instability occurs when $\Delta p = p_s(T_b) - p_v > \Delta p_c$, where the critical pressure drop Δp_c is proportional to $(1 + Bi)Ma_c(Bi)/K$, for a given fluid and depth h. For $Bi \gg 1$ (reaction-limited evaporation), it follows that $\Delta p_c \sim K$, while for $Bi \ll 1$ (heat-diffusion-limited evaporation), we have $\Delta p_c \sim K^{-1}$. Hence, the reference state is certainly stable both for large and small values of the phenomenological coefficient K, and there exists a range of values of K (hence, of the accommodation coefficient β) for which Marangoni convection can occur.

To summarize, the assumptions underlying the analysis presented in this section are the use of the phenomenological relation (3.240) to model the non-equilibrium mass flux, the one-sided model approximation (leading to decoupling of the dynamics of perturbations in the liquid and in the gas), the assumption that the surface deformation can be neglected (when the capillary number Ca is small, and the Galileo number Ga is large), and the assumption of linearity of the basic temperature profile induced by evaporation (the Peclet number is assumed small compared with unity). Furthermore, we also neglected the vapor recoil effect and the amount of kinetic energy imparted to gas molecules in the interfacial energy balance. Note that a direct comparison of neutral stability thresholds with a full numerical resolution of the two-layer problem including all the above effects has shown that the analysis presented here is indeed quantitatively valid in a large range of experimental conditions, for gas pressures that are not too low [253].

In conclusion, although the local equilibrium hypothesis at the interface may be valid for the calculation of the evaporation rate in the reference state (at large Bi), it might not be legitimate to neglect non-equilibrium effects when considering its stability against hydrodynamic fluctuations. Though this effect may be small, it is sufficient to allow Marangoni instability in some range of accommodation coefficients. In this sense, it would also be interesting to investigate the possible destabilization of the reference state by other small effects such as the hydrodynamic fluctuations of liquid and vapor pressure near the interface in the expression of the flux (3.274), and interfacial deformation. Clearly, the analysis should also be extended to non-zero values of the Peclet number Pe, which may be done on the basis of (3.279–3.282) and (3.292–3.294).

4 Weakly nonlinear pattern dynamics

4.1 Scope of nonlinear theories

In the previous chapter, various monotonic and oscillatory instability modes were identified. It was seen that above certain critical values of the dimensionless (or control) parameters characterizing the deviation from equilibrium, not all perturbations were damped but some of them were characterized by a positive growth rate. This indicates the loss of stability of the reference diffusive solution when the constraint is increased past its critical value. As the temporal growth rate of some of the normal modes becomes positive at that point, an exponential growth is predicted in the frame of linear theories, both for monotonic and for oscillatory modes (in the latter case the amplitude of the oscillation has an exponential growth). This unphysical behavior is connected to the linearization process, by which nonlinearities of various origins (e.g. advective terms in the basic equations, surface deformation, ...) were neglected. Further discussion of this point is presented in §4.1.1, where we also recall some basic results and definitions of elementary bifurcation theory.

Orientational degeneracy is an essential feature of isotropic systems such as those considered in this book. As all the horizontal directions are indistinguishable (rotational invariance), normal modes characterized by the same wavenumber but differing by their orientation in the horizontal plane are completely equivalent, and thus possess the same growth rate. Moreover, every superposition of plane wave solutions of the linear problem is itself a solution, reflecting the absence of interactions (or competitions) between eigenmodes of the linear problem. This infinite degeneracy is also greatly reduced by taking nonlinear interactions into account, in a way presented in §4.1.2 and detailed later on in this chapter. Note that even nonlinear theory will not allow prediction of the orientation of the nonlinear structure in a strictly isotropic horizontal plane. In a particular experiment, this orientation will ultimately depend on initial conditions, i.e. on unpredictable thermodynamic fluctuations, distant lateral walls or imperfections of the experimental set-up.

Another kind of degeneracy is the so-called "sideband degeneracy". It appears in laterally extended geometries, and a fortiori in laterally infinite systems, where the possible wavenumbers form a continuum. Just above the linear stability threshold, normal modes with wavenumbers within a band centered around the critical wavenumber k_c have a positive growth rate. Although some selection exists at the linear stage (because the linear growth rate depends on the wavenumber), the correct description of interactions between unstable modes once again requires the inclusion of nonlinear coupling and saturation effects. It will be seen in this chapter that nonlinear effects typically reduce the band of wavenumbers accessible to the system. Still, for strictly periodic structures in infinite domains, uncertainty remains as to which single wavelength will be selected within this narrowed band. This important question of wavelength selection or "realizability" of the stable nonlinear structures is still a subject of intensive research [19, 289]. Actually, wavelength selection appears to depend strongly upon several factors, such as the presence of lateral walls at large but finite distance, defects in the nonlinear structures, history of the system, ... The powerful method of envelope (Ginzburg–Landau) equations is introduced in §4.1.3, and used in the following sections to examine some of these effects.

Before embarking on detailed discussions, it is worth emphasizing the generality of weakly nonlinear theories. By opposition with the linear phenomena presented in the previous chapter, the nonlinear behaviors observed in the vicinity of critical points are in general much less dependent upon the specific mechanisms involved in the instability. At the linear stage, the microscopic peculiarities of a given problem (described in our case by thermo-hydrodynamic equations) are essential in the sense that they determine the appearance of instabilities, whose mechanism is intimately connected with the importance of various effects: heating from below or above, considering interfacial deformation, ... At the critical point however, many predictions can be made without this detailed knowledge, e.g. about the nature of the bifurcation, nonlinear competitions, secondary instabilities, ... These predictions depend on more general macroscopic features such as the nature of the instability (monotonic or oscillatory), the critical wavenumber (zero or finite) and symmetries (here, rotational, translational, mirror and time-translation invariances, or more specific symmetries). The role of symmetries relevant for Bénard instabilities will be discussed in §4.2.1. Given this limited amount of data, weakly nonlinear theories lead to some canonical forms (the normal forms) for the equations governing the dynamics of the amplitudes of near-critical modes, the study of which leads to predictions concerning the number and nature of bifurcating branches, dynamical regimes, ... This justifies that in the following, some of these bifurcations will first be studied without referring to any of the particular systems studied in Chapter 3.

However, the amplitude equations also generally depend on some non-rescalable coefficients (i.e. which cannot be eliminated by simple rescaling of amplitudes, and of space and time coordinates). Thus, the discussion of weakly nonlinear regimes should be achieved as a function of these coefficients. Contrary to the analytical form of the amplitude equations, coefficients depend on the particular physical system studied. Thus, among the possible behaviors associated with a certain normal form, only a subset will actually be observed for a given system. The calculation of the coefficients of amplitude equations is generally achieved via perturbation or projection methods, presented in §4.2, and developed for specific Bénard instabilities in subsequent sections.

In some sense, the possibility of establishing groups of physical systems exhibiting dynamically equivalent behaviors is similar to equilibrium phase transition phenomena, where some classes of systems are known to behave in similar ways (e.g. have the same scaling exponents) in the immediate vicinity of critical points (Universality Classes [52]). Other similarities exist between equilibrium and non-equilibrium transitions, some of which will be encountered in the following.

For non-equilibrium bifurcation phenomena considered here, three broad classes of instabilities may be defined [19], depending on the behavior of the growth rate in the neighborhood of the critical point (Fig. 4.1).

In the following, we will mainly be interested in case I, for which the critical wavenumber

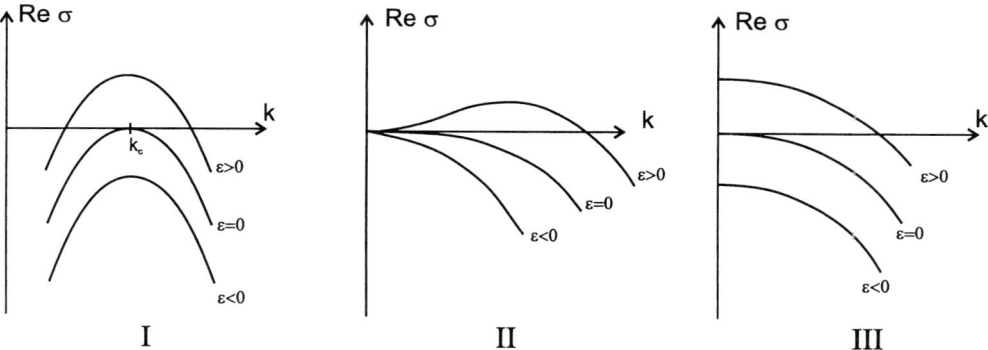

Figure 4.1: Schematic representation of the linear growth rate as a function of the wavenumber k, for several reduced control parameters ϵ (the reference state is stable for $\epsilon < 0$, while it is unstable for $\epsilon > 0$), defining instabilities of types I,II and III [19].

is finite (both monotonic and oscillatory cases will be studied, respectively in this chapter and in the next one). This indeed corresponds to most of the instabilities in finite-depth systems studied in the previous chapter. Long-wave instability modes (case II) are typically encountered when the upper and lower boundaries of the Bénard set-up are impervious to a diffusive quantity (see §3.3.1 and §3.7.1), or when surface deformation is allowed (see §3.6). These long-wave modes have been extensively studied both for Rayleigh and Marangoni–Bénard instabilities (see e.g. [215, 291, 292]) and will be discussed in §4.6. Note that the nonlinear equations derived for these long-wave modes (e.g. by Knobloch [215], or Shtilman and Sivashinsky [292]) can also be used as a model (simpler than the full system of hydrodynamic equations and boundary conditions), which can be reduced to case I in some particular cases. Finally, case III (where the homogeneous state can be amplified) is usually not encountered in Bénard problems, and will not be studied here. The interested reader is invited to consult the literature on reaction-diffusion systems [11, 19, 184] for the study of these type III modes.

4.1.1 Saturation of exponential growth – Landau's theory – Normal forms

To fix ideas, let us consider the Rayleigh–Bénard instability problem studied in §3.1.3 (stress-free boundaries). The control parameter is the Rayleigh number Ra, and the neutral stability curve is given by $Ra = (k^2 + \pi^2)^3/k^2$. The dispersion relation can here be calculated analytically [see Eq. (B.18) of Appendix B], and reads

$$(\sigma + k^2 + \pi^2)(Pr^{-1}\sigma + k^2 + \pi^2) = \frac{k^2}{k^2 + \pi^2}Ra \tag{4.1}$$

showing that two modes exist, whose growth rate σ can be calculated analytically [one of the modes is always stable, while the other one can bifurcate at the critical value $Ra = (k^2 + \pi^2)^3/k^2$, as seen in §3.1.2 and 3.1.3]. However, for our purpose in this section it will be sufficient to consider only the bifurcating mode in the vicinity of the critical point $k_c = \pi/\sqrt{2}$, $Ra_c = 27\pi^4/4$. Setting $k = k_c$, linearization of σ around $Ra = Ra_c$ yields

$$\sigma = \frac{3\pi^2}{2(1 + Pr^{-1})}\epsilon + O(\epsilon^2) \qquad (4.2)$$

where $\epsilon = (Ra - Ra_c)/Ra_c$ is a reduced supercriticality parameter.

According to the form (3.101) of normal modes ($\sim \exp[\sigma t]$), an exponential growth is predicted as soon as $\epsilon > 0$. This unbounded growth is clearly unrealistic after a time $\sim \epsilon^{-1}$, when perturbations become finite and get out of the linear range. In fact, in the linear stability studies of Chapter 3, it was seen that convective perturbations were determined up to an arbitrary multiplicative constant, say A, which could be set to a numerical value according to some normalization condition. This quantity actually represents the amplitude of the convective solution, as seen by rewriting the linear solution under the form

$$\begin{aligned} W &= A(t)\cos[k_c x]\sin[\pi z] \\ T &= A(t)\cos[k_c x]\sin[\pi z]/(\sigma + k_c^2 + \pi^2) \end{aligned} \qquad (4.3)$$

representing convective rolls (with axis parallel to y), satisfying stress-free heat-conducting boundary conditions $w = D^2 w = \theta = 0$ at horizontal boundaries $z = 0, 1$. Then, Eqs (3.99–3.100) are satisfied provided the amplitude $A(t)$ satisfies

$$\frac{dA}{dt} = \sigma A \qquad (4.4)$$

where σ is solution of the dispersion relation (4.1).

The first question to be answered by a *nonlinear theory* is the following: is it possible to achieve a description of the system above the convective threshold (at least in its vicinity) by considering a more general evolution equation than Eq. (4.4) for the time-dependent amplitude $A(t)$ of the convective perturbations? Later in this chapter, it will be seen that reintroducing nonlinear terms and applying perturbation theory in the vicinity of $\sigma = 0$ yields the so-called Landau equation

$$\frac{dA}{dt} = \sigma A - \beta A^3 \qquad (4.5)$$

where β is a constant coefficient. Note that more generally, the amplitude A is a complex quantity and $\{W, T\} \sim A\exp[ik_c x] + c.c.$, in which case A^3 must be replaced by $A|A|^2$ in Eq. (4.5). This will also be discussed later on.

Landau obtained Eq. (4.5) by the following considerations [10, 51, 210]. If the hypothesis is made that an equation of the form

$$\frac{dA}{dt} = f(A) \qquad (4.6)$$

holds in the vicinity of $\sigma = 0$ (excluding a possible time-dependence of f in view of the autonomous nature of the problem), and that the function f is analytical at $\sigma = 0$, a Taylor series limited to the cubic term gives

$$\frac{dA}{dt} = \sigma A + \delta A^2 - \beta A^3 \tag{4.7}$$

where coefficients σ, δ and β are constants independent of A. Now, for perturbations of the form (4.3), the coefficient of the quadratic term (and of every even-order term) should cancel for symmetry reasons. Consider the operation $A \to -A$. Applied to Eq. (4.3), this is seen to correspond to a horizontal translation $x \to x + \pi/k_c$ (i.e. *half* a horizontal period), or equivalently to a reversed sense of rotation of convective rolls. As the system is assumed to be infinite in horizontal directions, the translated solution is completely equivalent to the original, and the equation describing the evolution of $A' = -A$ must be identical to that valid for A. This is only possible if $\delta = 0$ [then Eq. (4.7) is invariant under $A \to -A$], and the Landau equation (4.5) indeed applies.

Consider first the case $\beta > 0$. For $\sigma < 0$, the only steady solution of Eq. (4.5) is the motionless state $A = 0$, which is linearly stable [linearization near $A = 0$ leads to Eq. (4.4)]. For $\sigma > 0$, $A = 0$ is an unstable solution, while two (equivalent) branches of steady convective states $A_s = \pm(\sigma/\beta)^{1/2}$ have emerged, which are stable [if $A = A_s + A'$, the evolution of the small perturbation A' is governed by $dA'/dt = -2\sigma A'$]. Thus, increasing σ (or ϵ) above zero leads to loss of stability of the motionless state, which is replaced by a convective state after a transient of duration $\sim \epsilon^{-1}$. The latter thus becomes infinitely long for $\epsilon \to 0$, a phenomenon called *critical slowing down*. A discontinuity of the derivative of the steady amplitude with respect to ϵ is observed at $\epsilon = 0$. This phenomenon is called a *supercritical, or pitchfork, bifurcation*, and the corresponding bifurcation diagram is represented in Fig. 4.2.

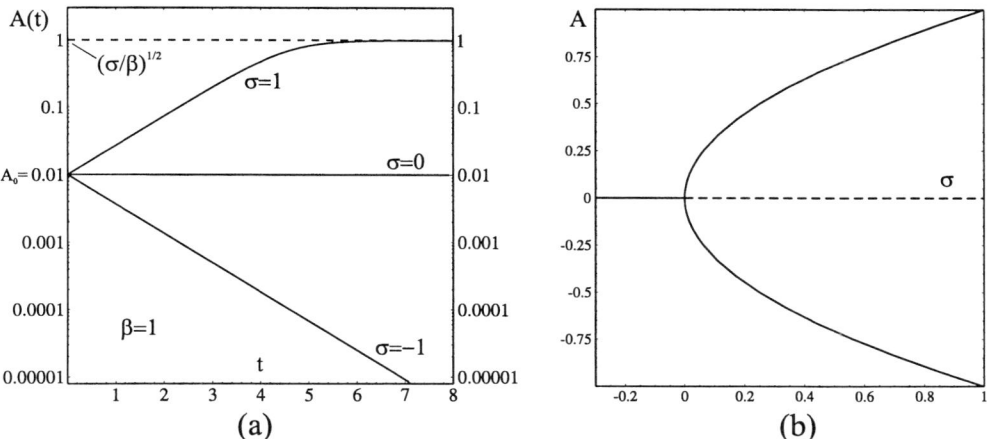

Figure 4.2: Supercritical bifurcation. (a): temporal evolution of the amplitude A starting from an initial fluctuation $A_0 = 0.01$, for several values of the linear growth rate σ. (b): steady states and their stability (full lines: stable solutions, dashed lines: unstable solutions).

Actually, terms of order higher than cubic should not be considered in Eq. (4.5) if $\beta > 0$, because the stable steady solutions ($A \sim \epsilon^{1/2}$) are small for $0 < \epsilon \ll 1$. For $\beta < 0$, the cubic term does not provide saturation of the linear exponential growth, and quintic order terms should be introduced, leading to

$$\frac{dA}{dt} = \sigma A - \beta A^3 - \gamma A^5 \qquad (4.8)$$

with $\gamma > 0$. This equation describes a *subcritical bifurcation*, whose bifurcation diagram is presented in Fig. 4.3(a). Note that if $\gamma < 0$, higher-order terms can be introduced, ...

For completeness, we also present in Fig. 4.3(b) the bifurcation diagram of Eq. (4.7) for $\delta > 0$, which will be encountered in the study of *hexagonal* convective patterns. In this case, A is the amplitude of a superposition of three sets of convective rolls rotated by $120°$ with respect to each other in the horizontal plane (§4.2.1). Then the solution is generally not invariant with respect to the transformation $A \rightarrow -A$, leading to $\delta \neq 0$. The result is a *transcritical bifurcation*, displaying a discontinuity of the steady amplitude A (and thus of convective quantities) at threshold. This is associated with a *hysteresis* phenomenon: once in a convective state (prepared by increasing σ above zero), the "constraint" σ must be decreased below a certain negative value ($\sigma_h = -\delta^2/4\beta$) before the system returns to the motionless state. Hysteresis is associated with *bistability*, i.e. there exist two stable states (the rest state $A = 0$ and the upper convective solution $A > 0$) for $\sigma_h < \sigma < 0$. Note that the stability of the convective branch is changed at the *turning point* $\sigma_h = -\delta^2/4\beta$, $A_h = \delta/2\beta$: introducing $\sigma = \sigma_h + \sigma'$ and $A = A_h + A'$ and keeping only the lowest-order terms in the small perturbation A', we get $dA'/dt = \delta(2\beta)^{-1}(\sigma' - \beta A'^2)$, describing a *saddle-node bifurcation*. A similar normal form describes the turning points of the subcritical bifurcation (see Fig. 4.3). More details about normal forms can be found e.g. in [4, 7, 51].

Strictly speaking, the analysis in terms of amplitude equations fails to describe quantitatively weakly nonlinear convective states in cases other than the supercritical bifurcation. Indeed, for subcritical and transcritical bifurcations, the stationary convective amplitudes are not small even at $\sigma \simeq 0$, and higher-order terms in amplitude equations cannot be neglected. However, the results are often assumed to be valid in a qualitative sense, and indeed apply to many experimental situations. Moreover, it will be seen later on in this chapter that for the transcritical bifurcation observed in hexagonal Marangoni–Bénard convection, the coefficient δ turns out to be numerically small, with the consequence that Eq. (4.7) describes the weakly nonlinear behavior with reasonable accuracy[1].

4.1.2   Orientational degeneracy – Rolls/Rhombs competition

For horizontally isotropic systems considered throughout this book, the parameter characterizing the horizontal dependency of perturbations at the linear stage is the wavenumber k (more precisely k^2). In fact, as seen in §3.3.1 for instance, the planform function $\phi(x, y)$ [defined by Eqs (3.101)] was found to satisfy the Helmholtz equation $\Delta_h \phi + k^2\phi = 0$ (where Δ_h is the horizontal Laplacian). Thus, even if linear theory provides the critical wavenumber k_c, it does not allow distinguishing between planform functions ϕ solutions of this equation. As

---

[1]Note that Eq. (4.7) with $\beta > 0$ is strictly valid for $\sigma \ll 1$ provided δ^2/β is at most of the order of σ. Then, the amplitude A scales as $A \sim (\sigma/\beta)^{1/2}$. Correspondingly, Eq. (4.8) with $\gamma > 0$ and $\beta < 0$ (subcritical bifurcation) is rigorous for small σ provided β^2/γ is at most of order σ, leading to $A \sim (\sigma/\gamma)^{1/4}$.

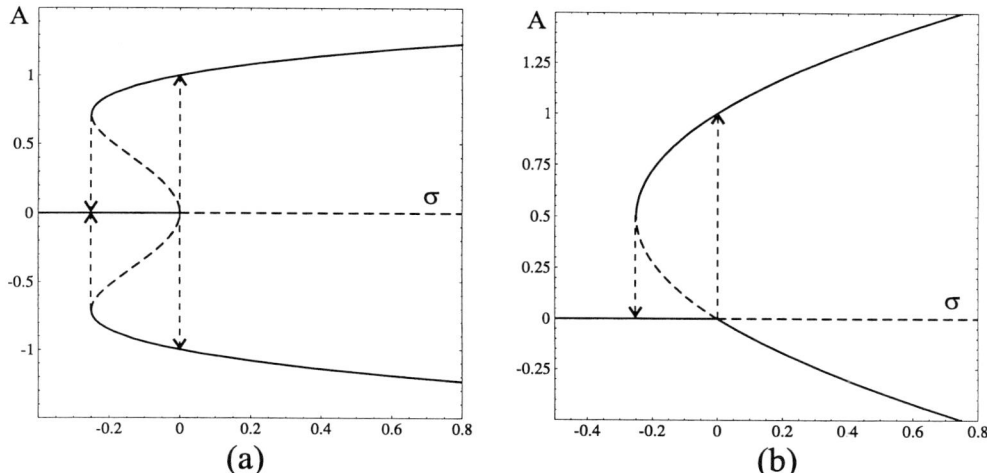

Figure 4.3: (a) Subcritical bifurcation: Eq. (4.8) with $\beta < 0$ and $\gamma > 0$. (b) Transcritical bifurcation: Eq. (4.7) with $\delta > 0$ and $\beta > 0$. Full lines: stable solutions, dashed lines: unstable solutions. Both cases (a) and (b) exhibit hysteresis: stable convective solutions exist in some range below the threshold, and can be reached either via finite amplitude perturbations, or by starting from a convective regime at $\epsilon > 0$ and decreasing ϵ below 0.

discussed earlier, the general solution of the Helmholtz equation can be written as the super-position of plane wave solutions $\exp[i\vec{k}.\vec{r}] = \exp[i(k_x x + k_y y)]$ (where $|\vec{k}|^2 = k_x^2 + k_y^2 = k_c^2$), with arbitrary amplitudes $A_{\vec{k}}$. This will be written

$$\phi(\vec{r} = x\vec{1}_x + y\vec{1}_y) = \sum_{\vec{k} \in P} A_{\vec{k}} \exp[i\vec{k}.\vec{r}] \tag{4.9}$$

where the set P contains an arbitrary number (possibly a continuum) of wavevectors $\vec{k}$ lying on the critical circle (Fig. 4.4).

Note however that wavevectors should always be included by conjugate pairs $\vec{k}$ and $-\vec{k}$, as required by the condition that the planform function ϕ is real (accordingly, we have $A_{-\vec{k}} = A_{\vec{k}}^*$, where the asterisk denotes the complex conjugate).

Now, the particular case considered in the previous section was a two-dimensional one, corresponding to

$$\phi = A \cos[k_c x] = \frac{A}{2}(\exp[ik_c x] + \exp[-ik_c x])$$

If the Landau equation (4.5) is valid in this case, it should also be valid for any orientation of $\vec{k}$ other than the x-axis (the solution then corresponds to rolls with axis perpendicular to the wavevector considered). A more general situation is obtained by superposing two of such

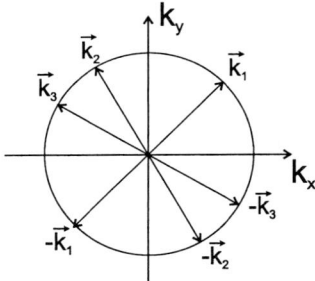

Figure 4.4: Possible combination of modes (set P) in the plane (k_x, k_y).

plane waves with different orientation. This reads

$$\phi(\vec{r}, t) = A(t) \exp[i\vec{k}_1.\vec{r}] + B(t) \exp[i\vec{k}_2.\vec{r}] + c.c. \tag{4.10}$$

where we allowed a dependence of complex amplitudes A and B on time t, and $\vec{k}_1$ and $\vec{k}_2$ form an angle θ (Fig. 4.5).

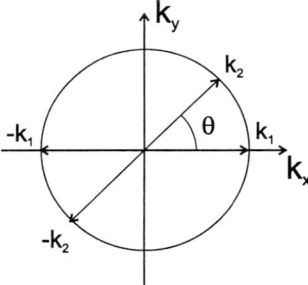

Figure 4.5: Interaction between rolls $\vec{k}_1$ and $\vec{k}_2$ (rolls/rhombs competition).

The planform ϕ is still a solution of the Helmholtz equation, but the nonlinear evolution equations for $A(t)$ and $B(t)$ cannot be expected to be of the type (4.5). Rather, we should consider more general (coupled) amplitude equations, namely

$$dA/dt = f_1(A, B)$$
$$dB/dt = f_2(A, B)$$

The explicit form of the functions f_1 and f_2 will be derived in the next sections. They may be obtained from symmetry arguments (see §4.2.1) or by direct calculation (perturbation or projection techniques, see §4.2.2 and §4.2.3), the latter methods also leading to numerical values of coefficients appearing in the Taylor expansions. For the moment, we just state the result up to order 3 in the amplitudes as

$$dA/dt = A \left(\sigma - \beta |A|^2 - \gamma(\theta)|B|^2 \right)$$
$$dB/dt = B \left(\sigma - \beta |B|^2 - \gamma(\theta)|A|^2 \right)$$

(4.11)

where the coupling constant $\gamma(\theta)$ generally depends on the angle θ between basic wavevectors. Clearly, a particular case of Eq. (4.11) is the Landau equation (4.5), when one of the amplitudes is equal to zero, and the other one is assumed real.

The analysis of the coupled amplitude equations (4.11) can be achieved as a function of the linear growth rate σ, the *self-interaction coefficient* β and the *coupling coefficient* $\gamma(\theta)$.

We will first discuss implications of Eqs (4.11) qualitatively. It is seen that each of the amplitudes evolves with an effective growth rate (the expressions between parentheses in both equations), determined by the linear growth rate σ and by the nonlinear interactions. Consider $\beta > 0$ and $\gamma > 0$. Then, the presence of one of the modes lowers the effective growth rate of the other one, i.e. competition takes place. If γ is large enough, this effect will oppose the formation of mixed states and it is likely that the evolution will tend to a preference of one of the modes over the other one (the choice of either mode depending on the initial conditions). Eventually, rolls or *stripes* ($A = 0$, $B \neq 0$ or $A \neq 0$, $B = 0$) are the final stable structure. Instead, if γ is small compared with β, this competition effect is weak and each mode evolves quasi-independently to a certain saturation value close to that of rolls. The final stage in this case will be a superposition (*mixed state*) of two sets of rolls with equal amplitude and orientations differing by an angle θ, called a rhombic pattern [see Fig. 4.6(b)].

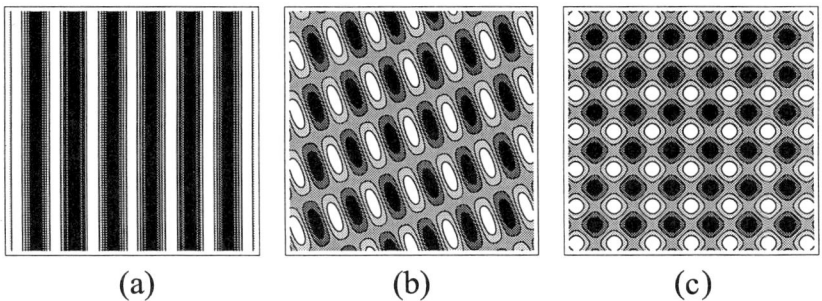

(a)   (b)   (c)

Figure 4.6: Patterns corresponding to the planform function (4.10). The contour levels of ϕ (corresponding e.g. to free surface isotherms for convection problems) are represented for (a) rolls (or stripes), (b) rhombs with $\theta = 40°$, (c) squares.

Note that apart from the particular case $\theta = \pi/2$ [squares, see Fig. 4.6(c)], mixed states of two sets of rolls are seldom observed in Bénard convection, an experimental fact which crucially depends on the detail of the nonlinear interactions quantified by β and $\gamma(\theta)$, as will be seen later.

The detailed discussion of Eqs (4.11) may be achieved by first separating modulus and phase for each of the complex amplitudes A and B. We thus substitute $A = a \exp[i\varphi]$ and

$B = b \exp[i\chi]$ in Eqs (4.11) and separate real and imaginary parts. Then,

$$\frac{da}{dt} = a\left(\sigma - \beta a^2 - \gamma b^2\right)$$
$$\frac{db}{dt} = b\left(\sigma - \beta b^2 - \gamma a^2\right) \tag{4.12}$$
$$\frac{d\varphi}{dt} = \frac{d\chi}{dt} = 0$$

indicating that the dynamics of a and b is independent of the phases φ and χ, which remain equal to their initial value. The physical meaning is clear, reconsidering the planform given by Eq. (4.10), as the phases φ and χ only determine the horizontal position of each set of rolls. Thus, phases have no role here as the system is invariant under any translation of the origin of coordinates in the horizontal plane.

An important property of the system (4.12) is to admit a *potential function*

$$V = -\frac{\sigma}{2}(a^2 + b^2) + \frac{\beta}{4}(a^4 + b^4) + \frac{\gamma}{2}a^2 b^2 \tag{4.13}$$

such that

$$\frac{da}{dt} = -\frac{\partial V}{\partial a}, \quad \frac{db}{dt} = -\frac{\partial V}{\partial b} \tag{4.14}$$

implying

$$\frac{dV}{dt} = \frac{\partial V}{\partial a}\frac{da}{dt} + \frac{\partial V}{\partial b}\frac{db}{dt} = -\left(\frac{\partial V}{\partial a}\right)^2 - \left(\frac{\partial V}{\partial b}\right)^2 \leq 0 \tag{4.15}$$

This indicates that the potential V (the *Lyapunov function*) is a monotonically decreasing function of time, which may only assume stationary values at critical points of $V(a, b)$ [where $\partial V/\partial a = \partial V/\partial b = 0$]. Note that this property is also valid for the more general system (4.11), and has important consequences on the dynamical regimes, such as the impossibility of periodic behaviors $V(t) = V(t + T)$ (this would imply that V increases during some part of the oscillation period), and of other more complicated time-dependencies. The dynamics described by such gradient systems is also said to be *relaxational* [51]. The fixed points (steady states) of the system (4.12) are stable if they correspond to local minima of V, and unstable otherwise (local maxima or saddles).

On the other hand, in the case $\beta > 0$, $\gamma > 0$, the trajectory of the system in the phase plane (a, b) cannot escape to infinity (because this would correspond to increasing V without bounds). For $\beta < 0$ however, or when $\beta > 0$ but $\gamma < -\beta$ (see Fig. 4.7), such divergences may occur, revealing the failure of the cubic system (4.11) in such cases (saturating quintic terms ought to be introduced as in §4.1.1). The equipotential curves (constant V) are illustrated in Fig. 4.7, for some typical situations. The trajectory in phase space is everywhere orthogonal to equipotential lines, and is indicated by arrows.

The steady states of Eqs (4.12) are:

$$- \text{Motionless state (O)} : \quad a = b = 0 \tag{4.16}$$

$$\text{– Rolls (Rl)} \qquad : \quad b = 0,\ a = \left(\frac{\sigma}{\beta}\right)^{1/2} \quad \text{or}\ \ a = 0,\ b = \left(\frac{\sigma}{\beta}\right)^{1/2} \tag{4.17}$$

$$\text{– Rhombs (Rh)} \quad : \quad a = b = \left(\frac{\sigma}{\beta + \gamma}\right)^{1/2} \tag{4.18}$$

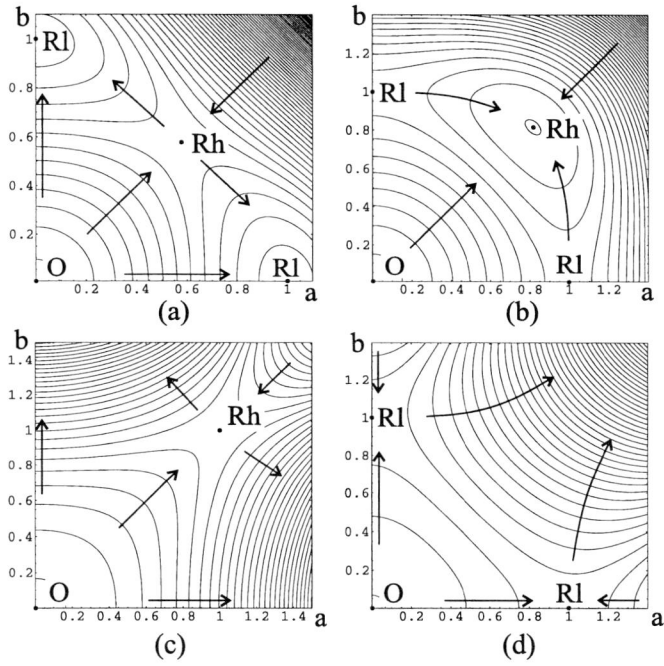

Figure 4.7: Level contours of V in the phase plane (a, b) for several values of σ, β, γ. (a) $\sigma > 0, 0 < \beta < \gamma$ (stable rolls, unstable rhombs); (b) $\sigma > 0, 0 < \gamma < \beta$ (stable rhombs, unstable rolls); (c) $\sigma > 0, \beta < 0, \gamma > -\beta$ (unstable rhombs); (d) $\sigma > 0$, $\beta > 0, \gamma < -\beta$ (unstable rolls). The origin is unstable in each case, and divergence occurs for cases (c) and (d).

The stability of the fixed points (4.16–4.18) is investigated by following the evolution of infinitesimal perturbations around a particular fixed point. This amounts to determine the eigenvalues of the Jacobian

$$J = -\begin{pmatrix} V_{aa} & V_{ab} \\ V_{ab} & V_{bb} \end{pmatrix} = \begin{pmatrix} \sigma - 3\beta a^2 - \gamma b^2 & -2\gamma ab \\ -2\gamma ab & \sigma - 3\beta b^2 - \gamma a^2 \end{pmatrix} \tag{4.19}$$

where indexes a and b refer to the corresponding derivative. Solving $Ju = \lambda u$ for each solution leads to the following eigenvalues and eigenvectors

$$- \text{Motionless state} : \quad \lambda_{1,2} = \sigma \ : \ u \text{ arbitrary (degeneracy)} \tag{4.20}$$

$$- \text{Rolls } (b = 0, \ a \neq 0) \ : \ \lambda_1 = -2\sigma, \ u_1 = (1,0); \ \lambda_2 = \sigma \left(1 - \frac{\gamma}{\beta}\right), \ u_2 = (0,1) \tag{4.21}$$

$$- \text{Rhombs} : \quad \lambda_1 = -2\sigma, \ u_1 = (1,1); \ \lambda_2 = -2\sigma \frac{\beta - \gamma}{\beta + \gamma}, \ u_2 = (1,-1) \tag{4.22}$$

Any of these solutions is linearly stable if both eigenvalues are negative. Fig. 4.8 illustrates the results of the discussion as a function of β and γ.

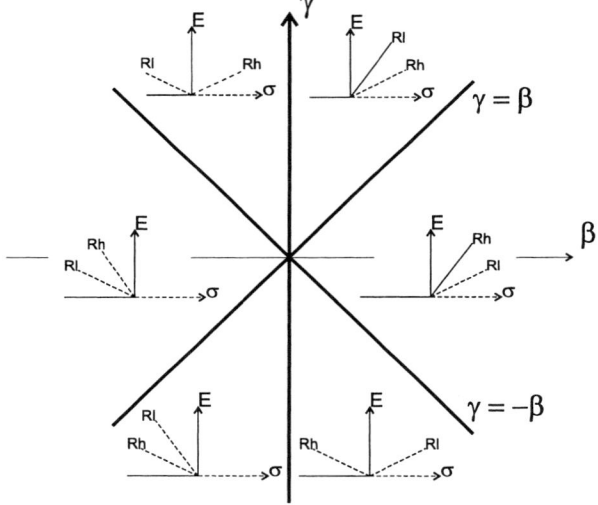

Figure 4.8: Bifurcation diagrams for the competition between rolls (Rl) and rhombs (Rh), as a function of β and γ. The quantity represented as a function of σ is the "energy" $E = a^2 + b^2$ of the convective flow, proportional to the convective heat transport.

The quantity $E = a^2 + b^2$ characterizes the intensity (or "energy") of the convective motions. In fact, it will be seen in the next sections that E is proportional to the increase of the heat transport through the layer by convection. More precisely, the Nusselt number $Nu = (\phi_{\text{cond}} + \phi_{\text{conv}})/\phi_{\text{cond}}$ is the ratio of the total heat transfer (conductive + convective) to the conductive heat transfer, and $Nu - 1 \sim E$. The quantity E may be calculated for each solution: this gives $E = 0$ for the rest state, $E = \sigma/\beta$ for rolls, and $E = 2\sigma/(\beta + \gamma)$ for rhombs. An important conclusion is that the heat transfer is maximal for the stable solution (see Fig. 4.8). This property was first suggested by Malkus [294], and stimulated an active research (see e.g. [23, 53, 296]). Unfortunately, this extremum principle is not general [23], as will be seen in the next sections. However, it holds for some Rayleigh–Bénard systems near the threshold of instability, in a qualitative sense, and provides a working criterion (though not universal) for explaining the competition between nonlinear structures above convective threshold.

As a conclusion, it may be inferred from the discussion made above that the orientational degeneracy of the linear problem will in general be greatly removed by the consideration of nonlinear effects. Indeed, among the infinite number of combinations (4.10) solutions of the Helmholtz equation, only those with $a = b \neq 0$ or $a = 0$, $b \neq 0$ ($a \neq 0$, $b = 0$) are actual steady solutions in the nonlinear regime. A further selection between these convective structures occurs due to their relative stability, which in the case studied above leads to a unique stable solution in the nonlinear regime (apart from equivalent solutions obtained by arbitrary translations, rotations or mirror reflections in the horizontal plane).

Finally we mention that the curve $\gamma(\theta)$, some examples of which will be presented in the following sections, generally presents a discontinuity (vertical asymptote) at $\theta = 60°$ and $\theta = 120°$. In fact, it will be seen that a *resonance* phenomenon occurs at these angles, at which the vectorial sum of some of the constitutive wavevectors [those belonging to P in Eq. (4.9)] also lie on the critical circle. As will be seen in §4.2.2, the nonlinear interactions of two modes with different wavevectors, say $\vec{k}_1$ and $\vec{k}_2$, leads to excitation of the modes with wavevectors $\pm\vec{k}_1 \pm \vec{k}_2$. When some of these modes lie on the critical circle (this occurs precisely at $\theta = 60°$ and $\theta = 120°$) and are thus linearly amplified, resonance occurs (energy injected by the nonlinearity is not linearly dissipated) and the analysis in terms of amplitude equations (4.11) fails. We will then have to consider a more general formulation (see §4.2.1) also including the mode generated by nonlinear interactions (and lying on the critical circle) as part of the set P. This "resonance" phenomenon will be studied in detail, as it is at the origin of the apparition of a generic pattern in nonlinear dissipative structures, namely the hexagonal pattern.

4.1.3 Sideband degeneracy – Modulation effects and envelope equations

When the control parameter is above its critical value (corresponding to some critical wavenumber k_c), all normal modes with wavenumbers within a band centered around k_c have a positive growth rate. In the previous chapter, it was seen that all the minima found for neutral stability curves were locally parabolic, i.e. the band of excited wavenumbers was found to grow as $\epsilon^{1/2}$ for $0 < \epsilon \ll 1$ [where $\epsilon = (Ra - Ra_c)/Ra_c$ for Rayleigh–Bénard, or $\epsilon = (Ma - Ma_c)/Ma_c$ for Marangoni–Bénard instabilities].

Contrary to what was assumed in the previous section, the structures observed above instability threshold, or at least their dynamical regimes, do in general involve superpositions of several wavelengths. As the excited wavelengths are close to each other, a beating spatial frequency phenomenon is observed, which leads to variations of the amplitude of convective structures on a large scale (compared with the basic wavelength $2\pi/k_c$ of the underlying pattern).

Accordingly, during experiments on Rayleigh–Marangoni–Bénard convection (as well as in other pattern-forming systems), perfectly regular ("ideal") patterns are seldom observed.

Indeed, many defects of various nature appear in the observed structures [106, 114]. The number and evolution of these defects are strongly influenced by the horizontal dimensions of the system, together with the shape of the vessel containing the system under investigation.

In this section, we recall some basic results about spatial modulations of a near two-dimensional convective structure (rolls). To illustrate the basic features of the methodology, we may begin as in the previous sections by superposing all near-critical perturbations, but rather than a sum we have to write an integral on the band $[k_1,k_2]$ of amplified wavenumbers: this gives, for the temperature perturbation for example

$$T = \int_{k_1}^{k_2} A_k(t) \exp[ikx]\theta_k(z)dk + c.c. \tag{4.23}$$

where $\theta_k(z)$ is the neutral stability function for temperature, and $c.c.$ denotes the complex conjugate. Near the critical point, the integration limits vary as $k_1 = k_c - \alpha\epsilon^{1/2}$ and $k_2 = k_c + \alpha\epsilon^{1/2}$ and we can change the variable of integration to $Q = (k - k_c)/\epsilon^{1/2}$. This gives, keeping only the lowest-order terms in ϵ

$$T = \exp[ik_c x]\theta_{k_c}(z) \int_{-1}^{+1} \alpha\epsilon^{1/2} A_{k_c}(t) \exp[i\alpha Q\epsilon^{1/2}x]dQ + c.c. + O(\epsilon) \tag{4.24}$$

which can be rewritten as

$$T = \left[A'(X = \epsilon^{1/2}x, t) \exp[ik_c x] + c.c. \right] \theta_{k_c}(z) \tag{4.25}$$

where the new amplitude $A'(X,t)$ depends on a slow length scale $X = \epsilon^{1/2}x$. Similarly, the relevant time scale is $T = \epsilon t$ and when transversal (i.e. along y) modulations of the amplitude need to be included, the relevant length scale is $Y = \epsilon^{1/4}y$. All these scalings may be justified by expanding the growth rate σ in the vicinity of $k_c\vec{1}_x$ [51, 297, 298]. Setting $\vec{k} = (k_c + \Delta k_\parallel)\vec{1}_x + \Delta k_\perp \vec{1}_y$ and keeping only the leading-order terms in the Taylor expansion of σ, we get

$$\sigma(\epsilon, \vec{k}) = \alpha \left[\epsilon - \xi_0^2 \left(\Delta k_\parallel + \frac{\Delta k_\perp^2}{2k_c} \right)^2 \right] + h.o.t. \tag{4.26}$$

where $h.o.t.$ denotes higher-order terms, and

$$\alpha = \frac{\partial\sigma}{\partial\epsilon}|_c > 0 \ , \quad \xi_0^2 = -\frac{1}{2\alpha}\frac{\partial^2\sigma}{\partial k^2}|_c > 0 \tag{4.27}$$

Eq. (4.26), which has been written in this form for later use, indeed shows that the range of unstable wavenumbers does not grow isotropically (see Fig. 4.9) around $\vec{k} = k_c\vec{1}_x$, and justifies the above scalings for space and time variables.

The amplitude equation satisfied by the spatially varying amplitude $A(X,Y,T)$ (omitting the prime) should be similar to the Landau equation presented in the last section, but may in general also include some spatial derivative terms. Symmetry considerations might be used

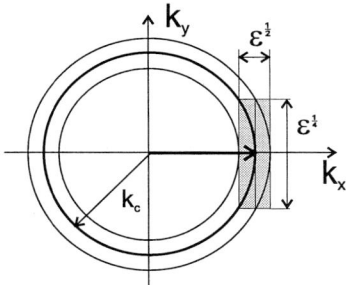

Figure 4.9: Anisotropic growth of the unstable interval around a particular wavevector $\vec{k} = k_c \vec{1}_x$.

for deriving the form of the equation (§4.2.1). Alternatively, a systematic perturbation method consists in expanding convective fields as

$$
\begin{pmatrix} W \\ T \end{pmatrix} = \epsilon^{1/2} A(X, Y, T) \exp[ik_c x] \begin{pmatrix} w_{k_c}(z) \\ \theta_{k_c}(z) \end{pmatrix} + c.c. + O(\epsilon)
\tag{4.28}
$$

It may then be shown [297, 298] that rescaling the variables adequately, we obtain the Newell–Whitehead–Segel envelope equation

$$
\frac{\partial A}{\partial T} = A + \left(\frac{\partial}{\partial X} + \frac{1}{2ik_c} \frac{\partial^2}{\partial Y^2} \right)^2 A - A|A|^2
\tag{4.29}
$$

which may also be written in terms of the original space and time variables, as

$$
\frac{\partial A}{\partial t} = \alpha \left[\epsilon A - \frac{\beta}{\alpha} |A|^2 A + \xi_0^2 \left(\frac{\partial}{\partial x} + \frac{1}{2ik_c} \frac{\partial^2}{\partial y^2} \right)^2 A \right]
\tag{4.30}
$$

Clearly, these equations can be transformed into one another by scalings $A \leftrightarrow (\alpha\epsilon/\beta)^{1/2} A$, $x \leftrightarrow \xi_0 \epsilon^{-1/2} X$, $y \leftrightarrow \xi_0^{1/2} \epsilon^{-1/4} Y$, $t \leftrightarrow (\alpha\epsilon)^{-1} T$.

Compared with Landau equations discussed in §4.1.1, it is seen that modulation effects lead to an extra *linear* term, whose form is thus directly linked to the growth rate σ found from linear stability theory. Assuming that the convective structure deviates slightly from a roll pattern with wavevector $k_c \vec{1}_x$, the expansion (4.26) is valid. Introducing the Fourier transform $T_{\vec{k}}$ of (say) the temperature field (4.28), we have

$$
A \sim \int T_{\vec{k}} \exp[i(\vec{k} - k_c \vec{1}_x).\vec{r}] d\vec{k}
\tag{4.31}
$$

where the integral is limited to half the Fourier space ($k_x > 0$). Taking into account that $\partial T_{\vec{k}}/\partial t = \sigma T_{\vec{k}}$ (each Fourier mode evolves independently at the linear stage), it can be shown that the linear part of (4.30) is indeed recovered[2].

The Newell–Whitehead–Segel equation (4.30) accounts for the possibility of spatial variations of A, hence *imperfect patterns* and *defects* in the convective structure, as studied later. The length $\xi = \xi_0 \epsilon^{-1/2}$ is called the *influence length*, by analogy with the *correlation length* of equilibrium phase transition phenomena, a quantity which diverges at the critical point. The two-dimensional version of this equation ($\partial/\partial y = 0$) is the real Ginzburg–Landau equation, and has extensively been applied in Rayleigh–Bénard convection, as well as in a variety of equilibrium [52] and non-equilibrium phase transitions [19]. In the following subsections, we recall some basic results about the real Ginzburg–Landau (GL) and Newell–Whitehead–Segel (NWS) equations, namely the influence of lateral boundaries, defects, phase diffusion and secondary instabilities.

Importantly, the functional form of GL and NWS equations does not depend on the particular mechanism responsible for instability. They hold in the vicinity of the threshold of any monotonic instability, provided the neutral stability curve presents a parabolic minimum at a critical wavenumber different from zero. Hence they are generic, as their rescaled form does not contain any parameter specific to the instability studied (k_c may also be eliminated by rescaling the Y variable). Details of the phenomenon considered only enter into some numerical coefficients determining the time, amplitude, and length scales of the observed patterns.

Distant lateral walls and defects

Linear analysis:

Segel [298] has considered the influence of *distant* lateral walls at $x = 0, L_x$ and $y = 0, L_y$. He used boundary conditions $A = 0$ at $x = 0, L_x$ and $A = \partial^2 A/\partial y^2 = 0$ at $y = 0, L_y$. From Eq. (4.28) it can be seen that the condition $A = 0$ on the boundaries is a reasonable approximation of realistic no-slip conducting walls. More details about higher-order boundary conditions can be found in [207]. Segel showed [298] that the basic conducting state $A = 0$ loses stability at $\epsilon = \Delta\epsilon$ (e.g. in Marangoni–Bénard convection, this corresponds to a positive shift $Ma_c\Delta\epsilon$ of the critical Marangoni number with respect to its value in infinite geometry), given by

$$\Delta\epsilon = \frac{\pi^2}{L_x^2}\xi_0^2 \tag{4.32}$$

showing that the wall-induced shift of the critical control parameter increases as L_x^{-2} when L_x decreases (stabilization due to lateral walls). This result is general and holds whatever the particular shape of the container [299], even though the coefficient of L_x^{-2} in (4.32) is dependent upon this shape. Owing to the assumptions of slow spatial variations of A however,

---

[2]A side remark is that Eq. (4.31) may be used for calculating the amplitude A corresponding to a roll-like experimental field whose Fourier transform is $T_{\vec{k}}$. This technique is called "demodulation" and may be efficiently used for comparing experimental data with amplitude equation theories (see e.g. [381]), e.g. localizing the defects, roll distortions, ... In case of patterns other than rolls, Eq. (4.31) may easily be generalized.

this is only valid for $L_x \gg k_c^{-1}, L_y \gg k_c^{-1}$. Note that although the result (4.32) does not depend on L_y, the corresponding neutral stability function does, since it is given [298] by

$$A_n(x, y) = A_0 \exp\left[-i\frac{q^2\pi^2}{2k_c L_y^2}x\right] \sin\left[\frac{\pi}{L_x}x\right] \sin\left[q\frac{\pi}{L_y}y\right] \tag{4.33}$$

where A_0 is an unknown amplitude factor, and $q = 1, 2, \ldots$ denotes a transversal wavenumber (which does not affect the threshold of instability at this order).

Another important result obtained from Eq. (4.32) is that rolls parallel to the shortest side of the container have a lower instability threshold than rolls parallel to the largest side, and are consequently preferred near the threshold. This fact has already been observed experimentally for Rayleigh–Bénard convection, and is also predicted theoretically by other means ([257], see also §3.3.4). It appears as general for every instability leading to a roll pattern. This is demonstrated in the following way: if L denotes the large dimension and S the small dimension of the box, rolls with axes parallel to the shorter side will be unstable for $\epsilon > \epsilon_S = \pi^2\xi_0^2/L^2$, while rolls with axes parallel to the longer side will be unstable for $\epsilon > \epsilon_L = \pi^2\xi_0^2/S^2 > \epsilon_S$.

Nonlinear results:

A finite-amplitude solution of Eq. (4.30) in the absence of transversal modulations (i.e. setting $\partial/\partial y = 0$) and for a *unique* rigid wall placed at $x = 0$ is the *kink* solution

$$A_s = \left(\frac{\alpha}{\beta}\epsilon\right)^{1/2} \tanh\left[\left(\frac{\epsilon}{2}\right)^{\frac{1}{2}}\frac{x}{\xi_0}\right] \tag{4.34}$$

displaying a boundary-layer profile in a region of thickness $\xi = \xi_0\epsilon^{-1/2}$ near the rigid wall $x = 0$, and tending to its saturation value $(\alpha\epsilon/\beta)^{1/2}$ in the bulk of the liquid ($x \to \infty$). Near the critical point, the correlation length is large (it diverges at $Ra = Ra_c$), and the effect of lateral walls is thus expected to propagate far into the bulk of the fluid. This has allowed precise experimental measurements of the correlation length scaling exponent $-1/2$ (and of other scaling laws such as $A \sim \epsilon^{1/2}$) in Rayleigh–Bénard convection [300, 301, 302]. Note also the divergence of the time scale for $\epsilon \to 0$ (the growth rate $\sigma \sim \epsilon$). As mentioned earlier, this phenomenon is called critical slowing down, by analogy with equilibrium phase transitions.

To simplify notations, the rescaled form (4.29) is used in the following. In the presence of lateral walls at scaled coordinates $X = 0, L$, and ignoring Y-modulations, we may obtain steady solutions of Eq. (4.29), by decomposing $A(X) = r(X)\exp[i\theta(X)]$, and separating real and imaginary parts: we obtain

$$\begin{aligned} r + r'' - r\theta'^2 - r^3 &= 0 \\ r\theta'' + 2\theta'r' &= 0 \end{aligned} \tag{4.35}$$

where the prime denotes a derivative with respect to X. The second of these equations implies that

$$(r^2\theta')' = 0 \tag{4.36}$$

so that the quantity $r^2\theta'$ should be independent of X. Moreover, the boundary conditions are $r = 0$ at $x = 0, L$, such that this quantity should be zero everywhere. This implies $\theta' = 0$ wherever $r \neq 0$, which indicates that the lateral boundaries exert a strong selection of the wavelength of the bulk convective structure. Indeed, the phase gradient θ' is a local (rescaled) wavenumber shift with respect to the critical wavenumber k_c of the convective structure [as can be seen by setting $A(X) = r(X)\exp[i\theta(X)]$ in Eq. (4.28)], and its vanishing for all x indicates that the wavenumber is constant (equal to k_c) everywhere in the bulk of the container. Note that this is only the lowest-order result: including next order corrections in Eq. (4.29) in fact leads to the conclusion that the selected wavenumber may deviate from the critical one by an amount $\sim \epsilon$ [51].

The amplitude $r(X)$ can then be obtained by solving the steady GL equation

$$r + r'' - r^3 = 0 \tag{4.37}$$

with boundary conditions $r(0) = r(L) = 0$.

Formally, Eq. (4.37) is equivalent to a Duffing oscillator [230] provided the spatial variable X is regarded as a fictive time variable. We may integrate it once after multiplying it by r', which gives

$$\frac{r'^2}{2} + V(r) = E \tag{4.38}$$

with the "potential energy" $V(r) = r^2/2 - r^4/4$, thus representing a nonlinear oscillator (force proportional to elongation $F = -k\,r$ with a non-constant elastic constant $k = 1 - r^2$). E is a constant of integration equivalent to the total energy.

The analysis of this system (see e.g. [230]) will only be summarized here. There exist solutions that satisfy the boundary conditions $r(0) = r(L) = 0$ when $0 < E < 1/4$, under the form of quasi-elliptical trajectories in the (r, r') phase plane (Fig. 4.10) for $E \ll 1$, that progressively deform into (heteroclinic) trajectories connecting the symmetric saddle points $(1, 0)$ and $(-1, 0)$, when E approaches $1/4$. This behavior, depicted in Fig. 4.10, is clearly due to the fact that the "force constant" $k = 1 - r^2$ vanishes at those points, so that the period of the cycle becomes infinitely large (in terms of patterns these near-heteroclinic trajectories correspond to large regions where the amplitude r is close to its saturation value 1, separated by sharp regions where r decreases to zero, corresponding to defects). Note that when $E \to 1/4$, it is possible to show that the solution $r(X)$ tends to the hyperbolic tangent kink solution (4.34) [after rescaling], which is thus the infinite period heteroclinic separatrix in the (r, r') phase plane. In some sense the lateral walls may thus be considered as defects, which should therefore also share the property of walls to impose a well-defined wavelength (equal to the critical one at the lowest order) in defect-free regions.

Thus, for $0 < E < 1/4$, the period of the cycle has to match the length L of the box we are considering. This provides a discretization of the possible values of E, obtained by expressing that $L = n\Delta x$ ($n = 1, 2, ...$), where the half period Δx of a cycle is calculated from Eq. (4.38). We obtain

$$L = n\Delta x = n \int_0^{r^*(E)} \frac{dr}{[E - V(r)]^{\frac{1}{2}}} \tag{4.39}$$

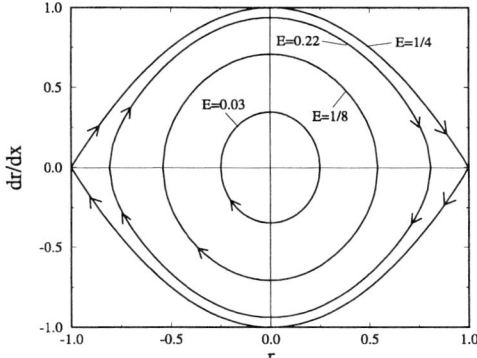

Figure 4.10: Phase portrait of the Duffing equation $r + r'' - r^3 = 0$, in the plane (r, r'). E = total energy. The trajectory is elliptical for $E \to 0$, and tends to a heteroclinic cycle for $E \to 1/4$.

where $r^*(E) = [1 - (1 - 4E)^{1/2}]^{1/2}$ is the value of r at the point where the trajectory in the (r, r') plane crosses the $r' = 0$ axis $[E = V(r^*)]$.

The equation (4.39) can be solved numerically for E, and the solutions obtained (for different $n = 1, 2, ..$ such that $n - 1$ represents the number of defects, i.e. the points where the amplitude decreases to zero in the bulk) are all equilibrium solutions of the steady nonlinear problem. However, as shown by Segel [298], only the $n = 1$ solution is found to be stable to infinitesimal perturbations. Thus, the result is that in two dimensions, no stable defect is possible, i.e. the envelope $A(x, t)$ cannot have nodes $A(x, t) = 0$ except at lateral boundaries (such that for L sufficiently large, A is close to the ideal saturation value in the bulk).

It is important to notice that the NWS equation (4.29) admits a *potential functional* [313]

$$V(A, A^*) = \iint dX \, dY \left[-|A|^2 + \left| \frac{\partial A}{\partial X} + \frac{1}{2ik_c} \frac{\partial^2 A}{\partial Y^2} \right|^2 + \frac{1}{2}|A|^4 \right] \tag{4.40}$$

so that $\partial A / \partial T = -\delta V / \delta A^*$ (functional derivative[3]), and

$$\frac{dV}{dT} = -2 \iint dX \, dY \left| \frac{\partial A}{\partial T} \right|^2 \leq 0,$$

thus again implying a purely relaxational behavior.

Siggia and Zippelius [313] and Pomeau, Zaleski and Manneville [314] have studied the motion of dislocations (point defects) in roll structures. An example of dislocation is shown in Fig. 4.11, where it is seen that this defect corresponds to a supplementary roll inserted at one point in an otherwise regular structure.

---

[3] If $\Phi[f(\vec{r})] = \int d\vec{r} \, \mathcal{L}[f(\vec{r})]$, where $\vec{r} \equiv (x, y)$ is the position vector, and $\mathcal{L}[f(\vec{r})]$ a certain function of $f(\vec{r})$ and its spatial derivatives, the functional derivative of Φ is formally defined [52] as $\delta \Phi / \delta f(\vec{s}) = \lim_{\epsilon \to 0} \epsilon^{-1} (\Phi[f(\vec{r}) + \epsilon \delta(\vec{r} - \vec{s})] - \Phi[f(\vec{r})])$. Then, if $\mathcal{L}[f(\vec{r})]$ depends, as in Eq. (4.40), on $f(\vec{r})$, $\partial_x f(\vec{r})$, and $\partial_{yy}^2 f(\vec{r})$, it is found using integration by parts that $\delta \Phi / \delta f = \partial \mathcal{L} / \partial f - \partial_x (\partial \mathcal{L} / \partial(\partial_x f)) + \partial_{yy}^2 (\partial \mathcal{L} / \partial(\partial_{yy}^2 f))$.

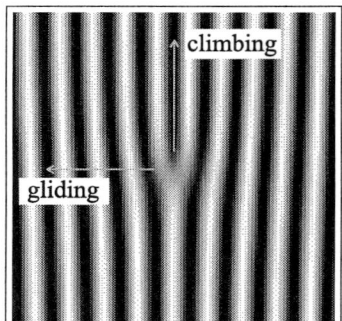 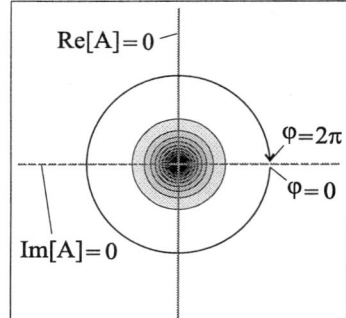

Figure 4.11: Dislocation in a roll structure. Left: grey-level plot of the function $T = A \exp[ix] + c.c.$, which may be interpreted as the temperature field in a Bénard experiment. The climbing motion allows a change of the wavenumber of the pattern, while gliding does not change the wavenumber. Right: contour plot of the modulus $|A|$ (thin lines), vanishing at the core of the defect, together with curves where $\text{Re}[A] = 0$ (thicker full curve) and $\text{Im}[A] = 0$ (thicker dashed curve). The phase $\varphi = \text{Arg}[A]$ undergoes a jump of 2π when following a contour encircling the dislocation.

The motion of the dislocation along the axis of the rolls ("climb" motion) provides a way either to create new rolls, or eliminate existing ones, thus allowing the wavelength of the structure to be optimized [i.e. minimizing the Lyapunov functional (4.40)]. It has been shown [313, 314] that the velocity of the defect vanishes when the roll pattern has the optimal wavenumber k_c at the lowest order [k_c in fact corresponds to a minimum of (4.40) as seen below] far from the defect core. In this case, the motion of the defect indeed does not provide a mechanism for diminishing V further. The force acting on the dislocation was also shown [314] to be similar to the Peach–Köhler force acting on dislocations in equilibrium crystals, and the motion of the defect was also studied in more general non-variational situations. Note that other wavelength selection mechanisms exist (e.g. lateral walls [207] or front propagation [303]). A detailed analysis and further references can be found in the review of Cross and Hohenberg [19], and in the book of Getling [289].

Secondary instabilities, Busse balloon and phase diffusion

Other important results obtained using the NWS equation (4.29) concern secondary instabilities, i.e. instabilities of the patterned regimes observed above the convective threshold. Some instabilities are indeed predicted and experimentally observed for structures that possess either too large a wavelength (*zig-zag instability*), or too small a wavelength (*Eckhaus instability*). These instabilities are universal and limit the range of stable wavelengths, thus providing a supplementary wavelength adjustment mechanism. However, as seen hereafter, these secondary instabilities merely limit the range of stable wavenumbers, but do not select any particular wavenumber inside this range.

In a laterally infinite domain, Eq. (4.29) admits *phase-winding solutions* [207]

$$A_Q = (1 - Q^2)^{1/2} \exp[iQX], \quad -1 < Q < +1 \tag{4.41}$$

which are constant modulus solutions, representing perfect rolls or stripes (in an infinite domain) with a shifted wavenumber $k_c + \epsilon^{1/2}Q/\xi_0$ (as can be seen by returning to the unscaled space variables). The parameter Q is often named *winding number*, and in our choice of scales, the limits $Q = \pm 1$ represent wavenumbers on the neutral stability curve at a given supercritical control parameter. Note that among the solutions (4.41), the structure with wavenumber $k = k_c$ ($Q = 0$) is the optimal one from the point of view of the Lyapunov functional (4.40). Indeed, direct substitution of (4.41) in Eq. (4.40) leads to the energy $-(1 - Q^2)^2/2$ per unit surface, which is indeed minimal at $Q = 0$.

The stability of the solutions (4.41) as a function of Q is investigated by inserting $A = A_Q + A'$ in Eq. (4.29) and linearizing with respect to A'. We then write $A' = \exp[iQX](u + iv)$, separate real and imaginary parts, and write u and v as normal modes

$$u = a \cos(\alpha X + \beta Y) \exp[\omega T]$$
$$v = b \sin(\alpha X + \beta Y) \exp[\omega T]$$

The compatibility condition of the system of equations obtained for a and b then leads to the eigenvalues (growth constants)

$$\omega_\pm = \pm \left[(1 - Q^2)^2 + p^2\right]^{1/2} - (1 - Q^2) - \alpha^2 - \frac{\beta^4}{4k_c^2} - Q\frac{\beta^2}{k_c} \tag{4.42}$$

where $p = \alpha(2Q + \beta^2/k_c)$.

While ω_- is always negative, ω_+ can become positive and lead to instability of the structure with winding number Q. Note first that ω_+ vanishes for $\alpha, \beta \to 0$ (while ω_- tends to a negative constant). Then, considering the case of purely longitudinal perturbations ($\beta = 0$), it is seen that ω_+ also vanishes at

$$\alpha^2 = \alpha_0^2 = 2(3Q^2 - 1) \tag{4.43}$$

which is only possible if $Q^2 > 1/3$. Thus, rolls with $Q < -1/3^{1/2}$ or $Q > 1/3^{1/2}$ are unstable to longitudinal perturbations with wavenumbers in the range $[0, \alpha_0]$ (Eckhaus instability).

Similarly, when only transversal perturbations are allowed ($\alpha = 0$), $\omega_+ = -\beta^4/4k_c^2 - Q\beta^2/k_c$, which is positive for β in the range $[0, 2(-k_cQ)^{1/2}]$ if $Q < 0$ (zig-zag instability).

Thus, these two universal instabilities (they do not depend on specific details of the pattern-forming system studied) limit the range of stable wavenumbers to the interval $0 < Q < 1/3^{1/2}$ (see Fig. 4.12, applied to the Marangoni–Bénard problem). When the local wavenumber is outside this interval, these instabilities induce an evolution of the pattern that generally allows it to return in the stable zone, via destruction of rolls (Eckhaus) when the local wavenumber is too large, or via undulation of the pattern along the axis of rolls (zig-zag) when the local wavenumber is too small (this instability occurs before the point where the pattern is able to create supplementary rolls via the Eckhaus instability).

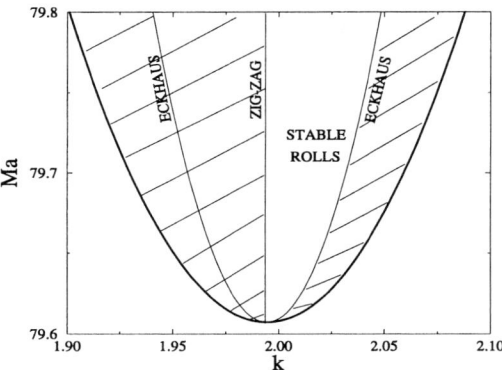

Figure 4.12: The stability of rolls as a function of their wavenumber in the case of Marangoni–Bénard convection. Rolls are stable in the undashed zone only. This diagram forms the lower part of the Busse balloon [315] delimitating the stability domain of rolls.

As said above, ω_- tends to a negative constant for $\alpha, \beta \to 0$, while ω_+ vanishes in the same limit. Expanding the latter for $0 < \alpha, \beta \ll 1$ and keeping only the lowest-order terms, we get

$$\omega_+ \simeq \frac{3Q^2 - 1}{1 - Q^2}\alpha^2 - Q\frac{\beta^2}{k_c} + h.o.t. \tag{4.44}$$

which indicates an (anisotropic) diffusive behavior in this limit. In fact, the quantity which "diffuses" is the phase $\varphi(X, Y)$ of the convective structure. This can be seen by perturbing the solution (4.41), now according to $A = [(1 - Q^2)^{1/2} + a]\exp[i(QX + \varphi)]$. A uniform shift of the phase φ thus corresponds to a translation of the pattern, leading to a completely equivalent solution (because of translational invariance). Hence, the solution (4.41) should be neutrally stable to such a uniform shift (corresponding to the eigenvalue $\omega_+ \to 0$ for $\alpha, \beta \to 0$). On the contrary, if the shift is slightly non-uniform (slow spatial variations of φ, i.e. $0 < \alpha, \beta \ll 1$), a slow evolution of the phase occurs, which is described at the lowest order by a *phase diffusion* equation [51, 19]

$$\frac{\partial \varphi}{\partial T} = D_{//}\frac{\partial^2 \varphi}{\partial X^2} + D_\perp\frac{\partial^2 \varphi}{\partial Y^2} \tag{4.45}$$

where $D_{//} = (1 - 3Q^2)/(1 - Q^2)$ and $D_\perp = Q/k_c$ are, respectively, longitudinal and transversal diffusion coefficients. A normal mode analysis of Eq. (4.45) indeed leads to (4.44), and shows that the pattern is unstable when either $D_{//} < 0$ (Eckhaus instability), or $D_\perp < 0$ (zig-zag instability). Detailed calculations of phase diffusion equations will be presented in §4.5.3 (hexagonal patterns) and §5.5 (waves).

4.2 Derivation of amplitude equations

As discussed in the previous sections, many problems arising in the study of weakly nonlinear instabilities can be solved using the method of amplitude equations. Given a certain set of near-critical normal modes, whose spatial dependency is known from the linear stability problem, we will be interested in the nonlinear system of coupled equations describing the evolution of their amplitudes. First, we will show how the analytical form of amplitude equations may be derived on the basis of symmetry arguments. As these macroscopic symmetries are common to many physical systems, the form obtained for the amplitude equations is universal, and does not depend on specific details of the problem. On the contrary, coefficients of the various terms appearing in the equations do depend on these details, and have to be computed from the full system of thermo-hydrodynamic equations and boundary conditions. This is typically achieved via perturbation methods (§4.2.2) or projection techniques (§4.2.3). Note that the methods presented in this chapter can also be used for oscillatory and multiple bifurcation problems, as seen in subsequent chapters.

4.2.1 Symmetries and amplitude equations

We here consider bifurcations from a steady reference state, such as those considered in Chapter 3. This solution has maximal symmetry (it has the symmetry group of the basic equations and boundary conditions), which is not the case for the patterned convective state, for which some symmetries are broken during the transition. For instance, the reference state is left unchanged by any translation in the plane of the layer (provided the latter is considered laterally infinite), which is clearly not the case once cellular convection has set in. However, translating the convective pattern, we should obtain an equally likely solution of the basic equations. Actually, this simple observation has important implications on the analytical form of amplitude equations, as shown in this section.

In a first stage, it will thus be useful to identify all the invariances enjoyed by the physical system considered. In a second stage, the implications on the form of amplitude equations and their solutions will be analyzed. Though more elaborate tools exist for these purposes (group theoretical methods [320]) and are commonly used in bifurcation theory, our presentation will be less formal, attempting to formulate our arguments on heuristic grounds.

Basic symmetries

Apart from *translational invariance* in the plane of the layer (continuous group), *rotational invariance* (another continuous group) is also part of the symmetry group of the problems considered in this book[4]. This means that given a particular pattern solution of the equations, any other convective pattern obtained from the original one by an arbitrary horizontal translation or a rotation around a vertical axis, is an equivalent solution which should evolve in the same manner as the original one, though in the transformed frame of reference. Note that another symmetry is the *parity invariance* (e.g. reflection with respect to a vertical line passing through the origin) but this is equivalent to a rotation of 180°. Now, the symmetry group of

---

[4]Note that we exclude here systems rotating around a vertical axis, for which translational and mirror invariances do not hold. However, such systems can be studied following the methods presented in this section.

the problem should also incorporate *mirror (reflection) invariance* with respect to any plane orthogonal to the plane of the layer.

The invariance of the physical set-up with respect to such transformations can be better understood when considering their effect on the basic equations directly. Equivalently, to simplify the presentation, we may examine their effect on some model equations, derived either phenomenologically (e.g. the Swift–Hohenberg equation presented hereafter), or rigorously in some limit (e.g. the long-wave Shtilman–Sivashinsky equation [292], or its variants derived by Knobloch [215], or by Golovin et al. [341]). A simple phenomenological model thought to capture most of the essential features of Rayleigh–Bénard convection was proposed by Swift and Hohenberg in 1977 [321]. Here, the evolution of the pattern (or planform function $\phi(\vec{r}, t) = \phi(x, y, t)$ is described by

$$\partial_t \phi = \epsilon \phi - (\Delta + 1)^2 \phi - \phi^3 \tag{4.46}$$

This equation, in which ϕ is real, ϵ is the supercriticality parameter, and Δ is the horizontal Laplacian, is clearly much simpler than the original set of thermo-hydrodynamic equations and boundary conditions. However, some of the properties of Rayleigh–Bénard problems are known to be independent of the detailed mechanisms leading to instability. These universal features should thus be shared by a number of different systems, said to belong to the same Universality Class. It is apparent that the Swift–Hohenberg equation has many important properties in common with Bénard problems considered in this book, which can be attributed to the fact that they belong to the same Class I (see Fig. 4.1), and have identical symmetry properties. Given a particular solution $\phi_0(\vec{r}, t)$ of the Swift–Hohenberg (SH) equation, consider first the translated solution $\phi(\vec{r}, t) = \phi_0(\vec{r} - \vec{r}_0, t)$, where $\vec{r}_0$ is an arbitrary horizontal vector. Clearly, the equation describing the evolution of the translated solution ϕ is identical to that describing the evolution of ϕ_0. Indeed, as ϕ_0 is a solution of Eq. (4.46), we have

$$\partial_t \phi_0(\vec{r}, t) = \epsilon \phi_0(\vec{r}, t) - (\Delta_{\vec{r}} + 1)^2 \phi_0(\vec{r}, t) - \phi_0(\vec{r}, t)^3 \tag{4.47}$$

where $\Delta_{\vec{r}}$ is the horizontal Laplacian with respect to $\vec{r}$-variables. Substituting $\vec{r}$ by $\vec{r} - \vec{r}_0$ in Eq. (4.47) directly leads to Eq. (4.46) for ϕ, as $\Delta_{\vec{r}-\vec{r}_0} = \Delta_{\vec{r}}$, indicating that the SH equation is invariant by translation.

Consider now the rotation symmetry, by transforming the particular solution $\phi_0(\vec{r}, t)$ onto a rotated solution $\phi(\vec{r}, t) = \phi_0(R_\theta \vec{r}, t)$, where R_θ is a rotation operator (with angle θ). Substituting $\vec{r}$ by $R_\theta \vec{r}$ in Eq. (4.47) leads to

$$\partial_t \phi(\vec{r}, t) = \epsilon \phi(\vec{r}, t) - (\Delta_{R_\theta \vec{r}} + 1)^2 \phi(\vec{r}, t) - \phi(\vec{r}, t)^3 \tag{4.48}$$

Now, using the matrix representation

$$R_\theta = \begin{pmatrix} \cos \theta & \sin \theta \\ -\sin \theta & \cos \theta \end{pmatrix} \tag{4.49}$$

and defining $\vec{r}\,' = R_\theta \vec{r}$, we have

$$
\Delta_{\vec{r}\,'} = \frac{\partial^2}{\partial x'^2} + \frac{\partial^2}{\partial y'^2} = \left(\cos\theta \frac{\partial}{\partial x} + \sin\theta \frac{\partial}{\partial y} \right)^2 + \left(-\sin\theta \frac{\partial}{\partial x} + \cos\theta \frac{\partial}{\partial y} \right)^2
$$
$$
= \frac{\partial^2}{\partial x^2} + \frac{\partial^2}{\partial y^2} = \Delta_{\vec{r}}
$$

$$(4.50)$$

and hence the equation satisfied by the rotated pattern ϕ reduces to the SH equation, which is thus invariant by rotation. From both cases of translation and rotation, it is seen that the condition of invariance of the SH equation with respect to a symmetry operator S acting on $\vec{r}$, is that the Laplacian operator is invariant with respect to this transformation, i.e. $\Delta_{S\vec{r}} = \Delta_{\vec{r}}$. It is straightforward to check that this is verified for the parity operator P (defined by $P\vec{r} = -\vec{r}$, in fact $R_\pi = P$), and for the mirror reflection

$$
M_\theta = \begin{pmatrix} \cos 2\theta & \sin 2\theta \\ \sin 2\theta & -\cos 2\theta \end{pmatrix}
$$

$$(4.51)$$

with respect to a line passing through the origin and forming an angle θ with the x-axis. Note that any mirror reflection with angle θ can be obtained as the composition of a mirror reflection with respect to x, followed by a rotation of angle -2θ (i.e. 2θ in the counter-clockwise direction).

For autonomous systems, the origin of time is arbitrary, so that another invariance to be satisfied is the time-translation $t \to t + \Delta t$ (which is trivially satisfied by the SH equation, but will be important for problems involving bifurcation of waves).

Moreover, the SH equation is invariant with respect to a change of the sign of ϕ (because it contains only odd terms in ϕ). As a consequence, the SH equation does not apply for the phenomenological description of problems in which the invariance with respect to the transformation $\phi \to -\phi$ is broken. Note that, as seen later in §4.3.5, such invariance continues to hold for purely buoyancy-driven convection in the Boussinesq approximation, even when boundary conditions at the top and bottom are asymmetric. It will actually be seen that the $\phi \to -\phi$ invariance may be broken e.g. by considering Marangoni effects, non-Boussinesq corrections, or surface deformations. In such cases, phenomenological modifications of the SH equation have been proposed, an example of which is the modified Swift–Hohenberg equation [322, 323]

$$
\partial_t \phi = \epsilon\phi - (\Delta + 1)^2 \phi + \delta\phi^2 - \phi^3
$$

$$(4.52)$$

which contains a supplementary quadratic term breaking the $\phi \to -\phi$ symmetry. For convection rolls, this is qualitatively unimportant, because the solution with reversed ϕ (i.e. reversed temperature perturbation and convective velocities) also corresponds to a translation of the original solution by half a horizontal period (i.e. one convective roll). On the contrary, it is not possible to find such kind of translation for the hexagonal pattern, where the upflows and the downflows do not form equivalent lattices in the horizontal plane (see Fig. 4.14 for illustrations of hexagonal patterns).

Now, let us determine the symmetry group of the systems of equations considered in this book. Typically, we will have to solve the Navier–Stokes equation coupled with an energy (or mass diffusion) equation, together with the continuity equation for an incompressible fluid. Using the Boussinesq approximation, the equations for the perturbations with respect to the motionless diffusive state are (see §3.3.1)

$$\vec{\nabla}.\vec{V} = 0 \tag{4.53}$$

$$\Delta\vec{V} - \vec{\nabla}p + Ra\vec{1}_z T = Pr^{-1}\left(\frac{\partial\vec{V}}{\partial t} + (\vec{V}.\vec{\nabla})\vec{V}\right) \tag{4.54}$$

$$\Delta T + W = \frac{\partial T}{\partial t} + (\vec{V}.\vec{\nabla})T \tag{4.55}$$

with appropriate boundary conditions at top and bottom boundaries.

Consider first the horizontal translation $\vec{r} \to \vec{r} - \vec{r}_0$ (hereafter, $\vec{r}$ will still denote the horizontal coordinate vector). It is clear that vertical derivative terms in Eqs (4.53–4.55) are unaffected, as well as the horizontal gradient and Laplacian terms. Thus, if

$$\left\{\vec{V}_0, T_0, p_0\right\}_{(\vec{r},z,t)} \tag{4.56}$$

is a particular solution of these equations, replacing $\vec{r}$ by $\vec{r} - \vec{r}_0$ shows that the same set of equations is satisfied by the translated solution

$$\left\{\vec{V}, T, p\right\}_{(\vec{r},z,t)} = \left\{\vec{V}_0, T_0, p_0\right\}_{(\vec{r}-\vec{r}_0,z,t)} \tag{4.57}$$

The same property holds for the boundary conditions (even including the Marangoni effect), and the Bénard system is thus translationally invariant. To conclude with translation, we mention that this invariance may be broken e.g. in systems where a ramp of supercriticality (e.g. this corresponds to a horizontal variation of Ra) is imposed, or when the system is submitted to uniform rotation around a vertical axis. These systems are not specifically investigated here.

Now, consider the rotational invariance $\vec{r} \to \vec{r}' = R_\theta \vec{r}$. It was seen above that the horizontal Laplacian Δ_h is invariant with respect to this transformation. Thus, Δ is also invariant. However, the horizontal derivatives are modified by the transformation $\vec{r} \to \vec{r}' = R_\theta \vec{r}$. In particular, for $\theta = \pi$ (parity), $(\partial_{x'}, \partial_{y'}) = -(\partial_x, \partial_y)$. More generally, the transformation of the horizontal derivatives is $(\partial_{x'}, \partial_{y'}) = R_\theta(\partial_x, \partial_y)$. However, imposing a simultaneous rotation of the horizontal velocity $\vec{V}_h \to R_{-\theta}\vec{V}_h$ (which is also intuitive on physical grounds), it is seen that the equations are unaffected. Thus, if

$$\left\{W_0, \vec{V}_{h0}, T_0, p_0\right\}_{(\vec{r},z,t)} \tag{4.58}$$

is a particular solution (W denotes the vertical velocity), the rotated solution

$$\left\{W, \vec{V}_h, T, p\right\}_{(\vec{r},z,t)} = \left\{W_0, R_{-\theta}\vec{V}_{h0}, T_0, p_0\right\}_{(R_\theta\vec{r},z,t)} \tag{4.59}$$

satisfies Eqs (4.53–4.55) and associated boundary conditions.

Similarly, it can be shown that the system is invariant to mirror reflections with respect to a vertical plane forming an angle θ with the x-axis. The transformed solution here reads

$$\left\{ W, \vec{V}_{\mathrm{h}}, T, p \right\}_{(\vec{r},z,t)} = \left\{ W_0, M_\theta \vec{V}_{\mathrm{h}0}, T_0, p_0 \right\}_{(M_\theta \vec{r},z,t)} \tag{4.60}$$

Note finally that for systems in uniform rotation around a vertical axis, this mirror symmetry does not hold. Indeed, in this case centrifugal and Coriolis forces [95] have to be incorporated in the Navier–Stokes equations, which breaks both translation and mirror symmetries (although invariance by rotation continues to hold).

These basic symmetries (translation, rotation, mirror reflection and time-translation) will be sufficient for our purposes, and it will be shown in the next section that they indeed allow determination of the analytical form of amplitude equations for near-critical modes, thereby strongly determining the nature of bifurcating branches appearing above convective threshold.

Influence of symmetry on the form of amplitude equations

Consider the instability of the motionless state to a monotonic mode with wavenumber $k_{\mathrm{c}} \neq 0$ (a typical situation for Bénard problems, as seen in Chapter 3). The solution of the linear problem at $Ra = Ra_{\mathrm{c}}$ can be written (see §3.3.1)

$$\begin{pmatrix} W \\ T \end{pmatrix} = \begin{pmatrix} W^0_{k_{\mathrm{c}}}(z) \\ T^0_{k_{\mathrm{c}}}(z) \end{pmatrix} \phi(\vec{r} = x\vec{1}_x + y\vec{1}_y) \tag{4.61}$$

where W and T are, respectively, vertical velocity and temperature perturbation, and $W^0_{k_{\mathrm{c}}}(z)$ and $T^0_{k_{\mathrm{c}}}(z)$ are the corresponding solutions of the neutral stability problem. According to the discussion of §4.1.2, the planform function ϕ is written

$$\phi(\vec{r},t) = \sum_{\vec{k} \in P} A_{\vec{k}}(t) \exp[i\vec{k}.\vec{r}] \tag{4.62}$$

where the set P contains an arbitrary number of wavevectors $\vec{k}$ lying on the critical circle $k = k_{\mathrm{c}}$ (see Fig. 4.4), and a (slow) time dependence of the amplitudes has been introduced. However, we ignore any spatial dependency of the amplitudes $A_{\vec{k}}$ in this section (this will be considered later on in §4.5). Remember that ϕ is real, thus requiring modes on the critical circle (in P) to be considered by conjugate pairs $(\vec{k}, -\vec{k})$, with $A_{-\vec{k}} = A^*_{\vec{k}}$.

It will be assumed that there exist evolution equations of the form

$$\frac{\partial A_{\vec{k}}}{\partial t} = f_{\vec{k}}(A_{\vec{k}'}) \quad \vec{k}, \vec{k}' \in P \tag{4.63}$$

which reflects the deterministic form of the basic equations, as well as their autonomous nature[5]. Moreover, the functions $f_{\vec{k}}$ will be assumed to be analytical at the origin $A_{\vec{k}} = 0$

[5] Note that none of the symmetries precludes a more general form of Eq. (4.63) including higher-order time derivatives. Though Eq. (4.63) is valid when one monotonic mode bifurcates (single zero eigenvalue), it might have to be generalized when simultaneous bifurcations of several modes occur (multiple zero eigenvalue). An example will be encountered in §5.4.6.

(motionless state). Clearly, we have $f_{\vec{k}} = 0$ at $A_{\vec{k}} = 0$, and assuming that the amplitudes remain small during their evolution (to be checked a posteriori), we may expand Eq. (4.63) in Taylor series

$$\frac{\partial A_{\vec{k}}}{\partial t} = \sum_{\vec{k}' \in P} \frac{\partial f_{\vec{k}}}{\partial A_{\vec{k}'}}|_0 A_{\vec{k}'} + \frac{1}{2!} \sum_{\vec{k}', \vec{k}'' \in P} \frac{\partial^2 f_{\vec{k}}}{\partial A_{\vec{k}'}, \partial A_{\vec{k}''}}|_0 A_{\vec{k}'} A_{\vec{k}''}$$
$$+ \frac{1}{3!} \sum_{\vec{k}', \vec{k}'', \vec{k}''' \in P} \frac{\partial^3 f_{\vec{k}}}{\partial A_{\vec{k}'}, \partial A_{\vec{k}''}, \partial A_{\vec{k}'''}}|_0 A_{\vec{k}'} A_{\vec{k}''} A_{\vec{k}'''} \tag{4.64}$$

limited to the cubic order. Note that this assumption can be relaxed if necessary, e.g. when the linear exponential growth is not saturated at cubic order.

In fact, it will appear that the matrix $\partial f_{\vec{k}}/\partial A_{\vec{k}'}$ of linear coefficients is equal to $\sigma \mathbf{I}$, where $\mathbf{I}$ is the identity matrix and σ is the linear growth rate. This is because in the linear regime, all eigenmodes are decoupled and have the same growth rate σ.

Bifurcation of rolls

Consider a single mode with wavenumber $\vec{k}_1$ and amplitude A_1 lying on the critical circle (still with its conjugate $-\vec{k}_1$ with associated amplitude A_1^*). At the lowest order, the planform function reads

$$\phi = A_1 \exp[i\vec{k}_1 . \vec{r}] + A_1^* \exp[-i\vec{k}_1 . \vec{r}] \tag{4.65}$$

Thus, according to Eq. (4.64), the general form of the amplitude equation for A_1 is (up to cubic order)

$$\dot{A}_1 = \sigma A_1 + \delta_1 A_1^2 + \delta_2 |A_1|^2 + \delta_3 A_1^{*2} + \beta_1 A_1^3 + \beta_2 A_1^2 A_1^* + \beta_3 A_1 A_1^{*2} + \beta_4 A_1^{*3} \tag{4.66}$$

where coefficients have been redefined, and the dot denotes the time-derivative. Now, this form may be greatly simplified by making use of the symmetry considerations. Consider the translational invariance $\vec{r} \rightarrow \vec{r} - \vec{r}_0$. If $\phi_0(\vec{r})$ is a solution, $\phi_0(\vec{r} - \vec{r}_0)$ must also be a solution. As seen from (4.65), this translated solution may also be obtained by the transformation $A_1 \rightarrow B_1 = A_1 \exp[-i\vec{k}_1 . \vec{r}_0]$ acting on the amplitude A_1. Thus, B_1 should satisfy the amplitude equation (4.66) if A_1 does. Substituting A_1 by B_1 in this equation, the result should be valid for all $\vec{r}_0$ and we see that all terms not proportional to $\exp[-i\vec{k}_1 . \vec{r}_0]$, after substitution, should cancel, i.e. we get $\delta_1 = \delta_2 = \delta_3 = \beta_1 = \beta_3 = \beta_4 = 0$. Redefining $\beta_2 = -\beta$, the equation (4.65) reduces to the Landau equation

$$\dot{A}_1 = \sigma A_1 - \beta A_1 |A_1|^2 \tag{4.67}$$

The only terms remaining are those multiplied by $\exp[i\varphi]$ under the transformation $A_1 \rightarrow A_1 \exp[i\varphi]$ (equivariant terms [320]). Note that the equivariant function $A_1(\sigma - \beta|A_1|^2)$ is the product of the equivariant function A_1 and the invariant function $\sigma - \beta|A_1|^2$ (i.e. it remains invariant under the transformation).

Now, although translational invariance has already highly simplified the amplitude equation, we have to consider the effect of other symmetries. The rotational invariance $\vec{r} \rightarrow \vec{r}' = R_\theta \vec{r}$ will not be useful here. According to Eq. (4.65), its action on the amplitude is $A_1 \rightarrow A_1 \exp[i\vec{k}_1.(R_\theta \vec{r} - \vec{r})]$, i.e. the phase generally depends on $\vec{r}$, a situation excluded in this section (though it will be allowed when considering spatial modulations in §4.5). In the case of parity $P\vec{r} = R_\pi \vec{r} = -\vec{r}$, the corresponding action on the amplitude is $A_1 \rightarrow A_1^*$. Thus, the equation (4.67) should be satisfied under this change, which is only possible if $\sigma = \sigma^*$ and $\beta = \beta^*$ (real coefficients). Finally, it can be seen that the mirror reflection symmetry faces the same problems as rotation. Strictly speaking, the fact that rotation and reflection invariances are not satisfied is a consequence of our arbitrary choice of a particular direction $\vec{k}_1$ on the critical circle.

The amplitude equation (4.67) for ideal (non-modulated) rolls may be solved by first separating modulus and phase $A = r \exp[i\varphi]$. This leads to $\dot{\varphi} = 0$ and to the Landau equation (4.5) for r (pitchfork bifurcation).

Bifurcation of 2 roll modes interacting at angle θ (rolls/rhombs competition)

We here derive the amplitude equations (4.11) for rolls interacting at angle θ, from symmetry considerations. The convective perturbations remain of the form (4.61), but the planform function is now given by

$$\phi = A_1 \exp[i\vec{k}_1.\vec{r}] + A_2 \exp[i\vec{k}_2.\vec{r}] + c.c. \tag{4.68}$$

with critical wavevectors $\vec{k}_1$ and $\vec{k}_2$ defined by Fig. 4.5. The unexpanded form of amplitude equations then reads

$$\begin{aligned} \dot{A}_1 &= f_1(A_1, A_1^*, A_2, A_2^*) \\ \dot{A}_2 &= f_2(A_1, A_1^*, A_2, A_2^*) \end{aligned} \tag{4.69}$$

Under the transformation $\vec{r} \rightarrow \vec{r} - \vec{r}_0$, Eq. (4.68) shows that the equivalent action on the amplitudes is $A_1 \rightarrow A_1 \exp[-i\vec{k}_1.\vec{r}] \equiv A_1 \exp[i\varphi]$ and $A_2 \rightarrow A_2 \exp[-i\vec{k}_2.\vec{r}] \equiv A_2 \exp[i\chi]$ with φ and χ arbitrary. The only linear terms equivariant under this transformation are A_1 in the equation for A_1 and A_2 in the equation for A_2. No equivariant quadratic term can be constructed (i.e. in the equation for A_1, none of the terms is multiplied by $\exp[i\varphi]$ under the above transformation). Thus, there are no quadratic terms in amplitude equations and a similar discussion of cubic terms finally yields

$$\begin{aligned} \dot{A}_1 &= \sigma A_1 - (\beta|A_1|^2 + \gamma|A_2|^2)A_1 \\ \dot{A}_2 &= \sigma' A_2 - (\beta'|A_2|^2 + \gamma'|A_1|^2)A_2 \end{aligned} \tag{4.70}$$

Consider now the parity invariance $\vec{r} \rightarrow -\vec{r}$. The action on the amplitudes is $A_1 \rightarrow A_1^*, A_2 \rightarrow A_2^*$, from which it is found that the reflected solution is a solution of the amplitude equations provided all coefficients in Eq. (4.70) are real.

As rotation invariance should be satisfied for any angle, we might in particular choose an angle equal to θ, i.e. the angle between $\vec{k}_1$ and $\vec{k}_2$. Substituting $\vec{r} \rightarrow R_\theta \vec{r}$ in ϕ leads to

$$\phi(R_\theta \vec{r}) = A_1 \exp[i\vec{k}_1.(R_\theta \vec{r})] + A_2 \exp[i\vec{k}_2.(R_\theta \vec{r})] + c.c. \tag{4.71}$$

While $\vec{k}_1.(R_\theta \vec{r}) = (R_{-\theta} \vec{k}_1).\vec{r} = \vec{k}_2.\vec{r}$, i.e. $\vec{k}_1$ is mapped onto $\vec{k}_2$ by the transformation (here, R_θ is the rotation in the clockwise direction), this is not the case for $\vec{k}_2$, which is transformed into a vector not included in the planform ϕ (if $\theta \neq \pi/2$). Once again, it is seen that the rotation invariance is broken by the particular choice of the set P of critical modes. It is thus not possible to find the action on A_1 and A_2 which would be equivalent to a rotation of angle θ. In this connection, we mention that the derivation of a rigorous set of equations invariant under the continuous group of rotation has not been achieved up to now [325], for the case of bifurcations with a finite critical wavenumber. A possibility allowing this ideal situation to be approached would be to choose an arbitrary number, say N, of critical modes regularly disposed on the critical circle, and then let $N \rightarrow \infty$. This is discussed in §4.2.4. Other approaches leading to rotationally invariant model equations are described in §4.5.5.

Finally, we consider the effect of the particular mirror reflection $M_{\theta/2}$, namely that with respect to the bisector of $\vec{k}_1$ and $\vec{k}_2$. Then, we have $\vec{k}_1.(M_{\theta/2} \vec{r}) = (M_{\theta/2} \vec{k}_1).\vec{r} = \vec{k}_2.\vec{r}$ and $\vec{k}_2.(M_{\theta/2} \vec{r}) = (M_{\theta/2} \vec{k}_2).\vec{r} = \vec{k}_1.\vec{r}$, such that this corresponds to the action $A_1 \rightarrow A_2, A_2 \rightarrow A_1$. To allow invariance with respect to this symmetry, the coefficients should thus satisfy $\sigma = \sigma'$, $\beta = \beta'$ and $\gamma = \gamma'$, and the final result is indeed given by Eqs (4.11). Note that the first two conditions $\sigma = \sigma'$ and $\beta = \beta'$ can also be obtained from rotational invariance (each roll mode, considered alone, should be equivalent). Hence, it results from symmetry considerations alone that the dynamics is purely relaxational, and the discussion of bifurcation diagrams presented in Fig. 4.8 holds. In particular, it was seen that stable small amplitude solutions may only exist above the threshold provided both rolls (pure solution) and rhombs (mixed solution) bifurcate supercritically. In this case, the stable solution is that which maximizes the energy $E = |A_1|^2 + |A_2|^2$ (proportional to the increase of heat transport by convection, as seen in §4.2.2).

Bifurcation of resonant triads of wavevectors (hexagons/rolls competition)

We now consider the important case of the resonant interaction of triads of steady modes with wavevectors $\vec{k}_1$, $\vec{k}_2$ and $\vec{k}_3$ oriented at 120° apart on the critical circle (Fig. 4.13).

The planform function is written

$$\phi = A_1 \exp[i\vec{k}_1.\vec{r}] + A_2 \exp[i\vec{k}_2.\vec{r}] + A_3 \exp[i\vec{k}_3.\vec{r}] + c.c. \qquad (4.72)$$

and the amplitudes satisfy equations of the form

$$\begin{aligned}
\dot{A}_1 &= f_1(A_1, A_1^*, A_2, A_2^*, A_3, A_3^*) \\
\dot{A}_2 &= f_2(A_1, A_1^*, A_2, A_2^*, A_3, A_3^*) \\
\dot{A}_3 &= f_3(A_1, A_1^*, A_2, A_2^*, A_3, A_3^*)
\end{aligned} \qquad (4.73)$$

A detailed discussion of the derivation of the cubic order amplitude equations is not presented here, as it proceeds mostly along the same lines as for other patterns. However, an important difference occurs due to the property $\vec{k}_1 + \vec{k}_2 + \vec{k}_3 = 0$, i.e. the basic wavevectors form a resonant triad, to employ the language of waves [326]. This difference is apparent when considering the occurrence of quadratic terms in the amplitude equations. While all these

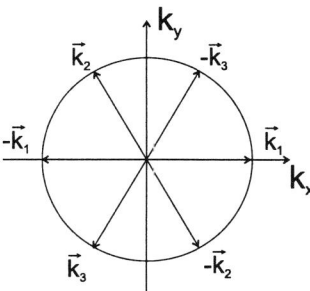

Figure 4.13: Resonant triad of wavevectors and their conjugates.

terms were eliminated in earlier sections, owing to the translational invariance $\vec{r} \to \vec{r} - \vec{r}_0$, this is not the case here because of the resonance condition $\vec{k}_1 + \vec{k}_2 + \vec{k}_3 = 0$. Indeed, the change $\vec{r} \to \vec{r} - \vec{r}_0$ is equivalent to the action $A_n \to A_n \exp[-i\vec{k}_n.\vec{r}_0]$, $n = 1, 2, 3$. Thus, in the equation for A_1 for example, the quadratic term $A_2^* A_3^*$ is equivariant (it is multiplied by $\exp[i(\vec{k}_2 + \vec{k}_3).\vec{r}_0] = \exp[-i\vec{k}_1.\vec{r}_0]$ under the translation, just as $\dot{A}_1$). This is the main difference with the preceding analyzes, and using the other symmetry invariances, we finally get

$$
\begin{aligned}
\dot{A}_1 &= \sigma A_1 + \delta A_2^* A_3^* - A_1 \left[\beta |A_1|^2 + \gamma(|A_2|^2 + |A_3|^2) \right] \\
\dot{A}_2 &= \sigma A_2 + \delta A_1^* A_3^* - A_2 \left[\beta |A_2|^2 + \gamma(|A_1|^2 + |A_3|^2) \right] \\
\dot{A}_3 &= \sigma A_3 + \delta A_1^* A_2^* - A_3 \left[\beta |A_3|^2 + \gamma(|A_1|^2 + |A_2|^2) \right]
\end{aligned}
\tag{4.74}
$$

where all the coefficients σ, δ, β and γ are real. Thus, a quadratic term occurs in each amplitude equation, whose origin is connected to the lack of symmetry with respect to the transformation $A_n \to -A_n$, $n = 1, 2, 3$ (corresponding to reversed fluid velocities and temperature perturbations). Indeed, if such a symmetry exists, Eqs (4.74) must be invariant with respect to the change of sign of each amplitude, which is not the case unless $\delta = 0$. This typically happens for pure Rayleigh–Bénard convection, but not for Marangoni–Bénard convection where $\delta \neq 0$ (see later).

Despite its more complicated form, the system (4.74) may still be put under a variational form. Indeed, a Lyapunov functional is

$$
\begin{aligned}
V = &-\sigma \left(|A_1|^2 + |A_2|^2 + |A_3|^2 \right) - \delta \left(A_1 A_2 A_3 + A_1^* A_2^* A_3^* \right) \\
&+ \frac{\beta}{2} \left(|A_1|^4 + |A_2|^4 + |A_3|^4 \right) + \gamma \left(|A_1|^2 |A_2|^2 + |A_1|^2 |A_3|^2 + |A_2|^2 |A_3|^2 \right)
\end{aligned}
\tag{4.75}
$$

such that $\dot{A}_n = -\partial V / \partial A_n^*$, $n = 1, 2, 3$ and $\dot{V} = -2 \sum_{n=1}^3 |\partial V / \partial A_n|^2 \leq 0$, thus implying a monotonic decrease of V up to one of its local minima. The behavior of the system (4.74) is thus severely restricted by this property: in particular, it excludes all non-monotonic behaviors such as oscillations, or chaotic dynamics. The existence of the potential function is thus a consequence of the symmetries of the original problem, and of the nature of bifurcating modes (monotonic with finite wavelength).

Separating moduli and phases as $A_n = r_n \exp[i\varphi_n]$, $n = 1, 2, 3$, and summing up the imaginary parts of amplitude equations, we get an equation for the sum of the phases $\varphi = \varphi_1 + \varphi_2 + \varphi_3$ as

$$\dot{\varphi} = -\delta \left(\frac{r_2 r_3}{r_1} + \frac{r_1 r_2}{r_3} + \frac{r_1 r_3}{r_2} \right) \sin \varphi \qquad (4.76)$$

On the other hand, the real parts of the amplitude equations give

$$
\begin{aligned}
\dot{r}_1 &= \sigma r_1 + \delta r_2 r_3 \cos \varphi - r_1 \left[\beta r_1^2 + \gamma (r_2^2 + r_3^2) \right] \\
\dot{r}_2 &= \sigma r_2 + \delta r_1 r_3 \cos \varphi - r_2 \left[\beta r_2^2 + \gamma (r_1^2 + r_3^2) \right] \\
\dot{r}_3 &= \sigma r_3 + \delta r_1 r_2 \cos \varphi - r_3 \left[\beta r_3^2 + \gamma (r_1^2 + r_2^2) \right]
\end{aligned}
\qquad (4.77)
$$

such that the dynamics is fourth-dimensional, as it only couples r_1, r_2, r_3 and φ.

Now, we look for the steady solutions of Eqs (4.76–4.77), and analyze their stability. From Eqs (4.76), it is seen that the only steady solutions for φ are $\varphi = 0$ and $\varphi = \pi$. Then, a steady solution is sought for r_1, r_2, r_3 from Eqs (4.77) in which φ is substituted either by $\varphi = 0$ or by $\varphi = \pi$. Thus, while the dynamics of the phases was unimportant for systems investigated in the previous sections, it appears here that a condition on the phases must be satisfied for the solution to be steady. This phenomenon, due to the quadratic coupling, is called *synchronization* (of the phases), and its role is particularly important as we shall see in later sections.

If $\delta > 0$, the solution $\varphi = 0$ is stable, as seen by linearizing Eq. (4.76) near a particular steady solution where $r_1, r_2, r_3 \neq 0$, while the solution $\varphi = \pi$ is unstable. The opposite holds for $\delta < 0$. In the following, we limit the discussion to the case $\delta > 0$ [in fact, if $\delta < 0$ we may define $B_n = -A_n$, for which the amplitude equations are identical to Eqs (4.74) except that the sign of δ is changed].

Considering the solution $\varphi = 0$ (stable to phase perturbations for $\delta > 0$), we substitute $\cos \varphi = 1$ in Eqs (4.77), and discuss their possible steady solutions. We can distinguish the following cases (without distinguishing symmetry-related solutions):

All amplitudes vanish:

$$r_1 = r_2 = r_3 = 0 \qquad\qquad \rightarrow \textbf{Reference state (O)} \qquad (4.78)$$

Two amplitudes vanish:

$$r_1 = r_2 = 0, r_3 = (\sigma/\beta)^{1/2} \qquad \rightarrow \textbf{Rolls (R)} \qquad (4.79)$$

One amplitude vanishes: According to Eqs (4.77), at least another amplitude must vanish and this case reduces to one of the first two.

All amplitudes are different from zero:

All amplitudes are different from each other: this case is impossible. Indeed, assume that r_1, r_2, r_3 are different from each other. Multiplying the first of Eqs (4.77) by r_1, the second by r_2, and subtracting we get $\beta(r_1^2 + r_2^2) + \gamma r_3^2 - \sigma = 0$. We also have $\beta(r_1^2 + r_3^2) + \gamma r_2^2 - \sigma = 0$. Subtracting these two relations, we get $r_2 = r_3$ if $\gamma \neq \beta$, thus contradicting the initial hypothesis.

2 of the 3 amplitudes are equal (say $r_1 = r_3$): assuming $\gamma \neq \beta$, we find that the only possibility is:

$$r_2 = \frac{\delta}{\gamma - \beta}, \quad r_1 = r_3 = \left[\frac{\sigma - \beta\delta^2/(\gamma - \beta)^2}{\gamma + \beta} \right]^{\frac{1}{2}} \quad \rightarrow \textbf{Mixed state (M)}$$

$$(4.80)$$

All amplitudes are equal (= r): we then have to solve $\sigma + \delta r - (\beta + 2\gamma)r^2 = 0$, leading to:

$$r_1 = r_2 = r_3 = \frac{\delta \pm \sqrt{\delta^2 + 4\sigma(\beta + 2\gamma)}}{2(\beta + 2\gamma)} \quad \rightarrow \textbf{Hexagons (H)}$$

$$(4.81)$$

For each of these solutions, permutations of amplitudes are possible, and are equivalent symmetry-related solutions (with the same stability properties). We may also obtain the existence conditions directly from Eqs (4.79–4.81).

The stability conditions of each solution may also be obtained analytically, and only the main results are reproduced here. Still assuming $\delta > 0$ and $\varphi = 0$ for each solution (and that the existence conditions are satisfied in each case), the eigenvalues of the Jacobian are:

Reference state (O): σ (6×). Thus O is stable if $\sigma < 0$, and unstable if $\sigma > 0$.

Rolls (R): $-2\sigma, 0, \sigma(1 - \gamma/\beta) \pm \delta(\sigma/\beta)^{1/2}$ (2×). The zero eigenvalue corresponds to neutral stability with respect to a shift of the phase of the rolls. Thus, subcritical rolls are always unstable, while supercritical rolls are always unstable near the threshold, and may become stable above $\sigma = \beta\delta^2/(\gamma - \beta)^2$ if $\gamma > \beta$. If not, rolls are always unstable.

Hexagons (H): We get two zero phase eigenvalues associated with translations of the hexagonal pattern in the horizontal plane. The eigenvalue associated with the sum of the phases φ is $-\delta r < 0$ [where r is given by Eq. (4.81)]. The three remaining eigenvalues are found from Eqs (4.77) as $-2r[\delta + (\beta - \gamma)r]$ (2×) and $r[\delta - 2r(\beta + 2\gamma)]$.

Mixed state (M): Again, we get 2 zero eigenvalues associated with horizontal translations, and the third phase eigenvalue is $-\delta(2r_2 + r_1^2/r_2) < 0$ [with r_1 and r_2 given by Eq. (4.80)]. One of the three remaining eigenvalues is $-2[\delta r_2 + (\beta - \gamma)r_1^2]$, and the other two have too cumbersome expressions to be reproduced here. It can be shown however that the mixed state is always unstable.

Up to now, we have only considered the solution $\varphi = 0$. The other case $\varphi = \pi$ gives branches of solutions which are phase unstable (for the case $\delta > 0$ considered here). Setting $\cos \varphi = -1$ in Eqs (4.77), it is seen that all solutions and eigenvalues given above are still valid provided δ is sign-changed. In particular, the phase eigenvalues of hexagons and mixed solutions become positive, i.e. the corresponding solutions are indeed unstable. In particular the hexagons with $\varphi = \pi$ are called down-hexagons (H–), by opposition with up-hexagons (H+) for which $\varphi = 0$. This denomination refers to the direction (down or up) of fluid motion at the center of hexagonal cells. These two physically different patterns are represented in Fig. 4.14. Note that down-hexagons are the preferred (stable) solution when $\delta < 0$.

$$\text{H+} \qquad\qquad\qquad \text{H-}$$

Figure 4.14: Free surface isotherms (e.g. for Marangoni–Bénard convection) for up-hexagons (H+) and for down-hexagons (H–). Dark zones correspond to cold fluid and white zones to hot fluid. The fluid motion at the surface is from hot to cold. H+ is stable (H– unstable) for $\delta > 0$, and vice versa for $\delta < 0$.

For moderate and high Prandtl number fluids, the up-hexagon solution is generally preferred (stable) near threshold in Marangoni–Bénard convection. Indeed, it will be seen later on that $\delta > 0$ for such situations. However, $\delta < 0$ (H– stable) is possible for low Prandtl number liquids.

According to the discussion given above, we may also draw complete bifurcation diagrams. This is done in Fig. 4.15 for some typical cases. It is seen that in case (a), which is the most common case in Marangoni–Bénard convection, the first bifurcation ($\sigma = 0$) occurs to up-hexagons (H+) which are stable in some range near the threshold (actually this range may be quite extended as we shall see later). Note that this first bifurcation is a *first-order* transition, i.e. there is a discontinuity of the steady state quantities when increasing the constraint past the threshold. As explained in §4.1.1, this is associated with a certain amount of *hysteresis*, i.e. σ must be decreased below a certain negative value $\sigma_h = -\delta^2/4(\beta + 2\gamma)$ for the system to return to the motionless state. On the other hand, above a certain value of σ (corresponding to a certain Marangoni number), a transition to rolls (R) occurs, since hexagons become unstable and the only remaining possibility for the system is to evolve into a roll structure (which remains the only stable solution for still higher constraints). Now, this hexagons/rolls transition is associated with a large hysteresis, since once in the roll state, the constraint must be significantly decreased for the system to switch back to the up-hexagonal pattern.

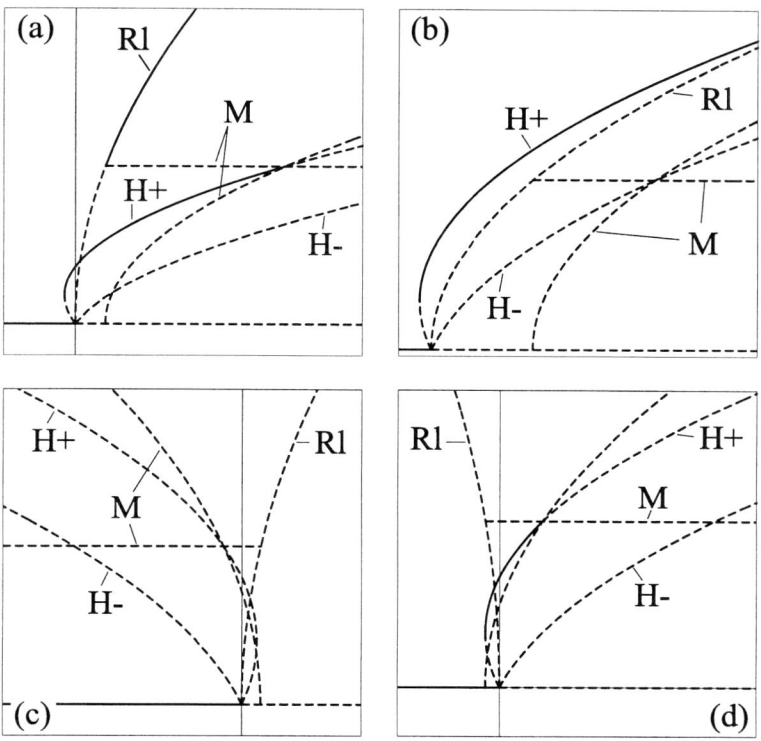

Figure 4.15: Typical bifurcation diagrams for the rolls/hexagons competition, in representative cases. (a) $\gamma > \beta > 0$, (b) $\beta > \gamma > 0$, (c) $\beta > 0$, $\gamma < -\beta$, (d) $\beta < 0$, $\gamma > -\beta$. Full lines correspond to stable states, and dashed lines to unstable solutions.

We end this section by commenting on the *zero eigenvalues* found for the above solutions. In fact, the occurrence of such phase eigenvalues is very general, as they are a direct consequence of the invariance by translation in the horizontal plane. Considering a certain pattern (say, hexagonal), any horizontal translation leads to an equivalent pattern. As there is no restoring mechanism which could act to force the pattern to return to the initial solution, the pattern is neutrally stable with respect to this translation. Such modes, which more generally occur whenever a continuous invariance is broken by the transition (from the disordered to the ordered phase), are named *Goldstone modes* [52] in Condensed Matter Physics. Returning to the case of translations, consider now a slightly non-uniform translation of an initially perfect pattern. We thus allow a slight distortion (non-zero wavenumber phase perturbation) of the pattern with respect to its initial "ideal" configuration. Here, it is likely that the pattern will acquire a certain dynamics, which may be assumed to be slow in view of the slow modulations. Thus, there should be a mode (more precisely two for the hexagons considered here) whose eigenvalue goes continuously to zero as the wavenumber of the perturbation tends to zero. The dynamics of such modes will be investigated later on. In particular, they are responsible for secondary *phase instabilities* of cellular structures (see §4.1.3 and §4.5.3). Other kinds of Goldstone modes will be discussed in §4.6.

4.2.2   Perturbation methods – Coefficients of amplitude equations

The amplitude equations determined by the symmetry considerations of the preceding sections still do not allow unambiguous assessment of the particular patterns occurring above the threshold in a specific convection problem. This is because they contain coefficients, whose calculation is dependent upon the specific details of the problem studied, i.e. the full system of thermo-hydrodynamic equations and boundary conditions. The calculation of these coefficients is typically achieved via perturbation methods, such as considered in this section, or projection methods (§4.2.3). We consider here a general operational formulation of the full system of equations, and establish useful expressions for the amplitude equations (for an arbitrary number of modes on the critical circle). The principle of the method described hereafter was first developed by Gor'kov [327], Malkus and Veronis [328] and later on by Schlüter, Lortz and Busse [199], and Segel [329]. Further details and references can be found in review papers of Busse [22, 23], Palm [27] and Normand, Pomeau and Velarde [25].

Consider that the problem may be written under the operational form

$$\mathcal{L}(U) = \mu M(U) + \Theta \left(\frac{\partial U}{\partial t} \right) + N(U, U) \tag{4.82}$$

where U is the vector of unknown perturbation fields $U = [\vec{V}, T, p]$, $\mathcal{L}$, M and Θ are linear operators [i.e. $\mathcal{L}(\lambda_1 U_1 + \lambda_2 U_2) = \lambda_1 \mathcal{L}(U_1) + \lambda_2 \mathcal{L}(U_2)$ for all U_1, U_2], $N(U, U)$ is a quadratic bilinear form (i.e. linear in the above sense in both its arguments), and μ is the control parameter (e.g. Ma, or Ra depending on the problem studied). The operators $\mathcal{L}$, M, Θ and N will typically contain the bulk equations, but it may also be convenient in some cases to include some of the boundary conditions in their definition. Eventually, the remaining boundary conditions (i.e. those which are linear and do not contain time-derivatives or control parameters) will be grouped into a set E to which the vector U is said to belong. Thus, we assume that the problem is well-posed by stating that $U \in E$ and U satisfies Eq. (4.82). In fact, all the problems studied in this book may actually be written in this form. In the case of interfacial deformations however, where nonlinearities are not only quadratic, a generalization could be made including higher-order nonlinear terms such as $N_3(U, U, U)$ [trilinear in U], which does not lead to specific difficulties.

The reference solution whose stability is studied is $U = 0$, for which Eq. (4.82) is identically satisfied. Note that boundary conditions in E are also satisfied provided the boundary conditions it includes are homogeneous (i.e. the set E is linear).

Though the method described here can be extended to multiple bifurcation problems (i.e. simultaneous bifurcation of several eigenvalues) and to the bifurcation of oscillatory modes (this is specifically done in the next chapter), we first consider here the case of the bifurcation at $\mu = \mu_c$ of a single eigenvalue with zero imaginary part (monotonic mode). The wavenumber k_c of the critical perturbations is assumed finite.

Thus, the problem

$$\mathcal{L}(U) = \mu_c M(U) \tag{4.83}$$

is assumed to admit non-trivial solutions

$$U_{\vec{k}} = \exp[i\vec{k}.\vec{r}]U_{0\vec{k}}(z) \tag{4.84}$$

with $|\vec{k}| = k_c$ and $U_{0\vec{k}} \in E_{\vec{k}}$, where $E_{\vec{k}}$ is a set of boundary conditions obtained by substituting horizontal gradients by $i\vec{k}$, horizontal Laplacian operators by $-k^2$, ... in the boundary conditions of E. We also define operators $\mathcal{L}_{\vec{k}}, M_{\vec{k}}, \Theta_{\vec{k}}$ by proceeding to the same substitutions in $\mathcal{L}, M$ and Θ. Thus,

$$\mathcal{L}_{\vec{k}}(U_{0\vec{k}}) = \mu_c M_{\vec{k}}(U_{0\vec{k}}) \tag{4.85}$$

which is the operational form of the differential problems (for z only) studied in linear neutral stability analyzes of Chapter 3.

As the problems considered in this book are homogeneous and isotropic in the horizontal directions, translational and rotational invariances apply and the eigenvalue which bifurcates at $\mu = \mu_c$ is infinitely degenerated (all modes with $\vec{k}$ on the critical circle bifurcate simultaneously). Adopting a certain normalization condition [typically one of the components of $U_{0\vec{k}}(z)$ is set to a constant value at a particular height z, e.g. the temperature perturbation $T_{0\vec{k}}$ can be set to 1 at the interface], the general solution of the linear stability problem (4.83), as discussed earlier, reads

$$U_0 = \sum_{\vec{k}\in P} A_{\vec{k}} \exp[i\vec{k}.\vec{r}]U_{0\vec{k}}(z) \tag{4.86}$$

where P is a set of modes with wavevectors on the critical circle, which rigorously should contain all of them, but in practice is always limited to a finite number of disturbances. Note that if the normalization condition is indeed $T_{0\vec{k}} = 1$ at the interface, the planform function

$$\phi_0 = \sum_{\vec{k}\in P} A_{\vec{k}} \exp[i\vec{k}.\vec{r}] \tag{4.87}$$

directly gives the surface temperature, which is generally representative of the organization of the convective motions.

U_0 being an exact solution of the linear problem, we are interested in the effect of nonlinear terms in the right-hand side (r.h.s.) of Eq. (4.82). Near threshold, the amplitude of convection is assumed to be small (which should be verified a posteriori). This smallness will be quantified by the smallness parameter ϵ, used to develop

$$U = \epsilon U_0 + \epsilon^2 U_2 + \epsilon^3 U_3 + ... \tag{4.88}$$

where U_0 is given by Eq. (4.86), and U_2, U_3 are yet undetermined vector fields. The reason for the choice of this asymptotic expansion is clearly related to the possibility of balancing the various powers of ϵ in Eq. (4.82), provided we also develop

$$\mu = \mu_c + \epsilon\mu_1 + \epsilon^2\mu_2 + ... \tag{4.89}$$

Such a relation between the control parameter μ and ϵ is one of the assumptions of the Gor'kov–Malkus–Veronis method [327, 328]. Clearly, it will be satisfactory if it is possible to find equations at each order, allowing the determination of μ_1, μ_2, ...

We will now discuss the scaling of the time-derivative term in Eq. (4.82). If the amplitudes $A_{\vec{k}}$ are functions of time t, then introducing Eqs (4.88) and (4.89) into Eq. (4.82) and taking into account Eq. (4.85) leads at the lowest order to $\Theta(\partial U_0/\partial t) = 0$, implying that the time scale of the evolution of the amplitudes should be long, i.e. $\partial/\partial t$ should at least be of order ϵ. More generally, using a multiscale perturbation technique [230], we could assume that the amplitudes $A_{\vec{k}}$ depend on several slow time scales $t_1 = \epsilon t$, $t_2 = \epsilon^2 t$, ... If, say, $A_{\vec{k}} = A_{\vec{k}}(t_1, t_2)$, we have

$$\frac{\partial A_{\vec{k}}}{\partial t} = \epsilon\frac{\partial A_{\vec{k}}}{\partial t_1} + \epsilon^2\frac{\partial A_{\vec{k}}}{\partial t_2} \tag{4.90}$$

It will turn out in the following that t_2 is the relevant time scale for supercritical bifurcation problems (the time derivative term will be in balance with other terms of the amplitude equations), and unless explicitly mentioned, we will adopt the slow time scale $t_2 = \epsilon^2 t$ only.

We may now insert Eqs (4.88–4.90) into Eq. (4.82) and identify the different powers of ϵ, leading to:

At order ϵ:

$$\mathcal{L}(U_0) - \mu_c M(U_0) = 0 \tag{4.91}$$

At order ϵ^2:

$$\mathcal{L}(U_2) - \mu_c M(U_2) = \mu_1 M(U_0) + N(U_0, U_0) \tag{4.92}$$

At order ϵ^3:

$$\mathcal{L}(U_3) - \mu_c M(U_3) = \mu_2 M(U_0) + \mu_1 M(U_2) + \Theta\left(\frac{\partial U_0}{\partial t_2}\right)$$
$$+ N(U_0, U_2) + N(U_2, U_0) \tag{4.93}$$

Thus, a sequence of linear inhomogeneous problems is generated, the first of which (homogeneous) being the linear neutral stability problem (4.83), whose solution is Eq. (4.86).

As the inhomogeneous part of Eq. (4.92) is known (apart from the undetermined constant μ_1, we may in principle proceed to its resolution. However, it will turn out that such a solution may not be found in all cases. To see this, let us rewrite its r.h.s. using Eq. (4.86) as

$$\mu_1 M(U_0) + N(U_0, U_0) = \mu_1 \sum_{\vec{k}\in P} A_{\vec{k}} \exp[i\vec{k}.\vec{r}] M_{\vec{k}}(U_{0\vec{k}})$$
$$+ \sum_{\vec{p},\vec{q}\in P} A_{\vec{p}} A_{\vec{q}} \exp[i(\vec{p}+\vec{q}).\vec{r}] N_{\vec{p},\vec{q}}(U_{0\vec{p}}, U_{0\vec{q}}) \tag{4.94}$$

where $M_{\vec{k}}$ has been defined above, and $N_{\vec{p},\vec{q}}$ can be defined in a similar way from $N(U_1, U_2)$, by replacing horizontal gradients by $i\vec{k}$, where $\vec{k}$ is the wavevector of the mode to which the gradient applies.

Examining the horizontal dependencies in Eq. (4.94), it is seen that the first term ($\sim \mu_1$) replicates the horizontal dependencies of critical modes, while the second (nonlinear) term generates all dependencies of the form $\exp[i(\vec{p} + \vec{q}).\vec{r}]$ with $\vec{p}, \vec{q}$ belonging to the critical set P. Depending on the choice of P, some of these nonlinearly generated modes might have wavevectors lying on the critical circle (this is the resonance phenomenon already encountered above for hexagons and associated with triads of wavevectors $\vec{k}_1$, $\vec{k}_2$ and $\vec{k}_3$ of P satisfying $\vec{k}_1 + \vec{k}_2 + \vec{k}_3 = 0$). If this is the case, attempts to solve Eq. (4.92) will generally lead to some incompatibility (i.e. U_2 contains non-physical secular terms [230]). However, this incompatibility may be removed by considering the nonlinearly generated resonant wavenumber as belonging to the set P. Then, it is seen that one term in the first sum of the r.h.s. of (4.94) has the same horizontal dependency as the resonant contribution generated by the other (double) sum. By an adequate choice of μ_1, it is then in general possible to solve for U_2 in terms of regular (periodic) functions. The condition allowing the determination of μ_1 is the solvability (Fredholm) condition of the problem (see Appendix E).

Considering a wavenumber $\vec{k}$ belonging to P, we may try to solve Eq. (4.92) for the contribution proportional to $\exp[i\vec{k}.\vec{r}]$ in the r.h.s. given by Eq. (4.94). For this contribution, we set

$$U_2 = U_{2\vec{k}}(z) \exp[i\vec{k}.\vec{r}] \tag{4.95}$$

Then, the problem to solve for $U_{2\vec{k}}(z)$ reads

$$\mathcal{L}_{\vec{k}}(U_{2\vec{k}}) - \mu_c M_{\vec{k}}(U_{2\vec{k}}) = \mu_1 A_{\vec{k}} M_{\vec{k}}(U_{0\vec{k}}) + \sum_{\vec{p},\vec{q}\in P} A_{\vec{p}} A_{\vec{q}}\, \delta_{\vec{k},\vec{p}+\vec{q}} N_{\vec{p},\vec{q}}(U_{C\vec{p}}, U_{0\vec{q}}) \tag{4.96}$$

where $\delta_{\vec{k},\vec{k}'} = 1$ if $\vec{k} = \vec{k}'$, otherwise $\delta_{\vec{k},\vec{k}'} = 0$. Thus, if $\tilde{U}_{0\vec{k}}(z)$ is a solution of the adjoint problem corresponding to the l.h.s. of Eq. (4.96), the Fredholm condition (see Appendix E) requires its orthogonality to the r.h.s. of (4.96), i.e.

$$\mu_1 A_{\vec{k}} \left\langle \tilde{U}_{0\vec{k}}, M_{\vec{k}}(U_{0\vec{k}}) \right\rangle + \sum_{\vec{p},\vec{q}\in P} A_{\vec{p}} A_{\vec{q}} \delta_{\vec{k},\vec{p}+\vec{q}} \left\langle \tilde{U}_{0\vec{k}}, N_{\vec{p},\vec{q}}(U_{0\vec{p}}, U_{0\vec{q}}) \right\rangle = 0 \tag{4.97}$$

where $<,>$ denotes a suitably defined scalar product. Thus, if there exist pairs of wavenumbers $\vec{p}, \vec{q}$ in P such that $\vec{k} = \vec{p} + \vec{q}$ (i.e. $\vec{p}, \vec{q}, \vec{k}$ form a resonant triad), and if none of the scalar products vanish, the condition (4.97) is a relation between μ_1 and the amplitudes $A_{\vec{k}}$. Strictly speaking, the condition (4.97) is then the sought leading-order amplitude equation. Eventually, a slower time scale t_1 may be introduced [see Eq. (4.90)], leading to a supplementary term

$$\frac{\partial A_{\vec{k}}}{\partial t_1} \left\langle \tilde{U}_{0\vec{k}}, \Theta_{\vec{k}}(U_{0\vec{k}}) \right\rangle \tag{4.98}$$

in Eq. (4.97). However, in this case the approach fails, because no stable finite amplitude solution is found at this order (due to the absence of cubic saturating terms, as examined below). Nevertheless, there exist conditions under which it is possible to complete this equation by cubic terms (see below), even in the case where resonance occurs. We thus have to distinguish between resonant and non-resonant cases.

Non-resonant set of critical modes

If no pair of wavevectors in P adds up to another critical wavevector, no resonance occurs and Eq. (4.97) is solved by

$$\mu_1 = 0 \tag{4.99}$$

Then, the solution at order ϵ^2 is

$$U_2 = \sum_{\vec{p},\vec{q}\in P} A_{\vec{p}} A_{\vec{q}} \exp[i(\vec{p}+\vec{q}).\vec{r}] U_{2\vec{p},\vec{q}}(z) \tag{4.100}$$

where z-dependencies are determined by the compatible problem

$$\mathcal{L}_{\vec{p}+\vec{q}}(U_{2\vec{p},\vec{q}}) - \mu_c M_{\vec{p}+\vec{q}}(U_{2\vec{p},\vec{q}}) = N_{\vec{p},\vec{q}}(U_{0\vec{p}}, U_{0\vec{q}}) \tag{4.101}$$

where $U_{2\vec{p},\vec{q}} \in E_{\vec{p}+\vec{q}}$.

The compatibility condition then has to be sought at order ϵ^3. Using Eqs (4.86) and (4.100), we may rewrite Eq. (4.93) as

$$
\mathcal{L}(U_3) - \mu_c M(U_3) =
$$
$$
\mu_2 \sum_{\vec{k}\in P} A_{\vec{k}} \exp[i\vec{k}.\vec{r}] M_{\vec{k}}(U_{0\vec{k}}) + \sum_{\vec{k}\in P} \frac{\partial A_{\vec{k}}}{\partial t_2} \exp[i\vec{k}.\vec{r}] \Theta_{\vec{k}}(U_{0\vec{k}})
$$
$$
+ \sum_{\vec{p},\vec{q},\vec{l}\in P} A_{\vec{p}} A_{\vec{q}} A_{\vec{l}} \exp[i(\vec{p}+\vec{q}+\vec{l}).\vec{r}] \left\{ N_{\vec{p}+\vec{q},\vec{l}}(U_{2\vec{p},\vec{q}}, U_{0\vec{l}}) + N_{\vec{l},\vec{p}+\vec{q}}(U_{0\vec{l}}, U_{2\vec{p},\vec{q}}) \right\}
$$
$$\tag{4.102}$$

The compatibility condition at this order will certainly contain nonlinear (cubic) resonant contributions (i.e. with a dependency $\exp[i\vec{k}.\vec{r}]$ where $\vec{k} \in P$). In fact, projecting Eq. (4.102) on $\exp[i\vec{k}.\vec{r}]\tilde{U}_{0\vec{k}}$ (Fredholm solvability condition) and solving for $\partial A_{\vec{k}}/\partial t_2$, we obtain the amplitude equations

$$\frac{\partial A_{\vec{k}}}{\partial t_2} = \mu_2 \alpha_{\vec{k}} A_{\vec{k}} - \sum_{\vec{p},\vec{q},\vec{l}\in P} A_{\vec{p}} A_{\vec{q}} A_{\vec{l}} \delta_{\vec{k},\vec{p}+\vec{q}+\vec{l}} Z_{\vec{p},\vec{q},\vec{k}} \tag{4.103}$$

with

$$\alpha_{\vec{k}} = -\frac{\left\langle \tilde{U}_{0\vec{k}}, M_{\vec{k}}(U_{0\vec{k}}) \right\rangle}{\left\langle \tilde{U}_{0\vec{k}}, \Theta_{\vec{k}}(U_{0\vec{k}}) \right\rangle} \tag{4.104}$$

and

$$Z_{\vec{p},\vec{q},\vec{k}} = \frac{\left\langle \tilde{U}_{0\vec{k}}, N_{\vec{p}+\vec{q},\vec{l}}(U_{2\vec{p},\vec{q}}, U_{0\vec{l}}) + N_{\vec{l},\vec{p}+\vec{q}}(U_{0\vec{l}}, U_{2\vec{p},\vec{q}}) \right\rangle}{\left\langle \tilde{U}_{0\vec{k}}, \Theta_{\vec{k}}(U_{0\vec{k}}) \right\rangle} , \quad \vec{l} = \vec{k} - \vec{p} - \vec{q} \quad (4.105)$$

where $U_{2\vec{p},\vec{q}}$ is determined from Eq. (4.101) with $U_{2\vec{p},\vec{q}} \in E_{\vec{p}+\vec{q}}$.

It can be shown [see Appendix E, Eq. (E.11)] that the linear coefficient $\alpha_{\vec{k}}$ is equal to the derivative of the growth rate σ with respect to the control parameter, evaluated at the threshold, i.e. the slope of the variation of σ with μ at $\mu = \mu_c$

$$\alpha_{\vec{k}} = \alpha = \frac{\partial \sigma}{\partial \mu}|_{\mu=\mu_c} \quad (4.106)$$

which is thus independent of the orientation of $\vec{k}$.

Now, consider the cubic terms in the amplitude equations (4.103). The triple sum may be simplified by considering the particular structure of the set P. For a given $\vec{k} \in P$, the problem is to find all triads of wavevectors $\vec{p}, \vec{q}, \vec{l} \in P$ whose sum is equal to $\vec{k} \in P$. By geometrical construction, it is seen that for this to be possible, at least one of the wavevectors $\vec{p}, \vec{q}, \vec{l}$ should be equal to $\vec{k}$. Then, taking into account that $A_{-\vec{k}} = A_{\vec{k}}^*$ for all $\vec{k}$, we obtain successively

$$\sum_{\vec{p},\vec{q},\vec{l} \in P} A_{\vec{p}} A_{\vec{q}} A_{\vec{l}} \delta_{\vec{k},\vec{p}+\vec{q}+\vec{l}} Z_{\vec{p},\vec{q},\vec{k}}$$
$$= A_{\vec{k}} \sum_{\vec{q} \in P} |A_{\vec{q}}|^2 Z_{\vec{k},\vec{q},\vec{k}} + \sum_{\vec{p} \neq \vec{k}} \sum_{\vec{q},\vec{l} \in P} A_{\vec{p}} A_{\vec{q}} A_{\vec{l}} \delta_{\vec{k},\vec{p}+\vec{q}+\vec{l}} Z_{\vec{p},\vec{q},\vec{k}}$$
$$= A_{\vec{k}} \sum_{\vec{q} \in P} |A_{\vec{q}}|^2 Z_{\vec{k},\vec{q},\vec{k}} + A_{\vec{k}} \sum_{\vec{p} \neq \vec{k}} |A_{\vec{p}}|^2 Z_{\vec{p},\vec{k},\vec{k}} + A_{\vec{k}} \sum_{\vec{p},\vec{q} \neq \vec{k}} A_{\vec{p}} A_{-\vec{q}} \delta_{\vec{q},-\vec{p}} Z_{\vec{p},\vec{q},\vec{k}}$$
$$= A_{\vec{k}} |A_{\vec{k}}|^2 Z_{\vec{k},\vec{k},\vec{k}} + A_{\vec{k}} \sum_{\vec{p} \neq \vec{k}} |A_{\vec{p}}|^2 (Z_{\vec{p},\vec{k},\vec{k}} + Z_{\vec{k},\vec{p},\vec{k}}) + A_{\vec{k}} \sum_{\vec{p} \neq \pm \vec{k}} A_{\vec{p}}|^2 Z_{\vec{p},-\vec{p},\vec{k}}$$
$$= \beta A_{\vec{k}} |A_{\vec{k}}|^2 + A_{\vec{k}} \sum_{\vec{p} \neq \pm \vec{k}} |A_{\vec{p}}|^2 (Z_{\vec{p},-\vec{p},\vec{k}} + Z_{\vec{p},\vec{k},\vec{k}} + Z_{\vec{k},\vec{p},\vec{k}})$$

$$(4.107)$$

where the self-interaction coefficient β is defined as

$$\beta = Z_{\vec{k},\vec{k},\vec{k}} + Z_{\vec{k},-\vec{k},\vec{k}} + Z_{-\vec{k},\vec{k},\vec{k}} \quad (4.108)$$

We may also define a cross-interaction coefficient $\gamma_{\vec{k},\vec{q}}$ as

$$\gamma_{\vec{k},\vec{q}} = Z_{\vec{q},-\vec{q},\vec{k}} + Z_{\vec{k},\vec{q},\vec{k}} + Z_{\vec{q},\vec{k},\vec{k}}$$
$$+ Z_{-\vec{q},\vec{q},\vec{k}} + Z_{\vec{k},-\vec{q},\vec{k}} + Z_{-\vec{q},\vec{k},\vec{k}} \quad (4.109)$$

Then, the final amplitude equations for the non-resonant case read

$$\frac{\partial A_{\vec{k}}}{\partial t_2} = \mu_2 \alpha_{\vec{k}} A_{\vec{k}} - A_{\vec{k}}[\beta |A_{\vec{k}}|^2 + {\sum}'_{\vec{q} \neq \vec{k}} \gamma_{\vec{k},\vec{q}} |A_{\vec{q}}|^2] \quad (4.110)$$

where $\sum'$ denotes a sum over half the set P (only one member of each conjugate pair is taken into account).

Note that $\gamma_{\vec{k},\vec{q}} = \gamma_{\vec{k},-\vec{q}}$ in view of Eq. (4.109), and $\gamma_{\vec{k},\vec{q}} = \gamma_{\vec{q},\vec{k}}$ (because of the mirror symmetry invariance with respect to the bisector of $\vec{q}$ and $\vec{k}$). In fact, $\gamma_{\vec{k},\vec{q}} = \gamma(\cos\theta)$ where θ is the angle between $\vec{q}$ and $\vec{k}$. This property does not hold for systems rotating around a vertical axis, where the reflection invariance is broken [19].

On the other hand, from Eqs (4.108) and (4.109), it can be seen that

$$\beta = \frac{1}{2}\lim_{\vec{q}\to\vec{k}}\gamma_{\vec{k},\vec{q}} = \frac{1}{2}\gamma(\theta = 0) \tag{4.111}$$

such that the curve of $\gamma(\theta)$ on the interval $[0, \pi/2]$ completely determines the cubic nonlinear terms in the amplitude equations, and should thus be sufficient to investigate the complete non-resonant pattern selection problem (see §4.2.4).

Resonant set of critical modes

When the critical set P contains resonant triads of wavevectors, it was seen that $\mu_1 \neq 0$ in general. More precisely, a non-trivial compatibility condition is found at order ϵ^2. Including the faster time scale t_1, we may rewrite it as

$$\frac{\partial A_{\vec{k}}}{\partial t_1} = \mu_1\alpha_{\vec{k}}A_{\vec{k}} + \sum_{\vec{p},\vec{q}\in P} A_{\vec{p}}A_{\vec{q}}\delta_{\vec{k},\vec{p}+\vec{q}}Z_{\vec{p},\vec{k}} \tag{4.112}$$

where $\alpha_{\vec{k}}$ is still given by Eq. (4.106), and the quadratic coefficient is

$$Z_{\vec{p},\vec{k}} = -\frac{\left\langle \tilde{U}_{0\vec{k}}, N_{\vec{p},\vec{q}}(U_{0\vec{p}}, U_{0\vec{q}}) \right\rangle}{\left\langle \tilde{U}_{0\vec{k}}, \Theta_{\vec{k}}(U_{0\vec{k}}) \right\rangle}, \quad \vec{q} = \vec{k} - \vec{p} \tag{4.113}$$

As mentioned above, these amplitude equations are not satisfactory, because the quadratic terms in general do not provide saturation of the linear exponential growth of amplitudes. For example, considering the resonant triad defined in Fig. 4.13, we get amplitude equations of the form

$$\frac{\partial a_1}{\partial t} = \alpha(\mu - \mu_c)a_1 + \delta a_2^* a_3^* \tag{4.114}$$

together with cyclic permutations of (1,2,3). Here we have multiplied Eq. (4.112) by ϵ^2, used Eq. (4.89) and defined rescaled amplitudes as $a_n = \epsilon A_n$, $n = 1, 2, 3$. The quadratic coefficient is defined as

$$\delta = Z_{-\vec{k}_2,\vec{k}_1} + Z_{-\vec{k}_3,\vec{k}_1} \tag{4.115}$$

The bifurcation described by the normal form (4.114) is represented in Fig. 4.16 (transcritical bifurcation) for $\alpha, \delta > 0$. The bifurcation diagram is a magnification of that of Fig. 4.15(a) near the origin, i.e. the leading-order approximation of the bifurcating branches for up- and down-hexagons.

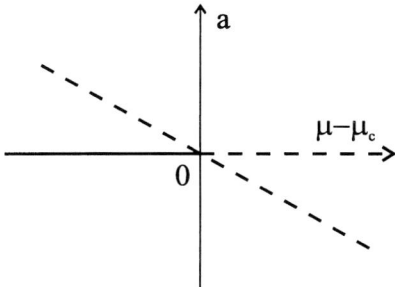

Figure 4.16: Transcritical bifurcation. No stable branch bifurcates at $\mu = \mu_c$. Here, $a = a_1 = a_2 = a_3$, and $a > 0$ corresponds to up-hexagons, while $a < 0$ corresponds to down-hexagons.

Thus, in the case where δ is of order unity, the analysis in terms of amplitude equations fails to describe the finite-amplitude convective regimes, because no stable small-amplitude solution exists above the threshold. However, in some circumstances, the quadratic coefficient computed from Eq. (4.115) is small, and cubic order amplitude equations may then be obtained under the assumption that $\delta = O(\epsilon)$.

This is the case when the symmetry with respect to the inversion of the fluid velocities (see §4.2.1) is only weakly broken. For instance, this is often assumed for weakly non-Boussinesq Rayleigh–Bénard convection [22, 332], e.g. when the viscosity slightly depends on temperature. Another possibility is the study of Marangoni–Bénard convection in the neighborhood of some values of the parameters for which $\delta = 0$ (such as when the Prandtl number $Pr \simeq 0.23$, as seen later), or Rayleigh–Marangoni–Bénard convection in liquid layers of relatively large depth where we may assume $Ma/Ra = 1/Bo_d = O(\epsilon)$. In any case, it will be seen in the next sections that the value of the quadratic coefficient δ is generally numerically small (compared with cubic coefficients) even for the pure Marangoni–Bénard instability [103, 104, 130, 151], such that the analysis presented below may certainly be expected to be qualitatively valid, with good accuracy.

The only rigorous case however is $\delta = O(\epsilon)$. Then, the quadratic term in Eq. (4.112) is small and enters at the next order (divided by ϵ), i.e. in the compatibility condition at order ϵ^3 [Eq. (4.110)]. The compatibility condition at order ϵ^2 is then identically satisfied by $\mu_1 = 0$ and $\partial A_{\vec{k}}/\partial t_1 = 0$, and the problem at order ϵ^2 is solved as for the non-resonant case. In this respect, note that when solving Eq. (4.101) for the cases where $\vec{k} = \vec{p} + \vec{q} \in P$, it is possible to include in $U_{2\vec{p},\vec{q}}$ the general solution of the corresponding homogeneous problem, say $B_{\vec{k}} U_{0\vec{k}}$ with arbitrary $B_{\vec{k}}$. This term might then lead to resonant contributions in the order ϵ^3 problem, by further interacting with terms in U_0. However, they will not contribute to the amplitude equation at this order because their coefficient is proportional to $\delta = O(\epsilon)$, i.e. it will appear at order ϵ^4. We therefore do not consider such terms here.

Finally, it can be seen that the discussion of the other cubic terms is unchanged compared with the non-resonant situation, such that the final amplitude equations for the weakly resonant case read

$$\frac{\partial A_{\vec{k}}}{\partial t_2} = \mu_2 \alpha A_{\vec{k}} + \epsilon^{-1} \delta A^*_{\vec{k}_+} A^*_{\vec{k}_-} - A_{\vec{k}} [\beta |A_{\vec{k}}|^2 + {\sum}'_{\vec{q} \neq \vec{k}} \gamma_{\vec{k},\vec{q}} |A_{\vec{q}}|^2] \qquad (4.116)$$

where $A_{\vec{k}_-}$ and $A_{\vec{k}_+}$ denote the amplitude of the modes with wavevectors at 120° apart from the wavevector $\vec{k}$ (i.e. $\vec{k} + \vec{k}_+ + \vec{k}_- = 0$) respectively in the clockwise and counter-clockwise directions on the critical circle. In Eqs (4.116), it should be remembered that $\delta = O(\epsilon)$, hence all terms are indeed of the same order.

We now briefly return to the case $\delta = O(1)$, for which solving problems (4.101) when $\vec{k} = \vec{p} + \vec{q} \in P$ faces more serious difficulties than in the case $\delta = O(\epsilon)$. Indeed, a direct resolution would lead to the incompatibility already mentioned above, and it appears that there is no asymptotically rigorous way to treat the problem. In principle, the only rigorous result is Eq. (4.112), which is unsatisfactory for the reasons mentioned above. However, it is still possible to obtain qualitatively valid cubic order amplitude equations by formally removing the resonant contribution of the r.h.s. of (4.101), and reintroducing it at order ϵ^3 (which should be satisfactory provided this resonant contribution is numerically small). Thus, for resonant problems (4.101), we will solve

$$\mathcal{L}_{\vec{p}+\vec{q}}(U'_{2\vec{p},\vec{q}}) - \mu_c M_{\vec{p}+\vec{q}}(U'_{2\vec{p},\vec{q}}) = N_{\vec{p},\vec{q}}(U_{0\vec{p}}, U_{0\vec{q}}) + Z_{\vec{p},\vec{q}} \Theta_{\vec{k}}(U_{0\vec{k}}) \qquad (4.117)$$

where $U'_{2\vec{p},\vec{q}} \in E_{\vec{p}+\vec{q}}$. This problem is compatible as can be checked by applying the Fredholm condition to its r.h.s. and using Eq. (4.113). Now, the resonant part $-Z_{\vec{p},\vec{q}} \Theta_{\vec{k}}(U_{0\vec{k}})$ must also be added (multiplied by $\epsilon^{-1} A_{\vec{p}} A_{\vec{q}}$) to the order ϵ^3 problem for the mode having wavevector $\vec{k} = \vec{p} + \vec{q}$. Finally, expressing the compatibility condition of this problem leads once again to Eqs (4.116), as in the case $\delta = O(\epsilon)$. The cubic coefficients $\gamma_{\vec{k},\vec{q}}$ are still given by (4.109), but the difference is that the coefficients $Z_{\vec{p},\vec{q},\vec{k}}$ with $\vec{p} + \vec{q}$ on the critical circle should be calculated by Eq. (4.105) with $U_{2\vec{p},\vec{q}}$ replaced by $U'_{2\vec{p},\vec{q}}$. Actually, $\gamma_{\vec{k},\vec{q}}$ contains this type of terms when the angle between $\vec{k}$ and $\vec{q}$ is either $\pi/3$ or $2\pi/3$, i.e. when $\vec{q} = \pm \vec{k}_+$ or $\vec{q} = \pm \vec{k}_-$. It is also checked that all these cases lead to the same γ. Accordingly, we define

$$\gamma' = \gamma_{\vec{k},\pm\vec{k}_+} = \gamma_{\vec{k},\pm\vec{k}_-} \qquad (4.118)$$

Note that the general form of the quadratic coefficient is

$$\delta = Z_{-\vec{k}_+,\vec{k}} + Z_{-\vec{k}_-,\vec{k}} \qquad (4.119)$$

which is independent of the orientation of $\vec{k}$, by isotropy.

Now, we may return to a non-rescaled form of Eqs (4.116) by multiplying it by ϵ^3, redefining amplitudes as $a_{\vec{k}} = \epsilon A_{\vec{k}}$, and taking into account that $t_2 = \epsilon^2 t$ and $\epsilon^2 \mu_2 = \mu - \mu_c$. Then, we get

$$\frac{\partial a_{\vec{k}}}{\partial t} = \alpha(\mu - \mu_c) a_{\vec{k}} + \delta a^*_{\vec{k}_+} a^*_{\vec{k}_-} - a_{\vec{k}} [\beta |a_{\vec{k}}|^2 + {\sum}'_{\vec{q} \neq \vec{k}} \gamma_{\vec{k},\vec{q}} |a_{\vec{q}}|^2] \qquad (4.120)$$

Finally, it is worth commenting on the general form and physical meaning of the cubic interaction coefficient $Z_{\vec{p},\vec{q},\vec{k}}$. According to the general expressions (4.101) and (4.105), the coefficient $Z_{\vec{p},\vec{q},\vec{k}}$ is the result of the nonlinear interaction of the mode $\vec{l} = \vec{k} - \vec{p} - \vec{q}$ with the mode nonlinearly generated by $\vec{p}$ and $\vec{q}$. The latter mode in fact results from a balance between the nonlinear interaction (injection of energy by interaction of $\vec{p}$ and $\vec{q}$) and the linear damping (dissipation of energy). The dynamics of the nonlinearly generated mode (in fact it is a harmonic of the fundamental modes belonging to P) is *adiabatically slaved* to the slow time variations of the generating modes. Indeed, the linear damping of generated modes is expected to occur on a $O(1)$ time scale compared with the slow $O(\epsilon^2)$ time scale of the critical modes. Then, damped modes instantaneously relax to their equilibrium value (i.e. to the *center manifold* [335, 51]) when the time scales are indeed well separated (i.e. in the limit $\epsilon \to 0$).

Now, a particularly important interaction is that generating the zero wavenumber (homogeneous) mode $\vec{k} = 0$. Indeed, this effect is at the origin of the modification by convection of spatially averaged characteristics, such as the increase of the heat transport due to convection (the Nusselt number), or mean flow effects (although it will be seen that for problems considered here, the homogeneous mode is purely thermal in the case of monotonic modes). According to Eqs (4.86), (4.88) and (4.101), the spatial average of the perturbation fields (the "mean field") reads, at the lowest order,

$$\bar{U} = \epsilon^2 \sum_{\vec{p} \in P} |A_{\vec{p}}|^2 U_{2\vec{p},-\vec{p}}(z) \tag{4.121}$$

where $U_{2\vec{p},-\vec{p}}(z)$ is determined from

$$\mathcal{L}_0(U_{2\vec{p},-\vec{p}}) - \mu_c M_0(U_{2\vec{p},-\vec{p}}) = N_{\vec{p},-\vec{p}}(U_{0,\vec{p}}, U_{0,-\vec{p}}) \tag{4.122}$$

with $U_{2\vec{p},-\vec{p}} \in E_0$. The explicit form of this system will be considered in some examples treated in the following sections.

Thus, the mean field is of order ϵ^2, and is a function of the intensities $|A_{\vec{k}}|^2$ only. In particular, as $U_{2\vec{p},-\vec{p}}$ is generally independent of the direction of $\vec{p}$, this implies that the Nusselt number (defined as the ratio of the total heat transport to the conductive heat transport) will verify

$$Nu - 1 \sim \epsilon^2 \sum_{\vec{p} \in P} |A_{\vec{p}}|^2 = \mathcal{E} \tag{4.123}$$

i.e. it is proportional to the "energy" $\mathcal{E}$ of the convective perturbations (which has already been encountered in §4.2.1 and used to illustrate some bifurcation diagrams). In the next section, we will see that this quantity is particularly important in determining stability properties, at least for the non-resonant case. Moreover, the nonlinear generation of a mean field temperature profile can be expected to play an important role in saturating the exponential growth of amplitudes. Indeed, it provides a direct modification of the cause of the instability (i.e. the imposed thermal gradient). In most cases, the distortion of this thermal gradient by nonlinear generation of a mean field will allow homogenization of the horizontally homogeneous temperature profile (stabilizing effect). This phenomenon will be studied later on, and in particular in Chapter 8 when treating highly nonlinear behaviors.

4.2.3   Projection methods – Galerkin–Eckhaus expansions

In this section, an alternative technique is presented, which also allows the systematic calculation of amplitude equations from the basic equations and boundary conditions. Rather than perturbation methods, we will here make use of expansions of the unknown fields in series of the eigenfunctions of the linear problem, followed by a projection of the equations on suitably defined functions, as in the Galerkin method. Here, the resulting system will be projected on the eigenfunctions of the adjoint linear problem, as proposed by Eckhaus [490]. This methodology has been used by several authors [132, 156, 149, 150, 151, 316, 345], and is very powerful, even for problems where separation of variables is not possible, when combined with appropriate numerical methods for resolution of spectral problems (e.g. spectral-Tau methods [152, 153]). It is also worth describing the method without restricting it to particular geometries such as layers, i.e. we will not attribute a particular role here to any of the spatial coordinates.

In order to simplify notations, we here rewrite the general problem (4.82) as

$$\Lambda(U) = \Theta \left(\frac{\partial U}{\partial t} \right) + N(U, U) \tag{4.124}$$

where $\Lambda = \mathcal{L} - \mu M$ is a linear operator now including the control parameter μ. Still, the perturbation vector $U(\vec{r}, t)$ is assumed to belong to a set E of sufficiently differentiable functions satisfying the boundary conditions. As in the last section, only those boundary conditions which are linear and do not involve time-derivatives should be included in E, the rest being included in Eq. (4.124). Note that the method presented below may also be easily generalized to nonlinearities of order higher than quadratic. Note also that in this section, $\vec{r}$ denotes the three-dimensional position vector.

Consider first the linear spectral problem

$$\Lambda(U) = \sigma \Theta(U), \quad U \in E \tag{4.125}$$

and assume that its solutions U_p with eigenvalue σ_p form a complete set[6], i.e. that every U in E may be represented as

$$U = \sum_p A_p(t) U_p(\vec{r}) \tag{4.126}$$

Then, it remains to determine evolution equations for the amplitudes $A_p(t)$. This can be achieved by inserting the expansion (4.126) in Eq. (4.124) and using Eq. (4.125), which yields

$$\sum_p \left(\frac{\partial A_p}{\partial t} - \sigma_p A_p \right) \Theta(U_p) = - \sum_{m,n} A_m A_n N(U_m, U_n) \tag{4.127}$$

Note that the summation signs used in this section have to be replaced by integrals when considering non-discrete spectra, i.e. typically when some of the dimensions of the system

---

[6]Completeness cannot be proven in general, except for the simplest situations like Rayleigh–Bénard convection with stress-free boundaries [489]. However, it turns out that near instability thresholds, methods based on this assumption compare very satisfactorily with more rigorous perturbation methods such as presented in §4.2.2.

considered are infinite. For instance, for layers of infinite lateral extension, the eigenvectors U_p are plane waves $\exp(i\vec{k}.\vec{r})u(z)$, where $\vec{k}$ is a horizontal wavevector. This should be kept in mind for the following, and in particular for §4.5.5.

We then define adjoint linear operators Λ^+ and Θ^+ as usual, i.e.

$$\begin{aligned}
\left\langle \Lambda^+(\tilde{U}), U \right\rangle &= \left\langle \tilde{U}, \Lambda(U) \right\rangle, \\
\left\langle \Theta^+(\tilde{U}), U \right\rangle &= \left\langle \tilde{U}, \Theta(U) \right\rangle, \quad \forall U \in E, \tilde{U} \in F
\end{aligned} \tag{4.128}$$

where $<,>$ is some suitably defined scalar product, usually consisting in integration over the fluid volume (or bounding surface) of the product of components of vector fields, and F denotes a set of boundary conditions that are part of the definition of adjoint operators Λ^+ and Θ^+. Examples of scalar products and calculations of adjoint problems are given in Appendices B and D.

Now, considering solutions of the adjoint spectral problem

$$\Lambda^+(\tilde{U}_l) = \sigma_l^* \Theta^+(\tilde{U}_l) \tag{4.129}$$

where an asterisk denotes the complex conjugate, it is readily shown using the definitions (4.128) that if U_p is a solution of Eq. (4.125) with eigenvalue σ_p, then either $\sigma_l \neq \sigma_p$ and $< \tilde{U}_l, \Theta(U_p) > = 0$, or $\sigma_l = \sigma_p$ and $< \tilde{U}_l, \Theta(U_p) > \neq 0$ in general. Then, projecting Eqs (4.127) on the adjoint eigenmodes $\tilde{U}_l$ and excluding degenerate cases[7] yields

$$\frac{\partial A_l}{\partial t} = \sigma_l A_l + \sum_{m,n} Z_{m,n,l} A_m A_n \tag{4.130}$$

with

$$Z_{m,n,l} = - \frac{\left\langle \tilde{U}_l, N(U_m, U_n) \right\rangle}{\left\langle \tilde{U}_l, \Theta(U_l) \right\rangle} \tag{4.131}$$

Thus, one is left with an infinite number of ordinary differential equations for the amplitudes $A_l(t)$, which is equivalent (provided the set of eigenmodes is indeed complete) to the original problem. Now, the structure of the spectrum of the eigenvalue problem (4.125) will allow important simplifications.

Throughout this chapter, we have been interested in near-critical situations of the type I in the classification of Cross and Hohenberg (see Fig. 4.1), where most of the modes are still damped. In a laterally infinite Bénard system, the growth rate of the most dangerous modes indeed has the behavior sketched in Fig. 4.17.

---

[7] Actually, even if degenerate eigenvalues exist, i.e. several eigenfunctions U_p correspond to the same eigenvalue σ_p, the linear part of Eqs (4.130) may still remain diagonal in some cases. For instance, for systems with periodic boundary conditions in some directions, this holds provided the scalar product is defined taking advantage of the usual orthogonality properties of trigonometric functions. This holds for instance for plane wave solutions with wavevectors of the same length but different orientation, or solutions of the Helmholtz equation in a cylinder with different azimuthal wavenumbers (see §3.3.4 and §7.1), ...

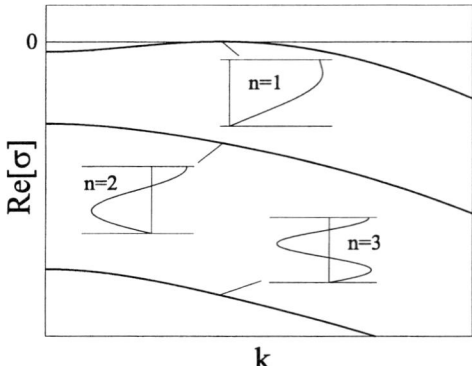

Figure 4.17: Typical variation of the real part of the most dangerous eigenvalues with the wavenumber k in a near-critical situation. The different bands correspond to vertical structures of increasing vertical wavenumber n (the insets represent typical z-dependencies of temperature perturbations for a Marangoni–Bénard problem).

For the moment, in order to avoid dealing with continuous bands of wavenumbers, we will require the system to be periodic in the horizontal directions, therefore selecting discrete values of the wavenumber k (see §3.3.4). Thus, it is possible to separate the set of solutions of Eq. (4.125) into a first set, say P, of modes with values of $\mathrm{Re}(\sigma)$ near zero, and a second set Q of modes with negative values of $\mathrm{Re}(\sigma)$, of order unity or higher. The modes belonging to P will be called "active", while it is natural to expect that the modes belonging to Q will be slaved to the dynamics of active modes, and hence will be called "passive".

Let us develop this *slaving* concept [51] in some detail, which will in particular allow to address the case where some eigenvalues are complex rather than purely real [148]. In the following, $B_q(t)$ will denote the amplitude of any of the passive modes. Its evolution equation will be limited to

$$\frac{\partial B_q}{\partial t} = \sigma_q B_q + \sum_{m,n \in P} Z_{m,n,q} A_m A_n \tag{4.132}$$

i.e. we neglect quadratic terms of Eq. (4.130) which involve damped modes. This is valid provided their amplitude remains small enough, which is guaranteed sufficiently near the primary (supposedly supercritical) bifurcation point. This is discussed at the end of this section. Then, Eq. (4.132) is formally integrated as

$$B_q(t) = \sum_{m,n \in P} Z_{m,n,q} \int_{-\infty}^{t} d\tau \, \exp[\sigma_q(t - \tau)] A_m(\tau) A_n(\tau) \tag{4.133}$$

where the choice of the lower integration bound at $-\infty$ is actually unimportant, since $\mathrm{Re}(\sigma_q) = O(1) < 0$ and the exponential factor shows that only values of τ within some time interval $\sim -1/\mathrm{Re}(\sigma_q)$ before t will contribute to the integral. This indicates that the amplitude of damped modes rapidly lose the memory of their past values. Now, if we assume further that the dynamics of all active modes is correctly described by $A_m(\tau) = \exp[i\,\mathrm{Im}(\sigma_m)\tau]a_m(\tau)$, where $a_m(\tau)$ is nearly constant on a time interval $\sim -1/\mathrm{Re}(\sigma_q)$, then $B_q(t)$ remains slaved to the value

$$B_q(t) = \sum_{m,n\in P} \frac{Z_{m,n,q}}{i\,\mathrm{Im}(\sigma_m + \sigma_n) - \sigma_q} A_m(t)A_n(t) \tag{4.134}$$

and in particular, if all eigenvalues are real, the result could have been obtained directly by assuming $\partial B_q/\partial t = 0$ in Eq. (4.132), i.e. that $\partial/\partial t$ can be neglected compared with $-\sigma_q$.

Separating the sums on active (P) and passive (Q) modes in Eq. (4.130), using Eq. (4.134) for all $q \in Q$, and neglecting terms of order higher than cubic, we get

$$\frac{\partial A_p}{\partial t} = \sigma_p A_p + \sum_{m,n\in P} Z_{m,n,p} A_m A_n + \sum_{m,r,s\in P} Z_{m,r,s,p} A_m A_r A_s \tag{4.135}$$

with

$$Z_{m,r,s,p} = \sum_{n\in Q} \frac{Z_{r,s,n}(Z_{m,n,p} + Z_{n,m,p})}{i\,\mathrm{Im}(\sigma_r + \sigma_s) - \sigma_n} \tag{4.136}$$

In this expression, the number of slaved modes to be considered in the sum should in principle be infinite, and in particular it should include all the stable bands corresponding to different vertical structures (see Fig. 4.17). However, the result has been found to converge rather rapidly for several Bénard convection problems [132, 151, 152, 345, 418]. Note that as the method described above has still not been restricted to a near-critical situation, the coefficients (4.131) and (4.136) generally depend on the control parameter μ. This in principle allows amplitude equations to be extrapolated further in the nonlinear regime, a possibility examined in §8.1.

Now, particularizing the method to laterally infinite (or periodic) Bénard systems, each mode p solution of Eq. (4.125) should be associated with some horizontal wavevector $\vec{k}_p$, i.e. it is a plane wave $\exp(i\vec{k}_p.\vec{r})u_p(z)$. Therefore, the scalar product should naturally include an average in the horizontal plane, which leads to the vanishing of a number of quadratic and cubic coefficients. Indeed, it is found that

$$Z_{m,n,p} \sim \delta_{\vec{k}_p, \vec{k}_m + \vec{k}_n}, \quad Z_{m,r,s,p} \sim \delta_{\vec{k}_p, \vec{k}_m + \vec{k}_r + \vec{k}_s} \tag{4.137}$$

Let us conclude this section by applying the results to the case considered in the previous section, namely to a near-critical situation for which the set P contains a certain distribution of *monotonic* modes $(\mathrm{Im}[\sigma_p] = 0, p \in P)$ with wavevectors all on the critical circle $|\vec{k}_p| = k_c = O(1)$, $p \in P$. A smallness parameter ε can then be used to measure the supercriticality $\mu - \mu_c = \varepsilon^2 \mu_2$, and a clear distinction is now possible between modes belonging to P (which all have the same growth rate $\sigma_p = \sigma \sim \varepsilon^2$) and slaved modes $(\mathrm{Re}[\sigma_q] = O(1) < 0)$. Then, the time variable should be scaled as $t \sim \varepsilon^{-2}$, and amplitudes of active modes as $A_p \sim \varepsilon$. Thus, according to Eq. (4.134), it is seen that $B_q \sim \varepsilon^2$ for slaved modes, which in fact justifies both Eq. (4.132) and the limitation to cubic terms in Eq. (4.135).

Using Eqs (4.137), and also Eq. (4.107) to simplify the cubic terms, the amplitude equations (4.135) reduce to the form (4.120) obtained via perturbation theory. Taking into account that the z-dependencies to be used in the calculation of coefficients (4.131) are given by the limit for $\varepsilon \to 0$ of the active eigenfunctions (i.e. the neutral stability functions), it is further seen that the numerical value of the quadratic coefficient δ has the same expression for both methods. However, it is not possible to show that the cubic coefficients are also identical. This fact, which should actually depend on the completeness of the set of eigenmodes used, has at present only been verified for particular cases, comparing for instance the value of some typical convective quantities at steady state. An excellent agreement is generally found for Bénard problems, and in particular for the range of hysteresis predicted in Marangoni–Bénard convection at the first bifurcation point [130, 132, 151, 345].

Finally, note that the limitation of amplitude equations (4.135) to cubic order is, as in the previous section, only valid provided the cubic terms are able to saturate the exponential growth of active amplitudes. Moreover, quadratic coefficients $Z_{m,n,l}, m, n, l \in P$ also need to be numerically small (of order ε) for all terms of Eqs (4.135) to be of the same order ε^3. In other cases, higher-order terms should be included in Eqs (4.135), which in principle does not lead to major difficulties, apart from the increasing computational effort in accurately evaluating higher-order coefficients. Examples of application of the method presented above, and its variants, may be found in [132, 149, 150, 151, 152, 345, 418] and in §7.1, dealing with Bénard convection in small aspect-ratio circular containers.

4.2.4   General non-resonant pattern selection problem

It is possible to obtain some general conclusions about the problem of the competition between an arbitrary large number, say N, of modes with wavevectors $\vec{k}_n$ ($n = 1, ..., N$) regularly disposed on the critical circle (the angle between two neighboring vectors is $\theta = \pi/N$). These modes (and their conjugates) form the set P defined in §4.2.2. The amplitude equations for this case have been obtained as Eq. (4.120), i.e.

$$\frac{\partial a_{\vec{k}}}{\partial t} = \alpha(\mu - \mu_c)a_{\vec{k}} + \delta a_{\vec{k}_+}^* a_{\vec{k}_-}^* - a_{\vec{k}} \left[\beta|a_{\vec{k}}|^2 + {\sum}'_{\vec{q} \neq \vec{k}} \gamma_{\vec{k},\vec{q}} |a_{\vec{q}}|^2 \right] \tag{4.138}$$

including the resonant quadratic interactions. However, in this section we will assume $\delta = 0$, which is valid either for specific problems (e.g. pure Rayleigh–Bénard convection, as will be seen in §4.3.5), or even for problems where $\delta \neq 0$ at a sufficiently large value of the constraint. Indeed, defining $\sigma = \alpha(\mu - \mu_c)$, we see that at large σ (in fact, if γ is of the same order as β, the condition is $\sigma \gg \delta^2/\beta$), the quadratic terms become negligible compared with the cubic ones [because the amplitudes grow as $(\sigma/\beta)^{1/2}$]. Although at large amplitudes, terms of order higher than cubic should be included, such an approximation can often be used to gain qualitative information about the selection of patterns at moderate supercriticality.

We thus rewrite the problem (4.138) as

$$\frac{\partial a_n}{\partial t} = \sigma a_n - a_n \left[\beta|a_n|^2 + {\sum}'_{m \neq n} \gamma_{m,n} |a_m|^2 \right] \tag{4.139}$$

where a_n stands for $a_{\vec{k}_n}$, and the cross-interaction coefficient $\gamma_{m,n}$ only depends on the non-oriented angle between the wavevectors $\vec{k}_m$ and $\vec{k}_n$.

Separating moduli and phases $a_n = r_n \exp[i\varphi_n]$ we get $\dot{\varphi}_n = 0$ (the dot denotes time derivative), and

$$\dot{r}_n = r_n \left[\sigma - \beta r_n^2 - \sum_{m \neq n} \gamma_{m,n} r_m^2\right] \tag{4.140}$$

Thus, the phases are not coupled (due to the absence of resonance), and the pattern selection problem should be entirely determined by the curve $\gamma(\theta)$. Note that $\beta = \gamma(0)/2$, see Eq. (4.111). By a rescaling of amplitudes $r_n \sim (\sigma/\beta)^{1/2}$, it turns out that pattern selection is independent of $\sigma > 0$, but this is not done here in order to keep usual notations.

Steady regular solutions

Seeking steady solutions of Eqs (4.140), it is seen that for every mode n, we may have the solution $r_n = 0$. If $r_n \neq 0$, we have to solve

$$\sigma = \beta r_n^2 + \sum_{m \neq n} \gamma_{m,n} r_m^2 \tag{4.141}$$

Thus, several solutions exist, obtained by setting to zero a certain set P_0 of amplitudes, and solving Eq. (4.141) for the non-zero amplitudes $n \in P_1 = P - P_0$. As a particular case, we consider hereafter the subset of regular solutions (i.e. those corresponding to a regular distribution of modes on the critical circle), for which it can be seen that the amplitudes are equal for each non-zero mode.

Let m be the number of non-zero amplitudes [i.e. $m = \dim(P_1)$] and denote by S_m the corresponding solution (with amplitude $r = R_m$). In particular, we have the patterns:

Rolls ($m = 1$)   : $S_1: R_1^2 = \frac{\sigma}{\beta}$ $E_1 = \frac{\sigma}{\beta}$

Squares ($m = 2$)   : $S_2: R_2^2 = \frac{\sigma}{\beta + \gamma_{\pi/2}}$ $E_2 = \frac{2\sigma}{\beta + \gamma_{\tau/2}}$

Hexagons ($m = 3$)   : $S_3: R_3^2 = \frac{\sigma}{\beta + \gamma_{\pi/3} + \gamma_{2\pi/3}}$ $E_3 = \frac{3\sigma}{\beta + \gamma_{\tau/3} + \gamma_{2\pi/3}}$

$$\cdot$$
$$\cdot$$
$$\cdot$$

Arbitrary m   : $S_m: R_m^2 = \frac{\sigma}{\beta + \sum_{p=1}^{m-1} \gamma_{p\pi/m}}$ $E_m = \frac{m\sigma}{\beta + \sum_{p=1}^{m-1} \gamma_{p\pi/m}}$

For every solution, we have also given the "energy"

$$E_m = \sum_{p=1}^{m} r_p^2 = m R_m^2 \tag{4.142}$$

We should also point out that for each of the solutions, the phases φ_n are in principle arbitrary (determined by initial conditions). This is strictly true as long as $\delta = 0$, but even if δ is small (or σ large), it is likely that a (slow) synchronization of the phases will occur for modes forming resonant triads. This is the case for hexagons ($m = 3$), as seen earlier, but also for dodecagons ($m = 6$), or more generally for each case where m is a multiple of 3. Some typical patterns are represented in Fig. 4.18.

Note that apart from the first three cases ($m \leq 3$) in Fig. 4.18, the higher order patterns are *quasi-periodic*, i.e. they are formed by the juxtaposition of periodic structures with incommensurate spatial frequencies. Although patterns of order higher than $m = 3$ are not common in convection problems (they are generally unstable), it is interesting to study their possibility in general, because they could appear in some other systems for which the nonlinear interactions between plane waves depend in some particular way on their angle of interaction [i.e. $\gamma(\theta)$ must assume a special form]. We also note that metastable quasi-periodic dodecagonal patterns have recently been predicted by Golovin et al. [341] (although in a small range of parameters) in large-scale Marangoni–Bénard convection with deformable interface (see also §4.6.3). Long ago, in his PhD dissertation, Busse [342] also considered quasi-periodic patterns.

Stability of regular solutions

Perturbing the amplitude equations by small disturbances around a particular steady solution (denoted by amplitudes r_n), the stability is determined by calculating the eigenvalues of a matrix J, the Jacobian, whose elements are

$$J_{n,m} = \delta_{n,m}(\sigma - 3\beta r_n^2 - \sum_{p \neq n} \gamma_{p,n} r_p^2) - 2(1 - \delta_{n,m})\gamma_{m,n} r_m r_n \tag{4.143}$$

Let us examine the structure of the Jacobian in some detail. For a mode (say, q) which has zero amplitude in the steady state considered (i.e. $r_q = 0$ and $q \in P_0$), Eq. (4.143) shows that $J_{q,m} = J_{m,q} = 0$ for $m \neq q$. Thus, the line q and the column q are zero except for the diagonal element $J_{q,q}$. Thus, this element directly gives an eigenvalue

$$\lambda_q^0 = \sigma - \sum_{p \in P_1} \gamma_{q,p} r_p^2 \qquad \text{for} \quad q \in P_0 \tag{4.144}$$

which characterizes the stability to "outer" disturbances (i.e. modes not present in the steady solution). The stability to "inner" disturbances has to be investigated by calculating the eigenvalues of the reduced matrix J' obtained by removing from J those lines and columns corresponding to elements of P_0. Thus, for S_1, we have a matrix with a single element $\sigma - 3\beta R_1^2 = -2\sigma$ (the remaining eigenvalue). For S_2, we have the 2×2 matrix

$$J_2' = -2R_2^2 \begin{pmatrix} \beta & \gamma_{\pi/2} \\ \gamma_{\pi/2} & \beta \end{pmatrix} \tag{4.145}$$

whose eigenvalues are -2σ and $-2\sigma(\beta - \gamma_{\pi/2})/(\beta + \gamma_{\pi/2})$, as already found in §4.1.2. In general for the regular solution S_m, it can be shown that the elements of the reduced matrix J' are given by

$$J_{p,q}' = -2R_m^2 [\beta \delta_{p,q} + \gamma_{p,q}(1 - \delta_{p,q})] \tag{4.146}$$

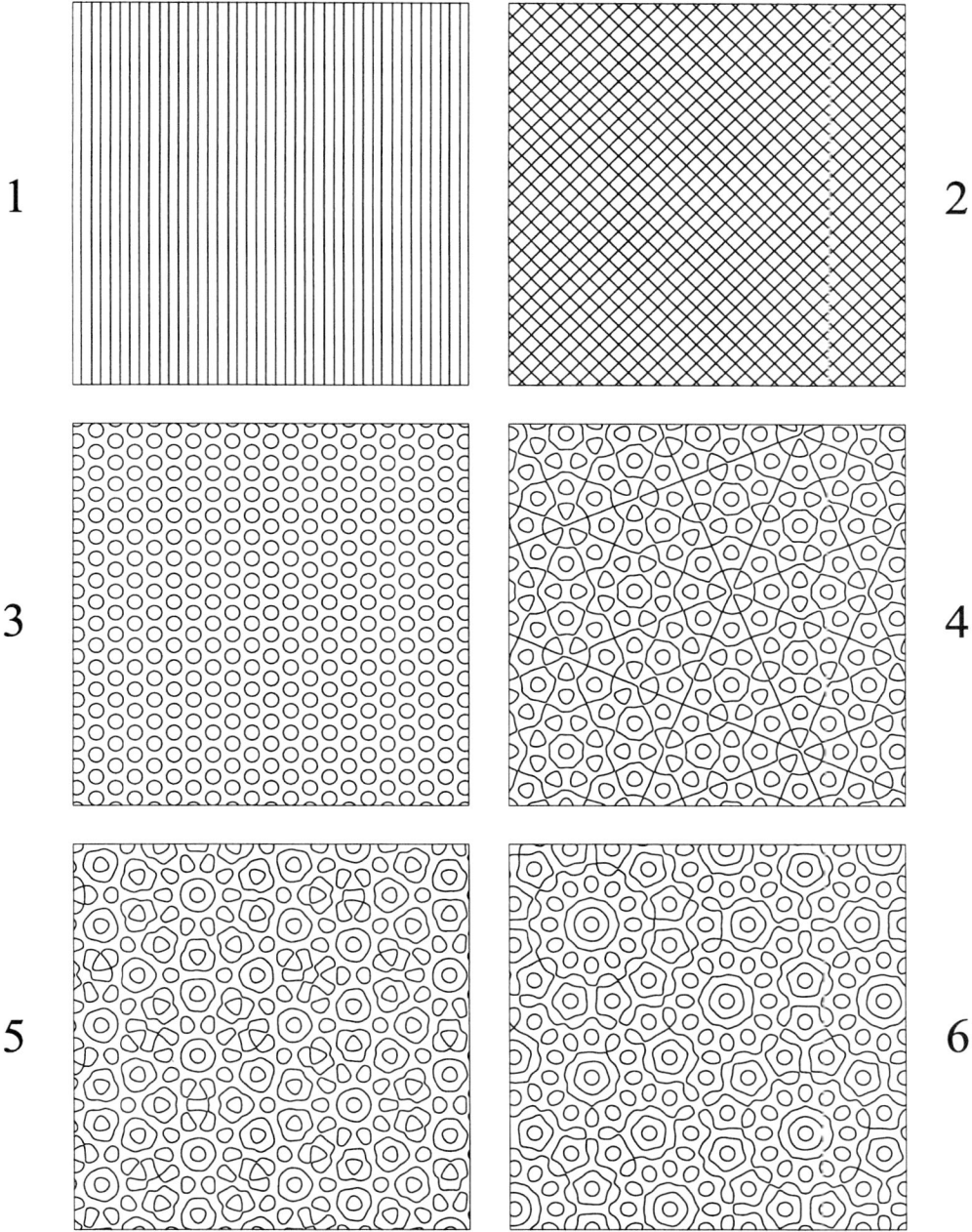

Figure 4.18: Zero contour line of the planform function for several patterns. Rolls ($m = 1$), squares ($m = 2$), hexagons ($m = 3$), octagons ($m = 4$), decagons ($m = 5$) and dodecagons ($m = 6$). For $m = 3$ and $m = 6$, a synchronization of the phases has been assumed.

with R_m given in the previous section. By using the symmetries $\gamma(\pi/2 + \theta) = \gamma(\pi/2 - \theta)$ and $\gamma(\theta) = \gamma(-\theta)$, it is possible to compute eigenvalues analytically, for relatively high-order solutions:

- S_3 : $-2R_3^2(\beta - \gamma_{\pi/3})$ (2×), $-2R_3^2(\beta + 2\gamma_{\pi/3})$

- S_4 : $-2R_4^2(\beta - \gamma_{\pi/2})$ (2×), $-2R_4^2(\beta - 2\gamma_{\pi/4} + \gamma_{\pi/2})$, $-2R_4^2(\beta + 2\gamma_{\pi/4} + \gamma_{\pi/2})$

- S_5 : $-2R_5^2(\beta + 2\gamma_{\pi/5} + 2\gamma_{2\pi/5})$, $-2R_5^2\left(\beta - \gamma_{\pi/5}\frac{1+\sqrt{5}}{2} - \gamma_{2\pi/5}\frac{1-\sqrt{5}}{2}\right)$ (2×),

  $-2R_5^2\left(\beta - \gamma_{\pi/5}\frac{1-\sqrt{5}}{2} - \gamma_{2\pi/5}\frac{1+\sqrt{5}}{2}\right)$ (2×)

- S_6 : $-2R_6^2(\beta + \gamma_{\pi/6} - \gamma_{\pi/3} - \gamma_{\pi/2})$ (2×), $-2R_6^2(\beta - \gamma_{\pi/6} - \gamma_{\pi/3} + \gamma_{\pi/2})$ (2×),

  $-2R_6^2(\beta - 2\gamma_{\pi/6} + 2\gamma_{\pi/3} - \gamma_{\pi/2})$, $-2R_6^2(\beta + 2\gamma_{\pi/6} + 2\gamma_{\pi/3} + \gamma_{\pi/2})$

- ...

Lyapunov function and extremum principles

The system (4.140) admits the potential

$$V = -\frac{\sigma}{2}\sum_n r_n^2 + \frac{\beta}{4}\sum_n r_n^4 + \frac{1}{4}\sum_m\sum_{n\neq m}\gamma_{m,n}r_m^2 r_n^2 \qquad (4.147)$$

from which it follows that $\dot{r}_p = -\partial V/\partial r_p$ and $dV/dt = -\sum_p(\dot{r}_p)^2 \leq 0$, thus implying a monotonic decrease of V up to one of its (possibly local) minima. This extremum principle was first established by Busse [53] who further showed that a potential function also exists in the general case where quadratic resonance occurs, as we shall see below.

It is interesting to examine the relation between the "energy" (or convective heat transport) $E = \sum_p r_p^2$ and V. Using Eqs (4.140) and (4.147) it can be shown that

$$\frac{1}{2}\frac{dE}{dt} = -\sigma E - 4V \qquad (4.148)$$

Thus, although it cannot be proven that the evolution of E will be monotonic, at steady state we must have

$$E = -\frac{4V}{\sigma} \qquad (4.149)$$

which implies that the steady solutions corresponding to higher E (or higher convective heat transport) are energetically more favorable than those of lower E (the discussion here refers to the supercritical case $\sigma > 0$). In particular, the absolute minimum of V corresponds to the absolute maximum of the heat transport E achievable in a steady state, and this solution should be absolutely stable (see also [53, 294, 296]). As also pointed out by Busse [53], every spatially averaged property other than the convective heat flux could also have served to obtain a physical interpretation of the maximum E criterion, such as the kinetic energy (also proportional to E), ...

The maximum E criterion provides a way to test the preference (from the point of view of the potential V) between the regular solutions established above. Indeed, among the regular solutions S_m ($m = 1, 2, 3, ...$), the energetically preferred one will be that corresponding to the maximum of the heat transport by convection. According to the above discussions, this quantity is proportional to

$$E_m = \frac{m\sigma}{\beta + \sum_{p=1}^{m-1} \gamma_{p\pi/m}} \tag{4.150}$$

which can be easily evaluated given the cross-interaction curve $\gamma(\theta)$. Note that it cannot be rejected that a non-regular solution could correspond to a higher heat transfer. In particular, the analysis made above may be easily extended to semi-regular solutions [53], i.e. those formed by superposing two identical regular solutions rotated by an arbitrary angle with respect to each other. More generally, the possibility of irregular distributions of modes on the critical circle should be checked by solving the system (4.141) for all steady solutions (e.g. numerically) and then comparing their energies $E = \sum_p r_p^2$.

It must also be emphasized that the maximum heat transfer criterion demonstrated here is not valid in the presence of resonant quadratic interactions (e.g. due to non-Boussinesq or thermocapillary effects [53, 23]). Although nothing can be said about the quantity E in these cases, a potential function still exists. Indeed, the general amplitude equations (4.120) may still be derived from the Lyapunov function

$$L = -\sigma \sum_p |a_p|^2 - \frac{\delta}{3} \sum_p \left(a_p^* a_{p+}^* a_{p-}^* + a_p a_{p+} a_{p-} \right)$$
$$+ \frac{\beta}{2} \sum_p |a_p|^4 + \frac{1}{2} \sum_p \sum_{q \neq p} \gamma_{p,q} |a_p|^2 |a_q|^2 \tag{4.151}$$

which implies $\dot{a}_n = -\partial L / \partial a_n^*$ and $dL/dt = -2 \sum_p |\partial L / \partial a_p|^2 = -2 \sum_p |\dot{a}_p|^2 \leq 0$.

4.3 Pattern selection in Bénard instabilities

We now turn to the explicit calculation of the linear and nonlinear coefficients appearing in the amplitude equations, for the particular case of Rayleigh–Marangoni–Bénard instabilities. For this purpose, we will use the general expressions established in §4.2.2. It is therefore necessary to obtain an operational form for differential equations and boundary conditions, as well as to discuss the nature of the mean field (homogeneous mode).

As an illustration, we will then consider the situation where a liquid layer is confined between two horizontal plates (which for simplicity will be considered as isothermal and stress-free) and heated from below. The corresponding linear stability has been studied in §3.3.1. For this classical Rayleigh–Bénard problem [22, 25, 27, 199, 327, 328, 329], it is possible to obtain the analytical form of the cubic coupling coefficient $\gamma(\theta)$ between modes whose orientation differs by θ on the critical circle. Implications for pattern selection and convective heat transfer are also briefly discussed. Then, in the next sections, more recent results are presented about the pattern selection problem for Rayleigh–Marangoni–Bénard instabilities, both for one-layer and two-layer systems (corresponding to the linear stability analyzes of §3.3.2 and §3.5.2, respectively).

Consider again the thermo-hydrodynamic equations in the Boussinesq approximation, under the form (see §3.3.1)

$$\vec{\nabla}.\vec{V} = 0 \tag{4.152}$$

$$\Delta\vec{V} - \vec{\nabla}p + Ra\vec{1}_z T = Pr^{-1}\left(\frac{\partial\vec{V}}{\partial t} + (\vec{V}.\vec{\nabla})\vec{V}\right) \tag{4.153}$$

$$\Delta T + W = \frac{\partial T}{\partial t} + (\vec{V}.\vec{\nabla})T \tag{4.154}$$

where $\vec{V} = W\vec{1}_z + \vec{V}_h$, T, p are respectively the velocity, temperature and pressure perturbations, $Pr = \nu/\kappa$ is the Prandtl number, and Ra is the Rayleigh number. The boundary conditions will not be explicitly needed here, and will only be introduced when calculating nonlinear coefficients.

We may eliminate the pressure p from the Navier–Stokes equations (4.153) by twice applying the curl operator. Then, using the relations $\vec{\nabla}\times\vec{\nabla}\times\vec{a} = -\Delta\vec{a}+\vec{\nabla}(\vec{\nabla}.\vec{a})$ and $\vec{\nabla}\times\vec{\nabla}f = 0$ valid for any $\vec{a}$ and f and the incompressibility relation (4.152), the projection of the result on the vertical $\vec{1}_z$ leads to

$$\Delta^2 W + Ra\Delta_h T = Pr^{-1}\frac{\partial}{\partial t}\Delta W - Pr^{-1}\vec{1}_z.\left[\vec{\nabla}\times\vec{\nabla}\times(\vec{V}.\vec{\nabla})\vec{V}\right] \tag{4.155}$$

where Δ_h is the horizontal Laplacian. Combining this equation with the energy equation (4.154) would lead to a closed problem for W and T if the horizontal velocity components were not appearing in the right-hand sides of both equations. Thus, we have to determine an equation for the horizontal velocity $\vec{V}_h$ by returning to the horizontal components of Eq. (4.153). We will use the following decomposition into an irrotational and a solenoidal part

$$\vec{V}_h = \vec{\nabla}_h\phi + \vec{\nabla}\times(\Psi\vec{1}_z) + \vec{V}_0(z,t) \tag{4.156}$$

where ϕ and Ψ are functions of x, y, z and t to be determined, while $\vec{V}_0(z,t)$ is the mean flow. This general representation is in fact equivalent to the decomposition of the solenoidal vector field $\vec{V}$ into a toroidal and a poloidal part (used e.g. in [316, 343])

$$\vec{V} = \vec{\nabla}\times(f\vec{1}_z) + \vec{\nabla}\times\vec{\nabla}\times(g\vec{1}_z) + \vec{V}_0(z,t) \tag{4.157}$$

which may indeed be written in the form $\vec{V} = W\vec{1}_z + \vec{V}_h$ [with $\vec{V}_h$ given by (4.156)] after the substitutions $W = -\Delta_h g$, $\phi = \partial g/\partial z$ and $\Psi = f$.

The horizontal velocity potential ϕ is related to the vertical velocity component W by

$$DW + \Delta_h\phi = 0 \tag{4.158}$$

obtained from the incompressibility condition (4.152). Now, the vertical component of the vorticity is given by

$$\omega_z = \vec{1}_z.\left[\vec{\nabla} \times \vec{V}\right] = -\Delta_h \Psi \tag{4.159}$$

for which an equation can be obtained by applying $\vec{1}_z.\vec{\nabla} \times$ to the horizontal component of Eq. (4.153). We obtain

$$\Delta\omega_z = Pr^{-1}\left\{\frac{\partial\omega_z}{\partial t} + \vec{1}_z.\vec{\nabla} \times [(\vec{V}.\vec{\nabla})\vec{V}_h]\right\} \tag{4.160}$$

Now, applying the horizontal divergence to the horizontal component of Eq. (4.153) leads to

$$\Delta\Delta_h\phi - \Delta_h p = Pr^{-1}\left\{\frac{\partial}{\partial t}\Delta_h\phi + \vec{\nabla}_h.[(\vec{V}.\vec{\nabla})\vec{V}_h]\right\} \tag{4.161}$$

Equations (4.154), (4.155), (4.156), (4.158), (4.160) and (4.161) in principle allow to determine the fields T, W, ϕ, Ψ (or ω_z) and p, while an equation for the mean flow $\vec{V}_0(z,t)$ still needs to be determined (see next section). Actually, important simplifications will be used in the following, by restricting consideration to supercritical bifurcations of monotonic modes.

4.3.1 Mean field dynamics

Of particular interest in nonlinear studies is the horizontal average of the perturbations (i.e. the "mean field", corresponding to $\vec{k} = 0$ Fourier components). As the continuity equation (4.152) reduces to $DW = 0$ for these modes, it is clear that the mean flow cannot have a vertical component in the presence of fixed and impervious horizontal boundaries (where $W = 0$). We may obtain an equation for the horizontal component $\vec{V}_0(z,t)$ of the mean flow, by averaging the horizontal Navier–Stokes equations (4.153). Assuming the fluctuations to remain bounded everywhere (e.g. periodic), we have

$$\frac{\partial^2\vec{V}_0(z,t)}{\partial z^2} = Pr^{-1}\left\{\frac{\partial\vec{V}_0(z,t)}{\partial t} + \frac{\partial}{\partial z}\left(\overline{W\vec{V}_h}\right)\right\} \tag{4.162}$$

where the overbar denotes the horizontally averaged value. It is interesting to put this equation in parallel with that governing the mean temperature field $T_0(z,t) = \overline{T}$. From Eq. (4.154), we similarly obtain

$$\frac{\partial^2 T_0}{\partial z^2} = \frac{\partial T_0}{\partial t} + \frac{\partial}{\partial z}\left(\overline{WT}\right) \tag{4.163}$$

Thus, at the linear stage, the dynamics of both the mean flow $\vec{V}_0(z,t)$ and the mean temperature perturbation $T_0(z)$ are purely diffusional. If at least one of the boundaries is rigid and heat conducting, the only steady solution will be $\vec{V}_0 = T_0 = 0$. When both boundaries are stress-free, a uniform mean flow solution is possible in principle. However, it is then possible to eliminate this component by working in a moving reference frame (Galilean invariance). On the other hand, if at both boundaries the heat flux is kept fixed, a uniform mean temperature shift is possible. This is also unimportant as long as no large scale modulations of these mean fields are allowed. In this case, the large-scale fields would acquire a slow dynamics, similarly to the Goldstone modes associated with uniform horizontal translations, and encountered in the study of amplitude equations (§4.1.3 and §4.5.3). Large scale modulations of the mean flow and vertical vorticity (see next section) may be generated by defects in cellular structures (see §4.5.6), and play an important role in pattern selection. Large scale modulations of the mean temperature is at the origin of the long-wavelength instabilities ($k_c = 0$) encountered in §3.3.1, and described by Sivashinsky-like equations [212, 215] in the nonlinear domain (see §4.5).

As we are not considering such kind of slow modulations for the moment, the solution of the linear problem (order ϵ in the notations of §4.2.2) will be characterized by a vanishing mean field [damping of the mean temperature field on a $O(1)$ time scale, and of the mean flow on a $O(Pr^{-1})$ time scale]. Now it is clear that the zero Fourier components may in general be excited by the nonlinear terms in Eqs (4.162–4.163). Thus, the solution at order ϵ^2 might contain a certain mean field component. However, even though this will always be the case for the mean temperature, mean flows in general do not occur at this order for defect-free patterns. Indeed, it is readily checked that the nonlinear term in Eq. (4.162) identically vanishes for any superposition of modes of the type (4.86). Accordingly, the mean flow is at least of order ϵ^3, and it is also checked that it will not affect amplitude equations at this order. If the bifurcation is supercritical, we may thus safely impose $\vec{V}_0 = 0$, keeping in mind that this assumption will need to be re-examined for imperfect patterns (see §4.5.6), and for weakly nonlinear waves (see Chapter 5).

Hence, for the general pattern selection problem considered here, the mean field is purely thermal at lowest order, and is obtained by solving Eq. (4.163) with $\partial/\partial t = 0$ (the time derivative enters at order ϵ^3). This gives

$$T_0 = \int_0^z dz \overline{WT}(z) + c_1 z + c_2 \tag{4.164}$$

with constants c_1 and c_2 to be determined from thermal boundary conditions. For instance, for isothermal boundaries at $z = 0, 1$, we have

$$c_2 = 0, \quad c_1 = - \int_0^1 \overline{WT} dz \tag{4.165}$$

As the dimensionless temperature profile in the reference state is given by $-z$, the Nusselt number is

$$Nu = 1 + \int_0^1 \overline{WT} dz \tag{4.166}$$

in agreement with Eq. (4.123), as both W and T are $O(\epsilon)$.

4.3.2 Vertical vorticity

Equation (4.160) shows that in the linear approximation, the vertical vorticity ω_z obeys a diffusion equation. This has to be solved with appropriate boundary conditions: for instance, at a rigid boundary, Eq. (4.159) shows that $\omega_z = 0$. Thus, in general, $\omega_z = 0$ is a solution at the linear stage, and the horizontal velocity field is irrotational. When both boundaries are stress-free (where $\partial \omega_z / \partial z = 0$), we will disregard the solution where ω_z is a constant (e.g. corresponding to dissipation-less uniform rotation around a vertical axis). Accordingly, it is possible to solve Eq. (4.159) by $\Psi = 0$ (which also satisfies boundary conditions at rigid or stress-free walls, as well as Marangoni boundary conditions). Thus, the linear solution is characterized by a purely potential horizontal velocity field $\vec{V}_h = \vec{\nabla}_h \phi$.

Just as for the mean flow, it can be shown that the nonlinear terms in Eq. (4.160) do not generate vertical vorticity at order ϵ^2. This is obvious for $Pr \to \infty$, as the equation for ω_z remains linear. When the Prandtl number is finite, we may still prove this result by computing the interaction of any two modes with wavevectors $\vec{k}_1$ and $\vec{k}_2$ on the critical circle. Denoting by $W_n(z)$ and $\phi_n(z) = k_c^{-2} D W_n(z)$ the neutral stability functions, the velocity components at the first order read

$$\vec{V}_1 = A_1 \exp[i\vec{k}_1.\vec{r}] \left(W_n \vec{1}_z + i\vec{k}_1 \phi_n \right), \quad \vec{V}_2 = A_2 \exp[i\vec{k}_2.\vec{r}] \left(W_n \vec{1}_z + i\vec{k}_2 \phi_n \right)$$

$$(4.167)$$

Then, the interaction term can be rearranged as

$$\vec{1}_z.\vec{\nabla} \times \left[(\vec{V}_1.\vec{\nabla})\vec{V}_{2h} \right] = A_1 A_2 \exp[i(\vec{k}_1 + \vec{k}_2).\vec{r}]\vec{1}_z.(\vec{k}_2 \times \vec{k}_1) \left[W_n D\phi_r - \vec{k}_1.\vec{k}_2 \phi_n^2 \right]$$

$$(4.168)$$

Now, in the calculation of terms proportional to $\exp[i(\vec{k}_1 + \vec{k}_2).\vec{r}]$ in the nonlinear term of Eq. (4.160) we must also consider the symmetric term $\vec{1}_z.\vec{\nabla} \times \left[(\vec{V}_2.\vec{\nabla})\vec{V}_{1h} \right]$, which (due to the cross-product) is just the opposite of Eq. (4.168). Consequently, the two contributions cancel each other, and no vertical vorticity is generated at order ϵ^2. We may then safely assume $\Psi = 0$ because the possible contributions generated at order ϵ^3 will not affect the calculation of the coefficients of amplitude equations at this order. Note that this result no longer holds when spatial modulations or defects occur in the patterns (see §4.5.6), or for waves (see Chapter 5).

4.3.3 Definition of linear and nonlinear operators

The above considerations allow important simplifications of the system of equations. In particular, the horizontal velocity potential can be readily obtained from Eq. (4.158) at each order of approximation, and Eq. (4.161) may be omitted, because it will merely serve to determine the pressure p (which does not appear in other equations, nor in boundary conditions if we assume boundaries to be undeformable). Thus, the full problem may be written under the operational form (4.82) with

$$\mathcal{L}(U) = \begin{pmatrix} \Delta^2 W \\ \Delta T + W \end{pmatrix}, \quad M(U) = \begin{pmatrix} -\Delta_{\mathrm{h}} T \\ 0 \end{pmatrix}, \quad \Theta(U) = \begin{pmatrix} Pr^{-1} \Delta W \\ T \end{pmatrix}$$

(4.169)

$$N(U_1, U_2) = \begin{pmatrix} -Pr^{-1}\vec{\imath}_z \cdot \left[\vec{\nabla} \times \vec{\nabla} \times (\vec{V}_1 \cdot \vec{\nabla})\vec{V}_2 \right] \\ (\vec{V}_1 \cdot \vec{\nabla}) T_2 \end{pmatrix}$$

(4.170)

and the control parameter is $\mu = Ra$. The form of the linear operators (4.169) is similar to those introduced in Appendices B and D. The subsequent definitions of the scalar product and of the adjoint linear problem are therefore valid, and may be used to evaluate the amplitude equation coefficients obtained in §4.2.2.

We also have to consider the boundary conditions forming the definition of the set E defined in §4.2.2. Some of the boundary conditions (such as the Marangoni stress boundary condition) are advantageously introduced as a supplementary component of the operators (see Appendix D). In each case, those linear boundary conditions not included in the operators will define the set E (i.e. the domain of the above operators).

The vector $U \in E$ can be written $U = [W, T]$ and it should be remembered that the associated horizontal velocity potential ϕ is determined by Eq. (4.158). Thus, for a vector $U_{\vec{k}} = [W_{\vec{k}}, T_{\vec{k}}]$ corresponding to a wavevector $\vec{k}$, the corresponding horizontal velocity is given by $\vec{V}_{\vec{k}h} = i\vec{k}k^{-2}DW_{\vec{k}}$. Accordingly, the bilinear operator (4.170) may be given a more explicit form. After some algebra, we finally get

$$N_{\vec{k},\vec{k}'}(U_{\vec{k}}, U_{\vec{k}'}) = \begin{pmatrix} Pr^{-1} \begin{pmatrix} \frac{(\vec{k}+\vec{k}')\cdot\vec{k}'}{k'^2} D \left[W_{\vec{k}} D^2 W_{\vec{k}'} - \frac{\vec{k}\cdot\vec{k}'}{k^2} DW_{\vec{k}} DW_{\vec{k}'} \right] \\ -(\vec{k}+\vec{k}')^2 \left[W_{\vec{k}} DW_{\vec{k}'} - \frac{\vec{k}\cdot\vec{k}'}{k^2} DW_{\vec{k}} W_{\vec{k}'} \right] \end{pmatrix} \\ W_{\vec{k}} DT_{\vec{k}'} - \frac{\vec{k}\cdot\vec{k}'}{k^2} DW_{\vec{k}} T_{\vec{k}'} \end{pmatrix}$$

(4.171)

For completeness, we also rewrite the linear operators in this representation:

$$\mathcal{L}_{\vec{k}}(U_{\vec{k}}) = \begin{pmatrix} (D^2 - k^2)^2 W_{\vec{k}} \\ (D^2 - k^2) T_{\vec{k}} + W_{\vec{k}} \end{pmatrix}, \quad M_{\vec{k}}(U_{\vec{k}}) = \begin{pmatrix} k^2 T_{\vec{k}} \\ 0 \end{pmatrix},$$

$$\Theta_{\vec{k}}(U_{\vec{k}}) = \begin{pmatrix} Pr^{-1}(D^2 - k^2) W_{\vec{k}} \\ T_{\vec{k}} \end{pmatrix}$$

(4.172)

4.3.4 Coefficients for the Rayleigh–Bénard instability

As a first illustration, let us use stress-free boundaries $W = D^2 W = T = 0$ at $z = 0, 1$. The linear stability problem has been treated in §3.3.1. Critical conditions were obtained as $Ra_c = 27\pi^4/4$, $k_c = \pi/\sqrt{2}$.

The neutral stability functions at $k = k_c$ (normalized such that the temperature perturbation is unity at mid-depth) read

$$U_{0k_c} = \sin[\pi z] \begin{pmatrix} \frac{3\pi^2}{2} \\ 1 \end{pmatrix} \tag{4.173}$$

where first and second components correspond to vertical velocity and temperature, respectively.

The adjoint problem may be found as in the Appendices B and D. In fact, the adjoint operators[8] are given by the transposed Eq. (4.172). Note that we will only be interested in the adjoint eigenvectors. These read

$$\tilde{U}_{0k_c} = a \sin[\pi z] \begin{pmatrix} 1 \\ -\frac{9\pi^4}{4} \end{pmatrix} \tag{4.174}$$

where the normalization factor a can be chosen for instance such that the denominator of the expressions (4.104), (4.105) and (4.113) is unity.

Then, the linear coefficient (4.104) is readily obtained as

$$\alpha = \frac{2}{9\pi^2(1 + Pr^{-1})} \tag{4.175}$$

which is indeed equal to $\partial\sigma/\partial Ra|_c$ [see Eq. (4.2)].

Now, we have to solve the second-order problem (4.101). In fact, it can be checked from Eq. (4.171) that the r.h.s. of this problem is proportional to $\sin[2\pi z]$. Thus, the adjoint vector (4.174) is orthogonal to this r.h.s. on the interval $[0, 1]$, and the problem is solvable (no resonance). Denoting by x the cosine of the angle between $\vec{p}$ and $\vec{q}$, we get

$$U_{2\vec{p},\vec{q}} = \sin[2\pi z] \begin{pmatrix} w_{2\vec{p},\vec{q}} \\ \theta_{2\vec{p},\vec{q}} \end{pmatrix} \tag{4.176}$$

with coefficients determined by

$$\begin{pmatrix} \pi^4(5+x)^2 & -\frac{27\pi^6}{4}(1+x) \\ 1 & -\pi^2(5+x) \end{pmatrix} \begin{pmatrix} w_{2\vec{p},\vec{q}} \\ \theta_{2\vec{p},\vec{q}} \end{pmatrix} = \begin{pmatrix} \frac{27\pi^7}{8}Pr^{-1}(x^2-1) \\ \frac{3\pi^3}{4}(1-x) \end{pmatrix} \tag{4.177}$$

Eq. (4.177) is unconditionally solvable, as expected. We may then proceed to the calculation of the cubic coefficients (4.105), though only the coefficients (4.108) and (4.109) are given here for conciseness. We obtain the self-interaction coefficient

$$\beta = \frac{9\pi^4}{8(1 + Pr^{-1})} \tag{4.178}$$

[8]Contrary to what is done in Appendix B, here we do not multiply the energy equation by $-k^2 Ra$, therefore not symmetrizing the linear operators. The adjoint eigenvector thus need not be identical to Eq. (4.173).

which is positive for all Pr, thus indicating that rolls always bifurcate supercritically. The general form of the cross-interaction coefficient $\gamma(x = \cos[\theta])$ is

$$\gamma = \frac{9\pi^4}{1 + Pr^{-1}} \left[\frac{\frac{1}{8}(318329 + 104295x^2 - 8112x^4 + 208x^6)}{(473 - 273x + 60x^2 - 4x^3)(473 + 273x + 60x^2 + 4x^3)} \right. }{}$$

$$\left. \qquad \frac{+(x^2 - 1)\left(3\frac{(44x^4 - 919x^2 - 2365)}{Pr^2} - 9\frac{(4x^4 + 333x^2 + 473)}{Pr}\right)}{(473 - 273x + 60x^2 - 4x^3)(473 + 273x + 60x^2 + 4x^3)} \right] \tag{4.179}$$

which is presented in Fig. 4.19, for various values of Pr. It is checked that γ tends to 2β for $x \to 1$ [verifying Eq. (4.111)], while in the limit $x \to 0$ ($\theta = \pi/2$, corresponding to squares), we obtain

$$\gamma_\perp = \frac{9\pi^4(673 + 72Pr^{-1} + 120Pr^{-2})}{3784(1 + Pr^{-1})} \tag{4.180}$$

which is always higher than β, indicating that squares are unstable to rolls whatever the Prandtl number. More generally, we may study the competition between rhombs and rolls (see §4.1.2 and §4.2.1). Accordingly, as $\gamma > \beta > 0$ for all Pr (see Fig. 4.19), rhombs of any angle are always unstable to rolls. From the results of §4.2.4, it can also be checked that hexagons S_3 and higher-order solutions are unstable.

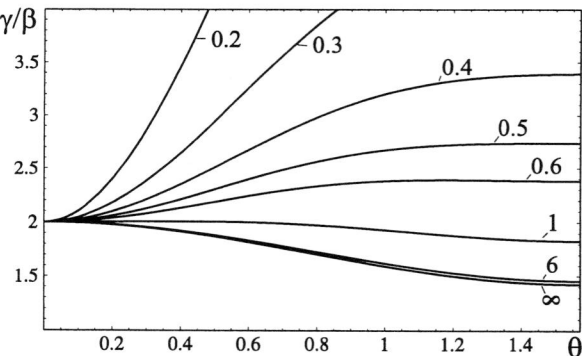

Figure 4.19: Rayleigh–Bénard convection between stress-free heat-conducting boundaries: ratio γ/β of the cross-interaction to the self-interaction coefficient as a function of the angle of interaction θ for various Prandtl numbers (from 0.2 to ∞).

In fact, it is possible to show that among the regular solutions (see §4.2.4), the absolutely stable one is the roll pattern, which corresponds to the highest Nusselt number. Indeed, using Eq. (4.173) and Eq. (4.166) valid for the present case, we get

$$Nu = 1 + \frac{3\pi^2}{4} \sum_{\vec{k} \in P} |a_{\vec{k}}|^2 \tag{4.181}$$

such that the Nusselt number for the pattern S_m is given by

$$Nu_m = 1 + \frac{3\pi^2}{2} m R_m^2 \tag{4.182}$$

with the values of R_m given in §4.2.4. In Fig. 4.20, we represent the quantity

$$\frac{\partial Nu_m}{\partial Ra} = \frac{m}{3\left(1 + Pr^{-1}\right)\left(\beta + \sum_{p=1}^{m-1} \gamma_{p\pi/m}\right)} \tag{4.183}$$

i.e. the initial slope of the heat transfer curve in a (Nu, Ra) diagram. As expected, the rolls are more effective in transporting energy than any other regular pattern.

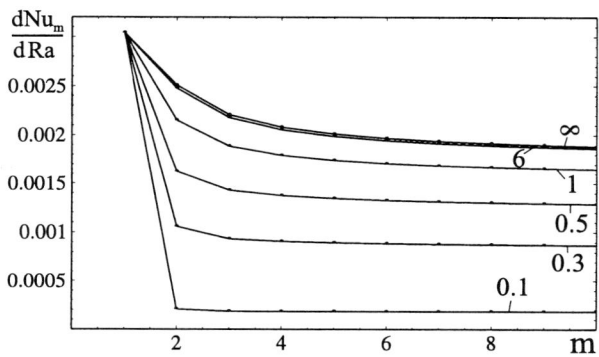

Figure 4.20: Rayleigh–Bénard convection between stress-free heat-conducting boundaries: slope of the Nusselt number versus the Rayleigh number at the bifurcation point, for various patterns and Prandtl numbers. m is the number of pairs of constitutive modes ($m = 1$: rolls, $m = 2$: squares, $m = 3$: hexagons, ...).

4.3.5 Self-adjointness, resonances and hexagonal patterns in Bénard problems

It was seen in §4.2.1 that quadratic terms in amplitude equations (4.120) always favor hexagonal patterns sufficiently near the threshold, whatever the detail of the cubic interaction coefficients[9]. In contrast, the Rayleigh–Bénard instability problem considered in the last section was characterized by a vanishing quadratic interaction coefficient δ. Hence, the amplitude equations were free of quadratic terms, and the pattern selection problem was shown to be governed by a principle of maximum "energy" $\sum_{\vec{k}} |a_{\vec{k}}|^2$ of convective perturbations.

What is the mathematical origin of the condition $\delta = 0$? We shall see that this result still holds for any of the boundary conditions considered in §3.3.1 (stress-free or rigid, heat-conducting or even partly conducting boundaries). Actually, top and bottom boundary conditions need not necessarily be symmetric. In fact, the condition $\delta = 0$ is a consequence of

---

[9]More precisely, these quadratic terms could also force dodecagons or higher-order resonant patterns where the number of constitutive modes is a multiple of 3, but this is not common in Bénard problems.

the self-adjoint nature of the linear problem. Indeed, it is shown in Appendix B that the pure Rayleigh–Bénard problem is self-adjoint, for any choice of boundary conditions considered in §3.3.1.

Self-adjointness of the linear problem implies that the solutions of the adjoint problem are equal to the neutral stability functions. Denoting these by $W_n(z)$ (vertical velocity) and $T_n(z)$ (temperature perturbation), we may directly compute the quadratic coefficient from Eqs (4.119), (4.113) and (4.171). This gives (omitting the denominator)

$$
Z_{-\vec{k}_+,\vec{k}} \sim \frac{Pr^{-1}}{2} \int_0^1 dz\, W_n D \left[W_n D^2 W_n + \tfrac{1}{2} DW_n DW_n \right]
$$
$$
- Pr^{-1} k_c^2 \int_0^1 dz\, W_n \left[W_n DW_n + \tfrac{1}{2} DW_n W_n \right] + \int_0^1 dz\, T_n \left[W_n DT_n + \tfrac{1}{2} DW_n T_n \right]
$$

$$(4.184)$$

and it is readily shown by integration by parts (using the boundary conditions $W_n = 0$ at $z = 0, 1$) that each integral vanishes identically. As $Z_{-\vec{k}_-,\vec{k}}$ yields the same result, it follows that $\delta = 0$ provided the nonlinearities are of advective origin, boundaries are undeformable and impervious, and the linear problem is self-adjoint. In fact, it may also be shown [96] that this result is a consequence of the equalities

$$
\int_V \vec{v}.(\vec{v}.\vec{\nabla})\vec{v}\, dV = \int_V T(\vec{v}.\vec{\nabla})T\, dV = 0
$$

$$(4.185)$$

valid for any solenoidal velocity field $\vec{v}$ and temperature field T provided the boundaries of the domain V are undeformable.

This holds in particular for a layer of fluid in contact with an inert gas at an undeformable interface with negligible Marangoni effect, and where the heat transfer is described by a Biot number. This will be confirmed in §4.4.3. Note that this result would probably not be valid when nonlinearities arising from non-Boussinesq effects or surface deformation are considered.

4.4 Pattern selection in Rayleigh–Marangoni instabilities

In this section, pattern selection is studied for Bénard instabilities driven by both surface tension and buoyancy. We consider a layer of fluid lying on a heated rigid plate and in contact with an inert gas phase at an undeformable interface. Linear stability analyzes and discussions of the validity of approximations used here are presented in Chapter 3.

Contrary to the pure Rayleigh–Bénard instability problems studied in the previous section, an important characteristic of problems involving thermocapillary effects is that they generally lead to resonant quadratic interactions between triads of modes 120° apart on the critical circle. As the Rayleigh–Marangoni problem could not be put under a self-adjoint form (see Appendix D), it is likely in view of the discussion of §4.3.5 that the quadratic interaction coefficient δ will not vanish. As shown in §4.2.1, hexagonal patterns then become the preferential form of convection directly above the threshold [a typical bifurcation diagram for the competition between rolls and hexagons is presented in Fig. 4.15(a)]. This is indeed the experimentally observed behavior [106, 114, 29]. The bifurcation to a resonant pattern such as hexagons is associated with hysteresis, i.e. the convective motions may be sustained even for moderate subcritical driving gradients. As nonlinear energy stability theories rule out the possibility of such behaviors for the pure Rayleigh–Bénard instability (see Appendix C), it is of interest to recall some basic results obtained when applying such energy stability theory to the case of Marangoni–Bénard convection. This is done in §4.4.1 and Appendix F.

As $\delta \neq 0$, the non-resonant pattern selection problem investigated in §4.2.4 will not be valid here. However, for small δ, it can be expected that results obtained for this non-resonant case will only be weakly affected by the resonances, at least for supercriticalities that are not too small. For instance, it is seen in Fig. 4.15(a) that above a certain secondary threshold, a transition from hexagons to rolls is predicted. If we assume that the cubic order amplitude equations may be extrapolated (as models) so far in the nonlinear regime, it is of interest to compute the cross-interaction coefficient $\gamma(\theta)$ to examine which conclusions can be obtained about the preference between non-resonant patterns such as rolls, squares, ... susceptible to appear when hexagons become unstable. In §4.4.2, values of relevant coefficients are given as a function of some dimensionless parameters, and the basic features of the pattern selection problem are investigated.

Then, the particular case of the competition between rolls, hexagons and square patterns is more extensively studied for several systems (including buoyancy effects, as well as a complete two-layer formulation) in §4.4.3. This is motivated by the recent experimental observation of a transition from hexagonal to square patterns (instead of rolls) by Nitschke–Eckert et al. [29, 30] and Schatz at al. [31], and squares very close to threshold in two-layer experiments of Tokaruk et al. [34]. In fact, it will be seen that this preference for squares over rolls appears as the rule rather than the exception when we consider more realistic descriptions than those most often adopted in the literature (namely a single layer with a zero free surface Biot number). These experimental findings have stimulated an active theoretical research [131, 132, 239], now explaining the hexagon-square transition with good qualitative agreement. The approach presented in §4.4.3 completes existing analyzes by the consideration of a more realistic full two-layer problem, including thermo-mechanical effects in the gas phase, and buoyancy effects in both layers.

4.4.1 Nonlinear energy stability theory

Energy stability theory has been applied to the Rayleigh–Marangoni–Bénard problem by Davis [347], Davis and Homsy [348], and Castillo and Velarde [349], also including other effects such as two-component fluids and interfacial deformation. We do not reproduce here all the results obtained so far, but rather focus on the pure Marangoni–Bénard problem in its simplest formulation (see §3.3.2). Further details about the method can be found in Appendix F.

While linear stability theory provides sufficient conditions for instability to infinitesimal perturbations (above the critical value $Ma_c = 79.6$), nonlinear energy stability provides sufficient conditions for stability to perturbations of finite amplitude. It is shown in Appendix F that a certain positive definite energy functional can be constructed, which is shown to decrease monotonically with time provided the Marangoni number is below a certain value $Ma_E < Ma_c$, determined by an extremum principle. It is also shown that the energy limit Ma_E is determined by the corresponding Euler–Lagrange problem. As this problem [Eqs

(F.7–F.11)] is linear and homogeneous, it is possible to solve it via normal modes analysis, just as for the linear problems of Chapter 3. The result is a compatibility condition $Ma_E(k, \mu)$, where k is the wavenumber and $\mu > 0$ is an auxiliary parameter used to make the energy limit Ma_E as close as possible to Ma_c.

The result of the maximization of $Ma_E(k, \mu)$ with respect to μ at given k is represented in Fig. 4.21, together with the neutral stability curve [Eq. (3.118) with $Bi = 0$]. Above the full curve (neutral stability), the motionless state is unstable to infinitesimal perturbations. Below the dashed one (energy limit), it is stable to perturbations of any amplitude. In between both curves, it is possible that convective states could be triggered by perturbations of sufficiently large amplitude.

The absolute minimum of the energy curve is $Ma_E = 56.77$ at $k_E = 2.224$ (corresponding to $\mu_E = 16.78$), as found in [347, 348, 349]. The corresponding z-dependencies of normal modes at the absolute minimum are presented in Fig. 4.22, to be compared with the neutral stability result (Fig. 3.9 of Chapter 3).

Thus, the possibility of subcritical motions exists as soon as surface tension effects determine the onset of convection. Their possible range of existence is rather large ($56.8 < Ma < 79.6$) but in the frame of this energy stability theory, nothing can be said about the actual size of the subcritical domain. While it is guaranteed that for $Ma < Ma_E$, no convective motions are possible, a more precise determination of the subcritical zone may only be achieved via direct resolution, or using amplitude equations.

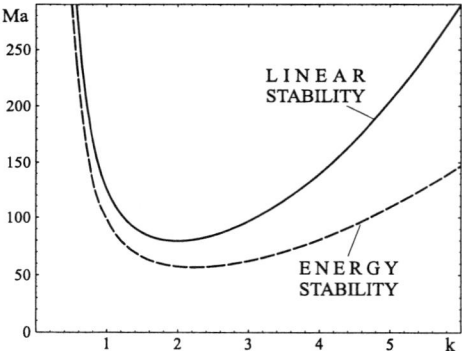

Figure 4.21: Linear stability threshold and energy stability limit for a pure Marangoni–Bénard problem (Pearson's formulation) with $Bi = 0$.

4.4.2 Hexagonal patterns in the Rayleigh–Marangoni instability

The amplitude equations to be considered here are the full cubic equations (4.120), now including quadratic resonances ($\delta \neq 0$). The calculation of the coefficients of these equations as a function of the governing parameters (such as the Prandtl number Pr, the dynamic Bond number $Bo = Ra/Ma$ and the Biot number Bi) will not be detailed here, as it strictly follows the method explained in §4.2.2. In particular, the coefficients are given by Eqs (4.104),

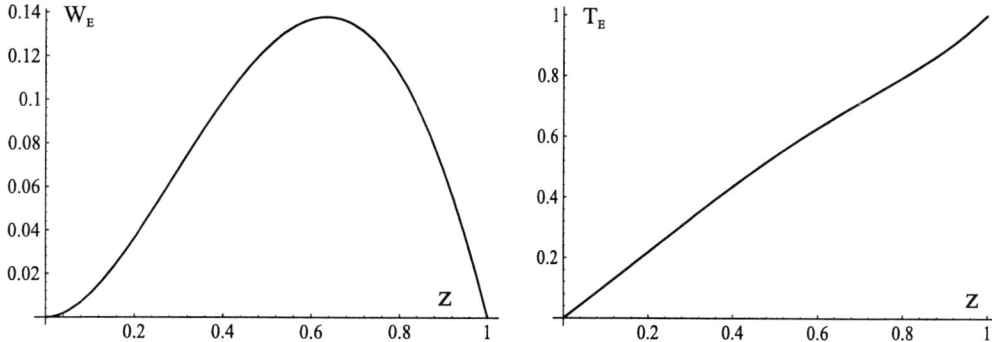

Figure 4.22: The z-dependencies of normal modes for the energy stability problem, at the minimum $Ma_E = 56.77$, $k_E = 2.224$ of the energy stability limit of Fig. 4.21.

(4.105), (4.108), (4.109), (4.113), (4.118) and (4.119). As the quadratic coefficient δ is a finite quantity for Marangoni–Bénard problems, the results should be considered as qualitatively valid (however, it is likely that reasonable quantitative agreement may be expected, because the numerical value of δ always remains small compared with the cubic coefficients β and γ).

The definition of the linear and nonlinear operators are as in §4.3.3. More precisely, we will add a supplementary component corresponding to the tangential stress boundary condition at the undeformable upper surface $z = 1$ (the remaining boundary conditions, i.e. $w = Dw = \theta = 0$ at $z = 0$ and $w = D\theta + Bi\theta = 0$ at $z = 1$, should be included in the definition of the domain E of the operators). Thus, more appropriate forms of the linear operators and of the scalar product are those given in Appendix D [Eqs (D.6) and (D.9)]. The nonlinear operator may still be put under the form (4.171), although with a third component equal to zero (the Marangoni condition is linear). The control parameter μ is now the Marangoni number Ma, and a convenient normalization condition of the neutral stability functions is $\theta(1) = 1$ (free surface temperature equal to unity). Note that the values of all the coefficients presented below are calculated at the wavenumber corresponding to the minimum of the neutral stability curve (critical conditions).

As the analytical form of the coefficients is too cumbersome to be reproduced here (in fact, it is more advantageous to calculate them numerically, e.g. using a shooting method for the resolution of linear inhomogeneous systems of ODEs), it is preferable to present them in the form of graphics. The linear coefficient α can be found from Fig. 4.23 as a function of Bo, for several values of Pr, and for $Bi = 0$ or $Bi = 0.8$. Remember that the linear growth rate is given by $\sigma = \alpha(Ma - Ma_c) = \sigma_0 \epsilon$, i.e. $\alpha = \sigma_0/Ma_c$.

The quadratic coefficient δ is given in Fig. 4.24(a) as a function of the Bond number Bo, for $Bi = 0$, and for several values of Pr. It is seen that its value is generally positive, except for small Prandtl numbers (liquid metals) where it is negative ($\delta \simeq 0$ near $Pr^* = 0.23$ [151, 350]). As expected, the value of δ tends to zero when Bo increases, because in this case buoyancy dominates, and no resonance occurs (the problem is self adjoint in the limit $Bo \to \infty$, i.e. $Ma = 0$ and $Ra > 0$, as seen in §4.3.5). Increasing Bi (increasing heat transfer coefficient at the upper free surface), the behavior of δ is not qualitatively modified,

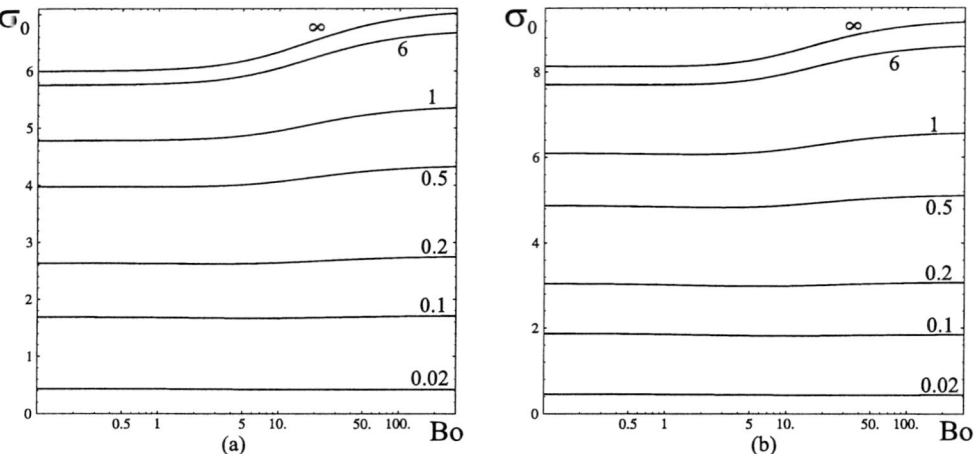

Figure 4.23: The linear coefficient σ_0 in the expression of the growth rate $\sigma = \sigma_0\epsilon$ [where $\epsilon = (Ma - Ma_c)/Ma_c = (Ra - Ra_c)/Ra_c$] as a function of the ratio of Rayleigh and Marangoni numbers (i.e. the dynamic Bond number $Bo = Ra/Ma$), for several values of Pr indicated on the figure. (a) corresponds to $Bi = 0$, (b) corresponds to $Bi = 0.8$.

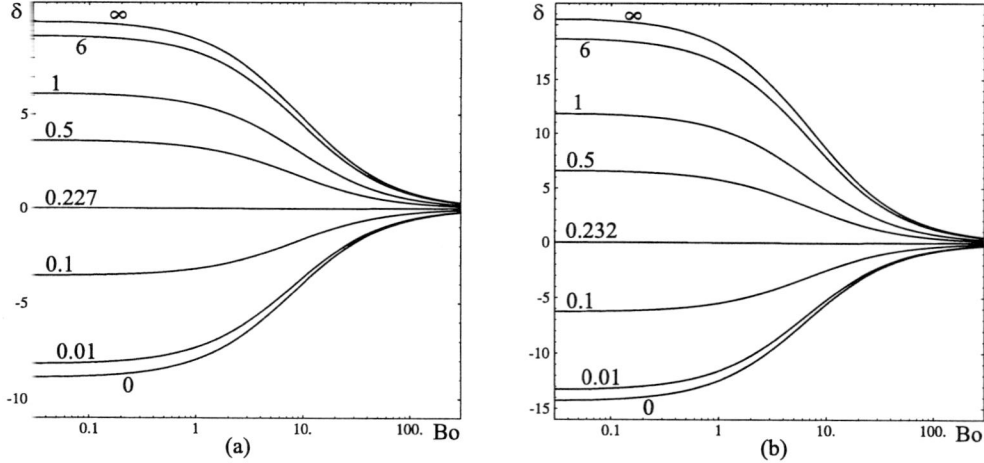

Figure 4.24: The quadratic interaction coefficient δ as a function of the ratio of Rayleigh and Marangoni numbers ($Bo = Ra/Ma$), for several values of Pr indicated on each curve. (a) corresponds to $Bi = 0$, (b) corresponds to $Bi = 0.8$. In both cases, δ changes sign for $Pr \simeq 0.23$.

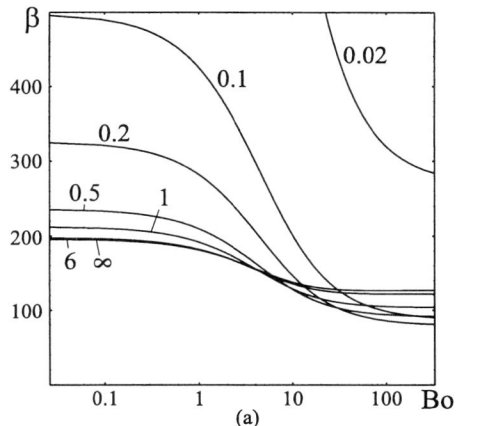
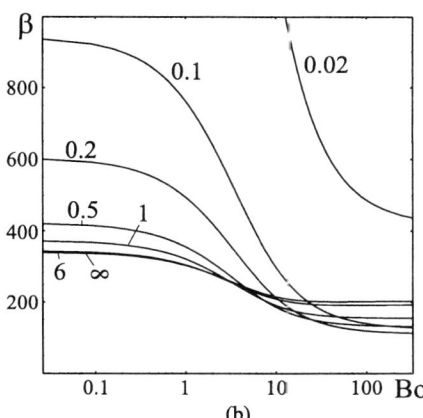

Figure 4.25: The self-interaction coefficient β as a function of the Bond number Bo, for several values of Pr indicated on the figure. (a) corresponds to $Bi = 0$, (b) corresponds to $Bi = 0.8$.

as seen in Fig. 4.24(b).

The cubic self-interaction coefficient β is depicted in Fig. 4.25, using the same representation as in Fig. 4.24. It is seen that β is always positive, with the consequence that the bifurcation to roll patterns is always supercritical.

The cross-interaction curve $\gamma(\theta)$ [normalized by β] is given in Fig. 4.26, for Marangoni–Bénard convection ($Bo = 0$), $Bi = 0$ and variable Prandtl number. An important difference with pure Rayleigh–Bénard convection (see Fig. 4.19) is that the coupling coefficient $\gamma(\theta)$ diverges at $\theta = \pi/3$, i.e. at the value for which nonlinear interactions generate modes on the critical circle (resonance). It is noteworthy that this divergence disappears at $Pr^* \simeq 0.23$, for which we also have $\delta = 0$. Thus, strictly speaking, the analysis of Marangoni–Bénard convection in terms of amplitude equations is only exact near this point (a fact which will be exploited later on in the study of modulated hexagonal patterns). Nevertheless, when $\delta = O(1)$, amplitude equations can still be expected to provide a reasonable approximation provided δ remains sufficiently small. It also appears that the validity of such models should be restricted to patterns with a small number of modes on the critical circle (due to the requirement that any interaction with angle close to $\pi/3$ should be avoided). In the following, we will thus limit our analysis to the bifurcation of low-order patterns such as rolls ($m = 1$), squares ($m = 2$) and hexagons ($m = 3$). Remember that for hexagons, care should be taken when calculating $\gamma' = \gamma(\pi/3)$ [see Eqs (4.117) and (4.118)]. The result is given in Fig. 4.27.

We may now evaluate the heat transfer characteristics. The expression (4.164) of the mean temperature field is still valid here, but constants c_1 and c_2 must be found from the boundary

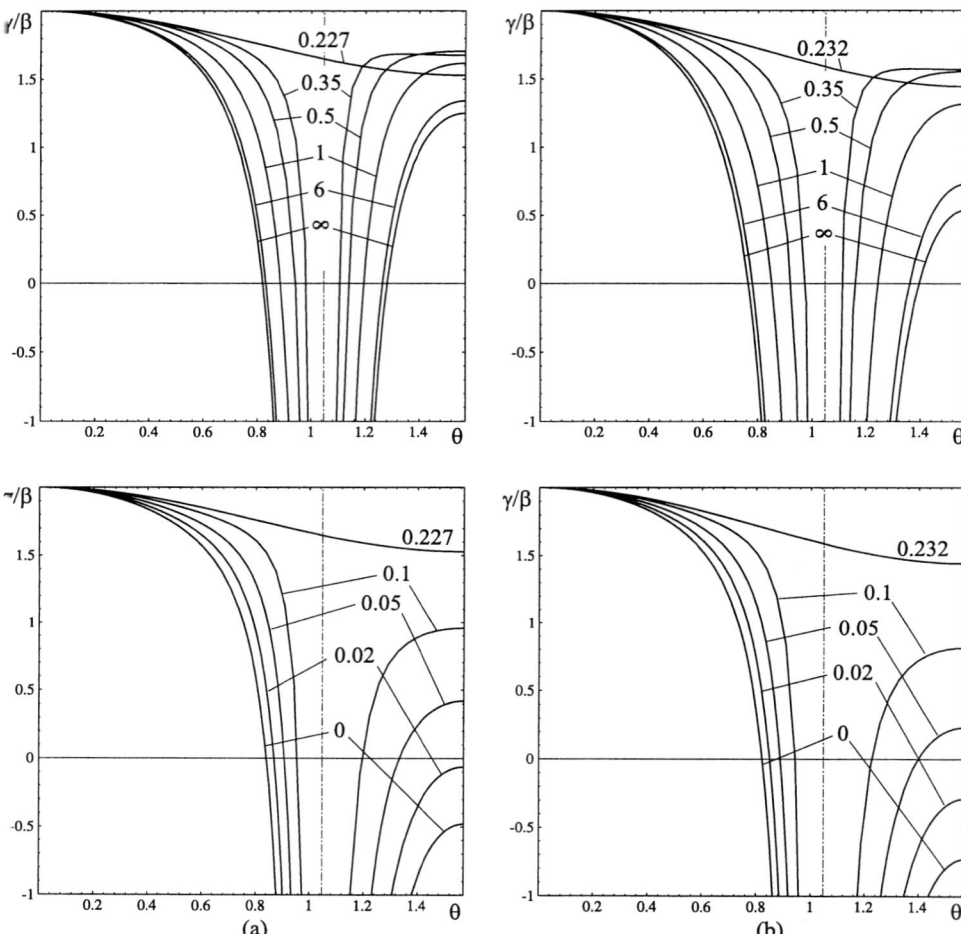

Figure 4.26: The coupling coefficient $\gamma(\theta)/\beta$ as a function of the angle of interaction θ, for several values of Pr indicated on each curve. (a) $Bi = Bo = 0$, (b) $Bi = 0.8$, $Bo = 0$. Resonance occurs at $\theta = \pi/3$, except for the particular value $Pr^* = 0.23$ [a difference exists between values of Pr^* corresponding to cases (a) and (b), but is practically negligible].

conditions $T_0 \mid_{z=0} = 0$ and $DT_0 + BiT_0 \mid_{z=1} = 0$. The Nusselt number then reads

$$Nu = 1 - DT_0(z = 1) = 1 + \frac{Bi}{1 + Bi} \int_0^1 dz \overline{WT} \tag{4.186}$$

Consequently, the Nusselt number remains equal to unity for $Bi = 0$, as indeed the heat flux through the free surface is held fixed in this limit ($DT = 0$ at $z = 1$). Accordingly, the heat

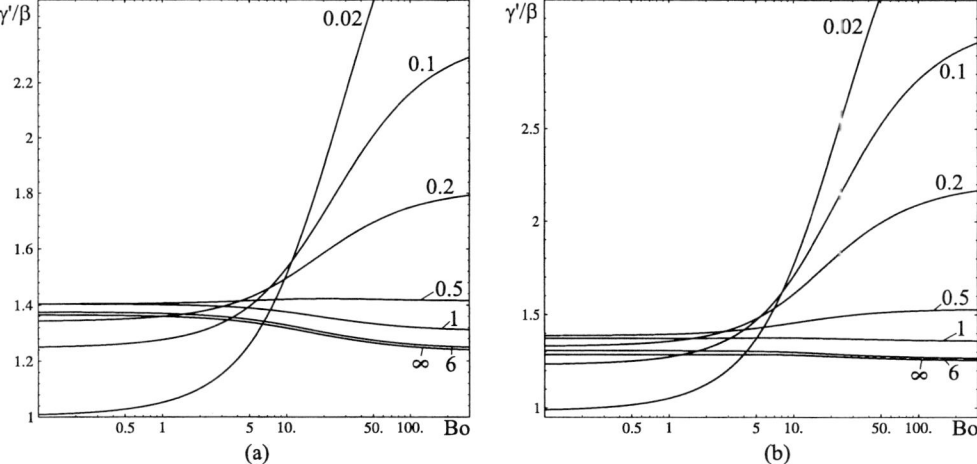

Figure 4.27: The modified coupling coefficient γ'/β for the angle $\pi/3$ as a function of the ratio of Rayleigh and Marangoni numbers (i.e. the Bond number Bo), for several values of Pr indicated on the figure. (a) corresponds to $Bi = 0$, (b) corresponds to $Bi = 0.8$.

transfer through the layer in a steady convective state is not affected (and remains equal to the reference value). For these situations, a better measure of the efficiency of heat transfer by convection is provided by the increase of the mean free surface temperature due to upward convective heat transport, i.e.

$$T_0(z = 1) = \frac{1}{1 + Bi} \int_0^1 dz \overline{WT} \tag{4.187}$$

This quantity is equal to the decrease Δ of the mean temperature difference between the hot bottom plate and the cold free surface, caused by the convective motions, and is thus representative of the fact that convection allows the transport of an imposed heat flux under a smaller temperature difference (increase of the apparent thermal conductivity by convection). The quantity (4.187) is also proportional to the "energy" of convective motions $E = \sum_{\vec{k}} |a_{\vec{k}}|^2$. Indeed, for patterns given at the first order by Eq. (4.86), Δ reads

$$\Delta = \frac{\int_0^1 dz \, w_{\mathrm{n}}(z)\theta_{\mathrm{n}}(z)}{1 + Bi} \sum_{\vec{k} \in P} |a_{\vec{k}}|^2 \tag{4.188}$$

In view of the discussion presented in §4.2.4, it can be expected that for non-resonant situations, the absolutely stable pattern will be that maximizing Δ. Actually, $\delta = 0$ is only satisfied at the particular value $Pr^* \simeq 0.23$ of the Prandtl number. The corresponding surface temperature increase Δ is represented in Fig. 4.28 for the lowest-order patterns.

It is seen in Fig. 4.28 that rolls are more efficient in transporting energy by convection, and thus correspond to the absolutely stable pattern (for $Pr = 0.23$). It is also seen that

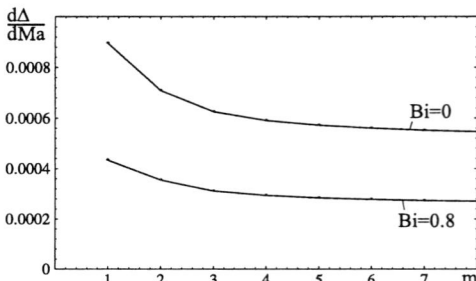

Figure 4.28: Variation of the free surface temperature increase Δ with Ma at the first bifurcation point in the non-resonant case $Pr = 0.23$ for various patterns : rolls ($m = 1$), squares ($m = 2$), hexagons ($m = 3$), ... Results are given for $Bi = 0$ and $Bi = 0.8$, in both cases for $Bo = 0$.

non-resonant hexagons generally lead to lower Δ than rolls and squares. It must be stressed here that this situation is specific to the case $Pr = 0.23$, and does not persist in a large range of Prandtl numbers, where squares can become preferable (especially at increasing Biot number, as seen later). Moreover, for other Prandtl numbers ($\delta \neq 0$), this maximum Δ principle fails for the reasons explained in §4.2.4, and *resonant* patterns such as hexagons will always be the stable pattern in some range near the threshold. This is visible in the bifurcation diagrams presented in Figs 4.29 and 4.30 for the case $Bi = 0$, $Bo = 0$ and $Pr \rightarrow \infty$ (corresponding to viscous liquids such as silicone oils, and also representative of moderate Prandtl number fluids). This diagram is obtained from the results of §4.2.1 for the competition between hexagons and rolls (thus excluding the possibility of square patterns for the moment, as for these values of Bi, Bo and Pr, they turn out to be unstable to rolls).

From Fig. 4.29, it can be seen that the stable pattern is not that maximizing Δ, due to quadratic resonances. In particular, there exists a bistability zone in a certain range of Marangoni numbers. Actually, the behavior near the first bifurcation point $\epsilon = 0$ cannot be seen in this figure. Figure 4.30 represents an enlarged view, making visible the (small) hysteresis effect typical of the bifurcation to resonant hexagons.

Returning to Fig. 4.29, we observe a rather large hysteresis loop associated with the transition from hexagons to rolls. Even though such behavior has never been experimentally observed in purely surface-tension-driven convection (probably because the effective Biot number is never zero, as discussed in the next section), we note that a competition between hexagons and rolls has only been reported in situations where the buoyancy effect cannot be neglected [115], and a fortiori in the absence of surface tension effects, when non-Boussinesq effects are important (as in the experiments of Bodenschatz et al. in a layer of gaseous CO_2 [134]). Such tendency to roll patterns at increasing buoyancy effect is apparent in Fig. 4.31, in a more convenient representation of the combined effect of surface tension and buoyancy (as in the Nield's diagram introduced in §3.3.3), again for $Bi = 0$ and $Pr \rightarrow \infty$. In this diagram, the trajectory described by the representative point of the system when increasing the injected heat flux (sufficiently slowly) is a line of slope $Bo = Ra/Ma$. Note that only stable patterns

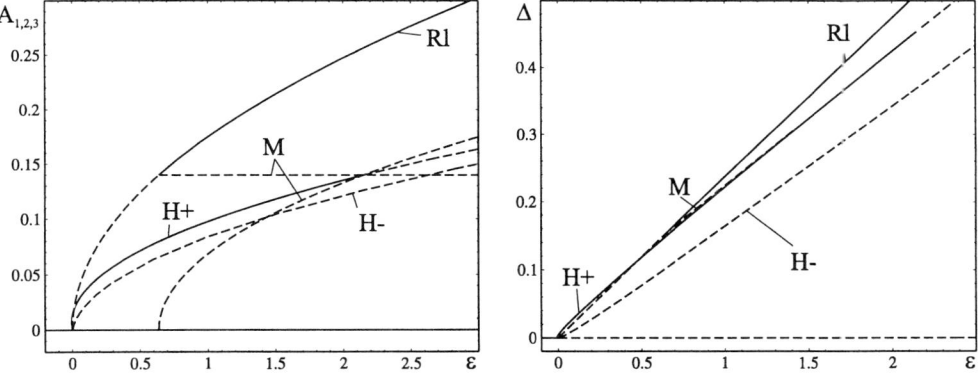

Figure 4.29: Bifurcation diagram for the pure Marangoni–Bénard problem ($Bo = 0$) with $Bi = 0$ and $Pr \to \infty$. Left: amplitudes as a function of the supercriticality parameter $\epsilon = (Ma - Ma_c)/Ma_c$. Right: increase of the mean surface temperature Δ versus ϵ. Full lines represent stable states, while dashed lines correspond to unstable states.

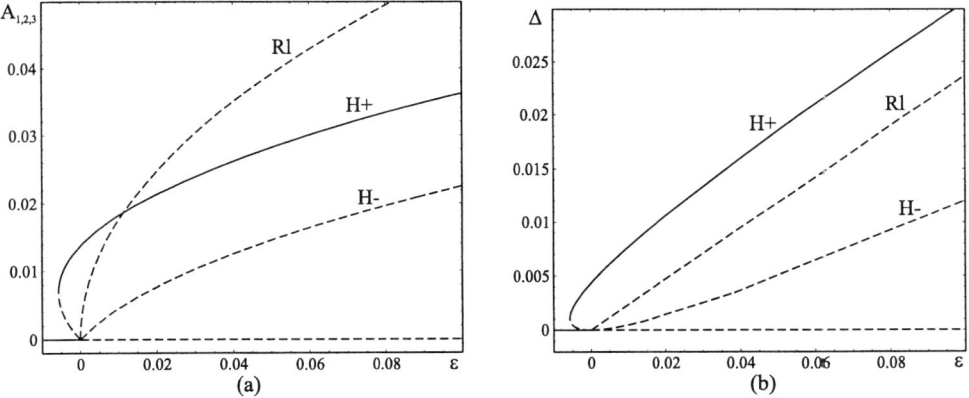

Figure 4.30: Magnification in the vicinity of the primary bifurcation point for the situation described in Fig. 4.29.

are indicated in this figure, as well as in the figures of the same type given in the next section. The symbols between parentheses denote patterns which are linearly stable, but unstable to perturbations of finite amplitude (metastable).

Note that until now, we have not distinguished between up- and down- hexagonal patterns. In fact, for $\delta > 0$ (i.e. $Pr > 0.23$), the preferred pattern near the threshold corresponds to up-hexagons (H+ in Fig. 4.31). For this pattern, the fluid motion is upwards at the center of hexagonal cells, corresponding to positive temperature perturbation at the free surface (free

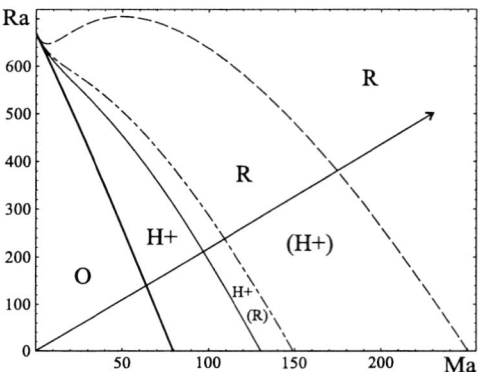

Figure 4.31: Competition between up-hexagons (H+) and rolls (R) for the combined Rayleigh–Marangoni–Bénard instability with $Bi = 0$ and $Pr \to \infty$. The symbol O denotes the motionless state. The arrow represents the path followed by the system when increasing the heat flux (slope $Bo = Ra/Ma$). Only stable patterns are indicated in each region. Symbols between parentheses denote metastable states (the dot-dashed line represents the locus of points where the Lyapunov functions of rolls and of hexagons are equal).

surface motions are always from hot to cold points for normal liquids).

In contrast, for $\delta < 0$ ($Pr < 0.23$), the free surface temperature is minimal at the center of hexagons, where the liquid flows downwards. Some examples of bifurcation diagrams involving stable down-hexagons (H–) will be encountered in the next section.

In Fig. 4.31, we have also represented the line (dot-dashed) where the potential function (4.75) is equal for rolls and hexagons. This line (lying in the bistability region) plays an important role in pattern selection in the presence of modulations of the amplitudes of the patterns (slow spatial variations of $a_{\vec{k}}$, as described in §4.1.3). Below this line, hexagons are energetically favorable [in the sense of the potential (4.75)] over metastable rolls, while the opposite holds above the line. In the presence of a front between a roll pattern and hexagons, it is likely that the front will propagate in a direction allowing the potential V to decrease (similarly to the motion of point defects in roll patterns [313, 314]). Eventually, the transition between patterns may be nucleated from other defects in the structure, such as those induced by lateral walls [134]. Thus, in the region between the thin full line (where rolls become stable) and the dot-dashed line, fronts will propagate towards the metastable roll pattern, allowing the hexagons to invade a larger portion of the set-up. When the driving flux is increased past the dot-dashed line, the fronts will move in order to allow rolls to occupy an increasing portion of the layer, at the expense of the metastable hexagonal pattern. Note that squares are always unstable to rolls in the situation considered here ($Bi = 0$, $Pr \to \infty$).

Let us conclude this section by giving [in Fig. 4.32(a)] the amount of hysteresis calculated for the first bifurcation (not visible in Fig. 4.31) to up-hexagons, as a function of the ratio between buoyancy and surface tension effects, for large Prandtl number fluids and for two representative Biot numbers. It is seen that although small in general, the hysteresis increases with the Biot number. For increasing buoyancy (i.e. increasing fluid depth), the hysteresis decreases and reaches zero in the limit of purely buoyancy-driven instability. There, the first bifurcation is to a roll pattern.

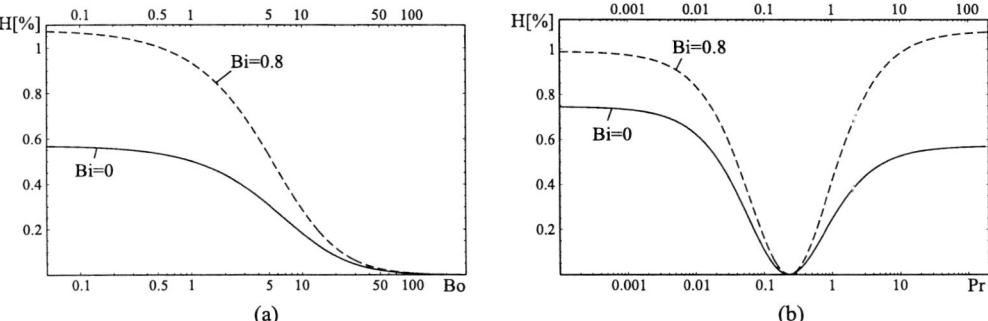

Figure 4.32: Hysteresis (given as a percentage of the critical heat flux) in the transition between the motionless state and the hexagonal pattern at the first bifurcation point. (a) Hysteresis as a function of the dynamic Bond number $Bo = Ra/Ma$, for $Pr \to \infty$ and different values of the Biot number Bi. (b) Hysteresis in the purely surface-tension-driven case ($Bo = 0$), as a function of Pr. The hysteresis vanishes at $Pr = Pr^* \simeq 0.23$, which depends only weakly on Bi and Bo.

In Fig. 4.32(b), we represent the variation of the hysteresis with the Prandtl number in the purely surface-tension-driven case ($Bo = 0$), again for $Bi = 0$ and $Bi = 0.8$. It is seen that the hysteresis is highest for high Prandtl number fluids. Owing to the vanishing of the resonant interaction at $Pr = Pr^* \simeq 0.23$, the hysteresis vanishes at this point. For values of Pr lower than Pr^* (liquid metals), the hysteresis increases again, but is associated with a down-hexagonal pattern. In fact, although overlooked in the above discussion, there is a slight variation of Pr^* with Bi and Bo. For example, at $Bi = 0$, Pr^* varies from 0.227 to 0.235 when Bo varies from 0 to ∞. At $Bo = 0$, $Pr^* = 0.227$ at $Bi = 0$ and $Pr^* = 0.232$ for $Bi = 0.8$.

4.4.3 Rolls, squares and hexagons

In the previous section, the interaction between rolls and hexagonal patterns was considered, using a system of three resonantly coupled amplitude equations [see Eq. (4.74)]. Although it will be seen that this is unimportant for the particular conditions ($Bi = 0$, $Pr \to \infty$) corresponding to Figs 4.29–4.31, by doing so we have explicitly excluded the possibility of square patterns. However, in view of the recent experimental results of Eckert (née Nitschke) et al. [29, 30] (see also Schatz et al. [31] and Tokaruk et al. [34]), it appears that conditions should exists for which square patterns become the preferred stable pattern, rather than rolls, above a certain value of the driving flux. As seen in Fig. 4.33, Eckert indeed carefully observed a distinct transition between a regular hexagonal pattern at small values of the constraint $\epsilon = (Ma - Ma_c)/Ma_c$, to a pattern strongly dominated by square convection cells at higher values of ϵ. In the intermediate stage of the transition from the hexagonal to the square planform, islands of square patterns coexist with hexagonal domains, the transition regions between them appearing as fronts of pentagons. In Fig. 4.33, the drastic change in the structure of the pattern when ϵ is increased is further confirmed by the azimuthal distribution function $Q(\varphi)$ obtained by integrating the power spectral density $P(k_x, k_y) = P(k, \varphi)$ over a range of wavenumbers k centered around the mean wavenumber of the pattern.

Actually, looking back to Fig. 4.26 representing the angular dependency of the coupling coefficient $\gamma(\theta)/\beta$, it is seen that the value for $\theta = \pi/2$ at $Pr \to \infty$ is 1.25 for $Bi = 0$, and 0.54 for $Bi = 0.8$. On the other hand, the discussion in §4.1.2 (summarized in Fig. 4.8) has shown that the condition for the stability of mixed (square) solutions was precisely $\gamma(\pi/2)/\beta < 1$ (we have $\beta > 0$ here). Thus, squares are predicted to be stable with respect to rolls in the second case $Bi = 0.8$, while rolls are stable if $Bi = 0$. Although the competition with hexagons remains to be investigated, this seems to indicate that the thermal conditions at the upper free surface of the liquid play a decisive role in the transition observed by Eckert et al. This is further reinforced by the fact that the air gap between the oil layer and the top (sapphire) cooling plate is only 0.4 mm, thereby strongly reducing the resistance to heat transfer. In fact, Eckert et al. [30] also predicted the transition experimentally observed from a full three-dimensional simulation of the governing equations in a single layer of fluid with a certain effective (and non-zero) Biot number.

We will now show that the transition is indeed predicted by an amplitude equation model using the coefficients calculated in the previous section for $Bi = 0.8$. This has also been recently predicted by Regnier et al. [132]. Still, in view of the discussions of §3.2.3 about the Biot number and the validity of the thermal boundary condition $DT + BiT = 0$ at a free surface, it appears more realistic to consider a two-layer formulation of the problem, taking into account equations governing the dynamics of both the thermal and the velocity field in the gas phase (although the main effect is thermal in general). This will be done later on. Golovin et al. [239] also recently considered a two-layer system without buoyancy effect, and obtained predictions in agreement with the experiment of Eckert et al. Motivated by the good agreement between amplitude equations and experiment (even in the moderate nonlinear regime), the analysis is here extended to more general conditions than those pertaining to the experiments of Eckert et al. Namely, we consider the effect of the Prandtl number and of the Bond number, measuring the relative importance of buoyancy and surface tension effects.

One-layer Rayleigh–Marangoni–Bénard instability

In this section, we begin by investigating the effect of Prandtl and Bond numbers on the selection between rolls, hexagons and squares. Note that instead of $Bo = Ra/Ma$, we will generally use the slope $\varphi = \arctan[Ra\,Ma_c/Ra_c\,Ma] = \arctan[0.12Bo]$ in a Nield's diagram.

We will first discuss the main features of the competition between rolls, squares and hexagons, and also provide analytical expressions for the various transition points. We refer to [53, 132, 239, 341] for a complete discussion of the relevant system of six coupled amplitude equations.

The selection between rolls and squares is determined by two cubic coefficients, namely β and $\chi = \gamma_{\pi/2}$. If $\chi > \beta$, rolls are stable and squares are unstable, while the opposite holds if

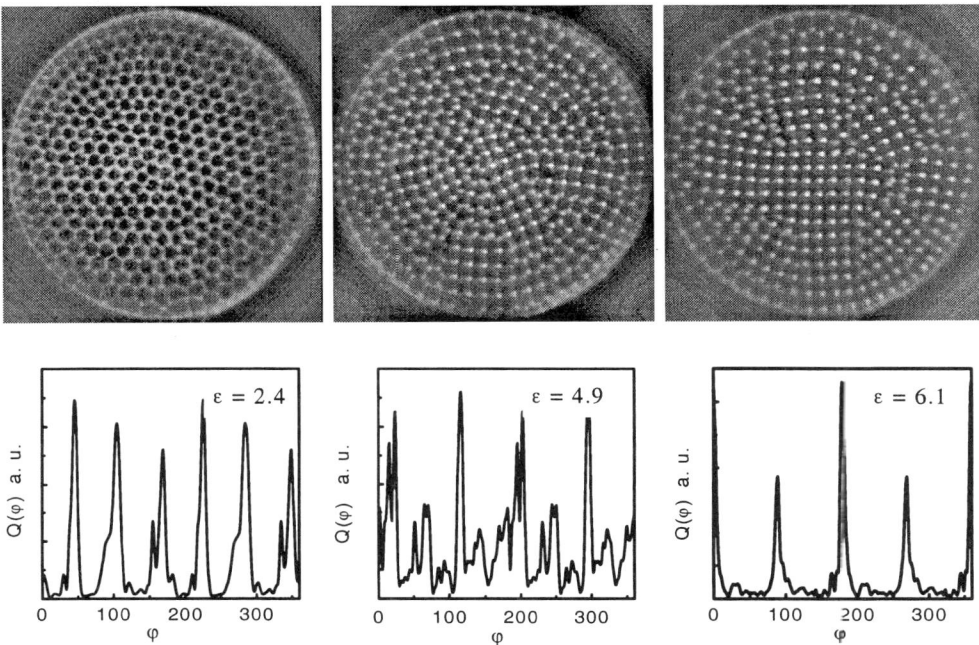

Figure 4.33: Hexagon-square transition in an experiment of Marangoni–Bénard convection by Eckert et al. [30]. Shadowgraph pictures at $\epsilon = 2.4$, $\epsilon = 4.9$ and $\epsilon = 6.1$, together with corresponding azimuthal distribution functions. The circular cell has a diameter 64 times larger than the 10 cSt silicone oil depth 1.41 mm. The air gap is 0.4 mm. Courtesy of K. Eckert and A. Thess.

$\chi < \beta$. It can also be shown that the Lyapunov functions of the rolls and of the hexagons are equal at $\chi = \beta$ ("second-order transition"). Thus, it appears that the selection between rolls and squares will be independent of the distance to the threshold ϵ (just as was the case for the general non-resonant pattern selection problem). In Fig. 4.34, we show the zones in the plane $(\log_{10} Pr, \varphi)$ where $\chi > \beta$ (stable rolls) and $\chi < \beta$ (stable squares) for a zero Biot number Bi. It is seen that apart for a region at small Prandtl number (liquid metals) and dominating surface tension effect, squares should in general be unstable to rolls for $Bi = 0$.

The competition of these two types of patterns with hexagons is more complicated due to quadratic resonances, quantified by δ. In particular, the selection here depends on the distance to the threshold, or equivalently on the growth rate σ. For small σ, transcritical hexagons are expected to be stable, while rolls and squares are unstable. For $\delta > 0$, the first bifurcation is to up-hexagons (for $\delta < 0$, to down-hexagons).

Increasing σ (or decreasing δ), the effects of resonance become less important, and results obtained for the non-resonant patterns should become increasingly valid. Keeping the notation $\gamma' = \gamma_{\pi/3}$, the discussion of §4.2.4 [valid for $\sigma \gg \delta^2/\beta$ if $\gamma' = O(\beta)$] indicates that hexagons should remain stable with respect to rolls for all σ when $\beta > \gamma'$. If $\beta < \gamma'$, it can be shown

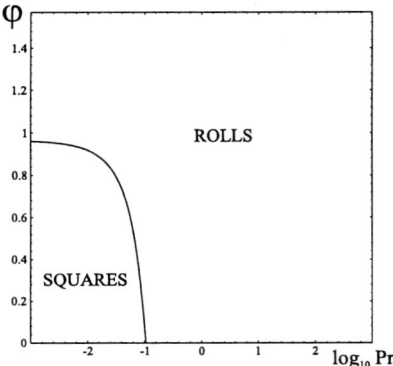

Figure 4.34: Competition between rolls and squares for $Bi = 0$, as a function of the Prandtl number and of the slope $\varphi = \arctan[0.12Ra/Ma]$ in a Nield's diagram.

that hexagons lose stability with respect to rolls for

$$\sigma > \sigma_{\mathrm{hr}} = \frac{\delta^2(\gamma' + 2\beta)}{(\gamma' - \beta)^2} \tag{4.189}$$

whi e rolls become stable with respect to hexagons for

$$\sigma > \sigma_{\mathrm{rh}} = \frac{\delta^2\beta}{(\gamma' - \beta)^2} \tag{4.190}$$

Note that even if Eq. (4.190) is satisfied, rolls may still be unstable with respect to squares (if $\chi < \beta$). Using the notation $\xi = \gamma_{\pi/6}$, it is seen from the results of §4.2.4 that hexagons should remain stable with respect to squares provided $\beta + 2\gamma' - \chi - 2\xi < 0$. If $\beta + 2\gamma' - \chi - 2\xi > 0$, hexagons lose stability with respect to squares for

$$\sigma > \sigma_{\mathrm{hs}} = \frac{\delta^2(\chi + 2\xi)}{(\beta + 2\gamma' - \chi - 2\xi)^2} \tag{4.191}$$

Now, for large σ, squares remain unstable with respect to hexagons when $\beta + \chi - \xi - \gamma' > 0$. If $\beta + \chi - \xi - \gamma' < 0$, squares gain stability with respect to hexagons at

$$\sigma > \sigma_{\mathrm{sh}} = \frac{\delta^2(\beta + \chi)}{(\xi + \gamma' - \chi - \beta)^2} \tag{4.192}$$

Thus, an important difference between the rolls-squares and both the hexagons-rolls and hexagons-squares transitions is that the last two are "first-order transitions", associated with bistability and hysteresis. Indeed, for the hexagons-rolls case for example, the point σ_{rh} at which rolls gain stability with respect to hexagons is different from the point $\sigma_{\mathrm{hr}} > \sigma_{\mathrm{rh}}$ where hexagons lose stability with respect to rolls. In between these two values of σ, both patterns are stable, although one of them is energetically favorable over the other (which is thus metastable). There also exists a point σ_{hr}^{M} (which will be called the Maxwell point by analogy with equilibrium phase transitions) where the Lyapunov function (4.151) is equal for both patterns. It can be calculated that

$$\sigma_{hr}^{M} = \frac{\delta^2 \left[\beta^2 + 3\beta\gamma' + \sqrt{2\beta(\beta + \gamma')^3} \right]}{2(\beta - \gamma')^2 (\beta + 2\gamma')} \tag{4.193}$$

while the Maxwell point of the hexagons-squares transition is given by

$$\sigma_{hs}^{M} = \frac{\delta^2 \left[(\beta + \chi)(5\beta + 12\gamma' - \chi) + \sqrt{(\beta + \chi)(3\beta + 4\gamma' + \chi)^3} \right]}{(\beta + 2\gamma')(\beta - 4\gamma' + 3\chi)^2} \tag{4.194}$$

Although only non-modulated patterns are considered here, the selection between patterns in the bistability region in general depends on modulation effects, e.g. fronts between the two patterns, defects, ... In the simplest descriptions, a more general Lyapunov functional still exists [extending Eq. (4.40) to several amplitudes], which includes supplementary energies (similar to a surface tension) from the gradients in the region of the front. The behavior of the system remains constrained by the requirement that this "free energy" should decrease with time. This leads to motion of the fronts in a direction tending to eliminate the energetically unfavorable pattern. Thus, it is likely that for $\sigma_{rh} < \sigma < \sigma_{hr}^{M}$, the hexagonal pattern will be preferred over the metastable rolls, while the opposite holds for $\sigma_{hr}^{M} < \sigma < \sigma_{hr}$. The first-order character of the transitions involving hexagons is further confirmed by the discontinuous change of the steady amplitudes (i.e. the order parameter) during the transition, in contrast with the second-order rolls-squares transition for which the order parameter varies continuously when passing the line $\chi = \beta$. Note that the above arguments should merely be taken in a *qualitative* sense, because in general, situations involving hexagonal patterns should actually be described by non-variational amplitude equations, as will be seen in §4.5.

Some of the bifurcation points (4.189–4.192) have already been represented in Fig. 4.31. In this figure and in the following ones, a thin full line represents a locus of points where rolls (or squares) gain stability to hexagons (i.e. σ_{rh} or σ_{sh}). The thin dashed line is the locus of points where hexagons lose stability either to rolls (σ_{hr}) or to squares (σ_{hs}). Finally, the dot-dashed line is the Maxwell line σ_{hr}^{M} for the hexagons/rolls transition (or σ_{hs}^{M} for the hexagons-squares transition). The threshold of instability of the motionless state O (first bifurcation) is depicted as a thicker line.

In Fig. 4.31, $Bi = 0$ and $Pr \to \infty$, such that squares are unstable everywhere (see Fig. 4.34), and the competition takes place between hexagons and rolls only. In particular, resonances vanish for $Bo \to \infty$ ($\varphi \to \pi/2$), such that all lines collapse on the threshold value $Ra_c = 669$, $Ma_c = 0$. Note that the small hysteresis loop associated with the first bifurcation is not visible in this graph.

Let us now consider progressively decreasing values of the Prandtl number Pr. The behavior is not qualitatively modified before reaching relatively low values of Pr (see Fig. 4.35 for $Pr = 1$). From Fig. 4.24, it is seen that δ decreases with Pr until $Pr^* \simeq 0.23$, where it changes sign. As this value of Pr^* is slightly dependent on φ, a resolution is necessary in the vicinity of this point. This is represented in Fig. 4.36 for $Pr = 0.229$, where it is seen that δ vanishes at $\varphi \simeq 0.52$. At the left of this point (i.e. for dominating surface tension effect), we are still in the situation of Figs 4.31 and 4.35, although the sequence of transitions occurs in a much more narrow interval of $\epsilon = (Ma - Ma_c)/Ma_c$. For $\varphi > 0.52$ (dominating buoyancy effect), the sequence of transitions is the same apart from the fact that up-hexagons H+ are replaced by down-hexagons H– (stable for $\delta < 0$). The inset of Fig. 4.36 represents a magnification of the behavior at negative ϵ (subcritical domain), i.e. a region where hexagons may coexist with the rest state. Note that we also represented the Maxwell line $\sigma_{oh}^M = -2\delta^2/9(\beta + 2\gamma')$ for this transition.

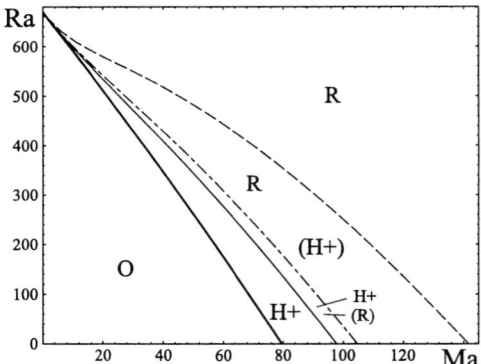

Figure 4.35: Competition between up-hexagons (H+) and rolls (R) for $Pr = 1$ and $Bi = 0$. The threshold is indicated as a thicker line. Only stable states are indicated. Symbols between parentheses are metastable states. O = motionless state. Analytical expressions for transition lines are given in the text (full = σ_{rh}, dashed = σ_{hr}, dot-dashed = σ_{hr}^M).

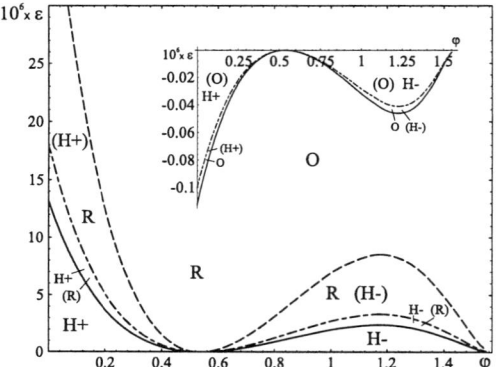

Figure 4.36: Resolution in the vicinity of Pr^* ($Pr = 0.229$), using the same notations as in Fig. 4.35. H– stands for the down-hexagonal pattern. The inset gives the region of bistability of the motionless state and of the hexagonal pattern.

For still smaller Pr, $\delta < 0$ for all φ and the stable hexagonal pattern at threshold is H–. However, looking back to Fig. 4.34, we see that for $Pr \leq 0.1$, there exists a small interval of φ where squares are stable with respect to rolls (and rolls unstable to squares). Thus, for dominating Marangoni effect, the picture changes [see Fig. 4.37(a)] near the axis $Ra = 0$ corresponding to the purely surface-tension-driven case. There, a transition from down-hexagons to squares occurs when increasing the Marangoni number. In fact, the sequence almost follows the usual scheme: first, squares turn from unstable into metastable at σ_{sh} while down-hexagons remain the preferred pattern. Then, at σ_{hs}^{M}, squares become energetically favorable, and down-hexagons turn into a metastable state. Finally, down-hexagons lose stability and the only remaining pattern is S. Note however that hexagons in fact lose stability first with respect to roll disturbances (the dashed line is σ_{hr}). As rolls are unstable themselves, the disturbances then evolve to a square pattern (the only stable solution). When $Pr \rightarrow 0$, the behavior is not qualitatively modified [Fig. 4.37(b)]. Only a distortion of the various transition lines takes place, indicating that down-hexagons should remain metastable over a much wider interval.

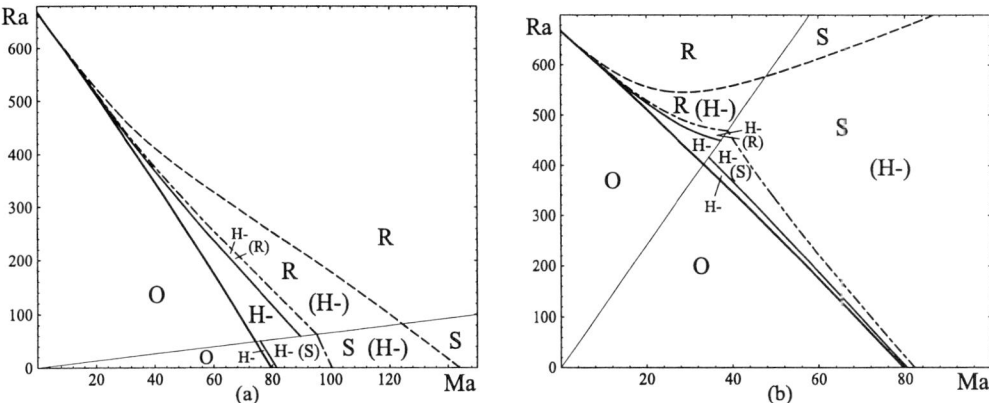

Figure 4.37: Behavior at low-Prandtl number for $Bi = 0$. (a) $Pr = 0.1$, (b) $Pr \rightarrow 0$. The line $\chi = \beta$ of slope φ^* passing through the origin separates a region where rolls are stable to squares (dominating buoyancy effect) from a region where squares are stable to rolls (dominating surface tension effect). Down-hexagons bifurcate first, and subsequent first-order transitions lead to rolls for ($\varphi > \varphi^*$) or to squares ($\varphi < \varphi^*$).

To conclude with the influence of the Prandtl number, it should be stressed that the results obtained for very low Prandtl number should be considered with some precautions. An important effect overlooked by amplitude equations is the influence of the mean flow (proportional to Pr^{-1}, see §4.3.1 and 4.3.2), whose importance at low Prandtl number is expected to be predominant [351], typically leading to time-dependence even in slightly supercritical convection. Moreover, transitions to inertial convection [375] might occur not far above threshold, as discussed in §4.5.6. Bestehorn [131] and Eckert et al. [30] have emphasized the role of the mean flow and vertical vorticity in acting as a "lubricant" allowing the rigidity of the convective structure to be reduced. In particular, their numerical simulations have shown an increased tendency towards more regular square convective cells, which arise even closer to the threshold. Moreover, mean flow effects (even at $Pr = 0.5$) are responsible for an intrinsic time-dependence which takes the form of oscillations between squares and roll patterns. In the following, we will thus restrict the analysis to moderate and large Prandtl number fluids.

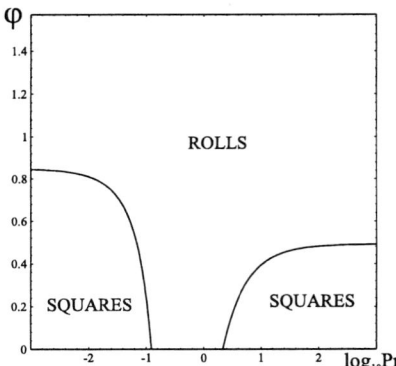

Figure 4.38: Competition between rolls and squares for $Bi = 0.8$.

Consider now an increase in the free surface Biot number to $Bi = 0.8$ (this is not too different from the value $Bi = 0.6$ used by Eckert et al. [30] for their numerical simulations). The competition between squares and rolls is summarized in Fig. 4.38. The region of squares at low Prandtl number [see Fig. 4.34] is still present, but a new region of squares has appeared at high Prandtl numbers (still for dominating surface tension effect).

Thus, the discussion made for low-Prandtl number fluids directly applies to the case of high Prandtl number fluids, apart from a preference for up-hexagons at threshold rather than down-hexagons. This behavior is depicted in Fig. 4.39 for $Pr \rightarrow \infty$, which appears in fairly good qualitative agreement with the experiment of Eckert et al. [29, 30].

Before presenting results relative to the two-layer instability, let us conclude this section by briefly examining the role of the heat transfer at the bottom rigid plate. This can be characterized by a Biot number Bi_0 using a mixed boundary condition of the form $DT - Bi_0 T = 0$ at $z = 0$, and generalizing the usual isothermal boundary condition $T = 0$ (corresponding to $Bi_0 \rightarrow \infty$). Note that the different sign of the condition with respect to that applying at the free surface (namely, $DT + BiT = 0$) follows from the reversed orientation of the normal to the boundary.

Consider the situation where $Bi = 0$ (poorly-conducting upper gas phase). When the thermal conductivity of the bottom wall decreases, Bi_0 decreases, such that we progressively tend to a situation where both boundaries are heat-insulating. In that case, the critical wavenumber of the pattern tends to zero, and the critical Marangoni number Ma_c tends to 48 (see e.g. [215, 292]). Similar behaviors were encountered in the case of the Rayleigh–Bénard instability (see §3.3.1 and Fig. 3.5). This situation is interesting from the theoretical point of view, because a separation between the vertical and the horizontal spatial scales occurs. The horizontal variation of temperature and velocity fields may then be rigorously described by long-wave equations [212, 215, 291, 292], as studied in §4.6. In §4.5.2, 4.5.3 and 4.5.4,

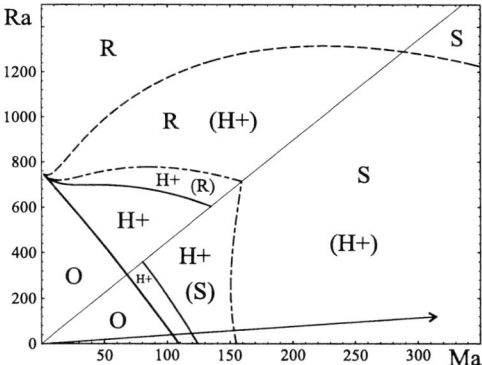

Figure 4.39: Behavior at high Prandtl number ($Pr \to \infty$) for $Bi = 0.8$. The behavior is qualitatively similar to that of Fig. 4.37, apart from a preference for up-hexagons H+ near threshold. The arrow describes the path followed when increasing ϵ in the experiment of Eckert et al. [Fig. 4.33].

we will use one of these Sivashinsky-like equations as a microscopic model replacing the full set of equations considered in the previous section. Results will then be obtained about the pattern selection problem in a way similar to that followed here, and it is interesting to check whether the results obtained here in the limit $Bi_0 \to 0$ tend to the results obtained by Golovin et al. [341] via the long-wave equation. This will also provides a check of the validity of the numerical procedure used to calculate cubic coefficients.

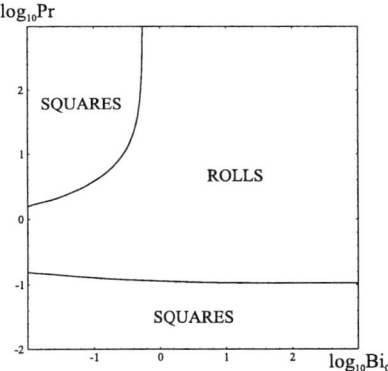

Figure 4.40: Competition between rolls and squares for $Bi = 0$, $\varphi = 0$, and for a variable bottom plate Biot number Bi_0.

Figure 4.40 illustrates the competition between squares and roll patterns as a function of the lower plate Biot number Bi_0 and of the Prandtl number Pr. In the limit $Bi_0 \to \infty$, we recover results of Fig. 4.34 at $\varphi = 0$ (pure Marangoni–Bénard instability). Squares only occur at low Prandtl ($Pr \leq 0.1$). When decreasing Bi_0, another region of squares appears (first at infinite Prandtl number) for $Bi_0 \leq 0.55$. In the limit $Bi_0 \to 0$, rolls are stable with respect to squares in the interval $0.21 < Pr < 0.82$ only, while squares are stable outside this interval.

Two-layer Rayleigh–Marangoni–Bénard instability

Although this system should present a much wider variety of qualitatively different behaviors[10], due to the large number of dimensionless parameters (ratios of fluid properties), we will here restrict the analysis to some situations which have been studied experimentally. Namely, we will first consider cases where the upper layer is a gas, and illustrate the role of the thermal conductivity ratio λ, together with the ratio a of the depths of the layers. This will allow to further discuss the experiment of Eckert [29]. Then, the recent experiments of Tokaruk et al. [34] in a Fluorinert-water system will be briefly considered, merely to show that squares can indeed be observed very close to threshold, as reported by these authors.

We first consider the conditions prevailing in the experiment of Nitschke–Eckert and Thess [29]. In Ref. [29], the ratio of layer thicknesses is $a = 0.258$, and the ratios of fluid properties (silicone oil/air) can be taken (using notations of §3.5) as $\lambda = 0.166$ and $\kappa = 188$ (in fact, we do not need to mention the ratios of viscosities, nor of thermal expansion coefficients, because the dynamical effects can here be neglected in the gas phase). For these parameters, the competition between squares and rolls is summarized in Fig. 4.41(a). We also considered that the Bond number (or angle φ) is a free parameter (actually, this would require varying the layer thicknesses proportionally because the results do depend on a). The Prandtl number Pr_1 of the oil is also varied, such a possibility being offered by the wide choice of viscosities available for silicone oils. In Fig. 4.41(b), we have also represented the case $a = 1$ (thicknesses equal for both layers) for comparison.

Note that in these plots and in the following one (Fig. 4.42), the range of φ has been extended to slightly negative values (inaccessible to experiments) in order to see how the large Prandtl region of squares disappears when increasing a. The "phase diagrams" corresponding to the situation $Pr_1 \to \infty$ are given in Figs 4.42(a,b). They are both qualitatively similar to Fig. 4.39, and in particular it is seen that the agreement between Fig. 4.42(a) and Fig. 4.39 is excellent, thus reflecting the fact that the one-layer problem works very satisfactorily in that case.

For increasing a, the thermal resistance to heat transfer is increased, and it is possible that for such conditions, the transition observed by Eckert et al. would have been to rolls rather than to squares. Note that the threshold for the transition to squares in the experiment is actually larger ($\epsilon \simeq 2.35$) than the Maxwell threshold [dot-dashed line of Fig. 4.42(a), leading to $\epsilon \simeq 0.6$]. The discrepancy might be attributed first to the qualitative validity of the amplitude equations at such large values of ϵ. A second explanation could be the following. As indicated by the authors, the transition to square patterns in their experiments appears to be mediated by defects (pentagon–heptagon pairs) in the hexagonal pattern. These defects induce the formation of fronts ("double lines" of pentagons) which allow squares to form via propagation of the fronts (which is coherent with the prediction of metastability – not

---

[10]As seen in §3.5, instability can occur for both directions of heating, and can also be oscillatory. Some aspects related to the onset of two-dimensional waves are treated in §5.4 for the buoyancy-driven case, and in §7.2 for the surface-tension-driven case.

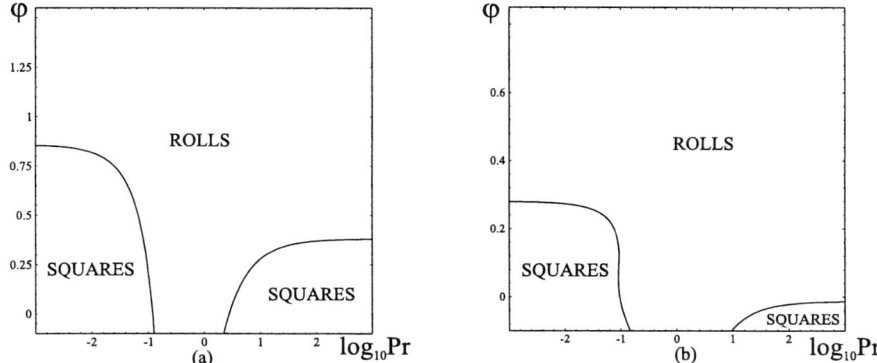

Figure 4.41: Silicone oil–air system: $\lambda = 0.166$, $\kappa = 188$. Competition between rolls and squares as a function of the Prandtl number Pr_1 of the oil and of the angle $\varphi = \arctan(0.12 Ra_1/Ma_1)$ in a Nield's diagram. (a) $a = 0.258$ (case of Eckert et al. [29]), (b) $a = 1$.

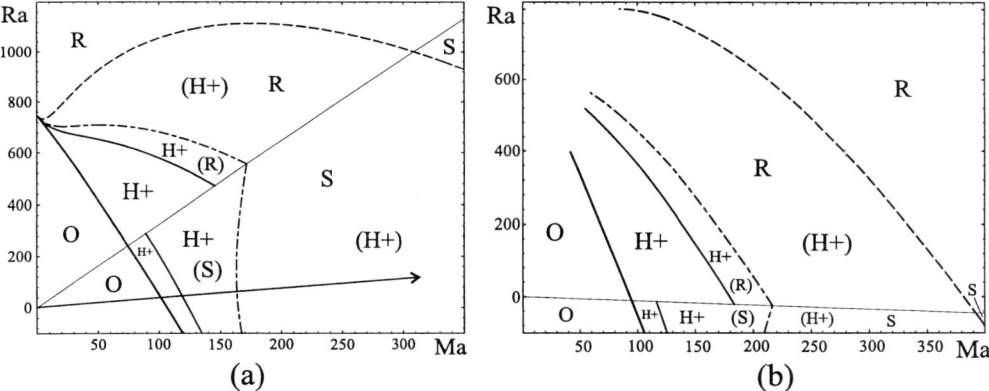

Figure 4.42: Behavior at high Prandtl number ($Pr \to \infty$) for the two-layer silicone oil–air system. (a) $a = 0.258$: the behavior is very similar to that of Fig. 4.39. The arrow describes the path followed when increasing ϵ in the experiment of Eckert et al. [29]. (b) $a = 1$: in that case, no transition to squares occurs, even for purely surface-tension-driven convection.

instability – of the hexagonal pattern). In fact, it may very well be that the higher value of the experimentally observed threshold is due to pinning effects, i.e. the interaction of fronts and defects with the underlying hexagonal structure, which are generally able to stop their motion, hence possibly delaying the appearance of squares.

Finally, we illustrate in Fig. 4.43 the effect of an increase in the thermal conductivity of the gas, by considering helium as the gaseous phase (the thermal conductivity of helium is approximately equal to that of oil). Equal thicknesses are assumed for oil and helium ($a = 1$).

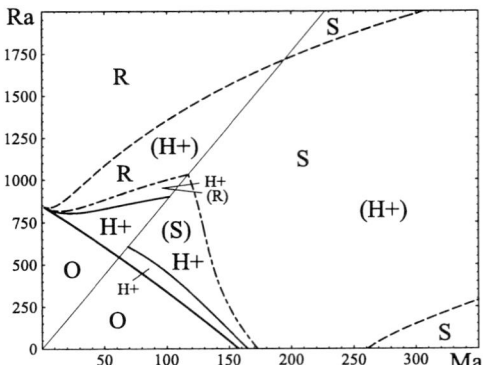

Figure 4.43: Behavior at high Prandtl number ($Pr \to \infty$) for the two-layer silicone oil–helium system with $a = 1$. The thermal conductivity ratio is $\lambda = 0.981$, and the thermal diffusivity ratio is $\kappa = 7480$.

To conclude this section, we briefly consider the recent experiments of Tokaruk et al. [34] in a *liquid/liquid system*. These authors report the careful observation of a square pattern close to onset, which undergoes a secondary transition to rolls at higher temperature differences. Typical shadowgraph images of the patterns are presented in Fig. 1.7 of Chapter 1. While the system considered in Fig. 1.7 is a layer of water on top of a layer of Fluorinert FC-104, we here consider the experiments reported in [34], for which a Fluorinert FC-75 (from 3M Inc.) was used instead. In this case, the ratios of fluid properties at 25 °C (using the notations of §3.5) are $\rho = 0.562$, $\mu = 0.629$, $\nu = 1.12$, $\kappa = 4.27$, $\lambda = 9.62$ and $\alpha = 0.184$. The Prandtl number of FC-75 is $Pr_1 = 23.5$.

The thickness ratio used in [34] is $a = 2.18$, while the total depth is $a_1 + a_2 = 0.406$ cm. For such moderate depths, both buoyancy and surface tension act in destabilizing the motionless state, and need to be incorporated in the analysis. Note that Tokaruk et al. *calculate* the surface tension variation $\sigma_T = -0.047$ dyn/cm K from the requirement that the linear theory reproduces the measured value of the critical temperature difference $\Delta T_c = 0.999 \pm 0.025$ °C (across the two layers). Adopting the same value of σ_T, and denoting the temperature drop across the lower FC-75 layer by ΔT_1, and its thickness measured in cm by a_1, we find using properties of the lower layer that Marangoni and Rayleigh numbers are respectively given by $Ma_1 = 9734\Delta T_1 a_1$ and $Ra_1 = 503446\Delta T_1 a_1^3$. Hence, the dynamic Bond number is $Bo_1 = Ra_1/Ma_1 = 51.7a_1^2$, showing that for the depth $a_1 = 0.128$, the Marangoni effect is actually dominant. Note that the ratio of local Rayleigh numbers in both layers is $Ra_2/Ra_1 = 0.09$, hence there is practically no buoyancy in the upper layer.

The linear stability analysis leads to the critical Marangoni number $Ma_c = 853.6$, while the critical wavenumber is $k_c = 2.377$, in perfect agreement with experiment [34]. Using the same notation as in the previous section, the nonlinear coefficients are calculated to be $\delta = 184.7$, $\beta = 15291$, $\chi = -5952$, $\gamma' = 16842$, $\xi = 17289$, while the linear growth rate is given by $\sigma = 12.27\epsilon$ (where $\epsilon = \Delta T/\Delta T_c - 1$).

As $\chi < \beta$, squares are preferred over rolls. Note that this fact will not be altered at higher

ϵ, such that we cannot hope to obtain a transition from squares to rolls in the frame of our weakly nonlinear theory. In fact, for the values of nonlinear coefficients given above, the sequence of events predicted at increasing ϵ is very similar to that described for the above analysis of the results of Eckert et al., though transitions typically occur at much smaller ϵ. As usual, hexagons are stable near threshold, and lose stability with respect to squares at $\epsilon = 0.192$. Squares become stable with respect to hexagons at $\epsilon = 0.042$, and the Maxwell point of the hexagons-squares transition is found from Eq. (4.194) at $\epsilon = 0.07$. This is slightly higher than the value $\epsilon = 0.05$, at which Tokaruk et al. have observed the first clear square pattern, and these authors also do not exclude that a hexagonal pattern could exist in the range $0 < \epsilon < 0.05$ [34].

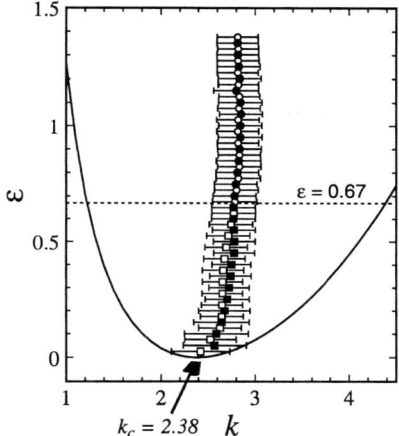

Figure 4.44: Mean wavenumber $< k >$ variation with the distance to the threshold ϵ in the experiment of Tokaruk et al. [34], for a Fluorinert FC-75–water system. The full line shows the linear stability boundary, while the bars indicate the width of the azimuthally averaged structure function. A square planform is found for $\epsilon < 0.67$, and rolls otherwise. Courtesy of W. Tokaruk and S.W. Morris (unpublished).

The Fluorinert–water system thus appears as a nice candidate to test weakly nonlinear theories, in view of the small values of ϵ needed for the various transitions to occur. In particular, the theory presented here could be refined to include spatial modulations, in order to account for the important role of defects in first-order transitions. As noted above, the transition to rolls observed by Tokaruk et al. at $\epsilon = 0.7$ cannot be predicted by the present theory. It is worth pointing out in this respect that the mean wavenumber of the square pattern was found to increase with ϵ, up to a value close to $k = 2.8$ at $\epsilon = 0.7$ (see Fig. 4.44). This effect is not easily incorporated in weakly nonlinear theories, and could be responsible for the preference of rolls at larger ϵ.

4.5   Imperfect patterns

In this section, we describe recent progress achieved in the study of spatially modulated cellular structures. As explained in §4.1.3 for rolls, slow spatial modulations of the amplitudes lead to amplitude equations containing spatial derivative terms. This in principle allows the description of patterns containing defects (e.g. point defects such as dislocations, or linear defects such as fronts or grain boundaries [51, 352] between patterns of different nature or orientation) and also allows to consider slow spatial variations of the wavenumber of the pattern. Furthermore, this in principle also permits the incorporation of the effect of lateral boundaries at large distance [298].

For hexagonal patterns, the most general amplitude equations are non-variational at the cubic order [346, 355], contrary to the situation prevailing for rolls. Hereafter, such amplitude equations are derived both from symmetry arguments, and from a direct calculation from a long-wave equation, describing the spatio-temporal dynamics of Bénard convection patterns in the limit of vanishing critical wavenumber (such as when both boundaries of the physical set-up are poor heat conductors [215, 238, 291, 292]). Then, we examine the phase stability of hexagonal patterns, which allows determination of the thresholds for secondary phase instabilities of hexagonal patterns with variable wavenumber inside the neutral stability boundary (see Fig. 4.12 for the case of roll patterns). Finally, we consider some important effects associated with the dynamics of point defects (namely pentagon-heptagon pairs) in the presence of non-variational quadratic terms.

4.5.1   Amplitude equations for modulated hexagonal patterns

In §4.2.1, the amplitude equations for non-modulated hexagonal patterns were obtained from symmetry arguments (without reference to any particular physical problem). It is possible to extend such analysis when the amplitudes are allowed to depend on some slow spatial coordinates X and Y. Actually, for the case of roll patterns the most generic result is an amplitude equation of the Newell–Whitehead–Segel [297, 298] type [see Eq. (4.29)]. This equation is obtained when the scalings of the longitudinal (i.e. along the basic wavevector $\vec{k}_1 = k_c \vec{1}_x$) and transversal (perpendicular to $\vec{k}_1$) variables [see Fig. 4.9] are taken as $X = \epsilon^{1/2}x$ and $Y = \epsilon^{1/4}y$ respectively (ϵ is the supercriticality parameter and x, y are the horizontal space variables). Note that these scalings and the corresponding scaling $T = \epsilon t$ of the time variable are natural consequences of the form (4.26) of the growth rate in the vicinity of the critical point. Now, if we keep $\vec{k} = (k_c + \Delta k_{/\!/})\vec{1}_x + \Delta k_\perp \vec{1}_y$, but instead of $\Delta k_\perp \sim \epsilon^{1/4}$, we use the scaling $\Delta k_\perp \sim \epsilon^{1/2}$ (corresponding to a slower transversal variable $Y = \epsilon^{1/2}y$), the transversal term in the Newell–Whitehead–Segel operator becomes of higher order and the resulting equation only contains a diffusive term in the longitudinal direction. A first consequence is that no zig-zag instability boundary (see §4.1.3) is predicted in this case.

For hexagonal or slightly non-equilateral hexagonal patterns (i.e. for which the angle between constitutive wavevectors is not exactly 120° [346]), it appears [239, 312, 318] that it is sufficient to consider this isotropic scaling (i.e. $\Delta k_{/\!/} \sim \Delta k_\perp \sim \epsilon^{1/2}$ or $X = \epsilon^{1/2}x$ and $Y = \epsilon^{1/2}y$), as the presence of resonant quadratic terms in the amplitude equations induces a synchronization of the phases of the three constitutive modes.

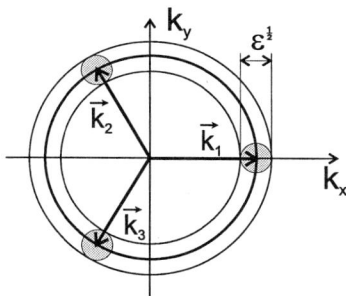

Figure 4.45: Band of amplified modes and isotropic scaling for hexagonal patterns.

In fact, it is seen from Fig. 4.45 that if we allow for some wavenumber shift of order $\epsilon^{1/4}$ in the transversal direction of the mode 1 for instance (e.g. a phase-winding $\varphi_1 \sim \epsilon^{1/4}y$ for the amplitude of the mode 1), the synchronization condition $\varphi_1 + \varphi_2 + \varphi_3 = 0$ cannot be satisfied unless a variation of order $\epsilon^{1/4} \gg \epsilon^{1/2}$ is allowed in the longitudinal wavevector shifts of modes 2 and 3. According to Eq. (4.26) such modes are linearly damped on a $O(\epsilon^{1/2})$ time scale, i.e. faster than the $O(\epsilon)$ time scale of amplified modes, with the consequence that the only allowed quasi-hexagonal patterns should have wavevector shifts inside the circular regions of Fig. 4.45. Moreover, a more rigorous analysis [364] shows that the zig-zag instability should indeed be much less dangerous for hexagonal patterns than it is for rolls.

It should also be emphasized that this argument is valid for patterns which are everywhere close to saturated hexagonal or slightly non-equilateral patterns, but breaks down in the vicinity of defects in the structure or near the lateral walls (where amplitudes locally vanish). In these cases, it might be necessary to re-consider equations containing Newell–Whitehead–Segel (NWS) operators.

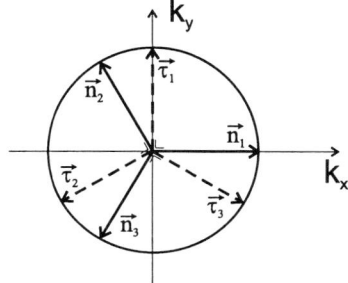

Figure 4.46: Longitudinal and transversal unit vectors.

In this section, we will consider the isotropic scaling only. Thus, the amplitudes a_1, a_2 and a_3 of the constitutive roll modes are assumed to depend on a slow horizontal variable $\vec{R} = \epsilon^{1/2}\vec{r}$. Consequently, the gradient operators $\vec{\nabla}$ are of order $\epsilon^{1/2}$ and it can be expected that quadratic terms involving a gradient operator should appear in the amplitude equations at order $\epsilon^{3/2}$ (the amplitudes are of order $\epsilon^{1/2}$). Note that cubic terms do not contain derivatives at this order, while linear terms should clearly be of the form already established for roll patterns [i.e. as in the NWS form (4.29) but without Y-derivatives].

Using the symmetry arguments of §4.2.1 (namely invariances to translations, rotations and reflections[11]), and defining unit vectors such as in Fig. 4.46 (vectors $\vec{n}_i$ and $\vec{\tau}_i$ respectively correspond to local longitudinal and transversal unit vectors for each amplitude $i = 1, 2, 3$), the amplitude equations up to order $\epsilon^{3/2}$ are obtained as

$$
\begin{aligned}
\dot{a}_1 =& \sigma a_1 + \xi(\vec{n}_1.\vec{\nabla})^2 a_1 - a_1 \left[\beta|a_1|^2 + \gamma(|a_2|^2 + |a_3|^2)\right] \\
&+ \delta a_2^* a_3^* + iB \left[a_2^*(\vec{n}_3.\vec{\nabla})a_3^* + a_3^*(\vec{n}_2.\vec{\nabla})a_2^*\right] \\
&+ iK \left[a_2^*(\vec{\tau}_3.\vec{\nabla})a_3^* - a_3^*(\vec{\tau}_2.\vec{\nabla})a_2^*\right]
\end{aligned}
\tag{4.195}
$$

together with cyclic permutations of $1, 2, 3$. Another consequence of the symmetry requirements is that all the coefficients σ, ξ, β, γ, δ, B and K are real. The new quadratic term proportional to B in Eq. (4.195) has been first obtained by Brand [355], while the form of the term proportional to K in Eq. (4.195) has first been obtained by Pontes [324] (although the corresponding coefficients B and K were zero in his case), and by Kuznetsov et al. [346] in the case of Rayleigh–Bénard convection. Later, the coefficients B and K were calculated for two-layer Marangoni–Bénard convection by Golovin et al. [239], and for one-layer Marangoni–Bénard convection by Bragard and Velarde [356], and Echebarría and Pérez-García [357, 358] (also including buoyancy).

It is important to note that the new quadratic terms appearing in Eq. (4.195) are non-potential, i.e. the amplitude equations cannot be written under a variational form [as was the case for unmodulated patterns, see Eq. (4.75), or even for Eq. (4.195) with $B = K = 0$]. Accordingly, the dynamics described by these equations is not restricted by "energetical" reasons anymore, and a possibility exists for more complex behaviors than in purely relaxational cases. Indeed, it has recently been shown by Bragard and Velarde [356], and by Golovin et al. [359, 360] that oscillatory behaviors occur in some circumstances. Other complex behaviors associated with motion of pentagon-heptagon defects (PHD) are described in §4.5.4.

4.5.2 Coefficients for long-wave Bénard instabilities

In this section, the coefficients of the amplitude equations (4.195) are calculated for a system of equations suggested by Knobloch [215], and modified by Shtilman and Sivashinsky [292]. This system governs long-wavelength convection patterns (see §4.6) and reads as

$$
\begin{aligned}
\dot{u} =& \epsilon^2 u - (1 + \nabla^2)^2 u + \vec{\nabla}.\left[(\vec{\nabla}u)^2\vec{\nabla}u\right] + \beta\vec{\nabla}.\left[\nabla^2 u\vec{\nabla}u\right] \\
&+ \delta\nabla^2 \left[(\vec{\nabla}u)^2\right] - \gamma\vec{\nabla}.\left[u\vec{\nabla}u\right] + p\left(\vec{\nabla}u \times \vec{\nabla}\Psi\right)_z,
\end{aligned}
\tag{4.196}
$$

---

[11]Note that amplitude equations limited to the lowest relevant order (here, $\epsilon^{3/2}$) are generally not invariant to *all* rotations and reflections. This will be further discussed in §4.5.5. Here, we consider only invariances to rotations with an angle multiple of $\pi/3$, and reflections with respect to the bisector of any pair of basic wavevectors, as is also done in §4.2.1.

$$\nabla^2 \Psi = \left(\vec{\nabla}\nabla^2 u \times \vec{\nabla}u\right)_z \qquad (4.197)$$

where the dot denotes the derivative with respect to time t, $u(x, y)$ and $\Psi(x, y)$ are respectively the scaled temperature and streamfunction of the mean flow, ϵ^2 is the supercriticality parameter, and β, γ, δ and p are constant coefficients which are specific for concrete pattern formation problems (note that these coefficients, as well as ϵ, are different from those of the previous section). For Marangoni convection, these coefficients were calculated in [215, 292, 341, 356]. In particular, the coefficient p is inversely proportional to the Prandtl number. Note that the vertical z component does not appear in the system (4.196–4.197), which is a consequence of the separation of scales occurring when the critical wavenumber of the neutral stability analysis is vanishingly small. This circumstance greatly simplifies the calculation of amplitude equations.

According to the above discussion, we use the multiscale expansions

$$u = \epsilon u_1 + \epsilon^2 u_2 + ..., \quad \Psi = \epsilon^2 \Psi_2 + \epsilon^3 \Psi_3 + ... \qquad (4.198)$$

$$\vec{\nabla} = \vec{\nabla}_0 + \epsilon\vec{\nabla}_1, \quad \frac{\partial}{\partial t} = \epsilon^2 \frac{\partial}{\partial \tau} \qquad (4.199)$$

where $\vec{\nabla}_0$ and $\vec{\nabla}_1$ are the gradients with respect to the fast and slow horizontal variables $\vec{r}_0$ and $\vec{r}_1$ respectively. The temperature field is written at the first order as

$$u_1 = a_1(\vec{r}_1, \tau)\exp[i\vec{n}_1.\vec{r}_0] + a_2(\vec{r}_1, \tau)\exp[i\vec{n}_2.\vec{r}_0] + a_3(\vec{r}_1, \tau)\exp[i\vec{n}_3.\vec{r}_0] + c.c. \qquad (4.200)$$

with the definition of the unit (critical) vectors of Fig. 4.46.

It is worth noting that inserting this expression into Eq. (4.197) leads to $\Psi_2 = 0$, i.e. there is no contribution to the vertical vorticity at this order, essentially for the same symmetry reasons as discussed in §4.3.1 and 4.3.2. Here again, we may set $\Psi = 0$, because the contribution to Ψ generated at order ϵ^3 will not affect the derivation of the amplitude equations.

Inserting the expressions (4.198) and (4.199) in Eq. (4.196) and collecting terms of order ϵ^2, we get an inhomogeneous problem for u_2 which contains resonant terms proportional to the quantity $\beta - \delta + \gamma$ (quadratic resonant coefficient). As in §4.2.2, we see that the amplitude equations should then rigorously be limited to this order. However, as no saturation can be expected from quadratic amplitude equations, we here also assume that $\beta - \delta + \gamma = O(\epsilon)$ and proceed to the next order after having solved the problem at order ϵ^2. Actually, the condition $\beta - \delta + \gamma = O(\epsilon)$ can be verified if we assume that the Prandtl number is near $1/5$ as seen below. If this is not the case, amplitude equations should be considered only as qualitatively valid.

We do not reproduce the intermediate calculations here, and directly state the compatibility conditions found at order ϵ^3 as

$$\begin{aligned}
\dot{a}_1 =\ &a_1 + 4(\vec{n}_1.\vec{\nabla}_1)^2 a_1 - \alpha a_1|a_1|^2 - \nu a_1\left(|a_2|^2 + |a_3|^2\right) \\
&+ Ca_2^* a_3^* + iB\left[a_2^*(\vec{n}_3.\vec{\nabla}_1)a_3^* + a_3^*(\vec{n}_2.\vec{\nabla}_1)a_2^*\right] \\
&+ iK\left[a_2^*(\vec{\tau}_3.\vec{\nabla}_1)a_3^* - a_3^*(\vec{\tau}_2.\vec{\nabla}_1)a_2^*\right]
\end{aligned} \qquad (4.201)$$

as well as the two other equations obtained by cyclic permutations of $1, 2, 3$. In Eq. (4.201), the quadratic coefficients are given by $C = \epsilon^{-1}(\beta - \delta + \gamma)$, $B = 2(\beta - \delta) + \gamma$, $K = -\sqrt{3}(\beta + \gamma)$, and the cubic coefficients by

$$\alpha = 3 + \frac{2}{9}(\beta + 2\delta + \gamma)(2\beta + 4\delta - \gamma), \quad \nu = 3 + \frac{3}{4}(3\delta - \gamma)(\beta + \delta + \gamma) \quad (4.202)$$

For completeness we mention that the cubic coefficient corresponding to the interaction of two modes with wavevectors forming an angle θ (rhombs) is given by

$$\chi(x = \cos\theta) = \frac{2}{(4x^2 - 1)^2}\left[\begin{array}{c} 1 + 4\beta\delta + 4\gamma\delta - 2\gamma^2 - 4\beta\gamma - 2\beta^2 \\ +32(1 + \beta\delta)x^6 + 8(\beta^2 + 4\delta^2 + \beta\gamma - 3\beta\delta)x^4 \\ +2(2\beta\delta - 3 - \beta^2 - 8\delta^2 - \beta\gamma)x^2 \end{array}\right]$$

$$(4.203)$$

and in particular for $\theta = \pi/2$ (squares), we have $\chi_{\pi/2} = 2 - 4(\beta + \gamma)(\beta + \gamma - 2\delta)$.

The system of amplitude equations (4.201) is asymptotically valid for $\epsilon \to 0$ provided $C = O(1)$, i.e. $\beta - \delta + \gamma = O(\epsilon)$, otherwise the amplitude of hexagons is not small even near the threshold, and Eq. (4.201) should be considered as a model. In the case of Marangoni–Bénard convection in a Boussinesq fluid adjacent to a layer of nearly-insulating gas [238], we have

$$\beta = -\frac{\sqrt{7}}{8}\left(1 + \frac{2}{Pr}\right), \quad \gamma = 0, \quad \delta = -\frac{3\sqrt{7}}{4}\left(1 + \frac{1}{6Pr}\right) \quad (4.204)$$

and it is seen that $\beta - \delta + \gamma$ vanishes at $Pr = Pr^* = 1/5$ (note that the result obtained in the general case was $Pr^* \simeq 0.23$ [151], see §4.4.2). Consequently, the analysis is rigorous if $Pr - Pr^* = O(\epsilon)$. Then, we also have $B = O(\epsilon)$, i.e. the Brand terms are negligible. This will be used in the following. Note that more generally [215], $\gamma \neq 0$ when non-Boussinesq effects are important, and $\beta = \delta$ when the problem is self-adjoint ($\beta = \delta = 0$ when the boundary conditions at the top and bottom plates are identical). According to the results presented above, it is seen that quadratic non-potential terms may occur even for pure Rayleigh–Bénard convection in a Boussinesq fluid, provided the boundary conditions are asymmetric. Then, $C = B = 0$ and $K \neq 0$ in general [346].

Using Eq. (4.204) and the above expressions of the coefficients, it can be seen that rolls are unstable to squares ($\chi_{\pi/2} < \alpha$) for

$$Pr < \frac{329 - 3\sqrt{4137}}{634} \simeq 0.215 \quad (4.205)$$

or

$$Pr > \frac{329 + 3\sqrt{4137}}{634} \simeq 0.823 \quad (4.206)$$

while rolls are stable inside this interval. We thus recover the numerical results of Fig. 4.40 [when the bottom plate Biot number Bi_0 tends to zero, which is just the condition necessary to derive the system (4.196–4.197)], as seen in §4.6. This provides a particularly convincing check of the calculations of the cubic coefficients, both in the full problem studied in §4.4.3 (numerical calculation of the coefficients) and in the long-wave situation considered here.

4.5.3 Phase dynamics and Busse balloon of hexagonal patterns

At the end of §4.2.1, it was seen that among the eigenvalues characterizing the stability of ideal (non-modulated) hexagonal planforms, two zero eigenvalues were found, associated with arbitrary uniform translations of the pattern in the horizontal plane. These eigenvalues reflect the absence of restoring forces in the case of uniform translations (generating a continuum of equivalent solutions). In case some long-wavelength modulations of the pattern are allowed, these two translational (phase) modes acquire a slow time dynamics, whose basic features are described in this section. The description of such phase dynamics, leading to secondary instabilities such as Eckhaus and zig-zag instabilities (see §4.1.3) is well known in the case of roll patterns (see e.g. [51, 314]), and has already been extended to hexagonal patterns [318, 319]. Here, we consider the extension of such analysis to the case of the most general amplitude equations (4.201), i.e. including non-potential quadratic terms. As discussed in the previous sections, the scaling used for horizontal variables is isotropic, with the result that NWS operators here reduce to longitudinal diffusion terms. This has the consequence that the zig-zag instability, corresponding to the vanishing of the transversal diffusion coefficient [51], is prevented here by the quadratic phase-synchronizing terms [318, 364]. The following analysis of secondary instabilities of the hexagonal pattern is thus restricted to the Eckhaus instability.

When $\alpha > 0$, it is possible to rescale amplitudes such as to rewrite Eq. (4.201) in the form

$$
\begin{aligned}
\dot{a}_1 =& \mu a_1 + 4(\vec{n}_1.\vec{\nabla})^2 a_1 - a_1|a_1|^2 - \nu a_1\left(|a_2|^2 + |a_3|^2\right) \\
&+ \delta a_2^* a_3^* + iK\left[a_2^*(\vec{\tau}_3.\vec{\nabla})a_3^* - a_3^*(\vec{\tau}_2.\vec{\nabla})a_2^*\right] \\
&+ iB\left[a_2^*(\vec{n}_3.\vec{\nabla})a_3^* + a_3^*(\vec{n}_2.\vec{\nabla})a_2^*\right]
\end{aligned}
\tag{4.207}
$$

We consider the linear stability of a stationary equilateral hexagonal pattern, still allowing for the possibility of some wavenumber shift q with respect to the critical wavenumber $k_c = 1$. Accordingly, the amplitudes will be written as

$$
a_p = (r_0 + \epsilon r_p)\exp[i(q\vec{n}_p.\vec{r} + \varphi_p)], \quad p = 1, 2, 3
\tag{4.208}
$$

Here, ϵ is the smallness parameter, r_0 is the amplitude of the unperturbed hexagonal pattern, r_p is the perturbation of the modulus, and φ_p that of the phase. When $r_p = \varphi_p = 0$, $p = 1, 2, 3$ (up-hexagonal pattern), we get the equation for r_0:

$$
\mu - 4q^2 + \delta r_0 + 2Bqr_0 - r_0^2(1 + 2\nu) = 0
\tag{4.209}
$$

It appears that the proper scalings of the horizontal coordinates and of time are given by

$$
r_p = r_p(\epsilon\vec{r}, \epsilon^2 t), \quad \varphi_p = \varphi_p(\epsilon\vec{r}, \epsilon^2 t)
\tag{4.210}
$$

and that the sum of the phases $\varphi_1 + \varphi_2 + \varphi_3$ should remain small (due to synchronization). Specifically, we set

$$
\varphi_1 + \varphi_2 + \varphi_3 = \epsilon^2\varphi(\epsilon\vec{r}, \epsilon^2 t)
\tag{4.211}
$$

In this scaling regime, the fast dynamics of the moduli r_p is adiabatically slaved to the dynamics of the slow phase variables [318, 325]. Indeed, inserting Eqs (4.208–4.211) in Eq. (4.207) and identifying the various powers of ϵ, the real part of the order ϵ result gives a linear inhomogeneous system of 3 equations for the 3 unknowns r_p, $p = 1, 2, 3$. The determinant of the matrix of this linear system is

$$4r_0^3(\delta + 2Bq - 2r_0 - 4\nu r_0)(\delta + 2Bq + r_0 - \nu r_0)^2 \tag{4.212}$$

In fact, the vanishing of this determinant corresponds to loss of stability of the regular hexagonal planform solution with respect to disturbances of the moduli r_p (the matrix mentioned above is the Jacobian of the system of equations governing the dynamics of the moduli r_p). According to Eq. (4.209) and Eq. (4.212), we see that this happens on two lines in the $(q\ \mu)$ plane. The first one is the existence boundary of hexagonal solutions, given by

$$\mu_{\text{ex}} = 4q^2 - \frac{(\delta + 2Bq)^2}{4(1 + 2\nu)} \tag{4.213}$$

while the second one is the line at which hexagons lose stability with respect to rolls, i.e. it is given by

$$\mu_{\text{roll}} = 4q^2 + \frac{\nu + 2}{(\nu - 1)^2}(\delta + 2Bq)^2 \tag{4.214}$$

Note that these two lines are tangent at the point $(q, \mu) = (-\delta/2B, \delta^2/B^2)$. Except in the neighborhood of the lines (4.213) and (4.214), we can thus solve the linear system obtained at order ϵ. This leads to

$$r_1 = a(\vec{n}_1 . \vec{\nabla})\varphi_1 + b\left[(\vec{n}_2 . \vec{\nabla})\varphi_2 + (\vec{n}_3 . \vec{\nabla})\varphi_3\right] + c\left[(\vec{\tau}_3 . \vec{\nabla})\varphi_3 - (\vec{\tau}_2 . \vec{\nabla})\varphi_2\right] \tag{4.215}$$

as well as its cyclic permutations leading to expressions of r_2 and r_3. In Eq. (4.215), $\vec{\nabla}$ denotes the gradient with respect to the slow variable $\vec{R} = \epsilon \vec{r}$, and the coefficients a, b, c are given by

$$a = \frac{r_0\left(-B\delta + 8q - 2B^2q + 8\nu q + 2B\nu r_0\right)}{[\delta + 2Bq - r_0(\nu - 1)][\delta + 2Bq - 2r_0(1 + 2\nu)]} \tag{4.216}$$

$$b = \frac{(8q - Br_0)(\delta + 2Bq) - 2r_0(Br_0 + 8\nu q)}{2[\delta + 2Bq - r_0(\nu - 1)][\delta + 2Bq - 2r_0(1 + 2\nu)]} \tag{4.217}$$

$$c = \frac{Kr_0}{2[\delta + 2Bq - r_0(\nu - 1)]} \tag{4.218}$$

Now, at order ϵ^2, we get the sought equations for the phases φ_p. Summing up over $p = 1, 2, 3$ and using Eq. (4.211), the time-derivative term is found to be $O(\epsilon^2)$ and may be omitted. The resulting equation may be solved for φ (slaving of the total phase). This expression is then reinjected in the individual phase equations. We do not present these intermediate steps here, but rather state the final result in terms of the two-dimensional phase vector

$$\vec{\phi} = [\Psi, \chi] = \left[\sqrt{3}(\varphi_2 + \varphi_3), \varphi_3 - \varphi_2 \right] \tag{4.219}$$

After some manipulations, we get the phase equation in the form

$$\partial_T \vec{\phi} = D_\perp \Delta \vec{\phi} + (D_{/\!/} - D_\perp) \vec{\nabla}(\vec{\nabla}.\vec{\phi}) \tag{4.220}$$

where the diffusion coefficients are given by

$$D_{/\!/} = 3 + \left(\frac{2q}{r_0} - \frac{B}{4} \right) \left(3a + 3b - \sqrt{3}c \right) + \frac{\sqrt{3}K}{4} \left(a + 5b + \sqrt{3}c \right) \tag{4.221}$$

$$D_\perp = 1 + \left(\frac{B}{4} + \frac{\sqrt{3}K}{4} - \frac{2q}{r_0} \right) (b - a + \sqrt{3}c) \tag{4.222}$$

We thus recover the form of the phase diffusion equation also obtained in [319], but with different values of the phase diffusion coefficients. The analysis of the stability of the reference hexagonal solution is straightforward: setting $\vec{\phi} \sim \exp[i\vec{k}.\vec{R}]$, the eigenvalues are found to be independent of the orientation of the disturbance wavevector $\vec{k}$, and given by $-D_{/\!/}k^2$ and $-D_\perp k^2$ (the eigenvectors do depend on the orientation of $\vec{k}$). Thus, the hexagonal pattern is stable provided $D_{/\!/} > 0$ and $D_\perp > 0$, and unstable otherwise.

The neutral stability conditions $D_{/\!/} = 0$ and $D_\perp = 0$, when combined with Eq. (4.209), yield lines delimitating the stability region ("Busse balloon") of hexagons in the plane (μ, q). It is worth noting that these two lines in general may intersect. At these points, the two eigenvalues are zero, thus allowing to expect that complex eigenvalues could appear (resulting from the interactions between the two phase modes) when considering other effects such as non-equilateral patterns, ... Moreover, the present analysis (derived under the so-called "phase approximation") does not allow the investigation of the stability to finite wavenumber disturbances, nor the calculation of the fastest growing modes (which should both depend on the orientation of the disturbance wavevector [318]). Several secondary instabilities can indeed be expected, as recently shown by Nuz et al. [360].

When $K = B = 0$, the results obtained above reduce to those obtained by Sushchik and Tsimring [318] for the potential case. This is represented in Fig. 4.47, for $\delta = 1, \nu = 2$. Then, it is readily seen from Eq. (4.209), Eqs (4.216–4.218) and Eqs (4.221–4.222) that the stability only depends on q^2 (and μ), i.e. the stability balloon is symmetric.

Note also that in contrast with the results found for roll patterns, the stability domain for hexagons is generally bounded from above (instability to rolls).

When $K \neq 0$ or $B \neq 0$, the stability domain is in general not symmetric. As an illustration, we will consider the case of the Marangoni–Bénard instability for which the coefficients have been calculated in the previous section. It should be recalled that unless the Prandtl number is near $Pr^* = 0.2$, the amplitude equations should merely be considered as a model.

When $Pr \simeq 0.2$, it can be shown (by a rescaling given in the next section) that the values of the coefficients should be taken as $\delta = 1$, $\nu = 1.118$, $B = 0$ and $K = 0.8425$. The corresponding stability region, represented in Fig. 4.48, is limited from above by the instability to rolls [Eq. (4.214)], and is now slightly asymmetric.

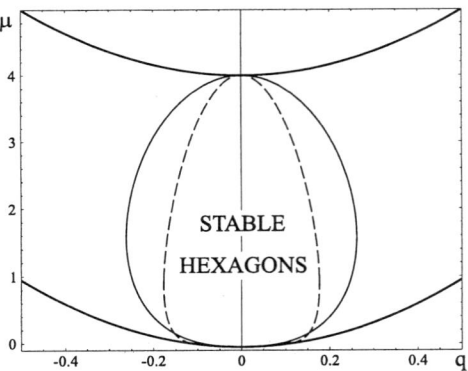

Figure 4.47: The stability region of up-hexagons in the potential case $K = B = 0$. Other coefficients are $\delta = 1$ and $\nu = 2$. The lower thicker full curve is the existence boundary (4.213) while the upper one indicates instability to rolls [Eq. (4.214)]. Thinner lines correspond to vanishing diffusion coefficients $D_{/\!/}$ (full) or $D_\perp$ (dashed).

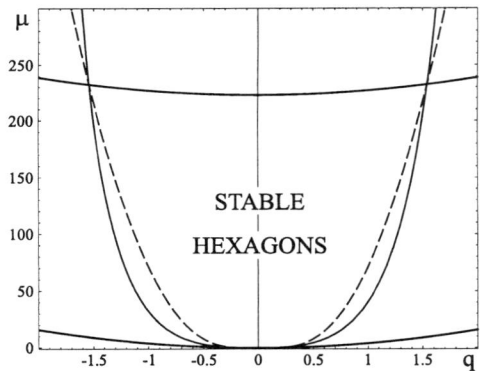

Figure 4.48: The stability region of up-hexagons in a non-potential case corresponding to Marangoni–Bénard convection between insulating boundaries at $Pr = 0.2$. The same lines are drawn as in 4.47. It has also been checked that the hexagons are locally stable to the cross-roll instability everywhere inside the Busse balloon obtained here.

To conclude this section, we apply the above results to the case of a fluid of higher Prandtl number. The stability diagrams are represented in Fig. 4.49 for $Pr = 0.3$ and $Pr \to \infty$.

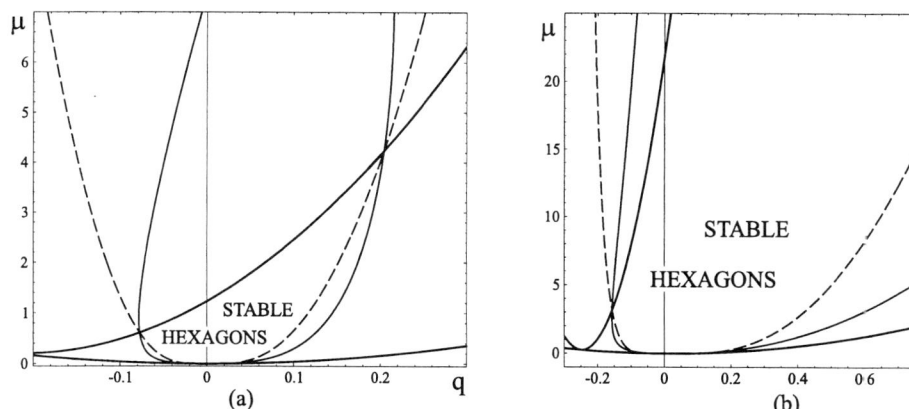

Figure 4.49: The stability region of up-hexagons in a non-potential case corresponding to Marangoni–Bénard convection between insulating boundaries at $Pr = 0.3$ (case a) and $Pr \to \infty$ (case b). The same lines are drawn as in Figs 4.47 and 4.48.

It is worth commenting that although the calculation of coefficients was performed from the Knobloch equation (4.196), i.e. for long-wavelength patterns ($k_c \to 0$), and even though the amplitude equations (4.201) must be seen has a model for Pr not close to $1/5$, the results obtained for large Prandtl number are in qualitative agreement with results obtained by Bestehorn [316], Echebarría and Pérez-García [357] and Bragard and Velarde [356] for short-wavelength convection patterns [$k_c = O(1)$]. In particular, the stability region of hexagons is seen to be displaced to the right (larger wavenumber) when increasing the control parameter μ, which is also in agreement with the experimental results obtained by Koschmieder and Switzer [108] in layers of silicone oil (see Fig. 4.50). Note finally that this asymmetry of the Busse balloon for hexagonal patterns is here entirely due to quadratic non-potential terms with coefficients B and K.

To summarize, the condition (4.214) represents the threshold of one of the *amplitude instabilities* of the hexagonal pattern, while the vanishing of either diffusion coefficient (4.221) or (4.222) defines the threshold of *phase instabilities* of perfect hexagonal patterns. Phase and amplitude instabilities of hexagonal patterns delimitate their *Busse balloon*, and have already been studied by different methods [316], also including the possibility of *cross-roll instabilities*. The latter (amplitude) instabilities correspond to the growth of disturbances with wavevectors orthogonal to one of the basic wavevectors of the hexagonal pattern (outer disturbances), but were not found [316] to affect the shape of the Busse balloon for hexagonal patterns (while it may indeed be more dangerous for rolls, as seen in Fig. 4.50). Using a system of *six* coupled amplitude equations of the form (4.207), written for two resonant triads rotated by $\pi/6$ with respect to each other, it is possible to calculate the eigenvalue corresponding to the cross-roll instability. While the latter is found to be maximal when the length of the cross-roll wavevector is k_c, it was indeed not found to be dangerous for hexagonal patterns, in the situations considered here (the threshold lies in the Eckhaus-unstable region).

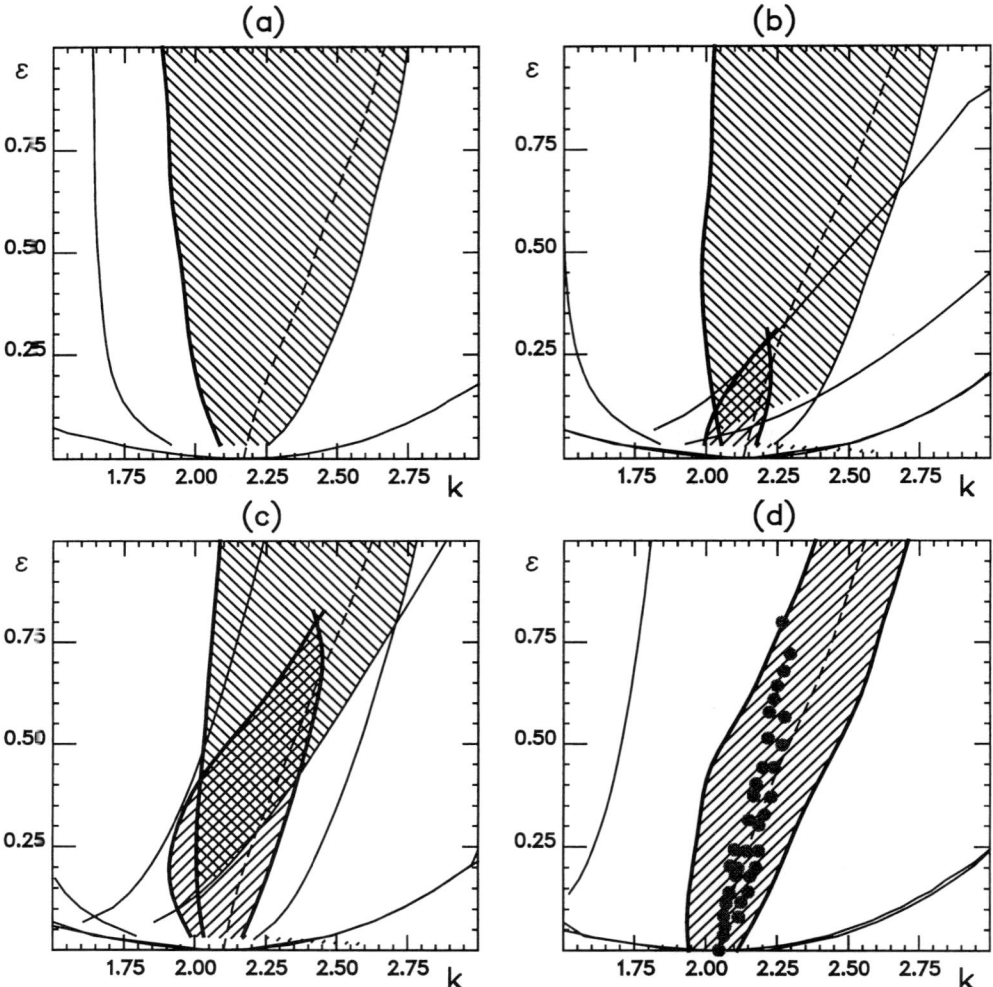

Figure 4.50: Stability regions of rolls (down-hatched zones) and hexagons (up-hatched zones) in Rayleigh–Marangoni–Bénard convection, for different angles ϕ of the physical line in a Nield's diagram ($\phi = 0$ and $\phi = \pi/2$ respectively correspond to purely buoyancy-driven and surface-tension-driven cases). Thin lines: amplitude instabilities (hexagon-rolls and cross-rolls); thick lines: phase instabilities; dashed lines: k_{max}, wavenumber which maximizes the linear growth rate. (a) $\phi = 0$, no stable hexagons, rolls are limited by zig-zag and cross-roll instabilities. (b) $\phi = 10°$, hexagons are stable in the small bubble around k_c. (c) $\phi = 20°$, large area of stable hexagons, confined by phase instabilities. (d) $\phi = 70°$, no stable rolls, hexagons stable in a narrow band around k_{max}. Large dots mark experimental data from Koschmieder and Switzer [108]. Courtesy of M. Bestehorn.

The reader is also referred to the more general analysis of Nuz et al. [360] of the stability of equilateral hexagons and slightly non-equilateral patterns, including in particular the possibility of short-wave instabilities. These authors also provided evidence a new kind of instability of roll patterns, which can generate steady non-equilateral hexagons, noise-sensitive transitions between roll patterns, or periodic oscillations. Clearly, these effects are linked with quadratic non-variational terms in Eqs (4.207), and would not be observed in a potential case.

4.5.4 Non-variational dynamics of pentagon-heptagon defects

In this section, we consider some aspects of the dynamics of pentagon-heptagon defects (PHD) in otherwise perfect hexagonal planforms, lying *within* the Busse balloon. The following results are obtained via numerical integration of the amplitude equations (4.201), with values of the coefficients appropriate to the pure Marangoni–Bénard instability problem (no buoyancy) in a Boussinesq layer of fluid confined between a bottom rigid insulating plate and an upper undeformable insulating free surface. The coefficients are then given by Eqs (4.202–4.204).

Multiplication of defects in hexagonal patterns

Here we restrict the analysis to the vicinity of the particular value $Pr^* = 0.2$ of the Prandtl number for which the amplitude equations were shown to be rigorously valid.

For $Pr - Pr^* = O(\epsilon)$, it is convenient to rescale the amplitude equations in the following manner. Rather than ϵ^2 in the linear term of the original equation (4.196), we could have written $\mu\epsilon^2$, where μ is some $O(1)$ parameter. Then, we would have obtained amplitude equations identical to Eq. (4.201) except for a coefficient μ multiplying the linear term a_1. Now, we substitute $a_p \rightarrow a_p \alpha^{-1/2}$ ($\alpha > 0$) and define ϵ such that the coefficient of the quadratic term $a_2^* a_3^*$ is unity. This gives

$$\epsilon = \frac{\beta - \delta}{\alpha^{1/2}} = O(Pr - Pr^*) \tag{4.223}$$

and the rescaled amplitude equations read as

$$\begin{aligned}
\dot{a}_1 =&\mu a_1 + 4(\vec{n}_1.\vec{\nabla}_1)^2 a_1 - a_1|a_1|^2 - \tilde{\nu}a_1\left(|a_2|^2 + |a_3|^2\right)\\
&+a_2^* a_3^* + i\tilde{K}\left[a_2^*(\vec{\tau}_3.\vec{\nabla}_1)a_3^* - a_3^*(\vec{\tau}_2.\vec{\nabla}_1)a_2^*\right]
\end{aligned} \tag{4.224}$$

where $\tilde{\nu} = \nu/\alpha$, $\tilde{K} = -\sqrt{3}\beta/\alpha^{1/2}$ and the term proportional to $B/\alpha^{1/2} = 2(\beta-\delta)/\alpha^{1/2}$ can be neglected in view of Eq. (4.223) in the limit $\epsilon \rightarrow 0$. An important point is that μ is a free $O(1)$ parameter (which will be considered as the control parameter in what follows), while the actual smallness parameter is defined by Eq. (4.223). Unless explicitly mentioned, all the simulations presented in the following correspond to the values $\tilde{\nu} = 1.118$, $\tilde{K} = 0.8425$ of the coefficients, obtained for the case $Pr = 1/5$. Note that the corresponding region of stability of the defect-free up-hexagonal solution is given in 4.48.

The numerical method is a second-order finite difference scheme with explicit Runge–Kutta fifth-order time integration [362] and adaptive time-step control ensuring a local relative error smaller than a given accuracy (generally 10^{-3}) at each time step. The simulation domain is rectangular and has dimensions (L_x, L_y), and the number of nodes in both directions will be denoted by (N_x, N_y). Some convergence tests will be presented in what follows. Validations of the code have been performed by imposing some constant phase gradients throughout the domain, and comparing the steady states with analytical results obtained in this case (this allows checking the discretization of all terms containing spatial gradients). The domain is taken sufficiently large, and the boundary conditions are selected such as to minimize the effect of lateral walls (although this appears to be too expensive in computer time for some simulations). The best choice appears to be "no-flux" conditions where the normal derivative of each amplitude is set to zero on each wall.

Figure 4.51 displays the main characteristics (amplitudes and reconstructed pattern) of a quasi-steady pentagon-heptagon defect (PHD) obtained for $\tilde{\nu} = 2, \tilde{K} = 0, \mu = 1$. Note that the reconstruction is calculated with $k_c = 1$ in Eq. (4.200), even though a rescaled wavenumber ($k_c \to \infty$ for $\epsilon \to 0$) should normally be used (for which the reconstructed defect would appear as a zone where a smooth transition occurs between hexagons and a nucleus of a roll pattern).

When $\tilde{K} = 0$, the amplitude equations (4.224) are free of non-potential quadratic terms, and it is known from results of Rabinovitch and Tsimring [311] that stable motionless PHDs (such as that of Fig. 4.51) are possible in this case (note that with the absorbing walls considered here, a weak attraction of the defects to the walls exists, likely to result in the disappearance of the defects after a sufficiently long time [353]).

A PHD is a bound state of two dislocations in two different roll subsystems (sublattices). Indeed, examining Fig. 4.51, it is seen that two of the constitutive amplitudes (a_2 and a_3) vanish locally, at the point where their real and imaginary parts are zero (intersection of full and dashed thick lines). The last amplitude a_1 is free of dislocation, although a variation of the amplitude occurs at the location of the defect (actually, as two of the amplitudes vanish there, the pattern locally resembles a nucleus of a roll pattern, for which the steady amplitude is different from that corresponding to the hexagonal pattern, i.e. far from the defect core).

If the phase of each amplitude is denoted by $\varphi_j, j = 1, 2, 3$, and if C denotes an arbitrary contour encircling the defect, the quantity

$$Q_j = \frac{1}{2\pi} \oint_C \vec{\nabla}\varphi_j . \vec{dl}, \quad j = 1, 2, 3. \tag{4.225}$$

is zero for $j = 1$ (the phase φ_1 is analytic), while it is equal to $+1$ or -1 for a_2 and a_3. Thus, the PHD is characterized by three different numbers Q_j, called topological charges. In the case of Fig. 4.51, the defect will be denoted by $(0, +1, -1)$.

Note that by suitably selecting the initial conditions, it is possible to create a $(0, 0, +1)$ defect for instance (one dislocation in a_3). This type of defect is not motionless in the case of hexagonal patterns, as shown in [311]. In this case, the synchronization of the pattern ($\varphi = \varphi_1 + \varphi_2 + \varphi_3 = 0$) induced by quadratic terms is not possible everywhere, and a thin region where $\varphi = \pi$ remains, originating at the core of the defect. For $\tilde{K} = 0$, the system of equation is potential, and this energetically unfavorable situation (the region where $\varphi = \pi$ corresponds to down-hexagons) is eliminated by a relatively fast motion of the defect in the direction of this "corridor" [311, 313, 314].

When two dislocations (in different sublattices) are initially located a certain distance apart from each other, the evolution depends on their relative topological charges. If their charges

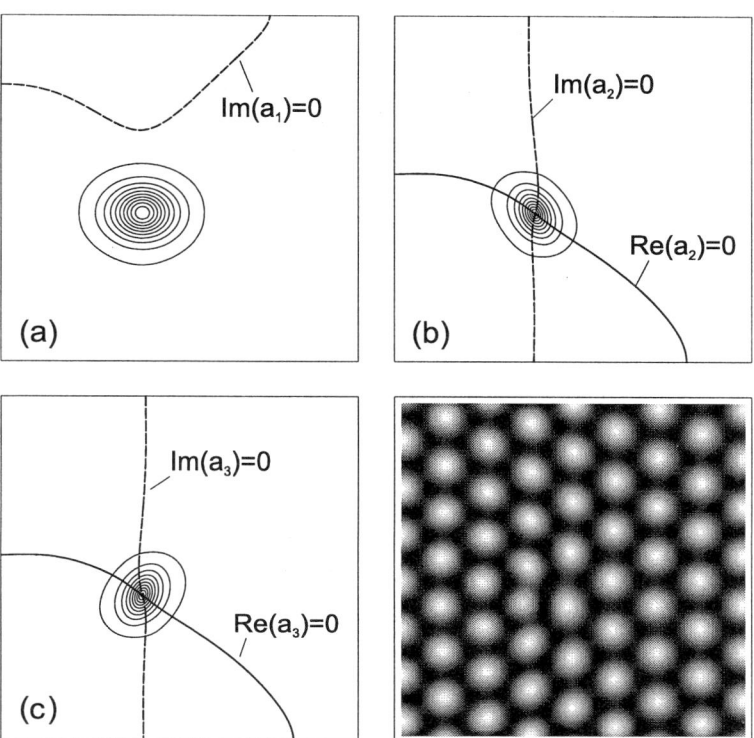

Figure 4.51: Stable quasi-motionless pentagon-heptagon defect. Plots (a), (b) and (c) represent the level lines of $|a_1|$, $|a_2|$ and $|a_3|$ respectively, together with the lines where the corresponding real parts (full lines) and imaginary parts (dashed lines) vanish. The last plot shows the full pattern reconstructed from the first-order temperature field (4.200). A slight deformation of the defect exists due to attraction by the walls.

Q_j have opposite signs, a region where $\varphi = \pi$ is created joining the two defects, therefore promoting their motion towards each other. When the two dislocations meet, the system cannot reduce the total energy anymore (the phases are synchronized everywhere except at the defect), and a stable PHD is created for which the two defects are bound a small distance apart from each other. When the two dislocations have the same charge, they mutually repel each other, up to their possible absorption by lateral walls. Finally, when the two dislocations are in the same sublattice, a similar evolution occurs. When they have opposite topological charge however, their motion towards each other is followed by annihilation of the two dislocations, thereby leaving a perfect hexagonal planform.

This picture is complicated by the presence of non-variational effects introduced by quadratic derivative terms ($\tilde{K} \neq 0$). In particular, the PHDs are generally no longer motionless, but acquire a velocity parallel to the wavevector $\vec{n}_j$ of the dislocation-free sublattice. A typical moving defect [corresponding to $(0, +1, -1)$] is represented in Fig. 4.52, where it is seen that the shape of the defect is now essentially asymmetric.

This behavior is similar to the influence of the phase gradients studied by Tsimring [363]. Similarly to the behavior of dislocations predicted in roll structures [313, 314] for variational systems, Tsimring has shown that a PHD is motionless in an optimal hexagonal pattern (i.e. when phase gradients vanish in the far-field of the defect). On the contrary, when such gradients exist (e.g. imposed externally), a constant velocity motion of the defect occurs, whose velocity and direction are determined by the phase gradients of the modes containing singularities, and by a certain mobility tensor [363]. Actually, extending some of the calculations of Tsimring, it can be seen that the quadratic non-variational terms can also be expected to produce a contribution to the force promoting defect motion, even in the absence of background phase gradients (i.e. even in an optimal structure). This is the situation considered in the following. Note however that the velocity of the defect is here found to depend on details of the core structure and cannot be calculated analytically.

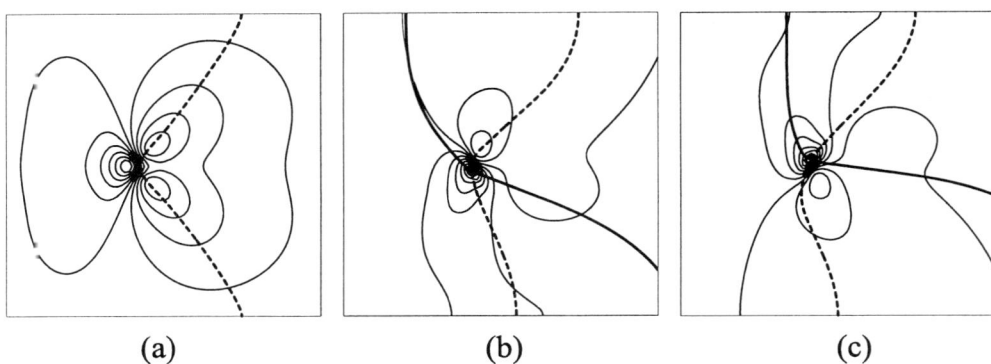

(a)   (b)   (c)

Figure 4.52: Stable moving $(0, 1, -1)$ defect for $\mu = 20$, $\tilde{\nu} = 1.118$, $\tilde{K} = 0.8425$. The PHD moves horizontally to the left. Plots (a), (b), (c) correspond to amplitudes a_1, a_2, and a_3 respectively, for which contour lines of the moduli are represented, together with lines where their real and imaginary parts vanish (thicker full and dashed curves respectively).

In Fig. 4.53, the velocity of the defect as a function of the parameter μ is depicted. It is seen that the velocity of the defect increases with μ. The scatter of the numerically obtained points is likely to result from the effect of lateral walls (the size of the box is not the same for all points), and possibly also from the fact that the velocity might not be completely stationary for each simulation. However, the linear tendency is rather clear, as well as the discontinuity occurring at $\mu \simeq 45$ (see below). A typical convergence test is represented in 4.54. Motion of defects has also been observed in different directions. For example, a PHD with charges $(+1, 0, -1)$ moves in the direction $\vec{n}_2$ (see Fig. 4.46), while a PHD with charges $(+1, -1, 0)$ moves with equal velocity (at a given μ) in the direction $\vec{n}_3$.

When the control parameter μ is increased past a critical value $\mu_c \simeq 45$, the moving PHD undergoes a nonlinear instability and splits into two new PHDs with different orientations. These PHDs then acquire motion in their respective directions, i.e. 60° apart from the direction of motion of the initial PHD. This basic process is represented in Fig. 4.55, for $\mu = 60$ in

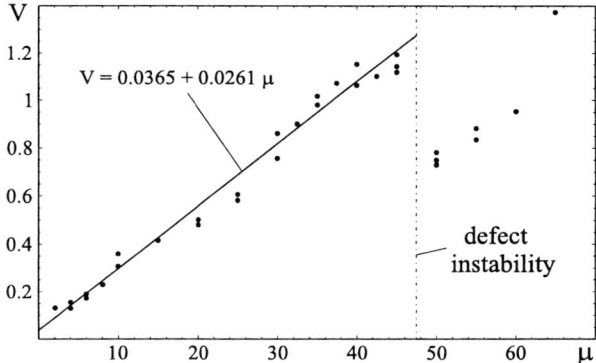

Figure 4.53: Velocity of the penta-hepta defect as a function of μ, with $\tilde{\nu} = 1.118$ and $\tilde{K} = 0.8425$).

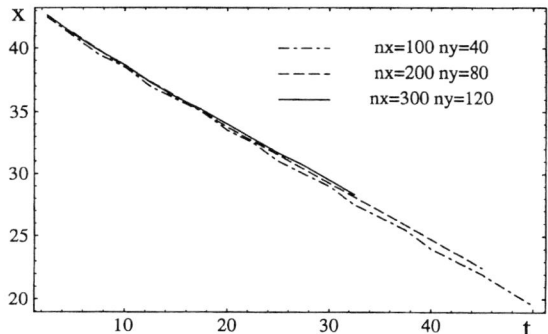

Figure 4.54: Convergence test (horizontal position of the defect versus time) in a box with dimensions $L_x = 50$, $L_y = 20$, for $\mu = 20$, $\tilde{\nu} = 1.118$, $\tilde{K} = 0.8425$, and different resolutions n_x, n_y. The velocity is here 0.46.

a box with lateral extension $L_x = L_y = 30$ (the evolution was convergence-tested up to $n_x = n_y = 300$, i.e. 10 nodes per unit length). The same representation is used as in Figs 4.51 and 4.52.

Figure 4.55 indeed shows that at time $t = 4$ (second line), two dislocations with charges $+1$ and -1 have appeared in the first sublattice (a_1). Note that the total charge is conserved during this event. In the absence of non-potential effects (or for lower values of μ), it is likely that such a situation would be eliminated by recombination of these two dislocations. However, in the situation considered here, the two newly created dislocations each recombine with the single dislocations already present in sublattices 2 and 3 to form two defects with different orientation. It appears that the forces promoting defect motion are strong enough to separate the two newly created PHDs, which may eventually become similar to the initial one

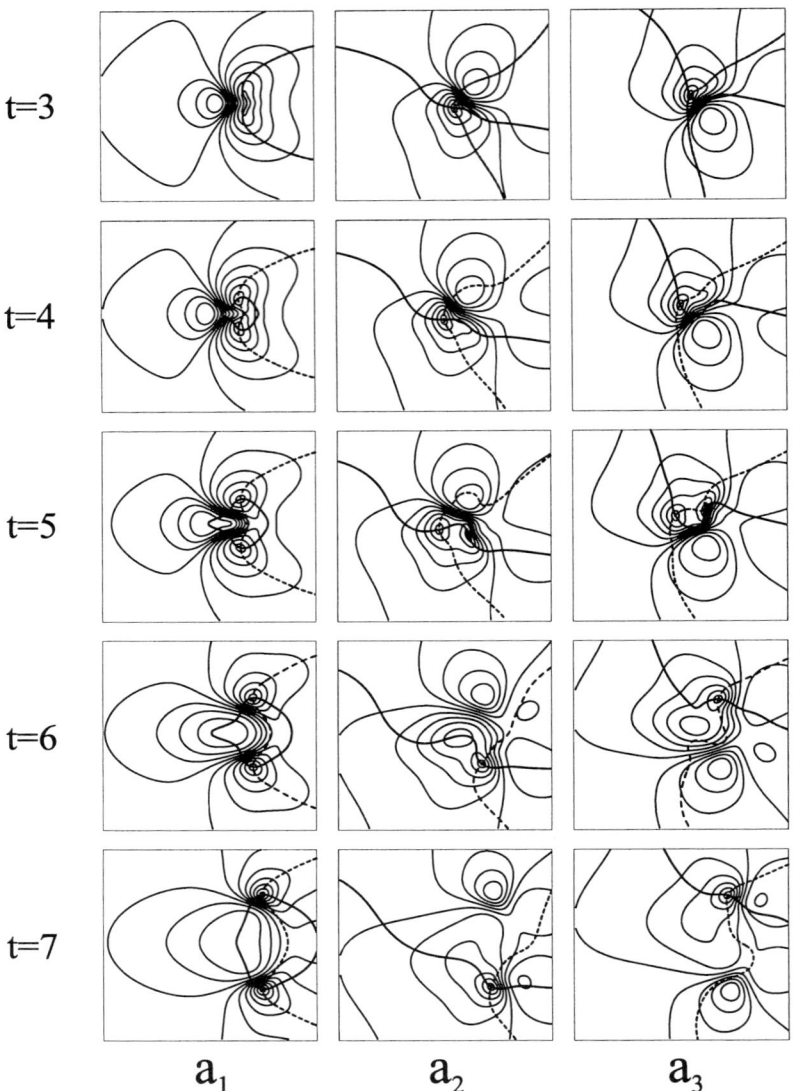

$$t=3$$

$$t=4$$

$$t=5$$

$$t=6$$

$$t=7$$

$$a_1 \qquad a_2 \qquad a_3$$

Figure 4.55: Evolution of an unstable penta-hepta defect $(0, +1, -1)$. The simulation is performed in a left-moving reference frame (velocity = 1.4) with $\mu = 60$, $L_x = L_y = 30$, $n_x = n_y = 300$. Only a subregion near the defect core is represented. The different rows correspond to times $t = 3, 4, 5, 6$ and 7 (from top to bottom). For each time, the 3 plots correspond to amplitudes a_1, a_2 and a_3, where the same representation is used as in Figs 4.51 and 4.52. During the process, the $(0, +1, -1)$ initial defect splits into two new penta-hepta defects with charges $(+1, 0, -1)$ and $(-1, +1, 0)$, moving in different directions.

after some time (see the last row, corresponding to $t = 7$).

This individual event is then able to repeat, as the new PHDs actually undergo a similar evolution, leading to four defects, which may then split again, and so on ... Hence, this eventually leads to a *multiplication of defects*, and to a non-trivial modification of the whole hexagonal pattern. The evolution of the system in the long-run is displayed in Fig. 4.56, and is seen to lead to quite complicated dynamics. It would be interesting to determine if the asymptotic state of the evolution ($t \to \infty$) remains time-dependent (even in a finite domain), which would then correspond to some form of spatio-temporal chaos (possibly similar in some respects to defect-mediated turbulence observed in other contexts, such as spiral defect chaos in reaction-diffusion systems and in the complex Ginzburg–Landau equation [366, 367], or in Rayleigh–Bénard convection in fluids with low Prandtl number [368]).

Such a question is presently unanswered however, as it appears that the lateral walls of the domain of integration always play an important role in the long-run evolution of the system. In a fixed reference frame (see Fig. 4.56), even though the number of defects could eventually increase up to about 26 PHDs (for a box size 100×100), the final stages appeared to be only weakly time-dependent, because most of the defects ultimately moved towards lateral walls and were absorbed there. Recombination of defects by pairs was also occasionally observed.

It is worth emphasizing the formation of strong longitudinal phase gradients (alternating full and dashed lines orthogonal to the directions of basic wavevectors) in the last plots of Fig. 4.56. These gradients correspond to a shift of the wavenumber of the pattern induced by the process of multiplicating defects (wavenumber selection). On the background of these phase gradients, it appears that the motion of the PHDs is significantly slowed down. For example, the single PHD remaining at time 150 still moves to the left, but with reduced velocity, and does not produce new dislocations until it reaches the region where phase gradients are lower (near the left wall). There, it re-accelerates and eventually split into other PHDs.

We also mention that the above results are not modified (at least up to $\mu = 60$) when including six amplitudes (i.e. another resonant triad of modes rotated by $\pi/6$ with respect to the original triad), hence allowing for hexagons/squares competition. Even though squares are stable with respect to rolls (which is indeed frequent for long-wave models like Eq. (4.196) [214, 215]), and may be locally stable to hexagons above a certain μ, hexagons have been found to remain locally stable to squares everywhere inside the corresponding Busse balloon given in Fig. 4.48, and limited by the Eckhaus instability boundaries (the cross-roll instability in fact appears within the Eckhaus-unstable region). To further validate the multiplication effect described above, some simulations were also performed including numerical noise (added at each time step), and indeed no significant effect was observed even for the evolution described in Fig. 4.55 (at $\mu = 60$). However, at larger μ, some simulations showed that the core of the PHDs (actually a nucleus of a roll pattern) can become unstable to squares, which leads to complicated dynamics.

To conclude this section and illustrate the generality of the multiplication effect described above, let us briefly consider direct numerical simulations of a non-variational model of pattern formation, namely the damped Kuramoto–Sivashinsky equation

$$u_t = \epsilon u - (1 + \Delta)^2 u - (\vec{\nabla} u)^2 \tag{4.226}$$

which is representative of several pattern-forming systems (e.g. combustion [211], solidification [213]), in addition to displaying rich dynamics for sufficiently large ϵ [496, 499].

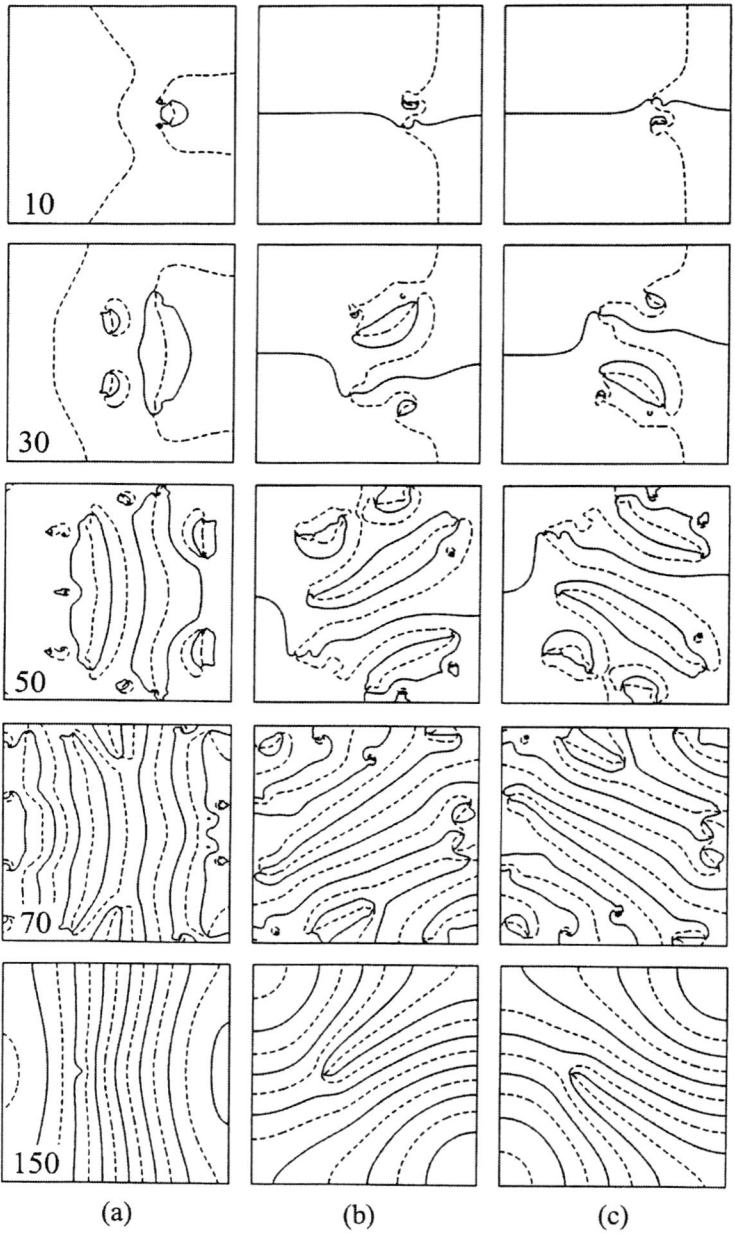

(a)   (b)   (c)

Figure 4.56: Evolution of the amplitudes a_1, a_2, and a_3 (columns a, b and c, respectively) for $\mu = 70$, in a fixed reference frame within a container with no-flux boundary conditions and dimensions $L_x = L_y = 100$ ($n_x = n_y = 300$). Only the lines where the real and the imaginary parts vanish are represented, respectively as full and dashed lines. Time is indicated for each group of 3 plots.

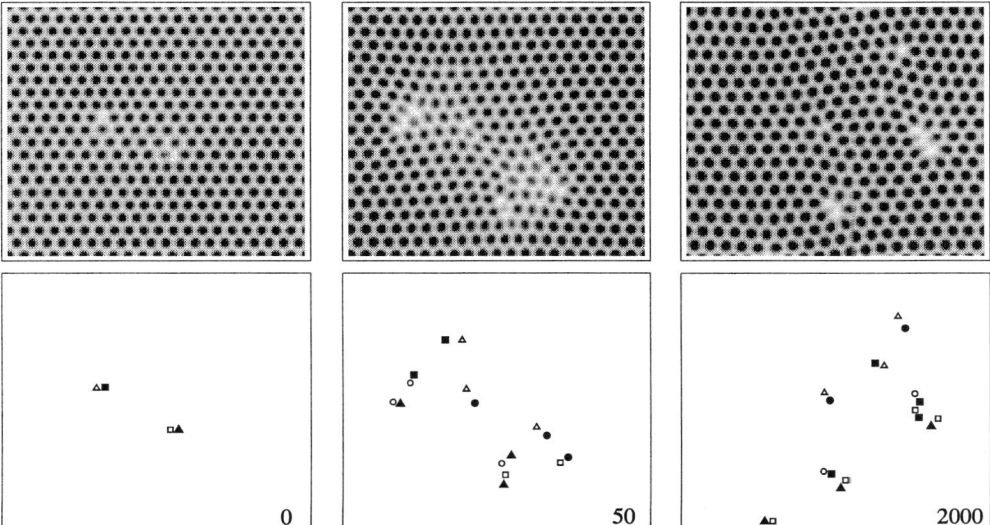

Figure 4.57: Direct 128×128 FFT-simulation of the damped KS equation, initiated by a hexagonal pattern with wavenumber $k = 10/9$ and two PHDs. Top: pattern evolution; Bottom: corresponding positions of dislocations calculated by a demodulation technique (circles, squares and triangles correspond to the sublattices in the 3 different directions, open and filled symbols to +1 and -1 topological charges). Time is indicated in the lower right corner.

Figure 4.57 displays some snapshots of a direct spectral Fast-Fourier-Transform simulation of Eq. (4.226), at $\epsilon = 0.1$ (i.e. before onset of spatio-temporal chaos), starting from a pattern with wavenumber $k = 10/9$ containing two PHDs (a single PHD would not satisfy periodic boundary conditions). Note that in contrast with amplitude equations, models such as (4.226) do capture non-adiabatic pinning effects, i.e. interaction of defects and fronts with the underlying small-scale structure [376, 377, 378].

While a pattern with $k = 1$ is found to be time-independent for the same value of ϵ (the motion of PHDs being prevented, probably due to strong pinning effects), a fast process of multiplication of PHDs is observed here, evolving to a quasi-steady structure with wavenumber closer to $k = 1$, and containing several defects still pinned by the hexagonal structure. A similar evolution occurs for $k < 1$. Hence, the initial wavenumber of the pattern also plays an important role, and even though the system is initially within the Busse balloon corresponding to perfect patterns, the presence of PHDs indeed allows re-adjustment of the global pattern wavelength. Note finally the bound states of several PHDs, remaining in the last picture at time $t = 2000$, and which seem not to evolve further over such a time scale.

Dynamics of the hexagons-squares transition

For other values of parameters than those considered above (e.g. using values of coefficients calculated for $Pr \to \infty$), it appeared from further simulations of the amplitude equations

(4.207), that at sufficiently large μ, the two dislocations forming the PHD can separate each other, therefore leading to a pair of non-bound (or free) dislocations. Once again, this behavior has to be attributed to non-potential quadratic terms, which appear to induce rather strong forces, acting against the energetic tendency of the system to form bound states of dislocations.

When considering a system of six amplitude equations, as already described above, the growing zone in between the two dislocations (where the phases are not synchronized) may be contaminated by a mode with wavevector orthogonal to the wavevector of the defect-free sublattice. In this region, whose growth is strongly anisotropic, squares are observed (the two sublattices containing dislocations acquire a very small amplitude), quite similarly to some of the behaviors reported by Eckert et al. [30, 33], and Thiele and Eckert [32]. In the latter work, Thiele and Eckert employ the stochastic geometry of polygonal networks, and provide a detailed study of the transition. They explain that a PHD found to be stable at low ϵ can indeed be a source of pentagon lines and squares at larger ϵ, via a process sketched in Fig. 4.58.

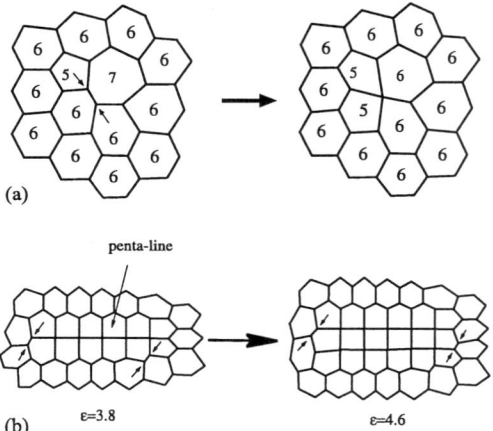

Figure 4.58: (a) Transformation of a PHD (left) into a 5-5-6-6 cluster (right) having a four-fold vertex at the center. (b) Transformation of a pentaline (left) into a patch of square cells with pentagonal edges (right). Schematic after shadowgraph images of the experiment. The arrows in (a) and (b) indicate vanishing sides. From [32]. Courtesy of U. Thiele.

In a first stage, the length of one of the sides of the heptagon forming the PHD decreases, and finally vanishes, leading to a 5-5-6-6 cluster (two pentagons and two hexagons). Then, other sides of the hexagons (and pentagons) again shrink to vanish, therefore increasing the number of four-fold vertices, and leading to patterns containing an increasing number of pentagons and squares. The reader is also referred to more recent works of Schatz et al. [31] and Eckert and Thess [33], for further details about the hexagons-squares transition. In the latter, Eckert and Thess calculate the position of dislocations using a demodulation technique, and indeed confirm the effect of separation of the two dislocations predicted by the amplitude equations. The underlying mechanism is also shown to consist in a combination of gliding and climbing motion of the two dislocations (see Fig. 4.11).

Still, amplitude equations considered throughout this section might not be completely appropriate to study this hexagon-squares transition, because of the absence of transversal diffusion. Indeed, remember that due to the explicit assumption of an isotropic scaling (see §4.5.1), the transversal derivatives in Newell–Whitehead–Segel operators have been dropped. While this assumption should be adequate for nearly-saturated hexagonal patterns, it has already been stressed that it might break down in the vicinity of defects or fronts in the hexagonal structures, and in zones (such as within islands of squares) where some of the modes forming resonant triads have small or vanishing amplitude.

As a conclusion, although this effect should be studied more systematically, non-potential amplitude equations (4.207) seem at least to indicate a tendency for dislocations to separate rather than to form bound states (depending on values of B and K), a possibility which was never described in earlier studies of modulated hexagonal patterns, and is in agreement with experiments. We also note in this respect that Afenchenko et al. [381] have also recently observed experimentally that the two dislocations forming a PHD are in general separated by a small distance (of the order of the basic wavelength), contrary to the result obtained in earlier studies of potential amplitude equations [312].

Clearly, another limitation of amplitude equations (potential or non-potential) is that they do not capture non-adiabatic pinning effects [376, 377, 378], i.e. the interaction of dislocations and fronts with the underlying periodic structure. Their role might not be negligible in some of the situations considered here. In addition to being able to stop the propagation of fronts and therefore to stabilize stripes of squares with pentagonal edges within hexagonal structures, they might also lead to bound states of several PHDs, such as seen in Fig. 4.57.

4.5.5 Rotationally invariant formulations

As shown in the previous sections, amplitude equation theories can be successfully applied to analyze competitions between patterns such as rolls, squares, hexagons, ... Moreover, typical defects of these structures may also be described, when including slow spatial modulations. Still, not all properties of real patterns may be accounted for properly by amplitude equations, because of their lack of rotational invariance. As amplitude equation theories imply the arbitrary selection of a certain set of "active" modes on the critical circle, the continuous rotational invariance of the original system of equations and boundary conditions is broken in a way discussed in §4.2.1, apart for some discrete values of the rotation angle. This appears to remain a major drawback of theories based on asymptotic expansions near the threshold of instability, and at present no amplitude equation formalism allows rigorous handling of this problem [19, 51, 419].

Important progress has been achieved in this direction by Cross and Newell [386, 387] (see also [19, 51]), within the formalism of phase equations. Even far from the threshold of instability, the phase of a convective structure (say, rolls) remains a slow variable compared with the relaxation time of the amplitude (due to translational invariance). Given a certain steady two-dimensional periodic structure $u(x, z) = u(x + 2\pi, z)$, one may define a generalized phase $\varphi(\vec{r}, t)$ and consider that the three-dimensional imperfect structure is described by $u(\varphi(\vec{r}, t), z)$ at the lowest order (the smallness parameter here is linked to the aspect ratio of the system, or to the typical size of large roll domains observed after sufficient time). It is then possible to obtain an equation for $\varphi(\vec{r}, t)$, or more precisely for the local wavevector $\vec{k} = \vec{\nabla}\varphi(\vec{r}, t)$, which is defined everywhere except at the core of the defects. It is beyond our scope here to describe the structure of such generalized phase equations, whose derivation can be found in the review of Cross and Hohenberg [19] and the book of Manneville [51].

Rather, we will focus hereafter on other more recent approaches, leading to rotationally-invariant *models* of pattern formation. For this purpose, we will first discuss some limitations of the Newell–Whitehead–Segel (NWS) equation (4.29), which we rewrite here in the form

$$\frac{\partial A}{\partial t} = \varepsilon A - A|A|^2 + \mathcal{D}_{\vec{r}}^2 A \tag{4.227}$$

where $\mathcal{D}_{\vec{r}}$ is the NWS operator

$$\mathcal{D}_{\vec{r}} = \partial_x - \frac{i}{2k_c}\partial_{yy}^2 \tag{4.228}$$

Note that the space variables have not been rescaled here, i.e. $\vec{r} = (x, y)$ is the usual horizontal position vector. Given a solution of this equation, say $A_0(\vec{r}, t)$, the planform function may be written

$$\phi_0(\vec{r}, t) = A_0(\vec{r}, t) \exp[ik_c x] + c.c. \tag{4.229}$$

Denoting the rotation operator with angle θ by R_θ and a rotated system of axes by $\vec{r}' = (x', y') = R_\theta \vec{r}$, the isotropy in horizontal directions implies that the rotated pattern

$$\phi(\vec{r}, t) = \phi_0(\vec{r}', t) = A_0(\vec{r}', t) \exp[ik_c x'] + c.c. \tag{4.230}$$

should be an equivalent solution of the original system, for any θ. Thus, the new amplitude

$$A(\vec{r}, t) = A_0(\vec{r}', t) \exp[ik_c(x' - x)] \tag{4.231}$$

should be a solution of Eq. (4.227). Taking into account that A_0 satisfies Eq. (4.227), the condition reads

$$\exp[-ik_c(x' - x)]\mathcal{D}_{\vec{r}}^2 \left\{ \exp[ik_c(x' - x)]A_0(\vec{r}', t) \right\} = \mathcal{D}_{\vec{r}'}^2 A_0(\vec{r}', t) \tag{4.232}$$

Though this relation is clearly not satisfied in general, it can be seen that it holds in the asymptotic sense $\varepsilon \to 0$, for sufficiently small rotation angles $\theta \sim \varepsilon^{1/4}$, i.e. just in the full range of transversal modulations allowed by the NWS equation. For instance, for a phase-winded solution of the form $A_0(x) = (\varepsilon - q^2)^{1/2} \exp(iqx)$, using $x' = x\cos\theta - y\sin\theta$ and taking into account the scaling $q \sim \varepsilon^{1/2}$, it can be calculated that the residue of the condition (4.232) is of the order ε^2, i.e. one order higher than the order $\varepsilon^{3/2}$ at which the NWS equation is obtained [297, 298].

Generalized Newell–Whitehead–Segel equations

In order to restore the full rotational symmetry, Gunaratne et al. [140, 142] have proposed to complete the NWS operator (4.228) by a second longitudinal derivative, i.e. the amplitude $A(\vec{r}, t)$ is assumed to satisfy

$$\frac{\partial A}{\partial t} = \varepsilon A - A|A|^2 + 4k_c^2 \square_{\vec{r}}^2 A \tag{4.233}$$

where

$$\Box_{\vec{r}} = \partial_x - \frac{i}{2k_c}(\partial^2_{xx} + \partial^2_{yy}) \tag{4.234}$$

Replacing $\mathcal{D}_{\vec{r}}$ by $\Box_{\vec{r}}$ in the condition (4.232), again for a solution of the form $A_0(x) = (\varepsilon - q^2)^{1/2} \exp(iqx)$, indeed shows that the latter is now identically satisfied for *every* rotation angle θ.

The term ∂^2_{xx} added to the NWS operator actually appears at the next orders in the perturbation scheme involving a scaling $\partial_x \sim \varepsilon^{1/2}$ of the longitudinal coordinate. When $\varepsilon \to 0$ *and for slow modulations* $\partial_x \sim q \sim \varepsilon^{1/2}$, the NWS equation is thus recovered. Gunaratne et al. have shown on a number of examples inspired from the Swift–Hohenberg equation (4.46), that spatial derivative terms appearing in NWS-like equations (eventually written for the hexagons-rolls competition [140, 142]) are always completed by higher-order terms such that only combinations of the kind (4.234) remain, i.e. full rotational symmetry is restored. Although not all terms in Eq. (4.234) are of the same order for $\varepsilon \to 0$, Gunaratne et al. [140, 142] have argued that it might be worth ignoring this scaling condition. given that an important symmetry of the basic system of equations can be fully restored at this cost.

Actually, the linear part of Eq. (4.233) can be readily obtained [143] for systems described at the linear stage by a Swift–Hohenberg form

$$\partial_t \phi(\vec{r}, t) = \left[\varepsilon - (k_c^2 + \Delta)^2\right] \phi(\vec{r}, t) \tag{4.235}$$

as can be seen by replacing $\phi(\vec{r}, t)$ by $A(\vec{r}, t) \exp[ik_c x]$ in this equation. Hence the operator (4.234) appears as the remainder of the operator $k_c^2 + \Delta$ when the basic structure $\exp[ik_c x]$ is extracted. Note that the linear part of the Swift–Hohenberg equation is universal sufficiently near the instability threshold, as seen later in this section.

However, Eq. (4.233) does not represent a universal generalization of the classical NWS equation. Indeed, in order to benefit from the full rotational symmetry of Eq. (4.233), phase variations of order unity must be assumed[12], i.e. the space variables should not be assumed to be slow. Therefore, several other terms might appear in the amplitude equations [141], such as cubic terms $\Box_{\vec{r}} A|A|^2$, ... Their occurrence depends on the nature of the problem considered. Indeed, for a system described by the Swift–Hohenberg equation [i.e. Eq. (4.235) completed by a saturating term $-\phi^3$], Eq. (4.233) is valid [143], but additional higher-order terms will probably need to be added when considering systems described by more complex dynamics, such as hydrodynamic systems. Still, rotationally invariant models obtained in this way have an important advantage over the NWS equation, namely the possibility of accounting for large changes of orientation of the patterns, such as seen in most experiments.

Derivation of rotationally invariant order-parameter models

Alternative models, aiming to describe the nonlinear dynamics of real systems by generalizations of the Swift–Hohenberg equation (4.46), have been proposed recently by Bestehorn

---

[12]Rotating a perfect pattern $\exp[ikx]$ by an angle θ corresponds to multiplying the amplitude by a phase factor $\exp[ik(x' - x)]$ where $x' = x \cos\theta - y \sin\theta$, see Eq. (4.231).

and Friedrich [147, 148], and are inspired by the earlier works of Haken [6]. Contrary to amplitude equations, Swift–Hohenberg-like equations discussed below are not only rotationally invariant, but also allow other essential properties of real patterns to be captured, namely pinning effects. Considering a constant phase shift $A(\vec{r}, t) \rightarrow \exp[i\alpha]A(\vec{r}, t)$, it is seen from Eq. (4.229) for instance that this corresponds to a translation of the periodic structure in the direction orthogonal to roll axes. The NWS equation (4.227) is unaffected by such a shift, which shows that no interaction occurs between the envelope $A(\vec{r}, t)$ and the underlying small-scale structure. However, it is known from experiments or direct numerical simulations of thermohydrodynamic equations that defects (spatial variations of the amplitude) do interact with the periodic structure. For example, the propagation of fronts, or motion of dislocations may even be prevented at small enough velocities. Clearly, the so-called non-adiabatic effects [19, 376, 377, 378] are not captured by amplitude equations.

The question to be addressed in the present section is whether it is possible to obtain the Swift–Hohenberg equation, or one of its generalizations, starting from the full system of hydrodynamic equations and boundary conditions under study. We outline in the following the essential steps of the recent theory of Bestehorn and Friedrich [147, 148], applying their study to the case where a single monotonic mode with wavenumber k_c becomes unstable at $\varepsilon = 0$.

Our starting point is Eq. (4.135), here written for a continuum of Fourier modes in the horizontal plane, i.e. we also take into account Eq. (4.137). Given that the active modes all belong to the first unstable band (i.e. the most dangerous vertical structure, see Fig. 4.17), each active mode is entirely characterized by its horizontal wavevector $\vec{k}$. Then, the result can be rewritten as

$$\frac{\partial A_{\vec{k}}}{\partial t} = \sigma(k^2)A_{\vec{k}} + \int Z^2_{\vec{k}_1, \vec{k}} A_{\vec{k}_1} A_{\vec{k}-\vec{k}_1} d\vec{k}_1 + \iint Z^3_{\vec{k}_1, \vec{k}_2, \vec{k}} A_{\vec{k}_1} A_{\vec{k}_2} A_{\vec{k}-\vec{k}_1-\vec{k}_2} d\vec{k}_1 \, d\vec{k}_2 \tag{4.236}$$

In §4.2.3, it was shown how the discrete version of this equation can be derived from hydrodynamic equations near the threshold of instability. Bestehorn and Friedrich [147, 148] apply the same method to a continuum of wavevectors. In short, the unknown fields are first expanded in series of the eigenmodes of the linear eigenvalue problem, equations are projected on adjoint eigenfunctions, and non-critical modes are then eliminated using adiabatic slaving. It is actually the latter assumption which allows reduction of the full three-dimensional dynamics to the dynamics of the most dangerous vertical structures, hence to a two-dimensional dynamics in the horizontal plane. The nonlinear coefficients Z^2 and Z^3 are in general complicated functions of the details of the problem investigated, and may in most cases only be evaluated numerically. In general, their dependence on the control parameter, say ε, may also be taken into account.

Now, define an order parameter field by

$$\phi(\vec{r}, t) = \int A_{\vec{k}}(t) \exp[i\vec{k}.\vec{r}]d\vec{k} \tag{4.237}$$

which is representative of the organization of the pattern. Given Eq. (4.236), let us now look for an evolution equation for $\phi(\vec{r}, t)$, i.e. in physical space.

Consider first the linear terms. The growth rate $\sigma(k^2)$ is assumed to be real and to depend only on k^2 (in addition to ε). Near the threshold $\varepsilon = 0, k = k_c$, it may be expanded as

$$\sigma(k^2, \varepsilon) = \alpha\varepsilon - \beta(k^2 - k_c^2)^2 + h.o.t. \tag{4.238}$$

with positive constants α and β. It is then found by Fourier transforming Eq. (4.236) that the linear part of the equation for $\phi(\vec{r}, t)$ reads

$$\partial_t \phi(\vec{r}, t) = \alpha\varepsilon\phi(\vec{r}, t) - \beta(k_c^2 + \Delta)^2 \phi(\vec{r}, t) \tag{4.239}$$

i.e. the Swift–Hohenberg form (4.235), up to a rescaling of space and time variables. This form is thus universal for type I monotonic instabilities (see Fig. 4.1) sufficiently near the threshold.

In principle, an equation in real space may be obtained provided the nonlinear coefficients Z^2 and Z^3 can be Fourier-transformed. Then, we define

$$\tilde{Z}^2(\vec{r}_1, \vec{r}_2) = \iint Z^2_{\vec{k}_1, \vec{k}_2} \exp[i\vec{k}_1.\vec{r}_1] \exp[i\vec{k}_2.\vec{r}_2] d\vec{k}_1 d\vec{k}_2 \tag{4.240}$$

and a similar expression for Z^3. The quadratic term in Eq. (4.236) leads to the real space expression

$$\iint \tilde{Z}^2(\vec{r}_2 - \vec{r}_1, \vec{r} - \vec{r}_2)\phi(\vec{r}_1, t)\phi(\vec{r}_2, t)d\vec{r}_1 d\vec{r}_2 \tag{4.241}$$

i.e. a non-local convolution-like term. This also happens for cubic (and higher-order) terms. Unfortunately, these equations cannot be solved analytically, and are also particularly time-consuming when handled numerically.

Is it possible to reduce nonlinear terms in the equation for the order parameter $\phi(\vec{r}, t)$ to a local form, involving only $\phi(\vec{r}, t)$ and its gradients at $\vec{r}$? To answer this question, we have to consider the dependency of coefficients Z^2 and Z^3 upon the interacting wavevectors. First, note that due to symmetries in horizontal plane[13], $Z^2(\vec{k}_1, \vec{k})$ should only depend on the length of wavevectors $\vec{k}_1$ and $\vec{k}_2 = \vec{k} - \vec{k}_1$ and on the angle between them (see also §4.2.3). Equivalently, rather than the angle we can give the length of the last side $\vec{k}$ of the triangle (Fig. 4.59), so that

$$Z^2_{\vec{k}_1, \vec{k}} = Z^2(k_1^2, k^2, k_2^2) \tag{4.242}$$

Now, Eq. (4.242) may be expanded around k_c with respect to each of its arguments:

$$Z^2(k_1^2, k^2, k_2^2) = \sum_{p,q,r} A_{p,q,r}(k_c^2 - k_1^2)^p (k_c^2 - k^2)^q (k_c^2 - k_2^2)^r \tag{4.243}$$

Note that taking into account that the Fourier transform $A_{\vec{k}}$ is mostly concentrated in some neighborhood of the critical circle $k = k_c$, the lowest-order terms in Eq. (4.243) should

---

[13]Here we restrict the analysis to systems satisfying translational, rotational and reflectional (or mirror) invariances. The reader is referred to [147, 148] for a generalization to systems lacking reflectional invariance (like rotating systems), or cases where several order parameters need to be considered.

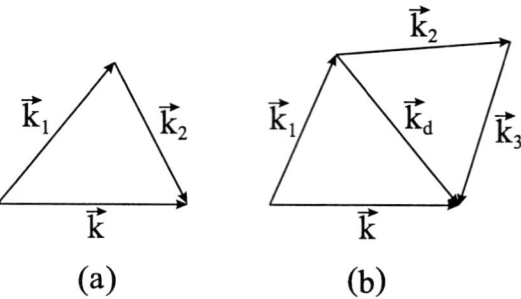

Figure 4.59: Quadratic (a) and cubic (b) interactions in Fourier space. (a) An arbitrary triangle is fixed by its three sides. (b) An arbitrary four-sided polygon needs its four sides plus the length of one diagonal.

usually dominate quadratic contributions in Eq. (4.236). However, formally we may retain an infinite number of terms in this expression. Indeed, note first that from Eq. (4.237), we obtain

$$(k_c^2 + \Delta)^p \phi(\vec{r}, t) = \int A_{\vec{k}}(t)(k_c^2 - k^2)^p \exp[i\vec{k}.\vec{r}]d\vec{k} \tag{4.244}$$

Then, inserting Eq. (4.243) in Eq. (4.236) and using Eq. (4.244), the quadratic contribution to $\partial_t \phi$ reads

$$\sum_{p,q,r} A_{p,q,r} \tilde{\Delta}^q \left[\tilde{\Delta}^p \phi \tilde{\Delta}^r \phi \right] \tag{4.245}$$

with $\tilde{\Delta} = k_c^2 + \Delta$.

Consider now the cubic term in Eq. (4.236). In view of Fig. 4.59, its dependency upon the interacting wavevectors should now be written

$$Z^3_{\vec{k}_1, \vec{k}_2, \vec{k}} = Z^3(k_1^2, k_2^2, k^2, k_3^2, k_d^2) \tag{4.246}$$

where $\vec{k}_3 = \vec{k} - \vec{k}_1 - \vec{k}_2$, and $\vec{k}_d = \vec{k} - \vec{k}_1 = \vec{k}_2 + \vec{k}_3$ is one of the diagonals of the four-sided polygon defined in Fig. 4.59. Taking into account that dominant contributions to the last term of Eq. (4.236) should correspond to values of k_1^2, k_2^2, k_3^2 and k^2 near k_c^2, we may again expand the expression (4.246) and possibly truncate the series at some order. However, such an assumption cannot be made for k_d^2, for which values within the range 0 to $4k_c^2$ should contribute to the integral. Bestehorn and Friedrich [147, 148] assume that the Taylor expansion of k_d^2 near 0 converges. Then, they expand

$$Z^3(k_1^2, k_2^2, k^2, k_3^2, k_d^2)$$
$$= \sum_{p,q,r,s,n} B_{n;p,q,r,s}(-1)^n(k_c^2 - k_1^2)^p(k_c^2 - k_2^2)^q(k_c^2 - k^2)^r(k_c^2 - k_3^2)^s k_d^{2n} \tag{4.247}$$

Transforming again to real space, and combining the result with Eqs (4.239) and (4.245), one finally gets

$$\partial_t \phi = \left[\alpha\varepsilon - \beta\tilde{\Delta}^2\right]\phi + \sum_{p,q,r} A_{p,q,r}\tilde{\Delta}^q\left[\tilde{\Delta}^p\phi\tilde{\Delta}^r\phi\right] + \\ \sum_{p,q,r,s,n} B_{n;p,q,r,s}\tilde{\Delta}^r\left[\tilde{\Delta}^p\phi\,\Delta^n\left(\tilde{\Delta}^q\phi\tilde{\Delta}^s\phi\right)\right] \tag{4.248}$$

where the linear part can also be made more complete, including other terms in the expansion (4.238). Equation (4.248) reduces to the modified Swift–Hohenberg equation (4.52) when considering only the lowest-order terms in the nonlinear interaction coefficients. For systems satisfying the $\phi \to -\phi$ symmetry, its quadratic term vanishes and the standard Swift–Hohenberg equation (4.46) is recovered.

Sufficiently near the threshold, the order parameter ϕ is dominated by Fourier modes with wavenumbers near k_c. Therefore, it can be expected, from Eq. (4.244), that $\tilde{\Delta}^p\phi$ will be much smaller than ϕ for all $p > 0$. This allows Eq. (4.248) to be approximated by

$$\partial_t \phi = \left[\alpha\varepsilon - \beta\tilde{\Delta}^2\right]\phi + \delta\phi^2 + \sum_n \gamma_n\phi\,\Delta^n\left(\phi^2\right) \tag{4.249}$$

with $\delta = A_{0,0,0}$ and $\gamma_n = B_{n;0,0,0,0}$. Interestingly, this model admits a Lyapunov functional

$$V = \int d\vec{r}\left[-\frac{\alpha\varepsilon}{2}\phi^2 + \frac{\beta}{2}(\tilde{\Delta}\phi)^2 - \frac{\delta}{3}\phi^3 - \frac{1}{4}\sum_n \gamma_n\phi^2\Delta^n(\phi^2)\right] \tag{4.250}$$

with $\partial_t\phi = -\delta V/\delta\phi$, thus implying a purely relaxational behavior, a result also predicted for near-critical perfect patterns (see §4.2.4).

Generalizations of the Swift–Hohenberg equation such as (4.248) or (4.249) have been used either as phenomenological models (see e.g. [51, 142, 378]), or as approximations of real Bénard systems [316, 317, 428]. In particular, the coefficients of nonlinear terms can be chosen on a case-by-case basis, in order to approximate the exact angular dependency of the coupling coefficients between plane waves, hence to arrange for a competition between rolls, squares, hexagons, ...

4.5.6 Low Prandtl number fluids, spiral patterns and inertial convection

Mean flow, vertical vorticity and spirals

The analysis of pattern selection presented in the last sections did not incorporate the effects of large-scale flows and vertical vorticity. As discussed in §4.3.1 and §4.3.2, this is valid for perfect patterns sufficiently near the threshold of monotonic instability, for which the horizontal flow field is purely potential. When defects exist in the convective structure, and when the Prandtl number is not too large, mean flows and vertical vorticity set up in a way described in this section, and generally play an important role in the overall pattern dynamics.

To discuss their influence, it is convenient to adopt a description based on modified Swift–Hohenberg equations [51, 144, 147, 322]. The approximation of real hydrodynamic systems by such rotationally invariant order parameter models was discussed in §4.5.5. We consider here that the temperature perturbation T and the vertical velocity W are given at the lowest order by

$$
\begin{pmatrix} W(\vec{r}, z, t) \\ T(\vec{r}, z, t) \end{pmatrix} = \begin{pmatrix} W_0(z) \\ T_0(z) \end{pmatrix} \varphi(\vec{r}, t) \tag{4.251}
$$

where $\varphi(\vec{r}, t)$ is the order parameter, and W_0 and T_0 are the neutral stability functions at the critical wavenumber k_c. The order parameter will be assumed to evolve according to a modified Swift–Hohenberg equation

$$
\partial_t \varphi = \alpha \varepsilon \varphi - \xi^4 (k_c^2 + \Delta)^2 \varphi - \delta \varphi^2 - \beta \varphi^3 \tag{4.252}
$$

where ε defines the supercriticality, and constants α, ξ, δ and β may in principle be chosen in order to approximate a particular convective problem. Other nonlinear terms involving the operator $k_c^2 + \Delta$ may be included, as described in §4.5.5.

However, we still need to modify Eq. (4.252) to incorporate the advection of the temperature field by the mean flow $\vec{V}_0(z, t)$ and vertical vorticity $\omega_z(\vec{r}, z, t)$, which need to be found from Eqs (4.160) and (4.162). Note that the expression (4.156) of the horizontal velocity also contains a potential part $\vec{\nabla}_h \phi$, whose effect is in principle already incorporated in the Swift–Hohenberg model (4.252). At moderate or large Prandtl number, this potential part actually represents the larger term in (4.156), the other terms being $O(Pr^{-1})$. Thus, the leading-order contribution to $\vec{V}_0(z, t)$ and $\omega_z(\vec{r}, z, t)$ may be found using $\vec{V} = W\vec{1}_z + \vec{\nabla}_h \phi$ in Eqs (4.160) and (4.162). Then, Eq. (4.160) can be rearranged as

$$
\Delta \omega_z = Pr^{-1} \frac{\partial \omega_z}{\partial t} + Pr^{-1} \vec{1}_z \cdot \left(\vec{\nabla}_h W \times D\vec{\nabla}_h \phi \right) \tag{4.253}
$$

Now, according to the continuity equation (4.158) and using Eq. (4.251), we have $\phi = DW_0(z)\chi(\vec{r}, t)$ where $\varphi + \Delta_h \chi = 0$. Thus, Eqs (4.253) and (4.162) can be rewritten as

$$
\Delta \omega_z = Pr^{-1} \frac{\partial \omega_z}{\partial t} + Pr^{-1} W_0(z) D^2 W_0(z) \vec{1}_z \cdot \left(\vec{\nabla}_h \chi \times \vec{\nabla}_h \Delta_h \chi \right) \tag{4.254}
$$

$$
D^2 \vec{V}_0 = Pr^{-1} \frac{\partial \vec{V}_0}{\partial t} - Pr^{-1} D \left[W_0(z) DW_0(z) \right] \overline{\Delta_h \chi \vec{\nabla}_h \chi} \tag{4.255}
$$

where the overbar denotes the horizontally averaged value. In principle, these equations can be solved using a Galerkin–Eckhaus method (see §4.2.3), i.e. expanding ω_z and $\vec{V}_0$ in series of eigenmodes of linear problems corresponding to Eqs (4.254) and (4.255), with appropriate boundary conditions. However, we will here consider a one-mode approximation, retaining only the vertical structure of ω_z and $\vec{V}_0$ corresponding to the most dangerous eigenvalue. Thus, referring to Eq. (4.159), we write $\omega_z = -f(z)\Delta_h \Psi(\vec{r}, t)$, where $f = 1$ for stress-free top and bottom plates, $f = \sin(\pi z)$ for rigid-rigid conditions, or $f = \sin(\pi z/2)$ for rigid bottom and free upper surface (even if subject to the Marangoni effect as the surface tension gradient affects only the potential part of the horizontal flow field). As the linear problem is self-adjoint, we project Eq. (4.254) on f, which leads to

$$\Delta_h \partial_t \Psi - Pr(\Delta_h - d^2)\Delta_h \Psi = a\vec{1}_z \cdot \left(\vec{\nabla}_h \varphi \times \vec{\nabla}_h \Delta_h \varphi\right) \tag{4.256}$$

where

$$a = \frac{\int_0^1 dz\, f W_0 D^2 W_0}{k_c^4 \int_0^1 dz\, f^2} \tag{4.257}$$

and d measures the viscous damping due to top and bottom walls ($d = 0$ for stress-free conditions, $d = \pi$ for rigid conditions, and $d = \pi/2$ for mixed rigid-free conditions).

Note that in Eq. (4.256), we also substituted $\chi \simeq \varphi/k_c^2$ which is indeed a solution of $\varphi + \Delta_h \chi = 0$ when the pattern is weakly modulated, i.e. the Fourier components of φ are concentrated near the circle of radius k_c. Note further that for perfect weak-amplitude patterns, $\Delta_h \varphi = -k_c^2 \varphi$, and the right-hand side of Eq. (4.256) vanishes, i.e. no vertical vorticity is generated. This is coherent with the discussion of §4.3.2.

For the mean flow, we equivalently set $\vec{V}_0 = f(z)\vec{v}_0(t)$, and the projected Eq. (4.255) reads

$$\partial_t \vec{v}_0 + d^2 Pr \,\vec{v}_0 = b \overline{\Delta_h \varphi \vec{\nabla}_h \varphi} \tag{4.258}$$

where

$$b = \frac{\int_0^1 dz\, f D(W_0 D W_0)}{k_c^4 \int_0^1 dz\, f^2} \tag{4.259}$$

and the mean flow also vanishes when $\Delta_h \varphi = -k_c^2 \varphi$ (see §4.3.1).

To close the problem, it remains to complete Eq. (4.252) by the advection of the structure induced by the non-potential part of the horizontal flow field. Considering only the temperature field, and according to Eqs (4.154), (4.156), and (4.251), we have to add a term $g\left[\vec{\nabla}_h \times (\Psi\vec{1}_z) + \vec{v}_0\right].\vec{\nabla}_h \varphi$ to the right-hand side of Eq. (4.252). Actually, the mean flow $\vec{v}_0$ will only transport the whole structure without distorting it, and can thus be eliminated working in a new reference frame $\vec{r}' = \vec{r} + g\int^t \vec{v}_0 dt$. Finally, the complete equation reads

$$\partial_t \varphi = \alpha\varepsilon\varphi - \xi^4(k_c^2 + \Delta)^2\varphi - \delta\varphi^2 - \beta\varphi^3 + g\vec{1}_z.(\vec{\nabla}_h \varphi \times \vec{\nabla}_h \Psi) \tag{4.260}$$

with $g = -\int_0^1 dz(\tilde{T}_0 T_0 f)/\int_0^1 dz(\tilde{T}_0 T_0)$, resulting from the projection on the adjoint temperature eigenfunction $\tilde{T}_0(z)$.

The equations (4.256) and (4.260) were first obtained by Manneville [51, 144]. Considering the case $\delta = 0$, Manneville has shown that at sufficiently low Pr, the vertical vorticity generated by defects leads to much longer transients than those observed for relaxational systems (e.g. at large Pr). In particular, turbulent bursts are generated by the nucleation of new dislocations in the roll structures, which also partly explains the permanent low-frequency regimes observed in low-Prandtl number weakly confined fluid layers [145, 146].

Later on, Bestehorn et al. [136, 147] showed using similar models that *spiral* and *target* patterns can be generated, as seen in Rayleigh–Bénard experiments [134, 135] with a layer of gaseous CO_2 ($Pr \simeq 0.9$), and in more recent direct numerical simulations of hydrodynamic equations [139]. As CO_2 shows some non-Boussinesq effects (the critical temperature difference is 29 °C in the experiments of Bodenschatz et al. [134, 135]), the first pattern observed near the threshold is hexagonal. This is also reproduced by Eq. (4.260) when $\delta \neq 0$. Above a certain value of the constraint, a transition to rolls is obtained, with little or no hysteresis [134]. In the roll state, and even during the transition, the rolls are curved and the system spontaneously forms stable rotating spirals, whose number of arms depends on ε. These features have been discussed by Bestehorn et al. [136], who also show that the lateral boundary conditions do have an influence on the spiral patterns. Rolls generally terminate perpendicularly to sidewalls when $\varphi = 0$ there, while they may arrange parallel to sidewalls if, say, some lateral heating is imposed, i.e. $\varphi > 0$. Eventually this effect leads to a unique n-armed spiral filling the entire circular container, as seen in experiments [135]. A further conclusion is that in the hexagonal state, the mean flow is weaker than for rolls, and the behavior is mostly relaxational.

Later experiments [137, 138] have shown that a transition to *spiral-defect chaos* occurs above a certain value of ε depending on the system size. Fig. 4.60 shows a snapshot of spiral defect chaos observed from direct simulations of the model equations (4.256) and (4.260), as also obtained by Bestehorn and Friedrich [147].

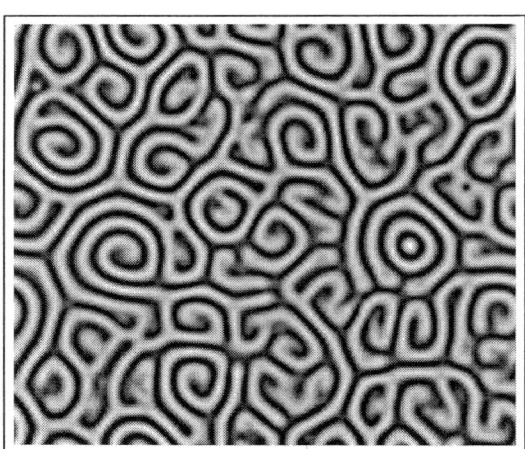

Figure 4.60: Spiral defect chaos observed by direct numerical resolution of model equations (4.256) and (4.260), in a rectangular domain with periodic boundary conditions. Spirals rotating in both directions form and annihilate in various parts of the domain. Occasionally, a target pattern is observed (such as in the right part of this figure), with a pulsating core emitting concentric waves.

Actually, for the simulation presented in Fig. 4.60, the Prandtl number has been assumed large enough for a dynamical slaving of Ψ to the dynamics of φ to occur, i.e. the time-derivative term is neglected in Eq. (4.256). Formally, a single equation remains for φ, after solving Eq. (4.256) with respect to Ψ (in Fourier space). The resulting equation for φ is therefore cubic, though the term linked with vertical vorticity is nonlocal, and leads to spiral formation, as studied by Bestehorn and Friedrich [147]. Note that these authors have also considered generalizations [147, 148] of Eq. (4.256), in the spirit of the method described in §4.5.5.

Le us conclude by mentioning that, to our knowledge, no experimental studies exist up to now on Marangoni–Bénard convection in low Prandtl number fluids and spatially extended systems. As a transition to squares rather than to rolls would generally be expected in this case (see §4.4.3), above a certain value of ε, it would certainly be interesting to examine whether large scale flows play an important role in the dynamics of the transition, and whether there is an analog of spiral-defect chaos for such systems. Already some answers to these questions are provided by Eckert et al. [30], whose direct numerical simulations of hydrodynamic equations have allowed the Prandtl number to be varied over a wide range. These authors have provided evidence on the role of the vertical vorticity in acting as a "lubricant" during the transition, favoring the motion of dislocations against pinning effects, and generally leading to more perfect square patterns. Note that this happens even at large (but not infinite) Prandtl numbers such as $Pr = 50$.

For low Prandtl numbers, the transition to squares is found to occur at lower ε (see also §4.4.3), and eventually squares may coexist with rolls (at $Pr = 0.5$ [30]). At these values of Pr, large scale flows are responsible for an intrinsic time-dependence in the form of oscillations between both types of cells. Note also that for very low Prandtl number fluids, some other types of time-dependence may be expected, which we now discuss.

Inertial Marangoni–Bénard convection and regularity of the zero-Prandtl limit

Boeck and Thess [203, 204] have recently studied surface-tension-driven Bénard convection in very low Prandtl number fluids such as liquid metals, both in two and three dimensions. We here describe some of their findings for two-dimensional laterally periodic domains with stress-free conducting bottom plate [203].

Just above the instability threshold (here, $Ma_c = 57.598$ and $k_c = 1.7003$ for zero free surface Biot number Bi, i.e. smaller than the usual value $Ma_c \simeq 80$ due to the no-slip condition at the bottom), Boeck and Thess find a supercritical regime of weak steady convection rolls, characterized by only slight deviations of isotherms from the basic conductive temperature profile. When increasing the Marangoni number, a transition is observed to another regime, which shows significantly deformed isotherms. This second regime corresponds to *inertial convection*, already encountered in Rayleigh–Bénard convection [205, 351], and corresponding to rolls acting like flywheels, i.e. the amount of energy dissipated per rotation of the rolls is much less than the total kinetic energy of the fluid motion.

To understand better the origin of such a transition, it is useful to come back to §3.2.2, and to discuss the regularity of the limit $Pr \to 0$, as done by Boeck and Thess. As the longest time scale is here the viscous one $\tau_{\text{visc}} = d^2/\nu$ (d is the liquid depth and ν the kinematic viscosity), it is natural to select it to write balance equations in dimensionless form. Note that we also use here a temperature scale $Pr\theta$, where θ is the imposed temperature difference between top and bottom, and a pressure scale $\rho\nu^2/d^2$ where ρ is the density. Then, Eqs (3.59) and (3.60) read, for perturbations,

$$\Delta \vec{V} - \vec{\nabla} p = \frac{\partial \vec{V}}{\partial t} + \left(\vec{V}.\vec{\nabla} \right) \vec{V} \tag{4.261}$$

$$\Delta T + W = Pr \left\{ \frac{\partial T}{\partial t} + \left(\vec{V}.\vec{\nabla} \right) T \right\} \tag{4.262}$$

where buoyancy has been neglected in Eq. (4.261). Note that the boundary conditions are unchanged using the viscous scaling, i.e. $W = D^2 W = T = 0$ at $z = 0$ and $W = DT + Bi\,T = D^2 W - Ma\Delta_h T = 0$ at $z = 1$.

Thus, for $Pr \simeq 0$, one would be tempted to neglect the right-hand side of Eq. (4.262), similarly to the usual neglect of the inertial terms in the Navier–Stokes equation for $Pr \to \infty$. Then, the temperature perturbation would adiabatically follow the dynamics of the velocity field, and the conductive profile would not be perturbed (at $Pr = 0$). From a computational point of view, such zero-Prandtl equations are much easier to simulate than the full equations with $Pr \simeq 0$, where the presence of two very different time scales usually makes time-stepping particularly slow. However, taking $Pr = 0$ in Eq. (4.262), it appears that only the weak convection regime found near the instability threshold is correctly represented. Note that even for this regime, the validity of this assumption decreases when Ma increases.

At larger Marangoni number ($Ma > Ma_i \simeq 73.4$ for a periodic box with dimensions $2\pi/k_c$), accurate (pseudo-spectral) numerical simulations of Eqs (4.261) and (4.262) with $Pr = 0$ and usual boundary conditions predict a catastrophic exponential growth of the kinetic energy with no tendency to nonlinear saturation, which finally leads to blow-up of the numerical code. During this unbounded growth, it has been observed by Boeck and Thess that the vorticity $\omega = \partial U/\partial z - \partial W/\partial x$ becomes nearly constant everywhere (*flywheel motion*), except near the boundaries. Combined with the incompressibility relation $\vec{\nabla}.\vec{V} = 0$, the constancy of ω implies that the nonlinear term in Eq. (4.261) is irrotational, and can thus be balanced by a pressure gradient. Thus, for $Pr = 0$, there is no nonlinear effect (except in some boundary layers) which could saturate the exponential growth of linear perturbations.

In reality, for small but finite Pr, i.e. when the thermal nonlinearity is re-incorporated in Eq. (4.262), the growth of perturbations observed above Ma_i does saturate to some value, though the corresponding velocity in this inertial regime is much higher because it has the thermal scale κ/d, where κ is the thermal diffusivity.

Boeck and Thess have also discussed the increase of the heat transport due to convection for both weak convection and inertial regimes. Actually, as also pointed out in §4.4.2, the classical definition of the Nusselt number (ratio of total to conductive heat flux) is not adequate for Marangoni–Bénard problems. Rather, the mean increase Δ of the surface temperature (measured in units of the basic temperature drop θ) appears as a more suitable measure of the convective heat transport. Similarly, Boeck and Thess use a modified Nusselt number Nu', defined as the ratio of the convective heat flux Q, to the heat flux Q_c that would occur with the fluid at rest, but with the free-surface temperature prescribed as in the convective state.

According to Eq. (4.187), the quantity Δ reads, in the viscous units used here

$$\Delta = \frac{Pr^2}{1 + Bi} \int_0^1 dz \overline{WT} \tag{4.263}$$

while $Q = \lambda\theta d^{-1}(1+Bi\Delta)$ and $Q_c = \lambda\theta d^{-1}(1-\Delta)$, such that the modified Nusselt number Nu' reads

$$Nu' = \frac{1+Bi\Delta}{1-\Delta} \simeq 1 + Pr^2 \int_0^1 dz\overline{WT} \tag{4.264}$$

Therefore, for weak convection, $W, T = O(1)$ at the lowest order in $Pr \to 0$, and both Δ and $Nu' - 1$ scale as Pr^2. In contrast, for inertial convection, it may be expected that $W, T \sim Pr^{-1}$, such that Δ and $Nu' - 1$ tend to a constant as Pr tends to zero. This fact was indeed observed by Boeck and Thess, and actually shows that the transition between weak and inertial convection becomes increasingly sharp as Pr is decreased. In fact, while the transition is found to be rather smooth at $Pr = 0.1$, it sharpens as Pr decreases, and may even be discontinuous (with little hysteresis) for $Pr < Pr_i$ with Pr_i between 0.01 and 0.02 [203].

At much larger Marangoni numbers, numerical simulations and physical arguments of Boeck and Thess show that a scaling regime is reached (up to $Ma = 40000$), where temperature and vorticity are nearly constant everywhere except within thin boundary layers near the top and bottom. Velocity V, typical temperature difference along the free surface ΔT_h, boundary-layer thickness δ and Nusselt number Nu are found to scale as $V \sim Ma^{2/3}$, $\Delta T_h, \delta \sim Ma^{-1/3}$, and $Nu \sim Ma^{-1/3}$, respectively. Note that these scalings apparently also hold for large Prandtl numbers fluids, as will be seen in §8.2 when considering large Marangoni number regimes.

Boeck and Thess have also studied the three-dimensional case, both for stress-free and realistic no-slip bottom boundary conditions [204]. At low Prandtl numbers, the first pattern at threshold consists in down-hexagons as expected (see §4.4.3 and [151, 350]). In the smallest possible periodic domain compatible with hexagonal cells (i.e. lateral dimensions are $L_x = 4\pi/k_c$ and $L_y = 4\pi/\sqrt{3}k_c$), they observe a variety of other patterns, some of them being time-dependent and eventually chaotic even at moderate Marangoni number such as $Ma = 120$ (stress-free case). However, the zero-Prandtl number equations do in this case provide a sufficient approximation of the behavior at finite but small Pr, as long as the system is in a three-dimensional state, and inertial convection is not observed for three-dimensional regimes. For a stress-free bottom wall, the flow always remain three-dimensional. For the no-slip case, two-dimensional rolls are observed in some interval of Marangoni numbers. Beyond a certain threshold, the flow becomes strongly intermittent because of the interplay between the flywheel effect and three-dimensional instabilities of the two-dimensional roll solutions (the exponentially growing roll solutions described above are shown to be unstable to three-dimensional disturbances).

4.6 Long-wave convection patterns

Significant progress in the understanding of instabilities and cellular patterns has been achieved via consideration of simplified nonlinear equations such as the Swift–Hohenberg model and its variants (see §4.5.5 and §4.5.6), or, as described in this section, long-wave order-parameter equations. Compared with the original system of three-dimensional thermo-hydrodynamic equations and boundary conditions, order-parameter equations may be written in the simplest case for a single scalar function of the horizontal spatial coordinates only, i.e. the vertical direction is eliminated. This can be achieved rigorously when the critical wavenumber k_c is vanishingly small, i.e. convection cells have a very large wavelength compared with the depth. As discussed below, this happens generally when particular types of boundary conditions are assumed at the top and bottom boundaries of the Bénard set-up. Despite this restriction, it turns out that the models obtained often describe, at least qualitatively, systems where the critical wavenumber is finite but not too large.

Our goal in this section is certainly not to achieve an exhaustive discussion of all possible cases treated in the literature, which of course extends much beyond the particular case of Bénard convection. Rather, we will explain the main steps in deriving such equations, and present some representative equations, which can also be used as starting models for developing amplitude equation analyzes.

4.6.1 Long-wave expansions, conservation laws and symmetries

Referring to Chapter 3, we note first that long-wave instabilities ($k_c = 0$) are typically encountered when the heat flux is held fixed at both top and bottom boundaries (or, equivalently, when the boundaries are heat-insulating, see §3.3.1). For multi-component mixtures, long-wave modes have been predicted when boundaries are impervious to species diffusion (see §3.7.1). Long-wave interface deformation modes encountered in §3.6 are another example where instability occurs first to vanishing wavenumbers, at least for sufficiently small depths of highly viscous liquids.

As already mentioned, such long-wave modes are in each case linked with a particular conservation law, and to its corresponding Goldstone mode. When both boundaries are heat-insulating (or impervious to mass diffusion), a uniform shift of the temperature (or concentration) is possible, and there is no force which acts in restoring the initial temperature (or concentration). Hence, the system is neutrally stable to such a uniform shift, and the eigenvalue problem should possess at least one zero eigenvalue. Note that for surface deformational modes, the corresponding Goldstone mode is a shift of the height of the interface, which is also neutrally stable as long as the shift is uniform.

Now, when the shift is non-uniform, i.e. it is slowly modulated along the horizontal coordinates, a slow dynamics can be expected, which may couple to convection flows and possibly lead to instability of the horizontally uniform state. Note that several Goldstone modes may exist simultaneously and be coupled. Remember also from §4.3 that when both boundaries are assumed stress-free, a uniform mean flow or vertical vorticity shift is possible, whose modulations play an important role especially at low Prandtl numbers (see §4.5.6).

Thus, denoting the control parameter by μ, in each case there exists an eigenvalue $\sigma(k, \mu)$ such that $\sigma(0, \mu) = 0$ for all μ. In this section, we further restrict attention to systems for which σ is real, and can be expanded as

$$\sigma(k, \mu) = k^2 \mu - k^4 + O(k^6) \tag{4.265}$$

where positive constants in front of the two terms have been set to unity without loss of generality (suitably rescaling time and μ). Systems obeying Eq. (4.265) belong to class II in the classification of Cross and Hohenberg (see Fig. 4.1). They may also be called "negative viscosity" systems [370]. Note that Eq. (4.265) is generally not valid when infinitely deep layers are considered [even though $\sigma(0, \mu) = 0$], and in this case σ might even not be analytical at $k = 0$ [370]. An example will be encountered in §8.1.

Now, there exist many systems in which the conservation law responsible for the Goldstone mode is only approximate. This happens for instance when top and/or bottom boundaries have a finite but small thermal conductivity, and even the homogeneous mode $k = 0$ is slightly dissipated. In such case, Eq. (4.265) can be completed by a small term $-\alpha$ with $\alpha > 0$.

The critical value of μ found from Eq. (4.265), complemented by a term $-\alpha$, is then $\mu = \alpha k^{-2} + k^2$, and has a minimum at $k^2 = k_c^2 = \alpha^{1/2}$ equal to $\mu_c = 2\alpha^{1/2}$ (see Fig. 4.61). For $\mu > 0$, the maximal growth rate occurs at $k_m^2 = \mu/2$ and its value is $\sigma_m = -\alpha + \mu^2/4$.

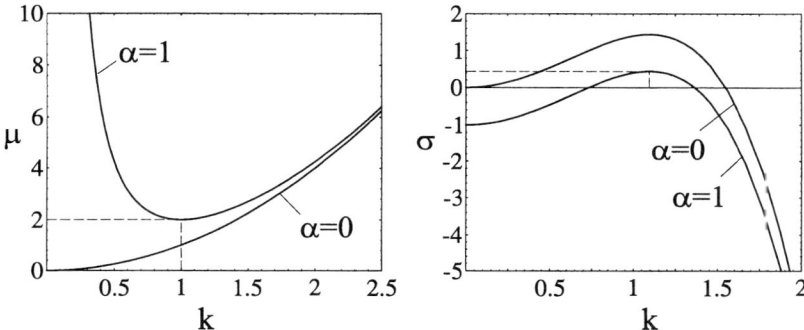

Figure 4.61: Linear stability results: (a) the neutral stability curve for $\alpha = 0, 1$; (b) the growth rate for $\mu = 2.4$ and $\alpha = 0, 1$.

If ϕ is the conserved (or quasi-conserved) scalar quantity, the order-parameter equation for ϕ may be written, consistently with Eq. (4.265) complemented by $-\alpha$, as

$$\partial_t \phi = -\alpha\phi - \mu\Delta\phi - \Delta^2\phi \qquad (4.266)$$

which is formally similar to the Swift–Hohenberg form, e.g. Eq. (4.235).

Above the instability threshold, Eq. (4.266) should be completed by nonlinear terms, whose form is restricted by symmetry considerations. A general discussion of one-dimensional long-wave order-parameter equations has been provided by Nepomnyashchy [370].

Possible nonlinear terms are also restricted by the scalings $\partial_t \sim \alpha \sim \mu\Delta \sim \Delta^2$, as all terms must be of the same order in Eq. (4.266). Therefore, if we assume $\mu \sim \varepsilon^2 \ll 1$, we have $\vec{\nabla} \sim \varepsilon$, and $\alpha \sim \partial_t \sim \varepsilon^4$. Note that no scaling is imposed yet on the amplitude of ϕ.

Given that terms such as ϕ^2 or ϕ^3 in general do not appear for Bénard systems[14], the lowest-order independent nonlinear terms respecting rotation, reflection and translation invariances are $(\vec{\nabla}\phi)^2$ and $\phi\Delta\phi$. If these are able to saturate the linear growth of ϕ, the scaling

---

[14]For instance, these terms may appear when spatially homogeneous perturbations of ϕ are not described by a linear damping $\partial_t \phi = -\alpha\phi$, e.g. when the law describing dissipation through top and bottom plates is nonlinear. The fact that Swift–Hohenberg models do contain such terms shows that they are not adapted to the description of large-scale modes far from the instability threshold. However, when the physical system studied is close to a phase transition between two homogenous states, Swift–Hohenberg equations may be used as a model. Indeed, the form

of the order-parameter is $\phi \sim \varepsilon^2$, and in this case, no other nonlinear term should be included. This is the case for instance for the damped Kuramoto–Sivashinsky equation

$$\partial_t \phi = -\alpha\phi - \mu\Delta\phi - \Delta^2\phi + (\vec{\nabla}\phi)^2 \qquad (4.267)$$

which has been used as a model to describe transitions from steady hexagonal patterns to oscillating and even chaotic structures (see e.g. [499]), and was originally derived in the context of front propagation [211, 213].

However, such terms do not appear in general for Bénard problems, unless some non-Boussinesq effects or surface deformation are considered (see below). For non-Boussinesq cases, but undeformable interfaces [215], they appear in the combination $\vec{\nabla}.(\phi\vec{\nabla}\phi) = (\vec{\nabla}\phi)^2 + \phi\Delta\phi$, and it appears that higher-order terms should be considered. By symmetry, next-order quadratic terms should contain at least four gradient operators, such that the balance with linear terms leads to $\phi \sim O(1)$. Hence, there is no longer any reason to consider quadratic terms before cubic ones, and there are many terms in general which are compatible with symmetry requirements. For a particular problem (see later), several of them may vanish, such that we will not attempt here to pursue the discussion in general terms, but rather focus on some particular cases. Such non-universality of long-wave order-parameter equations has also been pointed out by Nepomnyashchy [370].

For Bénard convection between poorly conducting undeformable boundaries, and negligible vertical vorticity effects, Knobloch [215] has shown that the general form of the order-parameter equation reads

$$\partial_t \phi = -\alpha\phi - \mu\Delta\phi - \Delta^2\phi + \kappa\vec{\nabla}.\left[(\vec{\nabla}\phi)^2\vec{\nabla}\phi\right] + \beta\vec{\nabla}.\left[\Delta\phi\vec{\nabla}\phi\right] +$$
$$\delta\Delta\left[(\vec{\nabla}\phi)^2\right] - \gamma\vec{\nabla}.\left[\phi\vec{\nabla}\phi\right] \qquad (4.268)$$

When $\beta \neq 0$, $\delta \neq 0$ and/or $\gamma \neq 0$, the reflection symmetry $\phi \to -\phi$, inherited from the reflection symmetry in the layer mid-plane, is broken. According to Knobloch [215], this is the case when top and bottom boundaries are not identical ($\beta \neq 0$, $\delta \neq 0$), or when non-Boussinesq effects are taken into account ($\gamma \neq 0$). Note that $\delta = \beta$ when the linear problem is self-adjoint, but $\delta \neq \beta$ otherwise. Knobloch has also provided a detailed analysis of pattern selection for $\alpha > 0$, $\kappa > 0$, as a function of coefficients β, γ and δ [215]. Note that coefficients of Landau equations derived from Eq. (4.268) near the instability threshold are provided in §4.5.2.

Shtilman and Sivashinsky [292] have then generalized Knobloch's equation to incorporate effects of the vertical vorticity, important at low Prandtl numbers.

4.6.2 Non-Boussinesq Rayleigh–Marangoni–Bénard convection

We now turn to a systematic derivation of the order-parameter equation from the thermohydrodynamic equations and boundary conditions, for a layer of infinite Prandtl number fluid lying on a rigid bottom plate and in contact with a gas. Both buoyancy and thermocapillarity

---

$\partial_t \phi = \varepsilon\phi - (1 + \Delta)^2\phi$ predicts an instability at $k = 0$ when $\varepsilon > 1$. This has been exploited [323, 398] to explain some particular patterns (coexistence between up and down hexagons, localized structures, ...) seen in experiments on Bénard convection near an equilibrium phase transition [395, 396, 397].

are considered. The interface is assumed to be undeformable, while both top and bottom boundaries are taken as poorly conducting. We will also consider viscosity variation with temperature, which is generally a dominant non-Boussinesq effect [268]. Specifically, we will consider that the dynamic viscosity, here denoted by η, can be linearized in the range of temperatures considered

$$\eta = \eta_0 \tilde{\eta}(T) = \eta_0 \left[1 + R_\eta (T - T_0)\right] \tag{4.269}$$

where η_0 is the dynamic viscosity at the temperature T_0 (taken as the temperature of the interface in the reference state). Both T and T_0 are taken dimensionless, by using as scale the temperature drop θ between the bottom and the free surface. $R_\eta = \theta \eta_0^{-1} (\partial \eta / \partial T)_0$ is a dimensionless viscosity-temperature coefficient.

Using the usual thermal scalings, Eqs (4.152–4.154) need to be modified to include viscosity variation with temperature, namely

$$\vec{\nabla}.\vec{V} = 0 \tag{4.270}$$

$$\vec{\nabla}.\left[\tilde{\eta}(T_{\text{ref}} + T)\, \mathbf{P}\right] - \vec{\nabla}p + RaT\vec{1}_z = 0 \tag{4.271}$$

$$\Delta T + W = \frac{\partial T}{\partial t} + (\vec{V}.\vec{\nabla})T \tag{4.272}$$

where $\vec{V} = v_i \vec{1}_i = \vec{V}_h + W\vec{1}_z$, T, p are respectively the velocity, temperature and pressure perturbations, $T_{\text{ref}} = 1 - z + T_0$ is the reference state temperature profile, and $\mathbf{P}$ is the dimensionless viscous stress tensor (see §2.1.2), whose components are $P_{i,j} = (\partial_j v_i + \partial_i v_j)$, according to Eq. (2.66). Note that $(\vec{\nabla}.\mathbf{P})_i = \partial_j P_{i,j}$. In Eq. (4.271), we have also taken $Pr = \eta_0 / \kappa \to \infty$.

The boundary conditions at the bottom plate $z = 0$ are

$$W = \vec{V}_h = DT - Bi_0 T = 0 \tag{4.273}$$

i.e. rigid, and poorly conducting when the Biot number Bi_0 is taken small. At the undeformable interface $z = 1$, we have

$$W = DT + Bi_1 T = 0 \tag{4.274}$$

where Bi_1 is another Biot number, also assumed to be small. Now, the tangential stress condition including viscosity dependence upon temperature reads

$$D\vec{V}_h + Ma\vec{\nabla}_h T + R_\eta T D\vec{V}_h = 0 \tag{4.275}$$

where, as before, $\vec{\nabla}_h$ is the horizontal gradient operator and D the z-derivative. We will consider the Marangoni number Ma as the control parameter, at a fixed dynamic Bond number $Bo = Ra/Ma$.

Apart from the fact that we consider a non-Boussinesq effect, the calculation of coefficients of Eq. (4.268) from the system (4.270–4.275) is very similar to those presented by

Gertsberg and Sivashinsky [212] for the Rayleigh–Bénard problem, by Knobloch [215] for Marangoni–Bénard convection at infinite Prandtl number, by Shtilman and Sivashinsky [292] for the same case but finite Prandtl number and mean flow effects (unfortunately, with some mistakes in the coefficients), by Hadji et al. [374] for the combined Rayleigh–Marangoni problem at infinite Prandtl number, and by Golovin et al. [341] for the two-layer Marangoni–Bénard case with surface deformation and mean flow effects (see below). In what follows, we will therefore only outline the main steps of the analysis.

The smallness parameter is taken as the slow scale of horizontal modulations, i.e. we scale $\nabla_h \sim \varepsilon \ll 1$, $D \sim 1$. According to the discussion of §4.6.1, also confirmed by a long-wave expansion of the linear dispersion relation corresponding to the problem (4.270–4.275), we should set $\partial_t = \varepsilon^4 \partial_\tau$, $Ma = Ma_c + \varepsilon^2 M_2$, and $(Bi_0, Bi_1) = \varepsilon^4 (a, b)$, while the order-parameter itself, i.e. $T(\vec{r}, z, t)$, should remain $O(1)$. According to Eqs (4.270), (4.271) and (4.275), we have $\vec{V}_h \sim \varepsilon$, $p = O(1)$ and $W \sim \varepsilon^2$ at the lowest order. Finally, in order to obtain the non-Boussinesq effect at the same order as the order-parameter equation (i.e. ε^4), we need to assume that it is small, i.e. $R_\eta = \varepsilon^2 g$.

Thus, we expand the various fields as

$$
\begin{aligned}
T &= \phi(\vec{R}, \tau) + \varepsilon T_1(z, \vec{R}, \tau) + \varepsilon^2 T_2(z, \vec{R}, \tau) + \dots \\
p &= p_0(z, \vec{R}, \tau) + \varepsilon p_1(z, \vec{R}, \tau) + \varepsilon^2 p_2(z, \vec{R}, \tau) + \dots \\
W &= \varepsilon^2 W_2(z, \vec{R}, \tau) + \varepsilon^3 W_3(z, \vec{R}, \tau) + \varepsilon^4 W_4(z, \vec{R}, \tau) + \dots \\
\vec{V}_h &= \varepsilon \vec{V}_{h1}(z, \vec{R}, \tau) + \varepsilon^2 \vec{V}_{h2}(z, \vec{R}, \tau) + \varepsilon^3 \vec{V}_{h3}(z, \vec{R}, \tau) + \dots
\end{aligned}
\tag{4.276}
$$

where $\vec{R} = \varepsilon \vec{r}$, $\tau = \varepsilon^4 t$, and from now on, $\vec{\nabla}$ will denote the gradient with respect to the slow horizontal coordinate $\vec{R}$. Note that the lowest-order temperature field $T_0 = \phi$ is independent of z, which directly follows from the lowest-order energy equation $D^2 T_0 = 0$ with boundary conditions $D T_0 = 0$ at $z = 0, 1$ (Goldstone mode).

Inserting the above relations in the problem (4.270–4.275), and identifying the successive powers of ε, one gets a hierarchy of linear inhomogeneous problems, whose resolution is cumbersome (high-order polynomials in z appear) but does not lead to major difficulties. At each order, a solvability condition is also obtained, which leads to the critical value Ma_c (at order ε^2) and finally to the order-parameter equation for ϕ (at order ε^4). At orders ε and ε^3, the solvability condition is trivially satisfied, and in fact it appears that we may set $T_1 = W_3 = \vec{V}_{h2} = p_1 = 0$ without loss of generality (at each order, it is possible to add a contribution identical to the solution of the lowest-order problem, but this does not affect the final result).

The lowest-order fields are found as a function of ϕ and its gradients, e.g.

$$
\begin{aligned}
p_0 &= \frac{Ma_c}{8}(8Boz - 12 - 5Bo)\phi , \\
\vec{V}_{h1} &= \frac{Ma_c\, z}{48}(24 + 6Bo - 36z - 15Boz + 8Boz^2)\vec{\nabla}\phi, \dots
\end{aligned}
\tag{4.277}
$$

At order ε^2, the solvability relation leads to

$$
\frac{Ma_c}{48} + \frac{Ra_c}{320} = 1 \quad \text{or} \quad Ma_c = \frac{960}{20 + 3Bo}
\tag{4.278}
$$

while at order ε^4, we find an order-parameter equation, which after some rescaling and possible rearrangement of terms, may be written as the Knobloch equation (4.268). The various coefficients are here given by

$$\alpha = 4\frac{\chi'}{\chi^2}(a+b), \quad \mu = 2, \quad \kappa = 1, \quad \gamma = -\frac{2g}{\chi}\sqrt{\frac{\chi'}{\delta_3}}, \quad \beta = \frac{\delta_1}{\sqrt{\chi'\delta_3}}, \quad \delta = \frac{\delta_2}{\sqrt{\chi'\delta_3}}$$

(4.279)

where

$$\chi = \frac{M_2}{Ma_c} - g\frac{36+7Bo}{4(20+3Bo)}, \quad \chi' = \frac{2(9240+3608Bo+261Bo^2)}{693(20+3Bo)^2},$$

$$\delta_1 = -\frac{12+Bo}{6(20+3Bo)}, \quad \delta_2 = -\frac{10080+1492Bo+21Bo^2}{42(20+3Bo)^2}$$

(4.280)

$$\delta_3 = \frac{40(864+252Bo+19Bo^2)}{63(20+3Bo)^2}$$

Note that $\chi > 0$ is required for the rescaling to hold. Then it can be checked that coefficients of Knobloch [215] and of Shtilman and Sivashinsky [292] are recovered for Marangoni–Bénard convection in the Boussinesq approximation ($Bo = 0, g = 0$). Moreover, for pure Rayleigh–Bénard convection ($Bo \to \infty$), it is found that $\delta = \beta \neq 0$, because the linear problem is self-adjoint, though boundary conditions at the top and bottom are not symmetric [215].

Non-Boussinesq effects affect not only nonlinear regimes (through the term $\vec{\nabla}.[\phi\vec{\nabla}\phi]$) but also cause a shift of the linear stability threshold (4.278). The linear part of Eq. (4.268) shows that instability occurs when the dissipation α decreases below unity, or equivalently when the Marangoni number exceeds the value

$$Ma'_c = \frac{960}{20+3Bo}\left[1+2\sqrt{\chi'(Bi_0+Bi_1)}+\frac{36+7Bo}{4(20+3Bo)}R_\eta\right]$$

(4.281)

which has been rewritten in terms of unscaled quantities. Hence, an increase (decrease) of viscosity with temperature stabilizes (destabilizes) the reference state. For instance, a liquid with $R_\eta = -0.1$ (i.e. a 10% difference between viscosity at the top and bottom in the reference state) would have its stability threshold decreased by 4.5% for pure Marangoni–Bénard convection, and 5.8% for pure Rayleigh–Bénard convection. Note that these figures are in good agreement with the linear stability analysis of Cloot and Lebon for pure Marangoni–Bénard convection [269], even though the latter is not restricted to small wavenumbers.

Now, pattern selection may be discussed using the expressions of the Landau coefficients obtained in §4.5.2, provided we consider the close vicinity of the instability threshold. In order to write Eq. (4.268) in the form (4.196), we thus set $Ma - Ma'_c = Ma_c\sqrt{\chi'(Bi_0+Bi_1)}\epsilon^2$, where ϵ is a *new* smallness parameter. Then, it is found that the coefficient γ to be considered in Eq. (4.196) can be rewritten

$$\gamma = -\frac{g}{\sqrt{\delta_3}(a+b)} = -\frac{R_\eta}{\sqrt{\delta_3}(Bi_0+Bi_1)}$$

(4.282)

while the values of β and δ_3 are still given by Eqs (4.279) and (4.280).

The result for the competition between rolls, squares and hexagons is summarized in Fig. 4.62 as a function of the angle $\varphi = \arctan(3Bo/20)$ of the physical line in a Nield's diagram, and of $r = R_\eta/\sqrt{Bi_0 + Bi_1}$. Note that, as in §4.4.3, squares or rolls will only appear above a certain supercriticality, while near the threshold, a hexagonal pattern is expected as long as the quadratic coefficient $\beta - \delta + \gamma$ (see §4.5.2) does not vanish. Actually, this happens on a line also represented in Fig. 4.62, and which separates a region of up-hexagons from a region of down-hexagons.

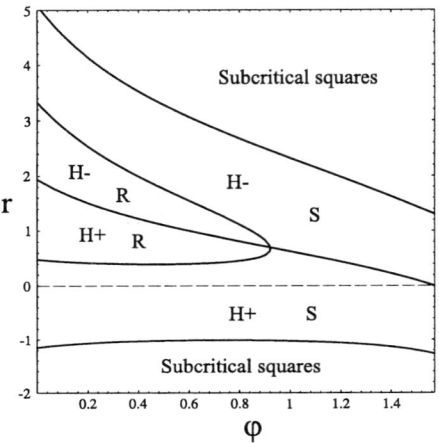

Figure 4.62: Competition between rolls (R), squares (S), up-hexagons (H+), and down-hexagons (H–) for Rayleigh–Marangoni convection between poorly-conducting boundaries, as a function of $\varphi = \arctan(3Ra/20Ma)$ and of a quantity r proportional to the variation of dynamic viscosity with temperature.

When non-Boussinesq effects become strong enough (increasing $|r|$), there is an increasing tendency to subcritical patterns. First, squares become subcritical, then hexagons, then finally rolls. The analysis in terms of cubic Landau equations is thus restricted to the region suitably close to the $r = 0$ axis. For liquids ($r < 0$ in general), up-hexagons are the first pattern at threshold (and are also called l-hexagons for this reason [22]), and are replaced by squares when the constraint is increased (there is a bistability region as in §4.4.3). The behavior is more interesting for the formal case $r > 0$ (increase of viscosity with temperature), showing in particular a point where all the patterns compete. This portion of the diagram is included as a mere illustration of the complexity of behaviors accessible to Eq. (4.268). Still, for gases ($r > 0$) between rigid plates ($Bo \to \infty$), the analysis confirms the preference of down-hexagons (or g-hexagons) near the instability threshold. Finally, we may again comment by comparison with results of §4.4.3 that there is a much greater tendency to square patterns when the layer is confined between poorly conducting boundaries, which was also observed experimentally [133].

4.6.3 Long-wave surface deformational modes

In this section, we consider another Goldstone-like mode, namely spatial modulations of the free surface height in a Marangoni–Bénard set-up. Large-scale surface deformations are coupled to thermal and velocity fields, and may become the most dangerous instability mode for sufficiently small Galileo number Ga and inverse capillary number Ca^{-1}, as shown in §3.6.1. Our goal here will not be to provide an extensive analysis of this instability, which has been the subject of numerous theoretical and numerical investigations [86, 35, 36, 237, 239, 341], but rather to show what type of order-parameter equations can be expected and to analyze the link between different approaches.

We will still consider the system (4.270–4.272), though here we neglect non-Boussinesq and buoyancy effects, i.e. $\tilde{\eta} = 1$ and $Ra = 0$. At the bottom plate $z = 0$, the boundary conditions are still taken as (4.273), except that we will now consider the more usual case of a conducting bottom plate $Bi_0 \rightarrow \infty$ (i.e. $T = 0$). At the moving interface $z = h(\vec{r}, t) = 1 + \xi(\vec{r}, t)$, the general boundary conditions were established in §2.2.2. For perturbations, neglecting evaporation ($J = 0$), and under dimensionless form, they can be rewritten in general as

$$-p + Ga\xi + \mathbf{P}.\vec{n}.\vec{n} - 2Ca^{-1}H = \mathbf{P}.\vec{n}.\vec{\tau}_i + Ma\frac{T_i + \xi_i(DT-1)}{\sqrt{1+\xi_i^2}} = 0 \ \ (i = x, y)$$

(4.283)

$$W - \xi_t - \left(\vec{V}_h.\vec{\nabla}_h\right)\xi = DT - \vec{\nabla}_h\xi.\vec{\nabla}_h T + NBi_1(T-\xi) + N - 1 = 0 \quad (4.284)$$

which all need to be expressed at the deformed interface $z = 1 + \xi(\vec{r}, t)$, and where $\mathbf{P}$ is the dimensionless viscous stress tensor, an index x, y or t means the corresponding derivative, $N = [1 + (\vec{\nabla}_h\xi)^2]^{1/2}$, and the unit normal $\vec{n}$ and tangents $\vec{\tau}_i$ are given by

$$\vec{n} = \frac{(-\xi_x, -\xi_y, 1)}{\sqrt{1+(\vec{\nabla}_h\xi)^2}}, \quad \vec{\tau}_x = \frac{(1, 0, \xi_x)}{\sqrt{1+\xi_x^2}}, \quad \vec{\tau}_y = \frac{(0, 1, \xi_y)}{\sqrt{1+\xi_y^2}}, \quad (4.285)$$

while the mean curvature H is

$$H = \frac{1}{2}\vec{\nabla}_h.\left(\frac{\vec{\nabla}_h\xi}{\sqrt{1+(\vec{\nabla}_h\xi)^2}}\right) \quad (4.286)$$

Note that we have included a free-surface Biot number Bi_1 in the last condition in Eqs (4.284), and that in general, the inverse capillary number Ca^{-1} should be written $Ca_0^{-1} - Ma(T-\xi)$ to include surface tension variation with temperature in the Laplace pressure term (in Ca_0, the surface tension to be used is that of the interface in the reference state). However, the latter effect turns out to be unimportant for the analyzes which follow.

The linearized version of this problem has been studied in §3.6.1. In the long-wave limit, the critical Marangoni number is given by Eq. (3.191). Note that below we will generally assume $Bi_1 = 0$, unless explicitly mentioned. In practice, Ga and Ca^{-1} are often very large, and in this section we will assume further that $Ca^{-1} \gg Ga^2$, which is achievable at sufficiently small depths of moderately viscous liquids [35]. Then, Eq. (3.191) leads to

$$Ma \simeq \frac{2}{3}Ga + \frac{2}{3Ca}k^2 \qquad (4.287)$$

Davis [86] and VanHook et al. [35] use the scaling $k \sim \varepsilon \ll 1$, $Ga \sim \varepsilon^{-1}$, $Ca \sim \varepsilon^3$, $Ma \sim \varepsilon^{-1}$, which, according to Eq. (4.287), allows a balance between hydrostatic pressure, Laplace pressure and thermocapillarity at the linear stage. It then appears that the relevant scalings are $\partial_t \sim \vec{\nabla}_{\mathrm{h}} \sim \varepsilon$, and the expansions of perturbation fields begin at orders $h \sim \vec{V}_{\mathrm{h}} = O(1)$, $p \sim \varepsilon^{-1}$ and $W \sim \varepsilon$. Identifying successive powers of ε in the system (4.270–4.272), (4.283–4.284), and solving the corresponding differential problems, one gets at the lowest order a pressure perturbation induced by gravity and surface curvature effects, and rewritten in terms of unscaled variables as

$$p = Ga(h - 1) - Ca^{-1}\Delta h \qquad (4.288)$$

i.e. independent of z. Hence, a horizontal (potential) flow field is generated, which may be found from leading-order expressions of horizontal Stokes equation and tangential stress condition as

$$\vec{V}_{\mathrm{h}} = Ma\,z\vec{\nabla}h + \left(\frac{z^2}{2} - zh\right)\left(Ga\vec{\nabla}h - Ca^{-1}\vec{\nabla}\Delta h\right) \qquad (4.289)$$

the first term of which represents the thermocapillary flow generated by the modification of surface temperature induced by the surface deformation (e.g. at local depressions, the interface is closer to the heated bottom, and therefore hotter), while the second term is the flow generated by the pressure perturbation (4.288).

Then, taking into account the continuity equation, the first condition in Eqs (4.284) (kinematic condition) may be rewritten

$$\partial_t h = -\vec{\nabla}.\int_0^h dz\vec{V}_{\mathrm{h}} \qquad (4.290)$$

i.e. minus the divergence of the total mass flux. This is a general feature of lubrication theories [237]. Combining Eqs (4.289) and (4.290), one finally gets the evolution equation for the surface height $h(\vec{r}, t)$ as

$$\partial_t h = -\vec{\nabla}.\left\{\frac{Ma}{2}h^2\vec{\nabla}h - \frac{Ga}{3}h^3\vec{\nabla}h + \frac{1}{3Ca}h^3\vec{\nabla}\Delta h\right\} \qquad (4.291)$$

which has been first obtained by Davis [86]. Note that this equation has recently been improved by VanHook et al. [35], by taking conduction in the gas phase into account (two-layer model). Actually, using the above scalings, it is seen that the thermal fields in both liquid and gas may be found solving the steady unidirectional heat equations $D^2 T_{\mathrm{l}} = D^2 T_{\mathrm{g}} = 0$, complemented by boundary conditions $T_{\mathrm{l}}(0) = T_{\mathrm{b}}$, $T_{\mathrm{l}}(h) = T_{\mathrm{g}}(h)$, $DT_{\mathrm{l}}(h) = \lambda DT_{\mathrm{g}}(h)$, $T_{\mathrm{g}}(h + a) = T_{\mathrm{b}} - (1 + a\lambda^{-1})$, where T_{b} is the bottom temperature, a is the ratio of gas to liquid depths in the unperturbed state, λ is the ratio of gas to liquid thermal conductivities, and all temperatures are reduced using the temperature drop across the liquid layer in the unperturbed state $h = 1$. Solving this problem leads to the leading-order liquid temperature field

$$T_1(\vec{r}, z, t) = T_b - \frac{z}{1 + F - Fh(\vec{r}, t)} \tag{4.292}$$

where $F = (1 - \lambda)/(a + \lambda)$ is a new parameter introduced by VanHook et al. [35], and which is generally positive. This equation better describes the effect of deformation on the surface temperature profile. Taking into account that the leading-order tangential stress condition must now be written $D\vec{V}_h(z = h) = -Ma\vec{\nabla}[T_1(z = h)]$, it can be shown that Eq. (4.291) is generalized to

$$\partial_t h = -\vec{\nabla} \cdot \left\{ \frac{Ma}{2} \frac{1 + F}{(1 + F - Fh)^2} h^2 \vec{\nabla} h - \frac{Ga}{3} h^3 \vec{\nabla} h + \frac{1}{3Ca} h^3 \vec{\nabla} \Delta h \right\} \tag{4.293}$$

While both equations (4.291) and (4.293) predict the formation of localized depressions (dry spots) seen in experiments [35, 36], only the two-layer model (4.293) leads to localized elevations (high spots), which occur typically for F higher than about 0.5. These states are also obtained experimentally (see Fig. 1.8 of Chapter 1), for increasing liquid depths (hence decreasing a and increasing F). As it corresponds to $F = 0$, the one-layer model does not predict high spots. VanHook et al. have also shown that the bifurcation described by Eq. (4.293) is always subcritical, with an unstable solution branch which never turns over into a stable branch (i.e. no saddle-node bifurcation occurs). Numerical integration also shows that no stable deformed states exist, but that the solution always blows up in finite time, which is coherent with the experimental observation of dry and high spots [35].

4.6.4 Interaction between deformational and cellular convection modes

In this section, we will briefly describe the progress in modeling the interaction between long-scale deformational modes and cellular convection. A first possibility is to consider a small bottom plate Biot number Bi_0 (still with small or vanishing free surface Biot number Bi_1), such that the critical wavenumber of the cellular convection mode is also small [336, 337, 339, 340, 341]. A second possibility is to consider a conducting bottom plate [238, 239, 369], but in this case the cellular mode, being triggered at wavenumbers of order unity, should be modeled using Ginzburg–Landau-like equations.

Poorly heat-conducting bottom and top boundaries

Consider the system (4.270–4.272) with $\tilde{\eta} = 1$ and $Ra = 0$. At the bottom plate $z = 0$, the boundary conditions are taken as (4.273), while at the upper free surface $z = 1 + \xi(\vec{r}, t)$, the boundary conditions (4.283–4.285) hold.

i) Weak surface deformations

First, we consider $Ca, Ga = O(1)$, a peculiar situation which could apply for a very thin liquid layer with very high viscosity, low surface tension, or low effective gravity. In this case, it may be shown from linear stability analysis at $k \sim \varepsilon \ll 1$, $(Bi_0, Bi_1) = (a, b)\varepsilon^4$ that the critical Marangoni number (up to a correction of order ε^2) is given by

$$Ma_c = \frac{48\,Ga}{72 + Ga} \tag{4.294}$$

which, interestingly, leads to $Ma_c = 48$ for $Ga \gg 1$ [see Eq. (4.278) with $Bo = 0$] and $Ma_c = 2Ga/3$ for $Ga \ll 1$ [i.e. Eq. (4.287) for $k \to 0$]. Now, for small surface deformations $\xi \sim \varepsilon^2$ and temperature perturbations $T \sim \varepsilon^2$, it can be shown [337] that at the leading order ε^4 we get the coupling $T = \varphi - Ga\xi/72 \sim \varepsilon^2$ between temperature and surface deformation fields[15], where φ is a spatially constant integration constant. Expressing the total volume conservation $< \xi > = 0$ (where $< . >$ denote the horizontal average), we see that $\varphi = < T >$. Note that both T and ξ (and φ) evolve on the slow time scale $\tau = \varepsilon^4 t$. At the same order ε^4, we also get the compatibility relation (4.294), and a deviation from this value is allowed according to $Ma - Ma_c \sim \varepsilon^2$, as in §4.6.2.

It is necessary to pursue the analysis up to order ε^6 to get the relevant evolution equation for the surface position ξ, which is finally found as

$$\partial_t \xi = -\alpha\xi - \mu\Delta\xi - \nu\Delta^2\xi - \gamma\vec{\nabla}.\left[\xi\vec{\nabla}\xi\right] + \gamma'\left\{(\vec{\nabla}\xi)^2 - \left\langle(\vec{\nabla}\xi)^2\right\rangle\right\} \tag{4.295}$$

i.e. a two-dimensional generalization of the equation first obtained by Garcia-Ybarra, Castillo and Velarde [336, 337], which in fact corresponds to a low-order truncation of the Knobloch equation (4.268), except for the Kuramoto–Sivashinsky-like contribution (last term). In this equation, the coefficients are given by

$$\alpha = \delta^{-1}\left[Bi_1 + \frac{Ga}{72}(Bi_0 + Bi_1)\right], \quad \mu = \left(1 + \frac{Ga}{72}\right)\delta^{-1}\frac{Ma - Ma_c}{Ma_c},$$
$$\nu = \delta^{-1}\left(\frac{1}{5} + \frac{1}{GaCa} + \frac{Ga}{1080}\right), \quad \gamma = \delta^{-1}\left(\frac{Ga}{36} - 1\right), \quad \gamma' = \frac{1}{2\delta} \tag{4.296}$$

in which $\delta = 3/8 + 3/Ga + Ga/72$. Note that the mean value of the temperature perturbation, i.e. $\varphi = < T >$, is found to evolve according to

$$\partial_t \varphi = -(Bi_0 + Bi_1)\varphi - \frac{1}{2}\left\langle(\vec{\nabla}\xi)^2\right\rangle \tag{4.297}$$

by expressing the volume conservation $< \xi > = 0$ [if $< \xi > = 0$ at $t = 0$, Eq. (4.295) shows that this remains the case for all $t > 0$, e.g. assuming periodic boundary conditions].

Although the nonlinear term $\vec{\nabla}.[\xi\vec{\nabla}\xi]$ cannot saturate the linear instability [338, 339], the second nonlinear term $(\vec{\nabla}\xi)^2$ not only prevents blow-up in finite time, but also generates quite complex chaotic dynamics (even in one dimension [338]), as for the damped Kuramoto–Sivashinsky equation (4.267). In this respect, it is worth mentioning that this nonlinearity here originates only from the term $N - 1$ in the interfacial energy balance, i.e. the last of Eq. (4.284). The latter term is linked to the increase of surface area due to deformation and

---

[15]For simplicity, we do not develop intermediate steps of the multiscale analysis here. Note however that for $\xi, T \sim \varepsilon^2$, the other scalings are found as $W \sim \varepsilon^4$, $\vec{V}_h \sim \varepsilon^3$, $p \sim \varepsilon^2$, while the horizontal gradient and time derivative respectively scale as $\vec{\nabla}_h \sim \varepsilon$ and $\partial_t \sim \varepsilon^4$.

its interaction with the basic temperature gradient. This effect therefore appears sufficient (at least for values of Ga that are not too large) to prevent the surface deformation becoming of order unity, which on the other hand could be expected for $Ca, Ga = O(1)$ (small stabilizing effects of gravity and capillarity).

It is beyond our scope here to discuss chaotic regimes of Eq. (4.295) in details (see [338] for one-dimensional regimes). The reader is also referred to [51, 496, 497, 498] for detailed analyzes of the damped Kuramoto–Sivashinsky equation (4.267). While steady cellular structures and in particular hexagonal patterns may be found right above threshold (see also Fig. 4.57), increasing the constraint further (or decreasing the dissipation α) leads to transitions to chaotic regimes, both for one-dimensional and two-dimensional situations. Note also that the one-dimensional version of Eq. (4.267) with $\alpha = 0$ is the equation describing the phase turbulence of traveling waves (see §5.5.7).

ii) Very weak surface deformations

We now consider a different asymptotics, applying when Ga and Ca^{-1} are large, specifically $Ga \sim \varepsilon^{-2}$ and $Ca \sim \varepsilon^4$, as done by Golovin et al. [341]. Thus, surface deformation is small ($\xi \sim \varepsilon^2 T$), and does not affect the linear stability threshold $Ma_c = 48$ (for one-layer convection). Actually, the expansions (4.276) used to obtain the Knobloch equation (4.268) remain valid here, while the surface deformation ξ should be expanded as $\xi = \varepsilon^2 \xi_2 + ...$. Note that on the slow time scale $\tau = \varepsilon^4 t$ describing the evolution of the temperature field [$\phi = O(1)$ at the lowest order], surface deformation is found to be slaved to the dynamics of ϕ, according to the non-local equation

$$\Delta(g\xi - c^{-1}\Delta\xi) = -\Delta\phi \qquad (4.298)$$

where $g \sim Ga$, $c \sim Ca$. In fact, Golovin et al. have developed their analysis for the more realistic case of two-layer convection, and the reader is referred to their paper [341] for explicit values of the coefficients. Here, we will merely state their main results qualitatively. Importantly, Golovin et al. also show that for two-layer convection, Marangoni convection is indeed long-wave only when the gas phase is sufficiently thin. For large gas depths, there is a short-wave mode of instability and the theory presented here is not applicable.

Note that their analysis is not restricted to infinite Prandtl numbers, such that vertical vorticity effects (see also §4.5.6) should be included. The streamfunction Ψ of the non-potential part of the horizontal velocity is also slaved, according to

$$\Delta\Psi = p^{-1}\vec{1}_z \cdot \left(\vec{\nabla}\Delta\phi \times \vec{\nabla}\phi\right) - q\vec{1}_z \cdot \left(\vec{\nabla}\xi \times \vec{\nabla}\phi\right) \qquad (4.299)$$

which, compared with Eq. (4.256), contains a supplementary term linked to surface deformation. Here, $p \sim Pr$ and $q = O(1)$. Accordingly, vertical vorticity is generated by surface deformations *even at infinite Prandtl number*, due to the non-zero vertical component of the velocity at the interface. Finally, the Knobloch equation describing the evolution of the order-parameter ϕ is found to be completed by the coupling with surface deformation and vertical vorticity, namely

$$\partial_t \phi = -\alpha\phi - 2\mu\Delta\phi + 2\Delta\xi - \Delta^2\phi + \vec{\nabla}.\left[(\vec{\nabla}\phi)^2\vec{\nabla}\phi\right] + \beta\vec{\nabla}.\left[\Delta\phi\vec{\nabla}\phi\right] +$$
$$\delta\Delta\left[(\vec{\nabla}\phi)^2\right] - \nu\vec{\nabla}.\left[\xi\vec{\nabla}\phi\right] - \vec{1}_z.\left(\vec{\nabla}\phi \times \vec{\nabla}\Psi\right) \tag{4.300}$$

with values of the coefficients also given in [341], for two-layer convection. Note that $\mu = (Ma - Ma_c)/Ma_c$ and α is proportional to the bottom plate Biot number.

Golovin et al. have also analyzed pattern selection in the system (4.298–4.300), varying Ga and Ca^{-1} over a large range (using properties of a water–air system). Note that for perfect patterns, vertical vorticity does not influence pattern selection, as also discussed in §4.5.6. For large Ga and Ca^{-1} (very weak surface deformation), the usual squares and up-hexagons are recovered. When decreasing Ga and Ca^{-1}, the effect of surface deformation increases and the competition between squares and rolls turns in favor of the latter. Up-hexagons near the threshold are then replaced by down-hexagons, at still smaller Ga and Ca^{-1}. Then, squares may again become preferred over rolls, and finally become subcritical at small Ga and Ca^{-1} (rolls become also subcritical at another threshold). Just before this transition, an interesting docecagonal pattern (see Fig. 4.18, $m = 6$) is found to be stable in a small region, though it does not correspond to an absolute minimum of the Lyapunov function of perfect patterns and hence is metastable.

Conducting bottom plate

When the bottom plate is a good heat-conductor, thermal dissipation prevents convective modes to be large-scale, and cellular convection then has a wavelength of the order of the liquid depth (short-wave mode). Still, there exists a large-scale surface deformational mode, such that the neutral stability curve may possess two minima, as in Fig. 3.33.

It is beyond our scope here to describe the interaction of these two modes in details. Rather, we will summarize the relevant scalings and some basic features of this rich problem, which has been studied in details by Golovin et al. [238, 239, 369].

Denoting by Ma_s the threshold Marangoni number of the short-wave mode (say, with critical wavenumber k_c), and by Ma_l that of the long-wave mode (with vanishing critical wavenumber), Golovin et al. [238] have developed a weakly nonlinear analysis for rolls in the vicinity of Ma_s, using the usual scaling for deriving Ginzburg–Landau equations (see §4.2.2), i.e. $Ma - Ma_s \sim \epsilon^2, \partial_t \sim \epsilon^2, f = \epsilon f_1 + \epsilon^2 f_2 + ...$, where f stands for temperature, velocity and pressure perturbations. At the lowest order,

$$f_1 \sim A(X = \epsilon x, \tau = \epsilon^2 t)\exp[ik_c x] + c.c. \tag{4.301}$$

defining the slowly varying amplitude $A(X, \tau)$ of the short-scale structure $\exp[ik_c x]$.

Golovin et al. [238] further make use of the strong surface tension limit $Ca \sim \epsilon^2$, such that surface deformation is small $\xi = \epsilon^2\xi_2 + \epsilon^3\xi_3 + ...$ The Galileo number Ga is taken to be of order unity (in practice, $Ga \sim 100$), which allows having either $Ma_s < Ma_l$ or $Ma_s > Ma_l$, as seen in §3.6.1.

At the lowest order, the classical Pearson's problem without surface deformation is recovered (§3.3.2), which in fact yields Ma_s and k_c. At order ϵ^2, the temperature field contains, in addition to terms $\sim \exp[\pm ik_c x], \exp[\pm 2ik_c x]$, slowly varying contributions

$$T_2(z, X, \tau) = |A|^2 \alpha(z) + \xi_2 \beta(z) \tag{4.302}$$

while the part of the pressure field independent of x reads

$$p_2(X, \tau) = q|A|^2 + Ga\xi_2 - c^{-1}\partial_X^2 \xi_2 \tag{4.303}$$

where $c = \epsilon^{-2}Ca = O(1)$ is the rescaled capillary number, and $q \sim Pr^{-1}$.

In Eq. (4.302), the first term represents the modification of the reference temperature gradient by the short-scale flow (homogenization), while the second one is due to the variation of the reference heat flux induced by surface deformation (note that β vanishes when $Bi = 0$, and Golovin et al. actually consider mass transfer rather than heat transfer, such that Bi and Pr are replaced by Sherwood and Schmidt numbers, respectively). The first term in Eq. (4.303) is linked to momentum advection terms in Navier–Stokes equations, while the other ones have already been encountered above, e.g. in Eq. (4.288).

At order ϵ^3, a modified Ginzburg–Landau equation is found for A, namely

$$\partial_\tau A = \mu A + \delta \partial_X^2 A - \lambda A|A|^2 + \gamma A \xi_2 \tag{4.304}$$

and an equation for ξ_2 is found from the kinematic condition at order ϵ^4, namely

$$\partial_\tau \xi_2 = \partial_X^2 \left(-a\xi_2 - b\partial_X^2 \xi_2 + d|A|^2 \right) \tag{4.305}$$

where constants δ, λ, γ, b and d are all positive, while μ and $a \sim (Ma_s - Ma_l)$ can change sign.

The coupling terms in Eqs (4.304) and (4.305) originate from the following effects: when the interface is deformed, a region of local elevation corresponds to a higher local Marangoni number, hence the spatially varying effective growth rate $\mu + \gamma \xi_2$ in Eq. (4.304). The term $d\partial_X^2|A|^2$ in Eq. (4.305) suppresses surface elevations, since as the short-scale motions are more intensive there, the surface temperature (or concentration of the surfactant) is higher due to more intensive homogenization [see e.g. Eq. (4.187)], inducing a surface-tension-driven outflow leading to surface flattening.

Golovin et al. [238, 369] have studied the system of Eqs (4.304) and (4.305) in four different cases, namely (i) $\mu > 0, a > 0$ (both modes unstable), (ii) $\mu > 0, a < 0$ (only the short-wave mode is unstable), (iii) $\mu < 0, a > 0$ (only the long-wave mode is unstable), and (iv) $\mu < 0, a < 0$ (both modes are damped). Nothing interesting is found in cases (ii) and (iv), where the stable states are the usual short-scale mode without surface deformation, and the rest state, respectively.

Cases (i) and (iii) are much more interesting. When both modes are unstable, i.e. case (i), the only steady solution is still $A \neq 0, \xi_2 = 0$, but it may become unstable to both monotonic and oscillatory disturbances, depending on the value of coupling parameters. Surface deformation then sets in, and modulates the amplitude of convective rolls on a large scale, either in the form of a steady surface relief, or traveling or standing waves. Higher in the supercritical region, the equations can also blow up in finite time [369]. In case (iii), the increasing surface deformations may turn on short-scale convection under regions of surface elevations, due to the local increase of the Marangoni number. This phenomenon has been observed experimentally for liquid depths in some range [35], as seen in Fig. 1.8 of Chapter 1. This convection then stabilizes the growth of the surface relief, and depending on coupling parameters, various states may be observed: although the equations may also blow up when the coupling is small, at higher coupling, standing waves, localized depressions, and even chaotic behavior are predicted. Finally, Golovin et al. [239] extended their analysis to three dimensions.

5   Weakly nonlinear waves

To what extent is it possible to achieve a rigorous description of weakly nonlinear regimes, when the rest state becomes unstable to oscillatory disturbances with finite wavenumber (Hopf bifurcation) ? This is the question we address in this chapter, using similar tools as for bifurcations of monotonic modes studied in Chapter 4. In two dimensions (one of them being along the vertical), it is known that such bifurcation results in a competition between traveling and standing waves [163]. After having considered these classical results in §5.1.1, the competition between waves is analyzed in the three-dimensional regime (§5.1.2), ignoring spatial modulations of the amplitudes of waves. Without reference to any particular system, general (operational) expressions for the coefficients of Landau equations, characterizing the nonlinear coupling between waves propagating at different angles in the horizontal plane, are obtained via perturbation theory. Some general results about patterns with particular symmetries (square and hexagonal) are also presented, before embarking on a discussion of the role of mean flow and vertical vorticity effects.

Then, in the next three sections, this theory is applied to some of the oscillatory regimes encountered in Chapter 3. In §5.2, internal and surface waves and their interaction are studied, both for two-dimensional and three-dimensional situations. Predictions are given about the nature of the Hopf bifurcation, expected wave patterns, and order of magnitude of oscillating temperature and velocity fields. In §5.3, double-diffusive oscillatory regimes are investigated. There, attention is focused on the effect of surface tension gradients at the upper free surface, which drastically affect the observed patterns, when compared with the more extensively studied case of buoyancy-driven convection. In both cases however, the oscillations may also be characterized by a low frequency, a situation for which the theory developed in §5.1 breaks down. Finally, in §5.4, two-layer systems are studied, and only the buoyancy-driven mechanism is considered, as it may induce a particular kind of coupling between the two layers, known as thermal coupling, as opposed to mechanical coupling. Oscillations between these modes are studied, and peculiarities of the low frequency limit are also discussed.

The chapter is concluded by stressing the importance of slow spatio-temporal modulations of wave patterns (§5.5). Spatial derivative terms in the amplitude equations are then included, and their effects are analyzed in some detail. Importantly, even by restricting the analysis to two spatial dimensions, it is shown that weak amplitude waves may be subject to phase instabilities, which in some cases result in weak turbulence very close to the primary instability threshold. The role of lateral boundaries, defects, and group velocity effects is also discussed. These results are obtained from general complex Ginzburg–Landau equations, and hence apply to most of the oscillatory systems considered in Chapter 3.

5.1 Hopf bifurcation with finite wavenumber

Here, the basic features of the complex Landau equations (without spatial modulations) are analyzed, i.e. an attempt is made to generalize the theory presented in §4.2 for monotonic modes. First, using symmetry properties, Landau equations are established and analyzed for a two-dimensional situation (traveling and standing wave patterns). Then, three-dimensional set-ups enjoying the usual symmetry properties of Bénard-like systems are considered. Even without spatial modulations, it will be seen that strong differences with monotonic modes exist, namely the absence of quadratic terms in Landau equations, and the occurrence of cubic phase-coupling terms which rule out the possibility of a Lyapunov function.

5.1.1 Two-dimensional waves (traveling and standing waves)

The analysis presented in this section, based on symmetry properties, closely follows the general discussion of §4.2.1, to which the reader is referred. The time-translation invariance $t \to t + \Delta t$ had no implications on the form of Landau equations in Chapter 4, where only monotonic modes were considered. For the bifurcation of oscillatory modes (overstability) however, it will be necessary to consider its effect in addition to that of other invariance properties, as shown below. We thus consider the situation encountered in the linear stability analyzes of Chapter 3, where a pair of complex conjugate eigenvalues crosses the imaginary axis at some critical value of the control parameter. If ω_c denotes the corresponding critical frequency, the general solution of the linear problem (in two spatial dimensions) reads

$$
\begin{pmatrix} W \\ T \end{pmatrix} = \{A_1 \exp[i(-k_c x + \omega_c t)] + A_2 \exp[i(k_c x + \omega_c t)]\} \begin{pmatrix} W^0_{k_c}(z) \\ T^0_{k_c}(z) \end{pmatrix} + c.c.
$$

$$(5.1)$$

Note that here, the solutions corresponding to $\vec{k} = \pm k_c \vec{1}_x$ (the linear problem only depends on k^2) are physically different, corresponding to right-traveling and left-traveling waves, with amplitudes A_1 and A_2 respectively (if the phases of A_1 and A_2 are independent of time, the phase velocity of the waves is ω_c / k_c).

The evolution equations satisfied by A_1 and A_2 are in general of the form (4.69), as for the competition between rolls and rhombs. Expanding up to cubic order, we may eliminate terms by applying the following symmetries. First, Eq. (5.1) shows that the translation $x \to x - x_0$ leads to the action $A_1 \to A_1 \exp[ik_c x_0]$, $A_2 \to A_2 \exp[-ik_c x_0]$ on the amplitudes. As the translated solution should be a solution of the amplitude equations, the above action should leave them unchanged, such that the only linear terms allowed (i.e. the equivariant terms, see §4.2.1) are A_1, A_2^* in the first equation and A_2, A_1^* in the second one. Quadratic terms are never equivariant, and the only equivariant cubic terms in the first equation are $A_1|A_1|^2$, $A_1|A_2|^2$, $A_1^2 A_2$, $A_2^*|A_1|^2$, $A_1^* A_2^{*2}$, $A_2 A_2^{*2}$ and $A_2^*|A_2|^2$. Thus, not so many terms are eliminated by translation compared with the case of the rolls/rhombs interaction. However, we now have to consider the time-translation invariance $t \to t + \Delta t$. According to Eq. (5.1), the action on amplitudes is now $A_1 \to A_1 \exp[i\omega_c \Delta t]$, $A_2 \to A_2 \exp[i\omega_c \Delta t]$. Among the linear terms, A_2^* and A_1^* are ruled out by this invariance requirement, and only the diagonal ones A_1 and A_2 remain. Among the cubic terms in the first equation, only the first two remain, and those in the second equation may be found similarly. Thus, the amplitude equations are found to be given by Eq. (4.70), as in the case of the rolls/rhombs competition.

A difference with the rolls/rhombs competition occurs however. Indeed, the only remaining invariance is parity $x \to -x$ (here corresponding to the action $A_1 \to A_2, A_2 \to A_1$), which implies $\sigma = \sigma', \beta = \beta', \gamma = \gamma'$. However, these coefficients may be complex in general, which was not the case for monotonic modes. Finally, the amplitude equations for the competition between left- and right-traveling waves read

$$
\begin{aligned}
\dot{A}_1 &= \sigma A_1 - (\beta|A_1|^2 + \gamma|A_2|^2)A_1 \\
\dot{A}_2 &= \sigma A_2 - (\beta|A_2|^2 + \gamma|A_1|^2)A_2
\end{aligned}
\tag{5.2}
$$

thus also similar to Eq. (4.11), but with complex coefficients in general.

The analysis of the system (5.2) turns out to be very similar to that of Eqs (4.11). Indeed, introducing moduli and phases $A_n = a_n \exp[i\varphi_n], n = 1, 2$, we get

$$
\begin{aligned}
\dot{a}_1 &= \sigma_R a_1 - (\beta_R a_1^2 + \gamma_R a_2^2)a_1 \\
\dot{a}_2 &= \sigma_R a_2 - (\beta_R a_2^2 + \gamma_R a_1^2)a_2
\end{aligned}
\tag{5.3}
$$

$$
\begin{aligned}
a_1\dot{\varphi}_1 &= \sigma_I a_1 - (\beta_I a_1^2 + \gamma_I a_2^2)a_1 \\
a_2\dot{\varphi}_2 &= \sigma_I a_2 - (\beta_I a_2^2 + \gamma_I a_1^2)a_2
\end{aligned}
\tag{5.4}
$$

in which suffixes R and I denote real and imaginary parts, respectively. Thus, the equations for the phases φ_1 and φ_2 are once again decoupled from those governing a_1 and a_2. Once a particular solution of Eqs (5.3) is considered, the evolution of the phases is obtained from Eqs (5.4). In particular, for steady solutions of Eqs (5.3), Eqs (5.4) imply a linear variation of φ_1 and φ_2 with t, i.e. a correction with respect to the threshold frequency ω_c.

The analysis of steady solutions of Eqs (5.3) and their stability properties completely parallels that developed for Eqs (4.11) when studying the competition between rolls and rhombs. Thus, the steady solutions are

$$
- \text{Motionless state (O)} \quad : \quad a_1 = a_2 = 0 \tag{5.5}
$$

$$
\begin{aligned}
- \text{Traveling waves (TW)} \quad : \quad & a_2 = 0, \ a_1 = \left[\frac{\sigma_R}{\beta_R}\right]^{\frac{1}{2}} \ (\text{right} - \text{TW}) \ \ or \\
& a_1 = 0, \ a_2 = \left[\frac{\sigma_R}{\beta_R}\right]^{\frac{1}{2}} \ (\text{left} - \text{TW})
\end{aligned}
\tag{5.6}
$$

$$
- \text{Standing waves (SW)} \quad : \quad a_1 = a_2 = \left[\frac{\sigma_R}{\beta_R + \gamma_R}\right]^{\frac{1}{2}} \tag{5.7}
$$

and the supercritical frequency correction for TW and SW may then be computed from Eqs (5.4). The stability properties of waves are then readily obtained from Fig. 4.8, substituting

Rl by TW (pure solution) and Rh by SW (mixed solution). This is done in Fig. 5.1. This classification has first been provided by Knobloch [163].

We may also define an energy $E = a_1^2 + a_2^2$, proportional to the time-averaged increase of the heat transport by convection (a specific example in a two-layer system will be studied in §5.4). Stable small amplitude solutions only exist if both TW and SW bifurcate supercritically, and the selected pattern is that which maximizes the time-averaged convective heat transport.

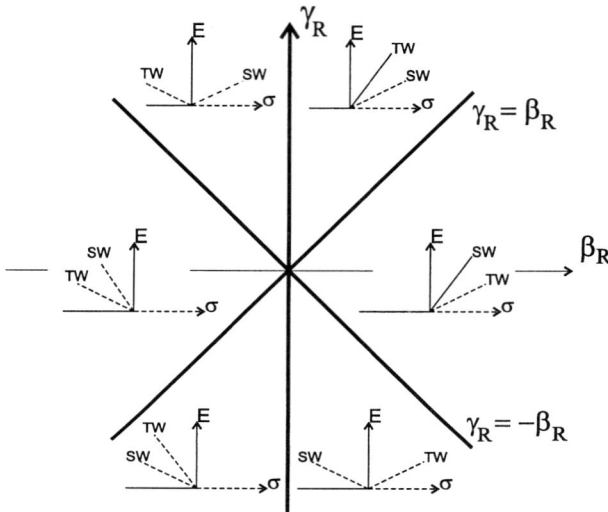

Figure 5.1: Summary of bifurcation diagrams for the competition between traveling waves (TW) and standing waves (SW), as a function of β_R and γ_R. The quantity represented as a function of σ is the "energy" $E = a_1^2 + a_2^2$ of the convective field, proportional to the time-averaged convective heat transport.

5.1.2 Three-dimensional waves – Amplitude equations coefficients

In this section, we apply perturbation theory to a Hopf bifurcation with finite wavenumber k_c and frequency ω_c, for a three-dimensional layered geometry. Our aim is to derive amplitude equations with explicit expressions for all their coefficients, following the methodology used in §4.2.2. In particular, Eqs (5.2) should be recovered as a special case.

The system of thermo-hydrodynamic equations is first written in operational form

$$\mathcal{L}(U) = \mu M(U) + \Theta \left(\frac{\partial U}{\partial t} \right) + N(U, U) \tag{5.8}$$

where U is the unknown perturbation vector, μ is the control parameter, and the various operators have the same definition as in §4.2.2. In particular, linear boundary conditions not included as particular components of Eq. (5.8) will enter in a set E, which, together with necessary continuity requirements, defines an abstract space to which U must belong.

At the Hopf threshold $\mu = \mu_c$, the linearized problem

$$\mathcal{L}(U) - \Theta\left(\frac{\partial U}{\partial t}\right) = \mu_c M(U) \tag{5.9}$$

with $U \in E$, is assumed to admit plane wave solutions

$$U_{\vec{k},\omega} = \exp[i(\vec{k}.\vec{r} + \omega t)]\, U_{0\,\vec{k},\omega}(z) \tag{5.10}$$

where $\vec{k}$ is a horizontal wavevector. Therefore, we have

$$\mathcal{L}_{\vec{k}}(U_{0\,\vec{k},\omega}) - i\omega\Theta_{\vec{k}}\left(U_{0\,\vec{k},\omega}\right) = \mu_c M_{\vec{k}}(U_{0\,\vec{k},\omega}) \tag{5.11}$$

and $U_{0\,\vec{k},\omega} \in E_{\vec{k}}$. The index $\vec{k}$ means that we have replaced the horizontal gradients by $i\vec{k}$, as in §4.2.2. Note also that as the original problem is real, we have $\mathcal{L}_{-\vec{k}} = \mathcal{L}_{\vec{k}}^*$, and the same property holds for other operators. Then, substituting $\vec{k} \to -\vec{k}, \omega \to -\omega$ in Eq. (5.11) and taking its complex conjugate shows that $U_{0\,-\vec{k},-\omega}(z) = U_{0\,\vec{k},\omega}^*(z)$.

Again in this section, we will consider rotationally invariant set-ups, which implies that the eigenvalues do not depend on the direction of $\vec{k}$, but only on its modulus. At criticality, only one value of $|\vec{k}| = k_c$ leads to a vanishing growth rate, all the other modes being damped. Note that due to invariance by parity $\vec{r} \to -\vec{r}$ (i.e. rotation by π), $U_{\vec{k},-\omega}$ is also solution of Eq. (5.9). Then, the general solution of the marginal problem may be written as

$$U_0 = \sum_{\substack{\vec{k}\in P,\\ \omega=\pm\omega_c}} A_{\vec{k},\omega} \exp[i(\vec{k}.\vec{r} + \omega t)]\, U_{0\,\vec{k},\omega}(z) \tag{5.12}$$

where the set P contains an arbitrary number of modes on the critical circle $|\vec{k}| = k_c$. Note that as in §4.2.2, we should always consider them by conjugate pairs $\{(\vec{k},\omega),(-\vec{k},-\omega)\}$, and that the corresponding amplitudes satisfy $A_{-\vec{k},-\omega} = A_{\vec{k},\omega}^*$ for U_0 to be real.

Past threshold but at a small supercriticality $\mu - \mu_c = \epsilon^2\mu_2$, assuming that all amplitudes $A_{\vec{k},\omega}$ depend only on the slow time scale $t_2 = \epsilon^2 t$, we may replace $\partial/\partial t$ by $\partial/\partial t_0 + \epsilon^2\partial/\partial t_2$ (where $t_0 = t$ is the fast time scale) in Eq. (5.8). Expanding $U = \epsilon U_0 + \epsilon^2 U_2 + \epsilon^3 U_3$, and identifying successive powers of ϵ in Eq. (5.8), shows that at order ϵ, we get the marginal problem for U_0, identically satisfied by Eq. (5.12), where t is replaced by t_0. The order ϵ^2 yields

$$\mathcal{L}(U_2) - \mu_c M(U_2) - \Theta\left(\frac{\partial U_2}{\partial t_0}\right) = N(U_0, U_0) \tag{5.13}$$

with $U_2 \in E$, while at order ϵ^3, we get

$$\mathcal{L}(U_3) - \mu_c M(U_3) - \Theta\left(\frac{\partial U_3}{\partial t_0}\right) = \mu_2 M(U_0) + \Theta\left(\frac{\partial U_0}{\partial t_2}\right) \\ + N(U_0, U_2) + N(U_2, U_0) \tag{5.14}$$

with $U_3 \in E$.

In fact, as in §4.2.2, the choice of the slow time scale t_2 and of $\mu - \mu_c = \epsilon^2 \mu_2$ is valid only provided Eq. (5.13) can be solved exactly. Thus, we set

$$U_2 = \sum_{\substack{\vec{k}_1, \vec{k}_2 \in P, \\ \omega_1, \omega_2 = \pm \omega_c}} A_{\vec{k}_1, \omega_1} A_{\vec{k}_2, \omega_2} \exp[i\{(\vec{k}_1 + \vec{k}_2).\vec{r} + (\omega_1 + \omega_2)t_0\}] U_{2\,\vec{k}_1, \vec{k}_2, \omega_1, \omega_2}(z) \tag{5.15}$$

with

$$\left[\mathcal{L}_{\vec{k}_1 + \vec{k}_2} - \mu_c M_{\vec{k}_1 + \vec{k}_2} - i(\omega_1 + \omega_2) \Theta_{\vec{k}_1 + \vec{k}_2} \right] (U_{2\,\vec{k}_1, \vec{k}_2, \omega_1, \omega_2})$$
$$= N_{\vec{k}_1, \vec{k}_2}(U_{0\,\vec{k}_1, \omega_1}, U_{0\,\vec{k}_2, \omega_2}) \tag{5.16}$$

and $U_{2\,\vec{k}_1, \vec{k}_2, \omega_1, \omega_2} \in E_{\vec{k}_1 + \vec{k}_2}$. This can certainly be solved if the kernel of the left-hand side operator is empty. This is indeed the case here, as $\omega_1 + \omega_2$ is either equal to $\pm 2\omega_c$ or to 0, and our initial hypothesis in this section is that the only possible marginal modes have frequencies $\pm \omega_c$. In fact, this is a major difference with the monotonic bifurcations treated in §4.2.2: even when $|\vec{k}_1 + \vec{k}_2| = k_c$, the fast time dependency of quadratic interactions between marginal waves never has the frequency $\pm \omega_c$ which would lead to resonance. Accordingly, no quadratic terms occur in amplitude equations.

Therefore, evolution equations for amplitudes have to be found at the next order. For this purpose, we may again define a suitable scalar product $<,>$, as in Appendices B, D and E. Then, adjoint operators satisfy the relationship

$$\left\langle \tilde{U}, (\mathcal{L}_{\vec{k}} - \mu_c M_{\vec{k}} - i\omega \Theta_{\vec{k}}) U \right\rangle = \left\langle (\mathcal{L}_{\vec{k}}^+ - \mu_c M_{\vec{k}}^+ + i\omega \Theta_{\vec{k}}^+) \tilde{U}, U \right\rangle \tag{5.17}$$

for all $U \in E_{\vec{k}}$, $\tilde{U} \in F_{\vec{k}}$, where $F_{\vec{k}}$ is a certain set of adjoint boundary conditions. Then, if $\tilde{U}_{\vec{k}, \omega}(z) \in F_{\vec{k}}$ denotes the solution of

$$(\mathcal{L}_{\vec{k}}^+ - \mu_c M_{\vec{k}}^+ + i\omega \Theta_{\vec{k}}^+) \tilde{U}_{\vec{k}, \omega} = 0 \tag{5.18}$$

we may find the evolution equation of $A_{\vec{k}, \omega}$ by projecting the right-hand side of Eq. (5.14) on $\tilde{U}_{\vec{k}, \omega}$ (Fredholm solvability condition). Taking into account Eqs (5.12) and (5.15), this leads to

$$\frac{\partial A_{\vec{k}, \omega}}{\partial t_2} = \mu_2 \alpha_\omega A_{\vec{k}, \omega} - \sum_{\substack{\vec{k}_{1,2,3} \in P \\ \omega_{1,2,3} = \pm \omega}} A_{\vec{k}_1, \omega_1} A_{\vec{k}_2, \omega_2} A_{\vec{k}_3, \omega_3} \delta_{\vec{k}, \vec{k}_1 + \vec{k}_2 + \vec{k}_3} \delta_{\omega, \omega_1 + \omega_2 + \omega_3} \underset{\substack{\omega_1, \omega_2, \omega}}{Z_{\vec{k}_1, \vec{k}_2, \vec{k}}} \tag{5.19}$$

where

$$\alpha_\omega = -\frac{\left\langle \tilde{U}_{\vec{k}, \omega}, M_{\vec{k}}(U_{0\,\vec{k}, \omega}) \right\rangle}{\left\langle \tilde{U}_{\vec{k}, \omega}, \Theta_{\vec{k}}(U_{0\,\vec{k}, \omega}) \right\rangle} \tag{5.20}$$

which does not depend on $\vec{k}$ owing to rotational invariance. The coefficients of cubic terms are given by

$$
Z_{\substack{\vec{k}_1,\vec{k}_2,\vec{k} \\ \omega_1,\omega_2,\omega}} = \frac{\left\langle \tilde{U}_{\vec{k},\omega}, N_{\vec{k}_1+\vec{k}_2,\vec{k}_3}(U_{2\,\vec{k}_1,\vec{k}_2,}, U_{0\,\vec{k}_3,\omega_3}) + N_{\vec{k}_3,\vec{k}_1+\vec{k}_2}(U_{0\,\vec{k}_3,\omega_3}, U_{2\,\vec{k}_1,\vec{k}_2,}) \right\rangle_{\omega_1,\omega_2}}{\left\langle \tilde{U}_{\vec{k},\omega}, \Theta_{\vec{k}}(U_{0\,\vec{k},\omega}) \right\rangle}
$$

(5.21)

where $\vec{k}_3 = \vec{k} - \vec{k}_1 - \vec{k}_2$ and $\omega_3 = \omega - \omega_1 - \omega_2$.

It remains to rearrange the cubic terms of the expression (5.19) in a more convenient form, as we did for monotonic modes. In particular, we may first develop the sum on wavevectors as for Eq. (4.107), except that now, $A_{-\vec{k},\omega} \neq A^*_{\vec{k},\omega}$ (if $\omega > 0$ say, $A_{-\vec{k},\omega}$ is the amplitude of the wave propagating in the direction of $\vec{k}$, which is not a priori related to $A_{\vec{k},\omega}$, the wave propagating in the direction opposite to $\vec{k}$). The summation on $\omega_1,\omega_2,\omega_3 = \pm\omega$ such that $\omega_1 + \omega_2 + \omega_3 = \omega$ reduces to three cases: using the notation $+$ for $+\omega$ and $-$ for $-\omega$, these are $(+,+,-), (+,-,+)$ and $(-,+,+)$. Finally, after lengthy calculation and using the relation $A_{-\vec{k},-\omega} = A^*_{\vec{k},\omega}$, we may write the amplitude equations in the form

$$
\begin{aligned}
\frac{\partial A_{\vec{k},\omega}}{\partial t_2} = & \mu_2 \alpha_\omega A_{\vec{k},\omega} - \beta_\omega A_{\vec{k},\omega} |A_{\vec{k},\omega}|^2 - \gamma_\omega A_{\vec{k},\omega} |A_{\vec{k},-\omega}|^2 \\
& - A_{\vec{k},\omega} \sum'_{\vec{q}\neq\vec{k}} \left(\gamma^+_{\vec{k},\vec{q},\omega} |A_{\vec{q},\omega}|^2 + \gamma^-_{\vec{k},\vec{q},\omega} |A_{\vec{q},-\omega}|^2 \right) \\
& - A_{\vec{k},-\omega} \sum'_{\vec{q}\neq\vec{k}} \chi_{\vec{k},\vec{q},\omega} A_{\vec{q},\omega} A^*_{\vec{q},-\omega}
\end{aligned}
$$

(5.22)

where $\sum'$ once again denotes the sum on half of the set P (i.e. only one wavevector of each conjugate pair is taken into account), and the cubic coefficients are defined as

$$
\begin{aligned}
\beta_\omega &= Z^{+++}_{\vec{k},\vec{k},\vec{k}} + Z^{+-+}_{\vec{k},-\vec{k},\vec{k}} + Z^{-++}_{-\vec{k},\vec{k},\vec{k}} \\
\gamma_\omega &= Z^{+-+}_{\vec{k},\vec{k},\vec{k}} + Z^{-++}_{\vec{k},\vec{k},\vec{k}} + Z^{+++}_{\vec{k},-\vec{k},\vec{k}} + Z^{-++}_{\vec{k},-\vec{k},\vec{k}} + Z^{+++}_{-\vec{k},\vec{k},\vec{k}} + Z^{+-+}_{-\vec{k},\vec{k},\vec{k}} \\
\gamma^+_{\vec{k},\vec{q},\omega} &= Z^{+++}_{\vec{k},\vec{q},\vec{k}} + Z^{+++}_{\vec{q},\vec{k},\vec{k}} + Z^{+-+}_{\vec{q},-\vec{q},\vec{k}} + Z^{+-+}_{\vec{k},-\vec{q},\vec{k}} + Z^{-++}_{-\vec{q},\vec{k},\vec{k}} + Z^{-++}_{-\vec{q},\vec{q},\vec{k}} \\
\gamma^-_{\vec{k},\vec{q},\omega} &= Z^{+-+}_{\vec{k},\vec{q},\vec{k}} + Z^{-++}_{\vec{q},\vec{k},\vec{k}} + Z^{-++}_{\vec{q},-\vec{q},\vec{k}} + Z^{+++}_{\vec{k},-\vec{q},\vec{k}} + Z^{+++}_{-\vec{q},\vec{k},\vec{k}} + Z^{+-+}_{-\vec{q},\vec{q},\vec{k}} \\
\chi_{\vec{k},\vec{q},\omega} &= Z^{-++}_{\vec{k},\vec{q},\vec{k}} + Z^{+-+}_{\vec{q},\vec{k},\vec{k}} + Z^{+++}_{\vec{q},-\vec{q},\vec{k}} + Z^{-++}_{\vec{k},-\vec{q},\vec{k}} + Z^{+-+}_{-\vec{q},\vec{k},\vec{k}} + Z^{+++}_{-\vec{q},\vec{q},\vec{k}}
\end{aligned}
$$

(5.23)

where a short-hand notation has been used, e.g.

$$
Z^{+-+}_{-\vec{q},\vec{q},\vec{k}} \equiv Z_{\substack{-\vec{q},\vec{q},\vec{k} \\ +\omega,-\omega,+\omega}}, \dots
$$

Note that from Eqs (5.23), it is seen that $\gamma^+_{\vec{k},-\vec{q},\omega} = \gamma^-_{\vec{k},\vec{q},\omega}$ and $\chi_{\vec{k},-\vec{q},\omega} = \chi_{\vec{k},\vec{q},\omega}$. On the other hand, owing to the invariance with respect to the action $A_{\vec{k},\omega} \to A^*_{\vec{k},-\omega}$ (parity), we have $\alpha_{-\omega} = \alpha^*_\omega, \beta_{-\omega} = \beta^*_\omega, \gamma_{-\omega} = \gamma^*_\omega, \gamma^+_{\vec{k},\vec{q},-\omega} = (\gamma^+_{\vec{k},\vec{q},\omega})^*, \dots$ While cubic coefficients coupling waves $\vec{q}$ and $\vec{k}$ should only depend on the angle between them (due to rotational invariance), it is further seen that if reflection invariance also holds, this angle should be non-oriented. Thus, pattern selection is entirely determined by the knowledge of two complex angular interaction curves, namely $\gamma^+(\theta) = \gamma^-(\pi - \theta)$ and $\chi(\theta) = \chi(\pi - \theta)$. Note finally that $\beta = \gamma(0)^+/2$ and $\gamma = \gamma^+(\pi) = \chi(0)$, as also checked from Eqs (5.23).

As expected, Eqs (5.22) reduce to Eqs (5.2) in the two-dimensional case $P = \{\vec{k}, -\vec{k}\}$. Note however that to be coherent with Eq. (5.1), we have to set $A_{\vec{k},+\omega_c} = A_2, A_{\vec{k},-\omega_c} = A_1^*$ The form (5.22) of amplitude equations was first obtained by Pismen [399], for long-wave convection problems (nearly-insulating boundaries, where $k_c \to 0$). While the angular interaction curves are found to depend rather simply on θ in this case [399], this is not so for $k_c = O(1)$, with important consequences on pattern selection, as seen in later sections.

We now consider some particular structures of the set P, namely hexagonal and square lattices, for which studies dealing with amplitude equations (5.22) exist.

Hexagonal lattice

Roberts et al. [401] have provided an analysis of the Hopf bifurcation on the hexagonal lattice $P = \{\pm\vec{k_1}, \pm\vec{k_2}, \pm\vec{k_3}\}$ with $\vec{k_1} + \vec{k_2} + \vec{k_3} = 0$. In order to use their results, we define $w_n = A_{\vec{k_n},+\omega_c}, z_n = A_{\vec{k_n},-\omega_c}^*, n = 1, 2, 3$, i.e. if $\omega_c > 0$, z_n and w_n represent, respectively, the amplitude of the wave propagating in the direction of $\vec{k_n}$, and opposite to it. Then, the amplitude equation for w_1 is found from (5.22) as

$$
\begin{aligned}
\dot{w}_1 = {}&\mu w_1 - \alpha_1 w_1 |w_1|^2 - \alpha_2 w_1 |z_1|^2 \\
&- \alpha_3 w_1 \left[|w_2|^2 + |w_3|^2\right] - \alpha_4 w_1 \left[|z_2|^2 + |z_3|^2\right] \\
&- \alpha_5 z_1^* \left[w_2 z_2 + w_3 z_3\right]
\end{aligned}
\tag{5.24}
$$

with $\alpha_1 = \beta, \alpha_2 = \gamma, \alpha_3 = \gamma^+(2\pi/3), \alpha_4 = \gamma^-(2\pi/3) = \gamma^+(\pi/3)$, and $\alpha_5 = \chi(2\pi/3) = \chi(\pi/3)$. The corresponding equation for z_1 is found by substituting z by w and vice versa in Eq. (5.24), and the other four equations are found by cyclic permutations of $(1, 2, 3)$.

Using group theoretical methods, Roberts et al. distinguish eleven different types of solutions of Eqs (5.24), listed in Table 5.1. As usual, any reflection, rotation or translation of each of these solutions is an equivalently valid solution. Note also that the system admits in general other types of solutions, i.e. the solutions classified by Roberts et al. are only those with maximal symmetry (i.e. only one of the z_n's is free, while the other z_n's and w_n's are either 0 or related to it). In order to find other less-symmetric solutions, the amplitude equations may for instance be integrated numerically.

Roberts et al. have also analyzed the stability of all solutions given in Table 5.1 as a function of coefficients $\alpha_i, i = 1, ...5$. Their results can be found in [401, 167, 220]. Renardy et al. have applied the theory of Roberts et al. to a number of realistic situations, such as interface deformational modes in two-layer buoyancy-driven convection [220, 402], oscillations between thermal and mechanical coupling in the same set-up ([178], see also §5.4), and buoyancy-driven double-diffusive instability [167]. In the last paper, Renardy also corrects a misprint in one of the stability conditions of twisted patchwork quilt and wavy rolls (II) solutions given by Roberts et al. Note that these stability conditions concern stability to inner

Table 5.1: The eleven solutions with maximal symmetry for the Hopf bifurcation on the hexagonal lattice [401].

Name	Fixed point set
Standing rolls	$z_1 = w_1, z_2 = w_2 = z_3 = w_3 = 0$
Standing hexagons	$z_1 = w_1 = z_2 = w_2 = z_3 = w_3$
Standing regular triangles	$z_1 = -w_1 = z_2 = -w_2 = z_3 = -w_3$
Standing patchwork quilt	$z_1 = w_1 = 0, z_2 = w_2 = z_3 = w_3$
Oscillating triangles	$z_1 = z_2 = z_3, w_1 = w_2 = w_3 = 0$
Wavy rolls (I)	$z_1 = z_3 = w_1 = -w_3, z_2 = w_2 = 0$
Twisted patchwork quilt	$z_2 = \exp[i2\pi/3]z_1, z_3 = \exp[i4\pi/3]z_1, w_p = z_p$
Wavy rolls (II)	$z_2 = \exp[i2\pi/3]z_1, z_3 = \exp[i4\pi/3]z_1, w_p = -z_p$
Traveling rolls	$w_1 = z_2 = w_2 = z_3 = w_3 = 0$
Traveling patchwork quilt (I)	$z_1 = z_3, z_2 = w_1 = w_2 = w_3 = 0$
Traveling patchwork quilt (II)	$z_1 = w_3, z_2 = z_3 = w_1 = w_2 = 0$

disturbances only, while stability to disturbances outside the set P is still not guaranteed, and may e.g. be checked numerically.

The traveling patchwork quilt (I) solutions are always unstable [401]. Also, when traveling rolls are stable, then standing rolls, oscillating triangles and traveling patchwork quilt (II) must all be unstable [167]. There are also some degenerate cases, for which the most dangerous eigenvalue of standing hexagons and regular triangles is exactly zero at the cubic order at which amplitude equations are truncated. Note that this may also happen for twisted patchwork quilts and wavy rolls (II). This degeneracy occurs because the cubic amplitude equations (5.24) are invariant under the change $w_p \to \exp[i\Psi]w_p, z_p \to z_p, p = 1, 2, 3$, which for $\Psi = \pi$, transforms standing hexagons into standing regular triangles and vice versa. Hence, these two solutions are equivalent at cubic order and have the same stability, a degeneracy which is removed when considering higher-order (quintic) terms [401]. The same degeneracy occurs for wavy rolls (II) and twisted patchwork quilts, as can also be seen from Table 5.1. For all other solutions, the shift $w_p \to \exp[i\Psi]w_p, z_p \to z_p, p = 1, 2, 3$ is unimportant, as it merely corresponds to a redefinition of space and time origins.

The eleven solutions, not represented here, are better visualized on a computer by animating the contour lines of the lowest-order planform function

$$\phi(\vec{r}, t) = \mathrm{Re} \left\{ \sum_{p=1}^{3} z_p \exp[i(\omega_c t - \vec{k}_p.\vec{r})] + \sum_{p=1}^{3} w_p \exp[i(\omega_c t + \vec{k}_p.\vec{r})] \right\} \qquad (5.25)$$

Note that all the solutions generated in this way can also be found in the book of Joseph and

Renardy [220]. In the next sections, we will only represent the solutions which will be found to be stable, when investigating some particular Bénard set-ups.

Finally, note that the case of the rhombic lattice where P contains two wavevectors at angle θ (plus their conjugates) can be seen as a particular case of the analysis of this section, corresponding e.g. to $z_3 = w_3 = 0$, and $\alpha_1 = \beta$, $\alpha_2 = \gamma$, $\alpha_3 = \gamma^+(\theta)$, $\alpha_4 = \gamma^-(\theta) = \gamma^+(\pi - \theta)$, and $\alpha_5 = \chi(\theta)$.

Square lattice

Considering now a set $P = \{\pm\vec{k_1}, \pm\vec{k_2}\}$ with $\vec{k_1}$ orthogonal to $\vec{k_2}$, it is readily seen, keeping the notation $w_n = A_{\vec{k}_n, +\omega_c}$, $z_n = A^*_{\vec{k}_n, -\omega_c}$, $n = 1, 2$, that the amplitude equation for w_1 reads

$$
\begin{aligned}
\dot{w}_1 = {} & \mu w_1 - \alpha_1 w_1 |w_1|^2 - \alpha_2 w_1 |z_1|^2 \\
& - \alpha_3 w_1 \left[|w_2|^2 + |z_2|^2 \right] - \alpha_5 z_1^* w_2 z_2
\end{aligned}
\tag{5.26}
$$

with $\alpha_1 = \beta$, $\alpha_2 = \gamma$, $\alpha_3 = \alpha_4 = \gamma^+(\pi/2)$, and $\alpha_5 = \chi(\pi/2)$. As before, other equations are found by exchanging w and z, and by the permutation of 1 and 2.

The analysis of this system has been achieved by Silber and Knobloch [400]. In addition to five solution branches directly found as a particular case of Table 5.1, Silber and Knobloch also consider the standing cross rolls $w_1 = z_1 \neq 0, w_2 = z_2 \neq 0, |w_1| \neq |w_2|$. All these solutions with their specific name are given in Table 5.2, while their stability conditions can be found in [400].

Table 5.2: Six possible solutions for the Hopf bifurcation on the square lattice [400, 430].

Name	Fixed point set
Standing rolls	$z_1 = w_1 \neq 0, z_2 = w_2 = 0$
Traveling rolls	$z_1 \neq 0, w_1 = z_2 = w_2 = 0$
Standing squares	$z_1 = w_1 = z_2 = w_2 \neq 0$
Alternating rolls	$z_1 = z_2 = w_1 = -w_2 \neq 0$
Traveling squares	$z_1 = z_2 \neq 0, w_1 = w_2 = 0$
Standing cross rolls	$w_1 = z_1 \neq 0, w_2 = z_2 \neq 0, |w_1| \neq |w_2|$

While standing and traveling rolls of Tables 5.1 and 5.2 correspond to the two-dimensional SW and TW solutions of §5.1.1, it is seen that standing squares, alternating rolls (also called anti-phase squares [399]) and traveling squares of Table 5.2 respectively correspond to standing patchwork quilts, wavy rolls (I), and traveling patchwork quilts (I or II) of Table 5.1, up to a change of the angle from $2\pi/3$ to $\pi/2$.

Some of these solutions will be represented later on in this chapter, for particular convective problems, when they will be found (numerically) to be stable against inner and outer disturbances.

General pattern selection problem

Owing to the absence of quadratic terms in the general amplitude equations (5.22), there is actually no reason (as was the case for monotonic patterns) for an eventual preference of wave patterns with hexagonal symmetry. Remember that this is due to our assumption $\omega_c = O(1)$, which should in particular not be valid in the neighborhood of a (codimension-2) point where $\omega_c \to 0$ and when some effects break the symmetry with respect to a change of sign of the perturbation fields. This is qualitatively discussed in §5.3.3.

Now, for $\omega_c = O(1)$ and small supercriticality, wave pattern selection depends on the detail of angular couplings $\gamma(\theta)^+$ and $\chi(\theta)$, and it is difficult to arrive at some generic conclusions, given the large number of solutions possible already for hexagonal or square lattices. Moreover, the amplitude equations (5.22) cannot in general be reduced to a potential form, due to cubic phase-coupling terms, i.e. the last ones in Eq. (5.22). If $\chi(\theta) = 0$, it can be shown after separation of the moduli and phases for each amplitude, that the evolution of the moduli (decoupled from that of the phases) is governed by a gradient system. This is also the case when $\chi(\theta) \neq 0$, provided that for each $\vec{k}$, only $A_{\vec{k},\omega}$ or $A_{\vec{k},-\omega}$ is different from zero (competition between traveling waves propagating in different directions).

In the general case, the evolution of phases and moduli cannot be decoupled, and no potential function can be found, such that the most satisfactory way to handle the general pattern selection problem is to integrate Eqs (5.22) numerically, with a large number of waves equispaced on the critical circle. Note in this respect that the angular interaction curves should in general not diverge at any angle, as was the case for monotonic patterns (see §4.4.2). Once again, this is linked to the absence of resonant contributions in the right-hand side of the order ϵ^2 problem (5.16), i.e. to the fact that none of the quadratically generated modes "replicates" the bifurcating ones. Some examples of interaction curves and numerical integrations of Eqs (5.22) will be presented in §5.2 and §5.3.

5.1.3   Mean-flow and vertical vorticity

To conclude this section, and before proceeding to the analysis of particular Bénard problems, let us briefly examine whether mean-flow and/or vertical vorticity could play a more important role here than for monotonic modes (see §4.3.1 and 4.3.2). In fact, while such velocity components were seen to cancel for symmetry reasons (or more exactly to be of order ϵ^3) in the case of monotonic modes, this will generally not be the case for waves, as we show below. Note that to keep within the framework set up in §5.1.2, it is convenient to include Eqs (4.160) and (4.162) from the outset in the definition of our operators $\mathcal{L}(U)$, $\Theta(U)$, $M(U)$ and $N(U,U)$, as well as associated boundary conditions in their domain E.

At the linear stage, as for monotonic patterns, the dynamics of vertical vorticity ω_z and mean flow $\vec{V}_m$ is purely diffusive, as seen from Eqs (4.160) and (4.162) (in this section, the mean flow is denoted by $\vec{V}_m$). Moreover, it is often decoupled from the dynamics of the thermal field and of the potential part of the horizontal velocity field, which are responsible for the instability of the motionless state. On a thermal time scale, the damping rate of ω_z and $\vec{V}_m$ is proportional to the Prandtl number Pr, and we here assume that at least one of the boundaries of the Bénard set-up is rigid, such that the bifurcating eigenmodes are characterized by $\omega_z = \vec{V}_m = 0$. According to Eqs (5.12), (4.156) and (4.158), the first-order velocity field may then be written as

$$
\vec{V}_0 = W_0 \vec{1}_z + \vec{V}_{h0} = \sum_{\vec{k} \in P} \left\{ \begin{array}{l} A_{\vec{k}+} \exp[i(\vec{k}.\vec{r} + \omega_c t_0)] \left(w_0(z)\vec{1}_z + \frac{i\vec{k}}{k_c^2} D w_0(z) \right) + \\ A_{\vec{k}-} \exp[i(\vec{k}.\vec{r} - \omega_c t_0)] \left(w_0^*(z)\vec{1}_z + \frac{i\vec{k}}{k_c^2} D w_0^*(z) \right) \end{array} \right\}
$$

(5.27)

where $w_0(z)$ is the vertical velocity component of the marginal eigenvector at wavenumber k_c and frequency $+\omega_c$, and is independent of the orientation of $\vec{k}$. According to Eq. (4.162), we see that $\vec{V}_{m2}$ (i.e. $\vec{V}_m$ at order ϵ^2) should contain a term independent of the fast time scale t_0, in addition to terms proportional to $\exp(\pm 2i\omega_c t_0)$. However, using Eq. (5.27), it turns out that

$$
\overline{W_0 \vec{V}_{h0}} = -\frac{i}{k_c^2} (w_0 D w_0^* - w_0^* D w_0) {\sum_{\vec{k} \in P}}' \vec{k} \left[|A_{\vec{k}+}|^2 - |A_{\vec{k}-}|^2 \right]
$$

(5.28)

where the overbar stands for the horizontally-averaged value, and $\sum'$ again denotes the sum on half of the set P. Thus, only the term independent of t_0 remains in the nonlinear term of Eq. (4.162), which can then be solved for the second-order mean flow

$$
\vec{V}_{m2} = Pr^{-1} v_m(z) {\sum_{\vec{k} \in P}}' \vec{k} \left[|A_{\vec{k}+}|^2 - |A_{\vec{k}-}|^2 \right]
$$

(5.29)

where $v_m(z)$ is a real function found as the solution of

$$
D^2 v_m = \frac{i}{k_c^2} D (w_0^* D w_0 - w_0 D w_0^*)
$$

(5.30)

with appropriate boundary conditions. Thus, a mean flow is generated whenever there is an asymmetry between left- and right-traveling waves (for any direction $\vec{k}$), and in particular it vanishes for standing waves.

Consider now Eq. (4.160) for the vertical vorticity ω_{z2} at order ϵ^2. Taking into account that the first-order horizontal velocity is given by $\vec{V}_{h0} = k_c^{-2} \vec{\nabla}_h D W_0$, we have

$$
\Delta \omega_{z2} - Pr^{-1} \frac{\partial \omega_{z2}}{\partial t_0} = Pr^{-1} k_c^{-2} \vec{1}_z . \left(\vec{\nabla}_h W_0 \times \vec{\nabla}_h D^2 W_0 \right)
$$

(5.31)

Replacing W_0 in this expression by the vertical component of Eq. (5.27), the right-hand side appears as a double sum on, say, $\vec{k}_1$ and $\vec{k}_2$, each term of which being proportional to $\vec{1}_z.(\vec{k}_1 \times \vec{k}_2)$, as in §4.3.2. Symmetrizing the double sum with respect to the permutation of $\vec{k}_1$ and $\vec{k}_2$, it is seen that terms proportional to $\exp[\pm 2i\omega_c t_0]$ cancel, and Eq. (5.31) can be rearranged as

$$
\Delta \omega_{z2} = \frac{w_0^* D^2 w_0 - w_0 D^2 w_0^*}{2 Pr k_c^2} \sum_{\vec{k}_{1,2} \in P} \vec{1}_z.(\vec{k}_1 \times \vec{k}_2) e^{i(\vec{k}_1 + \vec{k}_2).\vec{r}} \left\{ A_{\vec{k}_1 +} A_{\vec{k}_2 -} - A_{\vec{k}_1 -} A_{\vec{k}_2 +} \right\}
$$

(5.32)

Thus, both the mean flow $\vec{V}_{\mathrm{m}}$ and the vertical vorticity ω_z are found to be independent of the fast time scale t_0 at the lowest order ϵ^2. However, in contrast with monotonic modes, contributions independent of t_0 are generated, which affect the calculation of cubic coefficients (5.21), and hence influence wave pattern selection.

However, there exist cases where even the contributions independent of t_0 vanish at order ϵ^2. Looking back to Eqs (5.29), (5.30) and (5.32), it is seen that this is clearly the case for $Pr \gg 1$, but also when the argument of $w_0(z)$ is independent of z. Actually, it will be seen from particular examples that this can indeed happen, depending on the nature of the mechanism leading to wavy instability.

5.2 Weakly nonlinear internal and surface waves

The first application concerns the competition between internal and surface waves originating from the interaction between buoyancy and surface tension effects in a Bénard layer heated from above (see §3.4). In the present section, cubic coefficients and their angular dependency are explicitly calculated for this specific example. Owing to the absence of quadratic terms in the general amplitude equations (5.22), hexagonal patterns are indeed expected to be much less generic than for the monotonic instability case, such that the selection of wave patterns near the threshold depends much more critically on the details of the cubic interaction coefficients. Typically, the system studied here will be seen to be characterized by multistability between several spatially periodic and quasiperiodic wave patterns, with, however, indications that it might be necessary to pursue the analysis to orders higher than cubic in order to remove some degeneracies.

5.2.1 Two-dimensional waves

As seen in Fig. 3.22 of §3.4, for a fixed value of the slope $Bo = Ra/Ma$ in a Nield's diagram (remember that both Ra and Ma are negative here, corresponding to heating the layer from above), and at some critical $Ma = Ma_{\mathrm{c}}(Bo, Pr)$, a Hopf bifurcation occurs with finite critical frequency ω_{c} and wavenumber k_{c}, at least for values of Bo which are neither too small nor too large (see also Fig. 3.20). Therefore, we are under the conditions where the general analysis of §5.1 applies, and in particular, the normal form (5.2) is valid for the two-dimensional case, with values of nonlinear coefficients β and γ given by (5.23). The complex growth rate σ can be expanded near the threshold as $\sigma = \sigma_0\epsilon$, where $\epsilon = (Ma - Ma_{\mathrm{c}})/Ma_{\mathrm{c}}$ defines the supercriticality.

We will not specify the particular method used to compute β and γ here. In fact, owing to the high values of Rayleigh and Marangoni numbers involved, care has to be taken to avoid numerical inaccuracies in the resolution of the second-order problems (5.16) and when evaluating the integrals in (5.21). Specifically, the second-order problems can be solved analytically, though the resulting expressions are too cumbersome to be presented here. Then, scalar products in (5.21) can be obtained numerically. Actually, a useful estimate of the accuracy of the coefficients is also obtained by checking that the order ϵ^3 problem can indeed be solved (e.g. numerically, using a shooting method).

Critical parameters and corresponding values of the coefficients σ_0, β and γ are provided in Tables 5.3 and 5.4, for $Pr = 0.1$ and $Pr = 6$, and in both cases for a zero free surface Biot number (constant heat flux). Some values of Bo have been chosen in the vicinity of the part of Nield's marginal curve (see Fig. 3.22) yielding the lowest instability threshold (i.e. that corresponding to the optimal balance between buoyancy and surface tension effects). Note that the normalization condition adopted for the critical eigenfunctions is $T_c(z = 1) = 1$, which is responsible for the high numerical values of coefficients α and β (see below). It is seen that for all cases investigated, $\beta_R > 0$ and $-\beta_R < \gamma_R < \beta_R$, hence standing waves are the only stable solution above threshold (see Fig. 5.1). It also appears that the frequency of waves increases with increasing supercriticality ϵ.

Table 5.3: Critical parameters for zero free surface Biot number, and various Prandtl and dynamic Bond numbers Pr and Bo.

Pr	Bo	$-Ma_c \times 10^{-6}$	k_c	ω_c
6	3	0.4116	1.506	1405
6	4	0.3236	1.888	1612
6	5	0.3330	2.287	2017
0.1	5.2	0.1493	0.998	114.3
0.1	10	0.0554	1.501	116.4
0.1	15	0.0693	1.986	184.2

Table 5.4: Coefficients of Landau equations for situations considered in Table 5.3.

Pr	Bo	σ_0	$\beta \times 10^{-6}$	$\gamma \times 10^{-6}$
6	3	$20.4 + 689\,i$	$4.64 - 4.36\,i$	$0.80 + 3.83\,i$
6	4	$23.4 + 791\,i$	$3.50 - 2.11\,i$	$1.08 + 2.68\,i$
6	5	$26.0 + 991\,i$	$3.80 + 0.28\,i$	$1.66 + 1.17\,i$
0.1	5.2	$1.56 + 56.1\,i$	$0.373 - 0.306\,i$	$-0.0754 + 0.0714\,i$
0.1	10	$2.12 + 56.8\,i$	$0.0759 - 0.0079\,i$	$-0.0280 - 0.0481\,i$
0.1	15	$2.45 + 90.4\,i$	$0.0901 + 0.325\,i$	$-0.0574 - 0.379\,i$

The data of Table 5.4 also indicate that the amplitude of surface temperature fluctuations is usually very small for moderate ϵ, and hence difficult to measure in experiments.

Indeed, taking into account that the normalization of eigenfunctions has here been taken as $T_c(z = 1) = 1$, the amplitude of surface temperature oscillations for SW is given by $\Delta T_s = 4\Delta T_c[(\sigma_{0R}\epsilon/(\beta_R + \gamma_R)]^{1/2}$, where ΔT_c is the critical temperature difference. In §3.4, it was seen that for low-viscosity liquids such as water at about 50 °C, or acetone at room temperature, we had $\Delta T_c \sim 10$ °C. Thus, according to Table 5.4, ΔT_s should be of the order of $\epsilon^{1/2} \times 0.1$ °C, i.e. within the resolution of most thermocouples. Note that the temperature variations in the bulk are somewhat higher, though still difficult to measure. However, taking into account the large absolute values of critical Marangoni and Rayleigh numbers, it appears that the velocity fluctuations and in particular the surface velocities are easily measurable quantities. For the situation considered above, the maximal free surface velocity is indeed found to be of the order of $\epsilon^{1/2} \times 4$ mm/s.

5.2.2 Three-dimensional waves

Some examples of angular interaction curves are represented in Fig. 5.2, for two cases considered in Table 5.3, namely $Pr = 6, Bo = 4$, and $Pr = 0.1, Bo = 10$. Note that all curves are normalized by the corresponding value of β_R, i.e. the real part of the self-interaction coefficient. It should also be noted that, as expected from the analysis of §5.1.2 and in contrast with the corresponding curves for monotonic modes (e.g. Fig. 4.26), the curves of Fig. 5.2 do not diverge at any angle, i.e. no quadratic resonance occurs.

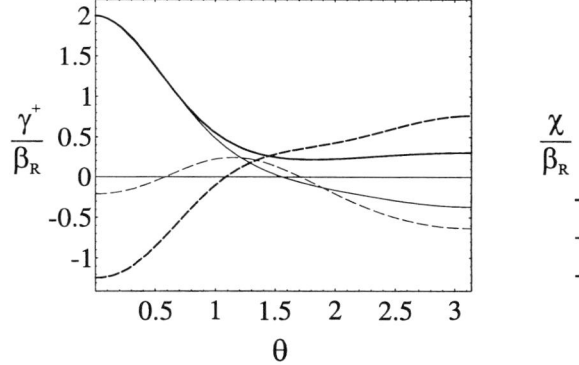
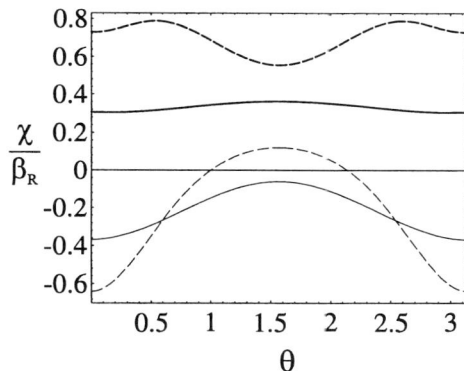

Figure 5.2: Real (full) and imaginary (dashed) parts of the angular interaction coefficients $\gamma^+(\theta)/\beta_R$ (left) and $\chi(\theta)/\beta_R$ (right). Thicker curves correspond to $Pr = 6, Bo = 4$, while thinner ones correspond to $Pr = 0.1, Bo = 10$.

Identifying the necessary coefficients, we may readily apply the analysis of solutions with hexagonal symmetry of Roberts et al. [401] to the problem considered here. On doing so, it is found that the eleven patterns of Table 5.1 are supercritical both for $Pr = 0.1, Bo = 10$ and $Pr = 6, Bo = 4$. For $Pr = 0.1, Bo = 10$, the only stable patterns among these solutions are standing hexagons and standing regular triangles. Unfortunately, this is just one of the degenerate cases mentioned above, for which both solutions are equivalent, due to an artificial symmetry of the cubic amplitude equations. Hence, one of their eigenvalues is necessarily 0 at this order, and we need to pursue calculations up to order five to remove the degeneracy and actually decide which of these two solutions will be selected. Note that for $Pr = 6, Bo = 4$, the same kind of degeneracy occurs, this time for wavy rolls (II) and twisted patchwork quilts.

However, the calculation of order five coefficients is a very cumbersome task, which will not be attempted here. Rather, we will limit ourselves to an illustration of some physical aspects of the phenomena which can be expected for the problem being considered. We will also consider hereafter a larger number of waves equi-spaced on the circle of radius k_c, thereby attempting to approach the ideal rotationally invariant pattern selection problem.

First, note that the standing waves which were found to be stable in two-dimensions are here unstable to three-dimensional perturbations. It is of interest to examine the dynamics of the wave amplitudes in this situation. In Fig. 5.3, we present the results of a direct numerical integration of the amplitude equations (5.22), considering a set of $N = 72$ waves on the critical circle, i.e. the minimal angle between waves is $5°$. The initial condition corresponds to standing waves in one direction, plus a random noise of amplitude 10^{-3} distributed among all the other waves.

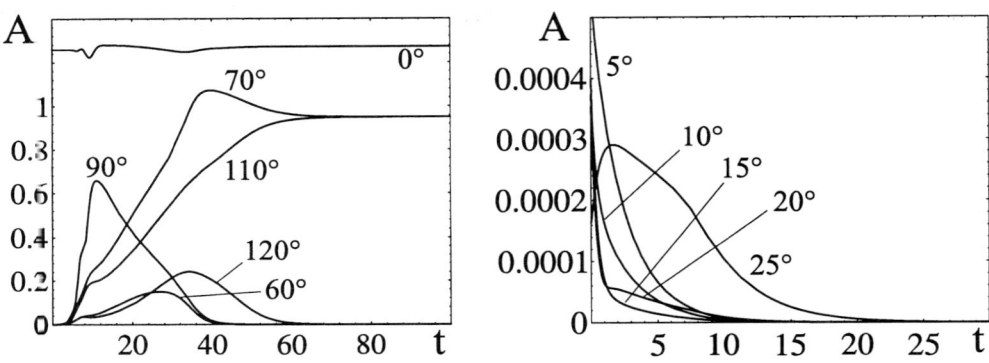

Figure 5.3: Evolution of several wave amplitudes for $Pr = 0.1, Bo = 10$, starting from a standing wave in a particular direction. Amplitudes are labeled by the angle they form with the direction of the initial standing wave (labeled 0). Left: some waves may grow and reach a finite-amplitude value, provided they develop at a sufficient angle from the initial wave. Right: wave amplitudes close to the initial wave are strongly damped.

Results of Fig. 5.3 may be qualitatively explained considering the strong damping effect a particular wave has on its nearest neighbors (right plot). However, this effect, quantified by the coefficient $\gamma^+(\theta)$, here appears to decrease rapidly with the angle θ, when compared with other physical systems (see e.g. §5.3). Eventually, $\gamma^+(\theta)$ may even become negative, indicating that a wave can help to excite waves propagating at a sufficient angle (left plot). Note that for the case considered here, $|A_{\vec{k}+}| \simeq |A_{\vec{k}-}|$ for all times and all $\vec{k}$, indicating a strong tendency to standing patterns, rather than propagating structures. The final stage of the evolution consists in a superposition of 3 standing waves at angles $0°$ (the initial one), $70°$ and $110°$, which in fact appears as a distorted type of standing hexagons or regular triangles (which would have been observed for the particular case of a hexagonal lattice), as discussed above. A snapshot of this pattern is represented in Fig. 5.4.

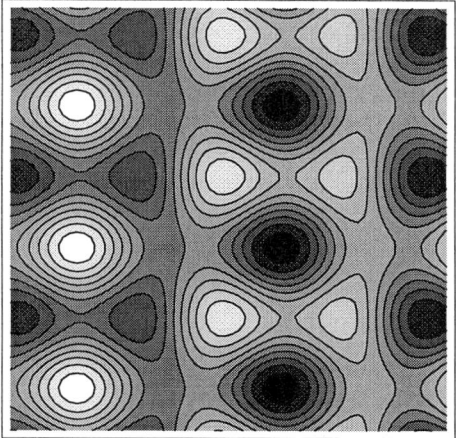

Figure 5.4: A snapshot of the standing wavy pattern reached for $t \to \infty$ in the simulation of Fig. 5.3 ($Pr = 0.1, Bo = 10$), intermediate between (distorted) standing hexagons and triangles (degenerate case). The pattern oscillates in time as a whole, with little variations of its shape and position of its nodes, similarly to a standing wave.

Actually, this is not the only pattern found to be stable for this problem. Indeed, for the case $Pr = 0.1, Bo = 10$, patterns consisting of five, six or even eight excited waves could be generated starting from different initial conditions, and found to be simultaneously stable. Some of these patterns look rather exotic, such as the one represented in Fig. 5.5. For $Pr = 6, Bo = 4$, not so many qualitatively different patterns are found, because of the stronger damping effect between waves propagating in different directions. Actually, it appears in this case that starting from random initial conditions, the final pattern often consists of anti-phase oscillating squares (an example of which is represented in Fig. 5.11 of the next section), or distorted versions of it, where the angle between basic standing waves is not exactly 90°.

To conclude, we emphasize that the results described above should be considered with caution, owing to the possible artificial degeneracies of the system of cubic amplitude equations, already encountered for the subset of solutions with hexagonal symmetry. These degeneracies might be removed at higher order, i.e. quintic in the perturbation scheme of §5.1.2. However, being small, this effect should manifest itself on a longer time scale, thus allowing the expectation that patterns of the kind found above might be observed as transient states in an experiment.

Remark finally that several patterns found here are spatially quasiperiodic, being composed of waves with incommensurate spatial frequencies. Therefore, these patterns cannot be predicted in numerical simulations assuming periodic boundary conditions. However, for sufficiently large domains (say, a box with aspect ratios $L_x \gg 1$ and $L_y \gg 1$), periodic patterns of increasing spatial complexity can be observed [405], looking similar to quasiperiodic ones,

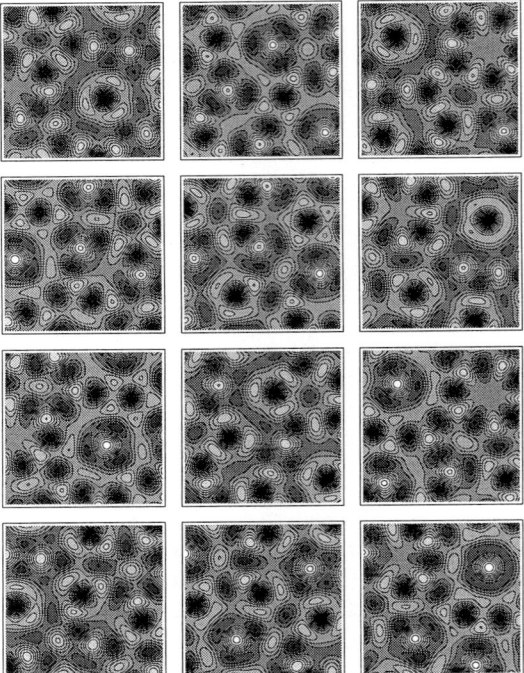

Figure 5.5: Twelve snapshots sampled over one period for one of the stable patterns found for $Pr = 0.1, Bo = 10$, from a direct simulation with $N = 48$ waves and random initial conditions. The evolution is from left to right, and top to bottom. The pattern consists in 5 waves, traveling at angles 0, 75, 142.5, 217.5, and 285°, with slightly different amplitudes and no phase-coupling.

as integer multiples of the fundamental wavenumbers $2\pi/L_x$, $2\pi/L_y$ can indeed be arbitrarily close to any wavenumber.

5.3   Double-diffusive oscillatory Marangoni convection

In this section, weakly nonlinear waves are studied for the Marangoni–Bénard problem in binary mixtures with Soret effect. The reader is referred to §3.7 for the corresponding linear stability analysis. It was seen there that an oscillatory onset is predicted for negative values of the separation ratio Ψ_{Ma}, i.e. when the thermal contribution is destabilizing, while the solutal contribution stabilizes the basic diffusive state. The weakly nonlinear regimes of this Hopf bifurcation are investigated in this section, using the methodology established in §5.1.2 to calculate the coefficients of complex Ginzburg–Landau equations.

As will appear in the following, various wave patterns are also predicted for this problem, depending on the governing parameters. Two-dimensional waves may develop either as standing waves, or traveling waves, though these 2D waves in general turn out to be unstable to 3D disturbances. For the hexagonal lattice, the stable wave patterns among those of Table 5.1 are oscillating triangles, wavy rolls (I), and traveling patchwork quilts (II). Multistability of these solutions is also possible, even with other solutions not on the hexagonal lattice such as oscillating squares.

Moreover, this section is concluded by presenting some recent results, based on "numerical experiments", i.e. direct numerical simulations of the hydrodynamic equations. These will show that weakly nonlinear theories presented in §5.1 do possess limitations, that are drastic in the case studied here. Namely, strong symmetry-breaking effects, here linked to the Marangoni effect at the free surface, will be shown to lead to waves and even steady patterns not described by amplitude equations derived in §5.1.

5.3.1 Two-dimensional waves

At the Hopf threshold $Ma = Ma_c$, a pair of complex conjugate eigenvalues $\sigma = \pm i\omega_c$ crosses the imaginary axis. The critical wavenumber k_c is finite and varies only slightly with Ψ_{Ma}, as seen in Fig. 3.39 of §3.7. The Landau equations (5.2) derived in §5.1.1 should thus be valid here for sufficiently large ω_c (i.e. large $|\Psi_{Ma}|$), as well as the discussion summarized in Fig. 5.1. As usual, we need to calculate the nonlinear coefficients β and γ, which can be achieved from Eqs (5.23). The complex growth rate σ can be expanded near the threshold as $\sigma = \sigma_0\epsilon$, where $\epsilon = (Ma - Ma_c)/Ma_c$ defines the supercriticality. The specific numerical method employed here is, as in §5.2, a semi-analytical method complemented by a check of the accuracy of coefficients by a numerical resolution of non-homogeneous systems of ODEs appearing at successive orders.

First of all, it turns out that the real parts of β and γ are positive for usual water–alcohol mixtures, in the absence of buoyancy ($Ra = 0$). Accordingly, the bifurcation is supercritical (see Fig. 5.1) leading either to TW or SW. This actually depends on the separation ratio Ψ_{Ma}, as can be seen from Fig. 5.6, obtained for $Le = 0.05$, $Ra = 0$, $Bi = 0.5$.

Note that the wavenumber is fixed to $k = 2$ in Fig. 5.6, a reasonable approximation in view of the corresponding Fig. 3.39 of §3.7. It is found that TW are preferred sufficiently close to the codimension-2 point $\Psi_{CTP} = -0.028$ while SW become stable for $\Psi_{Ma} < -0.095$. The reader is referred to §3.7.3 for representations of TW and SW, and for a discussion of physical mechanisms. Further details about two-dimensional steady and oscillatory regimes of double-diffusive Marangoni–Bénard convection in small aspect-ratio containers can be found in [280].

5.3.2 Three-dimensional waves

A typical angular dependency of cubic interaction coefficients is represented in Fig. 5.7.

When comparing Fig. 5.7 with Fig. 5.2 corresponding to the competition between internal and surface waves, it is seen that a stronger damping effect between neighboring waves should be observed here (the real part of γ^+ decreases less rapidly with the angle of interaction θ). However, for separation ratios closer to the codimension-2 point, where the basic frequency

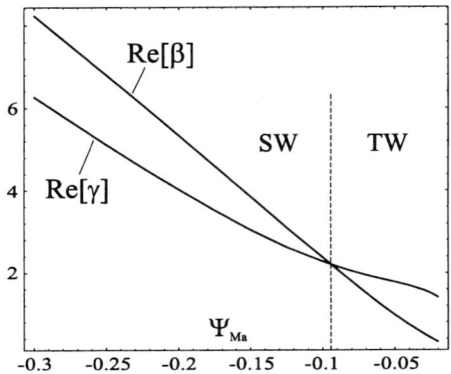

Figure 5.6: The real parts of cubic coefficients β and γ as a function of the Marangoni separation ratio Ψ_{Ma} for $Ra = 0$, $Pr = 6$, $Le = 0.05$, $Bi = 0.5$ and a fixed wavenumber $k = 2$.

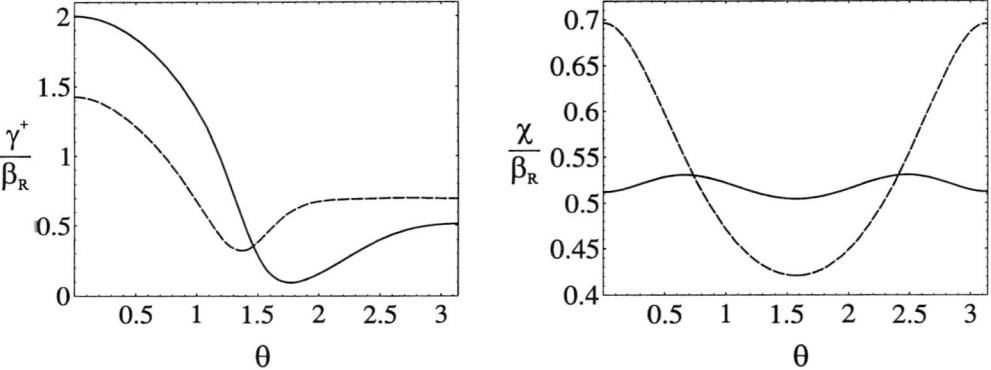

Figure 5.7: Real (full) and imaginary (dashed) parts of the angular interaction coefficients $\gamma^+(\theta)/\beta_{\mathrm{R}}$ (left) and $\chi(\theta)/\beta_{\mathrm{R}}$ (right), for pure Marangoni–Bénard convection ($Ra \simeq 0$), $k = 2$, $Le = 0.05$, $Pr = 6$, $Bi = 0$ and $\Psi_{\mathrm{Ma}} = -5$.

ω_{c} vanishes, resonances can be expected, as discussed in §5.1.2. The corresponding curves, which present increasingly sharp peaks at angles 60 and 120°, are not given here, as the validity of the theory presented in §5.1 in such degenerate cases is not guaranteed.

Hexagonal lattice

For perturbations with hexagonal symmetry, the discussion of Roberts et al. [401] of the eleven solutions with maximal symmetry is applicable. Here, we restrict attention to those patterns which are found to be stable for the Rayleigh–Marangoni–Bénard problem in binary mixtures with Soret effect. Owing to the large number of dimensionless parameters affecting the values of cubic coefficients, here we only examine a few representative cases.

We consider a set of parameter values typical of water–alcohol mixtures ($Pr = 6$, $Le = 0.05$) in contact with a poorly conducting gas ($Bi \simeq 0$). The separation ratio is varied over some range, such a possibility being offered by its strong variation with the initial composition of the mixture. For pure Marangoni–Bénard convection ($Ra = 0$) and large negative separation ratio ($\Psi_{Ma} = -5$), we find oscillating triangles (Fig. 5.8) to be the only stable supercritical solution of Table 5.1. The ten other solutions are found to be supercritical but unstable.

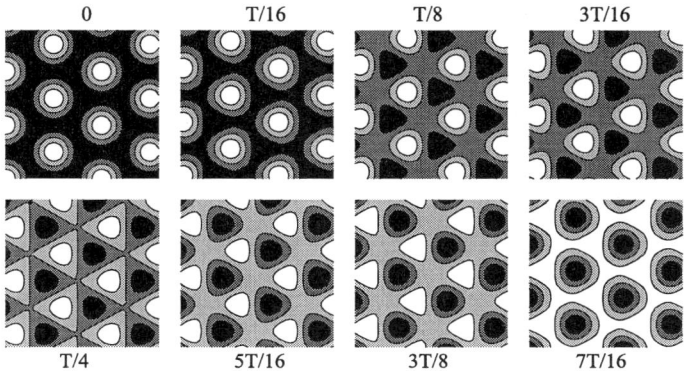

Figure 5.8: The oscillating triangles solution. Only one half of the period T is represented.

For decreasing $|\Psi_{Ma}|$, a region exists (e.g. for $\Psi_{Ma} = -2, -1, -0.3$) where the oscillating triangles are simultaneously stable with the wavy rolls (I), which are represented in Fig. 5.9. Therefore, pattern selection depends on the initial conditions in this case.

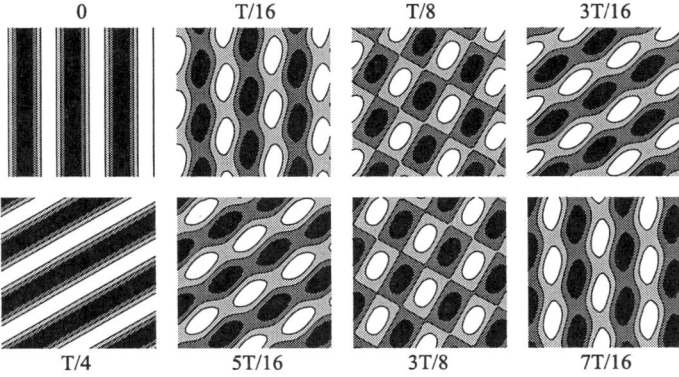

Figure 5.9: The wavy rolls (I) solution. Only one half of the period T is represented.

For smaller $|\Psi_{Ma}|$, wavy rolls (I) eventually become the only stable solution (such as for

$\Psi_{Ma} = -0.05$). For Ψ_{Ma} even closer to the codimension-2 point, some resonance effects occur due to the vanishing of the frequency (cubic coefficients diverge, see next section), and apparently destabilize all solutions. There, the validity of the amplitude equations breaks down, because they rely on the assumption that the frequency is asymptotically larger than the growth rate. For example, at $\Psi_{Ma} = -0.03$, traveling patchwork quilts (II), represented in Fig. 5.10, are found to be subcritical and unstable, and none of the eleven solutions classified in Table 5.1 is stable.

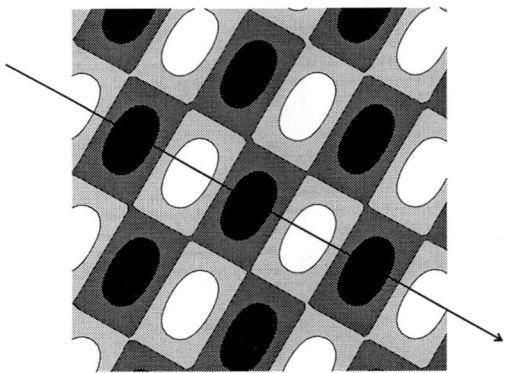

Figure 5.10: The traveling patchwork quilt (II) solution. The arrow indicates the direction of propagation.

There exist situations where traveling patchwork quilts (II) are stable. This was found for pure buoyancy-driven convection ($Ma = 0$) at $\Psi_{Ra} = -0.05$ and for $Bi = 0.5$. However, calculations of nonlinear coefficients indicate that there is a greater tendency to subcritical (unstable) patterns in the case of pure Rayleigh–Bénard convection. For larger $|\Psi_{Ra}|$, it appears indeed that 2D solutions may even be subcritical, which is in agreement with [166].

General pattern selection problem

As discussed in §5.1.2, no special role should be attributed to hexagonal patterns in the case of weak-amplitude waves, because amplitude equations do not contain quadratic phase-locking terms in general. Numerical integration of the full amplitude equations (5.22) with a large number of waves equi-spaced on the critical circle reveals that this is indeed the case for most of the situations described above. In particular, anti-phase oscillating squares (or alternating rolls, see Table 5.2 and Fig. 5.11) are generally found to be stable, as in previous studies of the long-wave purely buoyancy-driven case [399].

Eventually, they coexist with some solutions found above (oscillating triangles and wavy rolls I), therefore implying situations of bi-stability and even tri-stability. It is worth recalling here that the non-variational character of amplitude equations precludes any "energetic" preference for a particular pattern, as is the case for steady patterns. Front dynamics between different wavy three-dimensional patterns would thus be a particularly interesting subject of

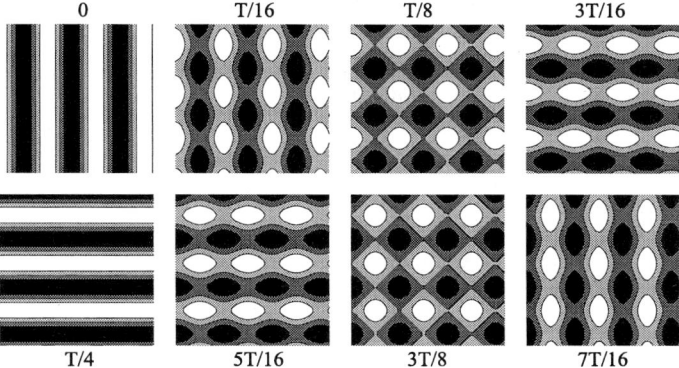

Figure 5.11: The anti-phase oscillating squares (alternating rolls) solution. Only one half of the period T is represented.

future investigation, although to the authors' knowledge, no rigorous analytical tool exists for this purpose (see also §5.5).

Finally, it is worth commenting that in the situation considered here, the basin of attraction of the anti-phase oscillating squares solution numerically appeared to be larger than that of other stable solutions with hexagonal symmetry. Indeed, starting from various sets of randomly chosen initial amplitudes, hexagonal solutions were very seldom observed.

5.3.3 Strong resonance and/or small frequency limitations

It is important to keep in mind that amplitude equations derived in §5.1 all rely on the basic assumption that the critical frequency remains of the order unity, while the amplitude of convection and the real growth rate are assumed to be small near the threshold. Such conditions are not met near a point where the frequency vanishes, e.g. when the oscillatory branch joins a neutral stability curve (codimension-2 point).

In the case of double-diffusive instabilities, such a degenerate case is known to occur at a sufficiently small separation ratio $|\Psi|$ (see e.g. Fig. 3.39 for surface-tension-driven double-diffusive instabilities, though qualitatively the same kind of diagram is found for the buoyancy-driven case [24]). In this section, some results of direct numerical simulations performed at moderate values of $|\Psi|$ are presented. Note that even though we will here stay at a reasonable distance from the codimension-2 point, the observed convective structures do not straightforwardly confirm predictions of the amplitude equation analyzes presented above.

Numerical experiments

An analysis of the Marangoni–Bénard problem in binary mixtures with Soret effect has recently been performed by Bestehorn and Colinet [465], in the region where waves with finite wavenumber are expected, i.e. at negative values of the separation ratio Ψ (see e.g. Fig. 3.39), us.ng properties of a typical water–alcohol mixture in contact with air ($Pr = 6$, $Le = 0.05$, $Bi \simeq 0$).

The basic equations considered in [465] and definitions of dimensionless parameters are given by Eqs (3.214–3.221) of §3.7. The method used is a pseudo-spectral method with a Fourier decomposition in the horizontal directions (periodic lateral boundary conditions), and finite differences for the vertical direction. The velocity field is first decomposed into a toroidal and a poloidal part [as in Eq. (4.157) of §4.3], whose governing equations can be obtained from Eqs (4.155) and (4.160). The mean flow can also be included, and calculated from Eq. (4.162), but turns out to be negligible (four orders of magnitude smaller than other velocity components) in the present case. In addition to the moderate value of the Prandtl number $Pr = 6$, a further reason for the smallness of the mean flow observed here might be the weak dependence of the phase of the critical velocity eigenmode on the vertical coordinate z (see the discussion of §5.1.3). More details about the numerical method may be found in [316, 465, 466].

The linear stability problem has been solved using the same numerical scheme for the vertical direction as for the full problem. A typical result is presented in Fig. 5.12, for a water–alcohol mixture, with a separation ratio taken in the range $-2 < \Psi_{Ma} < 0.1$.

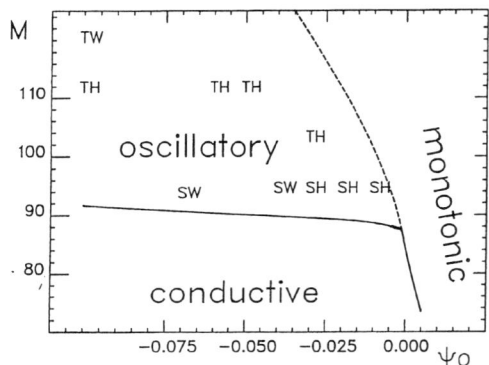

Figure 5.12: Linear stability results for $Pr = 6$, $Le = 0.05$, $Bi = 0.1$, in the region where the oscillatory branch joins the neutral stability curve. Results of some numerical simulations in the nonlinear regime are indicated (SH: time-steady hexagons, TH: traveling hexagons, SW: standing wave-like pattern, TW: traveling wave-like pattern). Note that $\Psi_0 = Le\Psi_{Ma}$ and M stands for Ma in this figure.

The nonlinear results which we now describe have been obtained in square periodic domains with dimensions 15×15 (in units of the layer depth), starting from random initial distributions with amplitude small compared with the one reached at saturation. Near the oscillatory line of Fig. 5.12 and for not too negative separation ratios (i.e. ω_c not larger than 1 in thermal units), steady hexagons are found, such as those shown in Fig. 5.13. These structures and their defects are typical for convection in a pure fluid, even though we are here in the oscillatory regime.

When increasing the Marangoni number [or $\epsilon = (Ma - Ma_c)/Ma_c$], a bifurcation to

Figure 5.13: Steady hexagons with point defects, observed at $\Psi_0 = -0.02$, $Ma = 95$ ($\epsilon = 0.1$).

traveling patterns is observed (see Fig. 5.14). These may either look like hexagonal structures of the previous figure (right plot), or be more asymmetric (left plot), depending on ϵ and Ψ_0.

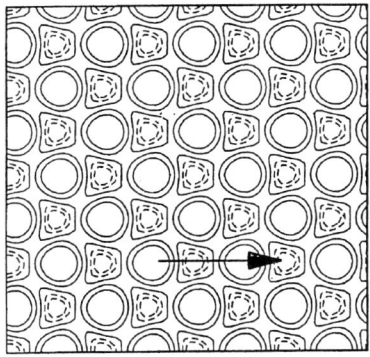
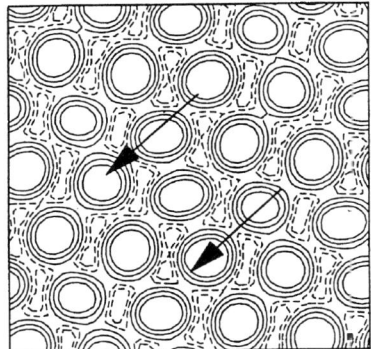

Figure 5.14: Left: traveling asymmetric regular structure with a strong square component observed at $\Psi_0 = -0.1$, $Ma = 112$ ($\epsilon = 0.3$). The structure travels with constant velocity to the right. Right: Traveling hexagons at $\Psi_0 = -0.03$, $Ma = 103$ ($\epsilon = 0.2$). The direction of motion is perpendicular to one of the three wavevectors.

At first sight, these results are rather surprising, given the amplitude equation analyzes presented in §5.1.2 and §5.3.2. First, we note a strong tendency to patterns with hexagonal symmetry, especially for low values of Ψ_0, while numerical integrations of amplitude equations with a large number of waves on the critical circle have typically shown a preference for solutions with square symmetry (see end of last section). Hence, it seems that quadratic resonance effects between triads of wavevectors $\vec{k}_1 + \vec{k}_2 + \vec{k}_3 = 0$ do enter the picture here. Actually, other direct numerical simulations at larger ϵ and/or ω_c (more negative separation ratio) have shown much less resonant structures, looking closer to the standing waves and traveling waves described in the previous sections (though spatial modulation effects might be important).

Moreover, even for perturbations on the hexagonal lattice, it is clear that the analysis of §5 1.2 cannot lead to steady solutions, nor can it predict the traveling hexagons of Fig. 5.14 (right). In both cases, it is easily seen that this would require the phase of some of the amplitudes to vary on a fast time scale, given the factor $\exp[i\omega_c t]$ in Eq. (5.25) for instance. Now, allowing the basic frequency ω_c to be small enough, it is clear that in principle, such solutions should be allowed. This has been thoroughly discussed by Renardy [470] and Renardy et al. [471, 485], who have indeed shown that the amplitude equations derivation need to be carefully reconsidered in this case, especially in the presence of strong quadratic resonance effects. These authors predict a variety of new patterns (including time-quasiperiodic ones), some of which being very similar to those observed in the direct simulations presented above, and to other ones found in [465]. They illustrate their approach on a two-layer buoyancy-driven Bénard system, whose two-dimensional regimes were studied theoretically by Colinet and Legros [177] (see also §5.4), and experimentally by Andereck et al. [180, 181].

Rather than amplitude equations analysis, Bestehorn and Colinet [465] have considered the role of quadratic nonlinearities using a generalization of the phenomenological (complex) Swift–Hohenberg model (see also §4.5.5). They also predict bifurcation sequences in good qualitative agreement with their direct numerical results based on hydrodynamic equations, which points to the generic nature of such types of patterns in the vicinity of a Hopf bifurcation with finite critical wavenumber, in the presence of strong resonance and/or small basic frequency effects.

Phenomenological model

We now turn to a simple phenomenological model which seems to capture the ingredients responsible for the breakdown of the amplitude equations description of §5.1, for small ω_c and/or strong symmetry-breaking effects. We will consider here the single complex ODE

$$\partial_t u = (\epsilon^2 + i\omega_c)u + \delta_1 u^2 + \delta_2 (u^*)^2 + \delta_3 |u|^2 - u|u|^2 \tag{5.33}$$

where u is a complex variable, and coefficients $\delta_{1,2,3}$ can also be taken to be complex in general. All the quadratic terms in this ODE break the symmetry $u \to -u$, as is often the case for hydrodynamic systems, e.g. when Marangoni or non-Boussinesq effects are significant. The ODE (5.33) is clearly much simpler than the full system of hydrodynamic equations and boundary conditions, but as we will see, it will help to understand some results presented in the previous sections.

For $\omega_c = O(1)$ and $\epsilon \ll 1$, we may in principle develop $u = \epsilon u_1 + \epsilon^2 u_2 + \ldots$ and $\partial_t = \partial_{t_0} + \epsilon^2 \partial_{t_2}$ as in §5.1.2, and identify terms at the successive powers of ϵ. At order ϵ, we find $(\partial_{t_0} - i\omega_c)u_1 = 0$, which is solved by $u_1 = a(t_2) \exp(i\omega_c t_0)$. After straightforward resolution of the order ϵ^2 problem for u_2, it is seen that the order ϵ^3 problem can be solved for u_3 in terms of periodic (i.e. non-secular) contributions, provided the resonant contributions cancel each other. This occurs when the amplitude equation for $a(t_2)$ satisfies

$$\partial_{t_2} a = a - \beta a |a|^2 \tag{5.34}$$

which is a complex Ginzburg–Landau equation with

$$\beta = 1 + \frac{i}{\omega_c}\left(-\delta_1\delta_3 + \frac{2}{3}|\delta_2|^2 + |\delta_3|^2\right) \tag{5.35}$$

This normal form describes the Hopf limit cycle observed in the vicinity of the instability threshold, i.e. locally near the origin $a = 0$, and is certainly valid at sufficiently small supercriticality ϵ. Note however that for ω_c tending to zero, the cubic coefficient β diverges, which indicates that an improved analysis might be needed under such conditions. Remark also that for the particular case $\mathrm{Im}[\delta_1\delta_3] = 0$, the limit $\omega_c \to 0$ reduces Eq. (5.35) to a nonlinear Schrödinger equation $\dot{a} = \pm ia|a|^2$, on a faster time scale $\omega_c^{-1}t_2$. Neglected dissipative corrections of order ω_c should then be re-incorporated to provide selection of the amplitude of oscillation.

Actually, coming back to the Bénard problem studied in the previous sections, it appears that the cubic coefficients β and γ not only diverge proportionally to ω_c^{-1}, but also become purely imaginary in the limit of small ω_c, as seen in Fig. 5.15.

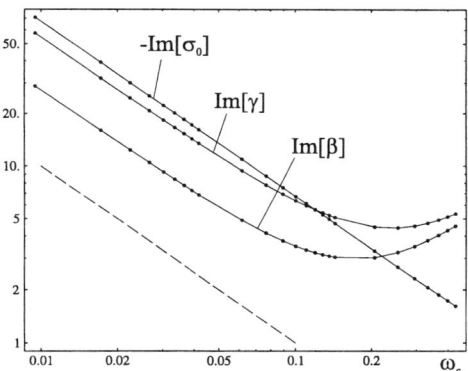

Figure 5.15: Imaginary parts of the cubic coefficients β and γ, for the Marangoni–Bénard instability in binary mixtures, in the limit of small Hopf frequency ω_c. Parameters used are those of Fig. 5.6, where the corresponding real parts are seen to tend to a constant for $\omega_c \to 0$ ($\Psi_{\mathrm{Ma}} \to \Psi_{\mathrm{CTP}}$). Also represented is the imaginary part of σ_0, i.e. the slope of the linear frequency versus ϵ at the critical point. The dashed line indicates a scaling ω_c^{-1}.

Now, even for $\omega_c = O(1)$, the limit cycle described by (5.34) may not be the only solution of Eq. (5.33), even at small ϵ. Indeed, for not too small values of δ_i (and not too large ω_c), this ODE can even have *fixed* points ($\partial_t u = 0$). For instance, for $\omega_c = 1$, $\epsilon = 0.02$, $\delta_1 = 2$, $\delta_2 = 0$ and $\delta_3 = 1$, there are two fixed points, one of them being stable and the other unstable (see Fig. 5.16).

Therefore, a preliminary conclusion drawn on the basis of the discussion of the last two sections, is that the weak-amplitude waves described by the general amplitude equations (5.22) are not the only solutions when strong symmetry-breaking effects occur or near a

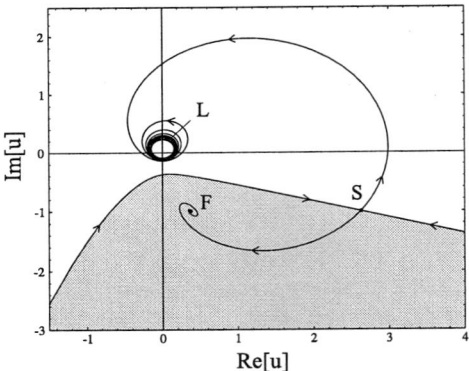

Figure 5.16: Phase portrait of the ODE $\partial_t u = (\epsilon^2 + i)u + 2u^2 + |u|^2 - u|u|^2$ just above the Hopf threshold $\epsilon = 0.02$. In addition to the limit cycle L, there is an (unstable) saddle-point S and a stable focus F. For initial conditions in the light grey region, the final state is F, while all other initial conditions lead to the limit cycle L.

codimension-2 point where the Hopf frequency is small. Other solutions with hexagonal symmetry, possibly time-independent, can be expected at $O(1)$ amplitudes, even for very small supercriticality. Recent analyzes have shown that the small-amplitude waves described by (5.22), although stable at very small ϵ (at least some of them, depending on cubic coefficients), very quickly become unstable when ϵ is increased, and indeed evolve to solutions with amplitudes of order unity. Note that a more general discussion of solutions with hexagonal symmetry expected in this situation has been provided by Renardy et al. [470, 471]. It also appears from these works and from the above results that near-Hamiltonian dynamics should be observed in this situation, i.e. the patterns may oscillate for a very long time before they finally settle into a stable state. Further discussions of regimes and transitions expected near a codimension-2 point where $\omega_c \to 0$ may be found in [163], and in the next section where another Bénard system is considered.

5.4 Two-layer Rayleigh–Bénard instability

In this section, a last example of a system involving oscillatory onset is considered, namely the competition between thermal and mechanical coupling modes in two-layer Rayleigh–Bénard convection. As this problem was not considered in Chapter 3, both linear and weakly nonlinear regimes will be addressed in this section, which will also serve to illustrate a different method based on Galerkin expansions (§5.4.2). As will be seen in §5.4.3, this problem is characterized by a Hopf bifurcation at onset when local Rayleigh numbers in both layers are close to each other. Moreover, a further condition is that the ratio of the thicknesses of the layers is near unity. Despite these restrictions, these waves have recently been experimentally observed by Andereck and co-workers [180, 181], who confirmed the prediction of Colinet and Legros [177] about the traveling nature of the waves. The latter authors have also shown that the range of existence of the waves is limited, and that a transition to steady mechanical coupling is generally observed when increasing the thermal constraint. In §5.4.5, some of their numerical results are described, which explain this transition in terms of the proximity of a codimension-2 point, a situation which bears some analogy with double-diffusive Bénard set-ups studied in the previous section.

5.4.1 Thermal versus mechanical coupling

Consider a system of two immiscible fluid layers, which are superimposed and confined between rigid isothermal plates kept at different temperatures. It is intuitive that mechanical and thermal continuity conditions at the interface will couple the individual behaviors in both layers. In particular, in contrast with one-layer Rayleigh–Bénard convection in pure fluid systems, two-layer systems may exhibit a number of qualitatively different behaviors. Among these, overstability is possible, as it was first shown by Gershuni and Zhukhovitskii [174], and by Rasenat, Busse and Rehberg [175]. As will be seen in next sections, these oscillations involve different mechanisms than those responsible for oscillatory "interfacial convection", associated with a non-vanishing deformation of the interface. This last kind of instability, which will not be addressed here, has already been studied by Wahal and Bose [406], Richter and Johnson [407], Richter and McKenzie [408], Renardy and Joseph [409], Y. Renardy and M. Renardy [176, 402, 410]. This mechanism is connected to a Rayleigh–Taylor kind of instability, and, as shown below, it is legitimate to neglect this interfacial mode provided that the density jump between both layers is large enough.

Besides the interesting theoretical aspects of these mechanisms for the understanding of bifurcation phenomena and pattern interactions in hydrodynamical systems, it is worth mentioning that research in the field of two-layer convection was triggered by studies of earth mantle convection: in this frame, bi-layered convection models have been constructed on the basis of the 700-km discontinuity in chemical composition, as opposed to "whole mantle" convection models (see [172, 407, 408, 413, 414, 415]).

Experimental and theoretical studies of Rasenat et al. [175], Nataf, Moreno and Cardin [171, 411, 412], and Prakash and Koster [179] have shown that stationary buoyancy-driven convection in bi-layered systems can be either mechanically coupled (MC), or thermally coupled (TC). The first case is more likely to appear, since except for peculiar cases (which are analyzed in this section), one of the layers reaches critical conditions faster (for a lower thermal constraint) than the other, which is in turn passively driven through the viscous coupling of horizontal velocities at the interface. Consequently, as seen on Fig. 5.17(a) in a two-dimensional case, convective cells that are facing each other in both layers rotate in opposite directions, and ascending motions in one layer face descending motions in the other one. The other kind of coupling, namely thermal coupling, requires both layers to be in supercritical conditions: this becomes clearer by looking at Fig. 5.17(b), which shows that two superposed cells are rotating in the same direction, and ascending motions in one layer face ascending motions in the other one. This mechanism occurs through diffusion of thermal fluctuations through the interface, as seen from the behavior of isotherms in Fig. 5.17(b). Both layers indeed need to be in supercritical conditions to overcome strong gradients of horizontal velocity near the interface.

Between these two extreme cases of mechanical and thermal coupling, each of them being

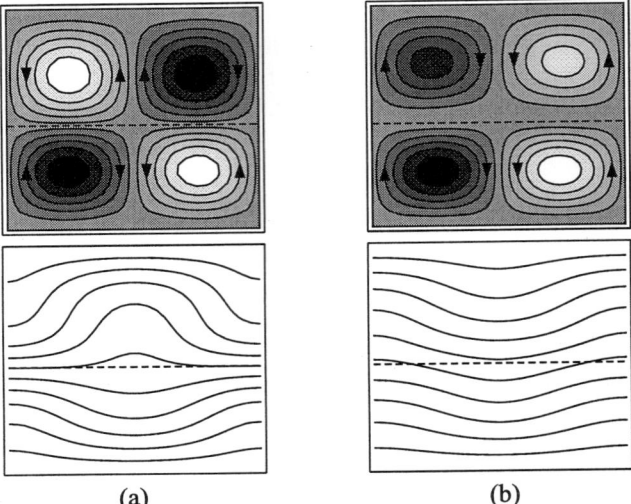

(a)   (b)

Figure 5.17: Streamlines and isotherms in the mechanical coupling situation (a), and in
the thermal coupling situation (b). The sense of rotation of convective cells is indicated
by arrows. The interface is indicated as a dashed line.

able to become the preferred mode of stationary convection for different values of dimensionless parameters (not less than seven are needed to characterize the system), the above-mentioned oscillatory mode can become the preferred one at criticality. These oscillations generally exhibit alternating thermal and mechanical coupling situations, at least when they grow as standing waves (SW). Nevertheless, one of the purposes of the following developments is to analyze if traveling waves (TW) would be preferred in a system with sufficient horizontal extension, and in this case to determine the corresponding behavior of the system.

These oscillations can arise when the effective Rayleigh numbers in both layers are approximately equal, and when the ratio of layer thicknesses is around unity. Many systems entering that category are close to the case of two identical liquids (with the same thicknesses), but need to differ from it by *balanced contrasts* in thermal diffusivity, kinematic or dynamic viscosity, thermal expansivity or thermal conductivity. By balanced contrasts, here it is meant that ratios of these properties in both layers must differ from unity, but be balanced in such a way that effective Rayleigh numbers in both layers are kept close to each other. In such a range of fluid parameters, slight shifts in the values of some parameters (such as the thickness ratio) can induce transitions between mechanical coupling, thermal coupling and oscillatory convection.

In §5.4.4, the competition between traveling and standing waves will be analyzed in a two-dimensional situation (which appears to be sufficient in view of recent results of Renardy [178]). Even though the structure of the amplitude equations will certainly be of the form (5.2) near the Hopf threshold, they are here re-derived from the hydrodynamic equations, using a Galerkin method. As this method is dependent upon some details of the problem studied, a first principle analysis is presented, which, we hope, will provide the reader with an alternative view on how to handle weakly nonlinear problems. Details and expressions of the amplitude equation coefficients can be found in Appendix G.

The validity of Eqs (5.2) is limited to a very close neighborhood of the bifurcation point. In §5.4.5, a numerical Galerkin procedure is then employed to derive further nonlinear results, such as the stability of the Hopf limit cycle, and the structure of other bifurcating solutions. Note that bifurcation diagrams will here be presented as a function of the Nusselt number, accounting for the heat transport due to convection. Hence, they can be expected to look similar to those of Fig. 5.1, at least sufficiently near the primary threshold.

The text presented from here on closely follows the work of Colinet and Legros [177], which was later extended to three-dimensional disturbances on the hexagonal lattice by Renardy [178]. In particular, Renardy confirmed the analysis presented here, in that she found three-dimensional patterns of Table 5.1 to be unstable to two-dimensional ones. An interesting experimental study of this Bénard set-up has also been achieved by Andereck et al. [180], and Degen et al. [181]. Three-dimensional pattern selection in two-layer systems, driven by both buoyancy and surface tension gradients, has also been studied by Tokaruk et al. [34]. Some of their results together with some predictions obtained from amplitude equations have been described at the end of §4.4.3. Two-layer thermocapillary convection will be further considered in §7.2.

5.4.2 Galerkin method

The geometry of the two-dimensional system is depicted in Fig. 5.18. Two rigid isothermal plates at constant temperatures are bounding a system of two superimposed immiscible fluids. A system of cartesian coordinates is chosen in such a way that the gravity vector is antiparallel to its z-axis. The lateral boundaries at $x = 0$ and $x = L$ are chosen to be free of tangential stress and thermally insulating (mainly for mathematical convenience [150], but also because this configuration allows easy switching to the case of infinite horizontal extent, as seen in §3.3.4).

Each fluid layer is characterized by its thickness a_i, thermal conductivity λ_i, thermal diffusivity κ_i, specific mass ρ_i, dynamical viscosity μ_i, kinematic viscosity $\nu_i = \mu_i/\rho_i$, and thermal expansion coefficient α_i ($i = 1$ refers to the lower layer, $i = 2$ to the upper layer). The scaling of the problem is performed by using physical properties of the lower layer: a_1 as length unit, a_1^2/κ_1 as time unit, $\rho_1 a_1^3$ as mass unit and the temperature drop ΔT_1 across the lower layer as temperature unit. Then, under the Boussinesq approximation (§3.1.4), the dimensionless equations describing Rayleigh–Bénard convection in such a system are:

$$\vec{\nabla}.\vec{V}_1 = 0 \tag{5.36}$$

$$\vec{\nabla}.\vec{V}_2 = 0 \tag{5.37}$$

$$\frac{\partial T_1}{\partial t} + (\vec{V}_1.\vec{\nabla})T_1 = \Delta T_1 + W_1 \tag{5.38}$$

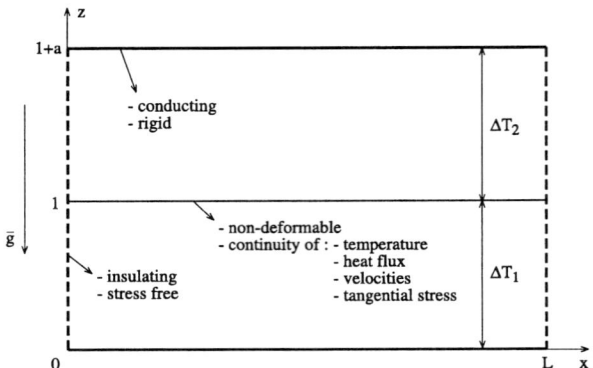

Figure 5.18: Geometry of the two-dimensional system and basic assumptions. Coordinates are scaled by the thickness of the lower layer. Idealized lateral walls (corresponding to periodic boundary conditions) are represented by dashed lines.

$$\frac{\partial T_2}{\partial t} + (\vec{V}_2.\vec{\nabla})T_2 = \kappa^{-1}\Delta T_2 + \lambda W_2 \tag{5.39}$$

$$\frac{\partial \vec{V}_1}{\partial t} + (\vec{V}_1.\vec{\nabla})\vec{V}_1 = Pr\Delta\vec{V}_1 - \vec{\nabla}p_1 + PrRaT_1\vec{1}_z \tag{5.40}$$

$$\frac{\partial \vec{V}_2}{\partial t} + (\vec{V}_2.\vec{\nabla})\vec{V}_2 = Pr\nu^{-1}\Delta\vec{V}_2 - \rho\vec{\nabla}p_2 + PrRa\alpha^{-1}T_2\vec{1}_z \tag{5.41}$$

where $\vec{\nabla}$ is the gradient operator; Δ is the Laplacian operator; $\vec{1}_z$ is the unit vector along z; $\vec{V}_i \equiv (U_i, 0, W_i)$, T_i and p_i the velocity, temperature and pressure perturbation fields in the ith layer. Ratios of fluid properties are $\kappa = \kappa_1/\kappa_2, \lambda = \lambda_1/\lambda_2, \alpha = \alpha_1/\alpha_2, \nu = \nu_1/\nu_2$. The ratio of thicknesses is $a = a_2/a_1$.

For mathematical convenience we consider the limiting case of an infinite Prandtl number $Pr = \nu_1/\kappa_1$ (which also implies the Prandtl number of the upper layer to be infinite, as ν and κ are taken to be of order unity). In fact, it will appear later on that contrary to several of the waves analyzed in Chapter 3, fluid inertia is not important here, therefore justifying this assumption at least for sufficiently viscous fluids (buoyancy-driven Stokes flow).

The only control parameter of the system of Eqs (5.36–5.41) is the Rayleigh number, defined as before, i.e. $Ra = g\alpha_1\Delta T_1 a_1^3/\nu_1\kappa_1$, where g is the acceleration of gravity.

The boundary conditions on the rigid isothermal plates are:

$$W_1(z = 0) = DW_1(z = 0) = T_1(z = 0) = 0$$
$$W_2(z = 1 + a) = DW_2(z = 1 + a) = T_2(z = 1 + a) = 0 \tag{5.42}$$

where D is the dimensionless z-derivative.

The model of interface we consider here is an undeformable one: this implies that we assume the density jump $\Delta\rho = \rho_1 - \rho_2$ to be large enough to stabilize the interfacial oscillatory mode, associated with a Rayleigh–Taylor kind of instability in part of the period of oscillation. Following Richter and Johnson [407], Richter and McKenzie [408], or later, Rasenat, Busse and Rehberg [175], we must have $\Delta\rho/\rho_1 \gg \alpha_i \Delta T_i$ ($i = 1, 2$). This hypothesis is not very restrictive, since the quantity $\alpha_i \Delta T_i$ must be small compared with unity for the Boussinesq approximation to be valid. Hence, the density jump at the interface must be large enough such that the density stratification of the system is not disturbed by density variations due to the imposed temperature difference. Then, continuity of temperature and heat flux leads to

$$
\begin{aligned}
T_1(z = 1) &= T_2(z = 1) \\
\lambda D T_1(z = 1) &= D T_2(z = 1)
\end{aligned}
\tag{5.43}
$$

while continuity of velocity at the undeformable interface implies

$$
\begin{aligned}
W_1(z = 1) &= W_2(z = 1) = 0 \\
D W_1(z = 1) &= D W_2(z = 1)
\end{aligned}
\tag{5.44}
$$

Furthermore, we disregard the Marangoni effect (this is valid for sufficiently thick layers, or low interfacial tension variation with temperature). In fact, it can be conjectured that the primary influence of the Marangoni effect will be to shift the preferred mode of instability to a mechanical coupling mode. We also neglect interfacial viscosity, even though the latter is thought to be important in case some surface contamination occurs. In fact, discrepancies between theoretical and experimental findings may exist due to this effect, as shown by Nataf, Moreno and Cardin [171, 411, 412]. Their studies show that, even when the predicted behavior is mechanical coupling (both linearly and numerically), the experimentally observed behavior is thermal coupling for their system. This is attributed to the presence of some interfacial viscosity whose origin is not well understood, but which makes the mechanical coupling pattern unobservable in their experiment.

Under these assumptions, the condition expressing the continuity of tangential stress is

$$
\mu D^2 W_1(z = 1) = D^2 W_2(z = 1)
\tag{5.45}
$$

As mentioned earlier, lateral walls are chosen to be free of tangential stress and thermally insulating, i.e.

$$
\begin{aligned}
U_1(x = 0) &= U_1(x = L) = U_2(x = 0) = U_2(x = L) = 0 \\
D^2 U_1(x = 0) &= D^2 U_1(x = L) = D^2 U_2(x = 0) = D^2 U_2(x = L) = 0 \\
\tfrac{\partial T_1}{\partial x}(x = 0) &= \tfrac{\partial T_1}{\partial x}(x = L) = \tfrac{\partial T_2}{\partial x}(x = 0) = \tfrac{\partial T_2}{\partial x}(x = L) = 0
\end{aligned}
\tag{5.46}
$$

For monotonic modes, this type of idealized walls is equivalent to periodic boundary conditions (see §3.3.4). This can also be expected for standing waves. Still, traveling waves are prevented by such walls (reflection occurs, exciting a counter-propagating wave), and their analysis will require to replace Eqs (5.46) by laterally periodic boundary conditions.

The system of partial differential equations and associated boundary conditions (5.36–5.46) is first transformed into an infinite system of ordinary differential equations, where the unknowns are the time-dependent amplitudes of a complete set of known spatial modes. Of course, such a Galerkin method is not the only way to derive the weakly nonlinear results that are presented in the next section (one could use the usual operational formulation of §5.1.2). But as we will see, the present formulation allows rather easy numerical integration, applied in §5.4.5 to obtain further nonlinear results.

Velocity and temperature fields are first expanded in series of spatial modes in which x and z dependencies are separated (this separation of variables is allowed by our choice of boundary conditions in $x = 0$ and $x = L$). This reads

$$
\begin{pmatrix} W_1 \\ W_2 \end{pmatrix} = \sum_{m,n} A_{mn}(t) e^{imk_0 x} \begin{pmatrix} W_{1mn}(z) \\ W_{2mn}(z) \end{pmatrix}
$$
$$
\begin{pmatrix} T_1 \\ T_2 \end{pmatrix} = \sum_{m,n} B_{mn}(t) e^{imk_0 x} \begin{pmatrix} T_{1mn}(z) \\ T_{2mn}(z) \end{pmatrix}
\tag{5.47}
$$

where the effective wavenumber k_0 is defined as

$$
k_0 = \frac{\pi}{L}
\tag{5.48}
$$

As the perturbation fields must be real, we also impose

$$
A_{mn}(t) = \bar{A}_{-mn}(t)
$$
$$
B_{mn}(t) = \bar{B}_{-mn}(t)
\tag{5.49}
$$

where the overbar denotes the complex conjugate [the functions $W_{imn}(z)$ and $T_{imn}(z)$ are taken to be real].

In the case of a system of infinite horizontal extent, we will choose k_0 as the critical wave number predicted by the linear theory and keep expansions (5.47), such that we explicitly neglect modulation effects associated with a finite bandwidth of wavenumbers (see §5.5): strictly speaking, this makes our results more appropriate for small aspect ratio boxes, or for patterns very near the convection threshold for which the correlation length is large (still, there are some conditions, as seen in §5.5).

The methodology is the following: we first choose a complete set $T_{imn}(z)$ of trial functions for temperature, which satisfy boundary conditions (5.42) and (5.43). We then insert the expansions (5.47) in Navier–Stokes equations (5.40) and (5.41) written for the z-component of velocity (i.e. after elimination of the pressure) in the approximation of an infinite Prandtl number. By writing

$$
A_{mn}(t) = Ra B_{mn}(t)
\tag{5.50}
$$

we can then solve these equations exactly by choosing $W_{imn}(z)$ as the unique solution of equations

$$(D^2 - m^2 k_0^2)^2 \begin{pmatrix} W_{1mn} \\ W_{2mn} \end{pmatrix} = m^2 k_0^2 \begin{pmatrix} T_{1mn} \\ \frac{\nu}{\alpha} T_{2mn} \end{pmatrix} \tag{5.51}$$

satisfying the remaining boundary conditions (5.42), as well as Eqs (5.44) and (5.45). Since all these boundary conditions are linear, they are also satisfied by expansions (5.47). Note that horizontal velocities in both layers are

$$\begin{pmatrix} U_1 \\ U_2 \end{pmatrix} = -Ra \sum_{m \neq 0, n} \frac{B_{mn}(t)}{imk_0} e^{imk_0 x} \begin{pmatrix} DW_{1mn} \\ DW_{2mn} \end{pmatrix} \tag{5.52}$$

satisfying the incompressibility relations and the lateral boundary conditions (5.46), provided that all coefficients $B_{mn}(t)$ are real. This last condition also ensures that the temperature functions satisfy insulating conditions on lateral walls. So,

$$B_{mn}(t) = \bar{B}_{mn}(t) \tag{5.53}$$

Clearly, this condition is relaxed if the system is laterally infinite, and in this more general case, complex amplitudes have to be taken into account.

As a summary, we have now satisfied all equations (5.36) to (5.46) except both nonlinear energy equations: these are used to find evolution equations for amplitudes $B_{mn}(t)$. We insert expressions (5.47) in these energy equations and obtain the evolution equation for a particular amplitude by projecting them on the corresponding spatial mode. It is worth noting here that we only imposed the system of temperature functions to be complete: the velocity field is considered to be a function of the temperature field (dynamical slaving), by choosing it to satisfy linear Stokes equations exactly. This would clearly not be possible if these equations were nonlinear, i.e. if the Prandtl number was finite, and our set of velocity functions would even not be complete in this case.

The resulting infinite system of ODEs can be written under its simplest form as

$$\dot{B} = AB + N(B, B) \tag{5.54}$$

where B is the vector whose components are the amplitudes $B_{mn}(t)$, A is the matrix of the linear problem and $N(u, v)$ is a real bilinear form producing nonlinear terms in the amplitude equations. The expressions of A and $N(u, v)$ are given as volume integrals in Appendix G.

Such a system of ODEs has already been obtained in different contexts and can even be used when lateral walls are realistic no-slip walls (using Chandrasekhar functions for the horizontal dependence [95]). For instance, it was successfully applied by Graham et al. [420] to the explanation of the diagonal oscillation formerly observed in Hele-Shaw cells [421].

5.4.3 Linear stability results

The problem is characterized by the values of seven dimensionless numbers, namely $\kappa, a, \nu/\alpha$ (only this combination enters the equations), μ, λ, L (or k_0) and Ra. In the following, we will neglect thermal conductivity stratification, and consequently assume $\lambda = 1$. Besides the fact that this simplification lowers the number of parameters characterizing the problem, another consequence is that the temperature field and its first z-derivative are now both continuous at the interface according to Eq. (5.43), which allows us to choose a very convenient orthonormal set of functions

$$T_{jmn}(z) = \left(\frac{2}{1+a}\right)^{\frac{1}{2}} \sin\left(\frac{n\pi z}{1+a}\right)$$

satisfying boundary conditions (5.42) and interface continuity conditions. When $\lambda \neq 1$, it is more difficult to determine a simple complete set of functions. Appendix G describes how a set of linearly independent polynomials can be constructed for this case. Since the completeness of this set has not been proved rigorously, we will only present results obtained for $\lambda = 1$.

The spectral problem corresponding to the system (5.54) is

$$A\zeta = \sigma\zeta \tag{5.55}$$

where σ is the growth rate. This problem admits non-trivial solutions for particular values of the Rayleigh number determined by

$$\det|A - \sigma I| = g(\sigma, \kappa, a, \nu\alpha^{-1}, \mu, k, Ra) = 0 \tag{5.56}$$

in which the effective wavenumber k_0 has been replaced by the wavenumber k, which can take any real value in an infinite layer. In general, the relation (5.56) is multi-valued for Ra. We thus denote its different roots by Ra_j and choose the ordering such that

$$Ra_1 < Ra_2 < ... < Ra_j(\kappa, a, \nu\alpha^{-1}, \mu, k, \sigma)$$

Furthermore, when lateral walls are included, the values of k are restricted to $k = \{nk_0\}, n = 1, 2, ...$, and the roots of (5.56) are

$$Ra_{ij} = Ra_j(\kappa, a, \nu\alpha^{-1}, \mu, ik_0, \sigma)$$

In this problem, exchange of stability cannot be assumed since the spectral problem is not self-adjoint (the real matrix A is not symmetric) in general. Hence, both neutral stability ($\sigma = 0$) and overstability ($\sigma = i\omega_0$) need to be analyzed. In the latter case, setting to zero the imaginary part of (5.56) allows determining the critical frequency ω_0.

Because of the high dimensionality of the parameter space, we restrict the analysis to model systems that are not too far from the case of two identical liquid layers with equal thicknesses ($a = \kappa = \nu\alpha^{-1} = \mu = 1$). For this peculiar system, overstability is not found (the linear problem is self-adjoint [178]), and the critical mode is always mechanical.

However, when varying fluid properties from this reference system, thermal coupling may become the preferred mode. This can happen in highly supercritical conditions when one layer has a much higher viscosity than the second, as studied by Cserepes et al. [172, 415]. Nevertheless, this is not a necessary condition for thermal coupling to occur at the threshold. For instance, Fig. 5.19 presents the case $\kappa = 2$, $\nu\alpha^{-1} = 0.5$, $\mu = 1$, $k_0 = 2.7$ (this value of the effective wavenumber corresponds to the critical value k_c in a laterally infinite layer for $a = 1$).

In Fig. 5.19, Ra_c is represented as a function of a : for $a < 0.92$, the critical mode is mechanical, while for $a > 1.067$, it is thermal (although when increasing a, it progressively turns back into a mechanical coupling mode via creation of small counter-rotating cells near the interface in the lower layer). Between these values of a, an oscillatory branch exists around the point where neutral stability curves would normally intersect. This is a generic phenomenon revealing the merging of both most dangerous real eigenvalues into a complex conjugate pair $\sigma = \pm i\omega_0$. Physically, this may occur when the effective Rayleigh numbers in both layers are approximately the same [175]. This requires the thickness ratio a to be about a value given by

$$a^* = \left(\frac{\alpha}{\lambda\nu\kappa}\right)^{\frac{1}{4}} \tag{5.57}$$

Furthermore, the corresponding oscillatory branch may become critical when $a^* \simeq 1$, which is the case for the system of Fig. 5.19. This system will be used in the following sections to illustrate nonlinear results. It is worth noting here that we have considered a fixed value $k_0 = 2.7$, but that we expect results to be valid for an infinite layer, since the critical wavenumber k_c is in fact almost constant along the overstable branch of Fig. 5.19. This has been done in order not to complicate the stability diagrams by discontinuities associated with the minimization process with respect to the wavenumber. On this figure, we have also represented the zone in which the growth constant σ, solution of the spectral problem (5.55), possesses a non-zero imaginary part. The nonlinear study that we present in the next sections is also aimed to compare this last zone with the zone of stability of the finite-amplitude Hopf cycle.

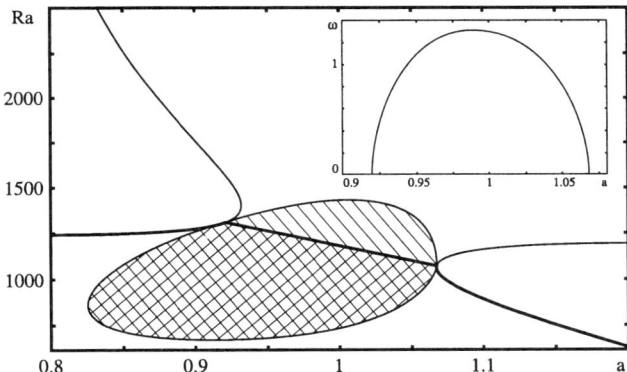

Figure 5.19: Threshold values of the Rayleigh number (thicker line) as a function of the thicknesses ratio a in the system $\kappa = 2$, $\nu a^{-1} = 0.5$, $\mu = \lambda = 1$, $k_0 = 2.7$. Overstability occurs for $0.92 < a < 1.067$. The corresponding critical frequency ω_0 is represented in the inset. The growth constant is complex with positive real part in the hatched area, with negative real part in the cross-hatched area, and real outside these areas. The critical mode is mechanical for $a < 0.92$, thermal for $a > 1.067$.

Finally, let us mention that a successful check of our results by comparison with previously published ones has been achieved in the case $a = \kappa = \nu\alpha^{-1} = \mu = 1$, and on the system $\kappa = \lambda = 1$, $\nu = \mu = \alpha = 4$, treated by Rasenat et al. [175]. Very similar behaviors as a function of c are observed for this system, and for the system presented in Fig. 5.19. These systems are in fact two examples of the balanced contrasts systems that we introduced in §5.4.1. Renardy has actually shown [178] that when $\alpha\kappa\mu/\nu = \alpha\kappa\rho = 1$, and at infinite Prandtl number, the linear problem is self-adjoint, hence it has only real eigenvalues. Therefore, Renardy recommends that this combination of parameters be taken as far as possible from unity, in order to maximize the region where oscillatory onset can occur. Note that $\alpha\kappa\rho = 4$ both for our case, and for the system of Rasenat et al. [175] mentioned above. Further examples are discussed by Renardy [178]. Moreover, a recent experimental analysis by Degen et al. [181] has allowed confirmation of the importance of the parameter $\alpha\kappa\rho$. For a silicone oil (top)–Fluorinert FC 70 (bottom) system, this parameter is 0.776 and oscillatory convection is only observed above the threshold of monotonic modes. In contrast, the oil (top)–water (bottom) system has $\alpha\kappa\rho = 0.375$ and indeed allows waves to be observed at the threshold. It also appears that the experimental work of Degen et al. confirms other theoretical predictions of Colinet and Legros [177] and Renardy [178], as seen below.

5.4.4   Weakly nonlinear analysis

In view of the similarities observed in the linear case (successive transitions from mechanical to oscillatory, then to thermal and back to mechanical coupling when increasing a), it is believed that the nonlinear study presented in this section is susceptible to possess some generality for other balanced contrasts systems.

The method we now describe is aimed at deriving a normal form for the Hopf bifurcation occurring at threshold, and to predict whether traveling or standing waves will be selected as a result of nonlinear interactions. Although the method of §5.1 could also be used for this purpose, the developments presented here (and in Appendix G) strictly follow the standard methodology of dynamical systems analysis [4], and should in our opinion provide some readers with an interesting alternative view of weakly nonlinear methods.

Since we search for time-periodic solutions $B(t) = B(t + T)$ of the system (5.54), we introduce a variable $s = \omega(\epsilon)t$, where $\omega(\epsilon) = 2\pi/T$ is the frequency and ϵ a small amplitude parameter characterizing the distance to the threshold. We then write $B(t) = v[s = \omega(\epsilon)t, \epsilon] = v[s + 2\pi, \epsilon]$ and choose [4] the expansions

$$\begin{pmatrix} v(s = \omega(\epsilon)t, \epsilon) \\ Ra(\epsilon) - Ra_c \\ \omega(\epsilon) - \omega_0 \end{pmatrix} = \sum_{n=1}^{\infty} \frac{\epsilon^n}{n!} \begin{pmatrix} v_n(s) \\ Ra_n \\ \omega_n \end{pmatrix} \tag{5.58}$$

By inserting these in Eq. (5.54) and identifying powers in ϵ, we get a hierarchy of linear inhomogeneous problems that can be solved sequentially. The first-order problem $J_0 v_1 = 0$ (see Appendix G) is homogeneous and corresponds to the spectral problem (5.55), which possesses four linearly independent solutions (two eigenvalues $\sigma = \pm i\omega_0$ of algebraic multiplicity two, corresponding to both directions of the wavevector $\pm k_0$). Then the solution is the linear superposition of all critical modes, i.e.

$$v_1(s) = c_+ e^{is} \zeta_{0+} + c_- e^{is} \zeta_{0-} + \bar{c}_+ e^{-is} \bar{\zeta}_{0-} + \bar{c}_- e^{-is} \bar{\zeta}_{0+} \tag{5.59}$$

where ζ_{0+} and ζ_{0-} represent solutions of the spectral problem $A_0(Ra_c)\zeta_0 = i\omega_0\zeta_0$ corresponding respectively to k_0 and $-k_0$. The coefficients c_+ and c_- represent respectively the amplitudes of the left and right propagating waves, as can be seen by coming back to physical quantities (5.47). Expression (5.59) also ensures that conditions (5.49) are satisfied, such that all physical quantities are real. In the case where lateral walls are included in the analysis, the condition (5.49) must be fulfilled, which implies $c_+ = c_-$ (standing waves).

The compatibility conditions at order ϵ^2 reduce to

$$i\omega_1 - Ra_1\sigma_{Ra} = 0 \tag{5.60}$$

where σ_{Ra} is the derivative of the growth rate σ with respect to the Rayleigh number, whose expression is given in Appendix G. It can numerically be seen that $\mathrm{Re}[\sigma_{Ra}] > 0$. The equation (5.60) then implies $Ra_1 = \omega_1 = 0$. It can also be shown [4] that higher-order compatibility conditions give

$$Ra_{2n+1} = \omega_{2n+1} = 0, \quad n = 0, 1, \ldots$$

and the Hopf bifurcation is a symmetric one.

The solution of the ϵ^2 problem is straightforwardly written as

$$v_2 = e^{2is}v_2^{I} + e^{-2is}\bar{v}_2^{I} + v_2^{III} \tag{5.61}$$

where v_2^{I} and v_2^{III} are quadratic functions of c_+ and c_-, given in Appendix G.

We can now proceed to the next order, whose compatibility conditions will lead to the amplitude equations for c_+ and c_-. By rescaling variables c_+ and c_- by a factor ϵ, and by assuming that their evolution is governed by a slow time scale $\epsilon^2 t$, we finally get

$$
\begin{aligned}
\dot{c}_+ &= c_+[\sigma_{Ra}(Ra - Ra_c) - i(\omega - \omega_0) + \alpha|c_+|^2 + \beta|c_-|^2] \\
\dot{c}_- &= c_-[\sigma_{Ra}(Ra - Ra_c) - i(\omega - \omega_0) + \alpha|c_-|^2 + \beta|c_+|^2]
\end{aligned}
\tag{5.62}
$$

These equations have also been obtained by Knobloch [163] in the case of Rayleigh–Bénard single-layer convection in binary fluids. As shown in §5.1.1, they can in fact be derived from symmetry arguments expressing the invariance of the infinite system to horizontal translations and reflection with respect to the z-axis. As usual, only the numerical values of the coefficients appearing in the amplitude equations do depend on the details of the problem, such as fluid properties, ... The discussion summarized in Fig. 5.1 is directly applicable here (replacing α by $-\beta$, and β by $-\gamma$).

Table 5.5 presents some values of α and β for the system of Fig. 5.19, while Table 5.6 contains transport properties for the same system, such as initial slopes Nu_{Ra} and ω_{Ra} of the time-averaged Nusselt number and frequency, respectively.

It is seen that traveling waves are predicted above the threshold and correspond to a better heat transport (higher Nusselt number), at least in the infinite system. Note that for other

Table 5.5: Coefficients of the amplitude equations for the system $\kappa = 2, \nu/\alpha = 0.5, \mu = \lambda = 1, k_0 = 2.7$ for three values of the thicknesses ratio a. Traveling waves (TW) are stable, while standing waves (SW) are unstable.

a	σ_{Ra}	Ra_c	ω_0	α	β
0.95	$0.0084 - 0.0037i$	1255	1.09	$-27.3 - 28.3i$	$-95.2 - 20.3i$
1.0	$0.0088 - 0.001i$	1173	1.29	$-46.9 - 12.4i$	$-138 + 27.8i$
1.05	$0.009 + 0.0003i$	1096	0.80	$-171 + 294i$	$-414 + 682i$

Table 5.6: Transport properties (slopes of the time-averaged Nusselt number Nu_{Ra} and of the frequency ω_{Ra} versus Ra at criticality) for the system $\kappa = 2, \nu/\alpha = 0.5, \mu = \lambda = 1, k_0 = 2.7$ for three values of the thicknesses ratio a. The heat transport is higher in the stable TW mode (for which the Nusselt number is time-independent).

a	TW		SW	
	Nu_{Ra}	ω_{Ra}	Nu_{Ra}	ω_{Ra}
0.95	$1.7\,10^{-3}$	$-1.2\,10^{-2}$	$7.6\,10^{-4}$	$-7.0\,10^{-3}$
1.0	$1.9\,10^{-3}$	$-3.3\,10^{-3}$	$9.7\,10^{-4}$	$-2.7\,10^{-4}$
1.05	$2.1\,10^{-3}$	$1.6\,10^{-2}$	$1.2\,10^{-3}$	$1.5\,10^{-2}$

systems standing waves could be preferred (and would lead to a better heat transport) even in infinite systems, because the values of α and β depend on fluid properties, layers depths, ...

In systems laterally confined by perfectly reflecting walls, since the condition $c_+ = c_-$ must be satisfied, standing waves are the only solution to the nonlinear problem. This conclusion should clearly be modified in the presence of modulation effects (i.e. a slow dependence of c_+ and c_- on the horizontal coordinate), that are neglected in our analysis. Indeed, several studies [424, 425, 427] have shown that spatially modulated traveling wave solutions indeed appear in relatively large aspect ratio containers (and generally do not occupy the whole container, but are concentrated mostly on one side). Other solutions also exist, under the form of traveling waves that exhibit an episodic reversal of their direction of propagation. The wave patterns may also present sources and sinks (basic defects of two-dimensional wave structures), as will be commented in §5.5.3. Note that the experimental works of Andereck et al. [180, 181] have essentially confirmed these expectations. In annular geometries (periodic boundary conditions), they observed sources and sinks of traveling waves both for oil–water and oil–Fluorinert combinations. In a rectangular cell (with non-perfectly reflecting end walls in general, see §5.5.3), both traveling waves and standing waves can be observed. In addition, at higher Rayleigh number and in annular containers, the authors report more complicated behaviors, such as temporal variations of the phase velocity of traveling waves, similar to a stick–slip motion.

5.4.5   Numerical results

The above predictions can be tested by direct numerical integration of the truncated system (5.54), in the case of real amplitudes (i.e. standing waves). A standard Runge–Kutta order 4 method turns out to be sufficient, with a control of the time step ensuring that the maximal relative error on any coefficient $B_{mn}(t)$ remains below 0.5% at each time step. The diagrams that are presented have been obtained by truncating the system to $m_{\max} = 4$, $n_{\max} = 12$ [upper bounds of the sums (5.47), corresponding to horizontal and vertical directions respectively], although satisfactory convergence is already obtained with $m_{\max} = 2, n_{\max} = 8$.

Direct numerical simulations also allow further nonlinear results to been obtained. When following the SW limit cycle while progressively increasing the Rayleigh number, it is observed that the time variation of transport properties becomes richer in harmonics, and that the basic frequency of the oscillation is decreasing. Above a value Ra^*, a transition to stationary convection (mechanical coupling) is predicted. This secondary bifurcation is accompanied by a sharp increase of the Nusselt number, although the associated hysteresis loop is small. This is seen in Fig. 5.20, which presents the global bifurcation diagram in the plane (Nu, Ra) for the case of Fig. 5.19 and Tables 5.5 and 5.6 with $a = 1$.

In fact, the frequency of the SW limit cycle decreases much more drastically than one could expect when extrapolating from the linear dependency existing at the leading order (the initial slope can even be positive, as for $a = 1.05$, see Table 5.6). When reaching Ra^* from below, the system begins to spend the largest part of the period of oscillation alternatively in the neighborhoods of two unstable saddle points, which makes the time-averaged Nusselt number approach the value calculated at those points (represented by the lower dashed branch of Fig. 5.20). At the limit Ra^*, the limit cycle ends in an heteroclinic orbit connecting both these unstable steady states. Increasing Ra above this value produces the observed transition to one of the two stable steady states (upper full branch of Fig. 5.20, labeled MC).

The similarity of behaviors between Fig. 5.20 and the results of Knobloch [163] is striking. His studies show that convection in one-layer binary fluid mixtures may exhibit such a type of dynamics near the codimension-2 point existing when the system is thermally destabilizing and solutally stabilizing (small negative separation ratio, defined as the ratio of solutal to thermal Rayleigh numbers). Near this point where a neutral stability branch and an overstable branch intersect (at the zero frequency point), the dynamics can be described by a Takens–Bogdanov [163, 420, 335] normal form. This is briefly discussed in the next section. Although in our case we are at a finite distance from both codimension-2 points of Fig. 5.19 ($a = 0.92$ and $a = 1.067$), it seems reasonable to infer that the behavior depicted in Fig. 5.20 (and subsequent bifurcation diagrams) is determined by the presence of these points.

As far as the traveling waves are concerned (complex vector B), numerical integration of the system (5.54) has only been performed for a low number of amplitudes ($m_{\max} = 2, n_{\max} = 5$). Although not really converged, we can expect the following results to be qualitatively valid. The TW limit cycle ends in a steady state bifurcation on the mechanical coupling branch. This happens for $Ra = Ra_t < Ra^*$, as also shown in Fig. 5.20. At this point the frequency of the TW limit cycle also vanishes. No hysteresis is expected in this case. Remember that Fig. 5.20 is valid for the laterally unbounded system, or at least for a sufficiently large one since modulation effects (envelope function) are expected to make the traveling wave fit in a realistic container. In the case of a strongly confined system (small

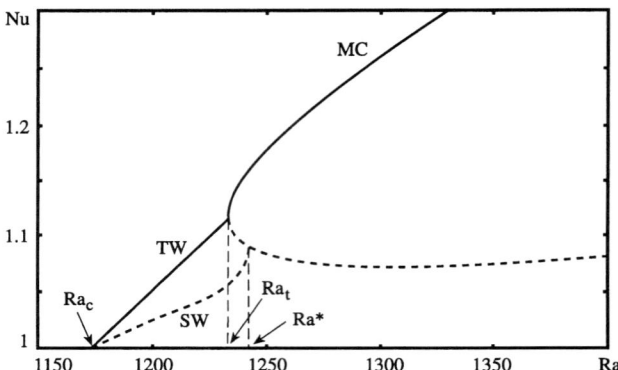

Figure 5.20: Bifurcation diagram (Nusselt number versus Rayleigh number) in the unbounded system for $\kappa = 2, \nu/\alpha = 0.5, \mu = \lambda = 1, k_0 = 2.7 (Ra_c = 1173)$. Full lines are stable states, dashed ones are unstable. In a system bounded laterally by perfectly reflecting sidewalls, the TW branch disappears, and the SW branch is stable. The heteroclinic orbit occurs at $Ra^* = 1241$, at which the frequency of the SW cycle vanishes. The turning point occurs at $Ra_t = 1232$.

aspect ratio), one expects the SW to be preferred over TW.

The physical mechanism of both kinds of oscillation can be investigated with the help of Figs 5.21 and 5.22. Figure 5.21 presents twenty streamline patterns, sampled during one period of oscillation of the standing waves. The succession is from left to right and top to bottom. As expected from linear theory [175], the system presents both thermal and mechanical coupling situations (twice) during one oscillation period. Transition from TC to MC occurs through generation of a mechanically coupled counter-rotating roll near the interface in the upper layer (this cannot be seen in Fig. 5.21, due to the large values of the streamfunction in the lower layer). This transition is clearly due to the fact that the system is unable to sustain large horizontal velocity gradients at the interface, since too strong viscous dissipation would occur there. The transition from MC to TC does not occur through generation of such a roll, and is due to diffusion of fluctuations of temperature through the interface, which, coupled to buoyancy, inverts the direction of rotation of convective cells in the lower layer (the fact that this inversion occurs in the lower layer is probably due to the thermal diffusivity contrast, and to the corresponding contrast of lifetime of temperature fluctuations).

Despite the fact that a more complete explanation of the oscillation mechanism should take into account the balanced contrasts of thermal diffusivity, viscosity, and other fluid properties, one can say that mechanical continuity conditions at the interface act as a restoring force when the system is in a thermally coupled situation, while thermal continuity conditions play this role when the system is mechanically coupled. Figure 5.22 presents the left-traveling wave pattern, in which a phase shift is apparent between both layers. Both basic kinds of couplings are present along the interface, and the corresponding restoring forces cooperate to create a uniform drift of the temperature and velocity fields to the left. By symmetry, the mirror state obtained by $x \rightarrow -x$ is also a solution, which can be generated with equal probability,

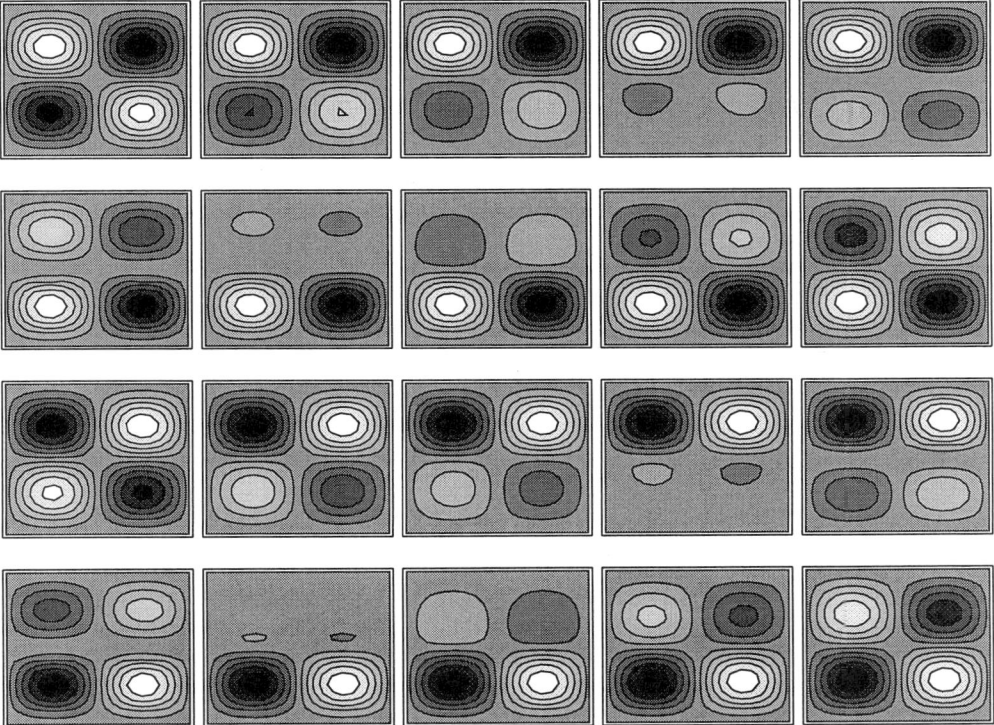

Figure 5.21: Streamlines in the standing wave mechanism: approximately one oscillation period ($T = 4.95$) is represented (the succession is from left to right and top to bottom). One half of the period is symmetrical to the other half (compare the two first lines with the two last ones). The oscillation is obtained for $Ra = 1190$, $\kappa = 2$, $\nu/\alpha = 0.5$, $\mu = \lambda = 1$, $k_0 = 2.7$.

depending on initial conditions.

It is also interesting to see what happens on the other sides of the codimension-2 bifurcation points, i.e. when the critical mode is the neutrally stable mode [1, 1]. This is the case for $a = 0.9$ for example [Fig. 5.23(a)], and for $a = 1.1$ [Fig. 5.23(b)].

For $a = 0.9$, the critical mode [1, 1] is stable (MC), and the mode [1, 2] is unstable. No secondary bifurcation occurs. For the more interesting case of $a = 1.1$, a closed loop between the critical mode [1, 1] (TC) and the mode [1, 2] exists, on which the Hopf bifurcation occurs. The corresponding limit cycle exists in a small range of values of Ra, whose size is decreasing with increasing a. The upper bound of this range of Rayleigh numbers is Ra^*, at which the limit cycle ends in a heteroclinic orbit (as in the case $a = 1$). When decreasing Ra very progressively and following the SW limit cycle, the system begins to spend a progressively larger part of the oscillation period near the origin, which makes the time-averaged Nusselt number decrease significantly. By adequately choosing the initial conditions, the trajectory of the system may eventually approach a pair of homoclinic trajectories that meet at the origin

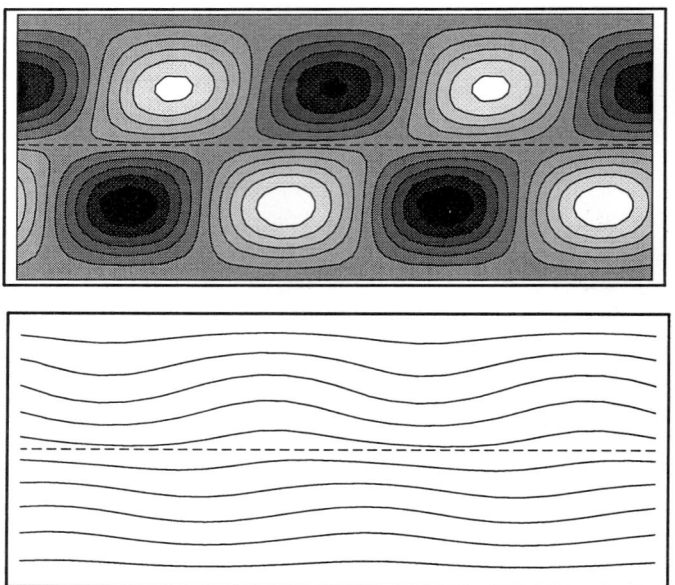

Figure 5.22: Streamlines and isotherms in the traveling wave mechanism. The pattern is drifting to the left with the velocity $c = \omega/k_0 = 1.86$. This pattern is obtained for $Ra = 1190$, $\kappa = 2$, $\nu/\alpha = 0.5$, $\mu = \lambda = a = 1$, $k_0 = 2.7$.

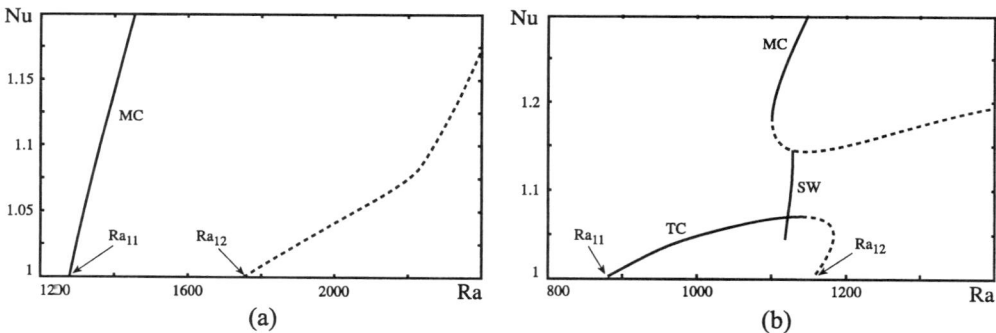

Figure 5.23: Bifurcation diagrams for $a = 0.9$ (a) and $a = 1.1$ (b). In case (a), no secondary bifurcation occurs in the range of the figure, while in case (b), the Hopf bifurcation occurs on the closed loop formed by the [1, 1] and [1, 2] branch. Only the corresponding standing waves are represented and exist for $1118 < Ra < Ra^* = 1127$. Around $Ra = 1118$, the SW limit cycle is close to a pair of homoclinic orbits. A heteroclinic cycle exists at $Ra^* = 1127$.

(saddle point). Decreasing Ra again causes the system to be attracted to the TC branch. The same transition to the MC stationary branch (as in the case $a = 1$) is observed above Ra^*. Consequently, a hysteresis loop behavior between TC and MC is seen in that case. For a increasing further, the hysteresis loop becomes larger, since both $[1,1]$ and $[1,2]$ branches split away from each other (the Hopf bifurcation remains on the $[1,1]$ branch, but occurs at larger distances from the threshold).

5.4.6   Three-dimensional case and Takens–Bogdanov dynamics

Renardy [178] has extended the weakly nonlinear analysis of the system described in this section by considering three-dimensional disturbances on the hexagonal lattice. She also considered the interaction between these bulk modes and interfacial oscillatory modes, which were extensively investigated earlier [176, 402, 410]. Contrary to the latter, which seem to lead to subcritical bifurcations in most cases [402], she showed that bulk modes generally bifurcate supercritically, in the range where oscillatory bifurcation occurs.

Among the eleven solutions with maximal symmetry classified in Table 5.1, Renardy finds that for some fluid systems in which the Hopf bifurcation occurs first, the only stable pattern consists in two-dimensional traveling waves. The ten other solutions are supercritical but unstable. However, in a later work [471], Renardy et al. mention that the wavy rolls (I) may also be stable, in addition to traveling waves. This solution was also found to be stable for Marangoni–Bénard convection in mixtures, and is represented in Fig. 5.9.

Now, the analysis of the previous section has shown that the proximity of codimension-2 points where the basic Hopf frequency vanishes complicates the overall bifurcation structure, in a way very similar to that described by Knobloch for oscillatory convection in binary mixtures [163] (see also [420] for another physical system). Recently, Renardy [470] and Renardy et al. [471] have extended his amplitude equation analysis to three-dimensional perturbations on the hexagonal lattice.

Let us succinctly illustrate the form of amplitude equations near a codimension-2 point (CTP), restricting the discussion to two-dimensional perturbations, and considering a fixed wavenumber k_0 as in Fig. 5.19. It is beyond our scope however to analyze the Takens–Bogdanov normal form in detail, which is done in [163, 164, 335, 472]. Rather, we will show how some of the behaviors obtained numerically in the last section can be predicted from the latter, at least qualitatively. Accordingly, we will not calculate the normal form coefficients for two-layer convection, but rather derive its general form from usual symmetry arguments (see §4.2.1 and §5.1.1).

In the vicinity of a point where ω_c vanishes, two convective modes are near-critical, and become identical at the CTP $\eta = \xi = 0$ (for definitions of the *unfolding* parameters ξ and η, see Fig. 5.24).

Actually, as the wavenumbers of both modes are assumed equal, the problem is a $1 : 1$ resonance[1] problem [472], and at the leading order, the planform function is

$$\phi(x,t) = a(t)\exp[ik_c x] + c.c. \tag{5.63}$$

---

[1]Chapter 7 is devoted to some other examples of codimension-2 points, namely $1 : 2$ resonances, where the wavenumbers of interacting modes are in the ratio $1/2$.

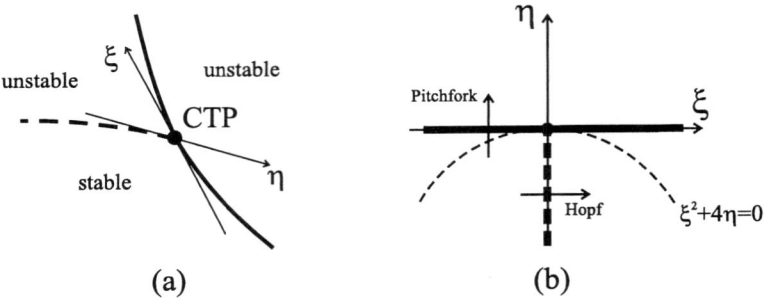

Figure 5.24: (a) Definition of a local system of coordinates at a codimension-2 point (CTP). (b) Thresholds of Hopf and steady (pitchfork) bifurcations close to the CTP in this representation.

since the differences between the two zero-eigenvalue modes only show up when ξ or η differ from zero, i.e. at higher order (see also [372]). In view of the similarity between Eqs (5.63) and (4.65), and because we consider here the usual invariance properties to translation $a \to a \exp[ik_c \Delta x]$ and parity $a \to a^*$, the discussion following Eq. (4.65) applies in principle, as well as Eq. (4.67). However, in all the discussions of monotonic modes in §4.2.1, we have explicitly assumed amplitude equations of the form $\dot{a} = f(a, a^*)$, i.e. a particular case of a more general relation $F(a, \dot{a}, \ddot{a}, ..., a^*, \dot{a}^*, \ddot{a}^*, ...) = 0$. Expanding the latter near the fixed point $a = 0$, we now get

$$0 = \alpha_0 a + \alpha_1 \dot{a} + \alpha_2 \ddot{a} + ... + \beta_0 a^* + \beta_1 \dot{a}^* + \beta_2 \ddot{a}^* + ... + \text{nonlinear terms} \quad (5.64)$$

where nonlinear terms are omitted for the moment. Translational invariance, i.e. invariance of (5.64) under $a \to a \exp[i\varphi]$ for all φ, requires $\beta_p = 0$ for all p. The discussion of nonlinear terms is similar to that achieved for deriving Eq. (4.67). In particular, quadratic terms are not allowed by translational invariance, and the cubic terms allowed are $a|a|^2$ as usual, but now also $\dot{a}|a|^2$, $a^2 \dot{a}^*$, and in fact all terms of the form $\partial_t^m a \, \partial_t^n a \, \partial_t^p a^*$. Note also that due to parity invariance $a \to a^*$, all coefficients need to be real.

Scaling considerations are now necessary to order the different terms. Different distinguished limits exist, depending on the nature of the bifurcation considered [which appears through the dependence of the various coefficients in Eq. (5.64) upon the control parameters]. First, for $\alpha_0 \sim \epsilon \ll 1$, and $\alpha_1 = O(1)$, a balance can occur between the first and second terms in Eq. (5.64), on the slow time scale $\partial_t \sim \epsilon$. Then, all other linear terms are of higher order, provided their coefficients are at most $O(1)$. This case corresponds indeed to the bifurcation of a single monotonic mode studied in §4.2.1. Choosing the scaling of the amplitude a such as to reach a balance with the largest nonlinear term $a|a|^2$, i.e. $a \sim \epsilon^{1/2}$, and neglecting terms of order higher than $\epsilon^{3/2}$, we can indeed rewrite the amplitude equation as Eq. (4.67).

For the double zero-eigenvalue considered here (Takens–Bogdanov bifurcation), we can in principle balance the three first terms in Eq. (5.64). This is achieved when $\alpha_0 \sim \epsilon^2$, $\alpha_1 \sim \epsilon$, $\alpha_2 = O(1)$, and $\partial_t \sim \epsilon$. Balancing these three terms with $a|a|^2$ leads to $a \sim \epsilon$. The normal form may then be rewritten (at leading order) $\ddot{a} = \eta a + \xi \dot{a} + \gamma a|a|^2$, whose linear part indeed reproduces Fig. 5.24(b). This corresponds to a situation where the growth rate is of the same order as the frequency and the amplitude. However, the term in $\dot{a}$ being the only dissipative term, the dynamics becomes in general quickly unbounded when $\xi > 0$, and higher-order terms should be included. It is more interesting to consider a case where near the Hopf line (i.e. $\eta < 0$), the growth rate is ϵ times smaller than the frequency and the amplitude, as done by Knobloch [163]. Then, after appropriate rescaling, the Takens–Bogdanov normal form becomes [163]

$$\ddot{a} = \eta a + \gamma a|a|^2 + \epsilon \left(\xi \dot{a} + \gamma' \dot{a}|a|^2 + \gamma'' a^2 \dot{a}^* \right) \tag{5.65}$$

where $O(\epsilon)$ dissipative corrections have been included, since at the lowest order, the dynamics is purely Hamiltonian. Note that a linear third-derivative term could also be included in the dissipative part, but it may then be replaced by the time derivative of the leading-order terms in Eq. (5.65), and hence only contributes to some renormalization of the coefficients.

The non-dissipative nature of the leading-order part in Eq. (5.65) can be recognized by looking for solutions of the form $a = r(t) \exp[i\varphi(t)]$. Then it is readily seen that if $r \neq 0$, the quantity $r^2 \dot{\varphi}$ is a constant, say c. Defining $v = \dot{r}$, we then find the relation between v and r as

$$\frac{v^2}{2} = \eta \frac{r^2}{2} + \gamma \frac{r^4}{4} - \frac{c^2}{2r^2} + E > 0 \tag{5.66}$$

where E is another integration constant. Thus, there is a two-parameter family (c and E) of solutions of the non-dissipative analog of Eq. (5.65), i.e. $\epsilon = 0$. Now, for $0 < \epsilon \ll 1$, we can in principle allow for a slow dissipative evolution of c and E (say, on a time scale $\tau = \epsilon t$), possibly leading to selection of a discrete set of values of c and E in the limit $\tau \to \infty$.

It is beyond the scope of our text to analyze the very rich weakly dissipative dynamics of Eq. (5.65), which the reader can find in [163, 164, 472]. From here, our goal will merely consist in addressing the structure of the phase space ($r, v = \dot{r}$) for $\epsilon = 0$, which can easily be done from Eq. (5.66).

First, consider $\eta < 0$ and $\gamma > 0$ (say, $\eta = -1$, $\gamma = 1$). For $c = 0$ (say, φ constant, i.e. standing waves), we recover the Duffing oscillator $\ddot{r} = -r + r^3$ represented in Fig. 4.10, apart for the substitution $x \to t$. Hence, SW have a small amplitude for $0 < E \ll 1$, and grow with E until they reach a heteroclinic trajectory connecting both unstable saddle-points $r = \pm 1$. Qualitatively, this corresponds to the SW branch of Fig. 5.20 (the relation between E and η, ξ has to be found for $\epsilon \neq 0$). Now, for $c \neq 0$, the term proportional to c^2 in Eq. (5.66) prevents the trajectory from reaching $r = 0$, and we can limit the discussion to $r > 0$ (the symmetric solutions for $r < 0$ correspond to a phase shift of π, i.e. half a period). This is represented in Fig. 5.25, where only the bounded solutions at fixed c and varying E are represented.

For $c^2 < 4/27$, there are two fixed points, a focus and a saddle, both corresponding to the family of TW ($a = r \exp[i\omega t]$ with constant r and $\omega = c/r^2$, $\omega^2 = 1 - r^2$). When increasing E past a certain threshold, a limit cycle bifurcates from the focus, and grows until a homoclinic trajectory of the saddle point is reached at higher E. These limit cycles are two-frequency waves (the period of the cycle, and the fluctuating phase speed $cr^{-2}k_c^{-1}$) called modulated waves (MW) by Knobloch [163]. Note that these solutions have not been observed for the particular case considered in the previous section, though they may be stable over some range for binary fluid convection [163].

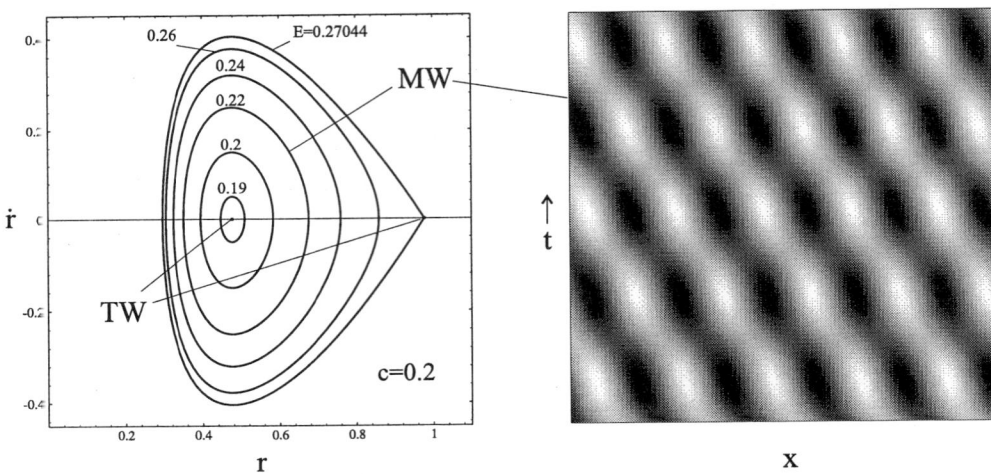

Figure 5.25: Phase portrait (left) of the non-dissipative equation $\ddot{r} = -r + r^3 + c^2/r^3$ at fixed $c = 0.2$, showing the possibility of bifurcation of a modulated wave (MW, right plot) from a traveling wave, and homoclinization of the latter on TW.

The situation $\eta > 0, \gamma < 0$ (say, $\eta = 1, \gamma = -1$) leads to qualitatively different bifurcation possibilities. For $c \neq 0$, there are only two symmetric foci $v = 0, r = \pm r_F$ (with $r_F > 1$), corresponding to TW. At increasing E and fixed c, two symmetric MW orbits bifurcate from these TW and progressively extend towards the origin, approaching it very close (at large E and/or small c) without ever reaching it. For $c = 0$, the origin becomes a saddle, and the foci are still present (now corresponding to steady states $r = \pm 1$). Increasing E above $-1/4$ leads to bifurcation of SW from these steady states. These standing waves are now asymmetric, since they bifurcate from a non-trivial steady state. At increasing E, they reach the origin (homoclinic gluing bifurcation) at $E = 0$, to form a single orbit surrounding the origin (now corresponding to symmetric SW) for $E > 0$. These solutions are not represented here, but this may easily be obtained from Eq. (5.66).

Although the stability of steady states, SW, TW and MW has not been addressed here, the analysis of the Hamiltonian dynamical system has shown some of their possible bifurcations. Stability of these solutions appears to depend sensitively on the values of ϵ, ξ, γ' and γ'' [163]. It cannot be guaranteed that the particular MW represented in Fig. 5.25 can be stable over some range of these parameters, but stable MW (possibly with smaller amplitude) have indeed been found by direct numerical integration of Eq. (5.65) at $\epsilon \neq 0$. It is worth emphasizing again that for very small ϵ, i.e. very close to the CTP, rather slow temporal evolutions of E and c occur as expected, and in practice, very long transients should be expected.

Renardy [470] and Renardy et al. [471] have recently extended the Takens–Bogdanov normal form (5.65) to the hexagonal lattice. The amplitude equations are then written for a resonant triad of wavevectors, and contain a number of additional terms, now including quadratic ones. Still, the amplitude equations have a Hamiltonian structure at the lowest order, and several of the properties of solutions found by these authors are shared by the two-dimensional case studied above. For instance, in addition to steady states like hexagonal patterns or rolls, there are possible bifurcations of periodic orbits from these steady states, including solutions of Table 5.1, as well as asymmetric SW (see also [164, 472]). Moreover, Renardy et al. [471] also find heteroclinic connections between steady hexagons, arising as a limit of oscillating triangles. Finally, by numerical integration of the complete amplitude equations including dissipation, several temporally quasiperiodic and possibly chaotic solutions are also found, in particular for the two-layer Rayleigh–Bénard problem considered throughout this section.

5.5 Modulations of two-dimensional waves

In this section, envelope equations are introduced for two-dimensional wave patterns. First, modulations of traveling waves (TW) are considered, using a single complex Ginzburg–Landau equation. As for monotonic modes (see §4.1.3), spatial modulations affect only the linear part of the amplitude equation at the relevant cubic order, and the coefficients multiplying the corresponding spatial derivative terms may thus be related to expansions of the linear growth rate in the vicinity of the bifurcation point.

The rigorous treatment of standing waves (SW) is more involved, due to the occurrence of $O(1)$ group velocity terms, which cannot be eliminated by a simple change of reference frame, as is the case for TW. Due to this circumstance, the correct amplitude equations applying to modulated standing waves should typically be non-local, unless the group velocity is small. Other results are presented which stress the importance of lateral boundary conditions, and allow the prediction of several of the states and defects found in experiment, such as chevron patterns, blinking and chaotic states, and wave sources and sinks.

Then phase diffusion equations are derived which allow studying the stability of traveling waves with wavenumber slightly different from the critical one. While in some cases there exists a certain range of wavenumbers for which the waves are stable, in some conditions this is not the case, and the system evolves to a state of phase turbulence as soon as the instability threshold is exceeded. In the former case, the nonlinear evolution of the phase obeys a dissipation-modified Korteweg–de Vries equation (near the Eckhaus stability limit), while in the latter, it evolves according to a Kuramoto–Sivashinsky (phase turbulence), or a modified version of it including dispersion, and known as the Kawahara equation.

5.5.1 Ginzburg–Landau equations for traveling waves

Consider first a state of pure (say, left) traveling waves. Note that this requires the system to be considered as laterally infinite (or periodic), which will be assumed in this section. Otherwise, the reflection of the wave on the left lateral wall would excite a counter-propagating wave [425]. This is considered in §5.5.3. For the moment, we write any perturbation field as a superposition of normal modes of the form

$$A_{\vec{q}}(t) \exp[i\vec{q}.\vec{r}] \exp[i(k_c x + \omega_c t)] U_{\vec{q}}(z) \tag{5.67}$$

where $\vec{r} = (x, y)$ is the horizontal position vector, $\vec{q}$ is a small wavevector shift with respect to the critical wavevector $k_c \vec{1}_x$, and $U_{\vec{q}}(z)$ is the eigenmode with wavevector $k_c \vec{1}_x + \vec{q}$. Denoting the linear complex growth rate by $\sigma(\mu, k^2)$, where μ is the control parameter and k the wavenumber, the linear evolution of $A_{\vec{q}}(t)$ should thus be governed by

$$
\frac{\partial A_{\vec{q}}}{\partial t} = \left\{ \sigma(\mu, k_c^2 + 2k_c q_x + q^2) - i\omega_c \right\} A_{\vec{q}}
$$
$$
\simeq \alpha(\mu - \mu_c) A_{\vec{q}} + (2k_c q_x + q^2) \left(\frac{\partial \sigma}{\partial k^2} \right)_c A_{\vec{q}} + \frac{1}{2}(2k_c q_x + q^2)^2 \left[\frac{\partial^2 \sigma}{\partial (k^2)^2} \right]_c A_{\vec{q}}
$$
(5.68)

where only the leading-order terms in the expansion of $\sigma(\mu, k^2)$ near the critical point $\mu = \mu_c$, $k^2 = k_c^2$ are kept, and $\alpha = (\partial\sigma/\partial\mu)_c$, which is now complex. We then define a complex envelope function by

$$
A(\vec{r}, t) = \int d\vec{q} A_{\vec{q}}(t) \exp[i\vec{q}.\vec{r}]
$$
(5.69)

Fourier-transforming Eq. (5.68), it is found that A evolves at the linear stage according to

$$
\frac{\partial A}{\partial t} = \alpha(\mu - \mu_c)A + c_g \frac{\partial A}{\partial x} - \frac{i}{2k_c} c_g \Delta A + \left(D + \frac{i}{2k_c} c_g \right) \left(\frac{\partial}{\partial x} - \frac{i}{2k_c} \Delta \right)^2 A
$$
(5.70)

where $\Delta \equiv \partial^2/\partial x^2 + \partial^2/\partial y^2$, and we have defined the group velocity $c_g = \text{Im}[(\partial\sigma/\partial k)_c]$ (the corresponding real part is zero at the critical point), and $D = -(\partial^2\sigma/\partial k^2)_c/2$.

Note that scaling conditions on the spatial variations of A have been ignored[2], which will allow us to discuss different cases in what follows. First, we may attempt the classical Newell–Whitehead–Segel (NWS) scaling $\mu - \mu_c = \epsilon^2\mu_2$, $t = \epsilon^{-2}t_2$, $x = \epsilon^{-1}X_1$, $y = \epsilon^{-1/2}Y_{1/2}$. On doing so, it is readily seen that some of the terms proportional to the group velocity c_g are of higher-order than the other ones in Eq. (5.70). While the term $c_g\partial A/\partial x$ can always be eliminated by working in a moving reference frame $x' = x + c_g t$, this is not the case for $\partial^2 A/\partial y^2$, such that another (longer) scaling of the transversal coordinate $y = \epsilon^{-1}Y_1$ might be required if $c_g = O(1)$ (which is the generic case).

For some problems, $c_g = O(\epsilon)$ (e.g. near a codimension-2 point [427, 435]), and redefining $v = c_g/\epsilon = O(1)$, we then obtain

$$
\frac{\partial A}{\partial t_2} = \alpha\mu_2 A + v\frac{\partial A}{\partial X_1} - \frac{i}{2k_c} v\frac{\partial^2 A}{\partial Y_{1/2}^2} + D \left(\frac{\partial}{\partial X_1} - \frac{i}{2k_c}\frac{\partial^2}{\partial Y_{1/2}^2} \right)^2 A
$$
(5.71)

However, as this is not a generic case, in what follows we will prefer to ignore transversal modulations, referring the reader to [19, 422, 423] for more detailed discussions. Focusing on purely two-dimensional regimes ($\partial/\partial y = 0$), eliminating the term $\partial A/\partial x$ by moving to a reference frame $x' = x + c_g t$, and re-incorporating nonlinearities as in Eq. (5.2), we obtain the complex Ginzburg–Landau (CGL) equation for $a = \epsilon^{-1}A$

$$
\frac{\partial a}{\partial t_2} = \alpha\mu_2 a + D\frac{\partial^2 a}{\partial X_1^2} - \beta a|a|^2
$$
(5.72)

---

[2] As no scaling assumption has yet been made for obtaining Eq. (5.70), spatial derivative terms always appear [140, 142] though the rotationally-invariant combination $\square \equiv \partial/\partial x - i(2k_c)^{-1}\Delta$, as in §4.5.5. As usual, rotational invariance is broken when slow spatial variations of A are assumed, because not all terms in $\square$ are of the same order of magnitude.

valid for any c_g and in the absence of lateral walls (or for periodic boundary conditions). Note that contrary to the real Ginzburg–Landau equation, the coefficients α, D and β are here complex. Their imaginary parts respectively describe linear frequency variation with μ, linear dispersion and nonlinear frequency detuning. Actually, the last two effects turn out to be sufficient to make the CGL equation non-variational and prone to very complex dynamics, as seen in the next sections.

5.5.2   Traveling and standing waves – group velocity effects

How can Eqs (5.2) be completed to incorporate slow spatial modulations of the amplitudes? As for pure TW, it is clear that the cubic Landau equations only need to be completed by linear spatial derivative terms. Note that we will here neglect transversal modulations from the outset. The leading-order physical fields (e.g. the temperature perturbation T) are first rewritten as

$$T(x, z, t) = A(x, t) \exp[i(k_c x + \omega_c t)]T_c(z) + B(x, t) \exp[i(k_c x - \omega_c t)]T_c^*(z) + c.c. \tag{5.73}$$

thus now including a right-traveling wave with amplitude B.

From Eq. (5.70) with $\partial/\partial y = 0$, we may apply the parity symmetry $x \to -x$ to find the corresponding equation for B. According to Eq. (5.73), this is equivalent to an action $A(x, t) \to B^*(-x, t)$ (where the asterisk denotes the complex conjugate), from which we readily obtain the linear (unscaled) evolution equation of $B(x, t)$ as

$$\frac{\partial B}{\partial t} = \alpha^*(\mu - \mu_c)B - c_g \frac{\partial B}{\partial x} + D^* \frac{\partial^2 B}{\partial x^2} \tag{5.74}$$

Now, it is clearly not possible in general to find a single reference frame eliminating group velocity terms of left- and right-TW simultaneously. We thus consider first the case where the group velocity is small, i.e. $c_g = O(\epsilon)$. Using the scalings $\mu - \mu_c = \epsilon^2 \mu_2$, $t = \epsilon^{-2} t_2$, $x = \epsilon^{-1} X_1$, $(A, B) = \epsilon(a, b)$, defining $v = c_g/\epsilon = O(1)$, and incorporating nonlinear terms as in Eqs (5.2), we obtain the system of coupled complex Ginzburg–Landau (CCGL) equations

$$\begin{aligned}
\frac{\partial a}{\partial t_2} &= \alpha \mu_2 a + v \frac{\partial a}{\partial X_1} + D \frac{\partial^2 a}{\partial X_1^2} - a \left(\beta|a|^2 + \gamma|b|^2 \right) \\
\frac{\partial b}{\partial t_2} &= \alpha^* \mu_2 b - v \frac{\partial b}{\partial X_1} + D^* \frac{\partial^2 b}{\partial X_1^2} - b \left(\beta^*|b|^2 + \gamma^*|a|^2 \right)
\end{aligned} \tag{5.75}$$

which has already been considered by various authors (see e.g. [427, 431, 432, 435]).

Before proceeding to the analysis of some solutions of the CGL and CCGL equations, we will consider the more general case where the group velocity is not small, i.e. $c_g = O(1)$. Clearly, its effect is larger than that of other terms, which requires either restricting attention to slower spatial modulations of the amplitudes (see end of next section), or reconsidering a more careful derivation of the equations. This has been achieved by Knobloch and co-workers [433, 435], who indeed emphasize that the CCGL system (5.75) should merely be considered as a phenomenological model in this case. The main steps of their analysis are outlined hereafter.

Equations (5.70) and (5.74) indicate that at $c_g = O(1)$, an intermediate (faster) time scale $t_1 = \epsilon t$ is relevant (keeping the same scalings of other variables). An expansion $(A, B) = \epsilon(c, b) + \epsilon^2(a_2, b_2)$ is also useful. After solving the linear (order ϵ) problem for μ_c, ω_c and k_c, solvability conditions of the order ϵ^2 are found as the leading-order contributions of Eqs (5.70) and (5.74), namely

$$\frac{\partial a}{\partial t_1} - c_g \frac{\partial a}{\partial X_1} = 0, \quad \frac{\partial b}{\partial t_1} + c_g \frac{\partial b}{\partial X_1} = 0 \tag{5.76}$$

whose solution is $a = a(\eta, t_2)$, $b = b(\xi, t_2)$, with $\eta = X_1 + c_g t_1$, $\xi = X_1 - c_g t_1$. Then, at order ϵ^3, the solvability conditions read

$$\frac{\partial a_2}{\partial t_1} - c_g \frac{\partial a_2}{\partial X_1} = -\frac{\partial a}{\partial t_2} + \alpha \mu_2 a + D \frac{\partial^2 a}{\partial X_1^2} - a\left(\beta|a|^2 + \gamma|b|^2\right),$$
$$\frac{\partial b_2}{\partial t_1} + c_g \frac{\partial b_2}{\partial X_1} = -\frac{\partial b}{\partial t_2} + \alpha^* \mu_2 b + D^* \frac{\partial^2 b}{\partial X_1^2} - b\left(\beta^*|b|^2 + \gamma^*|a|^2\right) \tag{5.77}$$

For a_2 and b_2 to remain of order unity on $O(\epsilon^{-2})$ time and length scales, it is necessary [433, 435] that the right-hand sides of Eqs (5.77) for a_2 and b_2 vanish when averaged over the comoving variables η and ξ respectively[3], which leads to

$$\frac{\partial a}{\partial t_2} = \left[\alpha \mu_2 - \gamma \left\langle |b|^2 \right\rangle_\xi\right] a + D \frac{\partial^2 a}{\partial \eta^2} - \beta a|a|^2,$$
$$\frac{\partial b}{\partial t_2} = \left[\alpha^* \mu_2 - \gamma^* \left\langle |a|^2 \right\rangle_\eta\right] b + D^* \frac{\partial^2 b}{\partial \xi^2} - \beta^* b|b|^2 \tag{5.78}$$

where $< \,.\, >_{\eta,\xi}$ denote averages over the corresponding variables. The resulting nonlocal system (5.78), called MFGL (mean-field Ginzburg–Landau equations), shows that left- and right-traveling waves are only coupled through their mean-field "energies", which is a consequence [435] of the fact that on the slow time scale t_2 considered, each wave "sees" the counter-propagating one moving through it very rapidly, i.e. on the scale t_1. The MFGL equations have been analyzed in details by Knobloch and co-workers [433, 434, 435]. Some of their results will be provided in the next sections.

5.5.3 Influence of lateral boundaries and defects

Coullet et al. [432] have identified the basic defects of wave structures on the basis of topological arguments, and studied their properties using CCGL equations (5.75), as well as two-dimensional generalizations of them including transversal modulations such as in Eq. (5.70). For CCGL equations (5.75), the basic point defects are of the kink-type, connecting a state of

---

[3]Consider for instance the first of Eqs (5.77). Changing variables as $a_2(X_1, t_1, t_2) = \tilde{a}_2(\xi, \eta, t_2)$, its left-hand side becomes $-2c_g \partial \tilde{a}_2 / \partial \xi$. Denoting the right-hand side by $f(\xi, \eta, t_2)$, a particular solution is found by direct integration over ξ, whose result remains bounded for large $|\xi|$ provided $\int_0^\xi f \, d\xi$ remains of order unity for $\xi \to \pm\infty$. In particular, for periodic boundary conditions, we get the conditions (5.78).

left-TW at $x \to -\infty$ to a state of right-TW at $x \to +\infty$, or vice versa. In the former case, the pattern appears as a source of TW propagating away from the defect, while in the latter, it consists in a sink towards which left- and right-TW converge. In both cases, amplitudes a and b are of order unity at the defect core. When viewed in a two-dimensional perspective, these defects appear as line defects orthogonal to the basic wavevectors. In addition, Coullet et al. [432] have considered line defects parallel to the basic wavevectors. These have been named "zipper" defects, as they actually separate regions of TWs propagating parallel and anti-parallel to the defect line. As for monotonic patterns (see §4.1.3 and §4.5.4), point dislocations are also found in two dimensions, which correspond to the insertion of an extra wavelength at some point (or more generally, N wavelengths), and actually look similar in some respects to dislocations found for monotonic patterns (vanishing of the corresponding order-parameter at the defect core, possible motion with respect to the basic structure, ...).

The incorporation of slow spatial modulations in principle also allows studying the influence of lateral walls at large distance (typically at $X_1 = \pm L/2$, such that the physical aspect ratio is L/ϵ). It is not our purpose here to discuss all possible regimes observed in confined oscillatory systems, which actually appear to be much more diverse than for monotonic modes. Rather, we will attempt to show how the effects of lateral walls may be studied within amplitude equation formalisms, referring the reader to the review of Cross and Hohenberg [19] for further details and more complete bibliography.

The effects of boundaries turn out to be very important for wave modes, mostly because of group velocity effects [19]. Because of terms proportional to v (say, $v > 0$) in Eqs (5.75), spatial inhomogeneities of a and b are propagated to the left and the right, respectively. Even for $\alpha_R \mu_2 > 0$, amplified perturbations can be propagated sufficiently fast, so that at a given location in a very large (or infinite) domain, a linear decay may rather be observed. In this case, the system is said to be *convectively* unstable. In contrast, the instability is called *absolute* when perturbations grow at every point in the laboratory frame. As the linear growth rate increases with μ, it can be expected that a transition from convective to absolute regimes will occur above some (finite) value of μ. Now, it appears that lateral *reflecting* boundaries may cause disturbances to grow locally even in the convective regime [19], which indeed justifies their careful study.

Actually, the boundary conditions to apply on a and b, e.g. in the CCGL equations, are not easily derived from a particular Bénard problem. Even for the simplest stress-free boundary conditions on top and bottom plates, and for monotonic buoyancy-driven convection, the exact calculation is rather involved [207]. When a TW propagates towards a rigid sidewall, exponentially decaying solutions are excited in some boundary-layer region, so as to satisfy the correct boundary conditions. In addition, a reflected wave is also excited, whose amplitude is related to that of the incoming wave via boundary conditions introduced by Cross [425]

$$b = r\,a \quad \text{at } X_1 = -L/2, \qquad a = r\,b \quad \text{at } X_1 = +L/2 \tag{5.79}$$

where r is a (complex) reflection coefficient. Cross has used these boundary conditions to analyze data of experiments on buoyancy-driven convection in mixtures [169], and calculated values of the reflection coefficient $|r|$. He showed (using stress-free top and bottom boundary conditions) that r tends to zero when approaching the codimension-2 point (proportionally to the Hopf frequency ω_c), i.e. just in the case where Eqs (5.75) are valid. In this case, and for small ratio of mass to heat diffusivities (i.e. small Lewis number), the four boundary conditions needed for the system (5.75) can be written [425]

$$a + \epsilon(\nu_1^* a_{X_1} + \nu_2^* b_{X_1}^*) = 0, \quad b + \epsilon(\nu_1^* b_{X_1} + \nu_2^* a_{X_1}^*) = 0 \text{ at } X = -L/2$$
$$a - \epsilon(\nu_1 a_{X_1} + \nu_2 b_{X_1}^*) = 0, \quad b - \epsilon(\nu_1 b_{X_1} + \nu_2 a_{X_1}^*) = 0 \text{ at } X = +L/2 \tag{5.80}$$

where a subscript X_1 denotes a derivative with respect to X_1, and ν_1 and ν_2 are two complex reflection coefficients, depending on the nature of the lateral walls, and also on L. From Eqs (5.75) and (5.80), Cross [425, 426] shows that the pattern found immediately above onset consists in a symmetric "chevron" pattern formed by right-TW strong in the right part of the container, and left-TW equally strong in the left part (for $v > 0$). At larger driving μ_2, this state may become asymmetric, one of the two waves becoming more pronounced on the corresponding side of the container. This state was called the "confined" state, because some fraction of the domain eventually remains very close to the rest state, while saturated traveling waves are observed in other parts. Actually, such states result from the interplay between propagation of a wave (say, the right one), excitation of the left-TW on the right wall, and nonlinear suppression of the left-TW in the region where the right-TW is saturated. At still larger μ, transitions to absolute instability may occur, resulting in waves filling the entire container. Actually, Cross and Hohenberg [19] emphasize that the system dynamics strongly depends on the system size L, on the group velocity v, as well as on the detail of the lateral boundary conditions (here quantified by ν_1 and ν_2).

Still from Eqs (5.75) and (5.80), Dangelmayr et al. [427] have explained other interesting dynamical regimes found in experiments [170] and in direct numerical simulations of governing equations for binary fluid convection between stress-free plates [424]. The so-called "blinking" states consist in traveling waves reversing their direction of propagation periodically, or even chaotically at larger Rayleigh number. Actually, other explanations have been proposed for these states. Instead of amplitude equations, Bestehorn et al. [428] used a complex order-parameter equation similar to those derived in §4.5.5, with coefficients calculated for convection in binary mixtures. They showed that blinking states may result from the nonlinear interaction of two linear eigenmodes with n and $n+1$ convective cells, in a large but finite box. It is beyond our scope here to discuss properties of blinking states in detail. However, it is worth mentioning that using the MFGL equations (5.78), even with the simplest type of boundary conditions (vanishing of the perturbation fields on lateral walls), Knobloch and De Luca [433] have also predicted blinking regimes. Interestingly, they showed that the pattern dynamics depends sensitively on the degree to which the sidewalls compress or expand the natural wavenumber k_c of the rolls, as also showed by Bestehorn et al. [428]. The relation between some of these studies has also been discussed by Knobloch [435].

To conclude this section, we briefly discuss another recent approach by Martel and Vega [436], and valid for $c_g = O(1)$, contrary to the approaches based on CCGL equations (5.75). Using a longer length scale $X_2 \simeq \epsilon^2 x$ (the other scalings are unchanged), it is possible to balance the group velocity term, e.g. in Eq. (5.74), with the other terms, while the linear diffusion/dispersion term becomes of higher order. Neglecting it, it appears that the dynamics of the phases of $A = \epsilon a$ and $B = \epsilon b$ may be decoupled from their moduli, for which a closed system is obtained. Using suitable $O(1)$ rescalings of time, length and amplitude, it can be shown [436] that we obtain a hyperbolic *real* system for $u = |a|^2$ and $v = |b|^2$, namely

$$\frac{\partial u}{\partial t_2} - \frac{\partial u}{\partial X_2} = 2u(\lambda - u - \alpha v)$$
$$\frac{\partial v}{\partial t_2} + \frac{\partial v}{\partial X_2} = 2v(\lambda - v - \alpha u) \tag{5.81}$$

where $\alpha = \gamma_R / \beta_R$. In principle, only two boundary conditions are now needed, which are obtained from Eqs (5.79) as

$$v = R u \quad \text{at } X_2 = -1/2, \qquad u = R v \quad \text{at } X_2 = +1/2 \tag{5.82}$$

where $R = |r|^2$ if $c_g > 0$, while $R = 1/|r|^2$ if $c_g < 0$. Note that the physical length of the system is here assumed $\sim \epsilon^{-2}$, i.e. ϵ^{-1} times larger than before, owing to the large propagation effects. From more general boundary conditions derived in earlier work, Martel and Vega also show that in order to avoid the formation of singularities, initial conditions $u_0(X_2) > 0$ and $v_0(X_2) > 0$ used to integrate (5.81) must satisfy the compatibility conditions

$$\partial v_0/\partial X_2 + R\,\partial u_0/\partial X_2 = 2(\alpha - 1)(R - 1)R\,u_0^2 \quad \text{at } X_2 = -1/2,$$
$$\partial u_0/\partial X_2 + R\,\partial v_0/\partial X_2 = 2(\alpha - 1)(1 - R)R\,v_0^2 \quad \text{at } X_2 = +1/2 \tag{5.83}$$

Martel and Vega [436] have shown that for $|\alpha| < 1$ (i.e. SW are preferred over TW in an infinite domain), all steady solutions are symmetric with respect to $x \to -x$, and correspond to stable symmetric chevron patterns, similar to those discussed above. The dynamics is actually much richer for $\alpha > 1$, i.e. when nonlinearities favor TW. For $\lambda_C = -\ln(R)/2 < \lambda < \lambda_{SB}$, the symmetric chevron pattern is still stable, while at $\lambda = \lambda_{SB}$, it undergoes a symmetry-breaking bifurcation leading to two asymmetric stable chevron patterns, looking in some cases (depending on R, α and λ) very similar to the confined states described above and also observed in experiments. At $\lambda = \lambda_H > \lambda_{SB}$, a Hopf bifurcation occurs, leading to oscillations around these asymmetric states (beating states). Martel and Vega also provide analytical and numerical values of λ_{SB} and λ_H, and show that while these bifurcations always occur, the dynamics at larger λ qualitatively depends on α and R, but always exhibits transitions to chaotic attractors that alternate with non-chaotic ones (period-doubling sequences, intermittency and quasiperiodic transitions). Their results are also shown to compare very satisfactorily with experimental measurements of Croquette and Williams [437] on the oscillatory instability of straight rolls in Rayleigh–Bénard convection.

Finally, Martel and Vega emphasize that their analysis should be restricted to Benjamin–Feir stable situations (see next sections). Otherwise, intermediate length scales X_1 are generated, which invalidates predictions obtained from Eqs (5.81–5.83).

In view of the complexity of behaviors observed even for one-dimensional waves, it is clear that an extension to realistic two-dimensional situations represents a huge task. To date, this problem has been tackled in a few works only. For instance, two-dimensional regimes of binary fluid convection in circular containers have been studied by Bestehorn et al. [429], who observe both zipper states and spatially confined waves. Pure Marangoni–Bénard double-diffusive oscillatory convection has also been recently studied [465] by direct numerical simulations of governing equations, both for periodic (see §5.3.3) and circular containers with idealized boundary conditions. In the latter case, defects generated and annihilated near the walls are shown to play an important role in the overall pattern dynamics. It is clear however that many open points remain, in particular in situations where the stable patterns predicted by unmodulated amplitude equations of §5.1.2 are built up by several plane waves propagating in different directions, and a fortiori when multistability occurs.

5.5.4   Phase instabilities of traveling waves

We now assume periodic lateral boundary conditions (which often applies satisfactorily to experiments in annular geometries[4]), and consider pure (left) TW described by the CGL equation (5.72), i.e. in a reference frame moving with the group velocity. After suitable rescalings (assuming $\beta_R > 0$, i.e. a supercritical bifurcation), this may be rewritten as

$$\dot{a} = \mu a + (1 + id)a'' - (1 + if)a|a|^2 \tag{5.84}$$

where dot and prime denote derivatives with respect to (slow) time t and length x, respectively, and $\mu = \alpha_R\mu_2$ is a control parameter (which can be assumed real, absorbing the imaginary part in the phase of a). In the following, we shall see that linear dispersion and nonlinear frequency detuning effects, quantified respectively by $d = D_I/D_R^{1/2}$ and $f = \beta_I/\beta_R$, play a fundamental role on the dynamical regimes predicted by Eq. (5.84). Note that as a particular case, we should recover results of §4.1.3, related to the real Ginzburg–Landau equation ($d = f = 0$).

As for the NWS equation (4.29), we will here be interested in the stability of the family of homogeneous states $a = a_s \exp[i(px + \omega t)]$, parameterized by the winding number p. Substituting in Eq. (5.84), we get $a_s^2 = \mu - p^2$ and $\omega = -f\mu + p^2(f - d)$. Thus, in our notations, the marginal curve is given by $\mu = p^2$, and at a given μ, p is allowed to vary between $-\mu^{1/2}$ and $\mu^{1/2}$. To examine the stability of these states, we perturb the modulus and the phase of a according to

$$a = \{a_s + r(t,x)\} \exp[i\{px + \omega t + \varphi(t,x)\}] \tag{5.85}$$

Inserting this expression in Eq. (5.84), linearizing with respect to r and φ, and separating real and imaginary parts, we get a linear homogeneous (real) system, which may be solved as usual by normal modes $(r, \varphi) \sim \exp[ikx + \sigma t]$. Then, the compatibility relation may be found as

$$\sigma^2 + 2\sigma\left(a_s^2 + k^2 + 2i\,k\,p\,d\right) + \\ 2k\,a_s^2\left\{k(1 + fd) + 2i\,p\,(d - f)\right\} + k^2(k^2 - 4p^2)(1 + d^2) = 0 \tag{5.86}$$

Thus, two modes exist, and at $k = 0$, one of them is damped on a $O(1)$ time scale (amplitude mode with $\sigma = -2a_s^2$), while the other has $\sigma = 0$ (phase mode). As usual, the Goldstone character of the phase mode is associated with invariance by translation, i.e. homogeneous perturbations of φ are marginally stable. Still, slow spatial modulations of φ may lead to instability, as seen below.

---

[4]Some examples of experiments in annular containers are presented in §6.1. Note that as we assume slow modulations, the analysis applies to long annular containers. A large radius is also necessary (though not sufficient) to neglect curvature effects, i.e. transversal modulations.

Because of complex coefficients in Eq. (5.86), it is difficult to achieve an analytical discussion of the sign of real parts of σ at arbitrary k. We thus rely on a numerical analysis, aimed at determining whether instability occurs first to long-wave modes $k \to 0$, or to short-scale ones $k = O(1)$, when $p/\mu^{1/2}$ is increased progressively from zero. The result is presented in Fig. 5.26.

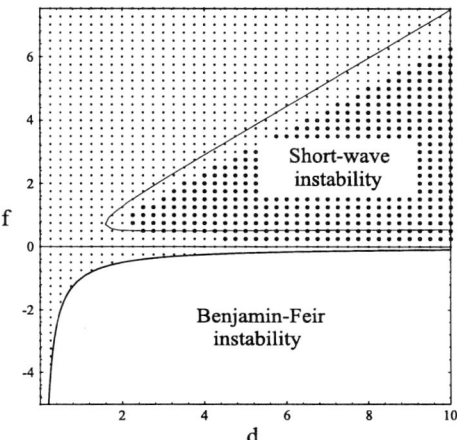

Figure 5.26: Numerical evaluation of the full dispersion relation (5.86). Regions of the plane (d, f) where TW are first unstable to the long-wave phase mode (small dots) or to a short-scale mode (big dots), when $p/\mu^{1/2}$ is progressively increased from zero. Within the thick curve $1 + df < 0$, TW are unstable for all p (Benjamin–Feir instability). The thin line corresponds to the simultaneous vanishing of $D_{/\!/}$ and of D_4 (codimension-2 line) in Eq. (5.87). The picture is symmetric with respect to the origin $d = f = 0$, as the original GL equation is invariant under the change $d \to -d$, $f \to -f, a \to a^*$.

Now, outside the short-scale instability regions, we expand the phase eigenvalue near $k = 0$, which leads to

$$\sigma = ikv_g - D_{/\!/}k^2 - i\Omega k^3 - D_4 k^4 + O(k^5) \tag{5.87}$$

with

$$\begin{aligned}
v_g &= 2p(f - d), \quad D_{/\!/} = 1 + df - 2p^2 a_s^{-2}(1 + f^2), \\
\Omega &= 2pa_s^{-4}(1 + f^2)(2fp^2 - da_s^2), \\
D_4 &= (1 + f^2)a_s^{-6}\{4p^4(1 + 5f^2) - 12d fp^2 a_s^2 + d^2 a_s^4\}/2
\end{aligned} \tag{5.88}$$

as obtained by Janiaud et al. [438]. According to Eq. (5.87), the linear phase diffusion equation [51, 431, 438] reads

$$\partial_t \varphi = v_g \partial_x \varphi + D_{/\!/}\partial_x^2 \varphi + \Omega \partial_x^3 \varphi - D_4 \partial_x^4 \varphi + \dots \tag{5.89}$$

When $D_4 > 0$ (which does not hold for all f and d, see Fig. 5.26), a long-wave instability occurs at $D_{/\!/} < 0$. Thus, if $1 + d\,f > 0$ and at a given $\mu > 0$, TW are stable within a range of wavenumbers (or winding numbers p) given by

$$p^2 < \mu \frac{1 + d\,f}{1 + d\,f + 2(1 + f^2)} \tag{5.90}$$

and are Eckhaus-unstable outside this interval. Note that for $d = f = 0$ (real GL equation), we indeed recover that rolls are stable within a range of phase-windings $p^2 < \mu/3$ (see §4.1.3). The evolution of the phase in the Eckhaus-unstable region will be examined in §5.5.6.

Now, when $1 + d\,f < 0$, the longitudinal diffusion coefficient $D_{/\!/}$ is negative in the whole range $p^2 < \mu$, and there is no stable traveling wave above the first instability threshold. The condition $1 + d\,f < 0$ is called the Newell criterion, and indicates the onset of phase turbulence in sufficiently large domains (see §5.5.7). Manneville [51] comments that this longitudinal instability is often named Benjamin–Feir instability, even though it was predicted first for surface water waves. Actually, for deep liquids, this corresponds to the conservative limit of the CGL equation (5.84), i.e. $\mu \to 0, f \to +\infty, d \to -\infty$, leading after some rescaling to a nonlinear Schrödinger equation $i\dot{a} = a'' + a|a|^2$.

An important locus of points has also been drawn in Fig. 5.26. This corresponds to values of d and f for which $D_{/\!/}$ and D_4 vanish simultaneously. Hence, in the neighborhood of this codimension-2 [431] line, we may have either sign of both $D_{/\!/}$ and D_4 when p is close to its threshold value (5.90). The value of D_4 calculated at this threshold [438] is

$$D_4^* = \frac{[1 + d\,f + 2(1 + f^2)][1 + 5f^2 + d^2(1 - 3f^2) + 4d\,f(f^2 - 1)]}{4(1 + f^2)^2} \tag{5.91}$$

Therefore, whenever this expression is negative and $1 + d\,f > 0$, short-wave instability occurs before the long-wave one when p^2/μ is increased, as confirmed by numerical results of Fig. 5.26. Still, the value of p^2/μ at which this happens can be very close to the Eckhaus threshold, as for the small dots remaining *inside* the zone $D_4^* < 0$. Note that nothing rigorous can be concluded when the expression (5.91) is positive, though Fig. 5.26 apparently indicates that the long-wave instability generally occurs first, apart from some region close to the $f = 0$ axis.

Before proceeding to the corresponding analysis of SW, it is worth commenting that here we considered the stability of TW with respect to internal perturbations only, i.e. we have not yet tested their stability with respect to waves propagating in the opposite (right) direction. Actually, denoting by g the cross-coupling coefficient ($g = \gamma_R/\beta_R$), it turns out that even if $1 + d\,f > 0$ and the condition (5.90) is satisfied, we must also have $g > 1$ and $p^2 < \mu(g-1)/g$ for TW of winding number p to be stable. Otherwise, counter-propagating waves develop (with growth rate maximal at the critical wavenumber k_c), and the system may end-up in a state of TW propagating in the other direction, with a more favorable wavenumber (or in a state of SW if $-1 < g < 1$, see below).

5.5.5   Phase instabilities of standing waves

The analysis of the modulational stability of SW solutions turns out to be more involved. As discussed in §5.5.2, the group velocity c_g is here expected to play an important role, and it is necessary to distinguish between $c_g = O(1)$ and $c_g = O(\epsilon)$.

Generic case $c_g = O(1)$

For $c_g = O(1)$, the relevant amplitude equations are the MFGL equations (5.73), which we rewrite here under the rescaled form

$$\dot{a} = \left[\mu - (g + ih)\langle|b|^2\rangle_\xi\right] a + (1 + id)\, a_{\eta\eta} - (1 + if)\, a|a|^2,$$
$$\dot{b} = \left[\mu - (g - ih)\langle|a|^2\rangle_\eta\right] b + (1 - id)\, b_{\xi\xi} - (1 - if)\, b|b|^2 \tag{5.92}$$

where we have used the same notation as in the previous section, while $g = \gamma_R/\beta_R$ and $h = \gamma_I/\beta_R$. The MFGL equations (as well as the CCGL [431, 434]) admit the phase-winded solutions

$$a = a_s \exp[i(p\eta + \omega_a t)], \quad b = b_s \exp[i(q\xi - \omega_b t)] \tag{5.93}$$

where $\omega_a = -fa_s^2 - dp^2 - hb_s^2$, $\omega_b = -ha_s^2 - fb_s^2 - dq^2$, and a_s, b_s are determined from $a_s^2 + gb_s^2 = \mu - p^2$ and $b_s^2 + ga_s^2 = \mu - q^2$. These solutions correspond in general to spatially quasiperiodic waves ($p \neq q$), or to standing waves ($p = q$). However, since we are dealing with periodic boundary conditions (say, an annulus containing $N \sim \epsilon^{-1}$ basic periods $2\pi/k_c$), we should in principle restrict the values of p and q to integer multiples of $\epsilon^{-1}k_c/N = O(1)$. To study the stability of these patterns, we write

$$a = \{a_s + r(t, \eta)\} \exp\left[i\left\{p\eta + \omega_a t + \varphi_a(t, \eta)\right\}\right],$$
$$b = \{b_s + s(t, \xi)\} \exp\left[i\left\{q\xi - \omega_b t + \varphi_b(t, \xi)\right\}\right] \tag{5.94}$$

and identify real and imaginary parts in Eqs (5.79). Linearizing with respect to r, s, φ_a and φ_b, and using normal modes $\exp[ikx + \sigma t]$, we see that the equations decouple, as observed by Knobloch [434], because spatial averages $< r >_\eta$ and $< s >_\xi$ vanish if $k \neq 0$ (the case $k = 0$ is analyzed below). Thus, in contrast to pure TW, there are now two decoupled phase modes, corresponding to space and time translations. We may analyze the system for r and φ_a alone (the corresponding system for s and φ_b is obtained by substitutions $r \leftrightarrow s$, $\varphi_a \leftrightarrow \varphi_b$, $p \leftrightarrow q$). In fact, replacing μ by $p^2 + a_s^2 + gb_s^2$ and ω_a by $-fa_s^2 - dp^2 - hb_s^2$, the system obtained for r and φ_a is identical to that obtained for pure TW. Hence, the compatibility relation (5.86) is also valid, together with Eqs (5.87–5.89). The only effect of the counter-propagating wave occurs through the value of $a_s^2 = [\mu(1 - g) - p^2 + gq^2]/(1 - g^2)$.

So, when the group velocity c_g is of order unity (i.e. the MFGL are valid), the Newell criterion $1 + df < 0$ still applies [433, 435, 434], and indicates that SW are unstable to both phase modes, whatever the values of p and q. For $1 + df > 0$, there exists in general a range of values of p and q where the homogenous states (5.93) are stable. When varying p^2 and q^2 from zero at given μ (or decreasing μ at fixed p and q), long-wave instability occurs to one of the phase modes (or both simultaneously if $p = q$), when one of the diffusion coefficients vanishes. These may be rewritten as

$$D_a = 1 + df - 2p^2 \frac{(1 + f^2)(1 - g^2)}{\mu(1 - g) - p^2 + gq^2},$$
$$D_b = 1 + df - 2q^2 \frac{(1 + f^2)(1 - g^2)}{\mu(1 - g) - q^2 + gp^2} \tag{5.95}$$

and this always occurs before one of the existence conditions $\mu(1 - g) - p^2 + gq^2 > 0$, $\mu(1 - g) - q^2 + gp^2 > 0$ fails to hold (see Fig. 5.27).

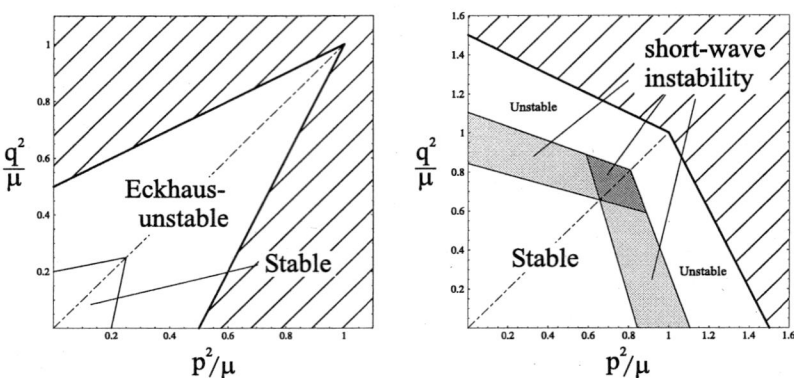

Figure 5.27: Two examples of stability diagrams for homogeneous quasiperiodic waves of winding numbers p and q. Left: $g = 0.5$, $d = 1$, $f = 1$. Right: $g = -0.5$, $d = 10$, $f = 2$. The thicker lines are existence boundaries (where either amplitude vanishes), while the thin lines correspond to vanishing of one of the longitudinal phase diffusion coefficients. In the right figure, for which the condition (5.96) holds, short-wave instability is found in the light grey (one mode) and dark grey (two modes) regions.

It is also easily shown that, as for TW, short-wave instability will always occur before the long-wave one when $1 + d f > 0$ and

$$1 + 5f^2 + d^2(1 - 3f^2) + 4d f(f^2 - 1) < 0 \tag{5.96}$$

while a numerical analysis of (5.86) is necessary in the opposite case.

Finally, we note that the results of this section are only valid provided the solutions (5.93) are stable to zero-wavenumber perturbations, i.e. spatially constant r, s, φ_a and φ_b. It is easy to show that this is the case when $-1 < g < +1$, independently of p and q. This is the usual condition for SW with $p = q = 0$ (see §5.1). When $g < -1$, the bifurcation to waves is subcritical, while for $g > 1$, one of the waves decays and the analysis of TW of the previous section is applicable.

Small group velocity $c_g = O(\epsilon)$

The analysis is more complicated in this case, because the phase equations for left- and right-propagating waves no longer decouple [431]. Even the situation $p = q = 0$ needs to be studied with some care, as shown by Knobloch [434, 435]. We also restrict the analysis to this case in the following, hence studying the modulational instability of homogeneous standing waves with critical wavenumber k_c.

Starting from the CCGL equations (5.75), using the same rescaling as in the previous sections (and defining $c = v/D_{\mathrm{R}}^{1/2} = \epsilon^{-1}c_{\mathrm{g}}/D_{\mathrm{R}}^{1/2}$), we seek solutions of the form

$$
\begin{aligned}
a &= \{a_{\mathrm{s}} + r(t,x)\}\exp[i\,\{\omega t + \varphi_a(t,x)\}],\\
b &= \{a_{\mathrm{s}} + s(t,x)\}\exp[i\,\{-\omega t + \varphi_b(t,x)\}]
\end{aligned}
\tag{5.97}
$$

with $\omega = -(f+h)a_{\mathrm{s}}^2$ and $\mu = (1+g)a_{\mathrm{s}}^2$. Then, a system of four coupled equations is obtained, for r, s, φ_a and φ_b. Using the phase approximation, Coullet et al. [431] have shown that after elimination of the amplitude modes, the system of linear phase equations may be written

$$
\begin{aligned}
\partial_t\varphi_+ &= c\,\partial_x\varphi_- + D_+\partial_{xx}^2\varphi_+,\\
\partial_t\varphi_- &= c\,\partial_x\varphi_+ + D_-\partial_{xx}^2\varphi_-
\end{aligned}
\tag{5.98}
$$

where we have defined $\varphi_+ = \varphi_a+\varphi_b$ and $\varphi_- = \varphi_a-\varphi_b$, and $D_+ = (1+df-g-hd)/(1-g)$, $D_- = (1+df+g+hd)/(1+g)$. Of course, this system is only valid at the leading order for slow spatial modulations $\partial_x \sim \epsilon$. Note that Eqs (5.98) actually combine terms which are not of the same order in ϵ, when $c = O(1)$. This is also seen by inserting normal modes $(\varphi_+,\varphi_-) \sim \exp[ikx + \sigma t]$, which leads to eigenvalues

$$
\sigma = -D_1 k^2 \pm \sqrt{D_2^2 k^4 - c^2 k^2}
\tag{5.99}
$$

with $D_1 = (D_+ + D_-)/2$ and $D_2 = (D_+ - D_-)/2$. For $k \to 0$ and $c = O(1)$, $\sigma \to -D_1 k^2 \pm ikc$, and instability (to both modes) occurs when $D_1 < 0$, or

$$
1 + d\,\frac{f - gh}{1 - g^2} < 0
\tag{5.100}
$$

as found by Coullet et al. [431]. Now, if $c^2 \ll k^2$, Eq. (5.99) gives $\sigma = -D_\pm k^2$, hence instability occurs when either or both diffusion coefficients D_+ and D_- becomes negative, or, using $1 \pm g > 0$ (stability to homogeneous perturbations)

$$
1 + df - (g + hd) < 0 \quad \text{or} \quad 1 + df + (g + hd) < 0
\tag{5.101}
$$

as found by Knobloch [435]. The origin of the discrepancy between conditions (5.100) and (5.101) stems from the non-commutativity of the limits $k \to 0$, $c \to 0$, or, equivalently, from the different time scales involved in Eqs (5.98). Knobloch further shows that the general case corresponds to $c \sim k$ (say, $c = c_1 k$) which allows matching of both results (5.100) and (5.101) in the limits $c_1 \to \infty$ and $c_1 \to 0$, respectively.

Note finally that none of the conditions (5.100) and (5.101) matches the Newell criterion $1 + df < 0$ for $c \to \infty$, though this was found above for $c_{\mathrm{g}} = O(1)$ from the MFGL equations. Again, this is because the limits $c \to \infty$ and $k \to 0$ do not commute. In fact, Eqs (5.98) are not valid in this case, and we need to come back to the full dispersion relation. Letting $c \to \infty$ before $k \to 0$ in the latter, Knobloch [434] has shown that the Newell criterion $1 + df < 0$ for modulational instability of SW at $c_{\mathrm{g}} = O(1)$ is indeed recovered.

5.5.6   Eckhaus instability of traveling waves

What is the nonlinear evolution of the system when TW with winding number p become Eckhaus-unstable, i.e. when p^2/μ is increased past the threshold defined by Eq. (5.90)? Here, we assume $1 + df > 0$, i.e. there does exist a stable range of TW. The analysis presented here closely follows the study of Janiaud et al. [438], which was also motivated by experiments on the secondary oscillatory instability of Bénard rolls in an annular geometry. Under some conditions, waves developing spontaneously on the (toroidal) convection rolls show compression pulses [438] propagating azimuthally, i.e. along the annulus.

Note that similar localized wavenumber modulations have been observed in quite different experiments on the dynamics of liquid columns falling from an annulus [307] (see also earlier experiments with non-periodic boundary conditions [126, 306]). Needless to say, it is not so likely that even our starting CGL equation (5.84) applies to this situation. In particular, some behaviors observed in the latter experiment clearly cannot be explained by the theory presented below, and are actually better described by the general analysis of Coullet and Iooss [303] of instabilities of periodic patterns (see also [309, 310]). Still, it is interesting to track the origin of possible generic characteristics in different physical systems.

Derivation of the phase equation

Let us now complete the linear phase diffusion equation (5.89) by adding nonlinear terms. This is expected to be asymptotically valid in the immediate vicinity of the Eckhaus threshold $D_{/\!/} = 0$, i.e. we assume $D_{/\!/} = -\eta\varepsilon^2$, which defines the smallness parameter $\varepsilon \ll 1$ and the unfolding parameter $\eta = O(1)$. According to Eqs (5.88), we have

$$a_s^2 = \mu - p^2 = \frac{2p^2(1+f^2)}{1+df}\left[1 - \frac{\eta\varepsilon^2}{1+df} + O(\varepsilon^4)\right] \tag{5.102}$$

i.e. we formally consider a_s^2 as the control parameter (we could also have used p or μ), and it will turn out that the other coefficients may be evaluated at $\eta = 0$, namely we define

$$\mu_1 = \Omega|_{\eta=0} = \frac{(f-d)(1+df)}{p(1+f^2)},$$
$$\delta_1 = -D_4|_{\eta=0} = -\frac{(1+df)[1+5f^2+d^2(1-3f^2)+4df(f^2-1)]}{4p^2(1+f^2)^2} \tag{5.103}$$

Thus, we should restrict the analysis to values of f and d such that $\delta_1 < 0$ (stabilizing fourth-order derivative). Now, the form of the linear growth rate (5.87) shows that the fastest growing modes have a scale $k \sim \varepsilon$ and growth rate $\mathrm{Re}[\sigma] \sim \varepsilon^4$. However, given that we may eliminate the group velocity term $\partial_x\varphi$ in Eq. (5.89) by a change of reference frame $x \to x + v_g t$, it is seen that the fastest time scale is provided by the dispersion term $\partial_x^3\varphi$. Accordingly, we set $\partial_t \sim \varepsilon^3$ and $\partial_x \sim \varepsilon$.

Still, it remains to chose the scale of the phase φ itself. The lowest-order nonlinear terms compatible with the invariance to homogeneous shifts of φ are $(\partial_x\varphi)^2$ and $\partial_x\varphi\partial_x^2\varphi$. Hence, a balance can occur between the first of these and dispersion if $\varphi \sim \varepsilon$. The phase equation can then be written [438]

$$\partial_T\phi = \mu_1\partial_X^3\phi + \mu_2(\partial_X\phi)^2 + \varepsilon\left(\delta_1\partial_X^4\phi - \eta\partial_X^2\phi + \delta_2\partial_X\phi\partial_X^2\phi\right) \tag{5.104}$$

where we have defined $T = \varepsilon^3 t$, $X = \varepsilon(x + v_g t)$ and $\phi = \varepsilon^{-1}\varphi$, and μ_2 and δ_2 are two nonlinear coefficients which still need to be calculated. Note that in the so-called Dissipation-Modified Korteweg–de Vries (DMKdV) equation, we have kept terms formally of higher-order (those multiplied by ε) because these turn out to be important on long time scales, as explained below.

Actually, there are several methods possible to calculate μ_2 and δ_2. First, following Janiaud et al. [438] (see also Coullet et al. [431]), we note that instead of Eq. (5.85), we could also have written

$$a = \{a_s(p + P) + s(t, x)\} \exp[i\{(p + P)x + \omega(p + P)t + \Psi(t, x)\}] \tag{5.105}$$

with $a_s(q) = (\mu - q^2)^{1/2}$ and $\omega(q) = -f\mu + (f - d)q^2$, i.e. Ψ is a phase perturbation (and s the associated amplitude) superposed on a TW of winding number $p + P$. Identifying Eqs (5.85) and (5.105), we see that $\Psi = \varphi - Px - [\omega(p + P) - \omega(p)]t$, which should satisfy the same equation as φ, though with new coefficients calculated at $p + P$. We thus have to reconsider the DMKdV equation in the unscaled form

$$\partial_t \varphi = v_g \partial_x \varphi + D_{//}\partial_x^2 \varphi + \Omega\partial_x^3 \varphi - D_4\partial_x^4 \varphi + \mu_2(\partial_x \varphi)^2 + \delta_2 \partial_x \phi \partial_x^2 \phi \tag{5.106}$$

Then, replacing φ by Ψ, and all coefficients by their value at $p + P$, and expanding in powers of $P \ll 1$ [because Eq. (5.104) is valid for slow modulations, i.e. it is *truncated* at some order], we can identify the remaining coefficients

$$\mu_2 = f - d, \quad \delta_2 = -\frac{(1 + df)[1 + df + 2(1 + f^2)]}{p(1 + f^2)} \tag{5.107}$$

A second method, which will only be summarized here because it is similar to the multi-scale perturbation techniques used to study long-wave modes in §4.6, consists in starting from the original CGL equation (5.84), and looking for solutions of the form (5.85). After separating the real from the imaginary part, we get a system of two equations for $r(t, x)$ and $\varphi(t, x)$. According to the above discussion of scalings, we then write

$$\begin{aligned} \varphi(t, x) &= \varepsilon\phi\left[T = \varepsilon^3 t, X = \varepsilon(x + v_g t)\right], \\ r(t, x) &= \varepsilon^2 R_2(T, X) + \varepsilon^3 R_3(T, X) + \dots \end{aligned} \tag{5.108}$$

and expand both equations in powers of ε. Then, at order ε^2, we get $R_2 = -p\,\partial_X\phi/a_s$ from the real part, while the imaginary part is then satisfied identically (thanks to the change of reference frame). At order ε^3, we get $R_3 \sim \partial_X^2\phi$, as well as a compatibility condition solved (at this order) by Eq. (5.102). Finally, the next two orders provide further corrections to r, in addition to the DMKdV equation (5.104), with coefficients (5.103) and (5.107).

Before discussing some properties of the DMKdV equation, it is useful to define the local wavenumber $u = \partial_X\phi$, and to rewrite the X-derivative of Eq. (5.104) under the form

$$\partial_T u = \mu_1 \partial_X^3 u + \mu_2 \partial_X(u^2) + \varepsilon\left[\delta_1 \partial_X^4 u - \eta\partial_X^2 u + \frac{\delta_2}{2}\partial_X^2(u^2)\right] \tag{5.109}$$

Nature of the Eckhaus instability

Using Eq. (5.104), it is possible to establish the conditions necessary for the Eckhaus bifurcation at $\eta = 0$ to be supercritical, hence leading to a modulated state of weak amplitude. More precisely, in a periodic domain (say, with fundamental wavenumber $k_0 = 2\pi/L$, where L is the domain size), instability of the homogeneous TW state $u = 0$ will occur when $\eta > -\delta_1 k_0^2$, as seen from the linear stability analysis of Eq. (5.109). Then, for $\eta = -\delta_1 k_0^2 + \xi^2$, where $\xi \ll 1$ is a new smallness parameter (a priori, not related to ε), we may attempt a weakly nonlinear analysis to determine the evolution of the unstable mode, expanding

$$\phi = \xi \left\{ A(\tau = \xi^2 T) \exp[ik_0(X - vT)] + c.c. \right\} + \xi^2 \phi_2(T, \tau, X - vT) + \dots \quad (5.110)$$

where v is an $O(\varepsilon)$ correction to v_g, to be determined in the course of the analysis. Omitting details, it can be shown [438] that this standard Ginzburg–Landau expansion leads at order ξ^3 to

$$\partial_\tau A = \varepsilon k_0^2 A + i \frac{k_0}{3} \frac{2\mu_2^2 + 3i\delta_2\mu_2\varepsilon k_0 - \varepsilon^2 k_0^2 \delta_2^2}{\mu_1 + 2i\delta_1\varepsilon k_0} A|A|^2 \quad (5.111)$$

For $\xi \ll \varepsilon \ll 1$, this leads to a nonlinear Schrödinger equation (the cubic coefficient is purely imaginary, and the linear growth rate vanishes). This is not too surprising, given the Hamiltonian character of the Korteweg–de Vries equation [Eq. (5.104) with $\varepsilon \to 0$]. Including the first order correction in ε, we then get a dissipation-modified nonlinear Schrödinger equation

$$\partial_\tau A = \frac{2ik_0\mu_2^2}{3\mu_1} A|A|^2 + \varepsilon k_0^2 \left\{ A + \frac{\mu_2}{3\mu_1^2}(4\delta_1\mu_2 - 3\delta_2\mu_1)A|A|^2 \right\} \quad (5.112)$$

whose dissipative part provides amplitude selection (on a longer time scale $\varepsilon\tau$, i.e. $\xi^2\varepsilon^4 t$ in the original time scale of the CGL equation), and saturation if

$$\frac{\mu_2}{\mu_1^2}(4\delta_1\mu_2 - 3\delta_2\mu_1) < 0 \quad (5.113)$$

which reads, in terms of f and d and for $1 + df > 0$,

$$d^2(1 - 6f^2) - (8 + f^2)(1 + 2df) > 0 \quad (5.114)$$

as obtained by Janiaud et al. [438]. If this condition is not satisfied, no saturation occurs and the approach based on Eq. (5.112) is not valid after a time of order $\varepsilon^{-1}k_0^{-2}$, where the amplitude A becomes $O(1)$. It is better in this case to reconsider the original CGL equation (5.84), which may account for large phase modulations, and associated strong deviations of the amplitude with respect to its saturation value a_s. Typically, in the regions where phase gradients (i.e. local wavenumber) become large, the amplitude of the wave is depressed, and may even lead to the formation of dislocations [points where $\mathrm{Re}(a) = \mathrm{Im}(a) = 0$] allowing phase slips and readjustment of the pattern wavelength (annihilation or creation of cells). An example of phase slip is presented in Fig. 5.28. As also observed by Janiaud et al., the region of amplitude depression may also be contaminated by the counter-propagating wave, and the theoretical description would in this case require consideration of a coupled system like the CCGL or MFGL.

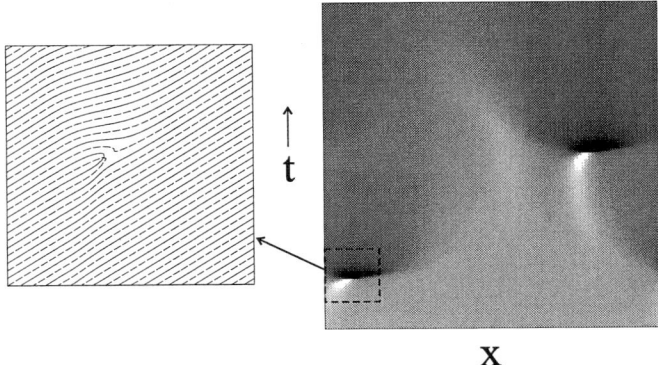

Figure 5.28: Direct numerical simulation of the CGL equation in a periodic domain, for $\mu = d = f = 1$. The initial condition corresponds to a traveling wave with winding number p slightly in the Eckhaus-unstable region. In the space–time plot (right) of the phase gradient $\partial_x \varphi$ (local wavenumber), two phase slips are visible, which allow the mean wavenumber to decrease within the stable region (increasing grey level). The left plot represents the lines $\mathrm{Re}(a) = 0$ (full) and $\mathrm{Im}(a) = 0$ (dashed) in the vicinity of a phase slip.

According to the above analysis, it is clear that in an experiment, amplitude modulations may grow very slowly when entering the Eckhaus unstable region, and it may be difficult in practice to observe stabilization to a saturated state. These long-lived transients have indeed been observed by Janiaud et al. in their experiments on the secondary oscillatory instability of convection rolls. The authors conclude that in their case, the instability may actually well be subcritical.

In fact, the condition (5.114) for stabilization of amplitude modulations on very long time scales is satisfied only in a small region of the (d, f) plane, most of it lying within the Benjamin–Feir unstable region (see Fig. 5.29). Still, for systems in which the nonlinear frequency detuning parameter $|f|$ is small, this may be observed for sufficiently large linear dispersion parameters $|d|$. Note that in this case, one should also take the possibility of short-wave and Benjamin–Feir instability into account, as Fig. 5.26 shows.

Other interesting comments are that even though multiplied by $\varepsilon \ll 1$, the second non-linearity $\partial_X \phi \partial_X^2 \phi$ does play an important role in the possible saturation of modulated states, and it would be wrong to neglect it. On the other hand, this is the only nonlinearity present for $d = f = 0$, i.e. for the real Ginzburg–Landau or NWS equation (4.29) applying to cellular steady patterns. Hence, for this case, the KdV part in (5.104) vanishes, and we recover the equation

$$\partial_T \phi = \varepsilon \left(-\frac{1}{4p^2} \partial_X^4 \phi - \eta \partial_X^2 \phi - \frac{3}{p} \partial_X \phi \partial_X^2 \phi \right) \tag{5.115}$$

describing the Eckhaus instability of non-propagative patterns [19, 439]. In this case, the bifurcation is always subcritical, as the criterion (5.114) also shows.

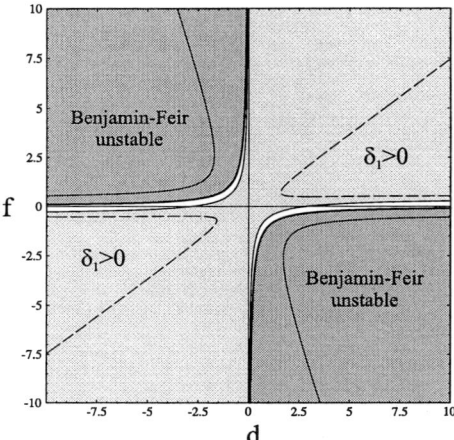

Figure 5.29: Regions of the plane (d, f) where traveling wave solutions of the CGL equation are unstable to the Benjamin–Feir mode (dark grey) or to a short-wave mode (inside the dashed line $\delta_1 = 0$, see also Fig. 5.26). In the light grey region, the Eckhaus instability of waves is subcritical, while it is supercritical only in the thin white region, also limited by the Benjamin–Feir instability.

Dissipation-modified KdV equation, solitons and cnoidal waves

In their experiments, Janiaud et al. have also observed that the growth of the disturbances following a quench in the Eckhaus-unstable region involves the excitation of several spatial scales, i.e. after some time, the evolution is not described by a sinusoidal perturbation such as (5.110). Rather, they observed that the phase gradient $u = \partial_X \phi$ becomes localized, i.e. pulse-shaped, and that an amplitude modulation also occurs at the same location. The next step in their theoretical analysis therefore consisted in looking for soliton-like solutions of the DMKdV equation (5.109). We will reproduce some steps of this analysis here, and also consider some other types of solutions of the Korteweg–de Vries equation, as some of these results might be needed later, for the consideration of surface waves in Marangoni–Bénard convection (§6.2). In addition to the partial results presented below, the reader is also referred to the more complete works of Christov and Velarde [447, 448], Nekorkin and Velarde [452], Velarde et al. [453], Bar and Nepomnyashchy [450], Rednikov et al. [451] and Kliakhander et al. [454] for more complete discussions and further references.

i) Solitons and cnoidal wave trains

The Korteweg–de Vries equation $\dot{u} = u_{xxx} + (u^2)_x$ is known (see e.g. [445]) to admit soliton solutions $u = 6b^2 \mathrm{sech}^2 z$, with $z = b(x + 4b^2 t)$. Owing to the absence of dissipation, there is no selection of the amplitude, and a family of solitons exists, here parameterized by b. The shorter the spatial scale b^{-1}, the larger the amplitude $\sim b^2$ and phase velocity $\sim b^2$. This solution is represented in Fig. 5.30, and appears as a homoclinic connection in the $(u, v = u_x)$

phase space, as can be seen by looking for solutions of the dynamical system

$$u_x = v \, , \quad v_x = E - cu - u^2 \tag{5.116}$$

where c is the phase velocity and E an integration constant. For $c^2 + 4E > 0$, this system has two fixed points $v = 0, u = (-c \pm \sqrt{c^2 + 4E})/2$, respectively a saddle $(-)$ and a focus $(+)$. Integrating (5.116) once yields $v^2 = 2(\alpha + Eu - cu^2/2 - u^3/3) > 0$. Given that a uniform shift of u amounts to a mere change of the velocity c (as seen from the KdV equation), we may place the saddle (soliton baseline) at $u = 0$, i.e. selecting $E = 0$. Then, it is seen that a two-parameter family of solutions exist, parameterized by α and c. The soliton family with amplitude $3|c|/2$ corresponds to $\alpha = 0$ (one recovers the sech^2 soliton after a further integration). In fact, the solitons correspond to the limit for $\alpha \to 0$ of a class of solutions of the KdV equation named cnoidal wave trains [445]. Their analytical form may be written in terms of elliptic functions [450, 451]. These are represented in Fig. 5.30, where we have fixed $c = -1$, and varied α from $\alpha = c^3/6 = -1/6$ (where a small amplitude orbit appears around the focus) to $\alpha = 0$ (where the periodic orbit collides with the saddle point in a homoclinic bifurcation).

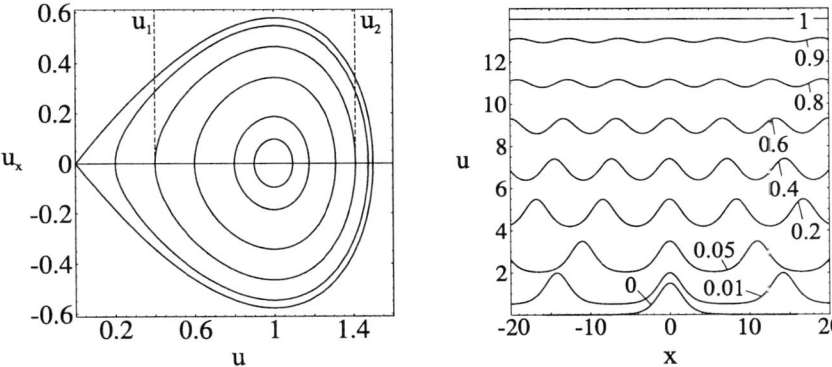

Figure 5.30: Cnoidal waves and soliton solutions of the KdV equation. Left: phase portrait (only bounded solutions are represented). Right: wave trains are labeled by the corresponding value of u_1, varying from $u = 1$ (zero amplitude) to $u = 0$ (soliton). The upwards shift of the wave trains is arbitrary.

ii) Weakly dissipative dynamics of solitons

Now, it can be expected that on a long time scale, the dissipative part of Eq. (5.109) will provide an energy input-output balance (if $\eta > 0$), hence leading to selection of the wave amplitude [448]. To examine this possibility, at least for solitons, we first rewrite the DMKdV equation (5.109) in a reference frame $y = X - Tc(T)$, and cast it in the form $L(u) = -\partial_T u + \varepsilon R(u)$, with $L(u) = -v\partial_y u - \mu_1 \partial_y^3 u - \mu_2 \partial_y(u^2)$ and $R(u) = \delta_1 \partial_y^4 u - \eta \partial_y^2 u + \delta_2 \partial_y^2(u^2)/2$, with $v = \partial_T(Tc)$, to be determined. The soliton $u_0(z) = 6\mu_1 b^2 \text{sech}^2 z/\mu_2$ with $z = by$ being a solution of $L(u_0) = 0$ with $v = -4b^2\mu_1$, we follow Janiaud et al. [438], allowing b to vary on a slow time scale $\tau = \varepsilon T$, and incorporating corrections to the soliton shape. Specifically, we consider

$$u = u_0(z = b(\tau)y) + \varepsilon u_1(z, \tau) + ..., \quad v = v(\tau) = -4\mu_1 b^2(\tau) \tag{5.117}$$

Expanding with respect to ε, the order ε yields $\mathcal{L}(u_1) = R(u_0) - (zu_0' + 2u_0)\dot{b}/b$, where $\mathcal{L}$ is the linearized operator corresponding to L, and the prime and dot respectively denote derivatives with respect to z and τ. The compatibility of this problem requires the inhomogeneity to be orthogonal to the solution of the adjoint problem, which is easily found as $\mathrm{sech}^2(z)$. After some algebra, an evolution equation for b is found, namely

$$\dot{b} = \frac{4}{3}b^3 \left[\frac{2}{5}\eta - \frac{8b^2}{35\mu_2}(6\delta_2\mu_1 - 5\delta_1\mu_2) \right] \tag{5.118}$$

which predicts a stabilization to finite amplitude when

$$6\delta_2\mu_1/\mu_2 - 5\delta_1 > 0 \tag{5.119}$$

Using Eqs (5.103) and (5.107), it can be calculated that for the CGL equation in the Benjamin–Feir stable case $1 + df > 0$, this condition is satisfied if

$$-67 + 5d^2 - 116df - 23f^2 - 39d^2f^2 - 28df^3 > 0 \tag{5.120}$$

Actually, this condition is fulfilled in a thin region close to the Benjamin–Feir instability line, within the region where the Eckhaus instability is supercritical [i.e. the condition (5.114) holds].

Note that a more general calculation of the stability of cnoidal waves and solitons in *infinite* domains has been performed by Bar and Nepomnyashchy [450], who show that both solitons and very small amplitude cnoidal waves are in fact unstable, while a range of stability exists between these two extremes. The instability of solitons can be understood from the fact that their exponentially decaying tails are themselves unstable, therefore leading to excitation of other pulses far enough from the initial one. As for small-amplitude cnoidal waves, the result of Bar and Nepomnyashchy actually does not contradict the analysis based on Eq. (5.112), as these authors consider more general disturbances which have a spatial period not necessarily commensurate with the period of the cnoidal wave train (quasi-wavenumbers), while we assumed a periodic box. Rednikov et al. [451] carried out an analysis similar to that described above, but for cnoidal waves as well, and also considering an additional dissipative term $-\varepsilon u$ in the DMKdV equation (5.109).

Janiaud et al. [438] have also performed direct numerical simulations of both DMKdV and CGL equations, which indeed show the existence of both cnoidal-like waves and pulses. However, they note that both criteria Eqs (5.114) and (5.120) appear to overestimate the region of the plane (d, f) where stable modulated solutions can be obtained, presumably due to finite size effects. Stable pulses have an amplitude which grows when p is further increased in the Eckhaus-unstable range. Above a certain value of p, the pulse no longer saturates and a phase slip occurs, leading to a homogeneous traveling wave solution in the stable range. They also point out that stable pulses can also be observed in the Benjamin–Feir unstable range, with a non-zero basin of attraction even for very large systems.

5.5.7 Phase turbulence of traveling waves

Up to now, we have only considered situations where the Newell criterion $1 + df < 0$ was not satisfied, i.e. there still existed a range of values of p corresponding to stable traveling waves. When $1 + df \to 0_+$, this range shrinks, and it appears that, after some rescaling, the dispersive and the second nonlinear term in the DMKdV equation (5.104) become negligible. The latter then reduces to the Kuramoto–Sivashinsky (KS) equation

$$\partial_t \phi = -\partial_X^2 \phi - \partial_X^4 \phi + (\partial_X \phi)^2 \qquad (5.121)$$

which is known to lead to chaotic dynamics at sufficiently large system size (phase turbulence), and has been the subject of numerous investigations [51, 290, 496, 497, 498]. Note that if when taking the limit $1 + df \to 0_+$, we assume $1 + df \sim \varepsilon^2 \to 0$, dispersion is not negligible and we rather get the Kawahara equation

$$\partial_t \phi = -\partial_X^2 \phi - \partial_X^4 \phi + (\partial_X \phi)^2 + g \partial_X^3 \phi \qquad (5.122)$$

Actually, the limit $1 + df \to 0_+$ is certainly singular, and it is better to restart from Eqs (5.84) and (5.85), in order to derive the phase equation applying past the Benjamin–Feir threshold $(1 + df \to 0_-)$. We will omit details here, just mentioning that the correct scalings are now $\partial_x \sim \varepsilon$, $\partial_t \sim \varepsilon^4$, $1 + df = -b\varepsilon^2$, $\varphi \sim \varepsilon^2$, $r \sim \varepsilon^4$, and we can also allow for some small wavenumber shift $p = \varepsilon q$. It is also convenient to introduce a reference frame translating with velocity $c = \varepsilon c_1 + \varepsilon^2 c_2 + ...$, in order to eliminate the term $\partial_X \varphi$. Pursuing calculations up to order ε^6, we indeed get the Kawahara equation (5.122) after some rescaling of time, length and amplitude, and the coefficient g is found as

$$g = -2 \operatorname{sgn}(qd) \left[1 + \frac{b r_s^2}{2 q^2 (1 + d^{-2})} \right]^{-\frac{1}{2}} \qquad (5.123)$$

with $r_s^2 = \mu$ at the leading order. Hence, for $q \neq 0$, dispersion should play an important role near the threshold of the Benjamin–Feir instability, and can be expected to "laminarize" [447] the turbulent solutions of the KS equation. Still, $|g| < 2$ from Eq. (5.123), so that we cannot reach the weakly dissipative dynamics characteristic of the DMKdV equation. However, according to the numerical results of Christov and Velarde [447], a transition to softer pulse-dominated regimes can be expected when g is increased past $g = 1/2$. This might actually explain why stable pulses could be observed by Janiaud et al. [438], even in the Benjamin–Feir unstable regime.

Let us conclude this section by emphasizing that these pulses do not possess the usual properties of KdV solitons. In particular, in addition to the selection of amplitude and velocity brought in by dissipation, their collisions are inelastic, i.e. the energy of two colliding pulses is not the same before and after their crossing. On the other hand, for $g = 1$, Christov and Velarde [447] have also observed (starting from pulse-like initial conditions) bound states of pulses, with tails exhibiting local minima. Nekorkin and Velarde [452] and Christov and Velarde [447] have also shown numerically how soliton bound states are formed, as aperiodic wave trains with pulses all traveling at the same velocity. Further properties of "dissipative solitons" will be discussed in §6.2. The reader is also referred to the paper of Bar and Nepomnyashchy [450], for a discussion of conditions of applicability and further bibliography on both DMKdV and Kawahara equations.

6 Solitonic and shock-like waves

This chapter describes further oscillatory convection and wave phenomena observed and predicted past an instability threshold, for some of the problems whose linear stability was considered in Chapter 3. This active subject of current research is mostly motivated by the experimental studies of Linde, Velarde, and collaborators. After a presentation of the some relevant experimental findings in heat- and mass-transfer systems, two different axes of theoretical investigation are illustrated. Note also that the theories presented generally refer to thermal problems, in the "heated from above" situation, thus we will ignore here specific effects like matter accumulation at the interface, energy barriers, ... This oversimplification will allow some understanding to be gained of transverse and longitudinal waves *separately*, and to assess their basic nonlinear properties.

First, long-wavelength capillary-gravity waves sustained by the Marangoni effect at the free surface of a liquid layer heated from above are considered. The nonlinear equation describing the evolution of the free surface position is a dissipation-modified Korteweg–de Vries equation, which includes nonlinearity, dispersion, instability and dissipation. Note that similar equations have been derived in other contexts, such as in film flows down inclined planes, Eckhaus instability of traveling waves, ... Typical solutions are solitons or periodic trains of waves propagating with constant velocity, which is in qualitative agreement with experiments.

Then, we turn to another type of waves also observed when the layer is heated from above, namely longitudinal, dilational, or sound-like modes. In contrast with transverse waves, longitudinal oscillatory motions are possible even in the absence of surface deformation, and can be sustained e.g. when curvature of the temperature profile is taken into account within the layer. This is the approach considered here, even though several other effects (interaction with surface deformation, surface accumulation, ...) could also help exciting these waves. Although oversimplified, the theory to be presented allows a nonlinear equation to be derived rigorously, even for waves that are not long compared with the depth of the layer. This is made possible by the absence of dissipation and dispersion at the leading order when the Marangoni number tends to infinity. The resulting new equation is non-local, and therefore more advantageously written in Fourier space. Some numerical solutions are presented, showing that the evolution of a wave packet typically leads to the formation of shocks (smoothed by dissipation) in the free surface velocity profile.

6.1 Some experimental findings

In stationary and quasi-stationary heat-transfer experiments, a square or annular glass container was filled with a liquid, for instance octane with a depth ranging from 3 to 8 mm. The bottom was cooled by a continuous flow of air or water at 20 °C. A quartz cover, which was placed 3 to 15 mm above the liquid surface, was heated to establish the temperature gradient in the gas/liquid system. The waves were visualized by a shadowgraph technique with a parallel beam passing bottom-up through the container. Surface deformation and spatial refractive-index variation due to an inhomogeneous temperature field yield a deflection of the transmitted light and the modified light intensity distribution was detected with a camera.

For gradients in the range of 10^2 K/cm or higher, periodic wave trains were observed. The lowest gradient defines the critical Marangoni number for oscillatory instability, which in stationary heat-transfer experiments (using the above mentioned octane–air or diphenyl as liquid and nitrogen or carbon dioxide as the gas phase), was estimated at about 10^4–10^5. Precision could not be obtained beyond a 50% error bar due to the uncertainties in determining the liquid layer thickness and the critical thermal gradients. However, cross-checking on these estimates was performed by calculating the critical Marangoni number for the case when the system is heated from below, in which Bénard cells could be observed for the same apparatus, using the same fluids, and hence the same material properties. Values of about 10^2 were found, as expected, thus supporting the relative reliability of the determination of the critical values for the oscillatory regime.

Close to the above-mentioned oscillatory threshold, regular patterns of periodic wave trains were observed. For instance, counter-traveling waves appeared aligned parallel to one pair of walls in a square container of length $L = 50$ mm when the heat-transfer was from nitrogen into diphenyl with the thermal gradients $(dT/dz)_{\text{liquid}} = 113$ K/cm in the liquid and $(dT/dz)_{\text{gas}} = 168$ K/cm in the gas (the system is heated from the top). The free surface temperature was $T_{\text{surface}} = 143$ °C ($d_{\text{liquid}} = 3$ mm and $d_{\text{gas}} = 15$ mm). The wave pattern of two counter-traveling periodic trains experience alternating head-on collisions and reflections back and forth between the opposite walls of the square container. The wave parameters were: wavelength $\lambda = 9.1$ mm, frequency $f = 1.84$ Hz, phase velocity $v = 1.67$ cm/s, and the time between the alternating collisions 0.54 s.

Depending on the parameters of the experiment, the waves can also interact at some angle. In this case, the crossing of wave crests lead to a change of trajectory and hence a so-called phase-shift which may be positive or negative depending on the incident angle. Interactions with negative phase-shift at angles of 80°–98°, and even 103° have been reported while a positive phase-shift occurs at angles of 90°–103°. Accordingly, the experimental "neutral" angle at which no change of trajectory was found is about 90°.

The above mentioned phase-shift gives evidence for the solitonic properties of these waves. Here, as the system is driven, it has been established that after a sudden deceleration at the collision, the crests re-accelerate until they reach the pre-collision velocity. This behavior has also been reported for mass-transfer experiments.

Figure 6.1 depicts the kind of setup used for the mass-transfer experiments in circular containers. A circular container permits, by using a suitable concentric ring, to have an annular channel filled with the absorbing liquid. This may result in waves propagating azimuthally, corresponding roughly to a one-dimensional situation with periodic boundary conditions (see

Fig. 6.2).

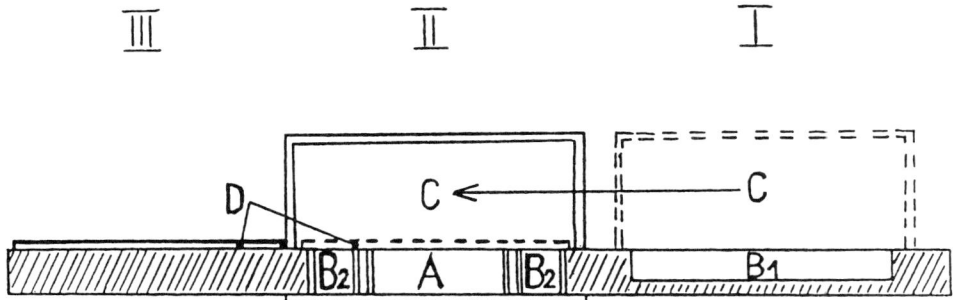

Figure 6.1: Set-up for mass-transfer experiments: liquid container A for absorbing liquid with circular or annular surface; reservoirs B_1 and B_2 for evaporating liquid; glass box C for vapor accumulation; and plate D to cover the container A and the reservoir B_2 before the experiment is started. Absorption starts after the glass box C, with the accumulated vapor, is moved from I to II on top of A and the cover plate D is removed from II to III.

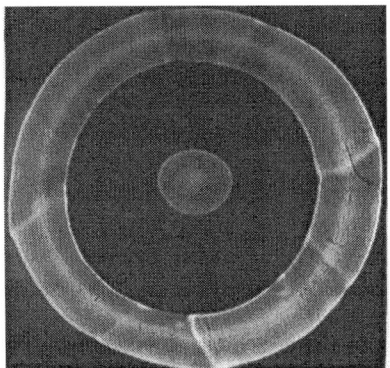

Figure 6.2: Typical long wave train observed in an annular (cylindrical) channel in a mass-transfer-driven system; absorption of pentane vapor by liquid toluene in a layer that is about 1 cm deep or less.

The reservoirs B_1 and/or B_2 were filled with a liquid with lower surface tension than the absorbing liquid in A. Two ways of doing the experiment were used. In the first approach the container A was filled, for instance, with toluene and the reservoirs B_1, and B_2 were filled with pentane. The container A together with the reservoir B_2 were closed with the cover D. The glass box C was placed on top of the reservoir B_1, and when the box C was full of pentane vapor, it was placed on top of the container A while the cover D was displaced. Throughout the whole duration of the experiment, pentane vapor was allowed to diffuse from the glass box and from the ring-shaped reservoir B_2 to the annular channel in A. The absorption process was able to start immediately with a very high concentration gradient leading to a very high driving force (Marangoni stress) and turbulent interfacial motions with irregularly traveling and breaking waves were observed. Then the concentration gradient spontaneously decreased with time to a moderate level, and more regular periodic wave trains were observed.

The other approach consisted in allowing vapor diffusion only from reservoir B_2 to the liquid in container A. After about 2 min this spontaneous process led only to moderate levels of driving force but generally enough to excite regular wave motions along the surface of the liquid in container A. Using xylene, nonane, trichloroethylene, or benzene instead of toluene and using hexane, acetone, or diethyl ether for the vapor phase, lead to qualitatively similar results.

Figure 6.3: The lateral transport of surface-active vapor from the reservoir B_2 compresses a film of an insoluble surfactant to the center of the container. Thereby, a quasi-annular channel of clean surface is spontaneously formed near the outer wall. The ridge indicates the borderline between the clean surface and the film-covered central pan. Images were taken 14, 24, and 34 s (from left to right) after commencing the experimental run, respectively. The third picture shows five simultaneous head-on collisions of two counter-traveling periodic wave trains, which deform the ridge and eventually break it, hence creating the wave guide that channels the two wave trains. Experiment of absorption of pentane vapor into toluene; diameter of the circular container, 43 mm; liquid depth, 18 mm.

As the absorbed vapor alters the parameters of the liquid, the experiments are unsteady by nature. Nevertheless, the observed periodic wave trains remained quasi-stationary as the waves traveled through the container many times without changing wavelength and frequency.

Figs 6.3 and 6.4 refer to a pair of synchronously colliding counter-rotating periodic wave trains that subsequently, when the driving force diminished, yielded to a single periodic wave train rotating either clockwise or counter-clockwise. A shadowgraph image of the latter is presented in Fig. 6.5. Finally, with low driving force the waves disappear, still having a well-defined wavelength and wave velocity.

When, for instance, radial transfer of the vapor occurs from the outer reservoir B_2, three stages of the initial process are shown in Fig. 6.3. Note the clean quasi annular surface in the periphery of the cylinder while the surfactant in the central part prevents Marangoni

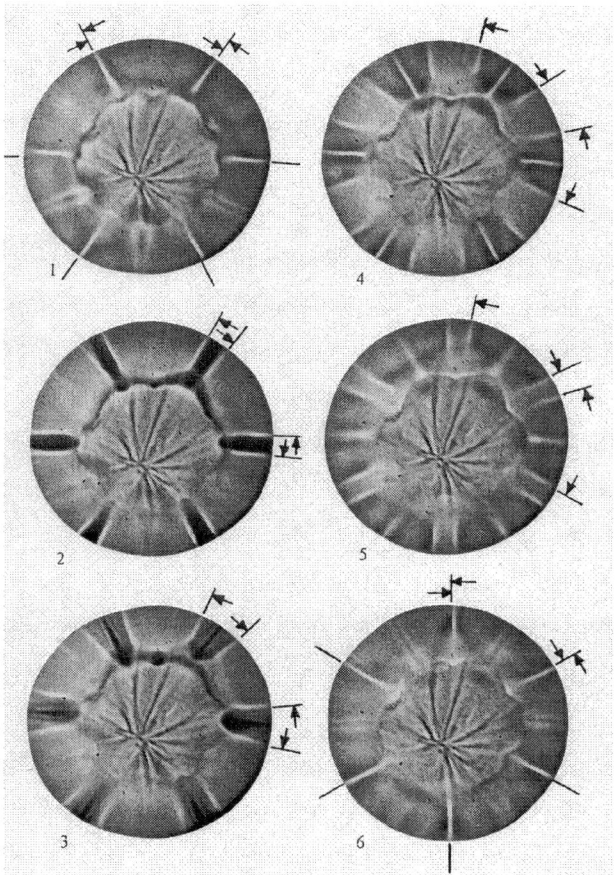

Figure 6.4: Two synchronously colliding counter-rotating trains of six waves each in a spontaneously created annular channel, formed by a concentric compressed film of insoluble surfactant in a circular container. The images were taken 70 s after commencing the experiment. The time interval between each image is 0.1 s.

convection. Two counter-rotating periodic wave trains appeared as shown in Fig. 6.4.

In the example of two counter-rotating trains, consisting of six waves each, shown in Fig. 6.4, the six images were taken consecutively with time intervals of 0.1 s, starting with the six head-on collisions of the wave crests in Fig. 6.4(1) until the next ones in Fig. 6.4(6). In Fig. 6.4(1) the collisions appear as bright sharp lines. Just after the collisions of the crests, broad dark lines are seen at the former collision spots in Fig. 6.4(2). In Fig. 6.4(3) the wave crests have separated further and still give a sharp line but with less contrast than the collision spots. The wave crests recover their form and velocity shortly after the collision in Fig. 6.4(3)–(5) and experience the next head-on collision in Fig. 6.4(6). This periodic behavior of the wave

Figure 6.5: Absorption of pentane vapor by liquid toluene: clockwise rotating periodic wave train with 20 wave crests in an annular channel of 57 mm and 37 mm outer and inner diameters, respectively, a few minutes after glass cover C is placed on top of A and diffusion proceeds from B_2 to A. Liquid depth 4 mm.

trains is repeated about 90 times over about 45 s. Note that though the insoluble film in the center of the container prevents the Marangoni waves there, radial lines were observed which may indicate longitudinal rolls below the surface formed by a flow instability.

The counter-rotating periodic wave trains in the quasi-annular channel formed by a central surfactant film in the center of a circular container behave similarly to the waves observed in a real annular channel. In annular channels with an inner wall, counter-rotating periodic wave trains also occur followed by a regime of only one wave train.

As in the above mentioned experiments the driving force decreases continuously due to diffusion, the integer number of crests in each wave train increases stepwise with corresponding stepwise wavelength decrease. The wave velocity also decreases, although during the interval of constant wavelength the wave velocity remains constant.

The regular behavior of counter-rotating periodic wave trains in the quasi-annular channel enables drawing of trajectories. Figure 6.6 shows a typical space–time plot of wave crests seen in Fig. 6.4.

In Fig. 6.6, already 0.2 s after the head-on collision, the crests move with constant velocity until they experience another collision which decelerates them. After the collision the crests suddenly decelerate and re-accelerate immediately past the collision event to attain their pre-collision velocity. In this manner the change of trajectory in the space–time plot of the head-on collision leads to a so-called negative phase-shift. In the case of Fig. 6.6 the spatial phase-shift is 17% of the wavelength, and the wave velocity right after collision is about 64% of the one before collision. Finally, Fig. 6.7 shows the corresponding space–time plot for a head-on collision of two single waves.

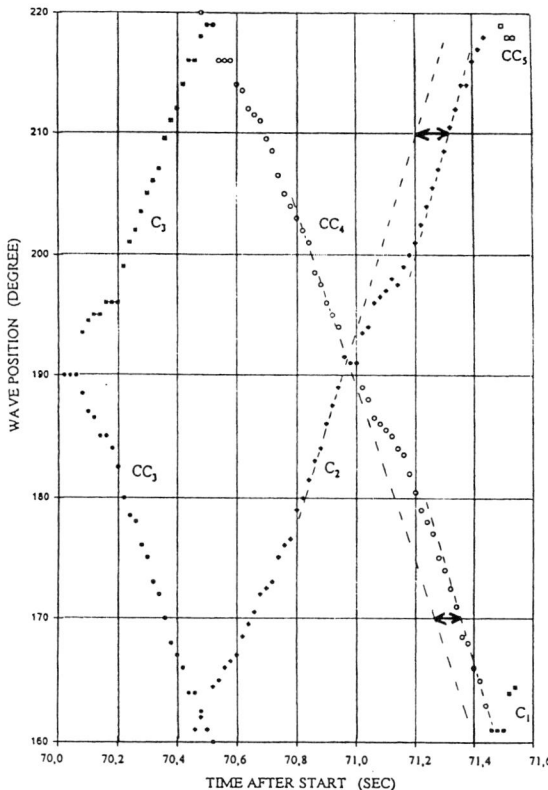

Figure 6.6: Space–time plot illustrating the phase shift due to transient deceleration, indicated with the double arrow, occurring in one of the head-on wave-crest collisions. C_i ($i = 1, 2, 3$) and CC_i, ($i = 3, 4, 5$) refer to crests of clockwise and counter-clockwise rotating periodic wave trains, respectively. The wavelength is 23 mm, the time lags of the phase-shifts are about 0.1 s, and the spatial phase-shifts are about 3.8 mm, which corresponds to 17% of wavelength. The wave velocities at the outer wall of the annular circular cylindrical channel before and after collision are 2.7 and 1.7 cm/s, respectively, corresponding to angular velocities of 71.4°/s and 45.7°/s. Thus, the wave velocity right after collision is about 64 ± 2% of the one before collision. About 0.2 s after collision the original wave velocity is recovered. The average velocities at the outer and inner wall, respectively, are 2.3 and 1.2 cm/s; the frequency is 1.0 Hz. All other synchronous head-on collisions between wave crests have identical phase shifts.

6.2 Dissipative Korteweg–de Vries description of transverse waves

In this section, we consider a multiscale expansion of the full nonlinear problem described by thermo-hydrodynamic equations. One foresees, however, some complications, for the dynam-

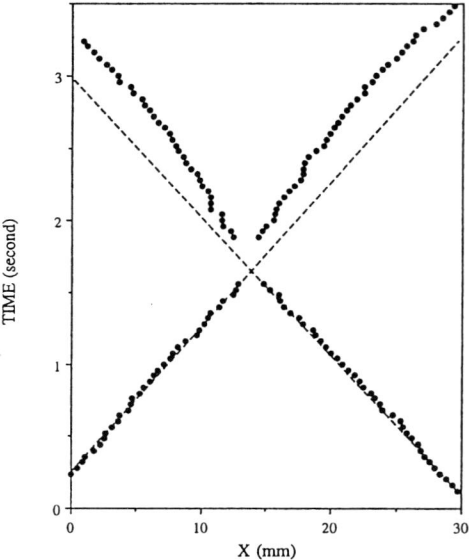

Figure 6.7: Space–time plot of the isolated head-on collision of two "solitary" waves in a mass-transfer-driven system, absorption of pentane vapor by liquid toluene in a layer about 1 cm deep or less.

ics of the liquid layer demands consideration of both the boundary layer in the upper surface open to the ambient air or gas phase and the boundary layer at the bottom where due account of the friction between liquid and solid must be considered. Furthermore, if theory is to be compared with experiments with thin liquid layers in small constrained geometries, as discussed in §6.1, then the full three-dimensional case ought to be considered together with the possible role played by menisci on lateral boundaries. In view of these difficulties, theoretical work has proceeded with drastic simplifications. Perhaps the minor one is considering a 2D geometry, that fits well with the experimental data earlier reported. Another simplification is to disregard, in a first approach, the two boundary layers mentioned above. Even more, consideration of a stress-free bottom greatly simplifies the problem albeit being too drastic a simplification. However, although one expects quantitative disagreement between theory and experiment, yet this qualitative description hopefully brings a significant understanding of the salient features found in experiments. Thus, for simplicity, we will consider a horizontal liquid layer, of thickness d, subject to a transverse vertical temperature gradient in a two-dimensional geometry. Let x and z denote the horizontal and vertical coordinates, respectively. With the horizontal and vertical velocity components (u, w), respectively, and ρ, p, T, and σ the density, pressure, temperature, and liquid–air interfacial tension, the evolution of the liquid layer is governed by the thermo-hydrodynamic equations in the Boussinesq approximation, which we recall for completeness and self-consistency in this chapter:

$$u_x + w_z = 0 \tag{6.1}$$

$$\rho_0 \left(u_t + u u_x + w u_z \right) = -p_x + \mu \left(u_{xx} + u_{zz} \right) \tag{6.2}$$

$$\rho_0 \left(w_t + u w_x + w w_z \right) = -p_z + \mu \left(w_{xx} + w_{zz} \right) - \rho g \tag{6.3}$$

$$T_t + u T_x + w T_z = \kappa \left(T_{xx} + T_{zz} \right) \tag{6.4}$$

$$\rho = \rho_0 \left[1 - \alpha \left(T - T_0 \right) \right], \quad \alpha = -\frac{(\partial \rho / \partial T)}{\rho_0} \tag{6.5}$$

$$\sigma = \sigma_0 + \gamma \left(T - T_0 \right), \quad \gamma = \frac{\partial \sigma}{\partial T} \tag{6.6}$$

where g denotes the gravitational acceleration, ρ_0 and σ_0 are some reference values, e.g. at the motionless quiescent state at temperature $T_0 = T(z = d)$, and subscripts t, x, and z refer to time and space derivatives. Note that generally α is positive (except for water around 4 °C) and γ, though generally negative, may according to the liquid become positive. Here μ is the dynamic viscosity and κ the thermal diffusivity of the liquid.

The above evolution equations must satisfy the following boundary conditions (b.c. in what follows).

At $z = d + \eta$, with $\eta(x, t)$ the time- and space-dependent interface deformation,

$$w = \eta_t + u \eta_x \tag{6.7}$$

$$p - p_{air} = \tfrac{2\mu}{N^2} \left[w_z + u_x \eta_x^2 - \eta_x \left(u_z + w_x \right) \right] - \\ - \tfrac{\eta_{xx}\sigma}{N^3} - \tfrac{\eta_{xx}}{N^5} \left(\bar{\kappa} + \bar{\epsilon} \right) \left(u_x + \eta_x w_x \right) \tag{6.8}$$

$$\mu \left(1 - \eta_x^2 \right) \left(u_z + w_x \right) + 2\mu \eta_x \left(w_z - u_x \right) = \\ = N \eta_x \sigma_z + N \sigma_x + \left[\left(\bar{\kappa} + \bar{\epsilon} \right) / N \right] \left(u_{xx} + \eta_x w_{xx} \right) + \\ + \left[\left(\bar{\kappa} + \bar{\epsilon} \right) / N^3 \right] \eta_{xx} \left(w_x - \eta_x u_x \right) \tag{6.9}$$

$$k \left(T_z - \eta_x T_x \right) = -k \beta N - h N \left(T - T_0 \right) \tag{6.10}$$

At $z = 0$

$$w = 0 \tag{6.11}$$

$$u_z = 0 \quad \text{(stress--free)} \tag{6.12}$$

and

$$T = T_\text{b} \quad \text{(constant temperature),} \quad \text{or } T_z = \text{const} \quad \text{(constant heat flux)} \tag{6.13}$$

The alternative to Eq. (6.12) is the more realistic case of a rigid boundary, $u = 0$. However, in order to simplify the mathematical description, we just retain (6.12) with the expectation that the essence of the problem is not qualitatively changed.

We have introduced $N^2 = 1 + \eta_x^2$ and $\bar{\kappa}$ and $\bar{\epsilon}$, the shear and dilational surface viscosities, respectively. The quantity h/k accounts for the liquid–air heat transfer characteristics, with $k = \kappa\rho c$ the thermal conductivity and c the specific heat of the incompressible liquid. The temperature gradient is $\beta = (T_\text{b} - T_0)/d$, taken to be positive when the heating is from the liquid side as in the earlier discussed traditional Bénard setup. As before, T_0 is a reference value taken at the undeformed liquid–air interface in the quiescent state and $T_\text{b} = T(z = 0)$. Note that for simplicity we have considered the air above the liquid layer to be weightless and dynamically passive with the pressure p_air coincident with that of the liquid surface at $z = d$.

The motionless quiescent steady state denoted with subscript s is a solution of the problem posed above. We have

$$T_\text{s}(z) = T_\text{b} - \beta z \tag{6.14}$$

$$\rho_\text{s}(z) = \rho_0 \left[1 + \alpha\beta (z - d)\right] \tag{6.15}$$

$$u_\text{s} = 0 \tag{6.16}$$

$$w_\text{s} = 0 \tag{6.17}$$

$$p_\text{s}(z) = p_\text{air} - \rho_0 g \left[(z - d) + (\alpha\beta/2)(z - d)^2\right] \tag{6.18}$$

Before embarking on the analysis we proceed to the adimensionalization of the problem and to an *a priori* estimate of its pertinent scales.

As usual, we choose units d for length, βd for temperature, κ/d for velocity, d^2/κ for time and $\mu\kappa/d^2$ for pressure, respectively. The dimensionless numbers characterizing the problem are

$$P = \nu/\kappa \qquad \text{(Prandtl number)}$$
$$Bi = hd/\kappa \qquad \text{(Biot number)}$$
$$K = \mu\kappa/\sigma_0 d \qquad \text{(capillary number)}$$
$$G = B/K \qquad \text{(Galileo number)}$$
$$B = \rho_0 g d^2/\sigma_0 \qquad \text{(Bond/``Eotvos'' number)}$$
$$Vi = (\bar{\kappa} + \bar{\epsilon})/\mu d \qquad \text{(surface viscosity number)}$$
$$R = \alpha g \beta d^4/\kappa\nu \qquad \text{(Rayleigh number)}$$
$$M = -\gamma\beta d^2/\kappa\mu \qquad \text{(Marangoni number or surface velocity Reynolds number)}$$

with $\nu = \mu/\rho_0$, the kinematic viscosity. Note that the notation of some of these dimensionless numbers has been modified compared with previous chapters, in order to simplify the writing of some expressions in the following discussions. Note also that $M/R = \Gamma$ is the *dynamic* Bond number, and is proportional to both α^{-1} and $(-\gamma)$, two quantities that although generally positive may under appropriate circumstances either vanish or change sign.

We now look for oscillatory motions, traveling waves of "long" wavelength and phase velocity C. A reasonable assumption is that the liquid–air heat transfer is low yet crucial. Introducing ϵ, a smallness parameter, we set $Bi = \epsilon^2$. Note that an expansion for Bi could also be written. For the remaining variables we shall use

$$\xi = \epsilon\,(x - Ct)\,, \quad \tau = \epsilon^3 t\,, \quad C = C_0 + \epsilon^2 C_2 + ...,$$
$$u = \epsilon^2\,(u_0 + \epsilon u_1 + ...)\,, \quad w = \epsilon^3\,(w_0 + \epsilon w_1 + ...)\,,$$
$$p - p_s = \epsilon^2\,(p_0 + \epsilon p_1 + ...)\,,$$
$$T - T_s = \epsilon^3\,(\theta_0 + \epsilon\theta_1 + ...)\,,$$
$$\eta = \epsilon^2\,(\eta_0 + \epsilon\eta_1 + ...)\,, \quad R = R_0 + \epsilon^2 R_2 + ...,$$
$$M = M_0 + \epsilon^2 M_2 + ...,$$
$$Vi = \epsilon^2\,\left(Vi_0 + \epsilon^2 Vi_2 + ...\right)\,.$$

Incorporating these scalings, inserting the above expansions in the evolution equations and b.c., and noticing that ϵ is an *ordering* parameter, from the original nonlinear problem we get a hierarchy of linear inhomogeneous equations that we shall solve sequentially.

To the lowest-order approximation, one finds

$$u_{0\xi} + w_{0z} = 0 \tag{6.19}$$

$$u_{0zz} = 0 \tag{6.20}$$

$$p_{0z} = 0 \tag{6.21}$$

$$w_0 + \theta_{0zz} = 0 \tag{6.22}$$

with the corresponding b.c. at $z = 1$ and $z = 0$, respectively,

$$w_0\,(1) + C_0\eta_{0\xi} = 0 \tag{6.23}$$

$$p_0\,(1) - G\eta_0 = 0 \tag{6.24}$$

$$u_{0z}\,(1) = 0 \tag{6.25}$$

$$\theta_{0z}(1) = 0 \tag{6.26}$$

$$w_0(0) = 0 \tag{6.27}$$

$$u_{0z}(0) = 0 \tag{6.28}$$

$$\theta_0(0) = 0 \quad \text{(conducting bottom)} \tag{6.29}$$

Note that $\eta = \eta(\xi, \tau)$. The solution at this level of approximation is the basic set of fields in the quiescent state.

In the subsequent order, i.e. to the first-order power in ϵ we obtain

$$u_{1\xi} + w_{1z} = 0 \tag{6.30}$$

$$u_{1zz} = -(C_0/P)\, u_{0\xi} + p_{0\xi} \tag{6.31}$$

$$p_{1z} = w_{0zz} + R_0 \theta_0 \tag{6.32}$$

$$w_1 + \eta_{1zz} = -C_0 \theta_{0\xi} \tag{6.33}$$

together with the corresponding b.c. at $z = 1$ and $z = 0$, respectively,

$$w_1(1) + C_0 \eta_{1\xi} = 0 \tag{6.34}$$

$$p_1(1) - G\eta_1 - 2w_{0z}(1) = 0 \tag{6.35}$$

$$u_{1z}(1) - M_0 \eta_{0\xi} = 0 \tag{6.36}$$

$$\theta_{1z}(1) - \eta_0 = 0 \tag{6.37}$$

$$w_1(0) = 0 \tag{6.38}$$

$$u_{1z}(0) = 0 \tag{6.39}$$

$$\theta_1(0) = 0 \quad \text{(conducting bottom)} \tag{6.40}$$

We clearly see the appearance, in particular, of the Marangoni and Rayleigh numbers as well as the structure of the hierarchy of linear inhomogeneous differential equations to be solved. According to the discussion of §3.1.4, the effect of buoyancy will here be considered in a qualitative sense, keeping in mind that the only situation coherent with the Boussinesq approximation is $R/G \to 0$.

Defining the vector $\psi = (u, w, p, \theta)^t$, with superscript t denoting transpose, the subsequent orders belong to the hierarchy of linear inhomogeneous problems:

$$
\begin{aligned}
L\psi_0 &= 0, \\
L\psi_1 &= \Phi_1(\psi_0; B, ...), \\
L\psi_n &= \Phi_n(\psi_0, ..., \psi_{n-1}; B, ...), \quad (n = 1, 2, ...),
\end{aligned}
\tag{6.41}
$$

where the operator L is the 4×4 matrix:

$$
\begin{pmatrix}
\frac{\partial}{\partial \xi} & \frac{\partial}{\partial z} & 0 & 0 \\
\frac{\partial^2}{\partial z^2} & 0 & 0 & 0 \\
0 & 0 & \frac{\partial}{\partial z} & 0 \\
0 & 1 & 0 & \frac{\partial^2}{\partial z^2}
\end{pmatrix}
$$

At each order of approximation, Fredholm's alternative must be used to ensure that the corresponding linear inhomogeneous problem can be solved. Here, in practice, it reduces to the integration of the horizontal Navier–Stokes equation over the liquid layer. The solvability condition determines C_0, the lowest-order coefficient in the expansion of the velocity C in powers of ϵ, which shows indeed the correction M_0 due to the Marangoni effect, though its value is to be found in the following order of approximation. Thus at order ϵ we obtain the first-order solutions and C_0 given by

$$
C_0 = (GP - PM_0)^{1/2}
\tag{6.42}
$$

We do not write the differential problem at the order ϵ^2. The main result is linked once more with the use of the solvability condition. In this case the horizontal Navier–Stokes equation integrated over the liquid layer leads to the following relationship:

$$
M_0/(-12) + R_0/30 = 1
\tag{6.43}
$$

which for vanishing buoyancy, $R_0 = 0$, reduces to $M_0 = -12$. The value $M_0 = -12$ is indeed very far from reality. However, theory unambiguously predicts that it is when the Marangoni number is negative that the possibility of exciting long capillary-gravity waves exists. The consideration of a no-slip b.c. at the bottom and its associated boundary layer is expected to bring the critical Marangoni number to a more realistic value (see §3.6.1). The contribution of the Rayleigh number, leading to $R_0 = 30$ when $M_0 = 0$, is maintained here for purely formal reasons. It also shows the spurious result, indeed never observed in experiments, that a long-wave instability may occur when heating a layer from below, within the Boussinesq approximation.

The next contribution is at order ϵ^3 and in this case the solvability condition brings in the time evolution of the leading term in the expansion of η, the interfacial deformation. As a result the following differential equation must be satisfied:

$$
\eta_{0\tau} + c\eta_{0\xi} + \alpha_1 \eta_0 \eta_{0\xi} + \alpha_2 \eta_{0\xi\xi\xi} = 0
\tag{6.44}
$$

which is just the Korteweg–de Vries (KdV) equation [48]. The coefficients c, α_1 and α_2 are

$$c = -C_2 - (6Bi - M_2/2 + BiR_0/30)A^{-1} \tag{6.45}$$

with $A = [(12 + G - 2R_0/5)/P]^{1/2}$

$$\alpha_1 = \frac{P^{1/2}(15B + 120K + KR_0)}{10K(12 + G - \frac{2}{5}R_0)^{1/2}} \tag{6.46}$$

and

$$\begin{aligned}
\alpha_2 = &-(126000P - 42000BP - 1411200KP - 201600BP^2- \\
&-2419200KP^2 + 46080KPR_0 - 80BP^2R_0 + 79680KP^2R_0+ \\
&+32KPR_0^2 + 32KP^2R_0^2)\{252000K[P(12 + G - \tfrac{2}{5}R_0)]^{1/2}\}^{-1}
\end{aligned} \tag{6.47}$$

As we are interested in the evolution equation obeyed by the surface deformation η when dissipation competes and/or cooperates with nonlinearity and dispersion, we ought to proceed to the next order approximation at order ϵ^4. We expect that this fourth-order approximation will provide an evolution equation for η_1 or more appropriately for $\eta = \eta_0 + \epsilon\eta_1$.

For η_1 we obtain

$$\eta_{1\tau} + c\eta_{1\xi} + \alpha_1(\eta_0\eta_1)_\xi + \alpha_2\eta_{1\xi\xi\xi} + \alpha_3\eta_{0\xi\xi} + \alpha_4\eta_{0\xi\xi\xi\xi} + \alpha_5(\eta_0\eta_{0\xi})_\xi = 0 \tag{6.48}$$

where α_1 and α_2 are the same coefficients that have already appeared in Eq. (6.44). Compiling results found to this order of approximation, we obtain the following evolution equation:

$$\eta_\tau + c\eta_\xi + \alpha_1\eta\eta_\xi + \alpha_2\eta_{\xi\xi\xi} + \epsilon[\alpha_3\eta_{\xi\xi} + \alpha_4\eta_{\xi\xi\xi\xi} + \alpha_5(\eta\eta_\xi)_\xi] = 0 \tag{6.49}$$

with

$$\alpha_3 = -\frac{P(360 + 15M_2 - R_0 - 6R_2 + 45Vi_0)}{90} \tag{6.50}$$

$$\begin{aligned}
\alpha_4 = &(653400K + 3979800KP + 504900BP^2+ \\
&+6058800KP^2 - 21120KR_0 - 111375KPR_0+ \\
&+220BP^2R_0 - 199320KP^2R_0 - 22KR_0^2- \\
&-112KPR_0^2 - 88KP^2R_0^2)/(1559250K)
\end{aligned} \tag{6.51}$$

$$\alpha_5 = \frac{2}{15}P(30 + R_0) \tag{6.52}$$

Equation (6.49) combines the Korteweg–de Vries equation with the Kuramoto–Sivashinsky equation [211, 213, 290]. It is a dissipation modified Korteweg–de Vries equation in the moving reference frame. A Galilean transformation to the laboratory reference frame provides the result

$$\begin{aligned}
\eta_t + C_{\exp}\eta_x + \alpha_1\eta\eta_x + \alpha_2\eta_{xxx} + \alpha_3'\eta_{xx}+ \\
+\alpha_4\eta_{xxxx} + \alpha_5(\eta\eta_x)_x = 0
\end{aligned} \tag{6.53}$$

where $\alpha_3' = \epsilon^2 \alpha_3$ and $C_{\exp} = C_0 + \epsilon^2 C_2 + \epsilon^2 c$ is the experimental velocity one would observe in the laboratory frame of reference. Note that C_0 and c are given by (6.42) and (6.45), respectively. Thus the velocity $C_{\exp}$ as well as $\epsilon^2 \alpha_3$ depend only on known or measurable quantities. Note that still another suitable Galilean transformation can be performed in order to eliminate the η_x term altogether. Consideration of the rigid boundary at the bottom and hence going away from the stress-free b.c. demands a more sophisticated treatment of the problem, which will not be attempted here. Still, we could phenomenologically add a dissipative term $\alpha_6 \eta$ to mimic it.

One of the formal findings is the relation

$$M_0/(-12) + R_0/30 = 1 \tag{6.54}$$

which delineates the instability threshold. Formally, Eq. (6.54) is the natural extension to oscillatory motions of a known relation for tightly coupled Marangoni and Rayleigh effects in long-wavelength steady convective instability, i.e.

$$M_0/48 + R_0/320 = 1 \tag{6.55}$$

Note, as already mentioned, that when buoyancy and surface deformation are considered together, a violation of the earlier assumed Boussinesq approximation is generally expected (§3.1.4). Hence, buoyancy is included here, merely to show that the relevant equation applicable to waves past the instability threshold is not qualitatively modified. Buoyancy will be neglected in what follows.

Now, let us comment on the (phase) velocity of the traveling wave. To the lowest-order approximation, and for a shallow layer without buoyancy effects, the dimensional result is

$$(C_0^2)_{\rm sw} \approx gd(1 + 12\mu\kappa/\rho_0 g d^3) \tag{6.56}$$

Thus (6.56) predicts a correction of order d^{-3}, which shows a departure from the ideal KdV evolution equation where the correction scales with d^{-1}. For a 10^{-2} cm liquid layer of silicone oil the predicted correction due to the dissipation is about 30%.

Eq. (6.53) deserves further consideration. Dissipation-modified Korteweg–de Vries equations have already been studied in the literature (see also §5.5.6). In particular, Eq. (6.53) with $\alpha_5 = 0$ has been extensively analyzed by Kawahara and collaborators [442]. On the other hand, Eq. (6.53) with $\alpha_2 = 0$ and nonzero α_5 is a modified Kuramoto–Sivashinsky equation also extensively studied. For the latter case the analysis shows that α_5 merely accelerates the cascade of expected bifurcating solutions when $\alpha_5 = 0$ [338]. Thus we expect that results obtained for Eq. (6.53) with vanishing α_5 suffice at this time to illustrate possible solutions of the equation for non vanishing α_5. When $\alpha_5 = 0$ Kawahara and collaborators [442] have shown that depending on the values of the coefficients α_i ($i = 1, ..., 4$) initial perturbations are followed by the formation of a row of soliton-like pulses, particularly in the strongly dispersive case, i.e., for relatively large values of α_2. They have also shown that the existence of a dispersive effect in the dissipation-modified KdV equation can bring about a kind of organization in the system that exhibits a turbulence-like behavior if the effect of dispersion is completely neglected (see also §5.5.7). Another conclusion is that the steady pulse solution plays an essential role in the evolution of the initial value problem under periodic boundary conditions. They have observed solutions consisting of distinct soliton-like pulses of equal amplitude when the interval of periodicity is sufficiently long in comparison with the characteristic scale determined by instability and dissipation, i.e. by the relative importance of α_3 and α_4. The interaction of two pulses in the unbounded region was also analyzed.

Kawahara further discussed in detail the effect of finite but small α_3 and α_4 [441, 442]. Because the growth term $(\alpha_3 \eta_{xx})$ is more important than the damping term $(\alpha_4 \eta_{xxxx})$ for small wave numbers in the dissipation-modified Korteweg–de Vries equation the width of a growing wave decreases and increases the relative effect of damping leading to a slowdown of the rate amplitude growth. For some fixed values of α_1 and α_4, Kawahara and Toh [442] have numerically and analytically illustrated some general features of periodic structures by means of a perturbation analysis of the cnoidal solution.

Finally, we note that with respect to the ideal KdV equation the linear terms with α_3 and α_4 provide wave velocity selection and thus uniqueness of the solitary wave solution of Eq. (6.53), as well as a wavy forerunner which is expected to permit easily formation of bound states, as shown by Nekorkin and Velarde [452], and Christov and Velarde [448].

The extension of the above given results to three dimensions was achieved by Nepomnyashchy and Velarde [446], and the more general approach to the problem, incorporating a realistic no-slip bottom and associated boundary layer, was considered by Rednikov and Velarde [449]. However, none of these asymptotic theories can account for the wave collision dynamics, anomalous (Mach–Russell) reflections, and other transient processes. Apparently, we have here a case where direct integration of the nonlinear thermo-hydrodynamic equations may be needed for a real understanding of the phenomena.

6.3   Dilational shock-like longitudinal waves

In this section, longitudinal waves resulting from a surface-tension-driven instability are studied. The physical system consists in a thin fluid layer, confined below by a rigid plane and above by a free surface with surface tension linearly depending on temperature (or concentration) and heated from above (or absorbing a surface-tension-lowering solute). In Chapter 3, dilational surface waves have been studied, though mostly in connection with other types of waves. Namely, their resonant interaction with internal waves due to buoyancy was considered in §3.4, while in §3.6, they were also shown to lead to mode-mixing and instability when interacting with capillary-gravity waves due to surface deformation. Nonlinear long-wave regimes of the latter problem were shown to be described by a dissipation-modified KdV equation in the previous section. Note also that earlier theories have shown that surfactant adsorption and accumulation at the interface may also trigger instability [235, 236, 241].

Now, for negligible buoyancy and surface deformation, it will be shown in this section that the possibility still exists to excite dilational surface waves in the set-up considered above. This even holds when dynamical effects in the gas phase are negligible (see §3.5 concerning two-layer systems), i.e. the gas phase is considered as inert (a fortiori, we will here consider a zero Biot number).

Rather, it will be shown that dilational surface waves can be excited by the sole consideration of the nonlinearity of the basic temperature (or concentration) profile. This may be created, e.g. in a transient situation (as in most experiments [44, 124, 46, 228]), or also when the thermal conductivity depends on temperature (or the diffusion coefficient depends on concentration). In the latter case, however, it should be assumed in addition that all other non-Boussinesq effects remain negligible. Interestingly, for a large Marangoni number, it is possible to achieve a rigorous asymptotic derivation of a nonlinear (non-local) evolution equation, some wave solutions of which are also presented in this section. Contrary to the derivation of the KdV-like equations of the last section, the analysis presented here does not require the long-wave assumption. Moreover, as we will see, dissipation here enters at the same order as dispersion and nonlinearity, and a numerical resolution of the non-local equation (more conveniently written in Fourier space) is necessary.

Still, a feature in common with the problem treated in §6.2 is the allowance for excitation of an extended range of length scales, accounting for the possibility of waves with distributed spatial frequencies and strongly non-harmonic shape, in contrast with weakly non-linear dissipation-dominated waves treated in Chapter 5.

6.3.1 Linear stability analysis and scalings

We consider a thin fluid layer (of depth now denoted by h) of infinite horizontal extent (or periodic), confined below by a rigid plane and open above to the ambient air. As usual, the surface tension is assumed to depend linearly on temperature. The layer is heated from above, but now both buoyancy and surface deformation are neglected.

The reference state is motionless with a quasi-stationary (nonlinear) temperature profile. The diffusive profile can indeed be regarded as quasi-stationary provided the time scale of phenomena considered is much smaller than the diffusive time scale h^2/κ, where κ is the thermal diffusivity. We will now refer exclusively to the thermal problem, keeping in mind the equivalence already exploited between thermal and concentrational problems. When such separation of time scales occurs, it is sufficient to consider the first two terms in a power series expansion of the diffusive temperature profile near the interface, i.e.

$$\vec{1}_z \cdot \vec{\nabla} T_{\text{ref}} = \beta + \gamma_{\text{d}} z , \quad \vec{1}_x \cdot \vec{\nabla} T_{\text{ref}} = 0 \tag{6.57}$$

It was seen in §3.4 and §3.6 that for dilational surface waves, the appropriate scalings of length L, temperature T, pressure p, velocity V and time t are respectively given by "thermocapillary" scales

$$[L] = h, \ [T] = \beta h, \ [p] = -\frac{d\sigma}{dT}\beta, \ [V] = \left(\frac{-\frac{d\sigma}{dT}\beta}{\rho}\right)^{1/2} , \ [t] = \left(\frac{\rho h^2}{-\frac{d\sigma}{dT}\beta}\right)^{1/2} \tag{6.58}$$

where σ is the surface tension and ρ is the fluid density. The dimensionless parameters of the problem are the Prandtl number $P = \nu/\kappa$ and Marangoni number $M = -(d\sigma/dT)\beta h^2/\mu\kappa$. Note that dissipative quantities (the thermal diffusivity κ and the kinematic viscosity ν) do not appear in scales (6.58), because as in §3.4 and §3.6, we will here consider the limit of large Marangoni number, for which the wave frequency is much larger than its damping/amplification rate. Assuming that $M \gg 1$ and $P = O(1)$, the appropriate smallness parameter of the problem is

$$\epsilon = \left(\frac{P}{M}\right)^{1/4} \ll 1 \tag{6.59}$$

The perturbations of pressure, horizontal velocity, vertical velocity and temperature are taken in the form

$$[p, u, w, T] \rightarrow \exp(\lambda t + ikx)\,[p(z), u(z), w(z), T(z)] \tag{6.60}$$

Using Eqs (6.57),(6.58) and (6.59), the linearized thermo-hydrodynamic equations reduce to

$$iku + w_z = 0 \tag{6.61}$$

$$\lambda u = -ikp + \epsilon^2(u_{zz} - k^2 u) \tag{6.62}$$

$$\lambda w = -p_z + \epsilon^2(w_{zz} - k^2 w) \tag{6.63}$$

$$\lambda T + (1 + \gamma z)\,w = \frac{\epsilon^2}{P}(T_{zz} - k^2 T) \tag{6.64}$$

where we have set $\gamma = \gamma_d h/\beta$.

The boundary conditions at the bottom ($z = -1$) are

$$u = w = T = 0 \tag{6.65}$$

while at the upper undeformable free surface ($z = 0$), we have

$$w = 0, \quad u_z + \frac{ik}{\epsilon^2}T = 0, \quad T_z = 0 \tag{6.66}$$

The second relationship is the Marangoni boundary condition (balance between Marangoni and viscous stresses at the free surface), while the third one expresses that the heat flux is fixed at the upper surface.

Leading-order dispersion relation

In this section, Eqs (6.61–6.66) are solved using an asymptotic expansion in powers of the ordering parameter ϵ. As also discussed in §3.4 and §3.6, it is necessary to divide the layer into three different regions: the upper boundary layer where the variables will be over-bared, the main bulk, and the lower boundary layer where the variables will be tilded. The solution is represented as

$$f = \epsilon f_1 + \epsilon^2 f_2 + o(\epsilon^2), \quad \tilde{f} = \epsilon \tilde{f}_1 + \epsilon^2 \tilde{f}_2 + o(\epsilon^2) \tag{6.67}$$

where f and $\tilde{f}$ stand for all the variables in the main bulk and the lower boundary layer, respectively. In the upper boundary layer, owing to the large Marangoni stress, we can start the expansion one order higher for the horizontal velocity field u. Thus the expansion in the upper boundary layer is given by:

$$\bar{u} = \bar{u}_0 + \epsilon \bar{u}_1 + \epsilon^2 \bar{u}_2 + o(\epsilon^2), \quad \bar{f} = \epsilon \bar{f}_1 + \epsilon^2 \bar{f}_2 + o(\epsilon^2) \tag{6.68}$$

where $\bar{f}$ stands for temperature, pressure, or vertical velocity.

At the leading order ϵ in the main bulk Eqs (6.61–6.64) become

$$iku_1 + w_{1z} = 0, \ \lambda u_1 = -ikp_1, \ \lambda w_1 = -p_{1z}$$
$$\lambda T_1 + (1 + \gamma z)w_1 = 0 \tag{6.69}$$

whose solution (potential flow) is

$$p_1 = C_1 \cosh(kz) + C_2 \sinh(kz)$$
$$u_1 = -\frac{ik}{\lambda}[C_1 \cosh(kz) + C_2 \sinh(kz)]$$
$$w_1 = -\frac{k}{\lambda}[C_1 \sinh(kz) + C_2 \cosh(kz)] \tag{6.70}$$
$$T_1 = \frac{(1+\gamma z)k}{\lambda^2}[C_1 \sinh(kz) + C_2 \cosh(kz)]$$

At the leading order in the lower boundary layer after making the change of variable $y = (z+1)/\epsilon$, Eqs. (6.61–6.64) become

$$\tilde{w}_{1y} = 0, \ \ \lambda \tilde{u}_1 = -ik\tilde{p}_1 + \tilde{u}_{1yy}, \ \ \tilde{p}_{1y} = 0$$
$$\lambda \tilde{T}_1 + (1 - \gamma)\tilde{w}_1 = P^{-1}\tilde{T}_{1yy} \tag{6.71}$$

After imposing boundary conditions (6.65) and the matching conditions with the bulk solution (6.70), i.e. $\tilde{f}_1(y \to +\infty) = f_1(z \to 0)$ at this order, we get

$$\tilde{p}_1 = C_1 \cosh(k) - C_2 \sinh(k)$$
$$\tilde{u}_1 = -\frac{ik}{\lambda}[C_1 \cosh(k) - C_2 \sinh(k)]\left[1 - \exp(-\sqrt{\lambda}y)\right]$$
$$\tilde{w}_1 = 0 \tag{6.72}$$
$$\tilde{T}_1 = 0$$
$$C_2 \cosh(k) - C_1 \sinh(k) = 0$$

where $\sqrt{\lambda}$ is taken as the square root with *positive* real part.

At the leading order in the upper boundary layer after making the change of variable $y = z/\epsilon$, Eqs (6.61–6.64) become

$$ik\bar{u}_0 + \bar{w}_{1y} = 0, \ \ \lambda \bar{u}_0 = \bar{u}_{0yy}, \ \ \bar{p}_{1y} = 0$$
$$\lambda \bar{T}_1 + \bar{w}_1 = P^{-1}\bar{T}_{1yy} \tag{6.73}$$

Solving these equations with boundary conditions (6.66), except for the Marangoni condition, and matching with the main bulk (6.70), the solution of Eqs (6.73) is

$$
\begin{aligned}
\bar{p}_1 &= C_1 \\
\bar{u}_0 &= C \exp\left(\sqrt{\lambda}\, y\right) \\
\bar{w}_1 &= \frac{ikC}{\sqrt{\lambda}} \left[1 - \exp\left(\sqrt{\lambda}\, y\right)\right] \\
\bar{T}_1 &= -\frac{ikC}{\lambda\sqrt{\lambda}} \left[1 + \frac{P}{1-P} \exp\left(\sqrt{\lambda}\, y\right) - \frac{P^{1/2}}{1-P} \exp\left(\sqrt{\lambda}P^{1/2}\, y\right)\right] \\
C_2 &= -i\,C\sqrt{\lambda}, \quad C_1 = -i \coth k\, C\sqrt{\lambda}
\end{aligned}
\tag{6.74}
$$

Imposing the Marangoni boundary condition at $y = 0$, i.e.

$$
\bar{u}_{0y} + ik\,\bar{T}_1 = 0 \tag{6.75}
$$

we get the leading-order dispersion relation

$$
\lambda_0^2 = \frac{-k^2}{1 + P^{1/2}} \tag{6.76}
$$

We observe that λ_0 is purely imaginary, which means that at the leading-order waves are neither damped nor amplified. Actually, such effects will enter the analysis at the next order in ϵ.

Damping/amplification rate

In this section the next order of approximation of the solution in the upper boundary layer is calculated. As we are interested in the first correction to the dispersion relation (6.76), it is not necessary to solve the equations (6.61–6.64) in the bulk at this order. Actually, this is also why it suffices to consider two terms in the development of the basic temperature gradient near the interface, as done for Eq. (6.57). Equations (6.61–6.64) for the upper boundary layer become

$$
ik\bar{u}_1 + \bar{w}_{2y} = 0\,, \quad \lambda\bar{u}_1 = -ik\bar{p}_1 + \bar{u}_{1yy}\,, \quad \lambda\bar{w}_1 = -\bar{p}_{2y} + \bar{w}_{1yy},
$$
$$
\lambda\bar{T}_2 + \gamma y\,\bar{w}_1 + \bar{w}_2 = P^{-1}\bar{T}_{2yy} \tag{6.77}
$$

The solution of Eqs (6.77) is, excluding the solution of the corresponding homogeneous problem[1]

$$
\begin{aligned}
\bar{p}_2 &= -ik\,C\sqrt{\lambda} \\
\bar{u}_1 &= -\frac{k \coth k\, C}{\sqrt{\lambda}} \\
\bar{w}_2 &= \frac{ik^2 \coth k\, C}{\sqrt{\lambda}}\, y \\
\bar{T}_2 &= \frac{ik^2 \coth k\, C}{\lambda\sqrt{\lambda}} \left[\frac{1}{P^{1/2}\sqrt{\lambda}} \exp\left(\sqrt{\lambda}P^{1/2}\, y\right) - y\right] + \\
&\quad \frac{ik\gamma C}{\lambda} \left\{ \frac{1-3P}{\lambda P^{1/2}(1-P)^2} \exp\left(\sqrt{\lambda}P^{1/2}\, y\right) + \right. \\
&\quad \left. \left[\frac{2P}{\lambda(1-P)^2} - \frac{P\,y}{(1-P)\sqrt{\lambda}}\right] \exp\left(\sqrt{\lambda}\, y\right) - \frac{y}{\sqrt{\lambda}} \right\}
\end{aligned}
\tag{6.78}
$$

---

[1] In fact, the homogeneous problem corresponding to Eqs (6.77) is given by Eqs (6.73). Thus, including its solution here would merely amount to redefine the amplitude of the leading-order solution (or to repeat the same calculations at each order), which of course does not modify the dispersion relation.

Applying the Marangoni boundary condition at $y = 0$, i.e.

$$\bar{u}_{0y} + \epsilon \bar{u}_{1y} + ik\,(\bar{T}_1 + \epsilon \bar{T}_2) = 0, \tag{6.79}$$

we obtain the dispersion relation

$$\lambda^2 + \frac{k^2}{1 + P^{1/2}} = \frac{\epsilon k^2}{\sqrt{\lambda} P^{1/2}} \left\{ k \coth k + \frac{\gamma(1 + 2P^{1/2})}{(1 + P^{1/2})^2} \right\} \tag{6.80}$$

which includes the first two orders of approximation

$$\lambda = \lambda_0 + \epsilon \lambda_1 \tag{6.81}$$

After some manipulations, one gets

$$\lambda_1 = -\frac{(1 \pm i)(1 + P^{1/2})^{3/4} |k|^{1/2}}{2\sqrt{2} P^{1/2}} \left\{ k \coth k + \frac{\gamma(1 + 2P^{1/2})}{(1 + P^{1/2})^2} \right\} \tag{6.82}$$

where the plus sign is for the left propagating wave. Contrary to the earlier leading order result, the real part of this expression generally does not vanish. Indeed, we have

$$\operatorname{Re}\lambda_1 = -\frac{(1 + P^{1/2})^{3/4} |k|^{1/2}}{2\sqrt{2} P^{1/2}} \left\{ k \coth k + \frac{\gamma(1 + 2P^{1/2})}{(1 + P^{1/2})^2} \right\} \tag{6.83}$$

The long-wavelength limit of this expression is

$$\lim_{k \to 0} \operatorname{Re}\lambda_1 = -\frac{(1 + P^{1/2})^{3/4}}{\sqrt{2} P^{1/2}} \left\{ 1 + \frac{\gamma(1 + 2P^{1/2})}{(1 + P^{1/2})^2} \right\} |k|^{1/2} \tag{6.84}$$

while in the short-wavelength limit, we always have a negative value

$$\lim_{k \to \infty} \operatorname{Re}\lambda_1 = -\frac{(1 + P^{1/2})^{3/4}}{2\sqrt{2} P^{1/2}} |k|^{3/2} \tag{6.85}$$

Hence, we may get a band of excited waves if Eq. (6.84) is positive, i.e. when

$$\gamma < \gamma_c = -\frac{(1 + P^{1/2})^2}{1 + 2P^{1/2}} \tag{6.86}$$

The above results also confirm that oscillatory instability is not possible, in our drastically simplified problem, when the basic temperature profile is linear, i.e. when $\gamma = 0$. Actually, this case corresponds to the Pearson's problem [101, 209] studied in §3.3.2, where we know that only monotonic instability can occur.

The real part of λ_1 versus k (wave number) for different values of γ is shown in Fig. 6.8, where one should remember that γ is proportional to the "concavity" of the temperature profile.

Let us further analyze the significant time scales of the problem, before embarking on a nonlinear description of the wave shape and evolution, applying when the condition (6.86) is fulfilled.

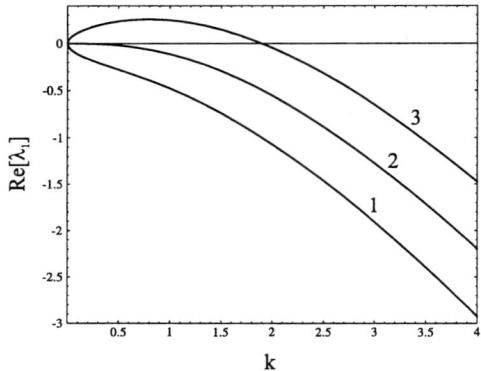

Figure 6.8: Leading-order growth rate of dilational, longitudinal waves. All the curves are drawn for $P = 6$, and correspond to $\gamma = 0$ (curve 1), $\gamma = \gamma_c = -2.017$ (curve 2) and $\gamma = -4$ (curve 3).

6.3.2 Separation of time scales

The analysis of the last section indicates that eventually, not just two, but rather three different time scales have to be taken into account. The longest one is the diffusion time scale over the depth of the layer. For $P = O(1)$, we may alternatively write it as h^2/κ or $h^2/\nu = \epsilon^{-2}[t]$, where $[t]$ has been defined in Eq. (6.58) and ϵ is given by Eq. (6.59). In contrast, the shortest time scale is given by $[t]$ itself, and is linked to the frequency of the waves, as indicated by the dimensionless result (6.76). Now, from §6.3.1, amplification and damping of the waves is seen to occur on an *intermediate* time scale $\epsilon^{-1}[t]$ [see Eqs (6.81) and (6.83)].

Hence, the ordering parameter $\epsilon \ll 1$ allows a clear (asymptotic) separation of time scales and associated physical effects. On a short time scale $[t]$, the diffusive profile has not significantly penetrated through the layer (only up to a depth $\sim \epsilon h$), and dilational waves propagate without significant dissipation or dispersion effect [indeed, Eq. (6.76) shows that the phase velocity $V = \omega_0/k = (1 + P^{1/2})^{-1/2}$ is independent of the wavenumber]. In the following, we shall concentrate on the evolution of a wave-packet on the intermediate time scale $\epsilon^{-1}[t]$, which should be influenced not only by dissipation and dispersion, but also by nonlinearity.

6.3.3 Nonlinear wave-packets

In the present section, the nonlinearities of the governing equations (Navier–Stokes and energy equations) are taken into account. At the leading order (6.76), the longitudinal waves are dispersion-less and dissipation-less, i.e. they behave as sound waves in a compressible gas. The next order correction (6.82) yields both dispersion and damping/amplification, which should thus be balanced by the effect of nonlinearity. This balance leads to the appropriate scaling of the wave amplitudes, and to the corresponding evolution equation.

As before, we separate the physical domain into three parts: the upper boundary layer where the variables are over-bared; the main bulk (perfect fluid at leading order); and the lower boundary layer where the variables are tilded. Assuming that the smallness parameter is still defined by Eq. (6.59), the solution in the upper boundary layer is sought as

$$\bar{u} = \epsilon \bar{u}_0 + \epsilon^2 \bar{u}_1 + o(\epsilon^2), \quad \bar{f} = \epsilon^2 \bar{f}_1 + \epsilon^3 \bar{f}_2 + o(\epsilon^3) \tag{6.87}$$

where f stands for temperature, pressure or vertical velocity. In the bulk and at the lower boundary layer, we have

$$f = \epsilon^2 f_1 + \epsilon^3 f_2 + o(\epsilon^3), \quad \tilde{f} = \epsilon^2 \tilde{f}_1 + \epsilon^3 \tilde{f}_2 + o(\epsilon^3) \tag{6.88}$$

where f can be any of the field variables. From the leading-order dispersion relation (6.76), we get the phase velocity

$$V = \left(1 + P^{1/2}\right)^{-1/2} \tag{6.89}$$

Let us consider a wave packet in the reference frame moving with the velocity (6.89) and introduce a slow time variable to adequately describe the evolution of the wave profile. Accordingly, the new independent variables for space and time (considering a wave traveling in the positive x direction) are

$$\xi = x - Vt, \quad \tau = \epsilon t \tag{6.90}$$

The thermo-hydrodynamic equations then read

$$u_\xi + w_z = 0 \tag{6.91}$$

$$\epsilon u_\tau - V u_\xi + u u_\xi + w u_z = -p_\xi + \epsilon^2 (u_{\xi\xi} + u_{zz}) \tag{6.92}$$

$$\epsilon w_\tau - V w_\xi + u w_\xi + w w_z = -p_z + \epsilon^2 (w_{\xi\xi} + w_{zz}) \tag{6.93}$$

$$\epsilon T_\tau - V T_\xi + u T_\xi + w T_z + (1 + \gamma z)\, w = \frac{\epsilon^2}{P}(T_{\xi\xi} + T_{zz}) \tag{6.94}$$

Note that the boundary conditions are the same as for the linear problem, i.e. Eqs (6.65) and (6.66).

Going to Fourier space, and hence limiting consideration to spatially periodic solutions, greatly helps solving Eqs (6.91–6.94). Moreover, periodic solutions are always of fundamental value by themselves, and are deemed to correctly represent the wave behavior far away from lateral boundaries. In addition, one of the standard geometrical configurations used in experiments with waves is a circular annular channel [124, 125], as also seen in Fig. 1.11 of Chapter 1. This naturally imposes spatial periodicity.

Let L be the (dimensionless) spatial period, while $k_0 = 2\pi/L$ is the fundamental wavenumber. Then the wavenumbers allowed in the system are $k = k_0 n$, n being an integer. All functions of ξ are represented in the form of Fourier series. For example for the velocity in the upper boundary layer, we can write

$$\bar{u}_0(\xi, z, \tau) = \sum_{k=-\infty}^{\infty} \bar{u}_{0k}(z)\, e^{ik\xi} \tag{6.95}$$

with

$$\bar{u}_{0k} = \frac{1}{L} \int_{-\frac{L}{2}}^{\frac{L}{2}} \bar{u}_0 \, e^{-ik\xi} \, d\xi \tag{6.96}$$

At the leading order in ϵ for the upper boundary layer we have

$$
\begin{aligned}
\bar{u}_{0k} &= C_k \exp\left(\sqrt{-iVk}\,y\right) \\
\bar{w}_{1k} &= \frac{ikC_k}{\sqrt{-iVk}} \left[1 - \exp\left(\sqrt{-iVk}\,y\right)\right] \\
\bar{T}_{1k} &= \frac{C_k}{V\sqrt{-iVk}} \left[1 + \frac{P}{1-P}\exp\left(\sqrt{-iVk}\,y\right) - \frac{P^{1/2}}{1-P}\exp\left(\sqrt{-iVk}P^{1/2}\,y\right)\right] \\
\bar{p}_{1k} &= -iC_k \coth k \sqrt{-iVk}
\end{aligned}
\tag{6.97}
$$

which is identical to the leading-order solution (6.74). The same is also true for the solution in the main bulk and in the lower boundary layer. The evolution equations for the quantities C_k will appear from the tangential stress condition at the interface ($y = 0$). At the leading order, this condition is satisfied identically, confirming that the phase velocity V is indeed defined by (6.39). It is at the next order that a meaningful equation for the coefficients C_k will emerge. Let us now outline the main steps of its derivation.

The tangential stress (Marangoni) condition (at $y = 0$) becomes

$$\frac{\partial \bar{u}_{1k}}{\partial y} + ik\,\bar{T}_{2k} = 0 \tag{6.98}$$

Thus, we need to evaluate $\bar{u}_{1k}$, $\bar{w}_{2k}$ and $\bar{T}_{2k}$. The equation for $\bar{u}_{1k}$ can be written as

$$\bar{u}_{1kyy} + iVk\bar{u}_{1k} = ik\,\bar{p}_{1k} + \partial_\tau \bar{u}_{0k} + \frac{ik}{2} \sum_{k'=-\infty}^{\infty} \bar{u}_{0k-k'}\,\bar{u}_{0k'} + \sum_{k'=-\infty}^{\infty} \bar{w}_{1k-k'}\,\bar{u}_{0k'y} \tag{6.99}$$

Here we have used a discrete version of the convolution integral. The homogeneous solution is the same as for the solution at leading order and we just calculate the particular solution corresponding to the four terms in the right-hand side (r.h.s.) separately. After the calculation of $\bar{u}_{1k}$ is completed, the next step consists in integrating the continuity equation in the upper layer, imposing that the vertical velocity is zero at the interface. We then get

$$\bar{w}_{2k} = -ik \int_{\zeta=0}^{\zeta=y} \bar{u}_{1k} \, d\zeta \tag{6.100}$$

where ζ is a dummy integration variable. Again we proceed to the integration of $\bar{u}_{1k}$ term by term. The last step in the calculation of the second order solution consists in integrating the equation for $\bar{T}_{2k}$

$$
\begin{aligned}
\bar{T}_{2kyy} + iVkP\bar{T}_{2k} = {}&\gamma P\,y\,\bar{w}_{1k} + P\,\bar{w}_{2k} + P\partial_\tau \bar{T}_{1k} + \\
&P\sum_{k'=-\infty}^{\infty} \bar{u}_{0k-k'}\,ik'\,\bar{T}_{1k'} + P\sum_{k'=-\infty}^{\infty} \bar{w}_{1k-k'}\,\bar{T}_{1k'y}
\end{aligned}
\tag{6.101}
$$

In Eq. (6.101) all terms in the r.h.s. have already been calculated. Therefore its solution can be found. Substituting all these terms into (6.98), we get an equation for each of the Fourier amplitudes, which can be written as

$$
\frac{\partial C_n}{\partial \tau} + a_n C_n + \frac{|k_0|}{2} i n \sqrt{-in} \sum_{n'=-\infty}^{+\infty} B(n-n',n') C_{n-n'} C_{n'} = 0 \tag{6.102}
$$

where the integer n subscripts are now used instead of $k = k_0 n$. The expressions for the coefficients a_n and $B(n - n', n')$ can be obtained analytically, though the latter turns out to be quite bulky [229]. In what follows, we restrict consideration to the case of large Prandtl numbers. This is a reasonable assumption for many liquids, not to mention the mass diffusion analog of the problem where P is replaced by the Schmidt number which is always about two orders of magnitude larger. Assuming $P \gg 1$ the coefficients become, after some rescaling,

$$
a_n = \left[\gamma_0 + \frac{nk_0 \coth(nk_0)}{2} \right] |k_0|^{1/2} \sqrt{-in} \tag{6.103}
$$

$$
B(n-n',n') = -\frac{1}{4} \left[\frac{1}{\sqrt{-in'}} + \frac{1}{\sqrt{-i(n-n')}} \right] + \frac{1}{2} \frac{1}{\sqrt{-in'}+\sqrt{-in}} +
$$
$$
\frac{1}{2} \frac{1}{\sqrt{-i(n-n')}+\sqrt{-in}} + \frac{\sqrt{-in}}{4\sqrt{-i(n-n')}\sqrt{-in'}} + \frac{1}{2\sqrt{-in}} \tag{6.104}
$$

Numerical integration of the evolution equation

In this section, some numerical simulations of the equations (6.102) are presented, considering a particular value of $\gamma_0 = -5$. The limiting (cut-off) wavenumber k_1 between the unstable and the stable modes is then $k_1 \simeq 10$. Consider first a fundamental wavenumber equal to $k_0 = 0.5$. This means that 19 modes are amplified. However, in order to get a numerically converged solution, a large number of modes $k_n = nk_0$, $n = 1, 2, ..., N$ must be taken into account, most of them being linearly damped. The integration of (6.102) is performed by using a standard 4th-order Runge–Kutta routine for a system of coupled ordinary differential equations (ODE). The time step is controlled such that a relative accuracy of 10^{-3} is ensured for each mode, and initial conditions are taken as a random "white-noise" of small amplitude.

A reconstruction of the horizontal velocity profile (6.95) in a space–time diagram is shown in Fig. 6.9, for which the evolution of 400 modes has been considered. In this graph, the time runs upwards and the spatial coordinate is horizontal.

From Fig. 6.9, we observe that as time proceeds, a single traveling wave is left, which reaches a stationary shape represented in Fig. 6.11. The final distribution of amplitudes is shown in Fig. 6.10, together with the corresponding spectra drawn for the waves with $k_0 = 1$ and $k_0 = 1.5$.

Note that the first unstable modes gather a large part of the "energy" of the system (here defined as $|C_k|^2$), but it is distributed continuously between all the modes. Note also that some (numerical) problems exist with wave modes at the large-wavenumber end of the ($N = 400$) spectrum considered, and more dramatically as we lower the value of k_0.

A snapshot of the space–time plot at the end of the simulation with $k_0 = 0.5$ is shown in Fig. 6.11, yielding the final profile of the nonlinear dilational wave. It has been checked [229]

Figure 6.9: Space–time plot of the surface velocity profile. Time is running upwards, and the spatial coordinate is represented horizontally. Small perturbations grow and form several waves which merge as time proceeds, leading to a single wave within the periodic domain (fundamental mode).

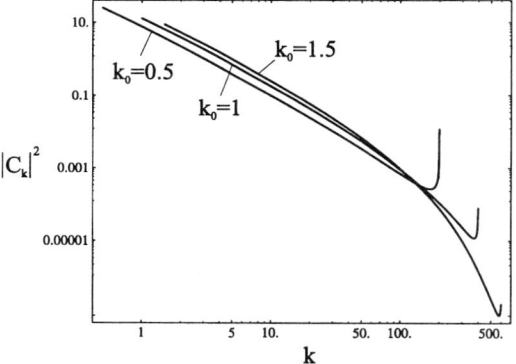

Figure 6.10: Distribution of the squared amplitudes versus wavenumber at steady state. Each energy spectrum corresponds to the final state of a simulation run with the corresponding value of k_0, maintaining the total number of modes $N = 400$ constant.

that an increase of N at given k_0 does not change the surface velocity profile significantly, while it strongly reduces the large-wavenumber inaccuracies in Fig. 6.10. It can be seen that the resulting wave is strongly nonlinear, very much appearing as a shock-like pulse. In fact when running simulations at smaller and smaller k_0 (and increasing N such as to reach convergence), the wave peak becomes more and more pronounced, possibly tending to some self-similar shape for $k_0 \to 0$ (while the amplitude diverges).

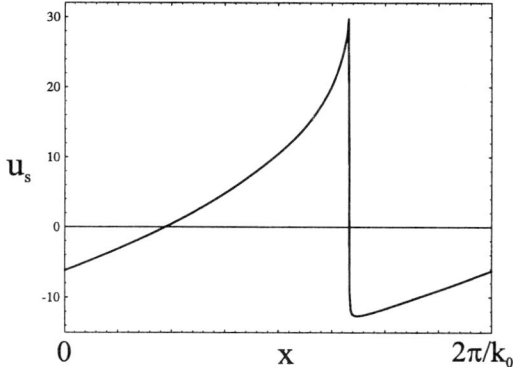

Figure 6.11: Surface velocity profile at the steady state for the simulation of Fig. 6.9.

The wavelength selection process described above has been cross-checked by three different methods. First, the time step has been decreased (increasing the required relative accuracy). However no appreciable variation in the results has been detected. Increasing the number of modes results in diminishing the numerical problem of the large-wavenumber modes. These modes will always be slightly out of the line, but when the number of modes is increased their relative weight becomes smaller and smaller, thus reducing the error when reconstructing the velocity profile.

A second line of investigation consisted in increasing the number of unstable modes, i.e. using a smaller k_0. As already said, this also requires to increase the number of stable modes (up to convergence). Note that good results are obtained when less than 5% of the modes are unstable. An unexpected result is that even when taking k_0 smaller and smaller the fundamental mode $k = k_0$ always remains predominant after a sufficient time. In other words, the wavenumber of the resulting nonlinear wave is always equal to k_0. Therefore, there is no intrinsic wavenumber selection in the system considered here. The wave always tends to occupy the whole externally imposed spatial period with a single peak, but at the same time does not get localized as a solitary wave.

A third check of the time evolution and stability of the constant-shape solutions of Eq. (6.102), motivated by the previous one, consisted in starting the simulation with an already formed wave (rather than initial random noise). The spatial period L was then doubled (or tripled) so that two (or three) periods of the initial wave fit into it. As the computation is run, it is found that the wave becomes unstable, even though it is a possible solution. The solution again ends up at a wave with the maximal possible spatial period, after some recombination process of wave crests as those observed in Fig. 6.9. Finally, this has also been confirmed by performing a numerical Floquet stability analysis of the final nonlinear wave. Nevertheless, the absence of an intrinsic wavenumber selection is probably a consequence of the oversimplified physico-chemical model used for the liquid layer rather than a genuine property of the longitudinal wave.

We cannot leave this chapter, however, without emphasizing how much theoreticians still have to accomplish to account for the wealth of results provided by the experimentalists [38,

39 44, 46, 47, 123, 122, 124, 125, 228]. However, it must be said that, unfortunately, the experiments so far reported provide mostly qualitative data, and very few quantitative findings due to the unsteady character of mass-transfer experiments. The need also exists for further and more reliable heat-transfer experiments.

7   Codimension-two bifurcations

While Chapters 4 and 5 were mostly concerned with bifurcations of a single mode, we focus in the present chapter on cases where two modes of different nature bifurcate quasi-simultaneously[1]. In addition to the usual control parameter (e.g. the Rayleigh or Marangoni number), an additional parameter will be needed here, in order to arrange for one of the two modes to bifurcate before the other, or vice versa. In such a case, we will speak of a codimension-2 bifurcation, i.e. two parameters are necessary to *unfold* the qualitatively different situations in the vicinity of the bifurcation point. Though this definition is rather intuitive, the reader is referred to classical textbooks on dynamical systems theory (e.g. [335]) for rigorous mathematical definitions of the codimension.

A classical example of codimension-2 point (CTP) in Bénard convection is encountered for binary mixtures, as studied in §3.7 and §5.3. In §5.4, the two-layer Bénard problem was also shown to exhibit such a degenerate situation under some conditions. In both cases, monotonic and oscillatory modes with almost identical wavenumbers interact in the vicinity of the CTP, which may result in complex dynamical regimes and transitions, as studied in §5.4. Assuming that the real system is correctly approximated by fixing the wavenumbers of both modes at a common value k_0 close to the actual critical wavenumber k_c, we have in fact a 1:1 resonance problem, for which the Takens–Bogdanov normal form applies, as partly discussed in §5.4.6. A more complete discussion of this CTP has been provided by Knobloch [163], Dangelmayr and Knobloch [164], and has been recently extended to perturbations on the hexagonal lattice by Renardy [470], and Renardy et al. [471]. All these works provide evidence that the dynamics in the vicinity of such a CTP may be quite complex.

We shall not study this 1:1 resonance problem further here, but rather focus on situations where the two nearly-critical modes have different wavenumbers. This includes bifurcations of monotonic modes with different spatial structure in confined geometries, a special case of which being the interaction of two monotonic modes with wavenumbers in the ratio 2:1. In §7.1, this will be studied for Bénard convection in circular cylinders, and the interacting modes will be characterized by different azimuthal wavenumbers $m = 1$ and $m = 2$. This problem is formally equivalent to that treated earlier by Proctor and Jones [417], apart from the fact that we will here also include an axisymmetric mode $m = 0$ in the dynamics, in view of both linear stability theory and experimental results of Johnson and Narayanan [43].

In §7.2, the resonance between waves and patterns will be studied, both for situations where their wavenumbers are not commensurate, and when they are locked in the ratio 2:1.

---

[1] Even though Chapters 4 and 5 dealt with interactions of several modes (e.g. plane waves with different orientation in the horizontal plane), consideration was restricted to cases where these modes were symmetry-related, hence bifurcated *simultaneously*. This is the actual difference with problems treated in this Chapter.

As strong resonances are also expected in the latter case, a much richer dynamics is predicted. Interestingly, it turns out that quadratic amplitude equations may be sufficient (under some conditions) to describe the dynamical regimes, including transitions to chaos.

Thus, our description of codimension-2 points is far from being exhaustive. In particular, as the normal forms applying in the vicinity of CTPs usually contain a number of non-rescalable coefficients, the complete analysis of possible regimes and transitions is typically quite complex (see e.g. [335, 164, 417]). Therefore, we have chosen to describe here some results illustrating this complexity, starting in each case from a particular physical set-up. Although the generic properties of the amplitude equations are emphasized, their coefficients are calculated from the basic governing equations and realistic boundary conditions, using the perturbative and projective methods described in the previous chapters.

7.1   Spatially resonant patterns in confined geometries

As described in Chapter 4, monotonic instabilities in containers of large horizontal extension most often lead to stationary regimes, such as hexagonal, square or roll-like convection cells. Even the competition between these patterns was experimentally observed to lead to only weakly time-dependent regimes, most often mediated by defects [28, 114, 118, 29, 34, 31].

However, when the liquid layer is more strongly confined (i.e. the horizontal extent is made comparable to the depth), convective patterns are not only strongly influenced by lateral walls [42, 152], but fascinating non-steady behaviors have also been reported. For instance, Ondarçuhu et al. [155] have observed three kinds of oscillating patterns in a square container with aspect ratio (width over height) 4.6. These dynamic behaviors are related by the authors to the presence of a Takens–Bogdanov bifurcation in the system.

Complex dynamics is also possible near instability threshold, when the aspect ratio(s) is (are) close to particular values for which two convective modes are equally susceptible to bifurcate (see §3.3.4). Recently, in a circular container with aspect ratio (radius over height) equal to 2.5, a dynamic mode switching between two two-cell patterns related by a $\pi/2$ rotation about the vertical symmetry axis has been observed by Johnson and Narayanan [43]. In their experiment, the fluid was lying below an air layer and its depth was of the order of millimeters, so that both buoyancy and thermocapillarity were active in the instability mechanism.

In the present section, we present a theoretical analysis of this mode switching. More precisely, we show that the dynamics of the system can be described by amplitude equations forming a low-dimensional system of ODEs whose bifurcations give rise to the same dynamical behavior as that seen in experiments [43]. More details about the analysis presented here can be found in the original paper of Dauby et al. [418].

For more complete theoretical studies of thermoconvective instabilities in confined domains, the reader is also invited to consult some earlier papers. The different works can be classified into two broad categories, depending on the type of boundary conditions used for the velocity along the sidewalls (see also §3.3.4 for linear stability analyzes). In [149, 150, 151, 468], the sidewalls of the containers are assumed to be "slippery" (for rectangular containers) or "vorticity-free" (for circular containers), while more realistic no-slip walls are considered for instance in [152, 154, 260] (and references therein).

The model of slippery/vorticity-free walls is quite convenient since it enables a complete separation of the vertical and horizontal space coordinates. Using the latter assumption, Echebarría, Krmpotić and Pérez-García [468] have described dynamic solutions in cylindrical containers. Their analysis has shown the possibility of non-stationary behaviors (rotating waves, modulated rotating waves and heteroclinic orbits) resulting from a 1:2 spatial resonance of two different convective modes. These modes differ by their azimuthal wavenumber ($m = 1$ and $m = 2$, respectively), though the analysis applies as well to two-dimensional patterns with wavenumbers k_1 and $k_2 = 2k_1$. This case was treated earlier by Proctor and Jones [416, 417], who also achieved a detailed analysis of the corresponding normal form.

Such analysis will now be extended to realistic no-slip circular containers, including the dynamics of the $m = 0$ axisymmetric mode, and comparing the theoretical predictions with available experiments.

7.1.1 Problem statement and linear stability results

We consider the basic system of equations (4.152–4.154), i.e.

$$\vec{\nabla}.\vec{V} = 0 \tag{7.1}$$

$$\Delta\vec{V} - \vec{\nabla}p + Ra\vec{1}_z T = Pr^{-1}\left[\frac{\partial\vec{V}}{\partial t} + (\vec{V}.\vec{\nabla})\vec{V}\right] \tag{7.2}$$

$$\Delta T + W = \frac{\partial T}{\partial t} + (\vec{V}.\vec{\nabla})T \tag{7.3}$$

where $\vec{V}, T, p$ are respectively the velocity, temperature and pressure perturbations, $Pr = \nu/\kappa$ is the Prandtl number, and Ra is the Rayleigh number. The components of $\vec{V}$ will be denoted by u, v and w in the radial, azimuthal and vertical directions, respectively.

The boundary conditions at the bottom of the box ($z = 0$) express that the wall is rigid and perfectly heat conducting. The upper free surface ($z = 1$) is assumed to remain undeformed while the surface tension is supposed to be a linear function of temperature. The heat transfer is modeled using a Biot condition, referring the reader to §3.2.3 and §3.6.1 for discussions about the validity of these assumptions. Hence, we have

$$u = T = 0 \text{ at } z = 0 \tag{7.4}$$

$$w = DT + BiT = 0 \text{ at } z = 1 \tag{7.5}$$

$$Du + Ma\frac{\partial T}{\partial r} = Dv + Ma\,r^{-1}\frac{\partial T}{\partial\phi} = 0 \text{ at } z = 1 \tag{7.6}$$

where r and ϕ are the horizontal polar coordinates, and Ma is the Marangoni number. Finally, the lateral sidewalls are rigid and insulating, i.e.

$$u = \frac{\partial T}{\partial r} = 0 \text{ at } r = a \tag{7.7}$$

where a is the aspect ratio of the container.

A general linear stability analysis for this problem was carried out in [154]. Only the results needed later are given in Fig. 7.1, where the critical non-dimensional temperature difference λ_c, defined as

$$\lambda_c = \frac{Ra_c}{669} + \frac{Ma_c}{79.6} \tag{7.8}$$

is plotted as a function of the aspect ratio a, for a in the neighborhood of 2.5 (Ra_c and Ma_c are the critical Rayleigh and Marangoni numbers).

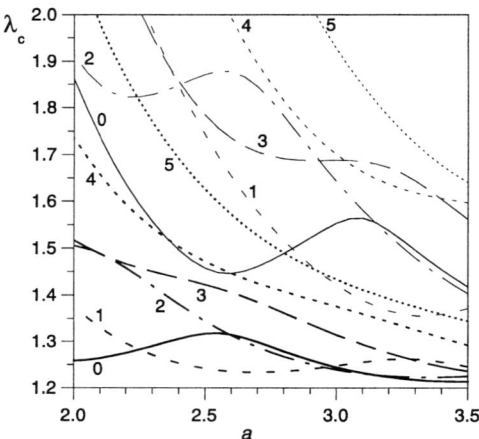

Figure 7.1: Critical non-dimensional temperature difference versus aspect ratio. The first and second eigenvalues are represented with thick and thin lines. The value of the azimuthal wavenumber m is indicated on each curve.

In the calculations leading to these curves, as well as in the other results presented in this section, the thickness of the liquid layer is taken as 5 mm (a is varied by changing the radius of the container). The viscosity of the fluid is 1 Stokes and the Prandtl number is equal to 1000. The air layer above the liquid has a depth equal to half the depth of the liquid layer so that the air layer can be considered as purely conductive [154] (see also §3.5.2). With conductivities for the air and for the liquid being respectively 0.026 and 0.16 $W\,m^{-1}\,K^{-1}$, an equivalent Biot number can be calculated as 0.43. The negative of the derivative of surface tension with respect to temperature is equal to $5 \times 10^{-5}\ N\,m^{-1}\,K^{-1}$ and the density of the liquid is 968 $kg\,m^{-3}$. The coefficient of thermal expansion is fixed to $9.6 \times 10^{-4}\ K^{-1}$. The above values for the different quantities are chosen in order to correspond to the experiments reported in [43], except that the thickness of the air layer is smaller in order to ensure that the upper gas remains mechanically passive [132].

The different curves in Fig. 7.1 correspond to different values of the azimuthal wave number m. For each m, thick and thin lines are used to represent the threshold of, respectively, the first and the second eigenmode. In the neighborhood of $a = 2.5$, the convective threshold is characterized by an $m = 1$ pattern. For $a = 2.5$, λ_c is equal to 1.24 while the critical

Rayleigh and Marangoni numbers are given by 292 and 64.1 respectively. A codimension-2 point occurs at $a = 2.961$. For this aspect ratio, the $m = 1$ and $m = 2$ critical curves intersect and the two modes are simultaneously critical. Note that the $m = 0$ mode is destabilized quite close to the instability threshold of the $m = 1$ and $m = 2$ patterns. It is important to note that these results are quite different from those obtained under the assumption of slippery lateral walls (see for instance Fig. 1 in [468]). In particular, the $m = 0$ curve for slippery walls is quite far away from the intersection of the other two modes. Moreover, the aspect ratio giving rise to the codimension-2 point is much smaller ($a \simeq 1.15$, see also Fig. 3.15) than the value found here.

7.1.2 Nonlinear regimes for $a = 2.5$

The method used for calculating amplitude equations and their coefficients consists in reducing the dynamics of the system to the dynamics of the most unstable modes. The procedure is very similar to that presented by Dauby and Lebon [152] for the study of rectangular no-slip containers and will only be summarized here (more details can be found in §4.2.3). The solution of the nonlinear PDE system is first expanded in a series of the eigenmodes of the linearized problem, using the linear growth rate as the eigenvalue. In the weakly nonlinear regime, the dynamics of the system is dominated by the dynamics of the most unstable modes of convection, i.e. those with the largest growth rates. A careful examination of the numerical data corresponding to Fig. 7.1 shows that for $a = 2.5$, the first $m = 0$, $m = 2$ and $m = 3$ modes (thick lines) become successively linearly unstable for supercriticalities $\epsilon = (\lambda - \lambda_c)/\lambda_c$ equal to 6.37×10^{-2}, 8.53×10^{-2} and 15.47×10^{-2}. All the other modes become unstable for still larger values of ϵ. Therefore we choose the $m = 0, 1$ and $m = 2$ modes as the most unstable or "active" modes, whose interactions determine the weakly nonlinear behavior of the fluid layer. Amplitude equations may then be obtained by a standard slaving principle (see for instance [152] and §4.2.3), for which the slaved modes are taken as being all the (non-active) eigenmodes with a linear growth rate larger than -50 and with an azimuthal wave number m between 0 and 4, while all the other modes are simply disregarded. The validity and convergence of this procedure will be discussed later. The system of equations obtained for the complex amplitudes A_0, A_1 and A_2 of the $m = 0, 1$ and $m = 2$ modes reads

$$
\begin{aligned}
\dot{A}_0 &= (l_0\epsilon + \sigma_0)\, A_0 + q_{000}A_0^2 + q_{011}\,|A_1|^2 + q_{022}\,|A_2|^2 + c_{0000}A_0^3 \\
&\quad + c_{0011}A_0\,|A_1|^2 + c_{0022}A_0\,|A_2|^2 + c_{0112}\left(A_1^2 A_2^\star + A_2 A_1^{\star\,2}\right) \tag{7.9}
\end{aligned}
$$

$$
\begin{aligned}
\dot{A}_1 &= (l_1\epsilon + \sigma_1)\, A_1 + q_{101}A_0 A_1 + q_{112}A_1^\star A_2 + c_{1001}A_0^2 A_1 \\
&\quad + c_{1012}A_0 A_1^\star A_2 + c_{1111}\,|A_1|^2\, A_1 + c_{1221}\,|A_2|^2\, A_1 \tag{7.10}
\end{aligned}
$$

$$
\begin{aligned}
\dot{A}_2 &= (l_2\epsilon + \sigma_2)\, A_2 + q_{202}A_0 A_2 + q_{211}A_1^2 + c_{2002}A_0^2 A_2 \\
&\quad + c_{2011}A_0 A_1^2 + c_{2112}\,|A_1|^2\, A_2 + c_{2222}\,|A_2|^2\, A_2 \tag{7.11}
\end{aligned}
$$

whose form may also be obtained from the usual symmetry arguments. Here, given that the lowest-order planform function (say, the surface temperature field) may be written as $T_s = A_0 f_0(r) + A_1 f_1(r) \exp[i\phi] + A_2 f_2(r) \exp[2i\phi] + c.c.$, rotational invariance (arbitrary shift of ϕ) requires the amplitude equations to be invariant under the action $A_p \to A_p \exp[i\, p\, \varphi], p =$

$0, 1, 2$, while mirror reflection ($\phi \rightarrow -\phi$) requires their invariance with respect to the action $A_p \rightarrow A_p^*$ (which can only be satisfied if all coefficients are real). Hence, amplitude equations (7.9–7.11) are generic and apply also to two-dimensional convection patterns, in the vicinity of a point where eigenmodes with wavenumbers 0, k_1 and $k_2 = 2k_1$ are near criticality.

Still, the values of the different coefficients do depend on the system studied, and in our case these are given in Table 7.1.

Table 7.1: Values of the coefficients of the amplitude equations (7.9–7.11).

A_0		A_1		A_2	
l_0	5.83	l_1	7.63	l_2	8.32
σ_0	-0.372	σ_1	0.00	σ_2	-0.710
q_{000}	1.89	q_{101}	0.742	q_{202}	-1.64
q_{011}	17.1	q_{112}	9.89	q_{211}	1.02
q_{022}	17.0				
c_{0000}	-26.4	c_{1001}	-24.5	c_{2002}	-9.99
c_{0011}	$-137.$	c_{1012}	-33.1	c_{2011}	-27.0
c_{0022}	$-180.$	c_{1111}	$-106.$	c_{2112}	$-164.$
c_{0112}	8.95	c_{1221}	$-106.$	c_{2222}	-71.8

When polar coordinates are used for the complex amplitudes, the rotational invariance of the physical system allows elimination of one of the two phase variables. The equilibrium solutions of this reduced system correspond for the real physical system either to stationary solutions or to traveling (here, rotating) waves [417, 468]. Results presented below have been obtained using the AUTO97 software. The bifurcation diagram is given in Fig. 7.2 where the full and dashed lines represent respectively the stable and unstable solutions.

At $\epsilon = 0$ (point B_1), the $m = 1$ mode bifurcates supercritically. Its nonlinear self-interaction generates the $m = 0$ and $m = 2$ modes so that the complete pattern is a superposition of the three modes. The convective pattern of this mixed mode solution M_1 is depicted in Fig. 7.3, where the vertical velocity at mid-depth of the container is plotted.

The primary bifurcations of the $m = 0$ and $m = 2$ modes at $\epsilon = 6.37 \times 10^{-2}$ (point B_0) and 8.53×10^{-2} (point B_2) do not give rise to stable solutions. A secondary bifurcation on the M_1-branch leads to another stable solution corresponding to a rotating wave pattern (RW). This RW exists only for a small range of ϵ, between 31.7×10^{-2} and 33.0×10^{-2}. The rotating pattern is also a superposition of the three elementary modes and is represented in Fig. 7.4.

The period of rotation is 126 (dimensionless thermal time) for $\epsilon = 32.8 \times 10^{-2}$. For the silicone oil used in the experiments [43], the dimensional period is then around 10 h, i.e. quite long. This RW-branch loses stability at $\epsilon = 33.0 \times 10^{-2}$ due to a Hopf bifurcation. This bifurcation is subcritical and no stable modulated wave (MW) can be observed. From this bifurcation point until $\epsilon = 53.5 \times 10^{-2}$, no stationary solution exists for the reduced system. Numerical integrations actually show that in this ϵ-range, a stable heteroclinic orbit exists. This heteroclinic orbit is due to the spatial nonlinear resonance of the $m = 1$ and $m = 2$ modes as already discussed by several authors [56, 417, 468]. For this solution, the role of

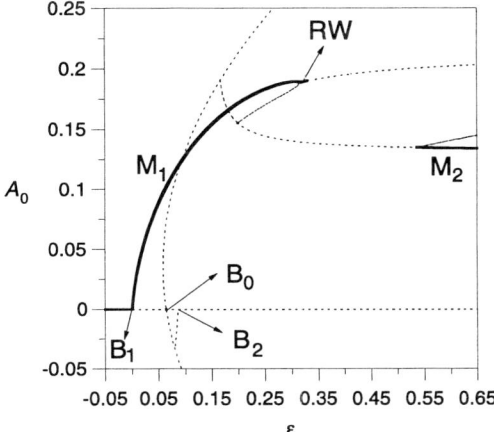

Figure 7.2: Bifurcation diagram for $a = 2.5$. The amplitude A_0 of the $m = 0$ mode is represented as a function of the relative distance to the threshold ϵ. Stable and unstable branches are represented by full and dashed lines. B_1, B_0 and B_2 indicate the primary bifurcation points for the $m = 1$, $m = 0$ and $m = 2$ modes.

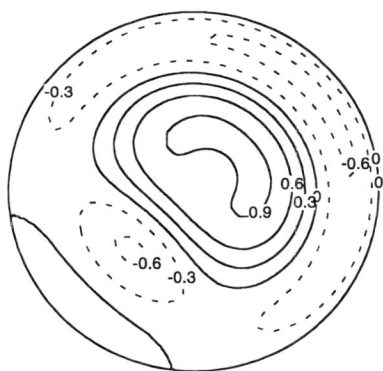

Figure 7.3: Vertical velocity at mid-depth for the M_1 solution ($\epsilon = 15 \times 10^{-2}$).

imperfections in the system is known to make the heteroclinic orbit periodic, with a period which decreases when imperfections become more important. In experiments, a possible imperfection originates from the fact that the container is never perfectly horizontal. For this reason, the $m = 1$ convective mode always has a small, but non-vanishing, amplitude. In the numerical calculations, these imperfections can easily be introduced by artificially adding a small constant term to the r.h.s. of Eq. (7.10). Note however that this term must not necessarily be introduced in the description since the numerical noise, which prevents this r.h.s. from becoming exactly zero, is in fact sufficient to give the heteroclinic orbit a finite

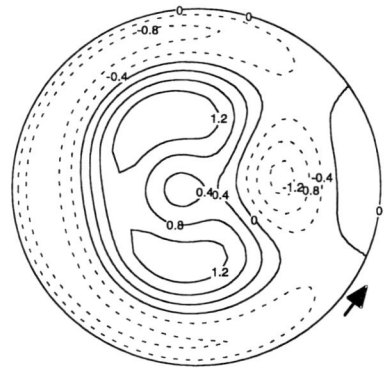

Figure 7.4: Vertical velocity at mid-depth for the rotating wave ($\epsilon = 32.8 \times 10^{-2}$).

period.

The periodic solution for the imperfect system is characterized by the alternation of two quasi-stationary patterns which are rotated by $\pi/2$ with respect to each other. For each of these patterns two convective cells are observed over very long time intervals. These cells result from a superposition of the $m = 2$ and $m = 0$ modes, while A_1 remains very small. The transition from one pattern to the other is induced by a sudden increase of A_1, which gives rise to a corresponding increase in size of one of the two cells. Then for a short time, the pattern consists of a unique cell that finally splits into the two cells of the rotated pattern. In Fig. 7.5, the time evolution of the amplitudes is shown (remember that the precise value of the period depends on the imperfections or on the precision of the calculations). Fig. 7.6 presents the time evolution of the convective pattern during one period.

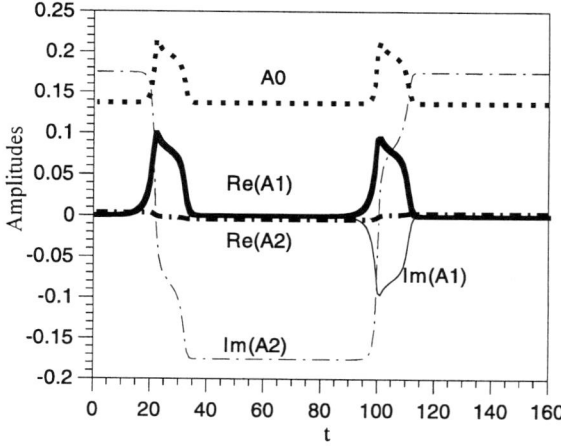

Figure 7.5: Time evolution of the amplitudes for the heteroclinic orbit ($\epsilon = 40 \times 10^{-2}$).

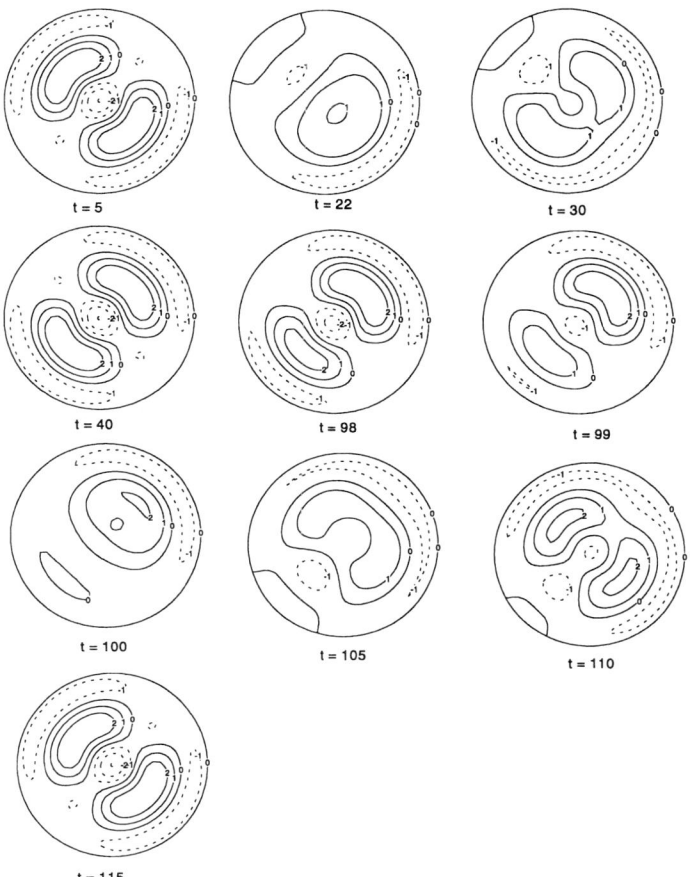

Figure 7.6: Vertical velocity at mid-depth for the heteroclinic orbit ($\epsilon = 40 \times 10^{-2}$). The time t corresponds to the abscissa in Fig. 7.5.

Finally, let us recall that if no constant term is added to Eq. (7.10) for the numerical calculations, the machine epsilon plays the role of the imperfection but its sign is undetermined. As a consequence, the transition between the two two-cell patterns has no fixed direction since any of the two cells can grow before splitting. This numerical solution with no explicit imperfection is thus not strictly-speaking periodic (noise-induced chaos).

The last stable solution of Eqs. (7.9–7.11) appears in the bifurcation diagram at $\epsilon = 53.5 \times 10^{-2}$. At this point, the quasi-stationary part of the heteroclinic orbit (the two-cell structure) becomes stable, with a zero value for A_1. This solution is a superposition of the $m = 2$ and $m = 0$ modes (mixed modes M_2) and is represented in Fig. 7.7.

To end this section, we briefly discuss the convergence of the bifurcation diagram when the number of modes taken into account in the description is increased. First, it has been checked that when the number of slaved modes is increased by decreasing the minimum linear growth

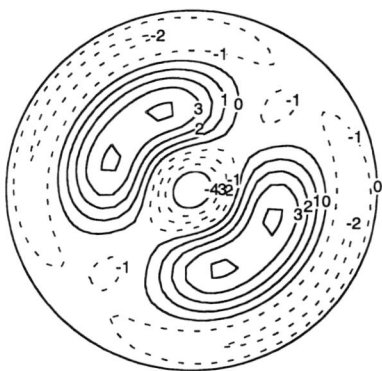

Figure 7.7: Vertical velocity at mid-depth for the M_2 solution ($\epsilon = 55 \times 10^{-2}$).

rate below -50, the values of the coefficients given in Table 7.1 indeed converge. The second important verification is carried out by writing amplitude equations not only for the three most unstable modes as in Eqs. (7.9–7.11), but also for modes with higher azimuthal wave numbers or for modes corresponding to the second or the third eigenvalue for a given m. In all cases, it has been observed that the bifurcation diagrams for the more complex system give rise to the same transitions with the same stable convective patterns.

7.13   Summary and discussion

The above analysis of nonlinear Rayleigh–Marangoni–Bénard thermoconvection in a circular container has been limited to an aspect ratio 2.5, in view of recent experimental findings of Johnson and Narayanan [43]. The theory predicts that just above the linear stability threshold, a one-cell solution occurs (mixed-mode M_1, Fig. 7.3). For larger values of the distance to the threshold, a rotating one-cell solution is stable (rotating wave RW, Fig. 7.4). Note however that the stability range on the ϵ-axis for this rotating pattern is very small. For this reason, this convective structure should be very difficult to observe in experiments (and is indeed not mentioned in [43]).

For the highest values of ϵ investigated here, a stable steady pattern can be observed which is made up by two symmetric convective cells (mixed modes M_2, Fig. 7.7). For distances to the threshold larger than the upper limit for the rotating wave and smaller than the lower limit for the M_2 solution, no stationary solution exists. The behavior of the fluid becomes dynamic and takes the form of a stable heteroclinic orbit (Fig. 7.6). This solution is characterized by a switching between two two-cell patterns rotated by $\pi/2$ with respect to each other. It is important to notice that this mode switching is completely similar to the experimental observations of Johnson and Narayanan [43], reproduced in Fig. 7.8.

In their work, these authors had felt that this dynamic behavior of the fluid had to be attributed to the presence of a codimension-2 point, for which the $m = 2$ and $m = 0$ modes are simultaneously unstable. The present analysis definitely shows that the codimension-2 point is actually present in the system but rather the $m = 1$ and $m = 2$ modes collaborate to

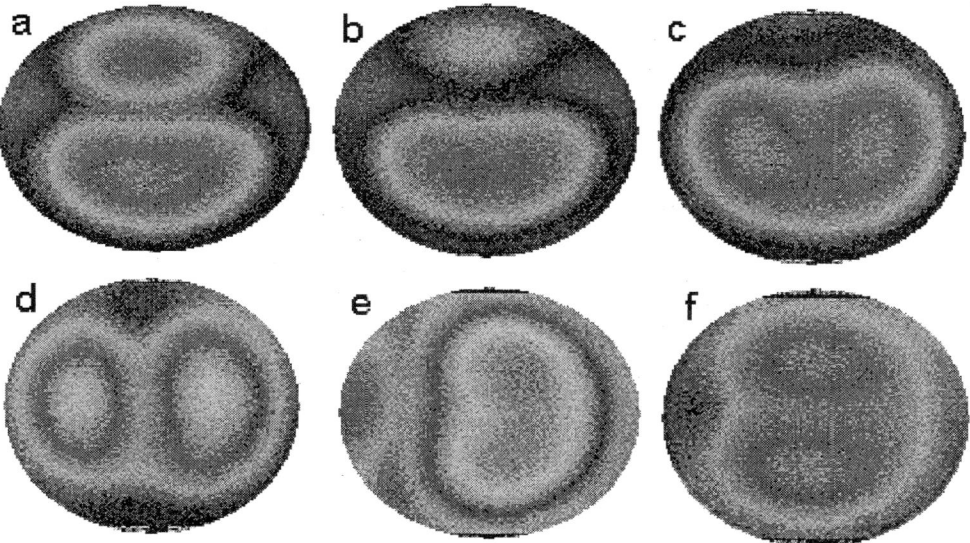

Figure 7.8: Infrared images of the free surface temperature showing the dynamic switching between two-cell patterns ($m = 2$) rotated by $\pi/2$ (a, d and f), via growth of the $m = 1$ mode (b, c and e). The experiment uses silicone oil 100 cSt, and the aspect ratio is $a = 2.5$, for a depth of 5 mm. Courtesy of D. Johnson and R. Narayanan.

generate the spatial resonance and the heteroclinic orbit.

This kind of heteroclinic behavior is important in the context of the study of the transitions to complex behaviors in dynamical systems and has already been considered by several authors in the past (see for instance [56, 416, 417, 468] and references therein). In particular, Echebarría, Krmpotić and Pérez-García have already described the possibility of such heteroclinic cycles in the case of slippery containers [468]. However, the aspect ratio they predict for this behavior is not the same as in experiments and the corresponding convective patterns are also different from the experimental structures, especially along the vertical walls. Note also that the MW which can be stable in slippery containers is here unstable. Still, using realistic boundary conditions, it has been shown that the heteroclinic orbit is still present. Hence, this seems to be a robust feature of the system investigated, and it is worth pointing out that, to our knowledge, Bénard convection in circular containers might be a unique example where both experiments and theoretical analysis show the presence of a stable heteroclinic connection.

7.2   Strong spatial resonances between waves and patterns

Another codimension-2 problem occurs when waves with wavenumber k_1 compete with stationary convection with wavenumber k_2. This possibility exists in two-layer systems, as shown by the linear stability analyzes of Chapter 3 (see the classification of Fig. 3.28). Moreover, systems of this class have also been identified recently in nonlinear optics [483], and seem to be particularly interesting, due to the possibility of observing patterns with complicated spatio-temporal dynamics near the instability threshold, especially in the case $k_2/k_1 = 2$. Strikingly, it can also be shown in this resonant case that some dynamical regimes may actually be described by quadratic – rather than the usual cubic – amplitude equations. This feature should make some of the results presented below rather generic, due to the small number of non-rescalable coefficients entering Ginzburg–Landau equations. In particular, even though the patterns considered in this section are one-dimensional, it turns out that some of the normal forms also apply for some two-dimensional optical patterns [483].

Specifically, after a calculation of typical coefficients for two-layer Bénard instabilities, non-resonant and resonant situations are analyzed separately. While the non-resonant case exhibits trivial dynamics leading to pure as well as mixed states (spatially quasiperiodic), the resonant dynamics appears much richer. In particular, in addition to asymmetric solutions and quasiperiodic regimes, the analysis of the resonant case focuses on a transition to chaotic dynamics via an infinite sequence of homoclinic bifurcations, which differs from the more classical period-doubling scenario in many respects. Reverse transitions involving an opposite sequence are also predicted, and are shown to be entirely determined by quadratic interactions.

The particular system considered here, of which no explicit use is made before the analysis of §7.2.2 apart for general symmetry properties, is a 2D Marangoni–Bénard system obtained by sandwiching two layers of immiscible fluids of infinite horizontal extent between two rigid conducting plates maintained at different temperatures (the linear stability of this system is presented in §3.5). No buoyancy effect is considered, and the interface is assumed to be un-deformable. For this system, a set of parameter values exists (most of them being fixed by the choice of liquids) for which waves of wavenumber k_1 and linear frequency ω_c are simultaneously critical with steady convection of wavenumber k_2. Accordingly, the deviation of the fields (temperature or vertical velocity) with respect to the diffusive solution can be written, at the lowest perturbative order, as a superposition of critical eigenmodes with complex amplitudes

$$U(x, z, t) = a_1(t)\exp[i(k_1 x + \omega_c t)]U_1(z) + a_2(t)\exp[i(k_1 x - \omega_c t)]U_2(z) + a_3(t)\exp[ik_2 x]U_3(z) + c.c.$$

$$(7.12)$$

where $U_2(z) = \bar{U}_1(z)$ and $U_3(z)$ is real (the overbar denotes the complex conjugate), x and z are the coordinates respectively parallel and perpendicular to the plane of the layers and t is the time. As usual, invariance properties of the physical system will first be used to derive the form of amplitude equations coupling the left- and right-traveling wave amplitudes a_1 and a_2, and the monotonic mode amplitude a_3. The analysis is restricted to non-zero values of k_1, k_2 and ω_c.

7.2.1 Derivation of the form of amplitude equations from symmetry properties

Following the method discussed in §4.2.1, the amplitude equations are first written in the form

$$\frac{da_i}{dt} = F_i(a_1, a_2, a_3, \bar{a}_1, \bar{a}_2, \bar{a}_3) \quad i = 1, 2, 3 \tag{7.13}$$

and Taylor-expanded around the origin $a_1 = a_2 = a_3 = 0$ (where the functions F_i vanish). Symmetry considerations are then used to simplify the resulting system. The physical set-up is invariant with respect to:

i) Time-translations $t \rightarrow t + \Delta t$ (autonomous system): according to Eq. (7.12), this is equivalent to the requirement that amplitude equations (7.13) are invariant under the transformation $\{a_1, a_2, a_3\} \rightarrow \{a_1 \exp[i\theta], a_2 \exp[-i\theta], a_3\}$ for every θ.

ii) Lateral translations $x \rightarrow x + \Delta x$: this implies that Eqs (7.13) are invariant under the transformation $\{a_1, a_2, a_3\} \rightarrow \{a_1 \exp[ik_1 \Delta x], a_2 \exp[ik_1 \Delta x], a_3 \exp[ik_2 \Delta x]\}$ for every Δx. When $k_2 = 2k_1$ (1:2 resonance case), this transformation becomes $\{a_1, a_2, a_3\} \rightarrow \{a_1 \exp[i\phi], a_2 \exp[i\phi], a_3 \exp[2i\phi]\}$ for every ϕ.

iii) Parity $x \rightarrow -x$ (left-right symmetry): this is equivalent to the invariance of Eqs (7.13) under the transformation $\{a_1, a_2, a_3\} \rightarrow \{\bar{a}_2, \bar{a}_1, \bar{a}_3\}$.

The combination of these fundamental invariance properties requires a number of terms in the Taylor expansions of Eqs (7.13) to vanish [163]. As an example, all linear coefficients except diagonal ones must be zero due to requirements i) and ii). The invariance iii) then requires the coefficients of linear terms in the equations for a_1 and a_2 to be complex conjugates of each other, and the linear coefficient in the equation for a_3 to be real. The discussion of nonlinear terms proceeds in the same way, although special attention has to be paid to quadratic terms, for which cases $k_2 = 2k_1$ and $k_2 \neq 2k_1$ are qualitatively different [417]. When the basic wavenumbers of the unstable modes are incommensurate, no resonance occurs, and the amplitude equations are free from phase-coupling terms. Among the possible $m : n$ resonance cases, a strong resonance occurs when k_1 and k_2 are in the ratio 1 : 2, and some quadratic phase-coupling terms now cannot be ruled out by considering the invariance ii).

After computing the possible cubic terms, and limiting Taylor expansions to this order, we arrive at the following system of coupled Ginzburg–Landau equations:

$$\begin{aligned}
\dot{a}_1 &= \mu a_1 + \delta a_3 \bar{a}_2 + a_1(\alpha|a_1|^2 + \beta|a_2|^2 + \gamma|a_3|^2) \\
\dot{a}_2 &= \mu a_2 + \bar{\delta} a_3 \bar{a}_1 + a_2(\bar{\alpha}|a_2|^2 + \bar{\beta}|a_1|^2 + \bar{\gamma}|a_3|^2) \\
\dot{a}_3 &= \mu' a_3 + \delta' a_1 a_2 + a_3(\alpha'|a_3|^2 + \gamma'|a_1|^2 + \bar{\gamma}'|a_2|^2)
\end{aligned} \tag{7.14}$$

with $\delta = \delta' = 0$ when $k_2 \neq 2k_1$. The dot denotes differentiation with respect to t. Note that the mirror symmetry iii) also requires the coefficients α' and δ' to be real.

An important feature of Eqs (7.14) is that they generally do not admit a potential function $\psi(a_i, \bar{a}_i)$ such that $\dot{a}_l = -\partial\psi/\partial\bar{a}_l$ for all l (giving $d\psi/dt = -2\Sigma_l|\partial\psi/\partial a_l|^2 \leq 0$). Indeed, expressing equalities of cross-derivatives of ψ leads to the necessary conditions $\alpha = \bar{\alpha}, \beta = \bar{\beta}, \gamma = \bar{\gamma} = \gamma'$, and $\delta = \bar{\delta} = \delta'$. Interestingly, the analysis of hexagonal convection [e.g. in one-layer Marangoni–Bénard instabilities [51, 151, 312, 316], see Eqs (4.74)] leads to equations similar to Eqs (7.14) for amplitudes of the three constitutive roll patterns. However, the conditions for existence of a potential are met in this case, such that the dynamics is purely relaxational [51, 312].

Owing to the large number of unknown coefficients, a general discussion of the possible solutions and stability properties of Eqs (7.14) appears to be outside the scope of the present text. As in [475], we thus focus our attention on a particular convective system for which coefficients can be calculated from the governing equations of fluid motion.

7.2.2 Coefficients for the two-layer Bénard instability

In each phase ($i = 1, 2$), the Boussinesq equations governing the velocity field $\vec{V}_i = (U_i, W_i)$ and the deviations of temperature T_i and pressure p_i with respect to the diffusive solution (constant temperature gradient $dT_i/dz = -\beta_i$) are

$$
\begin{aligned}
\vec{\nabla}.\vec{V}_i &= 0 \\
\kappa_i \Delta T_i + \beta_i W_i &= \dot{T}_i + (\vec{V}_i.\vec{\nabla})T_i \\
\mu_i \Delta \vec{V}_i - \vec{\nabla}p_i &= \rho_i \left[\dot{\vec{V}}_i + (\vec{V}_i.\vec{\nabla})\vec{V}_i \right]
\end{aligned}
\tag{7.15}
$$

where ρ_i, μ_i and κ_i are respectively the density, dynamic viscosity and thermal diffusivity.

On rigid conducting plates $z = -a_1$ and $z = a_2$ ($a_{i=1,2}$ are the layer thicknesses), the boundary conditions are

$$
\vec{V}_i = T_i = 0
\tag{7.16}
$$

while at the undeformable interface $z = 0$ the following conditions hold

$$
\begin{aligned}
W_1 = W_2 &= 0, U_1 = U_2, T_1 = T_2, \\
\lambda_1 \frac{\partial T_1}{\partial z} = \lambda_2 \frac{\partial T_2}{\partial z}, \mu_2 \frac{\partial U_2}{\partial z} &- \mu_1 \frac{\partial U_1}{\partial z} = \sigma_T \frac{\partial T_1}{\partial x}
\end{aligned}
\tag{7.17}
$$

where λ_i is the thermal conductivity and σ_T the interfacial tension variation with temperature.

This problem is put under dimensionless form using a_1 as the length unit, a_1^2/κ_1 as the time unit, and $\Delta T_1 = \beta_1 a_1$ as the temperature unit. Then, the linearized stability problem associated with Eqs (7.15–7.17) leads to the characteristic relation $\Delta(\sigma, Ma, k, a) = 0$ between the growth rate σ of the perturbations, the Marangoni number $Ma = -\sigma_T \Delta T_1 a_1/\mu_1 \kappa_1$, the wavenumber k, and the ratio of layer thicknesses $a = a_2/a_1$ (see §3.5). All other parameters (properties ratios) are determined by the choice of liquids: in the following, we use the values $\rho = \rho_2/\rho_1 = 0.893, \mu = \mu_2/\mu_1 = 1.02, \kappa = \kappa_2/\kappa_1 = 0.934, \lambda = \lambda_2/\lambda_1 = 0.698$, representative of the methanol (layer 1) – n-octane (layer 2) configuration. The numerically computed stability diagram for these parameters values is represented in Fig. 7.9.

Figure 7.9 is obtained for the particular value $a = a^* = 0.726$ of the thicknesses ratio, for which an asymptote exists in the neutral stability ($\sigma = 0$) limit at $k \simeq 5.4$, separating regions where the monotonous instability sets in either at low wavenumbers by heating from the

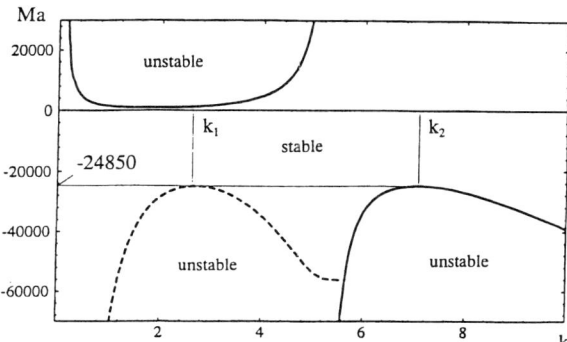

Figure 7.9: Linear stability results: Marangoni number Ma as a function of the wavenumber k for a system with parameters $\rho = \rho_2/\rho_1 = 0.893, \mu = \mu_2/\mu_1 = 1.02, \kappa = \kappa_2/\kappa_1 = 0.934, \lambda = \lambda_2/\lambda_1 = 0.698$ (methanol-octane), for a value $a = a^* = 0.726$ of the thicknesses ratio. Full curves: neutral stability, dashed curves: overstability (the critical frequency is $\omega_c = 64.7$).

methanol side ($Ma > 0$), or at high wavenumbers by heating from the octane side ($Ma < 0$). This behavior, already discussed in §3.5 (see the classification of Fig. 3.28), is connected with the fact that thermal diffusion times are similar for both liquid layers [476]. Furthermore, it is seen that an overstable ($\sigma = i\omega$) branch also exists when $Ma < 0$, and that its threshold coincides with the threshold of the monotonous mode at the critical Marangoni number $Ma_c = -24850$. This situation is occasional ($a = a^*$), and for increasing octane thickness $a > a^*$, the monotonous mode is first critical, while the oscillating mode becomes more dangerous when $a < a^*$. The point (Ma_c, a^*) thus defines a codimension-2 point (CTP) for this problem [51]. In order to investigate the behavior in the vicinity of this CTP, we use the real parameters μ and μ', defined by Eqs (7.14) and representing the real part of the growth constants of both modes. Thus, μ and μ' vanish at the CTP, and can be linearly approximated in its vicinity:

$$\begin{aligned} \mu &= s_1(Ma - Ma_c) + s_2(a - a^*) \\ \mu' &= s_1'(Ma - Ma_c) + s_2'(a - a^*) \end{aligned} \tag{7.18}$$

The values of linear and nonlinear coefficients are presented in Table 7.2, for both cases of non-resonant and resonant interactions.

The particular technique used to compute nonlinear coefficients of Table 7.2 will not be described in detail in this section, referring the reader to §4.2.2 and §4.2.3 for an explanation of possible methods. Actually, the coefficients have been cross-checked using two techniques. First, their values can be calculated by generalizing scalar products (4.105) and (4.113) to the problem considered here (also using the adjoint problem defined in Appendix D). Then, their accuracy can be tested by attempting to solve the corresponding linear inhomogeneous problems obtained at third order (e.g. using a shooting method). It was also checked that this method reproduced the values of coefficients calculated (using a different method) for

Table 7.2: Coefficients of the amplitude equations for non-resonant ($k_1 = 2.67, k_2 = 7.03 \neq 2k_1$) and resonant ($k_1 = 3, k_2 = 6 = 2k_1$) cases. The system parameters are $\rho = \rho_2/\rho_1 = 0.893, \mu = \mu_2/\mu_1 = 1.02, \kappa = \kappa_2/\kappa_1 = 0.934, \lambda = \lambda_2/\lambda_1 = 0.698$ (methanol–octane), and the CTP thickness ratio a^* and critical Marangoni number Ma_c are given for both cases.

	Non-resonant	Resonant		Non-resonant	Resonant
a^*	0.726	0.749	δ	0	$-14.7 - 18.4i$
Ma_c	-24850	-26895	δ'	0	0.643
ω_c	64.7	62.1	α	$-0.199 - 0.126i$	$-0.262 - 0.193i$
s_1	$-7.07\,10^{-4}$	$-6.9\,10^{-4}$	β	$0.270 - 0.393i$	$0.234 - 0.447i$
s_2	-34.3	-46.8	γ	$27.5 + 18.2i$	$-3.46 + 0.373i$
s_1'	$-1.48\,10^{-3}$	$-1.28\,10^{-3}$	α'	-14.6	-13.8
s_2'	111	278	γ'	$-0.997 + 3.74i$	$-0.235 + 1.34i$

wave interactions in the two-layer buoyancy-driven case treated in §5.4, and the value of the hysteresis of hexagons reported in Ref. [130] for the finite-depth case (see also §4.4.2).

7.2.3 Results and discussion

Non-resonant case

Substituting $a_l = r_l \exp[i\varphi_l]$ in the system (7.14) with $\delta = \delta' = 0$ leads to a system of equations for the moduli r_l, which is decoupled from the phases φ_l. The only possible solutions with stationary amplitudes are:

$$r_1 = r_2 = r_3 = 0 \qquad\qquad\qquad \text{:}\quad \text{diffusive rest state (O)} \qquad\qquad (7.19)$$

$$r_1 = r_2 = 0, r_3^2 = -\mu'/\alpha' \qquad\quad \text{:}\quad \text{steady convection (SC)} \qquad\qquad (7.20)$$

$$r_3 = r_{2(1)} = 0, r_{1(2)}^2 = -\mu/\alpha_R \qquad \text{:}\quad \text{traveling waves (TW)} \qquad\qquad (7.21)$$

$$r_3 = 0, r_1^2 = r_2^2 = -\mu/(\alpha_R + \beta_R) \quad \text{:}\quad \text{standing waves (SW)} \qquad\qquad (7.22)$$

$$r_{1(2)} = 0, r_{2(1)}^2 = \frac{\mu'\gamma_R - \mu\alpha'}{\alpha_R\alpha' - \gamma_R\gamma_R'}, \\ r_3^2 = \frac{\mu\gamma_R' - \mu'\alpha_R}{\alpha_R\alpha' - \gamma_R\gamma_R'} \qquad \text{:}\quad \text{mixed solution SC + TW (M}_2\text{)} \qquad (7.23)$$

$$r_1^2 = r_2^2 = \frac{\mu'\gamma_R - \mu\alpha'}{\alpha'(\alpha_R + \beta_R) - 2\gamma_R\gamma_R'}, \\ r_3^2 = \frac{2\mu\gamma_R' - \mu'(\alpha_R + \beta_R)}{\alpha'(\alpha_R + \beta_R) - 2\gamma_R\gamma_R'} \qquad \text{:}\quad \text{mixed solution SC + SW (M}_3\text{)} \qquad (7.24)$$

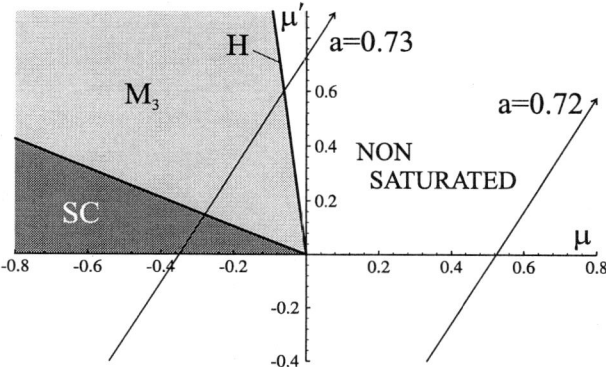

Figure 7.10: Stability map in the non-resonant case, as a function of the linear growth constants μ (of the waves) and μ' (of steady convection). Arrows indicate paths followed when the Marangoni number is increased, for two values of the thicknesses ratio a. Only stable convective solutions are indicated: SC = steady convection, M_3 = mixed (spatially quasiperiodic) mode. Non-saturated growth occurs for $u > 0$, and at the right of the Hopf bifurcation line H.

where an index R means the real part of a coefficient.

Existence conditions for each kind of solution are readily obtained from Eqs (7.20–7.24), by requiring the positiveness of squared amplitudes r_i^2. This leads to existence domains limited by lines going through the origin in the (μ, μ') unfolding plane (see Fig. 7.10). Stability conditions may also be obtained analytically, although results are not reproduced here for conciseness. Figure 7.10 summarizes the relevant results, i.e. the map of possible behaviors in the (μ, μ') plane for the particular system considered.

The corresponding bifurcation diagrams are represented in Fig. 7.11. For each solution, nonlinear corrections to the linear frequencies are computed from the imaginary part of amplitude equations:

$$\begin{aligned}
\dot{\varphi}_1 &= \omega_1 = \alpha_I r_1^2 + \beta_I r_2^2 + \gamma_I r_3^2 \\
\dot{\varphi}_2 &= \omega_2 = -\alpha_I r_2^2 - \beta_I r_1^2 - \gamma_I r_3^2 \\
\dot{\varphi}_3 &= \omega_3 = \gamma_I'(r_1^2 - r_2^2)
\end{aligned} \tag{7.25}$$

where $\omega_{i=1,2,3}$ are constant [and can be computed from Eqs (7.19–7.24) for each solution], such that $\varphi_i = \omega_i t + \varphi_{i0}$, and φ_{i0} is arbitrary.

Now, for symmetric solutions ($r_1 = r_2$), Eqs (7.25) lead to $\omega_2 = -\omega_1$ and $\omega_3 = 0$. According to Eq. (7.12), the perturbation fields are thus periodic with a single frequency $\omega = \omega_c + (\alpha_I + \beta_I)r_1^2 + \gamma_I r_3^2$. This is the case for both SW and M_3 solutions.

The SC solution also falls into this category, but the amplitudes of waves are zero, such that this solution is effectively steady. It is also checked that the TW solution is time-periodic, while the M_2 solution (dissymmetric) is the superposition of two traveling waves with incommensurate phase velocities $v = \omega/k_1 = (\omega_c + \alpha_I r_1^2 + \gamma_I r_3^2)/k_1$ and $v' = \omega'/k_2 = \gamma_I' r_1^2/k_2 \ll v$ (note that the direction of propagation is the same because $\gamma_I' > 0$, as seen in Table 7.2).

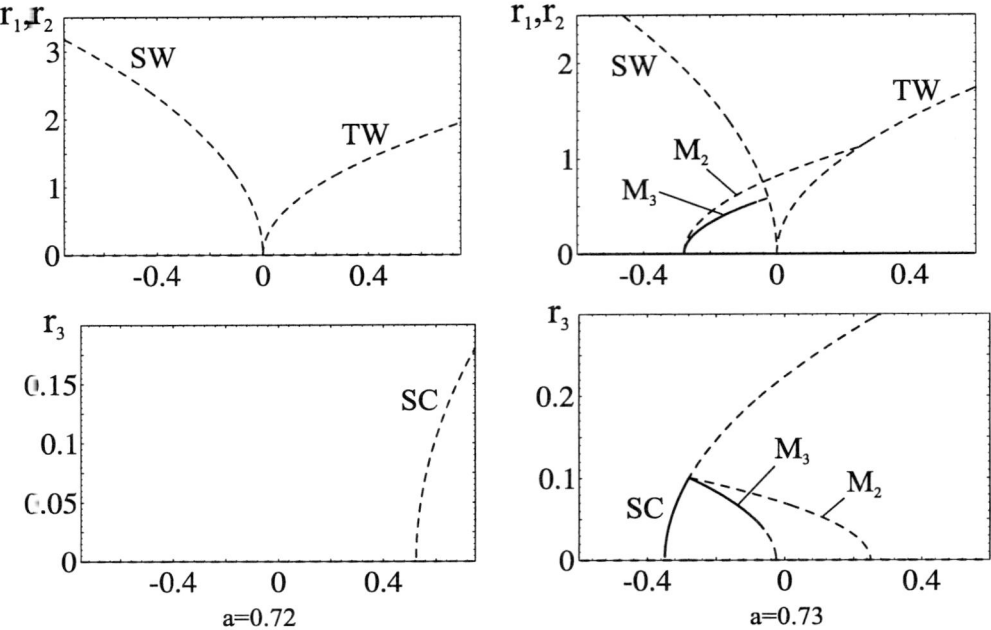

Figure 7.11: Bifurcation diagrams for the qualitatively different cases $a = 0.72 < a^*$ (left) and $a = 0.73 > a^*$ (right). The amplitudes of the waves $r_{1,2}$ and of steady convection r_3 are represented as a function of the growth constant μ of waves, for a displacement along the arrows labeled $a = 0.72$ and $a = 0.73$ of Fig. 7.10. Full curves represent stable states, while dashed curves represent unstable states. SC = steady mode, TW (SW) = traveling (standing) waves, M_2 = SC+TW, M_3 = SC+SW.

This M_2 solution is thus quasiperiodic both in time and in space (because the constitutive wavenumbers k_1 and k_2 are also incommensurate). However, the stability results indicate that it is always unstable for the present parameter values.

In Fig. 7.10, it is seen that in a large part of the diagram, no stable steady solution exists. In particular, as also seen in Fig. 7.11, the Hopf bifurcation occurring at $\mu = 0$ (for $a < a^*$, i.e. $u' < 0$) results in supercritical TW and subcritical SW, which are both unstable (see also §5.1.1). Since no saturation is obtained from cubic terms in this case (this is due to the fact that $\alpha_R + \beta_R > 0$), higher-order (quintic) terms should be included in amplitude equations. Such a procedure might result in standing waves of finite amplitude, existing in a certain subcritical region. Stable solutions exist when $a > a^*$, under the form of a SC supercritical solution (bifurcating from the rest state on the axis $\mu' = 0$, $\mu < 0$), undergoing a secondary bifurcation to the mixed solution M_3 when the constraint is increased (at $\mu'/\mu = \alpha'/\gamma_R = -0.531$). An interesting feature of this solution is its spatial quasi-periodicity (the constitutive modes have incommensurate wavenumbers k_1 and k_2). However, contrary to the M_2 solution, it has only one temporal frequency ω (which can be shown to be decreasing with increasing Marangoni number). This M_3 solution, represented in Fig. 7.12, finally undergoes a third

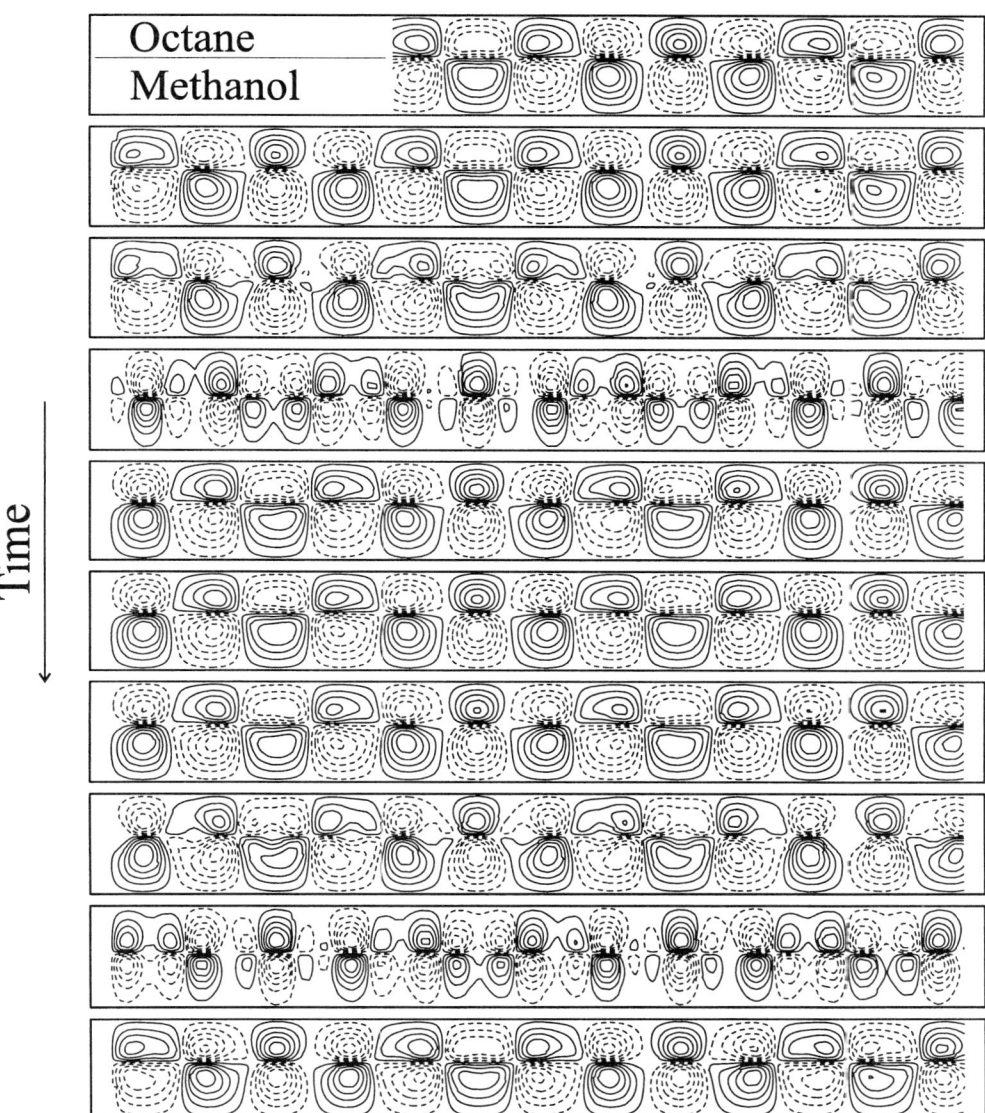

Figure 7.12: Representation of the stable mixed M_3 solution ($a = 0.73, Ma = -24760$): streamfunction in both layers. The system is quasiperiodic along the layers (wavenumbers $k_1 = 2.67$ and $k_2 = 7.03$), and oscillates in time with the pulsation $\omega \simeq \omega_c = 64.7$. The streamfunction is rescaled at each snapshot ($\Delta t = T/10 \simeq 9.710^{-3}$). Time runs downwards.

(Hopf) bifurcation when the constraint is increased, slightly before merging with the unstable subcritical SW branch at the point where the amplitude of the steady component r_3 vanishes [at $\mu'/\mu = 2\gamma'_R/(\alpha_R + \beta_R) = -28.1$]. As Figs 7.10 and 7.11 show, no stable saturated solution exists beyond the line H of the Hopf bifurcation, thus again requiring calculation of higher-order contributions.

Resonant case

For values of fluid properties selected in the last section, the condition $k_2 = 2k_1$ is clearly not satisfied (see Fig. 7.9). Rather than changing fluid properties in order to achieve exact resonance at the threshold (this is possible e.g. by increasing the viscosity ratios ν_2/ν_1 and μ_2/μ_1, and adjusting the thickness ratio $a = a_2/a_1$), we preferred to force resonance by arbitrarily selecting basic wavenumbers, say $k_1 = 3$ and $k_2 = 6$ (thus requiring the system to be periodic in the x-direction with the period $\Delta x = 2\pi/3$). In order for the thresholds of oscillatory and stationary modes to coincide, we have to adjust the value of a, such that the CTP now occurs for ($Ma_c = -26895, a^* = 0.7488$). Note that this somewhat artificial procedure might lead to some qualitative agreement with behaviors observed in a realistic system when the constraint is large enough, such that large bands of wavenumbers including k_1 and k_2 are unstable [this could be checked by direct numerical simulation of Eqs. (7.15–7.17)]. Moreover, some of the conclusions made in this section are quite general, as they are based on a discussion of the quadratic system, for which suitable rescalings eliminate most of the coefficients. Thus, in order to exploit some of the predictions made here for other resonant systems, only the quadratic coefficients δ and δ' need to be computed (which is much easier than the calculation of cubic coefficients). The new set of coefficients of Eqs (7.14) is also presented in Table 7.2.

A. Quadratic system

We will first concentrate on the role of quadratic nonlinearities. In this respect, it should be stressed that due to their perturbative origin, the strict validity of Eqs (7.14) may only be guaranteed in some neighborhood of the origin $a_1 = a_2 = a_3 = 0$. Since $\delta = |\delta| \exp[i\varphi_\delta]$ and δ' are finite quantities, the dynamics near this point should be governed by the quadratic system obtained by neglecting cubic nonlinearities in Eqs (7.14). This becomes clearer by introducing the change of scales

$$
\begin{aligned}
t &\to \mu^{-1}t \\
(a_1, a_2) &\to \left|\tfrac{\mu\mu'}{|\delta||\delta'|}\right|^{\frac{1}{2}} (a_1, a_2) \\
a_3 &\to \tfrac{\mu}{|\delta|} a_3
\end{aligned}
\tag{7.26}
$$

which allows the system (7.14) to be rewritten as

$$
\begin{aligned}
\dot{a}_1 &= a_1 + \exp[i\varphi_\delta]a_3\bar{a}_2 + a_1(\tilde{\alpha}|a_1|^2 + \tilde{\beta}|a_2|^2 + \tilde{\gamma}|a_3|^2) \\
\dot{a}_2 &= a_2 + \exp[-i\varphi_\delta]a_3\bar{a}_1 + a_2(\overline{\tilde{\alpha}}|a_2|^2 + \overline{\tilde{\beta}}|a_1|^2 + \overline{\tilde{\gamma}}|a_3|^2) \\
\dot{a}_3 &= \chi(a_3 + s\,a_1 a_2) + a_3(\tilde{\alpha}'|a_3|^2 + \tilde{\gamma}'|a_1|^2 + \overline{\tilde{\gamma}'}|a_2|^2)
\end{aligned}
\tag{7.27}
$$

where φ_δ is the phase of the quadratic coefficient δ, $\chi = \mu'/\mu$, $s = \mathrm{sgn}(\chi/\delta')$, and the tilded cubic coefficients are defined by

$$\tilde{\alpha} = s\frac{\alpha\mu'}{\delta'|\delta|}, \tilde{\beta} = s\frac{\beta\mu'}{\delta'|\delta|}, \tilde{\gamma} = \frac{\gamma\mu}{|\delta|^2}, \tilde{\alpha}' = \frac{\alpha'\mu}{|\delta|^2}, \tilde{\gamma}' = s\frac{\gamma'\mu'}{\delta'|\delta|} \tag{7.28}$$

These coefficients are proportional to μ or μ'. Near the CTP ($|\mu|, |\mu'| \ll 1$), it should thus be possible to neglect corresponding terms, provided that the amplitudes a_1, a_2 and a_3 remain bounded (i.e. of order unity). Most of this section is devoted to a discussion of the regions of the (μ, μ') unfolding plane where this condition is satisfied. In fact, cubic terms are essential in some limiting cases, as will be seen later. For the moment, neglecting them and separating amplitudes and phases in the usual manner $a_l = r_l exp[i\varphi_l]$, we arrive at the four-dimensional dynamical system

$$\begin{aligned}
\dot{r}_1 &= r_1 + r_2 r_3 \cos(\varphi_\delta - \varphi) \\
\dot{r}_2 &= r_2 + r_1 r_3 \cos(\varphi_\delta + \varphi) \\
\dot{r}_3 &= \chi(r_3 + s\, r_1 r_2 \cos\varphi) \\
\dot{\varphi} &= \frac{r_2 r_3}{r_1}\sin(\varphi_\delta - \varphi) - \frac{r_1 r_3}{r_2}\sin(\varphi_\delta + \varphi) - s\chi\frac{r_1 r_2}{r_3}\sin\varphi
\end{aligned} \tag{7.29}$$

where $\varphi = \varphi_1 + \varphi_2 - \varphi_3$ is the only quantity involving phases coupled to the amplitudes r_i. The only steady solutions of the system (7.29) are found to be:

- the symmetric solution (SS):

$$\begin{aligned}
r_1^2 &= r_2^2 = (s\cos\varphi_\delta)^{-1} \\
r_3^2 &= (\cos\varphi_\delta)^{-2}, \quad \varphi = n\pi \quad (n = 0, 1)
\end{aligned} \tag{7.30}$$

which exists provided $s\cos(\varphi_\delta) > 0$.

- the asymmetric solutions (AS):

$$\begin{aligned}
r_1^2 &= \frac{(1+q^2)(1+2\chi^{-1})}{s\cos\varphi_\delta q^2\left[1 \mp (1 + 2\chi^{-1}(1+q^2))^{1/2}\right]} \\
r_2^2 &= \frac{(1+q^2)(1+2\chi^{-1})}{s\cos\varphi_\delta q^2\left[1 \pm (1 + 2\chi^{-1}(1+q^2))^{1/2}\right]} \\
r_3^2 &= -\frac{(2+\chi)(1+q^2)}{2q^2}, \quad \tan\varphi = \pm\frac{1}{q}\left(1 + 2\chi^{-1}(1+q^2)\right)^{\frac{1}{2}}
\end{aligned} \tag{7.31}$$

where $q = \tan(\varphi_\delta)$. The AS solutions exist provided $s\cos(\varphi_\delta) > 0$ and $\chi < -2(1 + q^2)$. Contrary to the SS solution, the AS solutions are not individually obeying the left–right symmetry iii). However, they are mapped onto each other by applying the corresponding transformation.

A general discussion of the properties of the system (7.29) will not be attempted here. In the following, we restrict our analysis to cases where $\delta' > 0$ and $\cos(\varphi_\delta) < 0$ (see Table 7.2). Still, this will allow comparison with some results of Fujimura and Renardy [475] (denoted by F&R in what follows), as these conditions are also fulfilled in their analysis. Thus,

$s \cos(\varphi_\delta) > 0$ if $s = -1$, i.e. $\chi = \mu'/\mu < 0$ is a necessary condition for SS and AS to exist. Moreover, AS can only exist provided $\chi < -2(1 + q^2)$. The stability conditions may be obtained analytically for SS: computing eigenvalues of the Jacobian matrix leads to

$$\lambda^{SS}_{1,2} = \frac{\chi \pm \sqrt{\chi(\chi + 8)}}{2} \tag{7.32}$$

whose real part is always negative for $\chi = \mu'/\mu < 0$ (i.e. in the existence region of SS), and thus corresponds to stable directions when $\mu > 0$, and to unstable directions when $\mu < 0$ [because the eigenvalues have to be multiplied by μ in view of the scalings (7.26)]. The other two eigenvalues are

$$\lambda^{SS}_{+,-} = \frac{\chi + 4 \pm \sqrt{\chi^2 - 16q^2}}{2} \tag{7.33}$$

for which the discussion depends on whether $|q| > 1$ or $|q| < 1$, and is presented in Fig. 7.13.

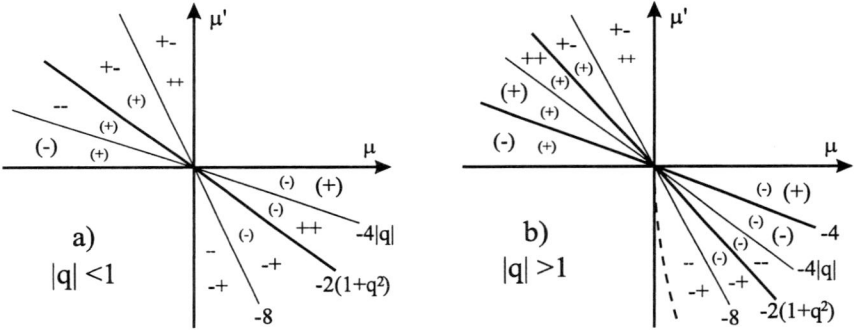

Figure 7.13: Stability diagram of the symmetric solution SS of the quadratic system for cases $| q | < 1$ (a) and $| q | > 1$ (b), and for $\delta' > 0$ and $\cos(\varphi_\delta) < 0$. On the axes, μ and μ' denote respectively the linear growth rates for waves and steady convection. In each sector (delimited by lines labeled by their slope $\chi = \mu'/\mu$), the signs of the real part of the four eigenvalues are given. A pair of complex conjugate eigenvalues is denoted by (+) if unstable, and by (−) if stable. Larger symbols correspond to eigenvalues (7.33), smaller ones to Eq. (7.32). The dashed curve is the parabola $\mu'^2/\mu = q^2 \delta' \delta_R/\alpha_R$ [Eq. (7.35)].

Note that $q = 1.25$ here and $q = -5.36$ in F&R, such that $| q | > 1$ in both cases, and we are in the situation sketched in Fig. 7.13(b). We now concentrate on the case $\mu > 0$ (and $\mu' < 0$), i.e. the only quadrant where SS may be stable. According to Eq. (7.33), when $| q | > 1$, the SS solution is stable in the range $-4 > \chi > -2(1 + q^2)$ $(= -5.13$ here, and -59.4 in F&R). This defines a sector of stability in Fig. 7.13(b). At $\chi = -4$, the SS solution undergoes a Hopf bifurcation, leading to time-dependent behaviors that diverge in the dynamics described by the quadratic system (7.29). Re-including cubic terms leads to 3-torus solutions (as discussed by F&R). However, these behaviors are not investigated here, as they are not fully governed by the quadratic system. At the opposite end of the stability interval, i.e. at $\chi = -2(1 + q^2)$, the SS solution undergoes a bifurcation to an AS solution (which starts to exist there and is stable in some range provided $|q| > 1$).

When $|\chi|$ is increased (χ is decreased), the stable AS solution may lose stability itself, or remain stable up to $\chi \to -\infty$. This again depends on the value of q. A numerical evaluation of eigenvalues characterizing the stability of AS shows that it remains stable for every $\chi <$ $-2(1 + q^2)$ provided $|q| > 2.297$, while it becomes unstable at a certain χ^* for $|q| < 2.297$. Thus, the F&R results differ from ours at this stage, because AS remains stable in their analysis (except for very large $|\chi|$ where the quadratic system is not valid, and cubic terms destabilize the asymmetric mode AS, leading to noise-sustained oscillations [475]). Here, the AS solution undergoes a Hopf bifurcation at $\chi^* = -5.65$. No stable fixed point of the quadratic system (7.29) exists beyond that point, such that time-dependent motions occur, that first take the form of limit cycles centered around both unstable AS foci [see Fig. 7.14(a)]. Note that these cycles are mirror images of each other with respect to the left–right symmetry (symbolized by the dot-dashed line in Fig. 7.14).

The scenario of transition to chaos that we now describe [obtained by numerical integration of the *quadratic* system (7.29) with a fourth-order Runge–Kutta procedure with relative accuracy 10^{-10} ensured by the timestep control] has already been encountered in Lorenz-like models of convection [477], and was first postulated by Arneodo et al. [478].

For larger $|\chi|$, both cycles of Fig. 7.14(a) grow symmetrically and approach the unstable SS solution (saddle point). At a certain χ (say χ_0^H), the periodic trajectories tend to homo-clinic orbits (saddle connections) of SS, approaching it infinitely close along one of its stable directions (the least stable), and leaving it along the unstable direction. At this point [see Fig. 7.14(b)], the period of each cycle is infinite, and the two cycles merge into a new "double-loop" cycle (subsequently denoted C-2), via a mechanism known as homoclinic gluing [479] bifurcation. Increasing $|\chi|$ again leads to a deformation of the symmetric C-2 cycle [see Fig. 7.14(c)], followed by a symmetry-breaking bifurcation (say, at $\chi = \chi_1^{SB}$), leaving the symmetric C-2 cycle unstable to two stable dissymmetric C-2 cycles [see Fig. 7.14(d)].

Increasing $|\chi|$ again leads to a new homoclinization of these stable cycles at the SS saddle point. At this moment [at $\chi = \chi_1^H$, see Fig. 7.14(e)], both C-2 cycles glue into a C-4 cycle (4 loops). The resulting cycle is symmetric in a first stage [Fig. 7.14(f)], but undergoes a symmetry-breaking bifurcation at χ_2^{SB}. At χ_2^H [Fig. 7.14(g)], both C-4 cycles glue into a C-8 cycle, and the process repeats, the values χ_n^{SB} (at which the C-2^n cycle becomes dis-symmetric) and χ_n^H (at which the two C-2^n cycles glue) converging geometrically to a value estimated at $\chi_\infty = -135$, which corresponds to the onset of chaos [the convergence rate is estimated at $(\chi_{n+1}^H - \chi_n^H)/(\chi_n^H - \chi_{n-1}^H) \to 2.6$]. Fig. 7.14(h) exhibits a magnification of the strange attractor (at $\chi = -150 < \chi_\infty$) near the SS saddle point. Note that this route to chaos actually differs from the more classical period-doubling sequence in many respects [477, 478, 479, 480] (divergence of the period of orbits at bifurcation points χ_n^H, dependence of the geometrical convergence rate on the saddle index leading to ratios different from the Feigenbaum ratio, and importance of symmetry and symmetry-breaking effects).

The behaviors just described have been obtained from the quadratic system (7.29), and are thus strictly valid near the CTP, i.e. in the limit $\mu, \mu' \to 0, \chi = O(1)$ [this has been checked from a simulation of the full cubic system (7.27)]. The result is that a transition to chaos occurs when $|\chi|$ is increased. It would thus be tempting to conclude that at the threshold of

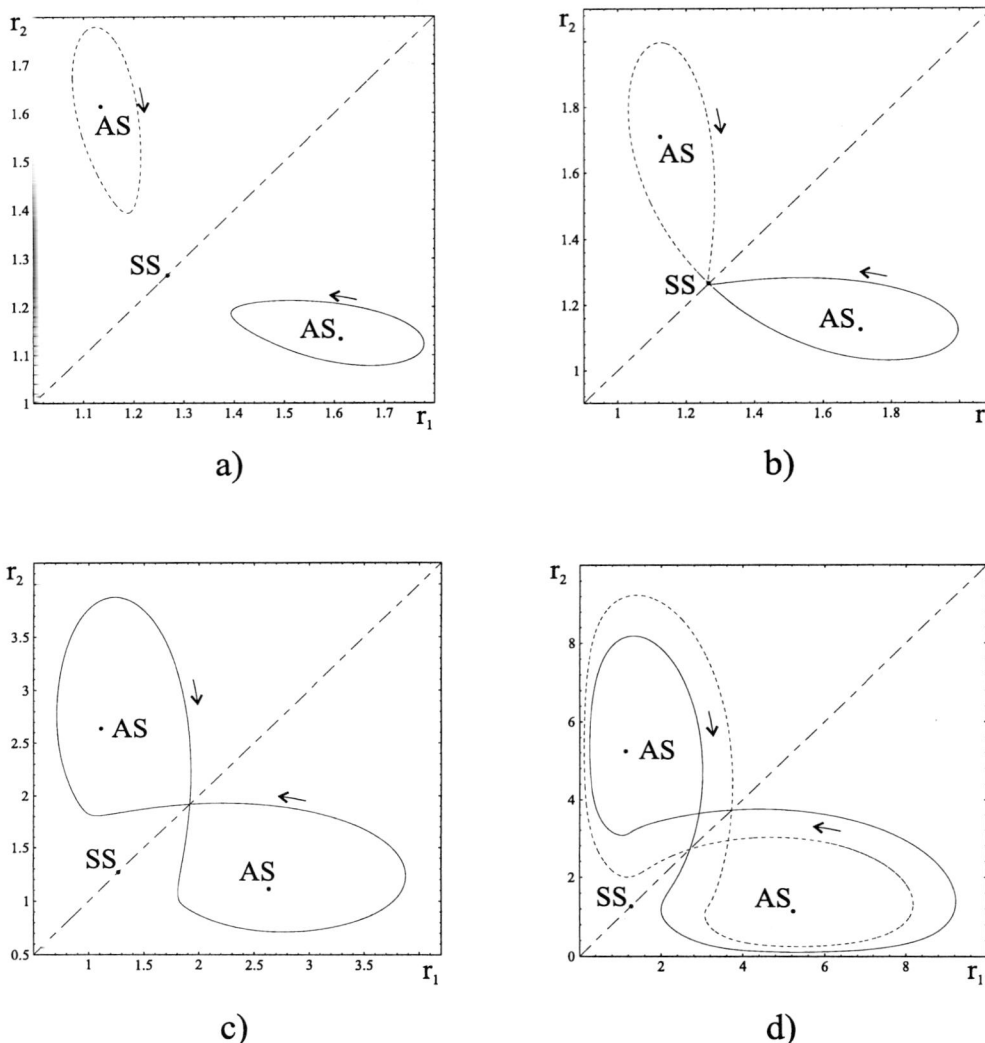

Figure 7.14: Time-dependent behaviors of the quadratic system occurring for $\chi < \chi^* = -5.55$, projected in the (r_1, r_2) plane. The left–right symmetry is represented as a dot-dashed line. Only stable cycles are represented. Fixed points (unstable) are the saddle point SS (symmetric mode), and the saddle-foci AS (asymmetric modes). a) Limit cycles (mirror images with respect to the left–right symmetry) for $\chi^* > \chi > \chi_0^H$. b) Saddle connections (homoclinic orbits) of the SS saddle point at $\chi = \chi_0^H = -6.09$. c) Symmetric double-loop cycle (C-2) for $\chi_0^H > \chi > \chi_1^{SB}$. d) Two stable dissymmetric C-2 cycles (mirror images) for $\chi_1^{SB} > \chi > \chi_1^H$.

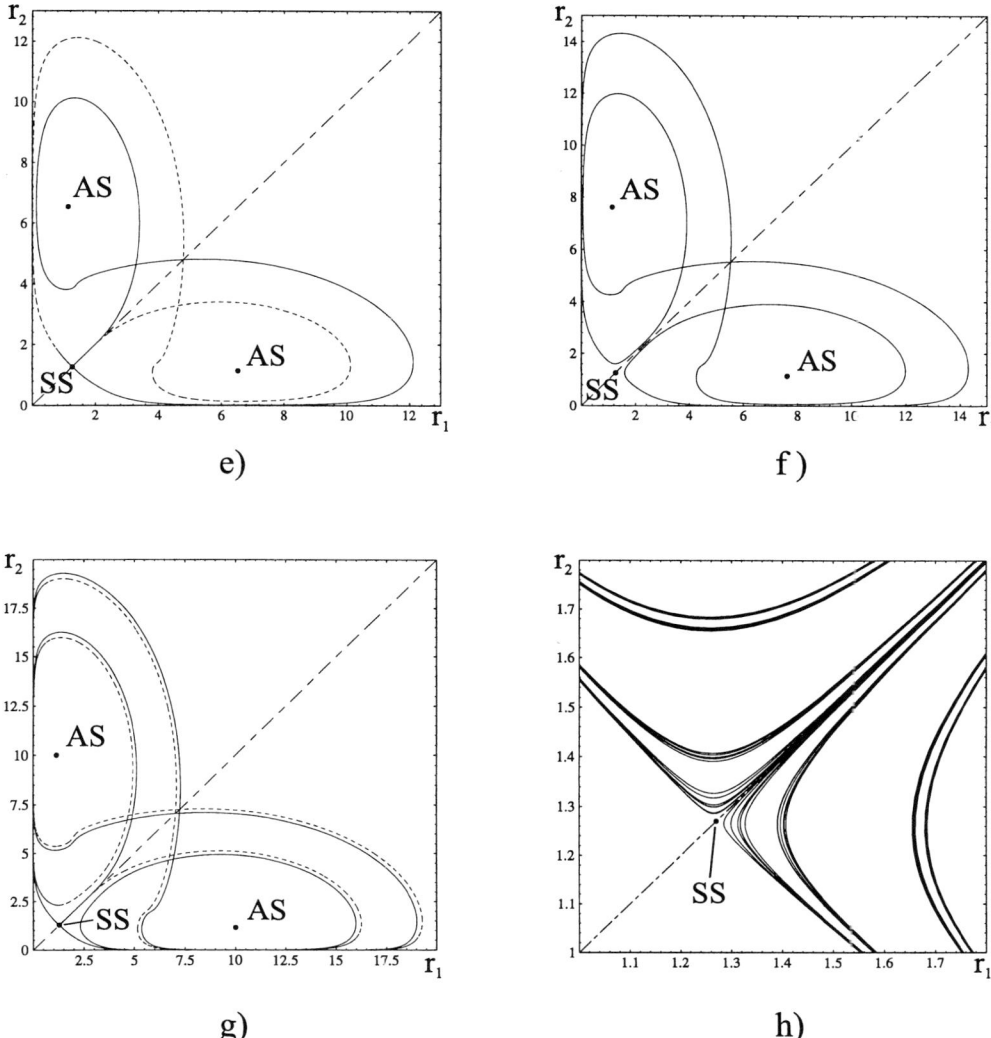

Figure 7.14: (continued): e) Homoclines of the saddle point SS at $\chi = \chi_1^H = -45.0$. f) Symmetric C-4 cycle for $\chi_1^H > \chi > \chi_2^{SB}$. g) Two dissymmetric C-4 cycles at the point $\chi = \chi_2^H = -101.5$ of homoclinic bifurcation. h) Trajectories in the vicinity of the saddle point in the chaotic regime ($\chi = -150$).

instability of the motionless state ($\mu = 0, \mu' < 0$), the system undergoes a direct transition to chaos. We will now show that this surprising behavior is a consequence of the truncation to the quadratic system (7.29), which may be resolved by reintroducing cubic terms. Thus, the following developments aim to determine what is the scenario of bifurcations leading from the motionless state to the above-described chaotic behaviors, when *increasing* the driving constraint, i.e. following arrows such as those represented in Fig. 7.10. It would be rather unusual that chaos occurs at the first bifurcation point, i.e. at $\mu = 0, \mu' = O(1) < 0$. In this limit ($\chi \to -\infty$), the quadratic system (7.29) is no longer valid. To show this, note that one of the components of the AS solutions (7.31) diverges [$r_{1(2)} \sim (-\chi)^{1/2}$] for $\chi \to -\infty$. Using (7.26) to return to the original variables, the corresponding physical solution turns out to be finite [in fact proportional to $\mu' = O(1)$] for $\mu \to 0$. Thus, cubic terms need to be reintroduced in this limit, and we must reconsider Eqs (7.27).

B. Cubic system

The full system (7.27) admits "pure solutions" (among others [475]), such as the TW solution

$$a_3 = a_2 = 0, \ |a_1|^2 = r_1^2 = -\tilde{\alpha}_R^{-1} \sim (\mu')^{-1} \tag{7.34}$$

which is identical to Eq. (7.21). Note that the phase of a_1 varies linearly with time as $\dot{\varphi}_1 = \tilde{\alpha}_I r_1^2 = -\tilde{\alpha}_I/\tilde{\alpha}_R = -\alpha_I/\alpha_R$ (nonlinear frequency shift). As (7.34) bifurcates from the trivial state at $\mu = 0$, there must be a region near the $\mu = 0$, $\mu' = O(1) < 0$ axis where cubic terms are essential (and where the TW solution has lower amplitude than the AS solution). On the contrary, the quadratic system has been shown to determine the dynamics when $\chi = O(1)$ and $\mu, \mu' \to 0$ (in which case the AS solution is smaller than TW). Thus, the limit of validity of the quadratic system should occur in a region where amplitudes of TW and AS are comparable. Using Eqs (7.31) and Eq. (7.34), it is found that equality between amplitudes of AS and TW occurs for $\chi/q^2 \cos \varphi_\delta = \tilde{\alpha}_R^{-1}$ (in the limit $\mu, \mu' \to 0$), which gives

$$\frac{\mu'^2}{\mu} = \frac{q^2 \delta' \delta_R}{\alpha_R} \tag{7.35}$$

i.e. a parabola $\mu = 0.0177\mu'^2$ in the (μ, μ') plane [dashed curve in Fig. 7.13(b)].

We now discuss the stability of TW. For this solution, analytical calculation of eigenvalues is possible [475] by perturbing the system (7.27) according to $a_1 = (r_1 + b_1) \exp[-i\alpha_I t/\alpha_R]$, $a_2 = b_2$, $a_3 = b_3 \exp[-i\alpha_I t/\alpha_R]$, and linearizing with respect to b_i. We first get $\dot{b}_1 = -(1 + i\alpha_I/\alpha_R)(b_1 + \bar{b}_1)$, decoupled from equations governing b_2 and b_3, and leading to eigenvalues -2 (stable) and 0 (perturbation along the limit cycle). The equations for b_2 and b_3 lead to two other eigenvalues λ (governing stability) which satisfy

$$\left(\lambda - 1 + \frac{\bar{\beta}}{\alpha_R}\right)\left(\lambda - \chi + \frac{\gamma'}{\alpha_R} - i\frac{\alpha_I}{\alpha_R}\right) + \frac{\chi\bar{\delta}\delta'}{\alpha_R\mu'} = 0 \tag{7.36}$$

Two different limiting cases of this dispersion relation are studied. The first one is $\mu \to 0$, $\mu' = O(1) < 0$, i.e. near the bifurcation point from the motionless state ($\chi \to -\infty$). It is found that eigenvalues tend to $\lambda_1 = \chi < 0$ (stable) and $\lambda_2 = 1 - \bar{\beta}/\alpha_R + \bar{\delta}\delta'/\alpha_R\mu'$. These eigenvalues reflect the fact that for $\chi \to -\infty$, the a_3 amplitude is adiabatically slaved to a_1 and a_2. Indeed, in the third of Eqs (7.27), $a_3 \to a_1 a_2$ when $s = -1$ and $\chi \to -\infty$. Substituting this expression in the first two equations (7.27), and neglecting $|a_3|^2$ (fourth order), we get a standard TW/SW competition (§5.1.1) with cubic interaction coefficient $\tilde{\beta}+\exp[i\varphi_\delta]$ and self-interaction coefficient $\tilde{\alpha}$. Therefore, as $\tilde{\alpha}_R < 0$ here, the supercritical TW solution is stable against SW if $\tilde{\alpha}_R - \tilde{\beta}_R - \cos(\varphi_\delta) > 0$ (equivalently $\lambda_{2R} < 0$), and unstable in the opposite case. Thus, TW is stable at the bifurcation point of waves if $0 > \mu' > \delta_R\delta'/(\beta_R - \alpha_R) = -19.1$. As we study the vicinity of the CTP, we assume in the following that this condition is satisfied.

The second limiting case is $\chi = O(1)$, and $\mu, \mu' \to 0$ [where the quadratic system (7.29) is valid, and possesses attracting solutions depicted in Fig. 7.14]. Eq. (7.36) shows that $\lambda_{1,2} = \pm(\chi\bar{\delta}\delta'/\alpha_R\mu')^{1/2}$, and one of these eigenvalues has positive real part. TW solutions have thus lost stability in between the two different regions of the (μ, μ') plane that we have investigated. The threshold may be found by substituting $\lambda = i\Omega$ in (7.36), which in fact exactly leads to the condition (7.35) for $\mu, \mu' \to 0$. Thus, on the parabola $\mu = 0.0177\mu'^2$ [see Fig. 7.13(b)], the TW can be guessed to undergo a bifurcation to an AS solution, which is indeed confirmed by direct numerical integration of Eqs (7.27).

Concluding results and summary

We will now summarize the developments made in this section for the quadrant $\mu > 0, \mu' < 0$, and supplement them by numerical results obtained from the full cubic system (7.27). The first bifurcation from the motionless state occurs at $\mu = 0$ to a pure stable TW solution (of the cubic system), which undergoes a secondary bifurcation to an AS solution at $\mu = 0.0177\mu'^2$ [dashed parabola of Fig. 7.13(b)]. For values of $\mu > 0.0177\mu'^2$ (e.g. increasing the Marangoni number Ma), the bifurcated AS solution is at first stable, then undergoes a tertiary Hopf bifurcation, leading to a situation similar to Fig. 7.14(a) (note that the unstable TW solution is not represented in this figure, and would in fact lie much further from the origin than the represented SS and AS). For increasing values of μ, a transition to chaos occurs via an infinite sequence of bifurcations qualitatively similar to that described in Fig. 7.14 (which corresponds to decreasing values of μ). However, contrary to this last sequence, the transitions observed here when increasing μ cannot be described by quadratic nonlinearities alone, and should thus be considered with some precautions, in view of the discussion made above. Thus, the sequence observed when increasing Ma should accumulate on a curve in the (μ, μ') plane, which might be of some parabolic form $\mu \sim \mu'^2$ (although this has not been checked). Beyond this line, the system is in a chaotic state similar to Fig. 7.14(h). Now, when increasing Ma again, a *reverse* transition represented in Fig. 7.14 occurs [from Figs 7.14(h) to 7.14(a)], the remaining two limit cycles of Fig. 7.14(a) finally collapsing on the AS solutions (at $\chi = \chi^* = -5.65$). As shown above, AS stay stable for a short range [up to $\chi = -2(1+q^2) = -5.13$], and both AS states in turn collapse on the SS solution, which becomes the only (stable) steady solution at this point. It was also seen that at $\chi = -4$, this SS state undergoes a Hopf bifurcation to a quasiperiodic state (3-torus) not investigated further here (see also [475]).

Finally, we briefly describe some results obtained via numerical integration of the full set of equations (7.27) in other quadrants, illustrating the wide variety of behaviors possible for this resonant system. We insist on the qualitative validity of these results, due to the $O(1)$ magnitude of the quadratic coefficients (see also [86]). In the quadrant $\mu < 0, \mu' > 0$ (where no stable fixed point exists), the (r_1, r_2, r_3, φ) phase space trajectory eventually describes a limit cycle (with frequency $\omega' \ll \omega_c$), thus again leading to temporal quasi-periodicity of the physical variables (3-torus). The small value of the frequency ω' is due to the fact that the system spends the largest part of the oscillation period near the $r_1 = 0, r_2 = 0$ coordinate line (where the system is in a quasi-steady state), and undergoes quick deviations (similar to relaxation oscillations) at moments that may eventually become irregularly spaced in time. During these high frequency (ω_c) relaxation periods, the amplitude of the steady mode changes sign (the phase undergoes a jump of π), such that the system may be viewed as regularly switching between two quasi-steady states with reversed convective velocities. When destabilizing the system with respect to the oscillating mode (quadrant $\mu > 0, \mu' > 0$), intricate phase space trajectories revealing deterministic chaos occur, the complicated phase space structure of which would certainly deserve further investigation.

8 Strongly nonlinear regimes

The analyzes of weakly nonlinear behaviors presented in Chapters 4, 5 and 7 focused on the interactions between a few near-critical modes, with well-defined length scales determined by their critical wavenumbers. As a rule, the dynamical regimes were described by low-dimensional systems of amplitude equations. Even when considering modulation effects of patterns (§4.1.3 and §4.5) and of waves (§5.5), only small wavenumber bands around the critical wavenumber were considered. Still, complex behaviors have been predicted in some cases, including various scenarios of transitions to chaotic regimes. However, it remains to determine what is the link between this kind of spatio-temporal complexity, and other forms of stochastic behaviors of deterministic systems, including fully-developed turbulence [55, 469]. This central question of Nonlinear Physics has not yet received a definitive answer, and it is our opinion that the analysis and comparison of quite different forms of turbulent-like phenomena should represent as many valuable steps in this direction.

A priori, what seems to distinguish spatio-temporal chaos from turbulence is the range of length and time scales involved in the dynamics [55]. Accordingly, the present chapter is devoted to recent attempts to model the interaction between a large number of unstable modes in Marangoni–Bénard convection, i.e. typically when the constraint is far above the threshold and large wavenumber bands are excited. This is also motivated by a number of experimental studies, undertaken during the 1960s, which provided evidence of various kinds of strongly time-dependent surface-tension-driven motions (see e.g. [38, 39, 40]) at sufficiently high Marangoni number. Owing to their importance in Chemical Engineering, these phenomena have most often been observed in mass (rather than heat) transfer systems, typically in situations where two unequilibrated fluid phases are initially put in contact and allowed to evolve autonomously towards a new state of thermal and chemical equilibrium.

Such chaotic interfacial flows have been interpreted as resulting from several kinds of hydrodynamic instabilities generated by the Marangoni effect [75], and affecting the unidimensional diffusive solutions generated by the heat/mass transfer across the interface (see Chapter 3). It was also shown that such convective instabilities are generally able to drive substantial increases of the overall transport efficiencies [81]. In the spirit of the original paper of Sternling and Scriven [75], a number of linear stability analyzes have been conducted, including deformability and rheology of the interface, insoluble surfactants, energy barriers to mass transfer [191, 241], evaporation [245], ... and were also extended by weakly nonlinear theories (see Chapter 4). However, only a few recent works [198, 334, 492, 494, 502, 503, 504] have attempted to explore strongly nonlinear regimes of this class of problems.

The two sections of this chapter concern the Marangoni–Bénard problem in its simplest formulation. Namely, the layer of fluid is taken to be infinitely deep, since, as described in the text, the resulting short-wave effects may be expected to be essential at a large constraint. Moreover, the interface is assumed to be undeformable, the Prandtl (or Schmidt) number is taken as infinite, and buoyancy effects are neglected. Such a formulation allows the analysis to be focused on the role of the surface tension dependency upon a scalar field (either temperature or concentration), and on the nonlinear mechanism of advection of this field by the flow it generates. To simplify the presentation, we will exclusively refer to the thermal situation in what follows. Note also that the Navier–Stokes equation is linear (Stokes flow) in the infinite Prandtl number limit, resulting in a dynamical slaving of the velocity field to the surface temperature distribution [198].

Even with these simplifications, no rigorous method exists to solve the full nonlinear problem. Direct numerical simulations are possible [198], but have not yet been conducted at sufficiently high Marangoni number to study turbulent behaviors in this high Prandtl number limit (while at low Prandtl number, chaotic behaviors have been found numerically [204, 494], in relation with inertial convection regimes, see §4.5.6). Therefore, the goal of the models described in what follows is merely to reach qualitative agreement with the exact results at moderate Marangoni number (some comparison with numerical simulations at steady state is presented in §8.1.5) or with experiments, and to allow identification of the main physical mechanisms responsible for the regimes observed.

The slaving of the velocity field to the surface temperature distribution at infinite Prandtl number has been recently exploited by Thess and co-workers [492, 502], who have proposed a model of infinite Marangoni number convection, in the form of a closed equation describing the evolution of the surface temperature field. Although this model does not take into account the energy input from the reference state temperature gradient, it predicts the formation of small-scale structures of the surface temperature field, a nonlinear feature shared by most experiments. An extension of this model is proposed in §8.2, which incorporates the possibility of boundary-layer instabilities, and leads to self-sustained chaotic regimes, also typical of experiments conducted at very high Marangoni number. Still, the progress accomplished in modeling these regimes does not yet allow anything to be concluded about their possible link either with fully-developed hydrodynamic turbulence, or with weaker forms of spatio-temporal chaos. Therefore, the so-called "interfacial turbulence" is here studied as a nonlinear phenomena in itself, whose generic features will certainly be subject to further analysis.

8.1 Finite-amplitude regimes of thermocapillary convection

In this first section, a model of thermocapillary instability is developed in the framework of the amplitude equations formalism. From a technical point of view, it will appear that the use of eigenfunctions at a given Marangoni number Ma (rather than those evaluated at the critical one) as a basis for the nonlinear problem, leads to amplitude equations with coefficients dependent upon Ma, which can be extrapolated rather high above the threshold. In particular, this will allow examination of the important wavelength selection problem between fastest growing modes (wavenumbers around $k_{max} \sim Ma^{1/2}$ for a zero free surface Biot number) and critical modes ($k_c = 0$ and $Ma_c = 0$). Actually, it will be seen that transient numerical integration of the equations leads to an unbounded growth of the mean wavelength, quite similarly to what was obtained for dilational waves in §6.2. Thus, at least in the range of

Marangoni numbers[1] considered in this first section (up to $Ma = 4000$), the system appears unable to generate a permanent regime with an intrinsic wavelength independent of additional data such as layer depth or lateral extension. Note already that this conclusion does not hold at still higher Marangoni numbers (for which the model developed here is not valid), as will be seen in §8.2.

8.1.1 Linear results

Figure 8.1 reproduces the neutral stability results of Pearson [101] for the Marangoni–Bénard problem, as well as results obtained by Scanlon and Segel [103] in the case of a layer of infinite depth (see §3.3.2). At a given Ma, perturbations with wavenumbers in the range lying above the neutral stability curve possess a positive amplification rate. There also exists a particular wavenumber k_{max} in this range which possesses the maximal amplification rate. In slightly supercritical conditions ($Ma \simeq Ma_c$), k_{max} is close to the critical wavenumber k_c, and does indeed predict the size of convective (hexagonal) cells observed experimentally. When Ma is increased, k_{max} is seen to increase, and the prediction of the selected wavelength becomes quite complicated, since it involves nonlinear competition between modes in the unstable range. For Ma increasing further, k_{max} can be made as large as desired compared with the critical wavenumber k_c (Fig. 8.1).

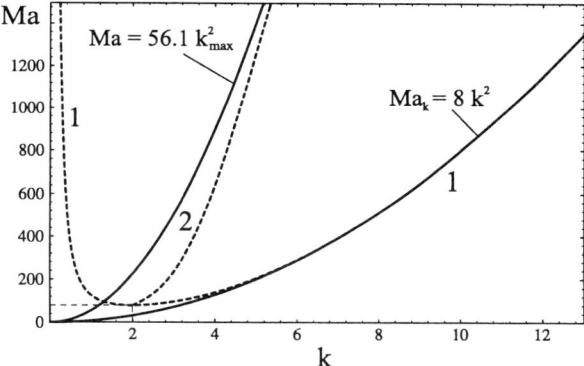

Figure 8.1: Linear stability results (curves 1: neutral stability boundaries; curves 2: loci of fastest growing wavenumbers) for the finite-depth case (dashed curves) and the infinite-depth case (full curves), for a zero Biot number. The critical conditions for the finite-depth case are defined by the critical Marangoni number $Ma_c = 79.6$ and the critical wavenumber $k_c = 1.99$. Formulas (infinite-depth case) are given for the neutral stability curve Ma_k [Eq.(8.14)] and for the fastest growing wavenumber k_{max} [Eq.(8.15)].

In the present section, our first goal will be to examine the possibility of some stable

[1] As the layer depth is here assumed large compared with the typical size of convective motions, the length scale considered in the definition of the Marangoni number is actually chosen as the lateral periodicity length, as seen later on in this section.

finite amplitude regime (steady or time-dependent) dominated by the fastest growing perturbations. After all transients have been damped out, will the convective structure be localized near the interface (as is the case for linear perturbations with $k \simeq k_{\max}$, see §3.3.2), or will it be depth-scaled ($k \simeq k_c$), as in the weakly nonlinear regime? In addition to its fundamental interest for turbulence[2], this question is related to the study of the similarities existing between non-equilibrium instabilities and equilibrium phase transitions [19]. Like some equilibrium phenomena, "non-equilibrium phase transitions" are sometimes characterized by intrinsic wavelengths, such as the Turing instability in reaction-diffusion systems [487], or the Kelvin–Helmholtz instability in Hydrodynamics [95]. For these instabilities, the stability boundary possesses a minimum of the control parameter at a finite wavenumber independent of the actual experiment size. It is seen in Fig. 8.1 that this is not the case for the Marangoni–Bénard instability (the critical wavenumber in a finite layer of depth h scales as $1/h$, just as for the Rayleigh–Bénard instability). However, we may conjecture that this does not rule out the possibility of an intrinsic wavelength, linked to the presence of fastest growing modes (as also seems to be the case for the Rayleigh–Taylor instability [488]). Note that the finite wavenumber of the fastest growing mode generally depends on the driving force amplitude (the thermal gradient in our case). Attempts to answer to the question of the preference of such modes at a given supercritical driving force obviously have to incorporate nonlinear effects in the analysis.

8.1.2   Problem formulation – Extrapolation of Landau equations

We will consider a semi-infinite viscous Boussinesq incompressible fluid in contact with an inert gas phase. The interface is located at the $z = 0$ coordinate plane of a cartesian reference frame with unit vectors $\vec{1}_i (i = x, y, z)$, and is assumed undeformable (this will allow analytical results to be obtained, and is justified since interfacial deformation is known to primarily affect long-wave modes, as seen in §3.6). The fluid is located in the domain $z < 0$, and a constant heat flux is injected in the system (a constant temperature gradient $-\beta$ is maintained at $z \rightarrow -\infty$). All equations and boundary conditions are scaled by a length scale d (arbitrary for the moment), a thermal time scale d^2/κ, a temperature scale βd and a pressure scale $\mu\kappa/d^2$. The Marangoni number $Ma = -\sigma_T \beta d^2/\mu\kappa$ is defined with respect to the length d, instead of the fluid thickness h ($h/d \rightarrow \infty$). Let $\vec{V} = \vec{V}_{\mathrm{h}} + W\vec{1}_z$ be the fluid velocity ($\vec{V}_{\mathrm{h}}$ is the horizontal velocity), T the temperature and p the pressure perturbations with respect to the purely conductive reference solution. A solution vector U is then defined by

$$U(\vec{r} = x\vec{1}_x + y\vec{1}_y, z, t) = \begin{pmatrix} \vec{V}_{\mathrm{h}} \\ W \\ p \\ T \end{pmatrix} \tag{8.1}$$

which is assumed to belong to a certain set, say E, of sufficiently derivable functions satisfying the boundary conditions of the problem: these are

$$\vec{V}_{\mathrm{h}}, W, DT, p \rightarrow 0 \quad \text{for } z \rightarrow -\infty \tag{8.2}$$

---

[2]For instance, phase turbulence in the Kuramoto–Sivashinsky equation (5.121) is known to be dominated by the scale of fastest growing modes of the linear problem [290], once the latter is much smaller than the system size.

$$W = DT + BiT = 0 \quad \text{for } z = 0 \tag{8.3}$$

where D is the dimensionless z-derivative and Bi is the free surface Biot number $Bi = \alpha d/\lambda$ (α is the free surface heat transfer coefficient and λ the thermal conductivity of the fluid).

As in §4.2.2, the system of partial differential equations for the solution vector U can be written under the general operational form

$$\mathcal{L}(U) = Ma\, M(U) + \Theta\left(\frac{\partial U}{\partial t}\right) + N(U, U) \tag{8.4}$$

where the linear part $\mathcal{L}(U)$ is given by

$$\mathcal{L}(U) = \begin{pmatrix} \Delta\vec{V}_h - \vec{\nabla}_h p \\ \Delta W - Dp \\ DW + \vec{\nabla}_h.\vec{V}_h \\ \Delta T + W \\ [D\vec{V}_h]_{z=0} \end{pmatrix} \tag{8.5}$$

the "evolution part" is defined as

$$\Theta(U) = \begin{pmatrix} 0 \\ 0 \\ 0 \\ T \\ 0 \end{pmatrix} \tag{8.6}$$

and the "constraint part" by

$$M(U) = \begin{pmatrix} 0 \\ 0 \\ 0 \\ 0 \\ [-\vec{\nabla}_h T]_{z=0} \end{pmatrix} \tag{8.7}$$

Finally, the bilinear form N is expressed as

$$N(U_1, U_2) = \begin{pmatrix} 0 \\ 0 \\ 0 \\ \vec{V}_1.\vec{\nabla}T_2 = \vec{V}_{1h}.\vec{\nabla}_h T_2 + W_1 DT_2 \\ 0 \end{pmatrix} \tag{8.8}$$

In the above relations, $\vec{\nabla}_h = \vec{1}_x \partial/\partial x + \vec{1}_y \partial/\partial y$ is the horizontal gradient, $\vec{\nabla} = \vec{\nabla}_h + \vec{1}_z D$ is the full gradient, and $\Delta = \vec{\nabla}^2$ is the Laplacian operator.

Apart from the fact that pressure is not directly eliminated from the equations, the problem (8.4) together with boundary conditions (8.2–8.3) is equivalent to the problem formulation of Scanlon and Segel [103]. Note also that we have included the Marangoni condition as the last component of (8.4), which has already been shown to simplify the process of deriving amplitude equations [103, 104, 130, 150, 151].

Derivation of amplitude equations

We first decompose U into Fourier modes

$$U(\vec{r}, z, t) = \int U_{\vec{k}}(z, t) \exp(i\vec{k}.\vec{r}) d\vec{k} \tag{8.9}$$

so that horizontal Fourier components $U_{\vec{k}}$ all belong to E [i.e. fulfill boundary conditions (8.2–8.3)] and satisfy

$$\mathcal{L}_{\vec{k}}(U_{\vec{k}}) = Ma\, M_{\vec{k}}(U_{\vec{k}}) + \Theta_{\vec{k}}\left(\frac{\partial U_{\vec{k}}}{\partial t}\right) + \int N_{\vec{k}', \vec{k}-\vec{k}'}(U_{\vec{k}'}, U_{\vec{k}-\vec{k}'}) d\vec{k}' \tag{8.10}$$

which is obtained by projecting (8.4) on $\exp(-i\vec{k}.\vec{r})$ and by replacing $\vec{\nabla}_{\mathrm{h}}$ by $i\vec{k}$ in linear operators (this is the meaning of the index $\vec{k}$). The bilinear form $N_{\vec{k}_1, \vec{k}_2}$ is defined in a similar way. Each Fourier mode is further decomposed as

$$U_{\vec{k}}(z, t) = A_{\vec{k}}(t) U_{\vec{k}}^\sigma(z) + U_{\vec{k}}^D(z, t) \tag{8.11}$$

where $U_{\vec{k}}^\sigma(z)$ is an eigenvector with eigenvalue σ_k ($k = |\vec{k}|$) of the linear spectral problem

$$\mathcal{L}_{\vec{k}}(U_{\vec{k}}^\sigma) - Ma M_{\vec{k}}(U_{\vec{k}}^\sigma) = \sigma_k \Theta_{\vec{k}}(U_{\vec{k}}^\sigma) \tag{8.12}$$

The resolution of (8.12), detailed in Appendix H, shows that for any $Ma > 0, 0 < k < Ma/2Bi$, an isolated eigenvalue σ_k exists (and is such that $\sigma_k + k^2 > 0$). This eigenvalue is the growth rate of the corresponding eigenmode $U_{\vec{k}}^\sigma(z)$, appearing in Eq. (8.11). For every value of Ma and k, there also exists a continuum of solutions of (8.12) that are bounded for $z \to -\infty$ (and which correspond to eigenvalues $\sigma \le -k^2$). This infinite set of solutions could eventually be used to develop the remainder term $U_{\vec{k}}^D(z, t)$ (the superscript D stands for "damped"), but it turns out to be simpler to compute $U_{\vec{k}}^D$ approximatively, by a method explained in Appendix H. As it turns out (see also [209]), eigenvalues σ_k are real and satisfy

$$Ma = \frac{2}{k}\left(-2k\sigma^{-2} + \frac{\sigma^{-1} + 2k^2\sigma^{-2} + 2kBi\sigma^{-2}}{Bi + \sqrt{\sigma + k^2}}\right)^{-1} \tag{8.13}$$

whose limit for $\sigma \to 0$ yields the neutral stability relation

$$Ma_k = 8k(k + Bi) \tag{8.14}$$

which is the asymptotic form ($k \to \infty$) of the neutral stability condition of Pearson [101], as also seen from Fig. 8.1. A relation between the maximal eigenvalue $\sigma_{\max}$, $k_{\max}$ and Ma may be found by differentiating (8.13) at constant Ma, setting $\partial\sigma/\partial k = 0$. For instance, at $Bi = 0$, this gives

$$\sigma_{\max} = 2(1 + \sqrt{2})k_{\max}^2 = \alpha Ma \quad \text{with } \alpha \simeq 0.086 \; (Bi = 0) \tag{8.15}$$

Although analytical results can also be obtained for $Bi \ne 0$, they are not reproduced here for conciseness. As commented by Scanlon and Segel [103], the minimum (critical) value of

Ma for which instability occurs is zero, due to the absence of stabilization by a rigid lower boundary of the modes with increasingly large wavelength. However, owing to their large inertia, the growth rate of these modes is vanishingly small for all Marangoni numbers. This is depicted in Fig. 8.2: it is seen that modes with wavenumbers between 0 and $k^* = (Ma/8)^{1/2}$ (for $Bi = 0$) are unstable, so that their amplitude $A_{\vec{k}}(t)$ in the decomposition (8.11) should grow exponentially with time, as long as nonlinear effects can be neglected.

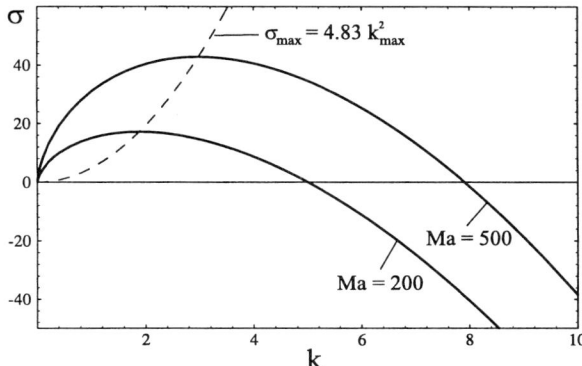

Figure 8.2: The growth rate σ as a function of the wavenumber $\vec{\kappa}$ for different Marangoni numbers Ma, and for $Bi = 0$. The dashed line represents the locus of the fastest growing perturbations, given by Eq. (8.15).

In fact, it is shown in Appendix H that the amplitudes $A_{\vec{k}}(t)$ obey evolution equations of the form

$$\frac{\partial A_{\vec{k}}}{\partial t} = \sigma_k A_{\vec{k}} + \int Z_{\vec{k}'\vec{k}} A_{\vec{k}'} A_{\vec{k}-\vec{k}'} d\vec{k}' + \iint Z_{\vec{k}'\vec{k}''\vec{k}} A_{\vec{k}'} A_{\vec{k}''} A_{\vec{k}-\vec{k}'-\vec{k}''} d\vec{k}' d\vec{k}''$$
$$(8.16)$$

These equations are strictly valid near the threshold. When the Marangoni number is increased, higher-order terms should be included. Equations (8.16) may then be considered as resulting from a truncated modified Galerkin scheme (see §4.2.3). Another hypothesis underlying the derivation of (8.16) is that the dynamics of damped modes (i.e. of $U_{\vec{k}}^D$) is determined by the evolution of the "primary" modes $A_{\vec{k}}(t)U_{\vec{k}}^{\sigma}$ (this amounts to neglecting time derivatives of damped modes). This slaving principle [51, 150, 151], strictly valid near the threshold, is here assumed to be qualitatively valid in the strongly nonlinear regime. This can be partly justified by the fact that damped modes cannot bifurcate ($\sigma < -k^2$), whatever the value of Ma (see also [489]).

Despite these assumptions, our model is expected to reflect physical reality even far from the threshold, provided that the eigenmodes $U_{\vec{k}}^{\sigma}$ are used for the Galerkin basis, rather than the neutral stability functions $U_{\vec{k}}^0$. In order to illustrate the differences between these different approaches, we now turn to the derivation of weakly nonlinear results [103, 104, 130, 150, 151], for which the latter option is sufficient.

Weakly nonlinear results

Making use of $U_{\vec{k}}(z,t) = A_{\vec{k}}(t)U_{\vec{k}}^0(z) + U_{\vec{k}}^D(z,t)$, instead of Eq. (8.11), and following a procedure similar to that described in Appendix H, we are left with amplitude equations identical to Eq. (8.16), although with different coefficients. As for classical perturbation expansions of §4.2.2, coefficients of the quadratic and cubic terms do not depend on Ma. It can also be shown that the coefficient of the linear term is the first term of a Taylor expansion of $\sigma_k(Ma)$ around $Ma = Ma_k$, i.e.

$$\sigma_k^0 = \frac{k(Ma - Ma_k)}{4(2k + Bi)} \tag{8.17}$$

where a superscript 0 subsequently denotes a value of a coefficient computed by using neutral stability functions.

From Eqs (8.9) and (8.11), it is seen that a roll mode with wavevector $\vec{k}_0$ is described by $A_{\vec{k}} = a_1(t)\delta(\vec{k} - \vec{k}_0) + \bar{a}_1(t)\delta(\vec{k} + \vec{k}_0)$, where δ is the Dirac-delta function, and an overbar denotes the complex conjugate. Substituting into (8.16) leads to the Ginzburg–Landau equation

$$\frac{\partial a_1}{\partial t} = \sigma_{k_0}^0 a_1 + (2Z_{111}^0 + Z_{-111}^0)a_1|a_1|^2 \tag{8.18}$$

where Z_{111}^0 and Z_{-111}^0 stand for $Z_{\vec{k}_0\vec{k}_0\vec{k}_0}^0$ and $Z_{-\vec{k}_0\vec{k}_0\vec{k}_0}^0$ respectively.

Defining a reduced distance to the threshold by

$$\epsilon = (Ma - Ma_{k_0})/Ma_{k_0} \tag{8.19}$$

it is seen that at $\epsilon = 0$, the rest state $a_1 = 0$ undergoes a pitchfork bifurcation to the steady amplitude

$$|a_{1s}| = \left[\frac{32(Bi + 3k_0)}{(Bi + k_0)(39Bi^2 + 248Bik_0 + 353k_0^2)} \epsilon \right]^{\frac{1}{2}} \tag{8.20}$$

obtained after evaluation of the cubic coefficients (see details in Appendix H).

Although strictly valid near the threshold, the limitations of this weakly nonlinear model for large ϵ are well known. Consider for example the temperature perturbation averaged in the horizontal direction (i.e. its $k = 0$ Fourier component):

$$\langle T \rangle = T_{\vec{k}=0}^D(z) = 2|a_1|^2 T_{01}^D(z) = O(\epsilon) \tag{8.21}$$

where $T_{01}^D(z)$ is the only non-zero component of $U_{01}^D(z)$ (Appendix H). The total averaged temperature profile is obtained by adding the reference profile $-z$ to (8.21), and is represented in Fig. 8.3, for several values of ϵ.

It is seen that in a region of depth $O(1/k_0)$ below the interface, the temperature profile is distorted (and somewhat homogenized) by Marangoni convection. It is also seen that unrealistic temperature distributions (strongly negative values of the mean temperature, leading to

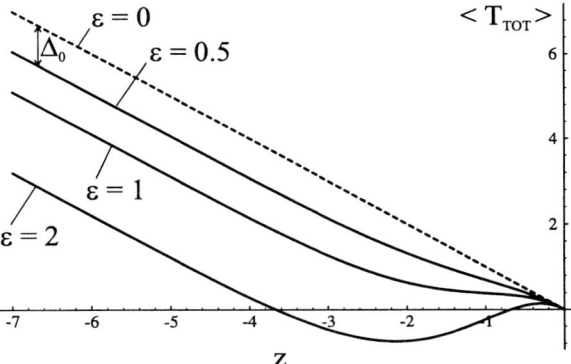

Figure 8.3: The total temperature $< T_{\mathrm{TOT}} >$ averaged in the horizontal plane as a function of the vertical coordinate z, as computed from weakly nonlinear results, and for different values of $\epsilon = (Ma - Ma_{\mathrm{c}})/Ma_{\mathrm{c}}$. The Biot number is $Bi = 0$, and the basic wavenumber is $k_0 = 1$. The distortion of the averaged temperature profile in the convective region near the interface $z = 0$ creates a decrease Δ_0 of the bulk temperature with respect to the purely conductive value (dashed line, $\epsilon = 0$). Δ_0 is here defined for $\epsilon = 0.5$.

large unrealistic cold spots in steady regimes) are obtained for ϵ superior to about 1.5. Defining Δ as the bulk temperature decrease with respect to its value in the conductive rest state, we may compute that

$$\Delta^0 = -2|a_{1s}|^2 \lim_{z \to -\infty} T_{01}^D(z) = \frac{32(Bi + 3k_0)(5Bi + 7k_0)}{k_0(39Bi^2 + 248k_0Bi + 353k_0^2)}\epsilon \qquad (8.22)$$

where the superscript 0 again denotes the weakly nonlinear result. The result (8.22) of course does not saturate for $\epsilon \to \infty$.

The importance of obtaining a better approximation of the bulk temperature decrease Δ is justified by the fact that it can be considered as equivalent to the classical Nusselt number Nu (more exactly to $Nu - 1$, which is also quadratic in the amplitudes near the instability threshold). Indeed, for systems in which the temperature difference is kept constant (such as Rayleigh–Bénard convection between conducting boundaries), Nu is defined as the dimensionless ratio of the total to the conductive heat flux, and therefore is a measure of the increase of the heat flux due to convection. For systems in which the heat flux is kept constant (as in the Marangoni–Bénard instability at $Bi = 0$), the decrease of the temperature difference between bulk and interface may also be perceived as an increase of the apparent thermal conductivity of the system, due to Marangoni–Bénard convection (see also the discussion of §4.3.1 and 4.4.2). In the next sections, it is shown that by using eigenvectors solution of (8.12) instead of neutral stability functions, a more realistic description of convective fields for large Ma can be obtained, including a saturation of the bulk temperature decrease at large Marangoni number.

8.1.3   Numerical results, mean-field approximation and physical interpretations

In this section, we will present results obtained by direct numerical integration of the set (8.16), for a two-dimensional domain of lateral length $L = 2\pi/k_0$ with periodic boundary conditions. The amplitude of the Fourier modes is given by

$$A_{\vec{k}}(t) = \sum_{n=-N, n\neq 0}^{+N} a_n(t)\delta(\vec{k} - n\vec{k}_0) \tag{8.23}$$

with $\bar{a}_n(t) = a_{-n}(t)$, and N sufficiently large to ensure numerical convergence. Substituting (8.23) in the relation (8.16) leads to

$$\frac{\partial a_m}{\partial t} = \sigma_m a_m + \sum_p Z_{p,m} a_p a_{m-p} + \sum_{p,q} Z_{p,q,m} a_p a_q a_{m-p-q} \tag{8.24}$$

where σ_m stands for σ_{mk_0}, $Z_{p,m}$ for $Z_{p\vec{k}_0,m\vec{k}_0}$ and $Z_{p,q,m}$ for $Z_{p\vec{k}_0,q\vec{k}_0,m\vec{k}_0}$, which are calculated as a function of the Marangoni number Ma.

In the following, we will take advantage of the fact that the length scale d of the problem is still arbitrary. We may thus choose $k_0 = 1$, which means that the dimensional length of the periodic box is $2\pi d$. From Eq. (8.14), the critical Marangoni number is given by $Ma_c = Ma_{k=1} = 8(1 + Bi)$.

The system of Eqs (8.24) has been integrated for a wide range of Marangoni and Biot numbers. Despite the large number of unstable modes in some cases (increasing with Ma), the long-term behavior appears to be surprisingly simple: a steady state is always reached, which is strongly dominated by the fundamental mode $n = 1$. Since the number of modes N needed to ensure convergence increases with Ma (see Fig. 8.2), it actually turns out to be too expensive in computer time to run simulations above $Ma = 500$ (for $Bi = 0$, and $N = 20$).

Still, it is possible to consider a simplified version of the system (8.24), which allows simulating the evolution of a larger number of amplitudes. This model is obtained by setting all cubic coefficients $Z_{p,q,m}$ with $p \neq m$ equal to zero. We then obtain

$$\frac{\partial a_m}{\partial t} = \left(\sigma_m - \sum_{q=1}^{N} S_{m,q}|a_q|^2\right) a_m + \sum_p Z_{p,m} a_p a_{m-p} \tag{8.25}$$

with $S_{m,q} = -2Z_{m,q,m}$. From Eq. (8.25), this quantity is seen to represent the strength with which the presence of the mode q lowers the effective growth rate of the mode m. The physical mechanism responsible for this stabilizing effect consists in the distortion of the mean temperature profile by convection (see Fig. 8.3), which lowers the destabilizing temperature gradient. A comparison of the time evolution and of the steady state values predicted by the full system (8.24) and the reduced set (8.25) reveals that the results differ only slightly (by less than 10% of the value of typical convective quantities at steady state, as shown in Fig. 8.7). In view of this rather good concordance, the mean temperature profile distortion by convection may be considered as a dominating effect in the nonlinear competition between unstable modes, which justifies the use of the "mean-field" system (8.25) to investigate larger

Marangoni number regimes. Note that such mean-field approximations have also been used in purely buoyancy-driven convection [23].

Again, even for Marangoni numbers as large as 4000 (for $Bi = 0$, i.e. $\epsilon \simeq 500$, and $N = 75$), the long-term behavior is not modified, independently of the initial conditions (here selected as a numerical "white noise", i.e. randomly chosen complex amplitudes of magnitude 10^{-8} to 10^{-3}) : the final state is still steady and dominated by the fundamental mode.

A sequence of a typical transient simulation is represented in Fig. 8.4. For sufficiently small initial perturbations, a convective structure dominated by the fastest linearly growing mode (the mode closest to $k_{\max}$) is observed after a relatively short time. This is the case as long as nonlinear effects can be neglected. At higher time intervals, this structure is progressively replaced by larger and larger wavelength structures, via a complex process of coalescence of neighboring convective cells. This evolution finally tends to the steady state with two convective cells occupying the entire domain (fundamental mode). Some properties of this steady state will be investigated in the next section.

It is interesting to compare Fig. 8.4 with the experimental observations made by Linde et al. [44] in mass transfer systems. Note that the latter systems actually correspond to high values of Bi, since the diffusion coefficients are generally much larger in the gas than in the liquid phase. However, our simulations were not found to be strongly dependent on Bi for large Ma. Linde observed a growth of the mean wavelength of the convective pattern, and interpreted it as an effect due to the non-stationary mass transfer occurring in his experiments. Indeed, after an experiment is started, for instance by contacting a gas phase containing a surface active solute with the liquid phase, the diffusion of this solute through the interface creates a growing diffusive boundary layer, which induces convective motions in the liquid, likely on a scale of the order of the boundary-layer thickness.

Since our formulation rather assumes that the basic temperature (or concentration) profile is already established throughout the entire layer before convection starts, the wavelength selection observed in Fig. 8.4 has to be intrinsically related to the nonlinear mechanism of heat (or mass) convective transport. This effect was indeed shown (see Fig. 8.3) to induce an homogenization of the temperature (or concentration) profile in a convective region located below the interface. This is also apparent in Fig. 8.5, which represents the temperature profile averaged along the horizontal direction corresponding to the evolution depicted in Fig. 8.4. It is seen that the temperature uniformization due to convection is more important at large times, when the penetration depth is large.

The growth mechanism can be explained by the following considerations: suppose that at one particular instant, the convective structure has a mean wavelength λ. Since the convective cells have to preserve a certain height/width ratio, temperature is almost homogenized in a region of depth λ below the interface (see Fig. 8.5). Modes with wavelengths smaller than λ may be considered as stable, since they see a nearly isothermal environment. On the contrary, modes with wavelengths larger than λ can penetrate sufficiently deep into the bulk of the liquid and bring hot fluid from the still conductive zone, to the interface. The effective growth rate of these modes remains nearly unchanged by the convective structure, so that these modes continue to grow (but slower and slower due to their growing inertia), and tend to replace smaller wavelength structures. The pattern wavelength λ may thus be expected to grow indefinitely, at least in an infinite system. In real experiments, the final wavelength will probably be determined by the actual depth of the experimental container (possibly near the

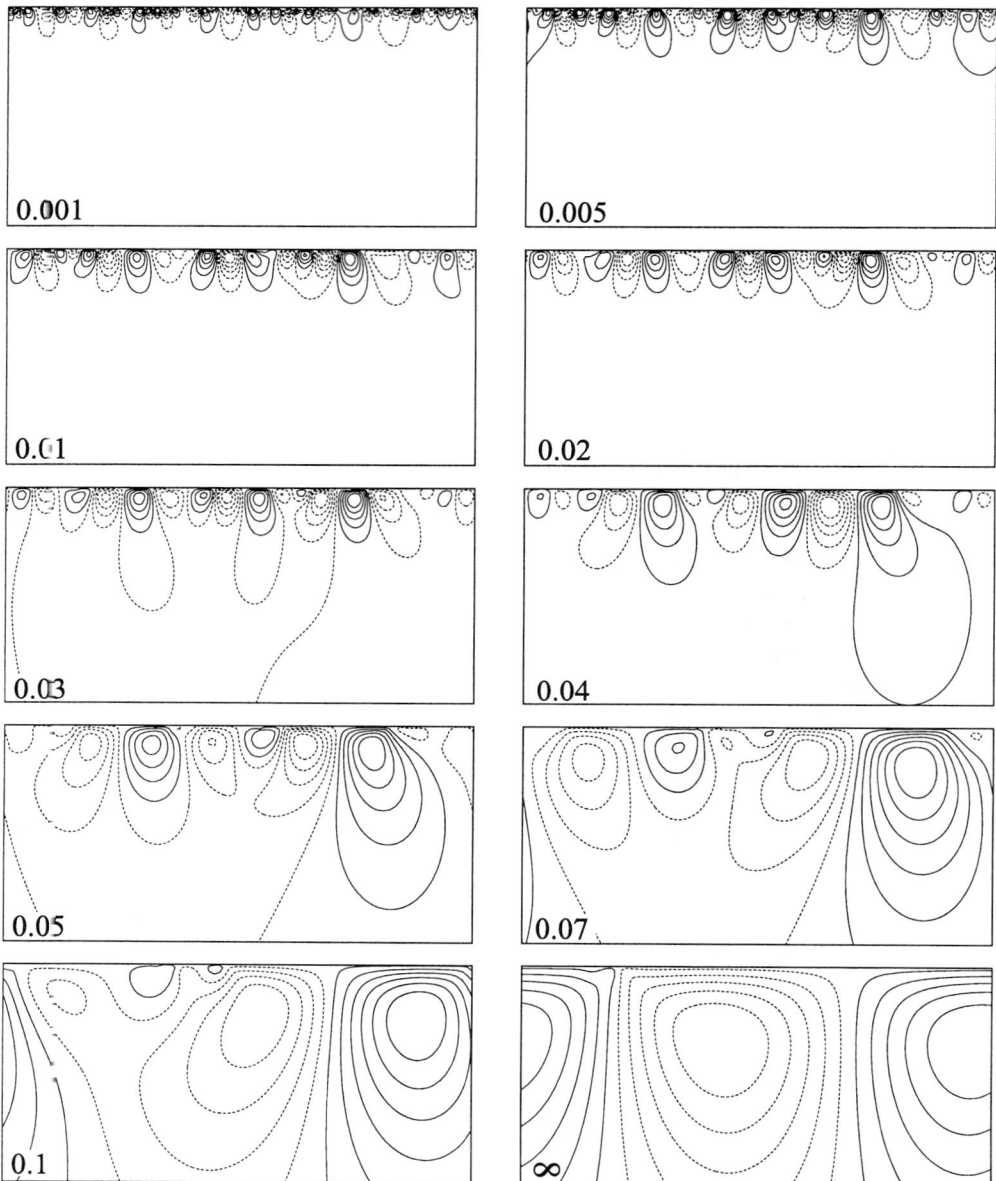

Figure 8.4: The evolution of the streamfunction pattern for $Ma = 3000$ ($\epsilon = 374$), $Bi = 0$, $N = 75$. Full closed curves correspond to clockwise motion. The initial condition was selected as a random noise of amplitude 10^{-5}. The streamfunction is rescaled at each snapshot (the reduced time is indicated). The fastest growing perturbation dominates for times $t < 0.03$. A continuous growth of the mean wavelength of the pattern is observed, the later stages of which tend to a steady state with two convective cells (1 period) in the simulation domain, after a time of order unity (in units $L^2/4\pi^2\kappa$, where L is the horizontal period).

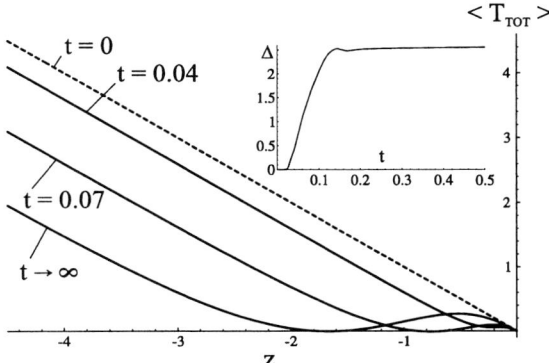

Figure 8.5: The evolution of the total temperature $< T_{\text{TOT}} >$ averaged in the horizontal plane as a function of the vertical coordinate z. The Biot number is $Bi = 0$. Several times are considered, which correspond to the simulation depicted in Fig. 8.4. Homogenization of the temperature occurs in a domain whose depth is growing with time, which results in a growth of the bulk temperature decrease Δ with time (see inset).

critical wavelength), indicating that an intrinsic wavelength is not likely to exist for the pure Marangoni–Bénard problem (at least up to $Ma = 4000$). Note that from the point of view of wavelength selection, the evolution described above presents some similarities with coarsening processes observed in a variety of physical systems (e.g. spinodal decomposition in binary mixtures [491]). These coarsening processes are usually characterized in terms of the scaling law of the wavelength versus time, which here appears close to $\lambda \sim t$, as is seen in the inset of Fig. 8.5 ($\Delta \sim \lambda$). To illustrate the robustness of the coarsening process described here, Fig. 8.6 presents a typical three-dimensional evolution, obtained from a generalization of Eqs (8.25) to two dimensions in Fourier space.

A last comment on Fig. 8.5 concerns the temperature profile near the interface. Since the vertical velocity is vanishing at $z = 0$, a thermal boundary layer is created there, in which the temperature gradient quickly recovers its bulk value. In Rayleigh–Bénard convection, boundary-layer effects are known to play a decisive role in the mechanisms of transition to turbulence (especially for high Prandtl number fluids [22]). However, even for the highest values of the Marangoni number considered here, boundary-layer instabilities have not been observed in our simulations, probably due to the different nature of these boundary layers (in particular the absence of no-slip condition for Marangoni–Bénard problems). However, it cannot be rejected that this kind of phenomena could appear for larger driving forces than those investigated in this section ($Ma = 4000$). Further discussion of boundary-layer instabilities is postponed to §8.2.

Note that a direct comparison of the results obtained from the present model with a finite difference resolution of the governing equations (see §8.1.5) has indicated a satisfactory agreement concerning the qualitative evolution of the system (i.e. the growth of the mean wavelength up to the final steady state with the largest wavelength). This confirms that the

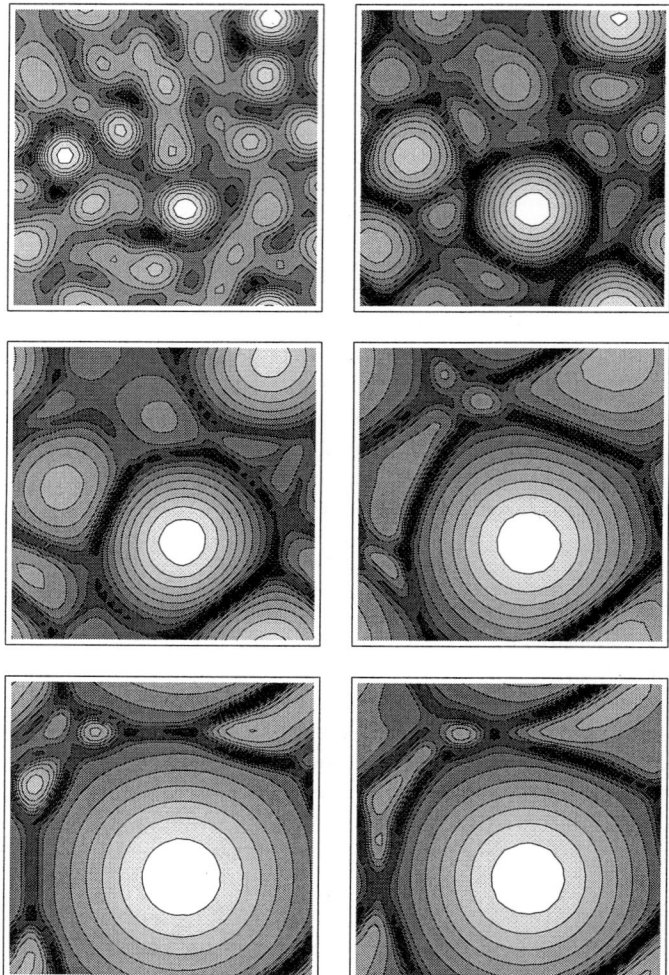

Figure 8.6: Typical 3D patterns observed from the amplitude equation model, for $Ma = 10000$. The 2D surface temperature field is represented (dark = cold, white = hot) for different times (the succession is from top to bottom and from left to right), starting from small "white-noise" random perturbations. Note the formation of small scale structures (sharp boundaries between neighboring convection cells).

most important ingredient responsible for this process is indeed the mean temperature profile distortion by convection. This in turn indicates that mean-field approximations [23], neglecting all cubic effects except the change in the mean temperature profile due to the convective heat transport, can lead to satisfactory approximations of highly supercritical behaviors.

8.1.4 Steady states at high Marangoni number

Bifurcation of rolls

Since the steady state reached by both full (8.24) and reduced (8.25) models is strongly dominated by the fundamental mode $n = 1$, we seek an approximate solution by setting to zero all
harmonics a_n with $n > 1$. The set (8.25) then reduces to the single equation

$$\frac{\partial a_1}{\partial t} = \left(\sigma - S_{11}|a_1|^2\right) a_1 \tag{8.26}$$

describing a pitchfork bifurcation similar to Eq. (8.18) but where the coefficients are now
computed from the eigenfunctions U_k^σ, and thus depend on Ma. After computation of these
coefficients, the steady convective solution of (8.26) can be found as

$$|a_1|^2 = \frac{\sigma}{S_{11}} = \frac{(Ma - Ma_c)\sigma^2(3 + \sqrt{1+\sigma})^3}{(1 + Bi)^2[512(Ma + \sigma^2) + Ma(3\sigma - 8)(3 + \sqrt{1+\sigma})^3]} \tag{8.27}$$

where the growth rate σ is solution of the dispersion relation (8.13), and thus also depends on
Ma. According to Eq. (8.26), the solution (8.27) is stable provided $\sigma > 0$, which is equivalent
to $Ma > Ma_c = 8(1 + Bi)$ (we have set $k_0 = 1$).

The decrease Δ of the bulk temperature due to Marangoni convection, as represented in
Figs 8.3 and 8.5, is expressed by

$$\Delta = \frac{8(Ma - Ma_c)(3 + \sqrt{1+\sigma})^3 \left[\sigma - 2 + \frac{8(Ma+\sigma^2)}{Ma(1+\sqrt{1+\sigma})^2}\right]}{512(Ma + \sigma^2) + Ma(3\sigma - 8)(3 + \sqrt{1+\sigma})^3} \tag{8.28}$$

This expression is represented in Fig. 8.7, together with results obtained from the integration of the full system (8.24) and of the reduced system (8.25). Another result found in
Fig. 8.7 is the expression (8.22), which reduces to $\Delta^0 = 672\epsilon/353$ for $Bi = 0$ and $k_0 = 1$.
Clearly this result is only valid near the threshold. On the contrary, the expression (8.28) for Δ
leads to the result $\Delta = 14/5\epsilon$ near the threshold [which is overestimated, due to the fact that
we have neglected the stabilizing coefficient Z_{-111}, see Eq. (8.18)]. Nevertheless, Eq. (8.28)
appears to be a better approximation of the bulk temperature decrease for large Marangoni
numbers (because Z_{-111} becomes negligible compared with Z_{111}). The corresponding mean
temperature profile can also be shown to be more realistic, since it does not exhibit cold spots
like those appearing in Fig. 8.3, but is rather similar to Fig. 8.5.

Hexagonal patterns

In view of the satisfactory agreement between the analytical result (8.28) and the results of the
numerical integration of (8.24), we reexamine the problem of the competition between three
sets of rolls forming angles of $60°$ with each other. We thus consider

$$A_{\vec{k}} = a_1(t)\delta(\vec{k} - \vec{k}_1) + a_2(t)\delta(\vec{k} - \vec{k}_2) + a_3(t)\delta(\vec{k} - \vec{k}_3) + \\ \bar{a}_1(t)\delta(\vec{k} + \vec{k}_1) + \bar{a}_2(t)\delta(\vec{k} + \vec{k}_2) + \bar{a}_3(t)\delta(\vec{k} + \vec{k}_3) \tag{8.29}$$

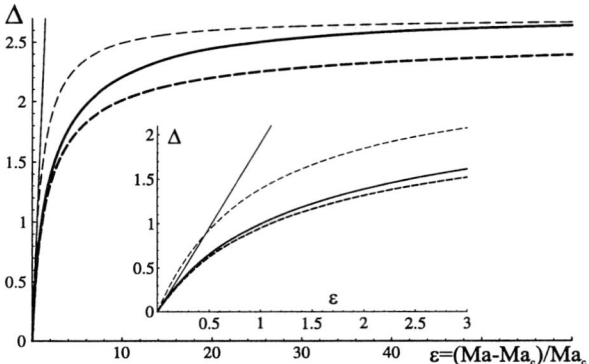

Figure 8.7: The bulk temperature decrease Δ as a function of the distance to the threshold $\epsilon = (Ma - Ma_c)/Ma_c$ for $Bi = 0$. Thick full curve: results of the numerical integration of the full system (8.25). Thick dashed curve: numerical integration of the "mean-field" system (8.24). Thin full line: the weakly nonlinear result $\Delta^0 = 672\epsilon/353$. Thin dashed curve: the analytical result for Δ given as Eq. (8.28) of the text (one-mode model). The inset represents a zoom of a region near the origin.

where the ordering of unit vectors $\vec{k}_i, |i| = 1, 2, 3$ is defined by Fig. 8.8.

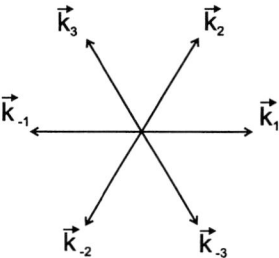

Figure 8.8: Definition of the basic wavevectors ($|\vec{k}_i| = 1$).

From Eq.(8.16), the corresponding amplitude equations are

$$\frac{\partial a_1}{\partial t} = \sigma a_1 + \delta a_2 \bar{a}_3 - \left[\alpha_1 |a_1|^2 + \alpha_2(|a_2|^2 + |a_3|^2)\right] a_1$$
$$\frac{\partial a_2}{\partial t} = \sigma a_2 + \delta a_1 a_3 - \left[\alpha_1 |a_2|^2 + \alpha_2(|a_1|^2 + |a_3|^2)\right] a_2 \qquad (8.30)$$
$$\frac{\partial a_3}{\partial t} = \sigma a_3 + \delta \bar{a}_1 a_2 - \left[\alpha_1 |a_3|^2 + \alpha_2(|a_1|^2 + |a_2|^2)\right] a_3$$

where

$$\delta = 2Z_{1,2} \qquad (8.31)$$

$$\alpha_1 = -(2Z_{1,1,1} + Z_{-1,1,1}) \tag{8.32}$$

$$\alpha_2 = -2(Z_{1,1,1} + Z_{2,1,1} + Z_{-2,1,1}) \tag{8.33}$$

and where $Z_{p,q,r}$ stands for $Z_{\vec{k}_p,\vec{k}_q,\vec{k}_r}$ (symmetry considerations have been used to minimize the number of coefficients to be calculated). The analysis of the gradient system (8.30) has been achieved in §4.2.1, and leads to steady solutions in the form of rolls (R), up-hexagons (U-H), down-hexagons (D-H), and other mixed solutions which are always unstable. Note that the possibility of square patterns has not been included here, even though this should be a natural extension of the present theory, in view of recent experimental results [29, 30, 31]. However, we may expect that the results presented below will be correct as far as the order of magnitude of main convective quantities is concerned, at least for Marangoni numbers that are not too large.

The analytical form of the coefficients δ, α_1, and α_2 (depending on Ma) is not written down here for conciseness. Rather, Fig. 8.9 presents their variation with the distance to the threshold ϵ for various Bi. Bifurcation diagrams are represented in Figs. 8.10 and Fig. 8.11.

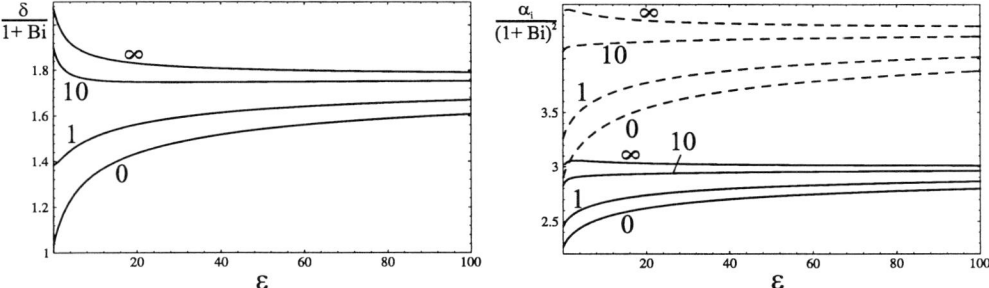

Figure 8.9: The coefficients $\delta/(1 + Bi)$ (left graph), $\alpha_1/(1 + Bi)^2$ (right graph, full curves) and $\alpha_2/(1 + Bi)^2$ (right graph, dashed curves) as a function of the distance to the threshold $\epsilon = [Ma - Ma_c(Bi)]/Ma_c(Bi)$, for various Biot numbers Bi (indicated on each curve).

Figure 8.10 represents the bulk temperature decrease Δ as a function of ϵ, for rolls, up-hexagons and down-hexagons. As expected[3], up-hexagons are the only stable solutions just above the threshold, and rolls become stable only at large ϵ. Down-hexagons are always unstable, because $\delta > 0$. Note that, although not visible on the figure, the first bifurcation to up-hexagons at $\epsilon = 0$ is slightly hysteretic: the depth of this subcritical region is 3.3% (for $Bi = 0$) in our model, slightly larger than the 2.3% value of Scanlon and Segel [103]. This is due to the fact that we have neglected the stabilizing action of "secondary" modes, i.e. those generated by quadratic interaction of the "primary" modes. Accordingly, the figures provided here cannot be expected to be exact, but at least they should provide fair approximations

---

[3]This is in qualitative agreement with the analysis of Scanlon and Segel [103], though there are quantitative differences: these authors predict that rolls should become stable above a value $\epsilon_1 = 64$ (our value is $\epsilon_1 \simeq 8$) of the constraint, while up-hexagons should become unstable above $\epsilon_2 = 196$ (our value is $\epsilon_2 \simeq 40$).

of main convective quantities, given that the harmonics of the fundamental modes appear to affect only weakly the value of typical convective quantities at moderate Ma (see Fig. 8.7).

Still, more accurate results may be obtained near the instability threshold by incorporating the amplitudes of these harmonics in Eq. (8.29), and finally eliminating them using adiabatic slaving [150, 151]. Doing so, the 2.3% value of the hysteresis is recovered. It is also possible to recover the result 0.56% of Bragard and Lebon [130], in the case of a layer of finite depth (the calculation of coefficients is then fully numerical). It is also apparent that the depth of the subcritical region increases with the Biot number, as already discussed in §4.4.2.

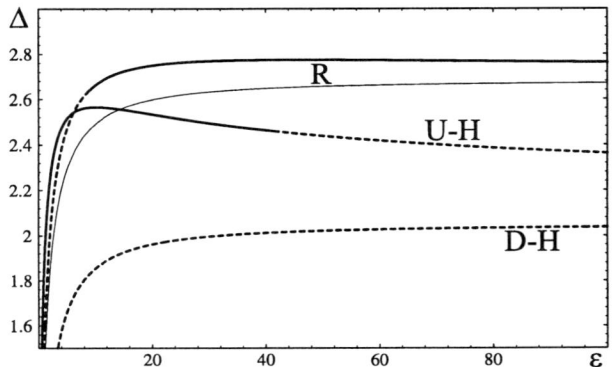

Figure 8.10: Bulk temperature decrease Δ of rolls (R), up-hexagons (U-H), and down-hexagons (D-H), as a function of the distance to the threshold ϵ, for $Bi = 0$. Full curves represent stable states, while dashed curves correspond to unstable states. The thin full line represents the analytical result given by Eq. (8.28) of the text for the bulk temperature decrease of rolls.

Lastly, we have represented the maximal surface velocity for the patterns considered here in Fig. 8.11, showing that this quantity is not strongly dependent on the particular planform selected.

Clearly, the previous figures show that the dependence of nonlinear coefficients upon the Marangoni number Ma allows for deviations from the weakly nonlinear scalings (velocity and temperature perturbations proportional to $\epsilon^{1/2}$) when Ma is increased. In particular, a saturation is observed for the bulk temperature decrease Δ. As shown by Colinet et al. [334], Δ calculated from the above model tends to 8/3 for rolls and to 2.08 for hexagons, for $Ma \to \infty$. These authors also find a scaling regime, where the amplitude of surface temperature fluctuations ΔT_s and the maximal surface velocity V_{max} behave as $\Delta T_s \sim Ma^{-2/3}$ and $V_{max} \sim Ma^{1/3}$, respectively. As mentioned by Colinet et al., these results should merely be considered as first approximations, rather than exact results. Indeed, later studies have shown (see next section) that although saturation of Δ indeed occurs, the correct scalings for other quantities appear to be $\Delta T_s \sim Ma^{-1/3}$ and $V_{max} \sim Ma^{2/3}$. Thus, even though this first model represents a considerable improvement of weakly nonlinear theory for $\epsilon = O(1)$, its quantitative validity should in fact be limited to $\epsilon < 10$, as seen later.

We conclude this section by commenting on an important mathematical aspect of the pure

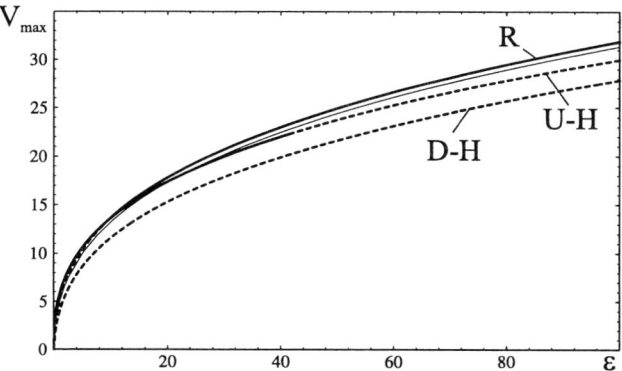

Figure 8.11: Maximal surface velocity V_{max} of rolls (R), up-hexagons (U-H), and down-hexagons (D-H), as a function of the distance to the threshold ϵ for $Bi = 0$. Full curves represent stable states, while dashed curves correspond to unstable states. The thin full line represents the analytical result for the maximal surface velocity of rolls.

Marangoni–Bénard instability. A particular feature of this problem is that the neutral stability condition provides Ma as a single-valued function of the wavenumber k (see also §3.3.2). This means that above the corresponding critical value and at given k, one and only one eigenmode is unstable, however large the value of Ma is. This has to be contrasted [489] with Rayleigh–Bénard instabilities, where n eigenmodes are linearly unstable above the value $Ra_n = (k^2 + n^2\pi^2)^3/k^2$ of the Rayleigh number [case of pure Rayleigh–Bénard convection between stress-free boundaries]. Clearly, the above analysis would require non-trivial modifications to account for interactions between these unstable vertical modes.

8.1.5 Comparison with full numerical simulations

The validity of the assumptions used in deriving the model amplitude equations (8.25) can be tested by direct comparison with a full numerical simulation of the two-dimensional governing equations. The latter uses a Fourier decomposition in the horizontal direction (periodic boundary conditions) and a finite difference discretization in the vertical direction. Realistic no-slip boundary conditions can then be accommodated on the bottom wall. However, as our main goal here will be to compare these exact numerical results with the model (8.25), the lateral size of the container will be taken to be smaller than its depth, ensuring that the lower plate indeed does not influence the results. Note that a stretched mesh has also been used to resolve the thin thermal boundary layer appearing near the interface at high Marangoni number.

As the Prandtl number is assumed infinite, the velocity field can be found analytically [198, 492, 502] from the Stokes equation $\Delta \vec{V} - \vec{\nabla} p = 0$, using the incompressibility condition $\vec{\nabla} \cdot \vec{V} = 0$ and appropriate boundary conditions at $z = 0$ and $z = -d$ (where d is here taken sufficiently large). As detailed in §8.2, this is most easily done in Fourier space, which finally leads to the velocity field everywhere in the fluid, as a function of Fourier coefficients of the surface temperature field. Therefore, it remains to iterate in time on the Fourier coefficients of the temperature field (using the energy equation written at each vertical mesh point), which can been achieved using a standard fourth-order Runge–Kutta routine. Once in the vicinity of a steady state, the latter may be determined more accurately, by switching to a Newton–Raphson technique.

In Fig. 8.12, some steady state characteristics are compared for the direct numerical simulations, for the model amplitude equations (8.25), and for another model developed in the following section.

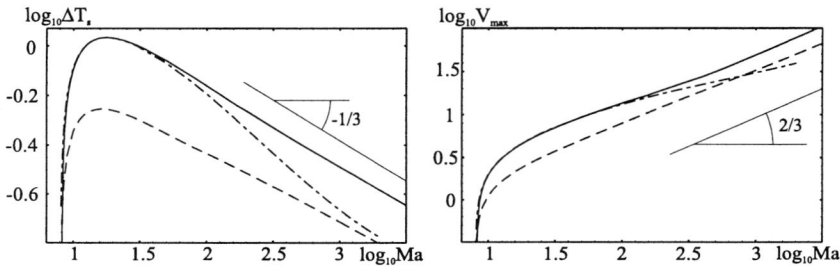

Figure 8.12: Comparison between the exact fully numerical results for 2D steady states (full lines), the amplitude equation model (8.25) [dot-dashed lines], and the boundary-layer model (8.43) [dashed lines]. Results are presented for the amplitude of surface temperature variations ΔT_s (left) and for the maximal free surface velocity V_max. While the amplitude equations provide an excellent quantitative agreement up to 10 times the critical Marangoni number, the boundary-layer model provides a better approximation of the scaling laws $\Delta T_\mathrm{s} \sim Ma^{-1/3}$ and $V_\mathrm{max} \sim Ma^{2/3}$ predicted above $Ma \simeq 400$.

For a zero Biot number, the critical Marangoni number is $Ma_\mathrm{c} = 8$, given that the fundamental wavenumber is taken as $k_0 = 1$. The amplitude equations (8.25) provide a nice approximation of exact results in the range 8–100, i.e. at least up to $\epsilon = 10$. For larger Marangoni numbers, the agreement becomes only qualitative, and in particular a scaling regime is found from direct simulations, which is not satisfactorily approximated by the amplitude equations. In the range $Ma = 400$–4000 (upper bound for direct numerical simulations), the surface temperature fluctuations are found to scale as $Ma^{-1/3}$, while the maximal free surface velocity grows as $Ma^{2/3}$. As will be seen in the next section, such a power-law regime is associated with a thermal boundary layer, whose thickness is found to decrease as $\delta \sim Ma^{-1/3}$. Note also that these scalings could well be valid over quite a large range of Prandtl numbers, as shown by the recent numerical results of Boeck and Thess for low Prandtl number fluids [20].

8.1.6 Extrapolation of the model at higher Marangoni number

Before proceeding to another model, it is worth attempting to simulate the amplitude equations (8.25) beyond their domain of quantitative validity. It should be emphasized however that at increasing Marangoni number, the number N of Fourier modes needed to reach convergence increases. Typically, spurious time-dependent regimes are observed when N is not large enough, and are replaced by a steady state at larger N. As studied by Thess and coworkers [198, 492], these high wavenumber harmonics are associated with sharp variations of the free surface temperature field (see Fig. 8.13).

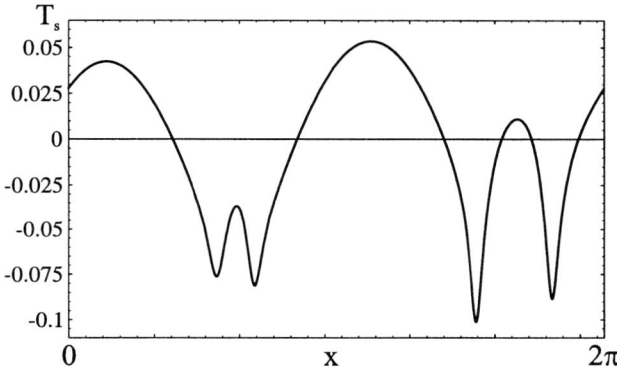

Figure 8.13: Typical 1D surface temperature field (transient state), illustrating the formation of cold ripples (sharp minima), separated by near-parabolic regions.

Such surface temperature fields can be explained [492] considering that fluid is accelerated from hot to cold zones at the free surface. The associated advection of temperature $(\vec{V}.\vec{\nabla}T)$ then generates small scales, i.e. hot zones expand and cold zones get narrower, resulting in a strong nonlinear energy transfer to large wavenumber harmonics. In the absence of heat diffusivity, this effect leads to finite-time blow-up [492] (the first and higher-order derivatives of the surface temperature field diverge) with associated self-similar behavior near the singularity (cold) point. Owing to heat dissipation however (as well as energy input at larger scales), a balance can occur leading to profiles such as in Fig. 8.13 (see also Fig. 8.6).

Now, above a certain value of the Marangoni number (which has not yet been determined accurately), it appears that an increase of N does no longer suppress time-dependent regimes of the model system (8.25). Numerical convergence is reached when the highest-wavenumber Fourier modes have very small amplitudes, and do not influence the dynamics any longer. An example of such chaotic states is presented in Fig. 8.14.

However, the accurate simulation of such chaotic regimes is particularly time-consuming. Moreover, in view of the discussion of the last section, it would be of interest to construct another model which provides better agreement with the scalings at steady state. Indeed, as explained in §8.2, secondary instabilities leading to time-dependent regimes appear to be strongly dependent upon the scaling of the boundary-layer thickness δ, just as in purely buoyancy-driven convection at large Prandtl numbers [23].

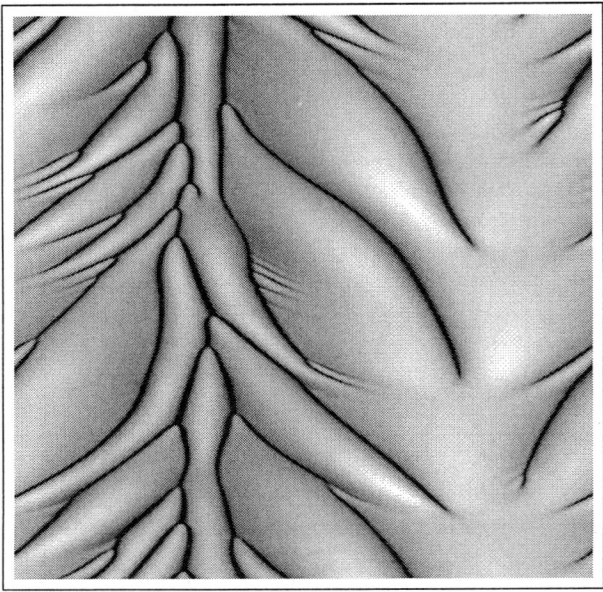

Figure 8.14: Space–time plot of the time-dependent behaviors predicted by the amplitude equation model (8.25), for $Ma = 30000$ and $N = 1000$, for a 2D situation. The 1D surface temperature field is represented (cold zones are dark, while hot zones are white) and time runs upwards.

8.2   Modeling of highly viscous interfacial turbulence

Instead of amplitude equations, our second model relies on the possibility of reducing the 3D Marangoni–Bénard instability problem to a 2D evolution equation for the surface temperature field. Surface advection of heat (nonlinear energy transfer) and small scale thermal dissipation should clearly be incorporated. Recent studies of Thess and co-workers [198, 492] have indeed shown the occurrence of cold ripples due to the latter effects, which is a feature common to many experiments. The infinite Marangoni number model of Thess can however be refined to incorporate an autonomous nonlinear term, providing both energy input (linear instability), and saturation (bulk temperature homogenization). This will also allow discussion of the role of the thermal boundary layer in triggering secondary instabilities and permanent chaotic regimes. Actually, contrary to the amplitude equations model considered in §8.1, the "boundary-layer model" developed below should be better adapted at very large Marangoni number, when the thermal boundary layer is indeed well-developed.

As in the previous section, we will consider a semi-infinite layer of infinite Prandtl number incompressible liquid in contact with an inert gas phase at an undeformable interface. In the absence of volume forces, the divergence-free velocity field $\vec{V}$ satisfies the Stokes equation $\Delta^2 \vec{V} = 0$ (Δ is the Laplacian). This is valid [198] provided the Reynolds number $Re = Ud/\nu$ is small (U and d are typical velocity and length scales, and ν is the fluid kinematic viscosity), or equivalently when the Peclet number $Pe = Ud/\kappa$ remains small compared with the Prandtl number $Pr = \nu/\kappa$ (κ is the thermal diffusivity). Then, the velocity equation is indeed linear and the only nonlinearity appears in the energy equation which reads, in units of d and $U = \kappa/d$

$$\partial T/\partial t = \Delta T + W - \vec{V}.\vec{\nabla}T \tag{8.34}$$

where T is the deviation of temperature with respect to the reference state linear profile $T_{\mathrm{ref}}(z) = -z$, and W is the vertical velocity. The temperature unit is $\theta = \beta d$, where $\beta > 0$ is the imposed gradient along the vertical z coordinate. We consider the usual thermocapillary boundary conditions at a thermally insulating undeformable interface $z = 0$. These are

$$W = DT = D^2W - Ma\Delta_{\mathrm{h}}T = 0 \quad \text{at } z = 0 \tag{8.35}$$

where $D = \partial/\partial z$, and Δ_{h} is the horizontal Laplacian operator. The driving parameter is the Marangoni number $Ma = -\sigma_T \beta d^2/\mu\kappa$, where σ_T is the surface tension variation with temperature and μ the dynamic viscosity. $\vec{V}$ and DT are assumed to vanish as $z \to -\infty$.

In §8.1, we described a model system of evolution equations for the complex Fourier coefficients $a_{\vec{k}}(t)$ of the free surface temperature field $T_{\mathrm{s}}(\vec{r}, t) = \sum_{\vec{k}} a_{\vec{k}}(t)exp[i\vec{k}.\vec{r}]$ where $\vec{r} = x\vec{1}_x + y\vec{1}_y$ and $\vec{k}$ are horizontal coordinate vectors and wavevectors. The equations were derived assuming adiabatic slaving of damped modes, including the mean-field temperature profile $T^D(z) =< T >$ where $< . >$ is the horizontal average. Moreover, amplitude equations were limited to cubic order in the amplitudes $a_{\vec{k}}(t)$ and could be written [see Eq. (8.25)]

$$\partial a_{\vec{k}}/\partial t = \tilde{\sigma}_k^{\mathrm{eff}} a_{\vec{k}} + \sum_{\vec{q}} Z_{\vec{q},\vec{k}} a_{\vec{q}} a_{\vec{k}-\vec{q}} \tag{8.36}$$

where $Z_{\vec{q},\vec{k}}$ is a quadratic coefficient, and $\tilde{\sigma}_k^{\mathrm{eff}}$ is an effective growth rate (depending on $\vec{k}$ only through its modulus $k = |\vec{k}|$) related to the intensity of convection by

$$\tilde{\sigma}_k^{\mathrm{eff}} = \sigma_k - \sum_{\vec{q}} S_{k,q} |a_{\vec{q}}|^2 \tag{8.37}$$

in which the linear growth rate σ_k is obtained from the linear eigenvalue problem, leading to the dispersion relation (8.13). For $Bi = 0$, the latter can be rewritten as $kMa - 2r_k(k+r_k)^2 = 0$ with $r_k = (\sigma_k + k^2)^{1/2}$. For a given Ma, modes with wavenumbers less than the cut-off $k^* = (Ma/8)^{1/2}$ are amplified (see Figs 8.2 and 8.15), while those with $k > k^*$ are damped [$\sigma_k \to -k^2$ for $k \to \infty$].

Note that the low wavenumber limit $k \to 0$ at fixed Ma is $\sigma_k \to (kMa/2)^{2/3}$, which is also the result obtained for $k = O(1)$ and $Ma \to \infty$. This is because in this problem, there is no other length scale than the scale of fastest growing modes $k_{\max} \sim Ma^{1/2}$. Therefore, a measure of supercriticality is the ratio of this wavenumber to an externally imposed wavenumber (the fundamental mode $k_0 = 1$, or the actual fluid depth).

As observed in the last section, thermocapillary convection at high Ma results in an intensive homogenization of the total mean temperature profile $T_{\mathrm{ref}} + T^D$ in the bulk [i.e. $DT^D \simeq 1$

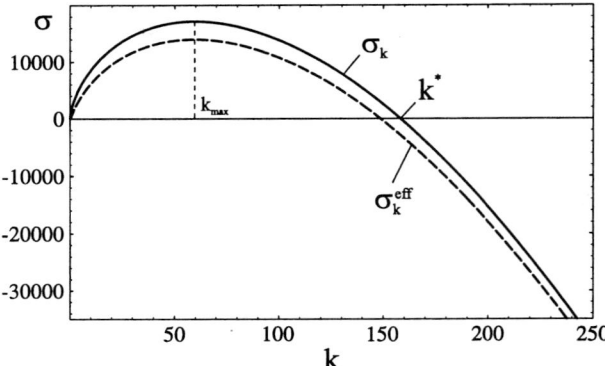

Figure 8.15: Linear growth rate σ_k (full line) versus wavenumber k at $Ma = 2\,10^5$ and effective growth rate σ_k^{eff} of the boundary-layer model (dashed line) at the time-averaged value of $\alpha = 19.7$ for the turbulent state of Fig. 8.17. Chaotic fluctuations around this value of α are about 12%.

there, and $D(T_{\mathrm{ref}} + T^D) \simeq 0$]. Actually, the second term in Eq. (8.37) was seen to represent the effect of such a homogenization on the effective growth rate of each mode (each mode "sees" the mean field temperature field induced by other modes). However, near the free surface (where $W = 0$) heat transport is purely conductive, and a boundary layer exists in which the mean vertical temperature gradient recovers its unperturbed value ($DT^D = 0$).

Actually, there are good reasons to think that secondary instabilities of this boundary layer may be expected when increasing Ma (similarly to what happens for Rayleigh–Bénard convection). In fact, the linear cut-off frequency $k^* \sim Ma^{1/2}$, such that if the boundary-layer thickness δ decreases more slowly than $Ma^{-1/2}$, some modes with wavenumbers $\delta^{-1} < k < k^*$ might still experience the unperturbed gradient (because their penetration depth is $\sim k^{-1}$ for a Stokes flow) and consequently be unstable as revealed by their positive *linear* growth rate. Using the same numerical code as in §8.1.5, the boundary-layer thickness is actually found to decrease as $\delta \sim Ma^{-1/3}$ (see also Fig. 8.16), i.e. not sufficiently fast to stabilize modes with wavenumber $k \sim Ma^{1/2}$.

As expected, secondary instabilities are observed from the amplitude equation model (8.36) when increasing Ma. Time-dependent regimes set in (see Fig. 8.14), whose complexity increases with Ma. However, as mentioned in §8.1.6, the numerical simulation of Eqs (8.36) is time-consuming, due to the large number of operations necessary to evaluate the convolution-like quadratic terms at each time step (increasing as N^2 with the number of Fourier modes N). On the other hand, N has to be taken sufficiently large, in order to avoid spurious time-dependent regimes. In the following, we will thus turn to another set of model evolution equations which avoid truncation to cubic terms (which is reminiscent of weakly nonlinear theories), and at the same time allow use of Fast Fourier Transform (FFT) techniques [198, 493], considerably speeding up the time integration [the number of floating-point operations per time step becomes $\sim N \log(N) \ll N^2$].

The derivation of the second model starts by evaluating Eq. (8.34) at the free surface

$z = 0$. This gives

$$\partial T_s/\partial t = [D^2 T]_{z=0} + \Delta_h T_s - \vec{V}_s.\vec{\nabla} T_s \tag{8.38}$$

where $\vec{V}_s = \vec{V}(z = 0)$ is the free surface velocity field which is a linear functional of the free surface temperature T_s in the limit $Pr \to \infty$ [198]. Keeping the notation $a_{\vec{k}}(t)$ for the Fourier amplitudes of $T_s(\vec{r}, t)$, the Stokes problem $\Delta^2 \vec{V} = 0$ with boundary conditions (8.35) indeed leads to

$$\vec{V}_s(\vec{r}, t) = Ma \sum_{\vec{k}} \exp[i\vec{k}.\vec{r}] a_{\vec{k}}(t) \frac{\vec{k}}{2ik} \tag{8.39}$$

The infinite Marangoni number model of Thess and co-workers [492, 502] rests on the hypothesis that in the limit $Ma \to \infty$, there exists a range of length scales where the first two (dissipative) terms in the right-hand side of Eq. (8.38) become negligible compared with the third one. In such "inertial" regime, numerical simulations predict the formation of small-scale cold ripples qualitatively similar to those observed in Fig. 8.13. These authors also propose to re-incorporate the horizontal Laplacian term in Eq. (8.38), which becomes important at large wavenumbers ("dissipative" range), and avoids blow-up after finite time via smoothing of the small-scale ripples.

Still, the first term in the r.h.s. of Eq. (8.38) is the only one providing coupling with the bulk temperature field, and its role is thus essential in the basic instability mechanism. Our goal in the following is to derive an approximate relation between this "energy input" term $[D^2 T]_{z=0}$ and T_s, which will allow the problem (8.38) to be closed, hence reducing the 3D problem to a 2D equation for T_s.

In fact, the boundary-layer model developed in the present section has a structure similar to Eqs (8.36) derived in §8.1. Namely, in addition to the quadratic term with coefficient $Z_{\vec{q},\vec{k}} = Ma\,\vec{q}.(\vec{q} - \vec{k})/2q$ found by Thess and co-workers, our model will incorporate an effective growth rate term, whose form will however be different from Eq. (8.37).

We first decompose the temperature field T into a mean $T^D(z) = <T>$ (horizontal average) and a fluctuating part T'. Using $<W> = 0$, and assuming the mean temperature profile to be quasi-stationary ($\partial T^D/\partial t = 0$), it can be shown using $\vec{\nabla}.\vec{V} = 0$ that averaging Eq. (8.34) and integrating the result once with respect to z leads to

$$DT^D = <WT'> \tag{8.40}$$

which satisfies $[DT^D]_{z=0} = 0$. Note that this assumption of quasi-stationarity should make the model better suited for regimes of well-developed interfacial turbulence, i.e. when the spatial average $<T>$ is equivalent to an ensemble average [57]. This high Ma limit is also coherent with the existence of a well-formed thermal boundary layer (see below). Subtracting the averaged Eq. (8.34) from itself, we get the fluctuating part

$$\partial T'/\partial t = \Delta T' + W(1 - DT^D) - \left\{ \vec{V}.\vec{\nabla} T' - <\vec{V}.\vec{\nabla} T'> \right\} \tag{8.41}$$

Now, the Fourier decomposition $T' = \sum_{\vec{k}} T_{\vec{k}}(z, t) \exp[i\vec{k}.\vec{r}]$ will be used. A second approximation is to assume that the vertical dependency of Fourier modes $T_{\vec{k}}(z, t)$ is determined by the most unstable eigenmodes of the spectral problem associated with Eq. (8.41), i.e.

$$\sigma_k^{\text{eff}} T_k = (D^2 - k^2) T_k + W_k f(z) \tag{8.42}$$

where $W_k = -k\,Ma\,z\exp[kz]\,T_k(0)/2$ (solution of the Stokes problem) and σ_k^{eff} again denotes an effective growth rate, here associated with the instantaneous nonlinear mean temperature gradient $f(z) = 1 - DT^D(z)$.

This hypothesis, which will allow the reduction to a 2D dynamics for these "interfacial modes", may be partly justified by the analysis of the case $f = 1$ ($DT^D = 0$) achieved in §8.1. In this case, the spectrum (at given k) is composed of the isolated eigenvalue σ_k (positive or negative), plus a continuum of Fourier-like modes whose growth rate is always negative for every Ma (note that strictly speaking, these Fourier modes are spatially oscillatory, and thus do not decay at $z \to -\infty$). These latter modes can thus only be excited by nonlinearities, and should merely act as passive modes. As this property apparently holds for $DT^D \neq 0$ too, we approximate $T_{\vec{k}}(z, t)$ by $a_{\vec{k}}(t)\theta_k(z)$ where $\theta_k(z)$ satisfies Eq. (8.42) and is normalized by $\theta_k(0) = 1$. Then, the spectral form (nonzero Fourier components) of Eq. (8.41) evaluated at $z = 0$ reads

$$\frac{\partial a_{\vec{k}}}{\partial t} = \sigma_k^{\text{eff}} a_{\vec{k}} - \{\vec{V}_s.\vec{\nabla}_s T_s\}_{\vec{k}} = \sigma_k^{\text{eff}} a_{\vec{k}} - Ma \sum_{\vec{p}} \frac{\vec{p}.(\vec{k} - \vec{p})}{2p} a_{\vec{p}} a_{\vec{k}-\vec{p}} \tag{8.43}$$

where $\{.\}_{\vec{k}}$ denotes the Fourier transform and Eq. (8.39) has been used for the second equality. Comparing Eq. (8.43) with the Fourier transform of Eq. (8.38), we may identify the energy input term $[D^2 T_{\vec{k}}]_{z=0}$ with $r_k^2 a_{\vec{k}}$ where $r_k^2 = \sigma_k^{\text{eff}} + k^2$ is obtained from the implicit dispersion relation associated with Eq. (8.42), i.e.

$$\frac{2r_k}{kMa} = \int_0^{-\infty} d\xi\, \xi f(\xi) \exp[(k + r_k)\xi] \tag{8.44}$$

In Eq. (8.44), $f(z) = 1- <WT'> = 1 + Ma \sum_{\vec{k}} kz\exp[kz]|a_{\vec{k}}|^2\theta_k(z)/2$ in which $\theta_k(z)$ also depends on $f(z)$. We are thus left with an integral equation for $f(z)$, for which no closed form solution could be found. Approximate solutions are possible, resting on the hypothesis that $f(z)$ is mainly determined by large scale flows. Then, a third-order differential equation is found for boundary-layer profiles of $f(z)$, which admits solutions $f(z) = h(z/\delta)$, where δ is the boundary-layer thickness. Note that this self-similarity [and the boundary-layer profile $h(y)$] is well verified by full numerical resolutions at steady state, for about one decade of Marangoni numbers (400–4000). Figure 8.16 represents a "collapse plot", where the numerically obtained data (by the mixed finite-difference/spectral method already mentioned in §8.1.5) for $f(z)$ are represented after stretching of the z-coordinate by a quantity α defined below.

The boundary-layer depth δ is directly related to free surface quantities as seen by evaluating the first derivative $f'(0) = h'(0)/\delta$ from Eq. (8.40). We get $\delta = h'(0)\alpha^{-1}$, where

$$\alpha = -D^2 T^D|_{z=0} = - <T'DW>|_{z=0} = \frac{Ma}{2} \sum_{\vec{p}} p|a_{\vec{p}}|^2 \tag{8.45}$$

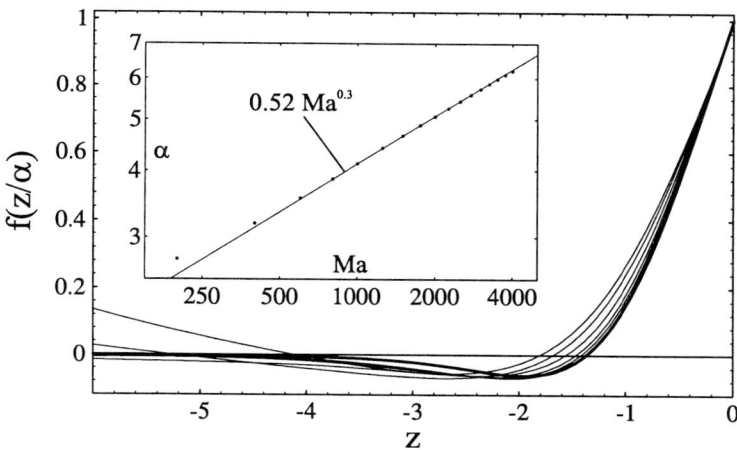

Figure 8.16: Collapse plot of the 2D numerical results at steady state (mixed spectral/finite-differences, stretched mesh), illustrating the self-similarity of the mean temperature gradient $f(z)$ for about one decade of Ma (400–4000). The boundary-layer depth is seen to scale as $\delta \sim \alpha^{-1} \sim Ma^{-1/3}$ approximately. The thicker line is the result found analytically.

Finally, inserting the boundary-layer solution $f(z) = h(z/\delta)$ in Eq. (8.44), a dispersion relation is found which provides the effective growth rate $\sigma_k^{\text{eff}}(k, Ma, \alpha)$ thus closing the model Eq. (8.43).

The "boundary-layer model" just described has been numerically investigated both for 1D and 2D surface temperature fields, using a standard Runge–Kutta order 5 method with adaptative time step control (ensuring an upper bound on the relative error on each mode) and FFT technique for the calculation of the second term in Eq. (8.43). Note that the latter method requires the number of Fourier modes to be taken as an integer power of 2. It consists in evaluating the nonlinear term $\vec{V}_{\mathrm{s}}.\vec{\nabla}_{\mathrm{s}} T_{\mathrm{s}}$ in physical space after calculating $\vec{V}_{\mathrm{s}}$ and $\vec{\nabla}_{\mathrm{s}} T_{\mathrm{s}}$ separately, using (inverse) Fast Fourier Transforms of $Ma\, \vec{k} a_{\vec{k}}/2ik$ and $i\vec{k} a_{\vec{k}}$ respectively. Then, the contribution to $\partial a_{\vec{k}}/\partial t$ is found by applying a (direct) FFT to the result. More details can be found in [198, 493].

Permanent chaotic regimes are obtained, a 1D example of which is depicted in Fig. 8.17. Note that this spatio-temporal dynamics appears to be in qualitative agreement with regimes predicted by the amplitude equation model (see Fig. 8.14). In both cases, sharp thermal ripples are created in a random way (as a result of the boundary-layer instability), and allow splitting of large convective cells into smaller ones. A higher degree of coherence is eventually observed at large scales, under the form of more regular oscillations.

To illustrate the large range of length scales involved, the corresponding time-averaged energy spectrum $E(k) = <\mid a_k \mid^2>$ is depicted in Fig. 8.18. A power-law domain $E \sim k^{-\nu}$ (similar to an inertial range) is observed, whose slope has not yet been systematically investigated, but here appears to lie in the range $\nu = 2$–2.3, slightly decreasing with Ma. A

Figure 8.17: Space–time plot of $T_s(x, t)$ in a typical chaotic state at $Ma = 2\,10^5$, from random initial conditions. Cold ripples appear as dark. Time runs upwards (total time interval = 0.011). Spatial resolution is 2048 Fourier modes.

dissipative range is also apparent for wavenumbers above a cut-off K.

It is worth noting that both K and $< \alpha >$ are found to scale as $Ma^{1/2}$ approximately, thus indicating that these regimes (in contrast with steady regimes observed at moderate Marangoni number) should be strongly influenced by fastest growing modes $k \sim Ma^{1/2}$. Thus, when increasing Ma, we may expect a transition (which has not yet been investigated in detail) be-

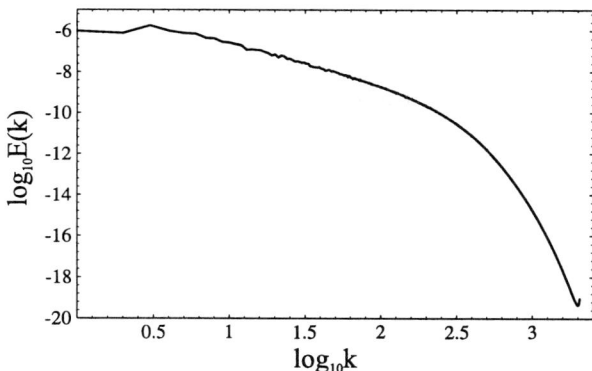

Figure 8.18: Time-averaged energy spectrum corresponding to Fig. 8.17. The "energy" $E(k) = <|a_k|^2>$ is represented as a function of the wavenumber k.

tween a steady regime dominated by the fundamental mode $n = 1$ (external scale) to a chaotic regime with spatial frequencies increasing as the fastest growing modes $k \sim Ma^{1/2}$ (intrinsic scale). It is tempting to conclude that a self-similar chaotic regime should be observed in the limit $Ma \to \infty$, though only very large scale simulations could really confirm or invalidate this conjecture. Indeed, it is not guaranteed a priori that the external scale, which appears to gather a large part of the energy (see Fig. 8.18), should have a vanishing influence in the limit $Ma \to \infty$. For instance, in hydrodynamic turbulence, the external scale at which energy injection occurs (e.g. the dimensions of a mesh placed in a high-speed flow) can affect even the exponents of power-law regimes in the inertial range, owing to intermittency effects. This is an example of incomplete self-similarity, several examples of which have been discussed in the book of Barenblatt [505].

Owing to the implicit assumption of a well-formed boundary layer, the above model appears unable to describe moderate Ma steady regimes. Other more phenomenological models might be used for this purpose. For instance, setting $f(z) = \exp(\alpha z)\cos(\beta z)$ (which closely resembles exact profiles when $\beta \to \alpha$), α is still given by Eq. (8.45), while β can be self-consistently determined by matching the second derivative of $f(z)$ at $z = 0$ with its exact value. Using Eq. (8.40), this gives $\beta^2 = \alpha^2 - Ma\sum_{\vec{p}} p^2 |a_{\vec{p}}|^2$ and the dispersion relation Eq. (8.44) finally leads to the value of the effective growth rate $\sigma_k^{\text{eff}}(k, Ma, \alpha, \beta)$. Apart from chaotic solutions (above $Ma \sim 10^4$) similar to those of Fig. 8.17, this model admits steady solutions for moderate Ma, characterized by scaling exponents in excellent agreement with those found from the full numerical resolution. Note however that this model is not quite satisfactory, as far as the recalculated shape of the boundary-layer function $f(z) = \exp(\alpha z)\cos(\beta z)$ is concerned (for $\beta^2 < -\alpha^2$, the latter diverges for $z \to -\infty$). Thus, other models (phenomenological or based on a more rigorous determination of the mean-field gradient f) should be developed to investigate the mechanism of transition from steady to chaotic regimes.

Still, a good approximation of steady regimes is also obtained by considering this phenomenological model with a single Fourier mode $k = 1$ with amplitude a_1. In this case, the "marginal stability" relation $\sigma_1^{\text{eff}} = 0$ leads to a relation between α, β and Ma (i.e. $|a_1|^2$ and Ma) at the steady convective state. Surprisingly, this relation accounts quite reasonably for the variation of convective quantities with Ma. Indeed, the asymptotics $Ma \to \infty$ gives $|a| \to 3^{1/6}2^{-1/3}Ma^{-1/3}$ and $\alpha, \beta \to (3/4Ma)^{1/3}$ leading to the above correct scalings for main convective quantities.

Finally, the following figures represent simulations of the boundary-layer model in 3D situations (2D surface temperature fields). These are in qualitative agreement with the corresponding simulations of the amplitude equation model presented in Fig. 8.6, although the number of Fourier modes could be taken much larger here, resulting in a better resolution of the small scale cold thermal ripples appearing at the boundaries of polygonal convective cells.

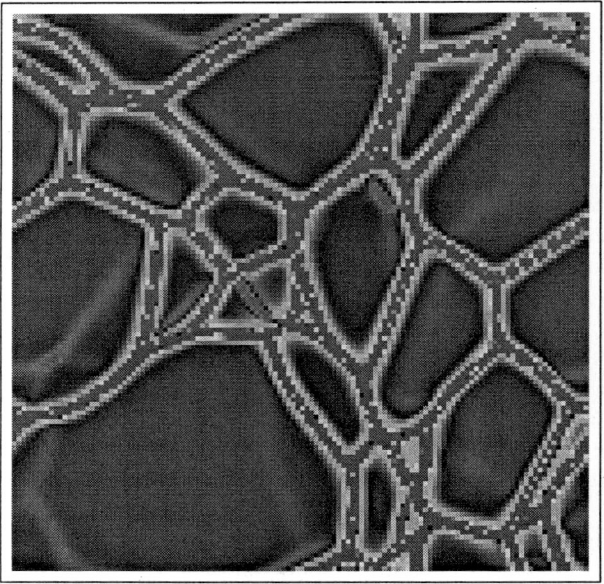

Figure 8.19: 2D surface temperature field resulting from a simulation of the boundary-layer model (resolution 256×256). The evolution of the pattern is slow and still appears to be influenced by the periodic boundary conditions.

Figure 8.19 is very similar to surface temperature fields obtained by Thess and Orszag [193, 503], from direct numerical simulation of the governing equations in finite-depth layers, and from the infinite-Marangoni number model [492, 502]. In Fig. 8.20, a 3D chaotic state is represented over a short time interval. It is seen that the pattern is basically composed of large convective cells growing at the expense of smaller ones. Owing to the instability of the thermal boundary layer however, these large cells split into smaller ones, while the process repeats in other parts of the container, thereby leading to persistent chaotic regimes.

An intensive investigation of scaling and spectral properties of the observed chaotic regimes would certainly appear to be the next step of the present analysis. This might allow understanding of peculiarities of thermal interfacial turbulence at high Marangoni number, when

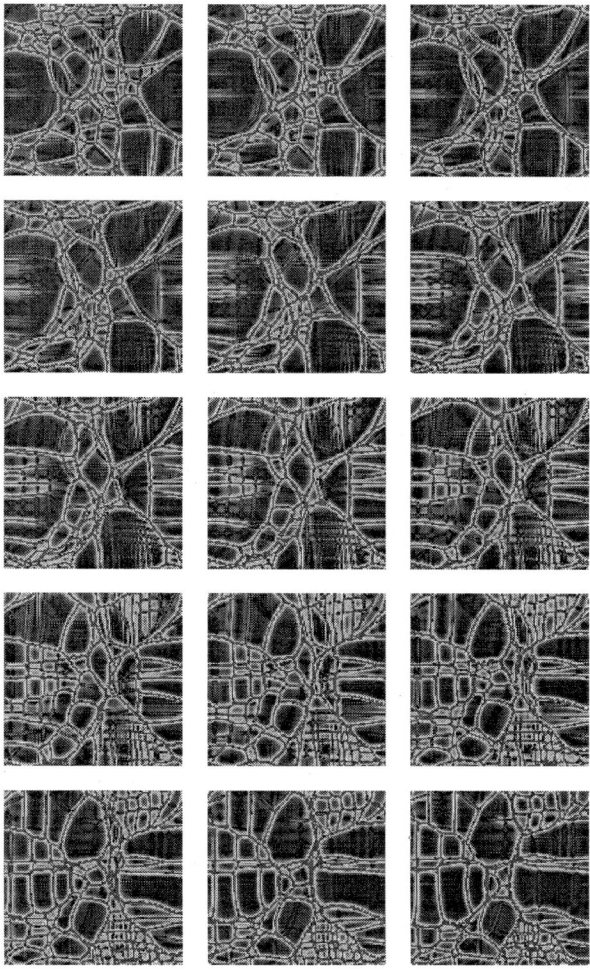

Figure 8.20: Typical chaotic evolution of a 2D surface temperature field at large Ma, with resolution 512×512. Small-scale structures are highlighted by applying the Laplacian operator to the actual surface temperature pattern [198].

compared with e.g. Navier–Stokes [55, 57, 505], or high Rayleigh number thermal turbulence [55, 59]. For this purpose, the model reduction of the problem dimensionality from 3D to 2D allows an important reduction of the computer cost of direct simulations. Another chief advantage of the model derived in this section is to provide a physically realistic stirring force term which is autonomous, i.e. an instantaneous function of the active scalar field whose turbulent dynamics is studied. In this sense, it generalizes earlier models of Thess et al. [492], and is thought to be applicable for describing generic features of interfacial turbulence in other configurations (e.g. two-layer systems, ...). We finally mention that the preliminary results presented above for time-dependent polygonal 2D surface temperature patterns indeed appear in good qualitative agreement with available experimental data [38, 39, 40], as well as with more recent experiments by the authors in highly volatile liquids (evaporative convection).

A Basic definitions and transport theorems

A.1 Particle paths, velocity and material derivative

The fluid is considered as a continuum, whose deformations can be characterized by the motions of fluid particles. The position of each of these particles is described [182] by a vector function $\vec{x}(\vec{\xi}, t)$ depending on the initial position $\vec{\xi}$ of the particle and on time t, and assumed to be continuous and single-valued. This function should also be invertible, such that $\vec{\xi}(\vec{x}, t)$ represents the position at time $t = 0$ of the particle which is at position $\vec{x}$ at time t. The function $\vec{x}(\vec{\xi}, t)$ thus describes the particle paths, along which any property of the fluid (such as density, or specific energy) can be followed. If such a quantity is denoted by f, the time variation of f can be defined in two ways. We may either be interested in the variation of f at a given fixed point $\vec{x}$:

- derivative at $\vec{x}$ constant: $\left(\dfrac{\partial}{\partial t} f \right)_{\vec{x}} = \dfrac{\partial f}{\partial t}$

or in the variation of f associated with a certain fluid particle $\vec{\xi}$ followed in its motion:

- derivative at $\vec{\xi}$ constant (material derivative): $\left(\dfrac{\partial}{\partial t} f \right)_{\vec{\xi}} = \dfrac{df}{dt}$

In particular, if $f = \vec{x}(\vec{\xi}, t)$ (i.e. the position of particle $\vec{\xi}$),

$$\frac{d\vec{x}}{dt} = \left[\frac{\partial \vec{x}(\vec{\xi}, t)}{\partial t} \right]_{\vec{\xi}} = \vec{v} \tag{A.1}$$

which defines the velocity. Thus, for any hydrodynamic quantity f, the material derivative can be expressed as

$$\begin{aligned}
\frac{df}{dt} &= \frac{df[\vec{x}(\vec{\xi}, t), t]}{dt} \\
&= \frac{\partial f}{\partial t} + v_i \frac{\partial f}{\partial x_i} = \frac{\partial f}{\partial t} + \vec{v}.\vec{\nabla}_{\vec{x}} f
\end{aligned} \tag{A.2}$$

where v_i denotes the ith component of velocity, and $\vec{\nabla}_{\vec{x}}$ the gradient with respect to the $\vec{x}$ variables. Note that the convention of summation on repeated indexes is adopted. Thus, the difference between derivatives of f at $\vec{x}$ and $\vec{\xi}$ constant is associated with the transport term $\vec{v}.\vec{\nabla}_{\vec{x}} f$, i.e. advection of f by the flow.

A.2 Streamlines

A flow is called steady if its velocity components v_i are independent of time. In general, even for unsteady flows, streamlines are defined at a given time as curves everywhere tangent to the local velocity vector. Streamlines are thus solutions $\vec{x}(s)$ of the equation

$$\frac{d\vec{x}}{ds} = v[\vec{x}(s), t]$$ (A.3)

at a given instant t, where s is a parameter along the streamline. From this definition, it is seen that streamlines coincide with particle paths at steady state only.

A.3 Streaklines

Streaklines of a given flow $\vec{x}(\vec{\xi}, t)$ are lines joining all particles that have gone through a given point $\vec{x}_0$ at some earlier time. Hence, they are the traces generated by a smoke or a dye continuously injected at the given point, disregarding diffusion. As $\vec{\xi}(\vec{x}_0, s), 0 < s < t$ defines a locus of points which all pass through the injection point $\vec{x}_0$ between times 0 and t, the position at time t of this set of points, i.e.

$$\vec{x} = \vec{x}(\vec{\xi}(\vec{x}_0, s), t) \quad 0 < s < t$$ (A.4)

defines the streakline at time t generated by injection of a dye at $\vec{x}_0$ from the initial time $t = 0$. At steady state, streamlines, streaklines and particle paths coincide.

A.4 Compressibility

Given some flow $\vec{x}(\vec{\xi}, t)$, consider a volume element $dV_0 = d\xi_1 d\xi_2 d\xi_3$ located at $\vec{\xi}$ at time $t = 0$. After a time t, this volume element is located at $\vec{x}(\vec{\xi}, t)$ and has a volume $dV = dx_1 dx_2 dx_3$ related to dV_0 by the coordinate transformation from $\vec{\xi}$ to $\vec{x}$, i.e.

$$dV = dx_1 dx_2 dx_3 = \frac{\partial(x_1, x_2, x_3)}{\partial(\xi_1, \xi_2, \xi_3)} d\xi_1 d\xi_2 d\xi_3$$ (A.5)

Thus, the Jacobian

$$J = \frac{dV}{dV_0} = \frac{\partial(x_1, x_2, x_3)}{\partial(\xi_1, \xi_2, \xi_3)} = \begin{vmatrix} \frac{\partial x_1}{\partial \xi_1} & \frac{\partial x_2}{\partial \xi_1} & \frac{\partial x_3}{\partial \xi_1} \\ \frac{\partial x_1}{\partial \xi_2} & \frac{\partial x_2}{\partial \xi_2} & \frac{\partial x_3}{\partial \xi_2} \\ \frac{\partial x_1}{\partial \xi_3} & \frac{\partial x_2}{\partial \xi_3} & \frac{\partial x_3}{\partial \xi_3} \end{vmatrix}$$ (A.6)

directly measures the dilatation or expansion of the fluid. Note that the material derivative of J is given [182] by

$$\frac{dJ}{dt} = \left(\frac{\partial v_1}{\partial x_1} + \frac{\partial v_2}{\partial x_2} + \frac{\partial v_3}{\partial x_3}\right) J = J\vec{\nabla}_{\vec{x}}.\vec{v}$$ (A.7)

or $d \ln J/dt = \vec{\nabla}_{\vec{x}}.\vec{v}$, i.e. the divergence of the velocity field represents the relative rate of change of the dilatation following a fluid element. This directly implies that for an incompressible flow, $dJ/dt = 0$, or equivalently

$$\vec{\nabla}_{\vec{x}}.\vec{v} = 0 \qquad (A.8)$$

A.5 Reynolds' Transport Theorem

We now turn to the transport of any quantity f, whose volume integral $F(t)$ at time t over a closed (time-dependent) material volume $V(t)$ is

$$F(t) = \iiint_{V(t)} f(\vec{x}, t) dV \qquad (A.9)$$

The convective derivative $dF(t)/dt$ can be easily computed by changing integration variables from $\vec{x}$ to $\vec{\xi}$, because in the latter system, the volume is the initial volume $V(t=0) = V_0$ which is fixed. Differentiating the integrand and using Eqs (A.2) and (A.7), we get

$$\frac{dF}{dt} = \iiint_{V_0} \left[J(\frac{\partial f}{\partial t} + (\vec{v}.\vec{\nabla}_{\vec{x}})f) + f(\vec{\nabla}_{\vec{x}}.\vec{v})J \right] dV_0 \qquad (A.10)$$

Returning to variables $\vec{x}$, we obtain the Reynolds' Transport Theorem

$$\frac{dF}{dt} = \iiint_{V(t)} \left(\frac{\partial f}{\partial t} + \vec{\nabla}.(f\vec{v}) \right) dV = \iiint_{V(t)} \frac{\partial f}{\partial t} dV + \iint_{S(t)} f\vec{v}.\vec{n} dS \qquad (A.11)$$

where Green's theorem has been used for the second identity [$S(t)$ is the surface of the volume $V(t)$, and $\vec{n}$ its outwards normal unit vector].

Thus, the rate of change of the quantity F inside a volume $V(t)$ is the instantaneous rate of change summed up at every point inside the volume, plus the net flux of f across the surface. As f can be any scalar or vector component, this theorem can be directly applied to express the conservation of mass, momentum and energy.

Another useful form of the Reynolds' Transport Theorem is obtained by combining Eqs (A.11) and (2.3) for a volumic quantity ρf (here f is a massic quantity). The Reynolds' Transport Theorem then reads

$$\frac{d}{dt} \iiint_{V(t)} \rho f dV = \iiint_{V(t)} \rho \frac{df}{dt} dV \qquad (A.12)$$

A.6 Deformation and rate of strain

If $\vec{v} = d\vec{x}/dt$ is the velocity, the components of the relative velocity of two nearby fluid elements $\vec{\xi}$ and $\vec{\xi} + d\vec{\xi}$ [such that at time t, their positions are respectively $\vec{x}(\vec{\xi}, t)$ and $\vec{x}(\vec{\xi} + d\vec{\xi}, t) = \vec{x}(\vec{\xi}, t) + d\xi_i \partial \vec{x}/\partial \xi_i + O(d\xi^2)$] is

$$dv_i = v_i(\vec{x} + d\vec{x}) - v_i(\vec{x}) = dx_k \frac{\partial v_i}{\partial x_k} + O(dx^2) \qquad (A.13)$$

in which

$$dx_k = d\xi_j \frac{\partial x_k}{\partial \xi_j} \tag{A.14}$$

where higher-order corrections are neglected. As dv_i and dx_k are components of cartesian vectors, Eq. (A.13) shows that the velocity gradients $\partial v_i / \partial x_k$ form a tensor called the velocity gradient tensor. This tensor can then be written as the sum of its symmetric and antisymmetric parts

$$\frac{\partial v_i}{\partial x_j} = \frac{1}{2}\left(\frac{\partial v_i}{\partial x_j} + \frac{\partial v_j}{\partial x_i}\right) + \frac{1}{2}\left(\frac{\partial v_i}{\partial x_j} - \frac{\partial v_j}{\partial x_i}\right) \tag{A.15}$$
$$= \quad e_{i,j} \quad + \quad \Omega_{i,j}$$

For rigid body translations, $\vec{v} = \vec{u}$ constant and $e_{i,j} = \Omega_{i,j} = 0$. If the motion is a rigid body rotation $\vec{v} = \vec{\omega} \times \vec{x}$ with constant rotation vector $\vec{\omega}$, it is easily checked that $e_{i,j} = 0$. Thus, the antisymmetric part $\Omega_{i,j}$ of the velocity gradient tensor corresponds to a rigid body rotation. As $e_{i,j}$ vanishes in the case of any rigid body motion (rotation + translation), it is called the deformation or rate of strain tensor and its vanishing is a necessary condition for the motion to be rigid.

B Self-adjointness and oscillatory modes

The linear stability problem will be taken in the form (3.104–3.105), which we rewrite here for convenience:

$$(D^2 - k^2)^2 w - k^2 Ra\theta = \sigma Pr^{-1}(D^2 - k^2)w \tag{B.1}$$

$$(D^2 - k^2)\theta + w = \sigma\theta \tag{B.2}$$

together with some set of boundary conditions which will be left unspecified for the moment, in order to examine their role on the properties of the adjoint problem. To simplify notation, we define a vector

$$U = \begin{pmatrix} w(z) \\ \theta(z) \end{pmatrix} \tag{B.3}$$

which belongs to a certain set E of such vectors satisfying the boundary conditions of the problem. As an example, for the stress-free heat conducting boundary conditions, we would define E as

$$E = \left\{ U = \begin{pmatrix} w(z) \\ \theta(z) \end{pmatrix} \; : \; w = D^2 w = \theta = 0 \; at \; z = 0, 1 \right\} \tag{B.4}$$

Thus, multiplying Eq. (B.2) by $-k^2 Ra$, the full linear stability problem may be written in the operational form

$$\mathcal{L}(U) - \sigma T(U) = 0 \quad \text{with} \quad U \in E \tag{B.5}$$

where the linear operators are defined by

$$\mathcal{L} \equiv \begin{pmatrix} (D^2 - k^2)^2 & -k^2 Ra \\ -k^2 Ra & -k^2 Ra(D^2 - k^2) \end{pmatrix} \tag{B.6}$$

$$T \equiv \begin{pmatrix} Pr^{-1}(D^2 - k^2) & 0 \\ 0 & -k^2 Ra \end{pmatrix} \tag{B.7}$$

Now, we define a scalar product in the form

$$\left\langle \tilde{U}, U \right\rangle = \int_0^1 dz(\tilde{w}^* w + \tilde{\theta}^* \theta) \tag{B.8}$$

where an asterisk denotes the complex conjugate, and by definition

$$\tilde{U} = \begin{pmatrix} \tilde{w}(z) \\ \tilde{\theta}(z) \end{pmatrix} \tag{B.9}$$

We also define the adjoint operators $\mathcal{L}^+$ and T^+ such that

$$\left\langle \tilde{U}, \mathcal{L}(U) \right\rangle = \left\langle \mathcal{L}^+(\tilde{U}), U \right\rangle \;,\; \left\langle \tilde{U}, T(U) \right\rangle = \left\langle T^+(\tilde{U}), U \right\rangle \tag{B.10}$$
$$\forall\, U \in E,\, \tilde{U} \in F$$

where F denotes a certain set of vectors to be defined later on, in a way similar to E. Repeated integration by parts of the left-hand side of the first of Eqs (B.10) leads to

$$\left\langle \tilde{U}, \mathcal{L}(U) \right\rangle$$
$$= \int_0^1 dz \left\{ \tilde{w}^* \left[(D^2 - k^2)^2 w - k^2 Ra\theta \right] + \tilde{\theta}^* \left[-k^2 Raw - k^2 Ra(D^2 - k^2)\theta \right] \right\}$$
$$= \int_0^1 dz \left\{ w \left[(D^2 - k^2)^2 \tilde{w}^* - k^2 Ra\tilde{\theta}^* \right] + \left[-k^2 Ra\tilde{w}^* - k^2 Ra(D^2 - k^2)\tilde{\theta}^* \right] \right\}$$
$$-2k^2 [\tilde{w}^* Dw]_0^1 + [\tilde{w}^* D^3 w]_0^1 - k^2 Ra[\tilde{\theta}^* D\theta]_0^1 + 2k^2 [D\tilde{w}^* w]_0^1$$
$$-[D\tilde{w}^* D^2 w]_0^1 + [D^2 \tilde{w}^* Dw]_0^1 - [D^3 \tilde{w}^* w]_0^1 + k^2 Ra[D\tilde{\theta}^* \theta]_0^1 \tag{B.11}$$

the first integral term of which is precisely $< \mathcal{L}^+(\tilde{U}), U >$ (with $\mathcal{L}^+ = \mathcal{L}$). Thus, it remains to cancel the boundary terms in this expression, using the boundary conditions for U (included in the definition of E) and for $\tilde{U}$, which will define the set F. The same procedure should be applied to the operator T. Similarly, we then obtain $T^+ = T$, with the supplementary boundary terms

$$Pr^{-1}[\tilde{w}^* Dw]_0^1 - Pr^{-1}[D\tilde{w}^* w]_0^1 \tag{B.12}$$

Now, it is readily seen that for all sets E of boundary conditions considered in §3.3.1, the boundary terms cancel provided $F = E$, i.e. the same boundary conditions are used for U and $\tilde{U}$. This completes the definition of self-adjointness for all the problems considered in §3.3.1. Note that this property persists if we adopt a mixed thermal boundary condition $D\theta + Bi\theta = 0$ (even with different Biot numbers at the top and bottom plates). Although not often used, a "partial-slip" boundary condition $D^2 w + Bi_v Dw = 0$ would also preserve the self-adjoint character of the linear operator.

Before examining the implications of self-adjointness, consider the case of Marangoni–Bénard convection (§3.3.2). Note that we will keep $Ra \neq 0$, so that the above developments are still applicable. However, now boundary conditions are $w = Dw = \theta = 0$ at $z = 0$, and $w = D\theta + Bi\theta = D^2 w + k^2 Ma\theta = 0$ at $z = 1$. Then the boundary terms may all be canceled provided the boundary conditions of $\tilde{U}$ are $\tilde{w} = D\tilde{w} = \tilde{\theta} = 0$ at $z = 0$ and $\tilde{w} = D^2 \tilde{w} = Ra(D\tilde{\theta} + Bi\tilde{\theta}) + MaD\tilde{w} = 0$ at $z = 1$. Here, the boundary conditions are different from those applying to the original stability problem, and the operators are in general not self-adjoint. The derivation of the adjoint problem just presented is however restricted to the case $Ra \neq 0$. The case $Ra = 0$ will be considered later on.

We may now establish some useful properties using the definitions (B.10). First, we may combine both these definitions to obtain the adjoint operator for the full problem. Taking into account Eq. (B.8), we have

$$\left\langle \tilde{U}, \mathcal{L}(U) - \sigma T(U) \right\rangle = \left\langle \mathcal{L}^+(\tilde{U}) - \sigma^* T^+(\tilde{U}), U \right\rangle \quad \forall\, U \in E, \tilde{U} \in F \tag{B.13}$$

Then, considering U_σ as the solution of the linear problem with eigenvalue σ, i.e. $\mathcal{L}(U_\sigma) - \sigma T(U_\sigma) = 0$, Eq. (B.13) yields

$$\left\langle \mathcal{L}^+(\tilde{U}) - \sigma^* T^+(\tilde{U}), U_\sigma \right\rangle = 0 \quad \forall\, \tilde{U} \in F \tag{B.14}$$

We may also rewrite this expression for $\tilde{U} = \tilde{U}_\lambda$, which is chosen as the eigenvector with eigenvalue λ of the adjoint problem, i.e. $\mathcal{L}^+(\tilde{U}_\lambda) - \lambda T^+(\tilde{U}_\lambda) = 0$. Thus, we obtain

$$(\lambda^* - \sigma) \left\langle T^+(\tilde{U}_\lambda), U_\sigma \right\rangle = 0 \tag{B.15}$$

which shows that either $\lambda = \sigma^*$, and $< T^+(\tilde{U}_{\sigma^*}), U_\sigma > = < \tilde{U}_{\sigma^*}, T(U_\sigma) > \neq 0$ in general, or $\lambda \neq \sigma^*$ and $< T^+(\tilde{U}_\lambda), U_\sigma > = < \tilde{U}_\lambda, T(U_\sigma) > = 0$ (orthogonality property). The latter property does not require the operators to be self-adjoint.

In the self-adjoint case, where the linear stability problem and its adjoint coincide, we may choose the eigenvalue λ equal to a certain eigenvalue σ of the original problem. Then, the eigenvectors also coincide (and are denoted as U_σ), and the identity (B.15) leads to

$$\mathrm{Im}[\sigma] \langle U_\sigma, T(U_\sigma) \rangle = 0 \tag{B.16}$$

This is very interesting insofar as the scalar product in this expression may be a definite positive function of U_σ (thus implying $\mathrm{Im}[\sigma] = 0$). However, this depends on the form of the operator T. As an example, for the self-adjoint Rayleigh–Bénard problems considered in §3.3.1, T is given by Eq. (B.7), and Eq. (B.16) yields

$$\mathrm{Im}[\sigma] \int_0^1 dz \left\{ Pr^{-1} \left(k^2 |w_\sigma|^2 + |Dw_\sigma|^2 \right) + k^2 Ra |\theta_\sigma|^2 \right\} = 0 \tag{B.17}$$

where we used the fact that $w = 0$ at $z = 0, 1$ for every set E of boundary conditions considered in §3.3.1. As the integral in Eq. (B.17) is clearly positive when $Ra > 0$, this shows that $\mathrm{Im}[\sigma] = 0$ for every eigenvector. This is the classical proof [20, 95, 210] of the principle of exchange of stability for the Rayleigh–Bénard problem (with all possible combinations of boundary conditions, even including Biot numbers, as seen above), stating that the loss of stability of the reference state may only occur through bifurcation of real eigenvalues (monotonic instabilities).

When $Ra < 0$, nothing can be predicted from Eq. (B.17), as there may exist cases for which the integral identically vanishes (and possibly $\mathrm{Im}[\sigma] \neq 0$). This is easily checked in the case of internal waves, at least in the simplest case of heat-conducting free-free boundary conditions [i.e. (B.4) is valid]. Solving the linear problem (B.5) leads to $w_\sigma = (\sigma + a^2) \sin[\pi z]$

and $\theta_\sigma = \sin[\pi z]$ with $a^2 = k^2 + \pi^2$ (considering only modes with vertical wavenumber $n = 1$), and the dispersion relation is given by

$$(\sigma + a^2)(a^2 + \sigma Pr^{-1}) = \frac{k^2 Ra}{a^2} \tag{B.18}$$

which is the dimensionless form of the relation (3.29). Now, the expression (B.17) is calculated as

$$\text{Im}[\sigma]\left\{Pr^{-1}a^2|\sigma + a^2|^2 + k^2 Ra\right\} = 0 \tag{B.19}$$

which is valid both for monotonic modes ($\text{Im}[\sigma] = 0$ and the factor between braces is different from zero), and oscillatory modes, in which case

$$\text{Re}[\sigma] = -a^2(1 + Pr)/2, \quad \text{Im}[\sigma] = \pm[-k^2 Ra Pr/a^2 + a^4(2Pr - Pr^2 - 1)/4]^{1/2} \tag{B.20}$$

from which it can be seen that the factor between braces in Eq. (B.19) vanishes.

Finally, we recall that nothing can be concluded about the nature of eigenvalues in the case where Marangoni effects are considered, because the linear problem is not self-adjoint.

C Linear energy stability and variational principles

Consider the linear Rayleigh–Bénard stability problem under the form (3.99–3.100) :

$$\Delta^2 w \, + \, Ra\Delta_{\mathrm{h}}\theta = Pr^{-1}\frac{\partial}{\partial t}\Delta w \qquad (C.1)$$

$$\Delta\theta + w = \frac{\partial\theta}{\partial t} \qquad (C.2)$$

For normal modes $(w,\theta) \sim \exp[i\vec{k}.\vec{r}]$, we first rewrite this system replacing $\Delta_{\mathrm{h}} \to -k^2$, $\Delta \to (D^2 - k^2)$.

Then, to construct an "energy" functional quantifying the strength of convective perturbations, we multiply Eq. (C.1) by w^*, Eq. (C.2) by $-k^2 Ra\theta^*$, integrate over the depth and add the resulting equations (also adding their complex conjugate). Integrating by parts as done in Appendix B (and for any of the sets of boundary conditions considered in §3.3 1), we obtain the energy

$$E = \frac{1}{2}\int_0^1 dz \left[Pr^{-1}\left(k^2|w|^2 + |Dw|^2\right) + k^2 Ra|\theta|^2 \right] \qquad (C.3)$$

which is positive definite for $Ra > 0$, and whose evolution is governed by

$$\frac{dE}{dt} = -I_{\mathrm{visc}} - k^2 Ra I_{\mathrm{th}} + 2k^2 Ra \left\{ \int_0^1 dz\theta^* w \right\}_{\mathrm{R}} \qquad (C.4)$$

where $\{.\}_{\mathrm{R}}$ denotes the real part, I_{visc} is the viscous dissipation integral

$$I_{\mathrm{visc}} = \int_0^1 dz \left[|D^2 w|^2 + 2k^2|Dw|^2 + k^4|w|^2 \right] \geq 0 \qquad (C.5)$$

and I_{th} is the thermal dissipation integral

$$I_{\mathrm{th}} = \int_0^1 dz \left[|D\theta|^2 + k^2|\theta|^2 \right] \, \geq \, 0 \qquad (C.6)$$

which allow a clear physical meaning to be attributed to Eq. (C.4), whose last term is the energy input from the basic gradient (work of buoyancy forces). The intensity or energy E

of the perturbations is seen to increase if the work of buoyancy forces overcomes viscous and thermal dissipation. Note that Eq. (C.4) is not valid if thermal boundary conditions are of the mixed type $D\theta + Bi\theta = 0$ (e.g. at a free surface), because in this case some boundary terms do not vanish.

For $Ra > 0$, no oscillatory modes are expected (see Appendix B) and we may consider only real perturbations. It is then possible to derive a variational principle defining the sufficient conditions for stability of the diffusive reference state. Stability is guaranteed if $dE/dt \leq 0$ for every perturbation (w, θ) satisfying the boundary conditions. Thus, the system is stable to any small perturbation if the Rayleigh number Ra is lower than Ra_E given by

$$\frac{1}{Ra_E} = \max_{\theta(z), w(z)} \frac{-k^2 I_{th} + 2k^2 \int_0^1 dz\theta w}{I_{visc}} \tag{C.7}$$

This energy bound depends on k if the maximization is done at fixed k. We may also obtain the lowest Ra_E (below which any perturbation of any k decays) by further maximizing Eq. (C.7) with respect to k. This leads to Eq. (3.111) of §3.3.1.

In fact, it may be rigorously shown [20] that Ra_E given by Eq. (C.7) is equal to the linear stability threshold Ra_c, above which perturbations are linearly amplified (i.e. the sufficient condition for instability). Indeed, the derivation of Euler–Lagrange equations for the maximization of the numerator of Eq. (C.7) with the constraint $I_{visc} = 1$ (normalization condition) leads to precisely the same differential problem as for neutral stability. Then, solving this problem leads to the optimal values of w and θ, and to the value of the Lagrange multiplier (assuming the role of a Rayleigh number) introduced to accommodate the normalization constraint $I_{visc} = 1$. Finally, substitution of the result into Eq. (C.7) then leads to $Ra_E = Ra_c$, i.e. the linear stability threshold and the energy bound coincide.

More importantly, this result continues to hold even when incorporating nonlinear advective terms in the basic equations (in fact they are canceled in the process of integrating over the volume, because of the incompressibility condition and by the use of Green's identity). This shows that for the full nonlinear Rayleigh–Bénard problem (with any of the sets of boundary conditions considered in §3.3.1), the sufficient condition for linear instability ($Ra > Ra_c$) and the sufficient condition for nonlinear stability ($Ra < Ra_E$) coincide ($Ra_E = Ra_c$), i.e. no subcritical instability is possible. Note that this result no longer holds in the presence of the Marangoni effect (in which case $Ma_E < Ma_c$, see §4.4.1 and Appendix F), thus indicating the possibility of subcritical convective motions.

D Adjoint problems for Rayleigh–Marangoni–Bénard instabilities

We define here the scalar products and give the adjoint problems for one-layer and two-layer Rayleigh–Marangoni–Bénard instabilities with undeformable interface.

D.1 One-layer Rayleigh–Marangoni–Bénard instability

The linear problem (3.99–3.100) reads, for normal modes,

$$(D^2 - k^2)^2 w - Rak^2\theta = Pr^{-1}\sigma(D^2 - k^2)w \tag{D.1}$$

$$(D^2 - k^2)\theta + w = \sigma\theta \tag{D.2}$$

with boundary conditions [see Eqs (3.114–3.115)]

$$w = Dw = \theta = 0 \quad at \ z = 0 \tag{D.3}$$

and

$$w = D\theta + Bi\theta = D^2 w + k^2 Ma\theta = 0 \quad at \ z = 1 \tag{D.4}$$

It is convenient to rewrite this problem under the operational form

$$\mathcal{L}(U) = MaM(U) + \sigma T(U) \tag{D.5}$$

with

$$\mathcal{L}(U) = \begin{pmatrix} (D^2 - k^2)^2 w \\ (D^2 - k^2)\theta + w \\ D^2 w|_{z=1} \end{pmatrix}, \quad M(U) = \begin{pmatrix} k^2 Bo\theta \\ 0 \\ -k^2\theta|_{z=1} \end{pmatrix},$$

$$T(U) = \begin{pmatrix} Pr^{-1}(D^2 - k^2)w \\ \theta \\ 0 \end{pmatrix} \tag{D.6}$$

where $Bo = Ra/Ma$ is the dynamic Bond number. We further define the domain of operators $\mathcal{L}$, M and T as the set of vectors

$$U = \begin{pmatrix} w \\ \theta \\ \theta|_{z=1} \end{pmatrix} \tag{D.7}$$

satisfying the boundary conditions $w(0) = Dw(0) = \theta(0) = 0$ and $w(1) = D\theta(1) + Bi\theta(1) = 0$. We then define the adjoint vectors (domain of adjoint operators) by

$$\tilde{U} = \begin{pmatrix} \tilde{w} \\ \tilde{\theta} \\ D\tilde{w}|_{z=1} \end{pmatrix} \tag{D.8}$$

satisfying the boundary conditions $\tilde{w}(0) = D\tilde{w}(0) = \tilde{\theta}(0) = 0$ and $\tilde{w}(1) = D^2\tilde{w}(1) = 0$. For every vectors $a = [a_1(z), a_2(z), a_3]$ and $b = [b_1(z), b_2(z), b_3]$, a scalar product is formed as

$$\langle a, b \rangle = \int_0^1 dz \left[a_1^* b_1 + a_2^* b_2 \right] + a_3^* b_3 \tag{D.9}$$

where an asterisk denotes the complex conjugate. Then, for every $U, \tilde{U}$ satisfying the above specified boundary conditions, it may be shown that

$$\begin{aligned} \left\langle \tilde{U}, \mathcal{L}(U) \right\rangle &= \left\langle \mathcal{L}^+(\tilde{U}), U \right\rangle \\ \left\langle \tilde{U}, M(U) \right\rangle &= \left\langle M^+(\tilde{U}), U \right\rangle \\ \left\langle \tilde{U}, T(U) \right\rangle &= \left\langle T^+(\tilde{U}), U \right\rangle \end{aligned} \tag{D.10}$$

with the adjoint operators

$$\mathcal{L}^+(\tilde{U}) = \begin{pmatrix} (D^2 - k^2)^2 \tilde{w} + \tilde{\theta} \\ (D^2 - k^2)\tilde{\theta} \\ -(D\tilde{\theta} + Bi\tilde{\theta})|_{z=1} \end{pmatrix}, \quad M^+(\tilde{U}) = \begin{pmatrix} 0 \\ k^2 Bo\tilde{w} \\ -k^2 D\tilde{w}|_{z=1} \end{pmatrix}, \tag{D.11}$$

$$T^+(\tilde{U}) = \begin{pmatrix} Pr^{-1}(D^2 - k^2)\tilde{w} \\ \tilde{\theta} \\ 0 \end{pmatrix} \tag{D.12}$$

Thus, the adjoint problem (associated with the above scalar product) reads

$$(D^2 - k^2)^2 \tilde{w} + \tilde{\theta} = Pr^{-1}\sigma^*(D^2 - k^2)\tilde{w} \tag{D.13}$$

$$(D^2 - k^2)\tilde{\theta} - k^2 Ra\tilde{w} = \sigma^*\tilde{\theta} \tag{D.14}$$

with boundary conditions

$$\tilde{w} = D\tilde{w} = \tilde{\theta} = 0 \quad at \ z = 0 \tag{D.15}$$

and

$$\tilde{w} = D^2\tilde{w} = D\tilde{\theta} + Bi\tilde{\theta} - k^2 MaD\tilde{w} = 0 \quad at \ z = 1 \tag{D.16}$$

Note that the adjoint problem is in general not unique. The above form is indeed dependent upon the choice of operators and/or scalar products. For example, we might have started from the equivalent problem (3.95–3.97) in which the horizontal velocity components and pressure have not yet been eliminated (this would have necessitated the definition of vectors U with more components), and this would have led to a different result for the adjoint problem (see Chapter 4). In fact, this is rather unimportant (the components of the adjoint vectors have no direct physical meaning), as long as the projection operations defined by the scalar products lead to the same physical results.

D.2 Two-layer Rayleigh–Marangoni–Bénard instability

Here, the linear problem [Eqs (3.141–3.148)] is

$$(D^2 - k^2)^2 w_1 - k^2 Ra_1 \theta_1 = \sigma Pr_1^{-1}(D^2 - k^2)w_1 \tag{D.17}$$

$$(D^2 - k^2)\theta_1 + w_1 = \sigma \theta_1 \tag{D.18}$$

$$(D^2 - k^2)^2 w_2 - k^2 \alpha \nu^{-1} Ra_1 \theta_2 = \sigma \nu^{-1} Pr_1^{-1}(D^2 - k^2)w_2 \tag{D.19}$$

$$(D^2 - k^2)\theta_2 + \beta \kappa^{-1} w_2 = \sigma \kappa^{-1} \theta_2 \tag{D.20}$$

with boundary conditions on bottom and top surfaces

$$w_1 = Dw_1 = \theta_1 = 0 \quad at\ z = 1 \tag{D.21}$$

$$w_2 = Dw_2 = \theta_2 = 0 \quad at\ z = a \tag{D.22}$$

and at the undeformable interface

$$w_1 = w_2 = 0,\ Dw_1 = Dw_2,\ \theta_1 = \theta_2,\ \lambda D\theta_2 = D\theta_1 \quad at\ z = 0 \tag{D.23}$$

$$\mu D^2 w_2 - D^2 w_1 = k^2 Ma_1 \theta_1 \quad at\ z = 0 \tag{D.24}$$

We may again rewrite this problem under the operational form

$$\mathcal{L}(U) = Ma_1\, M(U) + \sigma T(U) \tag{D.25}$$

now with

$$\mathcal{L}(U) = \begin{pmatrix} (D^2 - k^2)^2 w \\ (D^2 - k^2)\theta + \tilde{\beta}\tilde{\kappa}^{-1} w \\ \mu D^2 w_2 - D^2 w_1|_{z=0} \end{pmatrix},\ M(U) = \begin{pmatrix} k^2 \tilde{\alpha}\tilde{\nu}^{-1} Bo_1 \theta \\ 0 \\ k^2 \theta_1|_{z=0} \end{pmatrix} \tag{D.26}$$

$$T(U) = \begin{pmatrix} \tilde{\nu}^{-1} Pr_1^{-1} (D^2 - k^2) w \\ \tilde{\kappa}^{-1} \theta \\ 0 \end{pmatrix} \tag{D.27}$$

where an abbreviated notation has been used: the two first lines of each operator should be written for both layers (in the lower layer $-1 < z < 0$, all tilded quantities should be taken equal to unity, while in the upper layer $0 < z < a$, these quantities stand for the corresponding property ratios, i.e. $\tilde{\nu} = \nu$, $\tilde{\kappa} = \kappa$, ...). Note that $Bo_1 = Ra_1/Ma_1$ is the dynamic Bond number defined for layer 1 (the lower one).

Thus, the Marangoni condition (D.24) is included as the last component of the problem (D 25), while the remaining boundary conditions (D.21–D.23) should be satisfied by every vector U to which operators apply. Vectors U and $\tilde{U}$ are now defined by

$$U = \begin{pmatrix} w \\ \theta \\ \theta|_{z=0} \end{pmatrix}, \quad \tilde{U} = \begin{pmatrix} \tilde{w} \\ \tilde{\theta} \\ -D\tilde{w}|_{z=0} \end{pmatrix} \tag{D.28}$$

where the two first components should be considered as functions defined over the two layers (e.g. $w = w_1$ for $-1 < z < 0$ and $w = w_2$ for $0 < z < a$). We now define a scalar product as

$$\langle x, y \rangle = \int_{-1}^{a} dz \left[x_1^* y_1 + x_2^* y_2 \right] + x_3^* y_3 \tag{D.29}$$

Then it may be shown that the boundary conditions on $\tilde{U}$ are

$$\tilde{w}_1 = D\tilde{w}_1 = \tilde{\theta}_1 = 0 \quad at \; z = -1 \tag{D.30}$$

$$\tilde{w}_2 = D\tilde{w}_2 = \tilde{\theta}_2 = 0 \quad at \; z = a \tag{D.31}$$

$$\tilde{w}_1 = \tilde{w}_2 = 0, \; \mu D\tilde{w}_1 = D\tilde{w}_2, \; D^2\tilde{w}_1 = D^2\tilde{w}_2, \; \lambda\tilde{\theta}_1 = \tilde{\theta}_2 \quad at \; z = 0 \tag{D.32}$$

and the adjoint operators [still defined by Eq. (D.10)] read as

$$\mathcal{L}^+(\tilde{U}) = \begin{pmatrix} (D^2 - k^2)^2 \tilde{w} + \tilde{\beta}\tilde{\kappa}^{-1}\tilde{\theta} \\ (D^2 - k^2)\tilde{\theta} \\ D\tilde{\theta}_2 - D\tilde{\theta}_1|_{z=0} \end{pmatrix}, \quad M^+(\tilde{U}) = \begin{pmatrix} 0 \\ k^2 Bo_1 \tilde{\alpha}\tilde{\nu}^{-1}\tilde{w} \\ -k^2 D\tilde{w}_1|_{z=0} \end{pmatrix}, \tag{D.33}$$

$$T^+(\tilde{U}) = \begin{pmatrix} \tilde{\nu}^{-1} Pr_1^{-1} (D^2 - k^2) \tilde{w} \\ \tilde{\kappa}^{-1}\tilde{\theta} \\ 0 \end{pmatrix} \tag{D.34}$$

E   Fredholm solvability condition

Keeping notations of §4.2.2, let us attempt a resolution of the differential problem

$$\mathcal{L}(U_2) - \mu_c M(U_2) = \exp[i\vec{k}.\vec{r}]F(z) \tag{E.1}$$

with U_2 belonging to E [i.e. satisfying boundary conditions not included in (E.1)].
   Setting

$$U_2 = \exp[i\vec{k}.\vec{r}]U_{2\vec{k}}(z) \tag{E.2}$$

leads to the problem

$$\mathcal{L}_{\vec{k}}(U_{2\vec{k}}) - \mu_c M_{\vec{k}}(U_{2\vec{k}}) = F(z) \tag{E.3}$$

with $U_{2\vec{k}} \in E_{\vec{k}}$. Solving this inhomogeneous problem (IP) is possible in principle by the superposition of the General Solution of the Homogeneous Problem (GSHP) and of a Particular Solution of the Inhomogeneous Problem (PSIP). For the moment, imagine that we only solve the differential equations without expressing the boundary conditions. Then, PSIP is completely determined, but GSHP depends on a number of unknown integration constants equal to the order of the differential problem (say, n). Expressing the boundary conditions [both those included in (E.3) and those defining $E_{\vec{k}}$] will lead to a $n \times n$ linear inhomogeneous system for the unknown coefficients. If X is the vector of these n coefficients, this problem reads $AX = B$ where A is a $n \times n$ matrix and B is a non-zero vector of order n. However, without running into complicated notations, it is possible to see that the determinant of A precisely gives the compatibility condition of the neutral stability problem. This is because GSHP is a solution of the neutral stability problem [Eq. (E.3) with $F(z) = 0$], and the system $AX = 0$ admits non-trivial solutions only provided the neutral stability condition $\det(A) = 0$ is satisfied. Thus, if $|\vec{k}| \neq k_c$ (non-resonant case), $\det(A) \neq 0$ and the unknown constants for the IP may be found as $X = A^{-1}B$ (thus the problem is solved). On the contrary, if $|\vec{k}| = k_c$ (resonant case), $\det(A) = 0$ and the problem cannot be solved unless a condition is satisfied by B. This is the Fredholm solvability condition, which (if the compatibility condition is not degenerated, i.e. the rank of A is $n-1$), reads $\tilde{X}.B = 0$ where $\tilde{X}$ is the solution of the adjoint problem $A^+\tilde{X} = 0$. Here, A^+ is the adjoint (transposed conjugate) of A.
   Actually, it is not necessary to solve the system (E.3) to find the Fredholm solvability condition. Indeed, we may rather use the definition of the adjoint problem of the linear stability problem

$$\mathcal{L}_{\vec{k}}(U) - \mu_c M_{\vec{k}}(U) = 0 \tag{E.4}$$

with $U \in E_{\vec{k}}$.

According to the developments presented in Appendices B and D, it is possible to define a scalar product (dependent upon the problem studied) and an adjoint problem associated with Eq. (E.4). Denoting the scalar product as usual by $\langle a, b \rangle$, the adjoint problem of (E.4) then satisfies [see Eqs (B.10)]

$$\left\langle \tilde{U}, \mathcal{L}_{\vec{k}}(U) - \mu_{\mathrm{c}} M_{\vec{k}}(U) \right\rangle = \left\langle \mathcal{L}_{\vec{k}}^{+}(\tilde{U}) - \mu_{\mathrm{c}} M_{\vec{k}}^{+}(\tilde{U}), U \right\rangle \tag{E.5}$$

for all $U \in E_{\vec{k}}$ and $\tilde{U} \in F_{\vec{k}}$, where $F_{\vec{k}}$ is a certain set of boundary conditions.

In particular, if $\tilde{U}_{0\vec{k}}$ is a solution of the adjoint problem

$$\mathcal{L}_{\vec{k}}^{+}(\tilde{U}_{0\vec{k}}) - \mu_{\mathrm{c}} M_{\vec{k}}^{+}(\tilde{U}_{0\vec{k}}) = 0 \tag{E.6}$$

with $\tilde{U}_{0\vec{k}} \in F_{\vec{k}}$, then the condition (E.5) implies

$$\left\langle \tilde{U}_{0\vec{k}}, \mathcal{L}_{\vec{k}}(U) - \mu_{\mathrm{c}} M_{\vec{k}}(U) \right\rangle = 0 \tag{E.7}$$

for all $U \in E_{\vec{k}}$.

From this property, it is readily seen that the problem (E.3) with $U_{2\vec{k}} \in E_{\vec{k}}$ may be solved only if

$$\left\langle \tilde{U}_{0\vec{k}}, F(z) \right\rangle = 0 \tag{E.8}$$

i.e. only if F is orthogonal [in the sense of Eq. (E.8)] to every solution of the adjoint problem (Fredholm solvability condition). Actually, it is clear that this is only a necessary (but not a sufficient) condition. However, this will be sufficient for our purpose of finding the solvability conditions. In any case, a supplementary check of this solvability condition (and of the fact that it is sufficient) may be obtained by solving (E.3) directly as described above.

A first application of the definition of the adjoint problem is the following. Consider the spectral problem

$$\mathcal{L}_{\vec{k}}(U) - \mu M_{\vec{k}}(U) = \sigma \Theta_{\vec{k}}(U) \tag{E.9}$$

with $U \in E_{\vec{k}}$. This is the general form of the linear stability problems studied in Chapter 3, leading to the full dispersion relation $\sigma = \sigma(\mu, k)$ and to the corresponding eigenfunction $U = U_{\mu,\vec{k}}$. Now, consider a particular k (e.g. $k = k_{\mathrm{c}}$) and take the derivative of Eq. (E.9) with respect to μ evaluated at $\mu = \mu_{\mathrm{c}}$ (where $\sigma = 0$ and $U_{\mu_{\mathrm{c}},\vec{k}} = U_{0\vec{k}}$). This may be written in the form of a differential problem for $U'_{\vec{k}} = \partial U_{\mu,\vec{k}} / \partial \mu$, namely

$$\mathcal{L}_{\vec{k}}(U'_{\vec{k}}) - \mu_{\mathrm{c}} M_{\vec{k}}(U'_{\vec{k}}) = \frac{\partial \sigma}{\partial \mu}\big|_{\mu=\mu_{\mathrm{c}}} \Theta_{\vec{k}}(U_{0\vec{k}}) + M_{\vec{k}}(U_{0\vec{k}}) \tag{E.10}$$

whose compatibility condition [Eq. (E.8)] leads to

$$\frac{\partial \sigma}{\partial \mu}\big|_{\mu=\mu_{\mathrm{c}}} = -\frac{\left\langle \tilde{U}_{0\vec{k}}, M_{\vec{k}}(U_{0\vec{k}}) \right\rangle}{\left\langle \tilde{U}_{0\vec{k}}, \Theta_{\vec{k}}(U_{0\vec{k}}) \right\rangle} \tag{E.11}$$

thus providing an expression for the growth rate derivative with respect to the control parameter, in terms of scalar products involving neutral stability eigenfunctions only. This allows easy testing of the assumption of strict loss of stability $\partial\sigma/\partial\mu|_{\mu=\mu_c} > 0$ [4].

F   Energy stability theory for the Marangoni–Bénard problem

Rescaling the velocities as $\vec{V}' = Ma\vec{V}$ (and omitting the primes in what follows), the pure Marangoni–Bénard problem [see e.g. Eqs (4.152–4.154) with $Ra = 0$] reads

$$Ma\Delta\vec{V} - \vec{\nabla}p = Pr^{-1}Ma\left\{\frac{\partial\vec{V}}{\partial t} + Ma(\vec{V}.\vec{\nabla})\vec{V}\right\} \tag{F.1}$$

$$\Delta T + MaW = \frac{\partial T}{\partial t} + Ma(\vec{V}.\vec{\nabla})T \tag{F.2}$$

together with $\vec{\nabla}.\vec{V} = 0$ and the usual boundary conditions $W = DW = T = 0$ at $z = 0$, and $W = DT = D\vec{V}_{\mathrm{h}} + \vec{\nabla}_{\mathrm{h}}T = 0$ at $z = 1$.

We introduce the energy functional

$$E = \frac{1}{2}\int_V \left[\mu T^2 + MaPr^{-1}V^2\right]d\vec{r} \tag{F.3}$$

where μ is a yet undetermined positive constant ("optimizing" parameter [20]), Ma is taken as being positive, and the integral extends over the volume of the fluid (the horizontal integration is taken as the mean value).

Then, after integration by parts and use of the boundary conditions, it may be shown that dE/dt is negative (implying stability to disturbances of any amplitude) provided $Ma < Ma_{\mathrm{E}}$, where Ma_{E} is given by

$$\frac{1}{Ma_{\mathrm{E}}} = \max\left\{\frac{\mu\int_V TW\,d\vec{r} - \int_V(\vec{\nabla}\times\vec{V})^2 d\vec{r} - \oint_S\left[(\vec{V}_{\mathrm{h}}.\vec{\nabla}_{\mathrm{h}})T\right]_{z=1}\vec{dS}}{\mu\int_V(\vec{\nabla}T)^2}\right\} \tag{F.4}$$

and the maximization is taken over all solenoidal velocity fields $\vec{V}$ and temperature fields T satisfying the boundary conditions. Actually, we will also choose the auxiliary parameter μ such as to minimize the result of this maximization [20]. This is done with the intention of selecting an optimal form of the energy (F.3), leading to the sharpest possible energy threshold Ma_{E}, i.e. as close as possible to the linear stability threshold Ma_{c} (such as to minimize the region $Ma_{\mathrm{E}} < Ma < Ma_{\mathrm{c}}$ of possible instability to finite-amplitude disturbances).

It is important to note that the nonlinear terms in (F.1–F.2) do not contribute to the variational principle (F.4), because they vanish in the integration process according to the relations (4.185).

Thus, the problem amounts to determine the maximum of I_1/I_2 [where I_1 and I_2 denote the numerator and denominator of Eq. (F.4)] over fields satisfying the continuity constraint $\vec{\nabla} \cdot \vec{V} = 0$. This may be accommodated by using a Lagrange multiplier $\lambda(\vec{r})$, such that we will maximize the function

$$\phi = \frac{I_1}{I_2} + \int_V \lambda(\vec{r})\vec{\nabla}\cdot\vec{V}\,d\vec{r} \tag{F.5}$$

which is extremal provided $\delta\phi = 0$. Developing this condition using $Ma_{\mathrm{E}} = I_2^*/I_1^*$ (where I_1^* and I_2^* denote values of I_1 and I_2 at the extremum), and redefining $p(\vec{r}) = \lambda(\vec{r})I_2^*$, we get the condition

$$\delta I_1 - Ma_{\mathrm{E}}^{-1}\delta I_2 - \int_V \delta\vec{V}\cdot\vec{\nabla}p\,d\vec{r} = 0 \tag{F.6}$$

Developing this condition and integrating by parts (using the divergence theorem) such as to factorize δT and $\delta\vec{V}$ [the relation $\vec{\nabla}\cdot(\vec{a}\times\vec{b}) = (\vec{\nabla}\times\vec{a})\cdot\vec{b} - \vec{a}\cdot(\vec{\nabla}\times\vec{b})$ is also needed], we see that the Euler–Lagrange equations for the maximization of (F.5) reduce to the system

$$\vec{\nabla}\cdot\vec{V} = 0 \tag{F.7}$$

$$2\Delta\vec{V} - \vec{\nabla}p + \mu T\vec{1}_z = 0 \tag{F.8}$$

$$\frac{Ma_{\mathrm{E}}}{2}W + \Delta T = 0 \tag{F.9}$$

with boundary conditions

$$\vec{V} = T = 0 \quad \text{at } z = 0 \tag{F.10}$$

and

$$W = \vec{\nabla}_{\mathrm{h}}\cdot\vec{V}_{\mathrm{h}} - 2\mu Ma_{\mathrm{E}}^{-1}DT = 2D\vec{V}_{\mathrm{h}} + \vec{\nabla}_{\mathrm{h}}T = 0 \quad \text{at } z = 1 \tag{F.11}$$

Note that the boundary conditions at the interface $z = 1$ are the natural boundary conditions following from the cancelation of the boundary terms in (F.6). Actually, these differ from the usual conditions to be satisfied in the original problem ($W = DT = D\vec{V}_{\mathrm{h}} + \vec{\nabla}_{\mathrm{h}}T = 0$). However, we conjecture that this is unimportant because the physical boundary conditions can always be satisfied by a small change in the functions T and W solutions of Eqs (F.7–F.11) in the vicinity of the interface, thereby not affecting the value of the functional (F.4). This has been verified by numerical maximization (using normal modes of wavenumber k and polynomials of increasing order in the z-direction) of the functional (F.4), leading to numerical values of Ma_{E} within 3% of the exact value given in §4.4.1, even for relatively low-order polynomials (order 8). This also provides a confirmation that the extremum indeed corresponds to a maximum of (F.4).

G   Amplitude equation coefficients for the two-layer Rayleigh–Bénard problem

We give here the explicit form of the system of ODEs (5.54): it can be written as

$$\dot{B}_{m,n} = \sum_p \sigma_{n,m,p} B_{m,p} + Ra \sum_{p,q,r} \Psi_{n,m,p,q,r} B_{p,q} B_{m-p,r} \tag{G.1}$$

expressions for the coefficients of linear terms appearing in (G.1) are easily seen to be

$$\sigma_{n,m,p} = Ra\Theta_{n,m,p} + \Lambda_{n,m,p} \tag{G.2}$$

such that the matrix $A = Ra\Theta + \Lambda$ appearing in (5.54) is linear in the Rayleigh number.

Since $T_{i,m,n}(z)$ form a complete orthonormal set of temperature functions on $[0, 1+a]$, i.e.

$$\int_0^1 T_{1\,m,n} T_{1\,m,p} dz + \int_1^{1+a} T_{2\,m,n} T_{2\,m,p} dz = \delta_{m,p} \tag{G.3}$$

the expressions for Θ and Λ are

$$\Theta_{n,m,p} = \int_0^1 T_{1\,m,n} W_{1\,m,p} dz + \lambda \int_1^{1+a} T_{2\,m,n} W_{2\,m,p} dz$$
$$\Lambda_{n,m,p} = \int_0^1 T_{1\,m,n}(D^2 - m^2 k_0^2) T_{1\,m,p} dz + \kappa^{-1} \int_1^{1+a} T_{2\,m,n}(D^2 - m^2 k_0^2) T_{2\,m,p} dz$$

in which the velocity functions $W_{i,m,n}(z)$ are solutions of (5.51) satisfying the boundary conditions (5.42), (5.44) and (5.45).

Coefficients of the bilinear form $N(u,v)$ are given by

$$\Psi_{n,m,p,q,r} = \int_0^1 T_{1\,m,n}(\tfrac{m-p}{p} DW_{1\,p,q} T_{1\,m-p,r} - W_{1\,p,q} DT_{1\,m-p,r}) dz$$
$$+ \int_1^{1+a} T_{2\,m,n}(\tfrac{m-p}{p} DW_{2\,p,q} T_{2\,m-p,r} - W_{2\,p,q} DT_{2\,m-p,r}) dz$$

It can be seen that $\Psi_{n,-m,-p,q,r} = \Psi_{n,m,p,q,r}$, and that $\Psi_{n,m,0,q,r} = 0$ [since $W_{i,p,q}$ has a dependency $O(k^2)$ in the wavenumber k for $k \to 0$]. We will write the system (5.54) under a form that isolates the Rayleigh number, i.e.

$$\dot{B} = (Ra\Theta + \Lambda)B + RaM(B,B) \tag{G.4}$$

We define a scalar product by $< \xi, \eta >= \sum_{m,n} \xi_{mn} \bar{\eta}_{mn}$, and note $A_0 = Ra_c \Theta + \Lambda$, in which Ra_c is the critical value of the Rayleigh number corresponding to the critical frequency ω_c. A_0 is a block matrix, whose blocks are sub-matrices S_i [i corresponds to the horizontal wavenumber, and components of S_i are $(S_i)_{m,n} = Ra\Theta_{m,i,n} + \Lambda_{m,i,n}$]. The spectral problem will be written

$$A_0 \zeta_{0\pm} = i\omega_0 \zeta_{0\pm}$$
$$A_0 \bar{\zeta}_{0\pm} = -i\omega_0 \bar{\zeta}_{0\pm} \qquad (G.5)$$

where the overbar denotes the complex conjugate. ζ_{0+} and ζ_{0-} correspond respectively to positive and negative directions of the wavevector, i.e. $(\zeta_{0+})_{mn} = \delta_{m,1} \cdot \zeta_{0n}$ and $(\zeta_{0-})_{mn} = \delta_{m,-1} \cdot \zeta_{0n}$ where the vector ζ_0 is the solution of $S_1 \zeta_0 = i\omega_0 \zeta_0$. The orthonormalization conditions are

$$< \zeta_{0+}, \zeta_{0+}^* >=< \zeta_{0-}, \zeta_{0-}^* >= 1$$
$$< \zeta_{0+}, \zeta_{0-}^* >=< \zeta_{0-}, \zeta_{0+}^* >=< \zeta_{0\pm}, \bar{\zeta}_{0\pm}^* >=< \bar{\zeta}_{0\pm}, \zeta_{0\pm}^* >= 0 \qquad (G.6)$$

where the right vectors are the solutions of the adjoint problem

$$A_0^* \zeta_{0\pm}^* = -i\omega_0 \zeta_{0\pm}^*$$
$$A_0^* \bar{\zeta}_{0\pm}^* = i\omega_0 \bar{\zeta}_{0\pm}^* \qquad (G.7)$$

where A_0^*, defined by $< A_0 \xi, \eta >=< \xi, A_0^* \eta >$ for all ξ, η, is simply the transposed matrix of A_0. It can be computed [4] that the growth constant derivative with respect to the Rayleigh number (at criticality) can be written

$$\sigma_{Ra} =< \Theta \zeta_{0+}, \zeta_{0+}^* >=< \Theta \zeta_{0-}, \zeta_{0-}^* > \qquad (G.8)$$

Note that the real part of σ_{Ra} is positive in our problem, such that the hypothesis of strict loss of stability holds. In order to solve (G.4), we use expansions (5.58), after having defined the variable $s = \omega(\epsilon)t$. Identification of successive powers in ϵ gives $J_0 v_1 = 0$ at the first order, where J_0 is the linear operator defined by

$$J_0 \equiv -\omega_0 \frac{\partial}{\partial s} + A_0 \qquad (G.9)$$

which acts on 2π-periodic functions of s. We define the scalar product by

$$[a(s), b(s)] = \frac{1}{2\pi} \int_0^{2\pi} < a(s), b(s) > ds \qquad (G.10)$$

and the adjoint of J_0 is

$$J_0^* \equiv \omega_0 \frac{\partial}{\partial s} + A_0^* \qquad (G.11)$$

The solution of the first order problem $J_0 v_1 = 0$ is given by (5.59), which is the linear combination of independent vectors annihilated by J_0. The second order problem can be written

$$J_0 v_2 = 2 \left[\omega_1 \frac{\partial v_1}{\partial s} - Ra_1 \Theta v_1 - Ra_c M(v_1, v_1) \right] \qquad (G.12)$$

whose unique solution is given by (5.61), having taken into account Eq. (5.60). Indeed, this last equation expresses the compatibility conditions of (G.12), i.e. orthogonality of the second member of (G.12) with linearly independent solutions of the adjoint problem $J_0^* z^* = 0$ (which are $\exp[is]\zeta_{0\pm}^*$ and their complex conjugates). In Eq. (5.61), v_2^I and v_2^{III} are given by

$$
\begin{aligned}
(v_2^I)_{mn} &= c_-^2 \delta_{m,-2} u_{1n} + c_+^2 \delta_{m,+2} u_{1n} + c_+ c_- \delta_{m,0} \bar{u}_{0n} \\
(v_2^{III})_{mn} &= (|c_+|^2 + |c_-^2|) \delta_{m,0} u_{2n} + \bar{c}_+ c_- u_{3n} \delta_{m,-2} + c_+ \bar{c}_- u_{3n} \delta_{m,+2}
\end{aligned}
\tag{G.13}
$$

in which u_0, u_1, u_2, u_3 are determined by solving linear inhomogeneous problems obtained from the second order problem (G.12), written as $J_0 v_2 = -2 Ra_c M(v_1, v_1)$.

By assuming that c_+ and c_- depend on the slow time scale $t_2 = \epsilon^2 t$, the third order problem can be written

$$
J_0 v_3 = 3 \left\{ \omega_2 \partial v_1 / \partial s + 2 \partial v_1 / \partial t_2 - Ra_2 \Theta v_1 - Ra_c \left[M(v_1, v_2) + M(v_2, v_1) \right] \right\}
$$

whose compatibility conditions finally give the amplitude equations (5.62), in which $c_\pm$ has been rescaled as $\epsilon^{-1} c_\pm$.

The coefficients α and β are given by

$$
\begin{aligned}
\alpha &= \tfrac{1}{2} Ra_c \sum_{r,n,q} \bar{\zeta}_{0r}^* (\Psi_{r,1,1,n,q} \zeta_{0n} \bar{u}_{2q} + \Psi_{r,-1,1,n,q} \bar{\zeta}_{0n} u_{1q} \\
&\qquad\qquad\qquad + \Psi_{r,1,2,n,q} \bar{\zeta}_{0q} u_{1n}) \\
\beta &= \tfrac{1}{2} Ra_c \sum_{r,n,q} \bar{\zeta}_{0r}^* (\Psi_{r,1,1,n,q} \zeta_{0n} \bar{u}_{2q} + \Psi_{r,-1,1,n,q} \zeta_{0n} u_{3q} \\
&\qquad\qquad\qquad + \Psi_{r,1,2,n,q} \zeta_{0q} u_{3n} + \Psi_{r,1,1,n,q} \zeta_{0n} \bar{u}_{0q})
\end{aligned}
\tag{G.14}
$$

Orthonormal temperature functions:

We show here how an orthonormal set of temperature functions can be constructed in the case $\lambda \neq 1$. We first consider the set of linearly independent polynomial functions defined by

$$
\begin{aligned}
T_{1n}(z) &= z(a_{1n1} z^{n-1} + a_{1n2} z^n) \\
T_{2n}(z) &= (z - 1 - a)(a_{2n1} z^{n-1} + a_{2n2} z^n)
\end{aligned}
\tag{G.15}
$$

which satisfy the conducting conditions at $z = 0$ and $z = 1 + a$. We then fix two of the unknown coefficients (say a_{1n1} and a_{2n1}) such that interface conditions (5.43) are satisfied. Then, two linearly independent functions are obtained when $a_{1n2} = 0$ or $a_{2n2} = 0$. The orthonormalization is then achieved via a Gram–Schmidt procedure.

Since a rigorous proof of the completeness of the resulting set is lacking up to now, all the results presented in the text of §5.4 have been obtained in the case $\lambda = 1$, for which a trigonometrical complete basis of sine functions exists.

H   Extrapolation of amplitude equations

H.1   Derivation of model cubic amplitude equations

Using Eq. (8.11), we may write Eq. (8.10) as a differential problem for $U_{\vec{k}}^D(z,t)$:

$$
\begin{aligned}
\mathcal{L}_{\vec{k}}(U_{\vec{k}}^D) - Ma M_{\vec{k}}(U_{\vec{k}}^D) &= -A_{\vec{k}}\left[\mathcal{L}_{\vec{k}}(U_{\vec{k}}^\sigma) - Ma M_{\vec{k}}(U_{\vec{k}}^\sigma)\right] + \frac{\partial A_{\vec{k}}}{\partial t}\Theta_{\vec{k}}(U_{\vec{k}}^\sigma) \\
&+ \int d\vec{k}' A_{\vec{k}'} A_{\vec{k}-\vec{k}'} N_{\vec{k}',\vec{k}-\vec{k}'}(U_{\vec{k}'}^\sigma, U_{\vec{k}-\vec{k}'}^\sigma) + \int d\vec{k}' N_{\vec{k}',\vec{k}-\vec{k}'}(U_{\vec{k}'}^D, U_{\vec{k}-\vec{k}'}^D) \\
&+ \int d\vec{k}' A_{\vec{k}'}\left[N_{\vec{k}',\vec{k}-\vec{k}'}(U_{\vec{k}'}^\sigma, U_{\vec{k}-\vec{k}'}^D) + N_{\vec{k}-\vec{k}',\vec{k}'}(U_{\vec{k}-\vec{k}'}^D, U_{\vec{k}'}^\sigma)\right]
\end{aligned}
\tag{H.1}
$$

where the time-derivative of $U_{\vec{k}}^D$ has been canceled, as a result of our assumption to neglect the own dynamics of the damped modes (slaving principle). Note that because of Eq. (8.11), the boundary conditions can only be satisfied if $U_{\vec{k}}^D$ belongs to E. A very rough (and insufficient) model could be obtained at this stage by projection of Eq. (H.1) on some functions (generally the adjoint functions [490]), and assuming $U_{\vec{k}}^D = 0$. We would then obtain an equation of the form (8.16), but without stabilizing cubic terms. Rather, we will try to solve (H.1) for $U_{\vec{k}}^D$. This can be done only if (H.1) is compatible, which is not the case at $Ma = Ma_k$ (the kernel of the left-hand side operator is then not empty), except if the second member is orthogonal to the solution of the adjoint neutral stability problem (Fredholm's condition). This leads to the amplitude equations

$$
\frac{\partial A_{\vec{k}}}{\partial t} = \sigma_k A_{\vec{k}} + \int d\vec{k}' Z_{\vec{k}'\vec{k}} A_{\vec{k}'} A_{\vec{k}-\vec{k}'} - \int d\vec{k}' A_{\vec{k}'} \frac{\left\langle V_{\vec{k}}^*, N_{\vec{k}',\vec{k}-\vec{k}'}(U_{\vec{k}'}^\sigma, U_{\vec{k}-\vec{k}'}^D)\right\rangle}{\tau_k}
\tag{H.2}
$$

where Eq. (8.12) has been used, and it has been anticipated that the velocity (and pressure) components of $U_{\vec{k}}^D$ are zero. This will become apparent later, and is a consequence of the linearity of the equations of motion. In (H.2), $< V_{\vec{k}}^*, . >$ denotes the projection on the adjoint neutral stability solution (derived hereafter), $\tau_k = < V_{\vec{k}}^*, \Theta_{\vec{k}}(U_{\vec{k}}^\sigma) >$ is the normalization factor, and the quadratic coefficients are given by

$$
Z_{\vec{k}'\vec{k}} = -\frac{\left\langle V_{\vec{k}}^*, N_{\vec{k}',\vec{k}-\vec{k}'}(U_{\vec{k}'}^\sigma, U_{\vec{k}-\vec{k}'}^\sigma)\right\rangle}{\tau_k}
\tag{H.3}
$$

Now, by inserting the projected part (H.2) into the complete equation (H.1), we obtain

$$\mathcal{L}_{\vec{k}}(U_{\vec{k}}^D) - MaM_{\vec{k}}(U_{\vec{k}}^D) = \int d\vec{k'} A_{\vec{k'}} A_{\vec{k}-\vec{k'}} N_{\vec{k'},\vec{k}-\vec{k'}}^{NR}(U_{\vec{k'}}^\sigma, U_{\vec{k}-\vec{k'}}^\sigma) \\ + \int d\vec{k'} A_{\vec{k'}} N_{\vec{k'},\vec{k}-\vec{k'}}^{NR}(U_{\vec{k'}}^\sigma, U_{\vec{k}-\vec{k'}}^D) \tag{H.4}$$

where the "Non-Resonant" part of a term X is defined by

$$X^{NR} = X - \tau_k^{-1} \left\langle V_{\vec{k}}^*, X \right\rangle \Theta_{\vec{k}}(U_{\vec{k}}^\sigma),$$

such that $< V_{\vec{k}}^*, X^{NR} >= 0$ for every X. This ensures that (H.4) is compatible for every Ma.

Now, the form of equation (H.4) suggests an iterative series solution, starting with

$$U_{\vec{k}}^D = \int d\vec{k'} A_{\vec{k'}} A_{\vec{k}-\vec{k'}} U_{\vec{k}\vec{k'}}^D \tag{H.5}$$

in which $U_{\vec{k}\vec{k'}}^D$ is obtained from

$$(\mathcal{L}_{\vec{k}} - MaM_{\vec{k}})U_{\vec{k}\vec{k'}}^D = N_{\vec{k'},\vec{k}-\vec{k'}}^{NR}(U_{\vec{k'}}^\sigma, U_{\vec{k}-\vec{k'}}^\sigma) \tag{H.6}$$

Substituting this result in the r.h.s. of (H.4) leads to higher-order corrections [cubic terms for $U_{\vec{k}}^D$, generating quartic and higher-order terms in (H.2), which are not considered in the model of §8.1]. We limit the calculation of $U_{\vec{k}}^D$ to (H.5) and (H.6), such that the compatibility equations (H.2) reduce to the amplitude equations (8.16), with quadratic coefficients given by (H 3) and cubic coefficients given by

$$Z_{\vec{k'}\vec{k''}\vec{k}} = -\frac{\left\langle V_{\vec{k}}^*, N_{\vec{k'},\vec{k}-\vec{k'}}(U_{\vec{k'}}^\sigma, U_{\vec{k}-\vec{k'},\vec{k''}}^D) \right\rangle}{\tau_k} \tag{H.7}$$

H.2   Solution of the linear problem

Starting from Eq. (8.12), the neutral stability problem ($\sigma_k = 0$) can be written as

$$S_{\vec{k}}(U_{\vec{k}}^0) = \mathcal{L}_{\vec{k}}(U_{\vec{k}}^0) - Ma_k M_{\vec{k}}(U_{\vec{k}}^0) = \begin{pmatrix} (D^2 - k^2)\vec{V}_{r\vec{k}}^0 - i\vec{k}p_k^0 \\ (D^2 - k^2)W_k^0 - Dp_k^0 \\ DW_k^0 + i\vec{k}.\vec{V}_{r\vec{k}}^0 \\ (D^2 - k^2)T_k^0 + W_k^0 \\ \left[D\vec{V}_{r\vec{k}}^0 + i\vec{k}Ma_k T_k^0 \right]_{z=0} \end{pmatrix} = 0 \tag{H.8}$$

The solution of this problem which belongs to E [i.e. which satisfies boundary conditions (8.2) and (8.3)] reads

$$U_{\vec{k}}^0 = \begin{pmatrix} \vec{V}_{r\vec{k}}^0 \\ W_k^0 \\ p_k^0 \\ T_k^0 \end{pmatrix} = \exp[kz] \begin{pmatrix} -4i\vec{k}(k + Bi)(1 + kz) \\ -4k^2(k + Bi)z \\ -8k^2(k + Bi) \\ 1 - (k + Bi)z + k(k + Bi)z^2 \end{pmatrix} \tag{H.9}$$

where $k = |\vec{k}|$, and the normalization condition has been chosen such that $T_k^0(z = 0) = 1$. The compatibility condition leads to the neutral stability relation $Ma_k = 8k(k + Bi)$.

We can now define adjoint vectors

$$
V^* = \begin{pmatrix} \vec{V}_r^* \\ W^* \\ p^* \\ T^* \\ \vec{X}^* \end{pmatrix} \tag{H.10}
$$

belonging to a set E^* of adjoint boundary conditions to be defined later on.

A scalar product is introduced in the usual way (e.g. [103, 130]) by

$$
\langle V^*, S_{\vec{k}}(U) \rangle = \int_{-\infty}^{0} \left[\vec{V}_r^* . \vec{S}_1 + \overline{W^*} S_2 + \overline{p^*} S_3 + \overline{T^*} S_4 \right] dz + \left[\vec{X}^* . \vec{S}_5 \right]_{z=0} \tag{H.11}
$$

where S_i $i = 1, ..., 5$ are the components of (H.8), and the overbar denotes the complex conjugate. We also define the adjoint operator $S_{\vec{k}}^*$ of $S_{\vec{k}}$ by

$$
\langle V^*, S_{\vec{k}}(U) \rangle = \langle S_{\vec{k}}^*(V^*), U \rangle \tag{H.12}
$$

for all U belonging to E and V^* to E^*. Integration by parts leads to

$$
S_{\vec{k}}^*(V^*) = \begin{pmatrix} (D^2 - k^2)\vec{V}_r^* - i\vec{k}p^* \\ (D^2 - k^2)W^* - Dp^* + T^* \\ DW^* + i\vec{k}.\vec{V}_r^* \\ (D^2 - k^2)T^* \end{pmatrix} \tag{H.13}
$$

and the cancelation of the boundary term gives E^* as the set of sufficiently derivable functions satisfying

$$
\vec{V}_r^*, W^*, DT^*, p^* \to 0 \quad \text{for } z \to -\infty \tag{H.14}
$$

$$
W^* = D\vec{V}_r^* = \vec{X}^* + \vec{V}_r^* = DT^* + BiT^* + iMa_k\vec{k}.\vec{X}^* = 0 \quad \text{for } z = 0 \tag{H.15}
$$

The resolution of the adjoint problem $S_{\vec{k}}^*(V_{\vec{k}}^*) = 0$ (with $V_{\vec{k}}^* \in E^*$) gives

$$
V_{\vec{k}}^* = \frac{-k \exp[kz]}{4(2k + Bi)} \begin{pmatrix} i\vec{k}k^{-2}(1 - kz - k^2z^2) \\ z(1 - kz) \\ -4(1 + kz) \\ -8k \\ -i\vec{k}k^{-2} \end{pmatrix} \tag{H.16}
$$

where the normalization $\langle V_{\vec{k}}^*, \Theta_{\vec{k}}(U_{\vec{k}}^0) \rangle = 1$ has been adopted.

Finally, the eigenfunctions of the spectral problem (8.12) read

$$
U_{\vec{k}}^{\sigma} = \begin{pmatrix} \vec{V}_{r\vec{k}}^{\sigma} \\ W_{\vec{k}}^{\sigma} \\ p_{\vec{k}}^{\sigma} \\ T_{\vec{k}}^{\sigma} \end{pmatrix} = -4k^2(k+Bi) \begin{pmatrix} i\vec{k}k^{-2}\exp[kz](1+kz) \\ z\exp[kz] \\ 2\exp[kz] \\ \exp[kz]\left(2\frac{k}{\sigma^2}+\frac{z}{\sigma}\right) - \\ 2k\exp[(\sigma+k^2)^{1/2}z]\left(\frac{1}{k^2Ma}+\frac{1}{\sigma^2}\right) \end{pmatrix}
$$

$$\text{(H.17)}$$

where the corresponding eigenvalue σ is solution of the dispersion relation (8.13). Note that the derivation of (H.17) assumes $\sigma + k^2 > 0$ [which is verified provided that $k < Ma/2Bi$, as seen from Eq. (8.13)]. Otherwise, when $\sigma + k^2 < 0$, imaginary roots are obtained from the characteristic relation, and lead to a continuum of solutions bounded at $z \to -\infty$, but which do not satisfy boundary conditions (8.2).

H.3   Calculation of the mean temperature profile

For the roll mode $A_{\vec{k}} = a_1(t)\delta(\vec{k}-\vec{k}_0) + \bar{a}_1(t)\delta(\vec{k}+\vec{k}_0)$ [see Eq. (8.23)], the $k = 0$ Fourier component of the perturbation vector is found from (H.5) as

$$
U_{k=0}^D = 2|a_1|^2 U_{01}^D(z) \tag{H.18}
$$

where, according to (H.6), $U_{01}^D(z)$ is the solution of the inhomogeneous problem

$$
\mathcal{L}_0(U_{01}^D) = N_{1,-1}(U_1^0, U_{-1}^0) \tag{H.19}
$$

since $M_{k=0} = 0$, and the resonant part of $N_{1,-1}$ is zero, as $Z_{10} = 0$ from (H.3). Here again, the subscripts refer to the mode number (1 for $\vec{k}_0$, -1 for $-\vec{k}_0$). It is computed that the only non-zero component of $U_{01}^D(z)$ is $T_{01}^D(z)$, solution of

$$
D^2 T_{01}^D = W_1^0 DT_{-1}^0 + \vec{V}_{r1}^0 \cdot (-i\vec{k}_0 T_{-1}^0) = D(W_1^0 T_1^0) \tag{H.20}
$$

where the neutral stability functions are given by (H.9). Then, the solution belonging to E is

$$
T_{01}^D = (5Bi^2 + 12k_0Bi + 7k_0^2)\left(\exp[2k_0z]\left(\frac{1}{2k_0}-z\right)-\frac{1}{2k_0}\right) + \\ (Bi+k_0)^2 k_0 \exp[2k_0z]z^2(5-2k_0z) \tag{H.21}
$$

The cubic coefficient Z_{111}^0 can then be computed from (H.7) as

$$
\begin{aligned}
Z_{111}^0 &= -\langle V_1^*, N_{1,0}(U_1^0, U_{01}^D)\rangle = -\int_{-\infty}^0 T_1^* \left(W_1^0\right)^2 T_1^0 dz \\
&= -\frac{k_0^2(Bi+k_0)^2(3Bi+5k_0)}{2(Bi+2k_0)}
\end{aligned} \tag{H.22}
$$

A slightly longer but similar computation leads to the expression of the second cubic coefficient Z_{-111}^0, and then finally to Eq. (8.20). Equation (8.27) is also obtained in this way, but by using eigenfunctions (H.17) instead of the neutral stability functions (H.9).

Bibliography

[1] I. Prigogine, *Introduction to Thermodynamics of Irreversible Processes*, Interscience Publishers, Wiley, New York, 1961.

[2] P. Glansdorff and I. Prigogine, *Structure, Stabilité et Fluctuations*, Masson et Cie, Paris, 1971.

[3] R. Haase, *Thermodynamics of Irreversible Processes*, Dover, New York, 1990.

[4] G. Iooss and D.D. Joseph, *Elementary Stability and Bifurcation Theory*, Springer-Verlag, New York, 1990.

[5] G. Nicolis and I. Prigogine, *Self-organization in Nonequilibrium Systems, from Dissipative Structures to Order through Fluctuations*, Wiley, New York, 1977.

[6] H. Haken, *Advanced Synergetics*, Springer-Verlag, Berlin, 1987.

[7] G. Nicolis, *Introduction to Nonlinear Science*, Cambridge University Press, Cambridge, 1995.

[8] J.D. Murray, *Mathematical Biology*, Springer-Verlag, Berlin, 1989.

[9] G. Nicolis, "Physics of far-from-equilibrium systems and self-organization", in *The New Physics*, Paul Davies (Ed.), Cambridge University Press, Cambridge, 1989.

[10] P. Coullet, *Nonlinear Dynamics, Symmetries and Normal Forms*, lectures given at an International Summer School at El Escorial, Madrid, July 1999.

[11] D. Walgraef, *Spatio-temporal Pattern Formation. With examples from Physics, Chemistry and Materials Science*, Springer-Verlag, Berlin, 1997.

[12] H. Bénard, "Les tourbillons cellulaires dans une nappe liquide", *Rev. Gén. Sci. Pures Appl.* **11**, 1261–1271, 1309–1328 (1900).

[13] Q. Ouyang, J. Boissonade, J.C. Roux and P. de Kepper, "Sustained reaction-diffusion structures in an open reactor", *Phys. Rev. A* **134**, 282 (1989).

[14] Q. Ouyang and H.L. Swinney, "Transition from a uniform state to hexagonal and striped Turing patterns," *Nature* **352**, 610–612 (1991).

[15] J. Zierep and H. Oertel jr. (Eds), *Convective Transport and Instability Phenomena*, Braun-Verlag, Karlsruhe, 1982.

[16] D.T.J. Hurle (Ed.), *Handbook of Crystal Growth, Part B: Transport and Stability*, Elsevier, Amsterdam, 1993.

[17] D.T.J. Hurle, "On similarities between the theories of morphological instability of a growing binary alloy crystal and Rayleigh–Bénard convective instability", *J. Cryst. Growth* **72**, 738–42 (1985).

[18] A.C. Newell and J.V. Moloney, *Nonlinear Optics*, Addison–Wesley, Redwood City, 1992.

[19] M.C. Cross and P.C. Hohenberg, "Pattern formation outside of equilibrium", *Rev. Mod. Phys.* **65**, 851–1112 (1993).

[20] B. Straughan, *The Energy Method, Stability, and Nonlinear Convection*, Springer-Verlag, New York, 1992.

[21] J.S. Turner, *Buoyancy Effects in Fluids*, Cambridge University Press, Cambridge, 1973.

[22] F.H. Busse, "Nonlinear properties of thermal convection", *Rep. Prog. Phys.* **41**, 1929–1967 (1978).

[23] F.H. Busse, "Fundamentals of thermal convection", in *Mantle Convection, Plate Tectonics and Global Dynamics*, W.R. Peltier (Ed.), Gordon and Breach, 1989.

[24] J.K. Platten and J.C. Legros, *Convection in Liquids*, Springer-Verlag, Berlin, 1984.

[25] C. Normand, Y. Pomeau and M.G. Velarde, "Convective instability: a physicist's approach", *Rev. Mod. Phys.* **49**, 581–624 (1977).

[26] L.A. Segel, "Nonlinear hydrodynamic stability theory and its applications to thermal convection and curved flows", *J. Non-Equilib. Thermodyn.* **18**, 165–197 (1966).

[27] E. Palm, "Nonlinear thermal convection", *Annu. Rev. Fluid Mech.* **7**, 39–61 (1975).

[28] E.L. Koschmieder, *Bénard Cells and Taylor Vortices*, Cambridge University Press, Cambridge, 1993.

[29] K. Nitschke and A. Thess, "Secondary instability in surface-tension-driven convection", *Phys. Rev. E* **52**, 5772–5775 (1995).

[30] K. Eckert, M. Bestehorn and A. Thess, "Square cells in surface-tension-driven Bénard convection: experiment and theory", *J. Fluid Mech.* **356**, 155–197 (1998).

[31] M.F. Schatz, S.J. VanHook, W.D. McCormick, J.B. Swift and H.L. Swinney, "Time-independent square patterns in surface-tension-driven Bénard convection", *Phys. Fluids* **11**, 2577–2582 (1999).

[32] U. Thiele and K. Eckert, "Stochastic geometry of polygonal networks: An alternative approach to hexagon-square transition in Bénard convection", *Phys. Rev. E* **58**, 3458–3468 (1998).

[33] K. Eckert and A. Thess, "Nonbound dislocations in hexagonal patterns: pentagon lines in surface-tension-driven Bénard convection", *Phys. Rev. E* **60**, 4117–4124 (1999).

[34] W.A. Tokaruk, T.C.A. Molteno and S.W. Morris, "Bénard–Marangoni convection in two-layered liquids", *Phys. Rev. Lett.* **84**, 3590–3593 (2000).

[35] S.J. VanHook, M.F. Schatz, J.B. Swift, W.D. McCormick and H.L. Swinney, "Long-wavelength surface-tension-driven Bénard convection: experiment and theory", *J. Fluid Mech.* **345**, 45 (1997).

[36] S.J. VanHook, M.F. Schatz, J.B. Swift, W.D. McCormick and H.L. Swinney, "Long-wavelength instability in surface-tension-driven Bénard convection", *Phys. Rev. Lett.* **75**, 4397 (1997).

[37] M.F. Schatz, *personal communication*, 1997.

[38] A. Orell and W. Westwater, "Spontaneous interfacial cellular convection accompanying mass transfer: ethylene glycol–acetic acid–ethyl acetate", *A.I.Ch.E. J.* **8**, 350–356 (1962).

[39] H. Linde and K. Loeschcke, "Rollzellen und oszillation beim wärmeübergang zwischen gas und flüssigkeit", *Chem.-Ing.-Tech.* **39**, 65–74 (1966).

[40] J.C. Berg, A.A. Acrivos and M. Boudart, "Evaporative convection", *Adv. Chem. Eng.* **6**, 61–123 (1966).

[41] D. Schwabe, "The Bénard–Marangoni instability in small circular containers under microgravity : experimental results", *Adv. Space Res.* **24**, 1347–1356 (1999).

[42] E.L. Koschmieder and S.A. Prahl, "Surface-tension-driven Bénard convection in small containers", *J. Fluid Mech.* **215**, 571–583 (1990).

[43] D. Johnson and R. Narayanan, "Experimental observation of dynamic mode switching in interfacial-tension-driven convection near a codimension-two point", *Phys. Rev. E* **54**, 3102–3104 (1996).

[44] H. Linde, P. Schwartz and H. Wilke, "Dissipative structures and nonlinear kinetics of the Marangoni–Bénard instability", in *Dynamics and instability of fluid interfaces*, Lecture Notes in Physics **105**, T.S. Sorensen (Ed.), 75 (1979).

[45] H. Linde, S. Pfaff and C. Zirkel, "Strömungsuntersuchungen zur hydrodynamischen instabilität flüssig-gasförmiger phasengrenzen mit hilfe der kapillarspalt-methode", *Z. Phys. Chemie* **225**, 72–100 (1964).

[46] H. Linde, "Marangoni instabilities", in *Convective Transport and Instability Phenomena*, O. Zierep and H. Oertel (Eds), Braun-Verlag, Karlsruhe, 1982.

[47] P.D. Weidman, H. Linde and M.G. Velarde, "Evidence for solitary wave behavior in Marangoni–Bénard convection", *Phys. Fluids A* **4**, 921–926 (1992).

[48] A. N. Garazo and M.G. Velarde, "Dissipative Korteweg–de Vries description of Marangoni–Bénard oscillatory convection", *Phys. Fluids A* **3**, 2295–2299 (1991).

[49] A.Ye. Rednikov, P. Colinet, M.G. Velarde and J.C. Legros, "Two-layer Marangoni–Bénard instability and the limit of transverse and longitudinal waves", *Phys. Rev. E* **57**, 2872–2884 (1997).

[50] A.Ye. Rednikov, P. Colinet, M.G. Velarde and J.C. Legros, "Rayleigh–Marangoni oscillatory instability in a horizontal liquid layer heated from above: coupling and mode-mixing of internal and surface dilational waves", *J. Fluid Mech.* **405**, 57–77 (2000).

[51] P. Manneville, *Dissipative Structures and Weak Turbulence*, Academic Press, San Diego, 1990.

[52] P.M. Chaikin and T.C. Lubensky, *Principles of Condensed Matter Physics*, Cambridge University Press, Cambridge, 1995.

[53] F.H. Busse, "The stability of finite amplitude cellular convection and its relation to an extremum principle", *J. Fluid Mech.* **30**, 4, 625–649 (1967).

[54] E.N. Lorenz, "Deterministic nonperiodic flow", *J. Atmos. Sci.* **20**, 130 (1963).

[55] P. Tabeling and O. Cardoso (Eds), *Turbulence, A Tentative Dictionary*, NATO ASI Series, Series B: Physics Vol. 341, Plenum Press, New York, 1994.

[55] Ph. Holmes, J.L. Lumley and G. Berkooz, *Turbulence, Coherent Structures, Dynamical Systems and Symmetry*, Cambridge University Press, Cambridge, 1996.

[57] U. Frisch, *Turbulence*, Cambridge University Press, Cambridge, 1995.

[53] H.L. Swinney and J.P. Gollub (Eds), *Hydrodynamic Instabilities and the Transition to Turbulence*, Topics in Applied Physics, Vol. 45, Springer-Verlag, Berlin, 1981.

[59] E.D. Siggia, "High Rayleigh number convection", *Annu. Rev. Fluid Mech.* **26**, 137–168 (1994).

[60] E. Guyon, J.-P. Hulin et L. Petit, *Hydrodynamique Physique*, Savoirs Actuels, Editions du CNRS, Paris, 1994.

[61] F.J. Zuiderweg and A. Harmens, "The influence of surface phenomena on the performances of distillation columns", *Chem. Eng. Sci.* **9**, 89–103 (1958).

[62] F.J. Zuiderweg, "Marangoni effect in distillation of alcohol-water mixtures", *Chem. Eng. Res. Des.* **61**, 388–390 (1983).

[63] J.C. Berg and C.R. Morig, "Density effects in interfacial convection", *Chem. Eng. Sci.* **24**, 937–946 (1969).

[64] J.C. Berg and G.S. Haselberger, "Mass transfer during interfacial convection", *Chem. Eng. Sci.* **26**, 481–485 (1971).

[65] J. Berg, "Interfacial hydrodynamics: an overview", *Can. Metallurg. Q.* **21**, 121–136 (1982).

[66] P.L.T. Brian, J.E. Vivian and S.T. Mayr, "Cellular convection in desorbing surface-tension-lowering solutes from water", *Ind. Eng. Chem. Fundam.* **10**, 75–83 (1971).

[67] H.W. Hoogstraten, H.C.J. Hoefsloot, L.P.B.M. Janssen, "Marangoni convection in V-shaped containers", *J. Eng. Math.* **26**, 21–38 (1992).

[68] M.A. Mendes-Tatsis and E.S. Perez de Ortiz, "Spontaneous interfacial convection in liquid-liquid binary systems under microgravity", *Proc. R. Soc. London A* **438**, 389–396 (1992).

[69] L.J. Austin, W.E. Ying and H. Sawitowski, "Interfacial phenomena in binary liquid-liquid systems", *Chem. Eng. Sci.* **21**, 1109–1110 (1966).

[70] E.S. Perez de Ortiz and H. Sawitowski, "Interfacial stability of binary liquid-liquid systems", *Chem. Eng. Sci.* **28**, 2051–2069 (1973).

[71] L.M. Rabinovich, "Problems in modelling and intensification of mass transfer with interfacial instability and self-organisation", Chap. 4 in *Mathematical Modelling of Chemical Processes*, by L.M. Rabinovich, CRC Press, Boca Raton, 289–417, 1992.

[72] N. Kolev and K. Semkov, "On the evaluation of the interfacial turbulence (the Marangoni effect) in gas (vapour)-liquid mass transfer. Part II. Modelling of packed columns accounting for the axial mixing and Marangoni effects", *Chem. Eng. Process* **29**, 83–91 (1991).

[73] A.A. Golovin, "Mass transfer under interfacial turbulence: kinetic regularities", *Chem. Eng. Sci.* **47**, 2069–2080 (1992).

[74] J.C. Berg, M. Boudart and A.A. Acrivos, "Natural convection in pools of evaporating liquids", *J. Fluid Mech.* **24**, 721–735 (1966).

[75] C.V. Sternling and L.E. Scriven, "Interfacial turbulence : hydrodynamic instability and the Marangoni effect", *A.I.Ch.E. J.* **5**, 514–523 (1959).

[76] D.A. Edwards, H. Brenner and D.T. Wasan, *Interfacial Transport Processes and Rheology*, Butterworth–Heinemann, Boston (1991).

[77] Zh. Kozhoukharova and S.G. Slavtchev, "Influence of the surface deformability on Marangoni instability in a liquid layer with surface chemical reaction. Part I. Stationary convection", *J. Colloid Interface Sci.* **148**, 42 (1992).

[78] Zh. Kozhoukharova and S.G. Slavtchev, "Influence of the surface deformability on Marangoni instability in a liquid layer with surface chemical reaction. Part II. Overstability", *J. Colloid Interface Sci.* **152**, 473 (1992).

[79] M.A. Mendes-Tatsis and E.S. Perez de Ortiz, "Marangoni instabilities in systems with an interfacial chemical reaction", *Chem. Eng. Sci.* **51**, 3755 (1996).

[30] K. Eckert and A. Grahn, "Plume and finger regimes driven by an exothermic interfacial reaction", *Phys. Rev. Lett.* **82**, 4436–4439 (1999).

[31] T. Molenkamp, *Marangoni Convection, Mass Transfer and Microgravity*, PhD Dissertation, Rijksuniversiteit Groningen, 1998.

[32] O. Kabov, B. Scheid, I.V. Marchuk and J.C. Legros, "Free surface deformation by thermocapillary convection in a moving thin liquid layer locally heated", to appear in *J. of the Russian Academy of Sciences, Mechanics of Fluids and Gases* (2001).

[83] M.K. Smith and S.H. Davis, "Instabilities of dynamic thermocapillary liquid layers. Part 1. Convective instabilities. Part 2. Surface-wave instabilities", *J. Fluid Mech.* **132**, 119–162 (1983).

[84] F. Preisser, D. Schwabe and A. Scharmann, "Steady and oscillatory thermocapillary convection in liquid columns with free cylindrical surface", *J. Fluid Mech.* **126**, 545–567 (1983).

[85] Y. Kamotani, S. Ostrach and M. Vargas, "Oscillatory thermocapillary convection in a simulated floating-zone configuration", *J. Cryst. Growth* **66**, 83–90 (1984).

[86] S.H. Davis, "Thermocapillary instabilities", *Annu. Rev. Fluid Mech.* **19**, 403–435 (1987).

[87] Y. Shen, G.P. Neitzel, D.F. Jankowski and H.D. Mittelman, "Energy stability of thermocapillary convection in a model of the float-zone crystal-growth process", *J. Fluid Mech.* **217**, 639 (1990).

[88] G.P. Neitzel, K.-T. Chang, D.F. Jankowski and H.D. Mittelman, "Linear-stability theory of thermocapillary convection in a model of the float-zone crystal growth process", *Phys. Fluids A* **5**, 108 (1993).

[89] M. Wanschura, V.M. Shevtsova, H.C. Kuhlmann and H.J. Rath, "Convective instability mechanisms in thermocapillary liquid bridges", *Phys. Fluids* **5**, 912–925 (1995).

[90] V. Petrov, M.F. Schatz, K.A. Muehlner, S.J. VanHook, W.D. McCormick, J.B. Swift and H.L. Swinney, "Nonlinear control of remote unstable states in a liquid bridge convection experiment", *Phys. Rev. Lett.* **77**, 3779–3782 (1996).

[91] R. Savino and R. Monti, "Oscillatory Marangoni convection in cylindrical liquid bridges", *Phys. Fluids* **8**, 2906–2922 (1996).

[92] G.P. Neitzel and R.J. Riley, "Instability of thermocapillary-buoyancy convection in shallow layers. Part 1. Characterisation of steady and oscillatory instabilities", *J. Fluid Mech.* **359**, 143 (1998).

[93] S. Benz, P. Hintz, R.J. Riley and G.P. Neitzel, "Instability of thermocapillary-buoyancy convection in shallow layers. Part 2. Suppression of hydrothermal waves", *J. Fluid Mech.* **359**, 165 (1998).

[94] H.C. Kuhlmann, *Thermocapillary convection in models of crystal growth*, Springer-Verlag, Berlin, 1999.

[95] S. Chandrasekhar, *Hydrodynamic and Hydromagnetic Stability*, Dover, New York, 1961.

[96] G.Z. Gershuni and E.M. Zhukhovitskii, *Convective Stability of Incompressible Fluids*, Keterpress Enterprises, Jerusalem, 1976.

[97] I.B. Simanovsky and A.A. Nepomnyashchy, *Convective Instabilities in Systems with Interface*, Gordon and Breach, Amsterdam, 1993.

[98] P.G. Drazin and W.H. Reid, *Hydrodynamic Stability*, Cambridge University Press, Cambridge, 1981.

[99] Lord Rayleigh, "On convective currents in a horizontal layer of fluid when the higher temperature is on the under side", *Philos. Mag.* **32**, 529–546 (1916).

[100] M.J. Block, "Surface tension as the cause of Bénard cells and surface deformation in a liquid film", *Nature* **178**, 650–651 (1956).

[101] J.R.A. Pearson, "On convection cells induced by surface tension", *J. Fluid Mech.* **4**, 489 (1958).

[102] D.A. Nield, "Surface tension and buoyancy effects in cellular convection", *J. Fluid Mech.* **19**, 341 (1964).

[103] J.W. Scanlon and L.A. Segel, "Finite amplitude cellular convection induced by surface tension", *J. Fluid Mech.* **30**, 149–162 (1967).

[104] A. Cloot and G. Lebon, "A nonlinear analysis of the Bénard–Marangoni problem", *J. Fluid Mech.* **145**, 447–469 (1984).

[105] E.L. Koschmieder, "On convection under an air surface", *J. Fluid Mech.* **30**, 9–15 (1967).

[106] E.L. Koschmieder and M.I. Biggerstaff, "Onset of surface-tension-driven Bénard convection", *J. Fluid Mech.* **167**, 49–64 (1986).

[107] E.L. Koschmieder and S.G. Pallas, "Heat transfer through a shallow horizontal convecting fluid layer, *Int. J. Heat Mass Transfer* **17**, 991–1002 (1974).

[108] E.L. Koschmieder and D.W. Switzer, "The wavenumbers of supercritical surface tension driven Bénard convection", *J. Fluid Mech.* **240**, 533–548 (1992).

[109] J. Pantaloni, R. Bailleux, J. Salan and M.G. Velarde, "Rayleigh–Bénard–Marangoni instability: new experimental results", *J. Non-Equilib. Thermodyn.* **4**, 201–218 (1979).

[110] M.G. Velarde and Ch. Normand, "Convection", *Sci. Am.* **243**, 92 (1980).

[111] J. Pantaloni, P. Cerisier, R. Bailleux and C. Gerbaud, "Convection de Bénard–Marangoni: un pendule de Foucault?", *J. Phys. Lett.* **42**, 147–150 (1981).

[112] P. Cerisier, R. Occelli and J. Pantaloni, "Velocity and temperature fields in Bénard–Marangoni instability", *Physicochem. Hydrodyn.* **7**, 191–206 (1986).

[113] P. Cerisier, C. Pérez-García, C. Jamond and J. Pantaloni, "Wavelength selection in Bénard–Marangoni convection", *Phys. Rev. A* **34**, 1949–1952 (1987).

[114] P. Cerisier, R. Occelli, C. Pérez-García and C. Jamond, "Structural disorder in Bénard–Marangoni convection", *J. Phys.* **48**, 569–576 (1987).

[115] P. Cerisier, C. Jamond, J. Pantaloni and C. Pérez-García, "Stability of roll and hexagonal patterns in Bénard–Marangoni convection", *Phys. Fluids* **30**, 954–958 (1987).

[116] P. Cerisier, "Experimental study of the wavelength selection in a pentagonal vessel in Bénard convection", *Phys. Fluids A* **3**, 2061–2066 (1991).

[117] P. Cerisier, H. Nguyen Thi and B. Billia, "Disorder dynamics of arrays in Bénard convection", *Physica D* **61**, 113–118 (1992).

[118] M.F. Schatz, S.J. VanHook, J.B. Swift, W.D. McCormick and H.L. Swinney, "Onset of Surface-tension-driven Bénard convection", *Phys. Rev. Lett.* **75**, 1938–1941 (1995).

[119] M. Hennenberg, S.G. Slavtchev, M.A. Mendes-Tatsis and J.C. Legros, "Adsorption-desorption barriers and diffusional exchanges: linear surface instabilities of one and two layer systems", preprint (2001).

[120] S. Slavtchev, M. Hennenberg, J.C. Legros and G. Lebon, "Stationary solutal Marangoni instability in a two-layer system", *J. Colloid Interface Sci.* **203**, 354–368 (1998).

[121] J. Bragard, S.G. Slavtchev and G. Lebon, "Nonlinear solutal Marangoni instability in a liquid layer with an adsorbing upper surface", *J. Colloid Interface Sci.* **168** (1994).

[122] H. Linde, X.-L. Chu, M.G. Velarde and W. Waldhelm, "Wall reflections of solitary waves in Marangoni–Bénard convection", *Phys. Fluids A* **5**, 3162–3166 (1993).

[123] L. Weh and H. Linde, "Marangoni-stress driven solitonic (periodic) wave trains rotating in an annular container during heat transfer", *J. Colloid Interface Sci.* **187**, 159–165 (1997).

[124] H. Linde, M.G. Velarde, A. Wierschem, W. Waldheim, K. Loeschke and A.Ye. Rednikov, "Interfacial wave motions due to Marangoni instability. I. Traveling periodic wave trains in square and annular containers", *J. Colloid Interface Sci.* **188**, 16–26 (1997).

[125] A. Wierschem, M.G. Velarde, H. Linde and W. Waldheim, "Interfacial wave motions due to Marangoni instability. II. Three-dimensional characteristics of surface waves in annular containers" *J. Colloid Interface Sci.* **212**, 365–383 (1999).

[126] L. Limat, P. Jenffer, B. Dagens, E. Touron, M. Fermigier and J.E. Wesfreid, "Gravitational instabilities of thin liquid layers: dynamics of pattern selection", *Physica D* **61**, 166–182 (1992).

[127] P.G. Saffman and G.I. Taylor, "The penetration of a fluid into a porous medium or Hele-Shaw cell containing a more viscous liquid", *Proc. R. Soc. London Ser. A* **245**, 312–329 (1986).

[128] G.M. Homsy, "Viscous fingering in porous media", *Annu. Rev. Fluid Mech.* **19**, 271–311 (1987).

[129] A. Ye. Rednikov, Y.S. Ryazantsev and M.G. Velarde, "Active drops and drop motions due to nonequilibrium phenomena", *J. Non-Equilib. Thermodyn.* **19**, 95–113 (1994).

[130] J. Bragard and G. Lebon, "Nonlinear Marangoni convection in a layer of finite depth", *Europhys. Lett.* **21**, 831–836 (1993).

[131] M. Bestehorn, "Square patterns in Bénard–Marangoni convection", *Phys. Rev. Lett.* **76**, 46–49 (1996).

[132] V. Regnier, P.C. Dauby, P. Parmentier and G. Lebon, "Square cells in gravitational and capillary thermoconvection", *Phys. Rev. E* **55**, 6860–6865 (1997).

[133] P. Le Gal, A. Pocheau and V. Croquette, "Square versus roll pattern at convective threshold", *Phys. Rev. Lett.* **54**, 2501–2504 (1985).

[134] E. Bodenschatz, J.R. de Bruyn, G. Ahlers and D.S. Cannell, "Transitions between patterns in thermal convection", *Phys. Rev. Lett.* **67**, 3078–3081 (1991).

[135] E. Bodenschatz, D.S. Cannell, J.R. de Bruyn, R. Ecke, Y.-C. Hu, K. Lerman and G. Ahlers, "Experiments on three systems with non-variational aspects", *Physica D* **61**, 77–93 (1992).

[136] M. Bestehorn, M. Fantz, R. Friedrich and H. Haken, "Hexagonal and spiral patterns of thermal convection", *Phys. Lett. A* **174**, 48–52 (1993).

[137] S.W. Morris, E. Bodenschatz, D.S. Cannell and G. Ahlers, "Spiral defect chaos in large aspect ratio Rayleigh–Bénard convection", *Phys. Rev. Lett.* **71**, 2026–2029 (1993).

[138] Y. Hu, R.E. Ecke and G. Ahlers, "Transition to spiral-defect chaos in low Prandtl number convection", *Phys. Rev. Lett.* **74**, 391–394 (1995).

[139] W. Decker, W. Pesch and A. Weber, "Spiral defect chaos in Rayleigh–Bénard convection", *Phys. Rev. Lett.* **73**, 648–651 (1994); W. Pesch, "Complex spatio-temporal convection patterns", *Chaos* **6**, 348 (1996).

[140] G.H. Gunaratne, "Complex spatial patterns on planar continua", *Phys. Rev. Lett.* **71**, 1367–1370 (1993).

[141] A.A. Nepomnyashchy and L.M. Pismen, *Comment on* "Complex Spatial Patterns on Planar Continua", and *Gunaratne's reply*, *Phys. Rev. Lett.* **72**, 944–945 (1994).

[142] G.H. Gunaratne, Qi Ouyang and H.L. Swinney, "Pattern formation in the presence of symmetries", *Phys. Rev. E* **50**, 2802–2820 (1994).

[143] R. Graham, "Systematic derivation of a rotationally covariant extension of the two-dimensional Newell–Whitehead–Segel equation", *Phys. Rev. Lett.* **76**, 2185–2187 (1996).

[144] P. Manneville, "A two-dimensional model for three-dimensional convective patterns in wide containers", *J. Phys.* (Paris) **44**, 759–765 (1983); "Towards an understanding of weak turbulence close to the convection threshold in large aspect ratio systems", *J. Phys. Lett.* (Paris) **44**, 903–916 (1983).

[145] G. Ahlers and R.P. Behringer, "Evolution of turbulence from the Rayleigh–Bénard instability", *Phys. Rev. Lett.* **40**, 712 (1978).

[146] A. Pocheau, V. Croquette and P. Le Gal, "Turbulence in a cylindrical container of argon near threshold of convection", *Phys. Rev. Lett.* **55**, 1094 (1985).

[147] M. Bestehorn and R. Friedrich, "Natural patterns in nonequilibrium systems", in *Perspective Look at Nonlinear Media. From Physics to Biology and Social Sciences*, Springer-Verlag, Berlin, 31–44 (1998).

[148] M. Bestehorn and R. Friedrich, "Rotationally invariant order parameter equations for natural patterns in nonequilibrium systems", *Phys. Rev. E* **59**, 2642–2652 (1999).

[149] S. Rosenblat, G. M. Homsy and S. H. Davis, "Nonlinear Marangoni convection in bounded layers. Part 1. Circular cylindrical containers", *J. Fluid Mech.* **120**, 91–122 (1982).

[150] S. Rosenblat, G.M. Homsy and S.H. Davis, "Nonlinear Marangoni convection in bounded layers. Part 2. Rectangular cylindrical containers", *J. Fluid Mech.* **120**, 91–138 (1982).

[151] P.C. Dauby, G. Lebon, P. Colinet and J.C. Legros, "Hexagonal Marangoni convection in a rectangular box with slippery walls", *Q. J. Mech. Appl. Math.* **46**, 683–707 (1993).

[152] P.C. Dauby and G. Lebon, "Bénard–Marangoni instability in rigid rectangular containers", *J. Fluid Mech.* **329**, 25–64 (1996).

[153] C. Canuto, M.Y. Hussaini, A. Quateroni and T.A. Zang, *Spectral methods in Fluid Dynamics*, Springer, Berlin-Heidelberg, 1987.

[154] P.C. Dauby, G. Lebon and E. Bouhy, "Linear Bénard–Marangoni instability in rigid circular containers", *Phys. Rev. E* **56**, 520–530 (1997).

[155] T. Ondarçuhu, G.B. Mindlin, H.L. Mancini and C. Pérez-García, "Dynamical patterns in Bénard–Marangoni convection in a square container", *Phys. Rev. Lett.* **70**, 3892 (1993).

[156] D. Maza, B. Echebarría, C. Pérez-García and H. Mancini, "Bénard–Marangoni convection in small aspect ratio containers", *Phys. Scr.* **T67**, 82–85 (1996).

[157] G.B. Mindlin, T. Ondarçuhu, H.L. Mancini, C. Pérez-García and A Garcimartín, "Comparison of data from Bénard–Marangoni convection in a square container with a model based on symmetry arguments", *Int. J. Bifurcation Chaos* **4**, 1121–1133 (1994).

[158] T. Ondarçuhu, J. Millan-Rodrigues, H.L. Mancini, A. Garcimartín and C. Pérez-García, "Bénard–Marangoni convective pattern in small cylindrical layers", *Phys. Rev. E* **48**, 1051–1057 (1993).

[159] D. Johnson, R. Narayanan and P.C. Dauby, "The effect of air height on the pattern formation in liquid–air bilayer convection", in *Fluid Dynamics at Interfaces*, W. Shyy and R. Narayanan (Eds), Cambridge University Press, Cambridge, 1999.

[160] D. Johnson and R. Narayanan, "Marangoni convection in multiple bounded fluid layers and its application to materials processing", *Phil. Trans. R. Soc. London A* **356**, 885–898 (1998).

[161] D.T.J. Hurle and E. Jakeman, "Soret-driven thermosolutal convection", *J. Fluid Mech.* **47**, 667–687 (1971).

[162] J.C. Legros, J.K. Platten and P.G. Poty, "Stability of a two-component fluid layer heated from below", *Phys. Fluids* **15**, 1383–90 (1972).

[163] E. Knobloch, "Oscillatory convection in binary mixtures", *Phys. Rev. A* **34**, 1538–49 (1986).

[164] G. Dangelmayr and E. Knobloch, "The Takens–Bogdanov bifurcation with O(2)-symmetry", *Phil. Trans. R. Soc. London A* **322**, 243–279 (1987).

[165] W. Barten, M. Lucke, M. Kamps and P. Schmitz, "Convection in binary fluid mixtures. I. Extended traveling-wave and stationary states. II. Localised traveling waves", *Phys. Rev. E* **51**, 5636–80 (1995).

[166] W. Schopf and W. Zimmermann, "Convection in binary fluids: Amplitude equations, codimension-2 bifurcation, and thermal fluctuations", *Phys. Rev. E* **47**, 1739–64 (1993).

[167] Y.Y. Renardy, "Pattern selection for the oscillatory onset in thermosolutal convection", *Phys. Fluids A* **5**, 1376–89 (1993).

[168] J.K. Platten and G. Chavepeyer, "Oscillatory motion in Bénard cell due to the Soret effect", *J. Fluid Mech.* **60**, 305–319 (1973).

[169] R.W. Walden, P. Kolodner, A. Passner and C.M. Surko, "Traveling waves and chaos in convection in binary fluid mixtures", *Phys. Rev. Lett.* **55**, 496–9 (1985); P. Kolodner, A. Passner, C.M. Surko and R.W. Walden, *Phys. Rev. Lett.* **56**, 2621 (1986).

[170] J. Fineberg, E. Moses and V. Steinberg, *Phys. Rev. Lett.* **61**, 838 (1988); P. Kolodner and C. Surko, *Phys. Rev. Lett.* **61**, 842 (1988).

[171] H.C. Nataf, S. Moreno and Ph. Cardin, "What is responsible for thermal coupling in layered convection?", *J. Phys.* (Paris) **49**, 1707–1714 (1988).

[172] L. Cserepes, M. Rabinowicz and C. Rosemberg-Borot, "Three-dimensional infinite Prandtl number convection in one and two layers with implications for the earth's gravity field", *J. Geophys. Res.* **93**, 12009–25 (1988).

[173] R.W. Zeren and W.C. Reynolds, "Thermal instabilities in two-fluid horizontal layers", *J. Fluid Mech.* **53**, 305–327 (1972).

[174] G.Z. Gershuni and E.M. Zhukhovitskii, "Monotonic and oscillatory instabilites of a two-layer system of immiscible liquids heated from below", *Sov. Phys. Dokl.* **27**, 531–533 (1982).

[175] S. Rasenat, F.H. Busse and I. Rehberg, "A theoretical and experimental study of double-layer convection", *J. Fluid Mech.* **199**, 519–540 (1989).

[176] Y. Renardy and M. Renardy, "Perturbation analysis of steady and oscillatory onset in a Bénard problem with two similar liquids", *Phys. Fluids* **28**, 2699–2708 (1985).

[177] P. Colinet and J.C. Legros, "On the Hopf bifurcation occurring in the two-layer Rayleigh–Bénard convective instability", *Phys. Fluids* **6**, 2631–2636 (1994).

[178] Y.Y. Renardy, "Pattern formation for oscillatory bulk-mode competition in a two-layer Bénard problem", *Z. Angew. Math. Phys.* **47**, 567–590 (1996).

[179] A. Prakash and J.N. Koster, "Steady Rayleigh–Bénard convection in a two-layer system of immiscible liquids", *J. Heat Transfer* **118**, 366–373 (1996).

[180] C.D. Andereck, P.W. Colovas and M.M. Degen, "Observations of time-dependent behavior in the two-layer Rayleigh–Bénard system", in *Advances in Multi-Fluid Flows*, Y.Y. Renardy, A.V. Coward, D.T. Papageorgiou, S.M. Sun (Eds), SIAM (1995).

[181] M.M. Degen, P.W. Colovas and C.D. Andereck, "Time-dependent patterns in the two-layer Rayleigh–Bénard system", *Phys. Rev. E* **57**, 6647–6659 (1998).

[182] R. Aris, *Vectors, Tensors, and the Basic Equations of Fluid Mechanics*, Dover, New York, 1962.

[183] G. Nicolis, J. Wallenborn and M.G. Velarde, "On the validity of Gibbs entropy law in strongly coupled systems", *Physica D* **43**, 263–276 (1969).

[184] C. Vidal, G. Dewel and P. Borckmans, *Au-delà de l'équilibre*, Hermann, Paris, 1994.

[185] D. Jou, J. Casas-Vázquez and G. Lebon, *Extended Irreversible Thermodynamics*, 2nd Ed., Springer-Verlag, Berlin, 1996.

[186] J.P. Burelbach, S.G. Bankoff and S.H. Davis, "Nonlinear stability of evaporating/ condensing liquid films", *J. Fluid Mech.* **195**, 463–494 (1988).

[187] J.M. Delhaye, "Jump conditions and entropy sources in two-phase systems. Local instant formulation", *Int. J. Multiphase Flow* **1**, 395–409 (1974).

[188] S.R. de Groot and P. Mazur, *Non-Equilibrium Thermodynamics*, Dover, New York, 1984.

[189] R.B. Bird, W.E. Stewart and E.N. Lightfoot, *Transport Phenomena*, Wiley, Singapore, 1960.

[190] R. Defay, I. Prigogine, A. Bellemans and D.H. Everett, *Surface Tension and Adsorption*, Longmans, London, 1966.

[191] M.G. Velarde (Ed.), *Physicochemical Hydrodynamics. Interfacial Phenomena*, NATO ASI Series **174**, Plenum Press, New York, 1988.

[192] G. Fang and C.A. Ward, "Temperature measured close to the interface of an evaporating liquid", *Phys. Rev. E* **59**, 417–428 (1999).

[193] C.A. Ward and G. Fang, "Expression for predicting liquid evaporation flux: Statistical rate theory approach", *Phys. Rev. E* **59**, 429–440 (1999); G. Fang and Ch. Ward, "Examination of the statistical rate theory expression for liquid evaporation rates", *Phys. Rev. E* **59**, 441–453 (1999).

[194] M.J. Boussinesq, "Sur l'existence d'une viscosité superficielle, dans la mince couche de transition séparant un liquide d'un autre fluide contigu", *Annal. Chim Phys.* **8**, 349–357 (1913).

[195] M. Baus, "Elastic moduli of a liquid-vapor interface", *J. Chem. Phys.* **76**, 2003–2009 (1982).

[196] F.C. Goodrich, "The theory of capillary excess viscosities", *Proc. R. Soc. London A* **374**, 341–370 (1981).

[197] D.M. Anderson, G.B. McFadden and A.A. Wheeler, "Diffuse Interface Methods in Fluid Mechanics", *Annu. Rev. Fluid Mech.* **30**, 139 (1998).

[198] A. Thess and S.A. Orszag, "Surface-tension-driven Bénard convection at infinite Prandtl number", *J. Fluid Mech.* **283**, 201–230 (1995).

[199] A. Schlüter, D. Lortz and F. Busse, "On the stability of steady finite amplitude convection", *J. Fluid Mech.* **23**, 129–144 (1965).

[200] S.H. Davis and L.A. Segel, "Effects of surface curvature and property variation on cellular convection", *Phys. Fluids* **11**, 470–476 (1978).

[201] V.V. Pukhnachov, "Model of convective motion under low gravity", in *Proceedings of the VIIIth European Symposium on Materials and Fluid Sciences in Microgravity*, Brussels, Belgium, April 1992, pp. 157–160.

[202] P.S. Perera and R.F. Sekerka, "Nonsolenoidal flow in a liquid diffusion couple", *Phys. Fluids* **9**, 376–391 (1997).

[203] Th. Boeck and A. Thess, "Inertial Bénard–Marangoni convection", *J. Fluid Mech.* **350**, 149–175 (1997).

[204] Th. Boeck and A. Thess, "Bénard–Marangoni convection at low Prandtl number", *J. Fluid Mech.* **399**, 251–275 (1999).

[205] G. Veronis, "Large-amplitude Bénard convection", *J. Fluid Mech.* **26**, 49–68 (1966).

[206] P.C. Dauby, *Instabilités thermocapillaires et effets de confinement*, PhD Dissertation, Université de Liège, 1994.

[207] M.C. Cross, P.G. Daniels, P.C. Hohenberg and E.D. Siggia, "Phase-winding solutions in a finite container above the convective threshold", *J. Fluid Mech.* **127**, 155 (1983).

[208] A. Pellew and R.V. Southwell, "On maintained convective motion in a fluid heated from below", *Proc. R. Soc. London A* **176**, 312–343 (1940).

[209] A. Vidal and A. Acrivos, "Nature of the neutral state in surface-tension-driven convection", *Phys. Fluids* **9**, 615–616 (1966).

[210] L. Landau and E. Lifchitz, *Mécanique des Fluides*, 2nd Ed., Mir, Moscow, 1989.

[211] G.I. Sivashinsky, "Nonlinear analysis of hydrodynamic instability of laminar flames. Part 1. Derivation of basic equations", *Acta Astronautica* **4**, 1177–1206 (1977).

[212] V.L. Gertsberg and G.I. Sivashinsky, "Large cells in Rayleigh–Bénard convection", *Prog. Theor. Phys.* **66**, 1219–1229 (1981).

[213] G.I. Sivashinsky, *Physica D* **8**, 243 (1983).

[214] P. Le Gal, A. Pocheau and V. Croquette, "Square versus roll pattern at convective threshold", *Phys. Rev. Lett.* **54**, 2501–2504 (1985).

[215] E. Knobloch, "Pattern selection in long-wavelength convection", *Physica D* **41**, 450–479 (1990).

[216] M. Dubois, P. Bergé and J. Wesfreid, "Non-Boussinesq convective structures in water near 4 °C", *J. Phys* **39**, 1253–1257 (1978).

[217] M. Hennenberg, P.M. Bisch, M. Vignes-Adler and A. Sanfeld, "Mass transfer, Marangoni effect, and instability of interfacial longitudinal waves: 2. Diffusional exchange and adsorption-desorption processes", *J. Colloid Interface Sci.* **74**, 495–508 (1980).

[218] K.A. Smith, "On convective instability induced by surface tension gradients", *J. Fluid Mech.* **24**, 401–414 (1966).

[219] J. Reichenbach and H. Linde, "Linear perturbation analysis of surface-tension-driven convection at a plane interface (Marangoni instability)", *J. Colloid Interface. Sci.* **84**, 433–443 (1981).

[220] D.D. Joseph and Y.Y. Renardy, *Fundamentals of Two-Fluid Dynamics*, Springer-Verlag, New York, 1993.

[221] P. Welander, "Convective instability in a two-layer fluid heated uniformly from above", *Tellus* **16**, 349 (1964).

[222] G.Z. Gershuni and E.M. Zhukhovitskii, "On the instability of a horizontal layer system of immiscible fluids when heating from above", *Izv. AN SSSR, Mekh. Zhidk. i Gaza* **6**, 26 (1980), in Russian; *Fluid Dynamics* **15**, 816 (1980).

[223] I.B. Simanovsky, "Anticonvective, steady and oscillatory mechanisms of instability in multi-layer systems", *Physica D* **102**, 313–327 (1997).

[224] D. Johnson and R. Narayanan, "Geometric effects on convective coupling and interfacial structures in bilayer convection", *Phys. Rev. E* **56** (1997).

[225] H.A. Dijkstra, "On the structure of cellular solutions in Rayleigh–Bénard–Marangoni flows in small-aspect-ratio containers", *J. Fluid Mech.* **243**, 73–102 (1992).

[226] J.C. Earnshaw and A.C. McLaughin, "Waves at liquid surfaces: coupled oscillators and mode-mixing", *Proc. R. Soc. London A* **433**, 663–678 (1991).

[227] H. Lamb, *Hydrodynamics*, 6th Ed., Dover, New York, 1945.

[228] A. Wierschem, H. Linde and M.G. Velarde, "Internal waves excited by the Marangoni effect", *Phys. Rev. E* **62**, 6522–6530 (2000); A. Wierschem, H. Linde and M.G. Velarde, "Properties of surface wave trains excited by mass-transfer through a liquid surface", to appear in *Phys. Rev. E* (2001); H. Linde, M.G. Velarde, W. Waldhelm and A. Wierschem, "Interfacial wave motions due to Marangoni instability. III. Solitary waves and (periodic) wave trains and their collisions and reflections leading to dynamic network (cellular) patterns in large containers", to appear in *J. Colloid Interface Sci* (2001).

[229] J. Bragard, P. Colinet, A.Ye. Rednikov and M.G. Velarde, "Dilational Marangoni Waves" (unpublished).

[230] A.H. Nayfeh, *Introduction to Perturbation Techniques*, Wiley Classics, New York, 1993.

[231] E.B. Levchenko and A.L. Chernyakov, "Instability of surface waves in a nonuniformly heated liquid", *Sov. Phys. JETP* **54**, 102–105 (1981).

[232] P.L. Garcia-Ybarra and M.G. Velarde, "Oscillatory Marangoni–Bénard interfacial instability and capillary-gravity waves in single- and two-component liquid layers with and without Soret thermal diffusion", *Phys. Fluids* **30**, 1649–1655 (1987).

[233] L.E. Scriven and C.V. Sternling, "On cellular convection driven surface tension gradients: effects of mean surface tension and surface viscosity", *J. Fluid Mech.* **19**, 321–340 (1964).

[234] D.A. Goussis and R.E. Kelly, "On the thermocapillary instabilities in a liquid layer heated from below", *Int. J. Heat Mass Transfer* **33**, 2237–45 (1990).

[235] X.L. Chu and M.G. Velarde, "Sustained transverse and longitudinal waves at the open surface of a liquid", *Physicochem. Hydrodyn.* **10**, 727–737 (1988).

[236] X.L. Chu and M.G. Velarde, "Transverse and longitudinal waves induced and sustained by surfactant gradients at liquid-liquid interfaces", *J. Colloid Interface Sci.* **131**, 471–484 (1988).

[237] A. Oron, S.H. Davis and S.G. Bankhoff, "Long-scale evolution of thin liquid films", *Rev. Mod. Phys.* **69**, 931–980 (1997).

[238] A.A. Golovin, A.A. Nepomnyashchy and L.M. Pismen, "Interaction between short-scale Marangoni convection and long-scale deformational instability", *Phys. Fluids* **6**, 34–48 (1994).

[239] A.A. Golovin, A.A. Nepomnyashchy and L.M. Pismen, "Nonlinear evolution and secondary instabilities of Marangoni convection in a liquid-gas system with deformable interface", *J. Fluid Mech.* **341**, 317–341 (1997).

[240] M. Takashima, "Surface-tension-driven instability in a horizontal liquid layer with a deformable free surface. II Overstability", *J. Phys. Soc. Jpn.* **50**, 2751–2756 (1981).

[241] M. Hennenberg, X.L. Chu, A. Sanfeld and M.G. Velarde, "Transverse and longitudinal waves at the air-liquid interface in the presence of an adsorption barrier", *J. Colloid Interface Sci.* **150**, 7–21 (1992).

[242] J. Lucassen, "Longitudinal capillary waves", *Trans. Faraday Soc.* **64**, 2221–2230 (1968).

[243] J.C. Earnshaw, "Mode-mixing of surface waves", *J. Phys.: Condens. Matter* **8**, 9553–9558 (1996).

[244] A.Ye. Rednikov, P. Colinet, M.G. Velarde and J.C. Legros, "Oscillatory thermocapillary instability in a liquid layer with deformable open surface: capillary-gravity waves, longitudinal waves and mode-mixing", *J. Non-Eq. Thermodyn.* **25**, 0000 (2001).

[245] H. J. Palmer, "The hydrodynamic stability of rapidly evaporating liquids at reduced pressure", *J. Fluid Mech.* **75**, 487–511 (1976).

[246] F.J. Higuera, "The hydrodynamic stability of an evaporating liquid", *Phys. Fluids* **30**, 679–686 (1987).

[247] W.J. Bornhorst and G.N. Hatsopoulos, "Analysis of a liquid vapor phase change by the methods of Irreversible Thermodynamics", *Trans. ASME*, 840–846 (1967).

[248] I. Prigogine and R. Defay, *Thermodynamique Chimique*, Desoer, Liège, 1950.

[249] H.J. Palmer and J.C. Maheshri, "Enhanced interfacial heat transfer by differential vapor recoil instabilities", *Int. J. Heat Mass Transfer* **4**, 117–123 (1981).

[250] P. Gillon, P. Courville, A. Steinchen-Sanfeld, G. Bertrand and M. Lallemant, "Free convection and boundary layer instabilities in evaporating liquids under low pressure and/or microwave irradiation", *Physicochem. Hydrodyn.* **10**, 149–163 (1988).

[251] A. Prosperetti and M.S. Plesset, "The stability of an evaporating liquid surface", *Phys. Fluids* **27**, 1590–1602 (1984).

[252] E.H. Kennard, *Kinetic Theory of Gases*, McGraw-Hill, New York, 1938.

[253] B. Haut, "Contribution à l'étude du couplage entre l'évaporation et la convection thermocapillaire", Master thesis, Université Libre de Bruxelles (2000).

[254] V.C. Regnier and G. Lebon, "Time-growth and correlation length of fluctuations in thermocapillary convection with surface deformation", *Q. J. Mech. Appl. Math.* **48**, 57–75 (1995).

[255] G.K. Batchelor, *An Introduction to Fluid Dynamics*, Cambridge University Press, Cambridge, 1970.

[256] K.P. Chen, "Interfacial energy balance equation for surface-tension-driven Bénard convection", *Phys. Rev. Lett.* **78**, 4395–4397 (1997).

[257] S.H. Davis, "Convection in a box: linear theory", *J. Fluid Mech.* **30**, 465–478 (1967).

[258] K.H. Winters, T. Plesser and K.A. Cliffe, "The onset of convection in a finite container due to surface tension and buoyancy", *Physica D* **29**, 387–401 (1988).

[259] A.I. van de Vooren and H.A. Dijkstra, *Comput. Fluids* **17**, 467 (1989).

[260] H.A. Dijkstra, "Surface-tension-driven cellular patterns in three-dimensional boxes – linear stability", *Microgravity Sci. Technol.* **7**, 307–312 (1995); "Surface-tension-driven cellular patterns in three-dimensional boxes. II. A bifurcation study" *Microgravity Sci. Technol.* **8**, 70–76 (1995); "Surface-tension-driven cellular patterns in three-dimensional boxes. III. The formation of hexagonal patterns", *Microgravity Sci. Technol.* **8**, 155–162 (1995).

[261] J.K. Platten and D. Villers, "On thermocapillary flows in containers with differentially heated side walls", in *Physicochemical Hydrodynamics. Interfacial Phenomena*, M.G. Velarde (Ed.), NATO ASI Series, Series B: Physics, Vol. 174, 311–336, Plenum-Press, New York, 1988.

[262] B. Spindler, "Linear stability of liquid films with interfacial phase change", *Int. J. Heat Mass Transfer* **25**, 161–173 (1982).

[263] L.G. Badratinova, P. Colinet, M. Hennenberg and J.C. Legros, "On Rayleigh–Taylor instability in heated liquid-gas and liquid-vapor systems", *Russ. J. Eng. Thermophysics* **6**, 1–31 (1996).

[264] R. Perez-Cordon and M.G. Velarde, "On the (nonlinear) foundations of Boussinesq approximation applicable to a thin layer of fluid", *J. Phys.* (Paris) **36**, 591–601 (1975).

[265] M.G. Velarde and R. Perez-Cordon, "On the (nonlinear) foundations of Boussinesq approximation applicable to a thin layer of fluid (II) Viscous dissipation and large cell gap effects", *J. Phys.* (Paris) **37**, 177–182 (1975).

[266] P.C.T. de Boer, "Thermally-driven motion of strongly heated fluids", *Int. J. Heat Mass Transfer* **27**, 2239–2251 (1984).

[267] P.C.T. de Boer, "Thermally-driven motion of highly viscous fluids", *Int. J. Heat Mass Transfer* **29**, 681–688 (1986).

[268] R. Selak and G. Lebon, "Rayleigh–Marangoni thermoconvective instability with non-Boussinesq corrections", *Int. J. Heat and Mass Transfer* **40**, 785–798 (1997).

[269] A. Cloot and G. Lebon, "Marangoni instability in a fluid layer with variable viscosity and free interface, in microgravity", *Physicochem. Hydrodyn.* **6**, 453–462 (1985).

[270] S. Slavtchev and V. Ouzonov, "Stationary Marangoni instability in a liquid layer with temperature-dependent viscosity, in microgravity", *Microgravity Q.* **4**, 33–38 (1994).

[271] G. Lebon and A. Cloot, "Growth of fluctuations near the Bénard–Marangoni convective instability", *Acta Mech.* **48**, 43–55 (1983).

[272] R.S. Schechter, I. Prigogine and J.R. Hamm, "Thermal diffusion and convective stability", *Phys. Fluids* **15**, 379–386 (1972).

[273] M.G. Velarde and R.S. Schechter, "Thermal diffusion and convective stability. II. An analysis of the convected fluxes", *Phys. Fluids* **15**, 1707–1714 (1972).

[274] J.C. Legros, P. Goemaere and J.K. Platten, "Soret coefficient and the two-component Bénard convection in the benzene-methanol system", *Phys. Rev. A* **32**, 1903–1905 (1985).

[275] J.L. Castillo and M.G. Velarde, "Marangoni convection in liquid films with a deformable open surface", *J. Colloid Interface Sci.* **108**, 264–270 (1985).

[276] S. Van Vaerenbergh, P. Colinet and J.C. Legros, "The role of Soret effect on Marangoni–Bénard stability", Lecture Notes in Physics **386**, *Capillarity Today*, G. Pétré and A. Sanfeld (Eds), Springer-Verlag, Berlin, 1991.

[277] C.F. Chen and T.F. Su, "Effect of surface tension on the onset of convection in a double-diffusive layer", *Phys. Fluids A* **4**, 2360–2367 (1992).

[278] C.F. Chen and C.C. Chen, "Effect of surface tension on the stability of a binary fluid layer under reduced gravity", *Phys. Fluids* **6**, 1482–1490 (1994).

[279] A. Bergeon, D. Henry and H. Benhadid, "Marangoni–Bénard instability in microgravity conditions with Soret effect", *Int. J. Heat Mass Transfer* **37**, 1545–1562 (1994).

[280] A. Bergeon, D. Henry, H. Benhadid and L.S. Tuckerman, "Marangoni convection in binary mixtures with Soret effect", *J. Fluid Mech.* **375**, 143–177 (1998).

[281] L.E. Scriven, "Dynamics of a fluid interface: equation of motion for Newtonian surface fluids", *Chem. Eng. Sci.* **12**, 98–108 (1960).

[282] C. Pérez-García, J. Pantaloni, R. Occelli and P. Cerisier, "Linear analysis of surface deflection in Bénard-Marangoni instability", *J. Phys.* (Paris) **46**, 2047–2051 (1985).

[283] M.G. Velarde, P.L. Garcia-Ybarra and J.L. Castillo, "Interfacial oscillations in Bénard–Marangoni layers", *Physicochem. Hydrodyn.* **9**, 387–392 (1987).

[284] O. Dupont, M. Hennenberg and J.C. Legros, "Marangoni–Bénard instabilities under non-steady conditions. Experimental and theoretical results", *Int. J. Heat Mass Transfer* **35**, 3237–3244 (1992).

[285] M. Hennenberg, P.M. Bisch, M. Vignes-Adler and A. Sanfeld, "Mass transfer, Marangoni effect, and instability of interfacial longitudinal waves: 1. Diffusional exchange", *J. Colloid Interface Sci.* **69**, 128–137 (1979).

[286] P. Queeckers, J.C. Dupin and J.C. Legros, "Influence of surface viscosity on Marangoni–Bénard instabilities", *Heat Transfer D* **107**, 405–411 (1989).

[287] J.C. Earnshaw and A.C. McLaughin, "Waves at liquid surfaces II. Surfactant action and coupled oscillators", *Proc. R. Soc. London A* **440**, 519–536 (1993).

[288] M.G. Velarde and X.-L. Chu, "Dissipative hydrodynamic oscillators. I – Marangoni effect and sustained longitudinal waves at the interface of two liquids", *Il Nuovo Cimento* **11 D**, 709–716 (1989).

[289] A.V. Getling, *Rayleigh–Bénard Convection: Structures and Dynamics*, World Scientific, Singapore, 1998.

[290] H. Mori and Y. Kuramoto, *Dissipative Structures and Chaos*, Springer-Verlag, Berlin, 1998.

[291] L. Hadji, "Subcritical instability resulting from the low thermal conductivity of the boundaries", *Europhys. Lett.* **23**, 245–248 (1993).

[292] L. Shtilman and G. Sivashinsky, "Hexagonal structure of large-scale Marangoni convection", *Physica D* **52**, 477–488 (1991).

[293] A. De Wit, *Brisure de symétrie spatiale et dynamique spatio-temporelle dans les systèmes réaction-diffusion*, PhD Dissertation, Université Libre de Bruxelles, 1993.

[294] W.V.R. Malkus, "The heat transport and spectrum of thermal turbulence", *Proc. R. Soc. London A* **225**, 196 (1954).

[295] L.A. Segel, "The nonlinear interaction of a finite number of disturbances to a layer of fluid heated from below", *J. Fluid Mech.* **21**, 359–384 (1965).

[296] R.J. Donnelly, R. Herman and I. Prigogine Eds, *Non-Equilibrium Thermodynamics, Variational Techniques and Stability*, University of Chicago Press, Chicago, 1966.

[297] A.C. Newell and J.A. Whitehead, "Finite bandwidth, finite amplitude convection", *J. Fluid Mech.* **38**, 279–303 (1969).

[298] L.A. Segel, "Distant side-walls cause slow amplitude modulation of cellular convection", *J. Fluid Mech.* **38**, 203–224 (1969).

[299] Y.-Y. Chen, "Finite-size effects on linear stability of pure-fluid convection", *Phys. Rev. A* **45**, 3727–3731 (1992).

[300] M. Dubois and P. Bergé, "Experimental study of the velocity field in Rayleigh–Bénard convection", *J. Fluid Mech.* **85**, 641–653 (1978).

[301] J. Wesfreid, Y. Pomeau, M. Dubois, C. Normand and P. Bergé, "Critical effects in Rayleigh–Bénard convection", *J. Phys. Lett.* **39**, 725–731 (1978).

[302] J.C. Legros, J. Wesfreid and J.K. Platten, "Study of pretransitional induced Bénard convection by two-dimensional numerical experiments", *Int. J. Heat Mass Transfer* **22**, 976–978 (1979).

[303] G. Dee and J.S. Langer, "Propagating pattern selection", *Phys. Rev. Lett.* **50**, 383–386 (1983).

[304] S. Ciliberto, P. Coullet, J. Lega, E. Pampaloni and C. Pérez-García, "Defects in roll-hexagon competition", *Phys. Rev. Lett.* **65**, 2370–2373 (1990).

[305] L. Gil, G. Balzer, P. Coullet, M. Dubois and P. Bergé, "Hopf bifurcation in a broken-parity pattern", *Phys. Rev. Lett.* **66**, 3249–3252 (1991).

[306] F. Giogiutti, A. Bleton, L. Limat and J.E. Wesfreid, "Dynamics of a one-dimensional array of liquid columns", *Phys. Rev. Lett.* **74**, 538–541 (1995).

[307] C. Counillon, L. Daudet, T. Podgorski and L. Limat, "Dynamics of a liquid column array under periodic boundary conditions", *Phys. Rev. Lett.* **80**, 2117–2120 (1998).

[308] P. Coullet and G. Iooss, "Instabilities of one-dimensional cellular patterns", *Phys. Rev. Lett.* **64**, 866–869 (1990).

[309] F. Giorgiutti, L. Limat and J.E. Wesfreid, "Phase diffusion in the vicinity of an oscillatory secondary bifurcation", *Phys. Rev. Lett.* **80**, 2117–2120 (1998).

[310] L. Gil, "Secondary instability of one-dimensional cellular patterns: a gap soliton, black soliton and breather analogy", *Physica D* **147**, 300–310 (2000).

[311] M.I. Rabinovich and L.S. Tsimring, "Dynamics of dislocations in hexagonal patterns", *Phys. Rev. E* **49**, R35–38 (1994).

[312] L.M. Pismen and A.A. Nepomnyashchy, "Structure of dislocations in the hexagonal pattern", *Europhys. Lett.* **24**, 461–465 (1993).

[313] E.D. Siggia and A. Zippelius, "Dynamics of defects in Rayleigh–Bénard convection", *Phys. Rev. A* **24**, 1036–1049 (1981).

[314] Y. Pomeau, S. Zaleski and P. Manneville, "Dislocation motion in cellular structures", *Phys. Rev. A* **27**, 2710–2726 (1983).

[315] F.H. Busse and J.A. Whitehead, "Instabilities of convection rolls in a high Prandtl number fluid", *J. Fluid Mech.* **47**, 305–320 (1971).

[316] M. Bestehorn, "Phase and amplitude instabilities for Bénard–Marangoni convection in fluid layers with large aspect ratio", *Phys. Rev. E* **48**, 3622–3634 (1993).

[317] M. Bestehorn, "Pattern selection in Bénard–Marangoni convection", *Int. J. Bifurcation Chaos* **4**, 1085–1094 (1994).

[318] M.M. Sushchik and L.S. Tsimring, "The Eckhaus instability in hexagonal patterns", *Physica D* **74**, 90–106 (1994).

[319] J. Lauzeral, S. Métens and D. Walgraef, "On the phase dynamics of hexagonal patterns", *Europhys. Lett.* **24**, 707–712 (1993).

[320] J.D. Crawford and E. Knobloch, "Symmetry and symmetry-breaking bifurcations in fluid dynamics", *Annu. Rev. Fluid Mech.* **23**, 341–387 (1991).

[321] J.B. Swift and P.C. Hohenberg, "Hydrodynamic fluctuations at the convective instability", *Phys. Rev. A* **15**, 319 (1977).

[322] M. Bestehorn and H. Haken, "A calculation of transient solutions describing roll and hexagon formation in the convection instability", *Phys. Lett.* **99A**, 265–267 (1983).

[323] M' F. Hilali, S. Métens, P. Borckmans and G. Dewel, "Pattern selection in the generalized Swift–Hohenberg model", *Phys. Rev. E* **51**, 2046–2052 (1995).

[324] J. Da Rocha Miranda Pontes, "Pattern selection in spatially ramped Rayleigh–Bénard systems", PhD Dissertation, Université Libre de Bruxelles, 1992.

[325] A. Newell, T. Passot and J. Lega, "Order parameter equations for patterns", *Annu. Rev. Fluid Mech.* **25**, 399–453 (1993).

[326] K.M. Case and S.C. Chiu, "Three-wave resonant interactions of capillary-gravity waves", *Phys. Fluids* **20**, 742–745 (1977).

[327] L.P. Gor'kov, "Stationary convection in a plane liquid layer near the critical heat transfer point", *Sov. Phys. JETP* **6**, 311–315 (1957).

[328] W.V.R. Malkus and G. Veronis, "Finite amplitude cellular convection", *J. Fluid Mech.* **4**, 225–260 (1958).

[329] L.A. Segel, "The structure of non-linear cellular solutions to the Boussinesq equations", *J. Fluid Mech.* **21**, 345–358 (1965).

[330] S. Rosenblat, "Asymptotic methods for bifurcation and stability problems", *Stud. Appl. Math.* **60**, 241–259 (1979).

[331] D.J. Shipp, D.S. Riley and A.A. Wheeler, "On asymptotic methods for bifurcation and stability: confined Marangoni convection", *Q. J. Appl. Math.* **50**, 379–405 (1997).

[332] E. Palm, "On the tendency towards hexagonal cells in steady convection", *J. Fluid Mech.* **8**, 183–192 (1960).

[333] J.S. Vrentas, C.M. Vrentas, R. Narayanan and S.S. Agrawal, "Surface-tension-driven convection in a cylindrical geometry", *Chem. Eng. Comm.* **14**, 203–223 (1982).

[334] P. Colinet, J.C. Legros, Y. Kamotani, P.C. Dauby and G. Lebon, "Finite-amplitude regimes of the short-wave Marangoni–Bénard convective instability", *Phys. Rev. E* **52**, 2603–2616 (1995).

[335] J. Guckenheimer and Ph. Holmes, *Nonlinear Oscillations, Dynamical Systems, and Bifurcation of Vector Fields*, Springer-Verlag, New York, 1983.

[336] P.L. Garcia-Ybarra, J.L. Castillo and M.G. Velarde, "A nonlinear evolution equation for Bénard–Marangoni convection with deformable boundary", *Phys. Lett. A* **122**, 107–110 (1987).

[337] P.L. Garcia-Ybarra, J.L. Castillo and M.G. Velarde, "Bénard–Marangoni convection with a deformable interface and poorly conducting boundaries", *Phys. Fluids* **30**, 2655–2661 (1987).

[338] J.M. Hyman and B. Nicolaenko, "Coherence and chaos in the Kuramoto–Velarde equation", in *Partial Differential Equations*, M. Crandall (Ed.), Academic Press, New York, 1987.

[339] T. Funada and M. Kotani, "A numerical study of nonlinear diffusion equation governing surface deformation in the Marangoni convection", *J. Phys. Soc. Japan* **55**, 3857–3862 (1986).

[340] T. Funada, "Nonlinear surface waves driven by the Marangoni instability in a heat transfer system", *J. Phys. Soc. Japan* **56**, 2031–2038 (1987).

[341] A.A. Golovin, A.A. Nepomnyashchy and L.M. Pismen, "Pattern formation in large-scale Marangoni convection with deformable interface", *Physica D* **81**, 117–147 (1995).

[342] F.H. Busse, "Das Stäbilitetsverhalten der Zellularkonvektion bei endlicher Amplitude", PhD Dissertation, Munich (1962).

[343] R.M. Clever and F.H. Busse, "Convection in a fluid layer with asymmetric boundary conditions", *Phys. Fluids A* **5**, 99–107 (1993).

[344] J. Bragard, S.G. Slavtchev and G. Lebon, "Nonlinear solutal Marangoni instability in a liquid layer with an adsorbing upper surface", *J. Colloid Interface Sci.* **168** (1994).

[345] P.M. Parmentier, V.C. Regnier, G. Lebon and J.C. Legros, "Nonlinear analysis of coupled gravitational and capillary thermoconvection in thin fluid layers", *Phys. Rev. E* **54**, 411–423 (1996).

[346] E.A. Kuznetsov, A.A. Nepomnyashchy and L.M. Pismen, "New amplitude equations for Boussinesq convection and nonequilateral hexagonal patterns", *Phys. Lett. A* **205**, 261–265 (1995).

[347] S.H. Davis, "Buoyancy-surface tension instability by the method of energy", *J. Fluid Mech.* **39**, 347–359 (1969).

[348] S.H. Davis and G.M. Homsy, "Energy stability theory for free-surface problems: buoyancy thermocapillary layers", *J. Fluid Mech.* **98**, 527–553 (1980).

[349] J.L. Castillo and M.G. Velarde, "Buoyancy-thermocapillary instability: the role of interfacial deformation in one- and two-component fluid layers heated from below or above", *J. Fluid Mech.* **125**, 463–474 (1982).

[350] A. Thess and M. Bestehorn, "Planform selection in Bénard–Marangoni convection: l-hexagons versus g-hexagons", *Phys. Rev. E* **52**, 6358–6367 (1995).

[351] R.M. Clever and F.H. Busse, "Convection at very low Prandtl numbers", *Phys. Fluids A* **2**, 334–339 (1990).

[352] B.A. Malomed, A.A. Nepomnyashchy and M.I. Tribelsky, "Domain boundaries in convection patterns", *Phys. Rev. A* **42**, 7244–7263 (1990).

[353] Ph. de Doncker, "Etude de la sélection des structures convectives dans les instabilités de type Rayleigh-Marangoni–Bénard", Master Thesis, Université Libre de Bruxelles, 1996.

[354] K. Neuffer, "Quadratische Konvektionszellen in Flüssigkeiten", Master Thesis, Universität Stuttgart, 1996.

[355] H.R. Brand, "Envelope equations near the onset of a hexagonal pattern", *Prog. Theor. Phys. Suppl.* **99**, 442 (1989).

[356] J. Bragard and M.G. Velarde, "Bénard–Marangoni convection: planforms and related theoretical predictions", *J. Fluid Mech.* **368**, 165–194 (1998).

[357] B. Echebarría and C. Pérez-García, "Ginzburg–Landau equations for hexagonal patterns", communicated at the 4th SIAM conference on Applications of Dynamical Systems, Snowbird, Utah, USA, May 1997.

[353] B. Echebarría and C. Pérez-García, "Phase instabilities in hexagonal patterns", *Europhys. Lett.* **43**, 35–40 (1998).

[359] A.A. Golovin, A. Hari, A.A. Nepomnyashchy, A.E. Nuz and L.M. Pismen, "Non-potential effects in pattern formation", communicated at the 4th SIAM conference on Applications of Dynamical Systems, Snowbird, Utah, USA, May 1997.

[360] A.E. Nuz, A.A. Nepomnyashchy, A.A. Golovin, A.A. Hari and L.M. Pismen, "Stability of rolls and hexagonal patterns in non-potential systems", *Physica D* **135**, 233–262 (2000).

[36_] A.A. Golovin, B.A. Malomed, A.A. Nepomnyashchy, A.E. Nuz and L.M. Pismen, "Exotic planforms in non-equilibrium systems", in *Nonlinear Coherent Structures in Physics and Biology*, K.H. Spatschek and F.G. Mertens (Eds), Plenum Press, New York, 1994.

[362] W.H. Press, S.A. Teutolsky, W.T. Vetterling and B.P. Flannery, *Numerical Recipes in C*, 2nd Ed., Cambridge University Press, Cambridge, 1992.

[363] L.S. Tsimring, "Dynamics of penta-hepta defects in hexagonal patterns", *Physica D* **89**, 368–380 (1996).

[36₋] A.A. Nepomnyashchy, personal communication (1997).

[365] A.C. Newell and T. Passot, "Instabilities of dislocations in fluid patterns", *Phys. Rev. Lett.* **68**, 1846–1849 (1992).

[366] Q. Ouyang, "Transitions from spiral waves to chemical turbulence in a reaction-diffusion system", communicated at the 4th SIAM conference on Applications of Dynamical Systems, Snowbird, Utah, USA, May 1997.

[367] D. Egolf, "Using Finite-time Lyapunov dimensions to measure the dynamical complexity of topological defects", communicated at the 4th SIAM conference on Applications of Dynamical Systems, Snowbird, Utah, USA, May 1997.

[368] E. Bodenschatz, "Spatio-temporal chaos in Rayleigh–Bénard convection", communicated at the 4th SIAM conference on Applications of Dynamical Systems, Snowbird, Utah, USA, May 1997.

[369] D. Kazhdan, L. Shtilman, A.A. Golovin and L.M. Pismen, "Nonlinear waves and turbulence in Marangoni convection", *Phys. Fluids* **7**, 2679–2685 (1995).

[370] A.A. Nepomnyashchy, "Order parameter equations for long-wavelength instabilities", *Physica D* **86**, 90–95 (1995).

[371] L.M. Pismen, "Mean flow effects in defect dynamics", *Physica D* **61**, 217–226 (1992).

[372] D. Hefer and L.M. Pismen, "Long-scale thermodiffusional convection", *Phys. Fluids* **30**, 2648–2654 (1987).

[373] L. Hadji, "Nonlinear analysis of the coupling between interface deflection and hexagonal patterns in Rayleigh–Bénard–Marangoni convection", *Phys. Rev. E* **53**, 5982–5992 (1996).

[374] L. Hadji, J. Safar and M. Schell, "Analytical results on the coupled Bénard–Marangoni problem consistent with experiment", *J. Non-Equil. Thermodyn.* **16**, 343 (1991).

[375] W. Boos and A. Thess, "Cascade of structures in long-wavelength Marangoni instability", *Phys. Fluids* **11**, 1484–94 (1999).

[376] Y. Pomeau, *Physica D* **23**, 3 (1986); and in *Cellular Structures in Instabilities*, J.E. Wesfreid and S. Zaleski (Eds), Springer-Verlag, New York, 1984.

[377] D. Bensimon, B.I. Shraiman and V. Croquette, "Nonadiabatic effects in convection", *Phys. Rev. A* **38**, 5461–5464 (1988).

[378] C. Kubstrup, H. Herrero and C. Pérez-García, "Fronts between hexagons and squares in a generalized Swift–Hohenberg equation", *Phys. Rev. E* **54**, 1560–1569 (1996).

[379] I.B. Simanovsky, "Anticonvective, steady and oscillatory mechanisms of instability in multilayer systems", *Physica D* **102**, 313–327 (1997).

[380] B. Dionne, M. Silber and A.C. Skeldon, "Stability results for steady, spatially periodic planforms", *Nonlinearity* **10**, 321–353 (1997).

[381] O.V. Afenchenko, A.B. Ezersky, A.V. Nazarovsky and M.G. Velarde, "Experimental evidence of the structure and evolution of penta-hepta defect in hexagonal lattices due to Marangoni–Bénard convection", *Int. J. Bifurcation Chaos* (2001).

[382] A.A. Nepomnyashchy, "Large scale distortions of square patterns", *Int. J. Bifurcation Chaos* **4**, 1147–1154 (1994).

[383] R.B. Hoyle, "Long wavelength instabilities of square patterns", *Physica D* **67**, 198–223 (1993).

[384] L.M. Pismen and A.A. Nepomnyashchy, "Propagation of the hexagonal pattern", *Europhys. Lett.* **27**, 433–436 (1994).

[385] Y. Pomeau and P. Manneville, "Stability and fluctuations of a spatially periodic convective flow", *J. Phys. (Paris)* **40**, 609–612 (1979).

[386] M.C. Cross, "Phase dynamics of convective rolls", *Phys. Rev. A* **27**, 490–498 (1983).

[387] M.C. Cross and A.C. Newell, "Convection patterns in large aspect ratio systems", *Physica D* **10**, 299 (1984).

[388] J.E. Wesfreid, "Wave number variations of spatially damped Rayleigh–Bénard convection", *Phys. Fluids* **24**, 173–174 (1981).

[389] P.G. Drazin, "On the effects of side walls on Bénard convection", *J. Appl. Math. Phys.* **26**, 239–243 (1975).

[390] C. Normand, "Convective flow patterns in rectangular boxes of finite extent", *J. Appl. Math. Phys.* **32**, 81–96 (1981).

[391] P.G. Daniels, "The effect of distant sidewalls on the transition to finite amplitude Bénard convection", *Proc. R. Soc. London A* **358**, 173–197 (1977).

[392] P. Hall and I.C. Walton, "The smooth transition to a convective regime in a two-dimensional box", *Proc. R. Soc. London A* **358**, 199–221 (1977).

[393] V. Croquette, P. Le Gal and A. Pocheau, "Large-scale flow characterization in a Rayleigh–Bénard convective pattern", *Europhys. Lett.* **1**, 393–399 (1986).

[394] E. Bodenschatz, D.S. Cannell, J.R. de Bruyn, R. Ecke, Y.-C. Hu, K. Lerman and G. Ahlers, "Experiments on three systems with non-variational aspects", *Physica D* **61**, 77–93 (1992).

[395] G. Ahlers, L.I. Berge and D.S. Cannell, "Thermal convection in the presence of the first order phase change", *Phys. Rev. Lett.* **70**, 2399–2402 (1993).

[396] M. Assenheimer and V. Steinberg, "Rayleigh–Bénard convection near the gas-liquid critical point", *Phys. Rev. Lett.* **70**, 3888–3891 (1993).

[397] M. Assenheimer and V. Steinberg, "Observation of coexisting upflow and downflow hexagons in Boussinesq Rayleigh–Bénard convection", *Phys. Rev. Lett.* **76**, 756–759 (1996).

[398] G. Dewel, S. Métens, M'F. Hilali, P. Borckmans and C.B. Price, "Resonant patterns through coupling with a zero mode", *Phys. Rev. Lett.* **74**, 4647–4650 (1995).

[399] L.M. Pismen, "Selection of long-scale oscillatory convective patterns", *Phys. Rev. A* **38**, 2564–2572 (1988).

[400] M. Silber and E. Knobloch, "Hopf bifurcation on a square lattice", *Nonlinearity* **4**, 1063–1107 (1991).

[401] M. Roberts, J.W. Swift and D.H. Wagner, "The Hopf bifurcation on a hexagonal lattice", *Contemporary Mathematics* **56**, 283–318 (1986).

[402] M. Renardy and Y. Renardy, "Bifurcating solutions at the onset of convection in the Bénard problem for two fluids", *Physica D* **32**, 227–252 (1988).

[403] B. Dionne, "Planforms in two and three dimensions", *Z. Angew. Math. Phys.* **43**, 36–62 (1992).

[404] B. Dionne, M. Golubitsky, M. Silber and I. Stewart, "Time-periodic spatially periodic planforms in Euclidean equivariant partial differential equations", *Phil. Trans. R. Soc. London A* **352**, 125–168 (1995).

[405] A.C. Skeldon and M. Silber, "New stability results for patterns in a model of long-wavelength convection", *Physica D* **122**, 117–133 (1998).

[406] S. Wahal and A. Bose, "Rayleigh–Bénard and interfacial instabilities in two immiscible liquid layers", *Phys. Fluids* **31**, 3502–3510 (1988).

[407] F.M. Richter and C.E. Johnson, "Stability of a chemically layered mantle", *J. Geophys. Res.* **79**, 11 (1974).

[408] F.M. Richter and D.P. McKenzie, "On some consequences and possible causes of layered mantle convection", *J. Geophys. Res.* **86**, B7 (1981).

[409] Y. Renardy and D.D. Joseph, "Oscillatory instability in a Bénard problem of two fluids", *Phys. Fluids* **28**, 788–793 (1985).

[410] Y. Renardy, "Interfacial stability in a two-layer Bénard problem", *Phys. Fluids* **29**, 356–363 (1986).

[411] Ph. Cardin, H.C. Nataf and Ph. Dewost, "Thermal coupling in layered convection: evidence for an interface viscosity control from mechanical experiments and marginal stability analysis", *J. Phys. II* (Paris) **1**, 599–622 (1991).

[412] Ph. Cardin and H.C. Nataf, "Nonlinear dynamical coupling observed near the threshold of convection in a two-layer system", *Europhys. Lett.* **14**, 655–660 (1991).

[413] F.H. Busse, "On the aspect ratios of two-layer mantle convection", *Phys. Earth Planetary Interiors* **24**, 320–324 (1981).

[414] F.H. Busse, "Multiple solutions for convection in a two component fluid", *Geophys. Res. Lett.* **9**, 519–522 (1982).

[415] L. Cserepes and M. Rabinowicz, "Gravity and convection in a two-layer mantle", *Earth Planetary Sci. Lett.* **76**, 193–207 (1985/86).

[416] C. A. Jones and M.R.E. Proctor, "Strong spatial resonance and traveling waves in Bénard convection", *Phys. Lett. A* **121**, 5 (1987).

[417] M.R.E. Proctor and C.A. Jones, "The interaction of two spatially resonant patterns in thermal convection. Part 1. Exact 1:2 resonance", *J. Fluid Mech.* **188**, 301–335 (1988).

[418] P.C. Dauby, P. Colinet and D. Johnson, "Theoretical analysis of a dynamic thermoconvective pattern in a circular container", *Phys. Rev. E* **61**, 2663–2668 (2000).

[419] A.C. Newell, "The dynamics and analysis of patterns", in *Complex Systems, SFI Studies in the Sciences of Complexity*, D. Stein (Ed.), Addison-Wesley Longman Publishing Group Ltd., 107–173, 1989.

[420] M. D. Graham, U. Müller, P. H. Steen, "Time-periodic convection in Hele-Shaw slots: The diagonal oscillation", *Phys. Fluids A* **4**, 2382–93 (1992).

[421] U. Müller, "Bénard convection in gaps and cavities", in *Convective Transport and Instability Phenomena*, O. Zierep and H. Oertel (Eds), Braun-Verlag, Karlsruhe, 1982.

[422] H.R. Brand, P.S. Lomdahl, and A.C. Newell, "Evolution of the order parameter in situations with broken rotational symmetry", *Phys. Lett. A* **118**, 67 (1986).

[423] H.R. Brand, P.S. Lomdahl, and A.C. Newell, "Benjamin–Feir turbulence in convective binary fluid mixtures", *Physica D* **23D**, 345 (1986).

[424] A.E. Deane, E. Knobloch, J. Toomre, "Traveling waves in large-aspect-ratio thermosolutal convection", *Phys. Rev. A* **37**, 1817–20 (1988).

[425] M.C. Cross, "Traveling and standing waves in binary-fluid convection in finite geometries", *Phys. Rev. Lett.* **57**, 2935–38 (1986).

[426] M.C. Cross, "Structure of nonlinear traveling-wave states in finite geometries", *Phys. Rev. A* **38**, 3593 (1988).

[427] G. Dangelmayr, E. Knobloch and M. Wegelin, "Dynamics of traveling waves in finite containers", *Europhys. Lett.* **16**, 723–729 (1991).

[428] M. Bestehorn, R. Friedrich and H. Haken, "Modulated traveling waves in nonequilibrium systems: the blinking state", Z. Phys. B – Condensed Matter **77**, 151–155 (1989).

[429] M. Bestehorn, R. Friedrich and H. Haken, "Two-dimensional traveling wave patterns in nonequilibrium systems", Z. Phys. B – Condensed Matter **75**, 265–274 (1989).

[430] R.B. Hoyle, "Phase instabilities of oscillatory standing squares and alternating rolls", *Phys. Rev. E* **49**, 2875–2879 (1994).

[431] P. Coullet, S. Fauve and E. Tirapegui, "Large scale instability of nonlinear standing waves", *J. Phys. Lett.* **46**, 787–791 (1985).

[432] P. Coullet, C. Elphick, L. Gil and J. Lega, "Topological defects of wave patterns", *Phys. Rev. Lett.* **59**, 884–887 (1987).

[433] E. Knobloch and J. De Luca, "Amplitude equations for traveling wave convection", *Nonlinearity* **3**, 975–980 (1990).

[434] E. Knobloch, "Nonlocal amplitude equations", in *Pattern Formation in Complex Dissipative Systems*, S. Kai (Ed.), World Scientific, Singapore, 263–274, 1992.

[435] E. Knobloch, "Remarks on the Use and Misuse of the Ginzburg–Landau equation", in *Nonlinear Dynamics and Pattern Formation in the Natural Environment*, Pitman Research Notes in Mathematics, Longman, London, 1995.

[436] C. Martel and J.M. Vega, "Dynamics of a hyperbolic system that applies at the onset of the oscillatory instability", *Nonlinearity* **11**, 105–142 (1998).

[437] V. Croquette and H. Williams, "Nonlinear competition between waves on convective rolls", *Phys. Rev. A* **39**, 2765–2768 (1989); V. Croquette and H. Williams, "Nonlinear waves of the oscillatory instability on finite convective rolls", *Physica D* **37**, 300–314 (1989);

[438] B. Janiaud, A. Pumir, D. Bensimon, V. Croquette, H. Richter and L. Kramer, "The Eckhaus instability for traveling waves", *Physica D* **55**, 269–286 (1992).

[439] L. Kramer and W. Zimmermann, "On the Eckhaus instability for spatially periodic patterns", *Physica D* **16**, 221 (1985).

[440] A. La Porta and C.M. Surko, "Phase defects and spatiotemporal disorder in traveling-wave convection patterns", *Phys. Rev. E* **56**, 5351–5366 (1997).

[441] T. Kawahara, "Formation of saturated solutions in a nonlinear dispersive system with instability and dissipation", *Phys. Rev. Lett.* **51**, 381–383 (1983).

[442] T. Kawahara and S. Toh, "Nonlinear dispersive periodic waves in the presence of instability and damping", *Phys. Fluids* **28**, 1636–38 (1985); "Pulse interactions in an unstable dissipative-dispersive nonlinear system", *Phys. Fluids* **31**, 2103–11 (1988).

[443] S.M. Tobias, M.R.E. Proctor and E. Knobloch, "Convective and absolute instabilities of fluid flows in finite geometry", *Physica D* **113** 43–72 (1998).

[444] J. Martínez-Mardones, R. Tiemann, D. Walgraef and W. Zeller, "Amplitude equations and pattern selection in viscoelastic convection", *Phys. Rev. E* **54**, 1478–1488 (1996).

[445] J. Lighthill, *Waves in Fluids*, Cambridge University Press, Cambridge, 1978.

[446] A.A. Nepomnyashchy and M.G. Velarde, "A three-dimensional description of solitary waves and their interaction in Marangoni–Bénard layers", *Phys. Fluids* **6**, 187–198 (1994).

[447] C.I. Christov and M.G. Velarde, "Evolution and interactions of solitary waves (solitons) in nonlinear dissipative systems", *Physica Scripta* **T55**, 101–106 (1994).

[448] C.I. Christov and M.G. Velarde, "Dissipative solitons", *Physica D* **86**, 323–347 (1995).

[449] A.Ye. Rednikov and M.G. Velarde, "Nonlinear waves and (dissipative) solitons in thin liquid layers", in *Nonlinear Waves in Multi-Phase Flow*, H.-C. Chang (Ed.), Kluwer, Dordrecht, 2000, pp. 57–67.

[450] D.E. Bar and A.A. Nepomnyashchy, "Stability of periodic waves governed by the modified Kawahara equation", *Physica D* **86**, 586–602 (1995); D.E. Bar and A.A. Nepomnyashchy, "Stability of periodic waves generated by long-wavelength instabilities in isotropic and anisotropic systems", *Physica D* **132**, 411–427 (1999).

[451] A.Ye. Rednikov, M.G. Velarde, Yu. S. Ryazantsev, A.A. Nepomnyashchy and V.N. Kurdyumov, "Cnoidal wave trains and solitary waves in a dissipation-modified Korteweg–de Vries equation", *Acta Appl. Math.* **39**, 457–475 (1995).

[452] V.I. Nekorkin and M.G. Velarde, "Solitary waves, soliton bound states and chaos in a dissipative Korteweg–de Vries equation", *Int. J. Bifurcation Chaos* **4**, 1135–1146 (1994).

[453] M.G. Velarde, V.I. Nekorkin and A.G. Maksimov, "Further results on the evolution of solitary waves and their bound states of a dissipative Korteweg–de Vries equation", *Int. J. Bifurcation Chaos* **5**, 831–839 (1995).

[454] I.L. Kliakhander, A.V. Porubov and M.G. Velarde, "Localized finite-amplitude disturbances and selection of solitary waves", *Phys. Rev. E* **62**, 4959–4962 (2000).

[455] O. Lhost and J.K. Platten, "Large-scale convection induced by the Soret effect", *Phys. Rev. A* **40**, 6415–20 (1989); "Transitions between states, traveling waves, and modulated waves in the system water–isopropanol heated from below", *Phys. Rev. A* **38**, 3147–50 (1988); "Experimental study of the transition from nonlinear traveling waves to steady overturning convection in binary mixtures", *Phys. Rev. A* **40**, 4552–57 (1989).

[456] J.K. Platten and G. Chavepeyer, "An hysteresis loop in the two component Bénard problem", *Int. J. Heat Mass Transfer* **18**, 1071–1075 (1975).

[457] P. Kolodner, D. Bensimon and C.M. Surko, "Traveling-wave convection in an annulus", *Phys. Rev. Lett.* **60**, 1723–1726 (1988).

[458] E. Knobloch, M.R.E. Proctor and N.O. Weiss, "Heteroclinic bifurcations in a simple model of double-diffusive convection", *J. Fluid Mech.* **239**, 273–292 (1992).

[459] J.K. Platten and G. Chavepeyer, "Oscillatory motion in Bénard cell due to the Soret effect", *J. Fluid Mech.* **60**, 305–319 (1973).

[460] S. Van Vaerenbergh and J.C. Legros, "Soret coefficient of organic solutions measured in the microgravity SCM experiment and by the flow and Bénard cells", *J. Phys. Chem. B* **102**, 4426–4431 (1998).

[461] J.P. Larre, J.K. Platten and G. Chavepeyer, "Soret effects in ternary systems heated from below", *Int. J. Heat Mass Transfer* **40**, 545–55 (1997).

[462] C. Karcher and U. Muller, "Bénard convection in a binary mixture with a nonlinear density-temperature relation", *Phys. Rev. E* **49**, 4031–43 (1994).

[463] S. Van Vaerenbergh, P. Colinet, Ph. Géoris and J.C. Legros, "Preparatory results for a Marangoni–Bénard experiment with Soret effect under microgravity conditions", Proceedings of Spacebound'91, Ottawa, 1991.

[464] J.C. Legros, S. Van Vaerenbergh, Y. Decroly and P. Colinet, "Expériences en microgravité étudiant l'effet Soret: SCM, SCCO et MBIS", *Entropie* **184–185**, 38–43 (1994).

[465] M. Bestehorn and P. Colinet, "Bénard-Marangoni convection of a binary mixture as an example of an oscillatory bifurcation under strong symmetry-breaking effects", *Physica D* **145**, 84–109 (2000).

[466] M. Bestehorn, "Numerical results of 3D stationary convection in a binary mixture", *Phys. Lett. A* **174**, 43–47 (1993).

[467] E. Knobloch and J. Guckenheimer, "Convective transitions induced by a varying aspect ratio", *Phys. Rev. A* **27**, 408–417 (1983).

[468] B. Echebarría, D. Krmpotic and C. Pérez-García, "Resonant interactions in Bénard–Marangoni convection in cylindrical layers", *Physica D* **99**, 487–502 (1997).

[469] D. Ruelle, "The turbulent fluid as a dynamical system", in *New Perspectives in Turbulence*, L. Sirovich (Ed.), Springer-Verlag, New York, 1991.

[470] M. Renardy, "Hopf bifurcation on the hexagonal lattice with small frequency", *Adv. Diff. Eq.* **1**, 283–299 (1996).

[471] Y.Y. Renardy, M. Renardy and K. Fujimura, "Takens–Bogdanov bifurcation on the hexagonal lattice for double-layer convection", *Physica D* **129**, 171–202 (1999).

[472] K. A. Julien, "Strong spatial interactions with 1:1 resonance: a three-layer convection problem", *Nonlinearity* **7**, 1655–1693 (1994).

[473] A. De Wit, D. Lima, G. Dewel and P. Borckmans, "Spatiotemporal dynamics near a codimension-two point", *Phys. Rev. E* **54**, 261–271 (1996).

[474] A. Hill and I. Stewart, "Hopf-steady mode interactions with O(2) symmetry", *Dynam. Stab. Syst.* **6**, 149–171 (1991).

[475] K. Fujimura and Y.Y. Renardy, "The 2:1 steady/Hopf mode interaction in the two-layer Bénard problem", *Physica D* **85**, 25–65 (1995).

[476] Ph. Géoris, M. Hennenberg, I.B. Simanovsky, A. Nepomnyashchy, I.I. Wertgeim and J.C. Legros, "Thermocapillary convection in a multilayer system", *Phys. Fluids A* **5**, 1575–1582 (1993).

[477] D.V. Lyubimov and M.A. Zaks, "Two mechanisms of the transition to chaos in finite-dimensional models of convection", *Physica D* **9**, 52–64 (1983).

[478] A. Arneodo, P. Coullet and C. Tresser, "A possible new mechanism for the onset of turbulence", *Phys. Lett. A* **81**, 197–201 (1981).

[479] M.A. Zaks, "On the convergence rate of a sequence of homoclinic bifurcations in systems without a symmetry constraint", *Phys. Lett. A* **175**, 193–198 (1993).

[480] Y. Kuramoto and S. Koga, "Anomalous period-doubling bifurcations leading to chemical turbulence", *Phys. Lett. A* **92**, 1–4 (1982).

[481] A.S. Pikovsky, M.A. Zaks, U. Feudel and J. Kurths, "Singular continuous spectra in dissipative dynamics", *Phys. Rev. E* **52**, 285–296 (1995).

[482] P. Colinet, Ph. Géoris, J.C. Legros and G. Lebon, "Spatially quasiperiodic convection and temporal chaos in two-layer thermocapillary instabilities", *Phys. Rev. E* **54**, 514–524 (1996).

[483] B.Y. Rubinstein and L.M. Pismen, "Resonant two-dimensional patterns in optical cavities with a rotated beam", *Phys. Rev. A* **56**, 4264–4272 (1997).

[484] A. De Wit, G. Dewel and P. Borckmans, "Chaotic Turing-Hopf mixed mode", *Phys. Rev. E* **48**, R4191–4194 (1993).

[485] Y.Y. Renardy, "The Takens–Bogdanov bifurcation on the hexagonal lattice", communicated at the 4th SIAM conference on Applications of Dynamical Systems, Snowbird, Utah, USA, May 1997.

[486] Y. Kamotani, J. H. Lee, S. Ostrach and A. Pline, "An experimental study of oscillatory thermocapillary convection in cylindrical containers", *Phys. Fluids A* **4**, 955–962 (1992).

[487] G. Dewel, P. Borckmans, A. De Wit, B. Rudovics, J.-J. Perraud, E. Dulos, J. Boissonade, P. De Kepper, "Pattern selection and localised structures in reaction-diffusion systems", *Physica A* **213**, 181–198 (1995).

[488] M. Fermigier, P. Jenffer, L. Limat and J.E. Wesfreid, "Fluid-fluid interfacial instabilities induced by gravity", ESA SP-333, Proc. VIIIth European Symposium on Materials and Fluid Sciences in Microgravity, Brussels, Belgium, 1992.

[489] S. Rosenblat, G.M. Homsy and S. H. Davis, "Eigenvalues of the Rayleigh–Bénard and Marangoni problems", *Phys. Fluids* **24**, 11 (1981).

[490] W. Eckhaus, *Studies in Non-Linear Stability Theory*, Springer, New York, 1965.

[491] R. Lefever and D. Carati, "Intrinsic patterns in anisotropic multicomponent diffusive systems at equilibrium", *Physica A* **213**, 90–104 (1995).

[492] A. Thess, D. Spirn and B. Jüttner, "Viscous flow at infinite Marangoni number", *Phys. Rev. Lett.* **75**, 4614–4617 (1995).

[493] S.A. Orszag, "Numerical simulation of incompressible flows within simple boundaries. I. Galerkin (spectral) representations", *Stud. Appl. Math.* **L 4**, 293–327 (1971).

[494] Th. Boeck and A. Thess, "Turbulent Bénard-Marangoni convection: results of two-dimensional simulations", *Phys. Rev. Lett.* **80**, 1216–1219 (1998).

[495] A. Pumir and L. Blumenfeld, "Heat transport in a liquid layer locally heated on its free surface", *Phys. Rev. E* **54**, 4528–4531 (1996).

[496] H. Chaté and P. Manneville, "Transition to turbulence via spatiotemporal intermittency", *Phys. Rev. Lett.* **58**, 112–115 (1987).

[497] Ch. Misbah and A. Valance, "Secondary instabilities in the stabilised Kuramoto–Sivashinsky equation", *Phys. Rev. E* **49**, 166–183 (1994).

[498] K.R. Elder, J.D. Gunton and N. Goldenfeld, "Transition to spatiotemporal chaos in the damped Kuramoto–Sivashinsky equation", *Phys. Rev. E* **56**, 1613–1634 (1997).

[499] M. Paniconi and K.R. Elder, "Stationary, dynamical and chaotic states of the two-dimensional damped Kuramoto–Sivashinsky equation", *Phys. Rev. E* **56**, 2713–2721 (1997).

[500] J.P. Gollub and S.V. Benson, "Many routes to turbulent convection", *J. Fluid Mech.* **100**, 449–470 (1980).

[501] P. Bergé, M. Dubois and V. Croquette, "Approach to Rayleigh–Bénard turbulent convection in different geometries" in *Convective Transport and Instability Phenomena*, O. Zierep and H. Oertel (Eds), Braun-Verlag, Karlsruhe, 1982.

[502] A. Thess, D. Spirn and B. Jüttner, "A two-dimensional model for slow convection at infinite Marangoni number", *J. Fluid Mech.* **331**, 283–312 (1996).

[503] A. Thess and S.A. Orszag, "Temperature spectrum in surface tension driven Bénard convection", *Phys. Rev. Lett.* **73**, 541–544 (1994).

[504] P. Colinet, "Amplitude equations and nonlinear dynamics of surface-tension and buoyancy-driven convective instabilities", PhD Dissertation, Université Libre de Bruxelles, 1997.

[505] G.I. Barenblatt, *Scaling, Self-Similarity and Intermediate Asymptotics*, Cambridge University Press, Cambridge, 1996.

[506] C. Foias, O.P. Manley and R. Temam, "Physical estimates of the number of degrees of freedom in free convection", *Phys. Fluids* **29**, 3101–3103 (1986).

[507] B.I. Shrainan and E.D. Siggia, "Heat transport on high-Rayleigh-number convection", *Phys. Rev. A* **42**, 3650–3653 (1990).

[508] S. Ciliberto, S. Cioni and C. Laroche, "Large-scale flow properties of turbulent thermal convection", *Phys. Rev. E* **54**, 5901–5904 (1996).

[509] T. Takeshita, T. Segawa, J.A. Glazier and M. Sano, "Thermal turbulence in mercury", *Phys. Rev. Lett.* **76**, 1465–1468 (1996).

[510] K. Julien, S. Legg, J. Mc Williams and J. Werne, "Rapidly rotating turbulent Rayleigh–Bénard convection", *J. Fluid Mech.* **322**, 243–273 (1996).

Index